Student Solutions Manual
for

PHYSICAL CHEMISTRY
Seventh Edition

PETER ATKINS
CHARLES A. TRAPP
MARSHALL CADY
CARMEN GIUNTA

W. H. Freeman and Company
New York

Credits

Peter Atkins
Professor of Chemistry and Fellow of Lincoln College, Oxford

Charles A. Trapp
Professor of Chemistry, University of Louisville

Marshall Cady
Professor of Chemistry, Indiana University Southeast

Carmen Giuinta
Professor of Chemistry, Le Moyne College

ISBN: 0-7167-4388-4 (EAN: 9780716743880)

Printed in the United States of America

Fourth printing

Preface

This manual provides detailed solutions to all the end-of-chapter (**a**) Exercises, to a selection of slightly more than half the end-of-chapter Problems, and to half the new Box Questions. Solutions to Exercises and Problems carried over from the sixth edition have been reworked, modified, or corrected when needed.

The solutions to the Problems in this edition rely more heavily on the mathematical and molecular modelling software that is now generally accessible to physical chemistry students, and this is particularly true for many of the new Problems which request the use of such software for their solutions. But almost all of the Exercises and many of the Problems can still be solved with a modern hand-held scientific calculator. When a quantum chemical calculation or molecular modelling process has been called for, we have usually provided the solution with PC Spartan Pro^{TM} because of its common availability.

Most of the former Microprojects of the 6th edition have been incorporated into the end-of-chapter problems (a few as Box Questions) for this edition and the solutions to about half of them now appear in this manual rather than in a separate book as before. Their solutions have been rewritten to conform to the style of this manual. However, the book by M. P. Cady and C. A. Trapp, *A Mathcad Primer for Physical Chemistry*, Oxford University Press (1999) is still available and provides detailed instruction in the use of mathematical software for the solutions of Physical Chemistry problems.

In general, we have adhered rigorously to the rules for significant figures in displaying the final answers. However, when intermediate answers are shown, they are often given with one more figure than would be justified by the data. These excess digits are indicated with an overline.

We have carefully cross-checked the solutions for errors and expect that most have been eliminated. We would be grateful to any readers who bring any remaining errors to our attention.

We warmly thank our publishers for their patience in guiding this complex, detailed project to completion.

P. W. A.
C. A. T.
M. P. C.
C. G.

Contents

Part 1: Equilibrium

Part 1: Equilibrium

1 The properties of gases

Solutions to exercises

Discussion questions

E1.1(a) An equation of state is an equation that relates the variables that define the state of a system to each other. Boyle, Charles, and Avogadro established these relations for gases at low pressures (perfect gases) by appropriate experiments. Boyle determined how volume varies with pressure ($V \propto 1/p$), Charles how volume varies with temperature ($V \propto T$), and Avogadro how volume varies with amount of gas ($V \propto n$). Combining all of these proportionalities into one we find

$$V \propto \frac{nT}{p}$$

Inserting the constant of proportionality, R, yields the perfect gas equation

$$V = \frac{RnT}{P} \quad \text{or} \quad pV = nRT$$

E1.2(a) Consider three temperature regions:

(1) $T < T_B$. At very low pressures, all gases show a compression factor, $Z \approx 1$. At high pressures, all gases have $Z > 1$, signifying that they have a molar volume greater than a perfect gas, which implies that repulsive forces are dominant. At intermediate pressures, most gases show $Z < 1$, indicating that attractive forces reducing the molar volume below the perfect value are dominant.

(2) $T \approx T_B$. $Z \approx 1$ at low pressures, slightly greater than 1 at intermediate pressures, and significantly greater than 1 only at high pressures. There is a balance between the attractive and repulsive forces at low to intermediate pressures, but the repulsive forces predominate at high pressures where the molecules are very close to each other.

(3) $T > T_B$. $Z > 1$ at all pressures because the frequency of collisions between molecules increases with temperature.

E1.3(a) The van der Waals equation "corrects" the perfect gas equation for both attractive and repulsive interactions between the molecules in a real gas. See *Justification 1.1* for a fuller explanation.

The Bertholet equation accounts for the volume of the molecules in a manner similar to the van der Waals equation but the term representing molecular attractions is modified to account for the effect of temperature. Experimentally one finds that the van der Waals a decreases with increasing temperature. Theory (see Chapter 21) also suggests that intermolecular attractions can decrease with temperature. This variation of the attractive interaction with temperature can be accounted for in the equation of state by replacing the van der Waals a with a/T.

Numerical exercises

E1.4(a) Boyle's law [1.6] provides the basis for the solution.

Since $pV = \text{constant}$, $p_f V_f = p_i V_i$

Solving for p_f, $p_f = \frac{V_i}{V_f} \times p_i$

$$V_i = 1.0\,\text{L} = 10\overline{0}0\,\text{cm}^3, \qquad V_f = 100\,\text{cm}^3, \qquad p_i = 1.00\,\text{atm}$$

$$p_f = \frac{10\overline{0}0\,\text{cm}^3}{100\,\text{cm}^3} \times 1.00\,\text{atm} = 10 \times 1.00\,\text{atm} = \boxed{10\,\text{atm}}$$

E1.5(a)   (a) The perfect gas equation [1.12] is: $pV = nRT$

Solving for the pressure gives $p = \dfrac{nRT}{V}$

The amount of xenon is $n = \dfrac{131\,g}{131\,g\,mol^{-1}} = 1.00\,mol$

$$p = \frac{(1.00\,mol) \times (0.0821\,L\,atm\,K^{-1}\,mol^{-1}) \times (298.15\,K)}{1.0\,L} = \boxed{24\,atm}$$

That is, the sample would exert a pressure of 24 atm if it were a perfect gas, not 20 atm.

(b) The van der Waals equation [1.25a] for the pressure of a gas is $p = \dfrac{nRT}{V - nb} - \dfrac{an^2}{V^2}$

For xenon, Table 1.5 gives $a = 4.137\,L^2\,atm\,mol^{-2}$ and $b = 5.16 \times 10^{-2}\,L\,mol^{-1}$.
Inserting these constants, the terms in the equation for p become

$$\frac{nRT}{V - nb} = \frac{(1.00\,mol) \times (0.08206\,L\,atm\,K^{-1}\,mol^{-1}) \times (298.15\,K)}{1.0\,L - \{(1.00\,mol) \times (5.16 \times 10^{-2}\,L\,mol^{-1})\}} = 25.\overline{8}\,atm$$

$$\frac{an^2}{V^2} = \frac{(4.137\,L^2\,atm\,mol^{-1}) \times (1.00\,mol)^2}{(1.0\,L)^2} = 4.1\overline{37}\,atm$$

Therefore, $p = 25.\overline{8}\,atm - 4.1\overline{37}\,atm = \boxed{22\,atm}$

E1.6(a)   Boyle's law [1.6] in the form $p_f V_f = p_i V_i$ can be solved for either initial or final pressure, hence

$$p_i = \frac{V_f}{V_i} \times p_f$$

$$V_f = 4.65\,L, \qquad V_i = 4.65\,L + 2.20\,L = 6.85\,L, \qquad p_f = 3.78 \times 10^3\,Torr$$

Therefore,

(a) $p_i = \left(\dfrac{4.65\,L}{6.85\,L}\right) \times (3.78 \times 10^3\,Torr) = \boxed{2.57 \times 10^3\,Torr}$

(b)   Since 1 atm = 760 Torr exactly, $p_i = (2.57 \times 10^3\,Torr) \times \left(\dfrac{1\,atm}{760\,Torr}\right) = \boxed{3.38\,atm}$

E1.7(a)   Charles's law in the form $V = \text{constant} \times T$ [1.9] may be rewritten as $\dfrac{V}{T} = \text{constant}$ or $\dfrac{V_f}{T_f} = \dfrac{V_i}{T_i}$

Solving for T_f, $T_f = \dfrac{V_f}{V_i} \times T_i$, $V_i = 1.0\,L$, $V_f = 100\,cm^3$, $T_i = 298\,K$

$$T_f = \left(\frac{100\,cm^3}{1000\,cm^3}\right) \times (298\,K) = \boxed{30\,K}$$

E1.8(a)   The perfect gas law, $pV = nRT$ [1.12], can be rearranged to $\dfrac{p}{T} = \dfrac{nR}{V} = \text{constant}$, if n and V are

constant. Hence, $\dfrac{p_f}{T_f} = \dfrac{p_i}{T_i}$ or, solving for p_f, $p_f = \dfrac{T_f}{T_i} \times p_i$

Internal pressure = pump pressure + atmospheric pressure

$$p_i = 24\,\text{lb in}^{-2} + 14.7\,\text{lb in}^{-2} = 38.\overline{7}\,\text{lb in}^{-2}, \quad T_i = 268\,\text{K}\ (-5°\text{C}), \quad T_f = 308\,\text{K}\ (35°\text{C})$$

$$p_f = \frac{308\,\text{K}}{268\,\text{K}} \times 38.\overline{7}\,\text{lb in}^{-2} = 44.\overline{5}\,\text{lb in}^{-2}$$

Therefore, $p(\text{pump}) = 44.\overline{5}\,\text{lb in}^{-2} - 14.7\,\text{lb in}^{-2} = \boxed{30\,\text{lb in}^{-2}}$

Complications are those factors which destroy the constancy of V or n, such as the change in volume of the tyre, the change in rigidity of the material from which it is made, and loss of pressure by leaks and diffusion.

E1.9(a) The perfect gas law in the form $p = \dfrac{nRT}{V}$ [1.2] is appropriate. T and V are given; n needs to be calculated.

$$n = \frac{0.255\,\text{g}}{20.18\,\text{g mol}^{-1}} = 1.26 \times 10^{-2}\,\text{mol}, \qquad T = 122\,\text{K}, \qquad V = 3.00\,\text{L}$$

Therefore, upon substitution,

$$p = \frac{(1.26 \times 10^{-2}\,\text{mol}) \times (0.08206\,\text{L atm K}^{-1}\,\text{mol}^{-1}) \times (122\,\text{K})}{3.00\,\text{L}} = \boxed{4.20 \times 10^{-2}\,\text{atm}}$$

E1.10(a) The gas pressure is calculated as the force per unit area that a column of water of height 206.402 cm exerts on the gas due to its weight. The manometer is assumed to have uniform cross-sectional area, A.

Then force, $F = mg$, where m is the mass of the column of water and g is the acceleration of free fall. As in Example 1.2, $m = \rho \times V = \rho \times h \times A$ where $h = 206.402$ cm and A is the cross-sectional area.

$$p = \frac{F}{A} = \frac{\rho h A g}{A} = \rho h g$$

$$p = (0.99707\,\text{g cm}^{-3}) \times \left(\frac{1\,\text{kg}}{10^3\,\text{g}}\right) \times \left(\frac{10^6\,\text{cm}^3}{1\,\text{m}^3}\right)$$

$$\times\, (206.402\,\text{cm}) \times \left(\frac{1\,\text{m}}{10^2\,\text{cm}}\right) \times (9.8067\,\text{m s}^{-2})$$

$$= 2.0182 \times 10^4\,\text{Pa}$$

$$V = (20.000\,\text{L}) \times \left(\frac{1\,\text{m}^3}{10^3\,\text{L}}\right) = 2.0000 \times 10^{-2}\,\text{m}^3$$

$$n = \frac{m}{M} = \frac{0.25132\,\text{g}}{4.00260\,\text{g mol}^{-1}} = 0.062789\,\text{mol}$$

The perfect gas equation [1.12] can be rearranged to give $R = \dfrac{pV}{nT}$

$$R = \frac{(2.0182 \times 10^4\,\text{Pa}) \times (2.0000 \times 10^{-2}\,\text{m}^3)}{(0.062789\,\text{mol}) \times (773.15\,\text{K})} = \boxed{8.3147\,\text{J K}^{-1}\,\text{mol}^{-1}}$$

The accepted value is $R = 8.3145\,\text{J K}^{-1}\,\text{mol}^{-1}$.

Although gas volume data should be extrapolated to $p = 0$ for the best value of R, helium is close to being a perfect gas under the conditions here, and thus a value of R close to the accepted value is obtained.

E1.11(a) Since $p < 1$ atm, the approximation that the vapour is a perfect gas is adequate. Then (as in Exercise 1.10(b)),

$$pV = nRT = \frac{m}{M}RT. \text{ Upon rearrangement,}$$

$$M = \rho \left(\frac{RT}{p}\right) = (3.71\,\text{g}\,\text{L}^{-1}) \times \frac{(0.0821\,\text{L atm mol}^{-1}\,\text{K}^{-1}) \times (773\,\text{K})}{(699\,\text{Torr}) \times \left(\frac{1\,\text{atm}}{760\,\text{Torr}}\right)} = 256\,\text{g mol}^{-1}$$

This molar mass must be an integral multiple of the molar mass of atomic sulfur; hence

$$\text{number of S atoms} = \frac{256\,\text{g mol}^{-1}}{32.0\,\text{g mol}^{-1}} = 8$$

The formula of the vapour is then $\boxed{S_8}$

E1.12(a) The partial pressure of the water vapour in the room is: $p_{H_2O} = (0.60) \times (26.74\,\text{Torr}) = 16\,\text{Torr}$

Assuming that the perfect gas equation [1.12] applies, with $n = \frac{m}{M}$, $pV = \frac{m}{M}RT$ or

$$m = \frac{pVM}{RT} = \frac{(16\,\text{Torr}) \times \left(\frac{1\,\text{atm}}{760\,\text{Torr}}\right) \times (400\,\text{m}^3) \times \left(\frac{10^3\,\text{L}}{\text{m}^3}\right) \times (18.02\,\text{g mol}^{-1})}{(0.0821\,\text{L atm K}^{-1}\,\text{mol}^{-1}) \times (300\,\text{K})}$$

$$= 6.2 \times 10^3\,\text{g} = \boxed{6.2\,\text{kg}}$$

E1.13(a) (a) For simplicity assume a container of volume 1 L. Then the total mass is

$$m_T = n_{N_2}M_{N_2} + n_{O_2}M_{O_2} = 1.146\,\text{g} \tag{1}$$

Assuming that air is a perfect gas, $p_T V = n_T RT$, where n_T is the total amount of gas

$$n_T = \frac{p_T V}{RT} = \frac{(740\,\text{Torr}) \times \left(\frac{1\,\text{atm}}{760\,\text{Torr}}\right) \times (1\,\text{L})}{(0.08206\,\text{L atm K}^{-1}\,\text{mol}^{-1}) \times (300\,\text{K})} = 0.03955\,\text{mol}$$

$$n_T = n_{N_2} + n_{O_2} = 0.03955\,\text{mol} \tag{2}$$

Equations (1) and (2) are simultaneous equations for the amounts of gas and may be solved for them. Inserting n_{O_2} from (2) into (1) we get

$$(n_{N_2}) \times (28.0136\,\text{g mol}^{-1}) + (0.03955\,\text{mol} - n_{N_2}) \times (31.9988\,\text{g mol}^{-1}) = 1.146\,\text{g}$$

$$(1.2655 - 1.1460)\,\text{g} = (3.9852\,\text{g mol}^{-1}) \times (n_{N_2})$$

$$n_{N_2} = 0.02999\,\text{mol}$$

$$n_{O_2} = n_T - n_{N_2} = (0.03955 - 0.02999)\,\text{mol} = 9.56 \times 10^{-3}\,\text{mol}$$

The mole fractions are $x_{N_2} = \frac{0.02999\,\text{mol}}{0.03955\,\text{mol}} = \boxed{0.7583}$ $x_{O_2} = \frac{9.56 \times 10^{-3}\,\text{mol}}{0.03955\,\text{mol}} = \boxed{0.2417}$

The partial pressures are $p_{N_2} = (0.7583) \times (740\,\text{Torr}) = \boxed{561\,\text{Torr}}$

$$p_{O_2} = (0.2417) \times (740\,\text{Torr}) = \boxed{179\,\text{Torr}}$$

The sum checks, $(561 + 179)\,\text{Torr} = 740\,\text{Torr}$

(b) The simplest way to solve this part is to realize that n_T, p_T, and m_T remain the same as in part **(a)** as these are experimentally determined quantities. However, the simultaneous equations that need to be solved are modified as follows:

$$m_T = n_{N_2} M_{N_2} + n_{O_2} M_{O_2} + n_{Ar} M_{Ar} = 1.146\,g$$

$$n_T = n_{N_2} + n_{O_2} + n_{Ar} = 0.03955\,mol$$

Since $x_{Ar} = 0.0100$, $n_{Ar} = 0.0003955\,mol$
Solving the equations yields

$$n_{N_2} = 0.03084 \quad x_{N_2} = \boxed{0.7798}$$

$$n_{O_2} = 0.008314 \quad x_{O_2} = \boxed{0.2102}$$

The partial pressures are:

$$p_{N_2} = x_{N_2} p_T = 0.7798 \times 740\,Torr = \boxed{577\,Torr}$$

$$p_{O_2} = x_{O_2} p_T = 0.2102 \times 740\,Torr = \boxed{155\,Torr}$$

$$p_{Ar} = x_{Ar} p_T = 0.0100 \times 740\,Torr = \boxed{7\,Torr}$$

E1.14(a) This exercise uses the formula, $M = \rho \dfrac{RT}{p}$, which was developed and used in Exercises 1.10(b) and 1.11(a). Substituting the data, $M = \dfrac{(1.23\,g\,L^{-1}) \times (62.36\,L\,Torr\,K^{-1}mol^{-1}) \times (330\,K)}{150\,Torr}$

$$= \boxed{169\,g\,mol^{-1}}$$

E1.15(a) The easiest way to solve this exercise is to assume a sample of mass 1.000 g, then calculate the volume at each temperature, plot the volume against the Celsius temperature, and extrapolate to $V = 0$.
Draw up the following table

$\theta/°C$	$\rho/(g\,L^{-1})$	$V/(L\,g^{-1})$
-85	1.877	0.5328
0	1.294	0.7728
100	0.946	1.057

V versus θ is plotted in Fig. 1.1. The extrapolation gives a value for absolute zero close to $-273°C$. Alternatively, one could use an equation for V as a linear function of θ, which is Charles's law, and solve for the value of absolute zero. $V = V_0 \times (1 + \alpha\theta)$

At absolute zero, $V = 0$, then $\theta(\text{abs.zero}) = -\dfrac{1}{\alpha}$. The value of α can be obtained from any one of the data points (except $\theta = 0$) as follows.
From $V = V_0 \times (1 + \alpha\theta)$,

$$\alpha = \frac{\left(\frac{V}{V_0} - 1\right)}{\theta} = \frac{\left(\frac{1.057}{0.7728}\right) - 1}{100°C} = 0.003678(°C)^{-1}$$

$$-\frac{1}{\alpha} = -\frac{1}{0.003678(°C)^{-1}} = \boxed{-272°C}$$

which is close to the value obtained graphically.

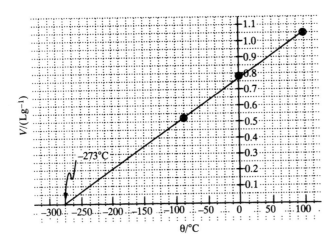

Figure 1.1

E1.16(a)   **(a)** $p = \dfrac{nRT}{V}$ [1.2]

$n = 1.0\,\text{mol}, \qquad T = 273.15\,\text{K (i)} \quad \text{or} \quad 1000\,\text{K (ii)}$

$V = 22.414\,\text{L (i)} \quad \text{or} \quad 100\,\text{cm}^3 \text{(ii)}$

(i) $\quad p = \dfrac{(1.0\,\text{mol}) \times (8.206 \times 10^{-2}\,\text{L atm K}^{-1}\,\text{mol}^{-1}) \times (273.15\,\text{K})}{22.414\,\text{L}} = \boxed{1.0\,\text{atm}}$

(ii) $\quad p = \dfrac{(1.0\,\text{mol}) \times (8.206 \times 10^{-2}\,\text{L atm K}^{-1}\,\text{mol}^{-1}) \times (1000\,\text{K})}{0.100\,\text{L}} = \boxed{8.2 \times 10^2\,\text{atm}}$

(b) $p = \dfrac{nRT}{V - nb} - \dfrac{an^2}{V^2}$ [1.25a]

From Table 1.5, $a = 5.507\,\text{L}^2\,\text{atm mol}^{-2}$ and $b = 6.51 \times 10^{-2}\,\text{L mol}^{-1}$. Therefore,

(i) $\quad \dfrac{nRT}{V - nb} = \dfrac{(1.0\,\text{mol}) \times (8.206 \times 10^{-2}\,\text{L atm K}^{-1}\,\text{mol}^{-1}) \times (273.15\,\text{K})}{[22.414 - (1.0) \times (6.51 \times 10^{-2})]\,\text{L}} = 1.00\overline{3}\,\text{atm}$

$\quad \dfrac{an^2}{V^2} = \dfrac{(5.507\,\text{L}^2\,\text{atm mol}^{-2}) \times (1.0\,\text{mol})^2}{(22.414\,\text{L})^2} = 1.1\overline{1} \times 10^{-2}\,\text{atm}$

and $p = 1.00\overline{3}\,\text{atm} - 1.1\overline{1} \times 10^{-2}\,\text{atm} = 0.992\,\text{atm} = \boxed{1.0\,\text{atm}}$

(ii) $\quad \dfrac{nRT}{V - nb} = \dfrac{(1.0\,\text{mol}) \times (8.206 \times 10^{-2}\,\text{L atm K}^{-1}\,\text{mol}^{-1}) \times (1000\,\text{K})}{(0.100 - 0.0651)\,\text{L}}$

$\qquad = 2.2\overline{7} \times 10^3\,\text{atm}$

$\quad \dfrac{an^2}{V^2} = \dfrac{(5.507\,\text{L}^2\,\text{atm mol}^{-1}) \times (1.0\,\text{mol})^2}{(0.100\,\text{L})^2} = 5.5\overline{1} \times 10^2\,\text{atm}$

and $p = 2.2\overline{7} \times 10^3\,\text{atm} - 5.5\overline{1} \times 10^2\,\text{atm} = \boxed{1.7 \times 10^3\,\text{atm}}$

Comment. It is instructive to calculate the percentage deviation from perfect gas behaviour for (i) and (ii).

(i) $\dfrac{0.99\overline{2} - 1.00\overline{0}}{1.000} \times 100\% = \overline{0.8}\%$

(ii) $\dfrac{(17 \times 10^2) - (8.2 \times 10^2)}{8.2 \times 10^2} \times 100\% = 10\overline{7}\%$

Deviations from perfect gas behaviour are not observed at $p \approx 1\,\text{atm}$ except with very precise apparatus.

E1.17(a) The three equations in [1.26] are used. The van der Waals parameters a and b and the gas constant R are substituted into the equations.

$$V_c = 3b = 3 \times (0.0226\,\text{L mol}^{-1}) = \boxed{6.78 \times 10^{-2}\,\text{L mol}^{-1}}$$

$$p_c = \frac{a}{27b^2} = \frac{0.751\,\text{L}^2\,\text{atm mol}^{-2}}{27 \times (0.0226\,\text{L mol}^{-1})^2} = \boxed{54.5\,\text{atm}}$$

$$T_c = \frac{8a}{27Rb} = \frac{8 \times (0.751\,\text{L}^2\,\text{atm mol}^{-2})}{27 \times (8.206 \times 10^{-2}\,\text{L atm K}^{-1}\,\text{mol}^{-1}) \times (0.0226\,\text{L mol}^{-1})} = \boxed{120\,\text{K}}$$

E1.18(a) The definition of Z is used $Z = \dfrac{pV_m}{RT}$ [1.20a] $= \dfrac{V_m}{V_m^{\circ}}$

V_m is the actual molar volume, V_m° is the perfect gas molar volume. $V_m^{\circ} = \dfrac{RT}{p}$. Since V_m is 12 per cent smaller than that of a perfect gas, $V_m = 0.88 V_m^{\circ}$, and

(a) $Z = \dfrac{0.88 V_m^{\circ}}{V_m^{\circ}} = \boxed{0.88}$

(b) $V_m = \dfrac{ZRT}{p} = \dfrac{(0.88) \times (8.206 \times 10^{-2}\,\text{L atm K}^{-1}\,\text{mol}^{-1}) \times (250\,\text{K})}{15\,\text{atm}} = \boxed{1.2\,\text{L}}$

Since $V_m < V_m^{\circ}$ attractive forces dominate.

E1.19(a) The amount of gas is first determined from its mass; then the van der Waals equation is used to determine its pressure at the working temperature. The initial conditions of 300 K and 100 atm are in a sense superfluous information.

$$n = \frac{92.4\,\text{kg}}{28.02 \times 10^{-3}\,\text{kg mol}^{-1}} = 3.30 \times 10^3\,\text{mol}$$

$$V = 1.000\,\text{m}^3 = 1.000 \times 10^3\,\text{L}$$

$$p = \frac{nRT}{V - nb} - \frac{an^2}{V^2}[1.25a] = \frac{(3.30 \times 10^3\,\text{mol}) \times (0.08206\,\text{L atm K}^{-1}\text{mol}^{-1}) \times (500\,\text{K})}{(1.000 \times 10^3\,\text{L}) - (3.30 \times 10^3\,\text{mol}) \times (0.0387\,\text{L mol}^{-1})}$$

$$- \frac{(1.352\,\text{L}^2\,\text{atm mol}^{-2}) \times (3.30 \times 10^3\,\text{mol})^2}{(1.000 \times 10^3\,\text{L})^2}$$

$$= (155 - 14.8)\,\text{atm} = \boxed{140\,\text{atm}}$$

E1.20(a) (a) The molar volume is obtained from

$$\rho = \frac{M}{V_m} = \frac{\text{molar mass}}{\text{molar volume}} \quad \text{or} \quad V_m = \frac{M}{\rho} = \frac{18.02\,\text{g mol}^{-1}}{133.2\,\text{g L}^{-1}} = \boxed{0.1353\,\text{L mol}^{-1}}$$

$$Z = \frac{pV_m}{RT}[1.20] = \frac{(327.6\,\text{atm}) \times (0.1353\,\text{L mol}^{-1})}{(0.08206\,\text{L atm K}^{-1}\,\text{mol}^{-1}) \times (776.4\,\text{K})} = \boxed{0.6957}$$

(b) The van der Waals equation is

$$p = \frac{RT}{V_m - b} - \frac{a}{V_m^2} \quad [1.25b]$$

Substituting this expression for p into Z [1.20] gives

$$Z = \frac{V_m}{V_m - b} - \frac{a}{V_m RT} = \frac{0.1353\,\text{L mol}^{-1}}{(0.1353\,\text{L mol}^{-1}) - (0.0305\,\text{L mol}^{-1})}$$

$$- \frac{5.464\,\text{L}^2\,\text{atm mol}^{-2}}{(0.1353\,\text{L mol}^{-1}) \times (0.08206\,\text{L atm K}^{-1}\,\text{mol}^{-1}) \times (776.4\,\text{K})}$$

$$= 1.291 - 0.633 = \boxed{0.658}$$

Comment. The difference is only about 5 per cent. Thus at this rather high pressure the van der Waals equation is still fairly accurate.

E1.21(a)   **(a)** $p = \dfrac{nRT}{V}[1.2] = \dfrac{(10.0\,\text{mol}) \times (0.08206\,\text{L atm K}^{-1}\,\text{mol}^{-1}) \times (300\,\text{K})}{4.860\,\text{L}} = \boxed{50.7\,\text{atm}}$

(b) $p = \dfrac{nRT}{V - nb} - a\left(\dfrac{n}{V}\right)^2 [1.25a]$

$$= \frac{(10.0\,\text{mol}) \times (0.08206\,\text{L atm K}^{-1}\,\text{mol}^{-1}) \times (300\,\text{K})}{(4.860\,\text{L}) - (10.0\,\text{mol}) \times (0.0651\,\text{L mol}^{-1})}$$

$$- (5.507\,\text{L}^2\text{atm mol}^{-2}) \times \left(\frac{10.0\,\text{mol}}{4.860\,\text{L}}\right)^2$$

$$= 58.4\overline{9} - 23.3\overline{2} = \boxed{35.2\,\text{atm}}$$

The compression factor is calculated from its definition [1.20] after inserting $V_m = \dfrac{V}{n}$.

To complete the calculation of Z, a value for the pressure, p, is required. The implication in the definition [1.20] is that p is the actual pressure as determined experimentally. This pressure is neither the perfect gas pressure, nor the van der Waals pressure. However, on the assumption that the van der Waals equation provides a value for the pressure close to the experimental value, we can calculate the compression factor as follows

$$Z = \frac{pV}{nRT} = \frac{(35.2\,\text{atm}) \times (4.860\,\text{L})}{(10.0\,\text{mol}) \times (0.08206\,\text{L atm K}^{-1}\,\text{mol}^{-1}) \times (300\,\text{K})} = \boxed{0.695}$$

Comment. If the perfect gas pressure had been used, Z would have been 1, the perfect gas value.

E1.22(a) $n = n(H_2) + n(N_2) = 2.0\,\text{mol} + 1.0\,\text{mol} = 3.0\,\text{mol} \qquad x_J = \dfrac{n_J}{n}\,[1.16]$

(a) $x(H_2) = \dfrac{2.0\,\text{mol}}{3.0\,\text{mol}} = \boxed{0.67} \qquad x(N_2) = \dfrac{1.0\,\text{mol}}{3.0\,\text{mol}} = \boxed{0.33}$

(b) The perfect gas law is assumed to hold for each component individually as well as for the mixture as a whole. Hence, $p_J = n_J \dfrac{RT}{V}$ [1.15]

$$\frac{RT}{V} = \frac{(8.206 \times 10^{-2}\,\text{L atm K}^{-1}\,\text{mol}^{-1}) \times (273.15\,\text{K})}{22.4\,\text{L}} = 1.00\,\text{atm mol}^{-1}$$

$$p(H_2) = (2.0\,\text{mol}) \times (1.00\,\text{atm mol}^{-1}) = \boxed{2.0\,\text{atm}}$$

$$p(N_2) = (1.0\,\text{mol}) \times (1.00\,\text{atm mol}^{-1}) = \boxed{1.0\,\text{atm}}$$

(c) $p = p(H_2) + p(N_2)[1.14] = 2.0\,\text{atm} + 1.0\,\text{atm} = \boxed{3.0\,\text{atm}}$

Question. Does Dalton's law hold for a mixture of van der Waals gases?

E1.23(a) Equations [1.26] are solved for b and a, respectively, and yield $b = \dfrac{V_c}{3}$ and $a = 27b^2 p_c = 3V_c^2 p_c$

Substituting the critical constants

$$b = \frac{1}{3} \times (98.7\,\text{cm}^3\,\text{mol}^{-1}) = \boxed{32.9\,\text{cm}^3\,\text{mol}^{-1}}$$

$$a = 3 \times (98.7 \times 10^{-3}\,\text{L mol}^{-1})^2 \times (45.6\,\text{atm}) = \boxed{1.33\,\text{L}^2\,\text{atm mol}^{-2}}$$

Note that knowledge of the critical temperature, T_c, is not required.

As b is approximately the volume occupied per mole of particles

$$v_{\text{mol}} \approx \frac{b}{N_A} = \frac{32.9 \times 10^{-6}\,\text{m}^3\,\text{mol}^{-1}}{6.022 \times 10^{23}\,\text{mol}^{-1}} = 5.46 \times 10^{-29}\,\text{m}^3$$

Then, with $v_{\text{mol}} = \dfrac{4}{3}\pi r^3$, $r \approx \left(\dfrac{3}{4\pi} \times (5.46 \times 10^{-29}\,\text{m}^3)\right)^{1/3} = \boxed{0.24\,\text{nm}}$

E1.24(a) The Boyle temperature, T_B, is the temperature at which $B = 0$. In order to express T_B in terms of a and b, the van der Waals equation must be recast into the form of the virial equation.

$$p = \frac{RT}{V_m - b} - \frac{a}{V_m^2} \quad [1.25b]$$

Factoring out $\dfrac{RT}{V_m}$ yields $p = \dfrac{RT}{V_m}\left\{\dfrac{1}{1 - b/V_m} - \dfrac{a}{RT V_m}\right\}$

So long as $b/V_m < 1$, the first term inside the brackets can be expanded using $(1 - x)^{-1} = 1 + x + x^2 + \cdots$, which gives

$$p = \frac{RT}{V_m}\left\{1 + \left(b - \frac{a}{RT}\right) \times \left(\frac{1}{V_m}\right) + \cdots\right\}$$

We can now identify the second virial coefficient as $B = b - \dfrac{a}{RT}$

Since at the Boyle temperature $B = 0$, $T_B = \dfrac{a}{bR} = \dfrac{27T_c}{8}$

(a) From Table 1.6, $a = 6.260\,\text{L}^2\,\text{atm mol}^{-2}$, $b = 5.42 \times 10^{-2}\,\text{L mol}^{-1}$. Therefore,

$$T_B = \frac{6.260\,\text{L}^2\,\text{atm mol}^{-2}}{(5.42 \times 10^{-2}\,\text{L mol}^{-1}) \times (8.206 \times 10^{-2}\,\text{L atm K}^{-1}\,\text{mol}^{-1})} = \boxed{1.41 \times 10^3\,\text{K}}$$

(b) As in Exercise 1.23(a), $v_{\text{mol}} \approx \dfrac{b}{N_A} = \dfrac{5.42 \times 10^{-5}\,\text{m}^3\,\text{mol}^{-1}}{6.022 \times 10^{23}\,\text{mol}^{-1}} = 9.00 \times 10^{-29}\,\text{m}^3$

$$r \approx \left(\frac{3}{4\pi} \times (9.00 \times 10^{-29}\,\text{m}^3)\right)^{1/3} = \boxed{0.59\,\text{nm}}$$

E1.25(a) The reduced temperature and pressure of hydrogen are calculated from the relations

$$T_r = \frac{T}{T_c} \quad \text{and} \quad p_r = \frac{p}{p_c} \quad [1.28]$$

$$T_r = \frac{298 \text{ K}}{33.23 \text{ K}} = 8.96\overline{8} \quad [T_c = 33.23 \text{ K}, \text{ Table } 1.5]$$

$$p_r = \frac{1.0 \text{ atm}}{12.8 \text{ atm}} = 0.078\overline{1} \quad [p_c = 12.8 \text{ atm}, \text{ Table } 1.5]$$

Hence, the gases named will be in corresponding states at $T = 8.96\overline{8} \times T_c$ and at $p = 0.078\overline{1} \times p_c$.

(a) For ammonia, $T_c = 405.5 \text{ K}$ and $p_c = 111.3 \text{ atm}$ (Table 1.5), so

$$T = (8.96\overline{8}) \times (405.5 \text{ K}) = \boxed{3.64 \times 10^3 \text{ K}}$$

$$p = (0.078\overline{1}) \times (111.3 \text{ atm}) = \boxed{8.7 \text{ atm}}$$

(b) For xenon, $T_c = 289.75 \text{ K}$ and $p_c = 58.0 \text{ atm}$, so

$$T = (8.96\overline{8}) \times (289.75 \text{ K}) = \boxed{2.60 \times 10^3 \text{ K}}$$

$$p = (0.078\overline{1}) \times (58.0 \text{ atm}) = \boxed{4.5 \text{ atm}}$$

(c) For helium, $T_c = 5.21 \text{ K}$ and $p_c = 2.26 \text{ atm}$, so

$$T = (8.96\overline{8}) \times (5.21 \text{ K}) = \boxed{46.7 \text{ K}}$$

$$p = (0.078\overline{1}) \times (2.26 \text{ atm}) = \boxed{0.18 \text{ atm}}$$

E1.26(a) The van der Waals equation [1.25b] is solved for b, which yields

$$b = V_m - \frac{RT}{\left(p + \frac{a}{V_m^2}\right)}$$

Substituting the data

$$b = 5.00 \times 10^{-4} \text{ m}^3 \text{ mol}^{-1} - \frac{(8.314 \text{ J K}^{-1} \text{ mol}^{-1}) \times (273 \text{ K})}{\left\{(3.0 \times 10^6 \text{ Pa}) + \left(\frac{0.50 \text{ m}^6 \text{ Pa mol}^{-2}}{(5.00 \times 10^{-4} \text{ m}^3 \text{ mol}^{-1})^2}\right)\right\}}$$

$$= \boxed{0.46 \times 10^{-4} \text{ m}^3 \text{ mol}^{-1}}$$

$$Z = \frac{pV_m}{RT} [34] = \frac{(3.0 \times 10^6 \text{ Pa}) \times (5.00 \times 10^{-4} \text{ m}^3)}{(8.314 \text{ J K}^{-1} \text{ mol}^{-1}) \times (273 \text{ K})} = \boxed{0.66}$$

Comment. The definition of Z involves the actual pressure, volume, and temperature and does not depend upon the equation of state used to relate these variables.

Solutions to problems

Solutions to numerical problems

P1.1 Boyle's law in the form $p_f V_f = p_i V_i$ is solved for V_f: $V_f = \dfrac{p_i}{p_f} \times V_i$

$$p_i = 1.0\,\text{atm}$$

$$p_f = p_{ex} + \rho g h [4] = p_i + \rho g h = 1.0\,\text{atm} + \rho g h$$

$$\rho g h = (1.025 \times 10^3 \, \text{kg m}^{-3}) \times (9.81 \, \text{m s}^{-2}) \times (50\,\text{m}) = 5.0\overline{3} \times 10^5 \, \text{Pa}$$

Hence, $p_f = (1.0\overline{1} \times 10^5 \, \text{Pa}) + (5.0\overline{3} \times 10^5 \, \text{Pa}) = 6.0\overline{4} \times 10^5 \, \text{Pa}$

$$V_f = \frac{1.0\overline{1} \times 10^5 \, \text{Pa}}{6.0\overline{4} \times 10^5 \, \text{Pa}} \times 3.0\,\text{m}^3 = \boxed{0.50\,\text{m}^3}$$

P1.3 Since the Neptunians know about perfect gas behaviour we may assume that they will write $pV = nRT$ at both temperatures. We may also assume that they will establish the size of their absolute unit to be the same as the °N, just as we write $1\,\text{K} = 1°\text{C}$. Thus

$$pV(T_1) = 28.0\,\text{L atm} = nRT_1 = nR \times (T_1 + 0°\text{N})$$

$$pV(T_2) = 40.0\,\text{L atm} = nRT_2 = nR \times (T_1 + 100°\text{N})$$

or $T_1 = \dfrac{28.0\,\text{L atm}}{nR}$, $T_1 + 100°\text{N} = \dfrac{40.0\,\text{L atm}}{nR}$

Dividing, $\dfrac{T_1 + 100°\text{N}}{T_1} = \dfrac{40.0\,\text{L atm}}{28.0\,\text{L atm}} = 1.42\overline{9}$ or $T_1 + 100°\text{N} = 1.42\overline{9}T_1$, $T_1 = 233$ absolute units

As in the relationship between our Kelvin scale and Celsius scale $T = \theta -$ absolute zero (°N) so absolute zero (°N) = $\boxed{-233°\text{N}}$

Comment. To facilitate communication with Earth students we have converted the Neptunians' units of the pV product to units familiar to humans, that is L atm. However, we see from the solution that only the ratio of pV products is required, and that will be the same in any civilization.

Question. If the Neptunians' unit of volume is the lagoon (L), their unit of pressure is the poseidon (P), their unit of amount is the nereid (n), and their unit of absolute temperature is the titan (T), what is the value of the Neptunians' gas constant (R) in units of L, P, n, and T?

P1.5 Solving for n from the perfect gas equation [1.12] yields $n = \dfrac{pV}{RT}$ and $n = \dfrac{m}{M}$, hence $\rho = \dfrac{m}{V} = \dfrac{Mp}{RT}$

Rearrangement yields the desired relation, that is $\boxed{p = \rho\dfrac{RT}{M}}$, or $\dfrac{p}{\rho} = \dfrac{RT}{M}$, and $M = \dfrac{RT}{p/\rho}$

Draw up the following table and then plot $\dfrac{p}{\rho}$ versus p to find the zero pressure limit of $\dfrac{p}{\rho}$ where all gases behave ideally.

$$\rho/(\text{g L}^{-1}) = \rho/(\text{kg m}^{-3});$$

$$1\,\text{Torr} = (1\,\text{Torr}) \times \left(\frac{1\,\text{atm}}{760\,\text{Torr}}\right) \times \left(\frac{1.013 \times 10^5 \, \text{Pa}}{1\,\text{atm}}\right) = 133.3\,\text{Pa}$$

p/Torr	91.74	188.98	277.3	452.8	639.3	760.0
$p/(10^4\,\text{Pa})$	1.223	2.519	3.696	6.036	8.522	10.132
$\rho/(\text{kg m}^{-3})$	0.225	0.456	0.664	1.062	1.468	1.734
$\left(\dfrac{p}{\rho}\right)/(10^4\,\text{m}^2\,\text{s}^{-2})$	5.44	5.52	5.56	5.68	5.81	5.84

$\dfrac{p}{\rho}$ is plotted in Fig. 1.2. A straight line fits the data rather well. The extrapolation to $p = 0$ yields an intercept of $5.40 \times 10^4\,\text{m}^2\,\text{s}^{-2}$. Then

$$M = \frac{RT}{5.40 \times 10^4\,\text{m}^2\,\text{s}^{-2}} = \frac{(8.314\,\text{J K}^{-1}\,\text{mol}^{-1}) \times (298.15\,\text{K})}{5.40 \times 10^4\,\text{m}^2\,\text{s}^{-2}}$$

$$= 0.0459\,\text{kg mol}^{-1} = \boxed{45.9\,\text{g mol}^{-1}}$$

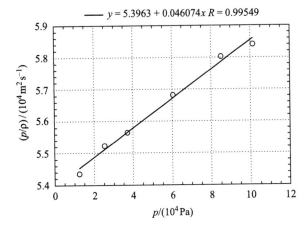

$y = 5.3963 + 0.046074x \quad R = 0.99549$

Figure 1.2

Comment. This method of the determination of the molar masses of gaseous compounds is due to Cannizarro who presented it at the Karlsruhe conference of 1860 which had been called to resolve the problem of the determination of the molar masses of atoms and molecules and the molecular formulas of compounds.

P1.8 $\dfrac{p}{T} = \dfrac{nR}{V} = \text{constant}$, if n and V are constant. Hence, $\dfrac{p}{T} = \dfrac{p_3}{T_3}$, where p is the measured pressure at temperature, T, and p_3 and T_3 are the triple point pressure and temperature, respectively. Rearranging, $p = \left(\dfrac{p_3}{T_3}\right) T$.

The ratio $\dfrac{p_3}{T_3}$ is a constant $= \dfrac{50.2\,\text{Torr}}{273.16\,\text{K}} = 0.183\overline{8}\,\text{Torr K}^{-1}$. Thus the change in p, Δp, is proportional to the change in temperature, ΔT. $\Delta p = (0.183\overline{8}\,\text{Torr K}^{-1}) \times (\Delta T)$

(a) $\Delta p = (0.183\overline{8}\,\text{Torr K}^{-1}) \times (1\,\text{K}) = \boxed{0.184\,\text{Torr}}$

(b) Rearranging, $p = \left(\dfrac{T}{T_3}\right) p_3 = \left(\dfrac{373.16\,\text{K}}{273.16\,\text{K}}\right) \times (50.2\,\text{Torr}) = \boxed{68.6\,\text{Torr}}$

(c) Since $\dfrac{p}{T}$ is a constant at constant n and V, it always has the value $0.183\overline{8}$ Torr K^{-1}, hence

$$\Delta p = p_{374.15\,K} - p_{373.15\,K} = (0.183\overline{8}\,\text{Torr}\,K^{-1}) \times (1\,K) = \boxed{0.184\,\text{Torr}}$$

P1.11 From definition of Z [1.20] and the virial equation [1.22], Z may be expressed in virial form as

$$Z = 1 + B\left(\frac{1}{V_m}\right) + C\left(\frac{1}{V_m}\right)^2 + \cdots$$

Since $V_m = \dfrac{RT}{p}$ [assumption of perfect gas], $\dfrac{1}{V_m} = \dfrac{p}{RT}$; hence upon substitution, and dropping terms beyond the second power of $\left(\dfrac{1}{V_m}\right)$

$$Z = 1 + B\left(\frac{p}{RT}\right) + C\left(\frac{p}{RT}\right)^2$$

$$Z = 1 + (-21.7 \times 10^{-3}\,\text{L}\,\text{mol}^{-1}) \times \left(\frac{100\,\text{atm}}{(0.0821\,\text{L}\,\text{atm}\,K^{-1}\text{mol}^{-1}) \times (273\,K)}\right)$$

$$+ (1.200 \times 10^{-3}\,\text{L}^2\,\text{mol}^{-2}) \times \left(\frac{100\,\text{atm}}{(0.0821\,\text{L}\,\text{atm}\,K^{-1}\,\text{mol}^{-1}) \times (273\,K)}\right)^2$$

$$Z = 1 - (0.0968) + (0.0239) = \boxed{0.927}$$

$$V_m = (0.927) \times \left(\frac{RT}{p}\right) = (0.927) \times \left(\frac{(0.0821\,\text{L}\,\text{atm}\,K^{-1}\,\text{mol}^{-1}) \times (273\,K)}{100\,\text{atm}}\right) = \boxed{0.208\,\text{L}}$$

Question. What is the value of Z obtained from the next approximation using the value of V_m just calculated? Which value of Z is likely to be more accurate?

P1.12 As indicated by eqns 1.21 and 1.22 the compression factor of a gas may be expressed as either a virial expansion in p or in $\left(\dfrac{1}{V_m}\right)$. The virial form of the van der Waals equation is derived in Exercise 1.24(a) and is $p = \dfrac{RT}{V_m}\left\{1 + \left(b - \dfrac{a}{RT}\right) \times \left(\dfrac{1}{V_m}\right) + \cdots\right\}$

Rearranging, $Z = \dfrac{pV_m}{RT} = 1 + \left(b - \dfrac{a}{RT}\right) \times \left(\dfrac{1}{V_m}\right) + \cdots$

On the assumption that the perfect gas expression for V_m is adequate for the second term in this expansion, we can readily obtain Z as a function of p.

$$Z = 1 + \left(\frac{1}{RT}\right) \times \left(b - \frac{a}{RT}\right) p + \cdots$$

(a) $T_c = 126.3\,K$

$$V_m = \left(\frac{RT}{p}\right) \times Z = \frac{RT}{p} + \left(b - \frac{a}{RT}\right) + \cdots$$

$$= \frac{(0.08206\,\text{L}\,\text{atm}\,K^{-1}\,\text{mol}^{-1}) \times (126.3\,K)}{10.0\,\text{atm}}$$

$$+ \left\{(0.0387\,\text{L}\,\text{mol}^{-1}) - \left(\frac{1.352\,\text{L}^2\,\text{atm}\,\text{mol}^{-2}}{(0.08206\,\text{L}\,\text{atm}\,K^{-1}\text{mol}^{-1}) \times (126.3\,K)}\right)\right\}$$

$$= (1.036 - 0.092)\,\text{L}\,\text{mol}^{-1} = \boxed{0.944\,\text{L}\,\text{mol}^{-1}}$$

$$Z = \left(\frac{p}{RT}\right) \times (V_{\text{m}}) = \frac{(10.0\,\text{atm}) \times (0.944\,\text{L mol}^{-1})}{(0.08206\,\text{L atm K}^{-1}\,\text{mol}^{-1}) \times (126.3\,\text{K})} = 0.911$$

(b) The Boyle temperature corresponds to the temperature at which the second virial coefficient is zero, hence correct to the first power in p, $Z = 1$, and the gas is close to perfect. However, if we assume that N_2 is a van der Waals gas, when the second virial coefficient is zero

$$\left(b - \frac{a}{RT_{\text{B}}}\right) = 0, \quad \text{or} \quad T_{\text{B}} = \frac{a}{bR}$$

$$T_{\text{B}} = \frac{1.352\,\text{L}^2\,\text{atm mol}^{-2}}{(0.0387\,\text{L mol}^{-1}) \times (0.08206\,\text{L atm K}^{-1}\,\text{mol}^{-1})} = 426\,\text{K}$$

The experimental value (Table 1.5) is 327.2 K. The discrepancy may be explained by two considerations.

1. Terms beyond the first power in p should not be dropped in the expansion for Z.
2. Nitrogen is only approximately a van der Waals gas.

When $Z = 1$, $V_{\text{m}} = \dfrac{RT}{p}$, and using $T_{\text{B}} = 327.2\,\text{K}$

$$= \frac{(0.08206\,\text{L atm K}^{-1}\text{mol}^{-1}) \times 327.2\,\text{K}}{10.0\,\text{atm}}$$

$$= \boxed{2.69\,\text{L mol}^{-1}}$$

and this is the ideal value of V_{m}. Using the experimental value of T_{B} and inserting this value into the expansion for V_{m} above we have

$$V_{\text{m}} = \frac{0.08206\,\text{L atm K}^{-1}\,\text{mol}^{-1} \times 327.2\,\text{K}}{10.0\,\text{atm}}$$

$$+ \left\{0.0387\,\text{L mol}^{-1} - \left(\frac{1.352\,\text{L}^2\text{atm mol}^{-2}}{0.08206\,\text{L atm K}^{-1}\,\text{mol}^{-1} \times 327.2\,\text{K}}\right)\right\}$$

$$= (2.68\overline{5} - 0.012)\,\text{L mol}^{-1} = \boxed{2.67\,\text{L mol}^{-1}}$$

and $Z = \dfrac{V_{\text{m}}}{V_{\text{m}}^{\circ}} = \dfrac{2.67\,\text{L mol}^{-1}}{2.69\,\text{L mol}^{-1}} = 0.992 \approx 1$

(c) $T_{\text{I}} = 621\,\text{K}$ [Table 3.2]

$$V_{\text{m}} = \frac{0.08206\,\text{L atm K}^{-1}\,\text{mol}^{-1} \times 621\,\text{K}}{10.0\,\text{atm}}$$

$$+ \left\{0.0387\,\text{L mol}^{-1} - \left(\frac{1.352\,\text{L}^2\text{atm mol}^{-2}}{0.08206\,\text{L atm K}^{-1}\,\text{mol}^{-1} \times 621\,\text{K}}\right)\right\}$$

$$= (5.09\overline{6} + 0.012)\,\text{L mol}^{-1} = \boxed{5.11\,\text{L mol}^{-1}}$$

and $Z = \dfrac{5.11\,\text{L mol}^{-1}}{5.10\,\text{L mol}^{-1}} = 1.002 \approx 1$

Based on the values of T_B amd T_I given in Tables 1.5 and 3.2 and assuming that N_2 is a van der Waals gas, the calculated value of Z is closest to 1 at T_I, but the difference from the value at T_B is less than the accuracy of the method.

P1.14   (a) $$V_m = \frac{\text{molar mass}}{\text{density}} = \frac{M}{\rho} = \frac{18.02 \text{ g mol}^{-1}}{1.332 \times 10^2 \text{ g L}^{-1}} = \boxed{0.1353 \text{ L mol}^{-1}}$$

(b) $$Z = \frac{pV_m}{RT}[1.20b] = \frac{(327.6 \text{ atm}) \times (0.1353 \text{ L mol}^{-1})}{(0.08206 \text{ L atm K}^{-1} \text{ mol}^{-1}) \times (776.4 \text{ K})} = \boxed{0.6957}$$

(c) Two expansions for Z based on the van der Waals equation are given in Problem 1.12. They are

$$Z = 1 + \left(b - \frac{a}{RT}\right) \times \left(\frac{1}{V_m}\right) + \cdots$$

$$= 1 + \left\{(0.0305 \text{ L mol}^{-1}) - \left(\frac{5.464 \text{ L}^2 \text{ atm mol}^{-2}}{(0.08206 \text{ L atm K}^{-1} \text{ mol}^{-1}) \times (776.4 \text{ K})}\right)\right\}$$

$$\times \frac{1}{0.1353 \text{ L mol}^{-1}} = 1 - 0.4084 = 0.5916 \approx 0.59$$

$$Z = 1 + \left(\frac{1}{RT}\right) \times \left(b - \frac{a}{RT}\right) \times (p) + \cdots$$

$$= 1 + \frac{1}{(0.08206 \text{ L atm K}^{-1} \text{ mol}^{-1}) \times (776.4 \text{ K})}$$

$$\times \left\{(0.0305 \text{ L mol}^{-1}) - \left(\frac{5.464 \text{ L}^2 \text{ atm mol}^{-2}}{(0.08206 \text{ L atm K}^{-1} \text{ mol}^{-1}) \times (776.4 \text{ K})}\right)\right\} \times 327.6 \text{ atm}$$

$$= 1 - 0.2842 \approx \boxed{0.72}$$

In this case the expansion in p gives a value close to the experimental value; the expansion in $\frac{1}{V_m}$ is not as good. However, when terms beyond the second are included the results from the two expansions for Z converge.

Solutions to theoretical problems

P1.17 $$Z = \frac{pV_m}{RT} = \frac{1}{\left(1 - \frac{b}{V_m}\right)} - \frac{a}{RTV_m} \quad \text{[see Exercise 1.24(a).]}$$

which upon expansion of $\left(1 - \frac{b}{V_m}\right)^{-1} = 1 + \frac{b}{V_m} + \left(\frac{b}{V_m}\right)^2 + \cdots$ yields

$$Z = 1 + \left(b - \frac{a}{RT}\right) \times \left(\frac{1}{V_m}\right) + b^2\left(\frac{1}{V_m}\right)^2 + \cdots$$

We note that all terms beyond the second are necessarily positive, so only if

$$\frac{a}{RTV_m} > \frac{b}{V_m} + \left(\frac{b}{V_m}\right)^2 + \cdots$$

can Z be less than one. If we ignore terms beyond $\frac{b}{V_m}$, the conditions are simply stated as

$$Z < 1 \quad \text{when} \quad \frac{a}{RT} > b \qquad Z > 1 \quad \text{when} \quad \frac{a}{RT} < b$$

Thus $Z < 1$ when attractive forces predominate, and $Z > 1$ when size effects (short-range repulsions) predominate.

P1.19 The Dieterici equation of state is listed in Table 1.6. At the critical point the derivatives of p w/r/t V_m equal zero along the isotherm for which $T = T_c$. This means that $(\partial p/\partial V_m)_T = 0$ and $(\partial^2 p/\partial V_m^2)_T = 0$ at the critical point.

$$p = \frac{RT\,e^{-\frac{a}{RTV_m}}}{V_m - b} \qquad \left(\frac{\partial p}{\partial V_m}\right)_T = p\left\{\frac{aV_m - ab - RTV_m^2}{V_m^2(V_m - b)(RT)}\right\}$$

$$\left(\frac{\partial^2 p}{\partial V_m^2}\right)_T = \left(\frac{\partial p}{\partial V_m}\right)_T \left\{\frac{aV_m - ab - RTV_m^2}{V_m^2(V_m - b)(RT)}\right\}$$

$$+ p\frac{(-2aV_m^2 + 4V_m ab + RTV_m^3 - 2ab^2)}{\left[V_m^3\left[(V_m - b)^2(RT)\right]\right]}$$

Each of these equations is evaluated at the critical point giving the three equations:

$$p_c = \frac{RT_c\,e^{-\frac{a}{RT_c V_c}}}{V_c - b} \qquad aV_c - ab - RT_c V_c^2 = 0$$

$$-2aV_c^2 + 4V_c ab + RT_c V_c^3 - 2ab^2 = 0$$

Solving the middle equation for T_c, substitution of the result into the last equation, and solving for V_c yields the result: $V_c = 2b$ or $b = V_c/2$ (The solution $V_c = b$ is rejected because there is a singularity in the Dieterici equation at the point $V_m = b$.) Substitution of $V_c = 2b$ into the middle equation and solving for T_c gives the result: $T_c = a/4bR$ or $a = 2RT_c V_c$. Substitution of $V_c = 2b$ and $T_c = a/4bR$ into the first equation gives:

$$p_c = \frac{1}{4}\left(\frac{a}{b^2}\right)e^{-2}$$

The equations for V_c, T_c, p_c are substituted into the equation for the critical compression factor (eqn 1.20) to give: $Z_c = p_c V_c/RT_c = 2e^{-2} = 0.2707$. This is significantly lower than the critical compression factor that is predicted by the van der Waals equation (eqn 1.27): $Z_c(\text{vdW}) = p_c V_c/RT_c = 3/8 = 0.3750$. Experimental values for Z_c are summarized in Table 1.4 where it is seen that the Dieterici equation prediction if often better.

P1.20 The critical point corresponds to a point of zero slope which is simultaneously a point of inflection in a plot of pressure versus molar volume. A critical point exists if there are values of p, V, and T which result in a point which satisfies these conditions.

$$p = \frac{RT}{V_m} - \frac{B}{V_m^2} + \frac{C}{V_m^3}$$

$$\left.\begin{aligned}
\left(\frac{\partial p}{\partial V_m}\right)_T &= -\frac{RT}{V_m^2} + \frac{2B}{V_m^3} - \frac{3C}{V_m^4} = 0 \\[2mm]
\left(\frac{\partial^2 p}{\partial V_m^2}\right)_T &= \frac{2RT}{V_m^3} - \frac{6B}{V_m^4} + \frac{12C}{V_m^5} = 0
\end{aligned}\right\} \text{ at the critical point}$$

That is, $\left.\begin{aligned} -RT_c V_c^2 + 2BV_c - 3C &= 0 \\ RT_c V_c^2 - 3BV_c + 6C &= 0 \end{aligned}\right\}$

which solve to $V_c = \boxed{\dfrac{3C}{B}}$, $\boxed{T_c = \dfrac{B^2}{3RC}}$

Now use the equation of state to find p_c

$$p_c = \frac{RT_c}{V_c} - \frac{B}{V_c^2} + \frac{C}{V_c^3} = \left(\frac{RB^2}{3RC}\right) \times \left(\frac{B}{3C}\right) - B\left(\frac{B}{3C}\right)^2 + C\left(\frac{B}{3C}\right)^3 = \boxed{\frac{B^3}{27C^2}}$$

It follows that $Z_c = \dfrac{p_c V_c}{RT_c} = \left(\dfrac{B^3}{27C^2}\right) \times \left(\dfrac{3C}{B}\right) \times \left(\dfrac{1}{R}\right) \times \left(\dfrac{3RC}{B^2}\right) = \boxed{\dfrac{1}{3}}$

P1.21 $\dfrac{pV_m}{RT} = 1 + B'p + C'p^2 + \cdots$ [1.21]

$\dfrac{pV_m}{RT} = 1 + \dfrac{B}{V_m} + \dfrac{C}{V_m^2} + \cdots$ [1.22]

whence $B'p + C'p^2 + \cdots = \dfrac{B}{V_m} + \dfrac{C}{V_m^2} + \cdots$

Now multiply through by V_m, replace pV_m by $RT\{1 + (B/V_m) + \cdots\}$, and equate coefficients of

powers of $\dfrac{1}{V_m}$: $B'RT + \dfrac{BB'RT + C'R^2T^2}{V_m} + \cdots = B + \dfrac{C}{V_m} + \cdots$

Hence, $B'RT = B$, implying that $\boxed{B' = \dfrac{B}{RT}}$

Also, $BB'RT + C'R^2T^2 = C$, or $B^2 + CR^2T^2 = C$, implying that $\boxed{C' = \dfrac{C - B^2}{R^2T^2}}$

P1.24   The critical temperature is that temperature above which the gas cannot be liquefied by the application of pressure alone. Below the critical temperature two phases, liquid and gas, may coexist at equilibrium; and in the two-phase region there is more than one molar volume corresponding to the same conditions of temperature and pressure. Therefore, any equation of state that can even approximately describe this situation must allow for more than one real root for the molar volume at some values of T and p, but as the temperature is increased above T_c, allows only one real root. Thus, appropriate equations of state must be equations of odd degree in V_m.

The equation of state for gas A may be rewritten $V_m^2 - \dfrac{RT}{p}V_m - \dfrac{RTb}{p} = 0$, which is a quadratic and never has just one real root. Thus, this equation can never model critical behavior. It could possibly model in a very crude manner a two-phase situation, since there are some conditions under which a quadratic has two real positive roots, but not the process of liquefaction.

The equation of state of gas B is a first-degree equation in V_m and therefore can never model critical behavior, the process of liquefaction, or the existence of a two-phase region.

A cubic equation is the equation of lowest degree which can show a cross-over from more than one real root to just one real root as the temperature increases. The van der Waals equation is a cubic equation in V_m.

Solutions to applications

P1.26   Avogadro's principle states that equal volumes of gas contain equal numbers of molecules. Consequently, the ratio of the masses of equal volumes of two gases is the ratio of their molecular masses and hence their molar masses. Thus, the density of a gas (mass per unit volume) is proportional to its molar mass (mass per mole), and ratios of densities are equal to ratios of molar masses. So the molar mass of water relative to that of hydrogen is

$$\frac{0.625}{0.0732} = \boxed{8.54}$$

For oxygen, we need the reaction stoichiometry: the mass of a unit volume of water vapour consists of the mass of a unit volume of hydrogen plus the mass of half a unit volume of oxygen. Thus, the density of oxygen would be expected to be $2 \times (0.625 - 0.0732)$, and its molar mass relative to hydrogen is

$$\frac{2 \times (0.625 - 0.0732)}{0.0732} = \boxed{15.1}$$

Comment. The differences from the modern values are not unexpected.

P1.28   The virial equation is

$$pV_m = RT \left(1 + \frac{B}{V_m} + \frac{C}{V_m^2} + \cdots \right) \quad \text{or}$$

$$\frac{pV_m}{RT} = 1 + \frac{B}{V_m} + \frac{C}{V_m^2} + \cdots$$

(a) If we assume that the series may be truncated after the B term, then a plot of $\dfrac{pV_m}{RT}$ vs $\dfrac{1}{V_m}$ will have B as its slope and 1 as its y-intercept. Transforming the data gives

p/MPa	V_m/L (mol^{-1})	pV_m/RT	$(1/V_m)/(\text{mol L}^{-1})$
0.4000	6.2208	0.9976	0.1608
0.5000	4.9736	0.9970	0.2011
0.6000	4.1423	0.9964	0.2414
0.8000	3.1031	0.9952	0.3223
1.000	2.4795	0.9941	0.4033
1.500	1.6483	0.9912	0.6067
2.000	1.2328	0.9885	0.8112
2.500	0.98357	0.9858	1.017
3.000	0.81746	0.9832	1.223
4.000	0.60998	0.9782	1.639

A plot of the data in the third column against that of the fourth column is shown in Fig. 1.3. The data fit a straight line reasonably well, and the y-intercept is very close to 1. The regression yields $B = \boxed{-1.32 \times 10^{-2}\, \text{L mol}^{-1}}$

(b) A quadratic function fits the data somewhat better (Fig. 1.4) with a slightly better correlation coefficient and a y-intercept closer to 1. This fit implies that truncation of the virial series after the term with C is more accurate than after just the B term. The regression then yields

$$B = \boxed{-1.51 \times 10^{-2}\, \text{L mol}^{-1}} \quad \text{and} \quad C = \boxed{1.07 \times 10^{-3}\, \text{L}^2\, \text{mol}^{-2}}$$

P1.30 $n = \dfrac{pV}{RT}$ [1.2], $V = \dfrac{4\pi}{3} r^3 = \dfrac{4\pi}{3} \times (3.0\,\text{m})^3 = 11\overline{3}\,\text{m}^3 = $ volume of balloon

$p = 1.0\,\text{atm},$ $T = 298\,\text{K}$

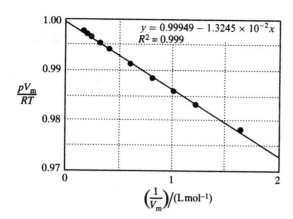

$$y = 0.99949 - 1.3245 \times 10^{-2}x$$
$$R^2 = 0.999$$

Figure 1.3

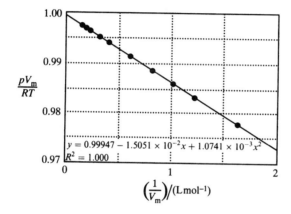

$$y = 0.99947 - 1.5051 \times 10^{-2}x + 1.0741 \times 10^{-3}x^2$$
$$R^2 = 1.000$$

Figure 1.4

(a) $n = \dfrac{(1.0\,\text{atm}) \times (11\overline{3} \times 10^3\,\text{L})}{(8.206 \times 10^{-2}\,\text{L atm K}^{-1}\,\text{mol}^{-1}) \times (298\,\text{K})} = \boxed{4.6\overline{2} \times 10^3\,\text{mol}}$

(b) The mass that the balloon can lift is the difference between the mass of displaced air and the mass of the balloon. We assume that the mass of the balloon is essentially that of the gas it encloses.

Then $m(\text{H}_2) = nM(\text{H}_2) = (4.6\overline{2} \times 10^3\,\text{mol}) \times (2.02\,\text{g mol}^{-1}) = 9.3\overline{3} \times 10^3\,\text{g}$

Mass of displaced air $= (11\overline{3}\,\text{m}^3) \times (1.22\,\text{kg m}^{-3}) = 1.3\overline{8} \times 10^2\,\text{kg}$

Therefore, the payload is $(13\overline{8}\,\text{kg}) - (9.3\overline{3}\,\text{kg}) = \boxed{1.3 \times 10^2\,\text{kg}}$

(c) For helium, $m = nM(\text{He}) = (4.6\overline{2} \times 10^3\,\text{mol}) \times (4.00\,\text{g mol}^{-1}) = 18\,\text{kg}$

The payload is now $13\overline{8}\,\text{kg} - 18\,\text{kg} = \boxed{1.2 \times 10^2\,\text{kg}}$

P1.32 Avogadro's principle states that equal volumes of gases represent equal amounts (moles) of the gases, so the volume mixing ratio is equal to the mole fraction. The definition of partial pressures is

$$p_\text{J} = x_\text{J} p$$

The perfect gas law is

$$pV = nRT \quad \text{so} \quad \frac{n_\text{J}}{V} = \frac{p_\text{J}}{RT} = \frac{x_\text{J} p}{RT}$$

(a) $\dfrac{n(CCl_3F)}{V} = \dfrac{(261 \times 10^{-12}) \times (1.0\,\text{atm})}{(0.08206\,\text{L atm K}^{-1}\text{mol}^{-1}) \times (10 + 273)\,\text{K}} = \boxed{1.1 \times 10^{-11}\,\text{mol L}^{-1}}$

and $\dfrac{n(CCl_2F_2)}{V} = \dfrac{(509 \times 10^{-12}) \times (1.0\,\text{atm})}{(0.08206\,\text{L atm K}^{-1}\text{mol}^{-1}) \times (10 + 273)\,\text{K}} = \boxed{2.2 \times 10^{-11}\,\text{mol L}^{-1}}$

(b) $\dfrac{n(CCl_3F)}{V} = \dfrac{(261 \times 10^{-12}) \times (0.050\,\text{atm})}{(0.08206\,\text{L atm K}^{-1}\text{mol}^{-1}) \times (200\,\text{K})} = \boxed{8.0 \times 10^{-13}\,\text{mol L}^{-1}}$

and $\dfrac{n(CCl_2F_2)}{V} = \dfrac{(509 \times 10^{-12}) \times (0.050\,\text{atm})}{(0.08206\,\text{L atm K}^{-1}\text{mol}^{-1}) \times (200\,\text{K})} = \boxed{1.6 \times 10^{-12}\,\text{mol L}^{-1}}$

P1.33 We want the height h at which $p(O_2) = \dfrac{1}{9}p(N_2)$. According to the barometric equation

$$p(O_2) = p_0(O_2)e^{-\left(\frac{M_{O_2} gh}{RT}\right)} \quad \text{where } p_0(O_2) = 0.20\,\text{bar}$$

and

$$p(N_2) = p_0(N_2)e^{-\left(\frac{M_{N_2} gh}{RT}\right)} \quad \text{where } p_0(N_2) = 0.80\,\text{bar}$$

$$p(O_2) = \frac{1}{9}p(N_2)$$

$$p_0(O_2)e^{-\left(\frac{M_{O_2} gh}{RT}\right)} = \frac{1}{9}p_0(N_2)e^{-\left(\frac{M_{N_2} gh}{RT}\right)}$$

$$e^{\left(\frac{(M_{N_2} - M_{O_2})gh}{RT}\right)} = \frac{p_0(N_2)}{9 p_0(O_2)}$$

Solving for h,

$$h = \frac{RT}{(M_{N_2} - M_{O_2})g} \ln\left(\frac{p_0(N_2)}{9 p_0(O_2)}\right)$$

$$= \frac{(8.31447\,\text{J K}^{-1}\,\text{mol}^{-1}) \times (298.15\,\text{K}) \times \ln\left(\frac{0.80\,\text{bar}}{9(0.20\,\text{bar})}\right)}{(0.02802\,\text{kg mol}^{-1} - 0.03200\,\text{kg mol}^{-1}) \times (9.80665\,\text{m s}^{-2})}$$

$$= 5.15 \times 10^4\,\text{m} = \boxed{51.5\,\text{km}}$$

The pressure at 51.5 km can be obtained by using the barometric equation twice, once for O_2 and again for N_2

$$p(O_2) = 0.20 \times e^{-\left(\frac{0.03200\,\text{kg mol}^{-1} \times 9.8067\,\text{m s}^{-2} \times 5.15 \times 10^4\,\text{m}}{8.314\,\text{J K}^{-1}\,\text{mol}^{-1} \times 298\,\text{K}}\right)}$$

$$= 2.9 \times 10^{-4}\,\text{bar}$$

$$p(N_2) = 0.80 \times e^{-\left(\frac{0.02802\,\text{kg mol}^{-1} \times 9.8067\,\text{m s}^{-2} \times 5.15 \times 10^4\,\text{m}}{8.314\,\text{J K}^{-1}\,\text{mol}^{-1} \times 298\,\text{K}}\right)}$$

$$= 2.66 \times 10^{-3}\,\text{bar}$$

$$p = p(O_2) + p(N_2) = \boxed{3.0 \times 10^{-3}\,\text{bar}}$$

2 The First Law: the concepts

Solutions to exercises

Discussion questions

E2.1(a) Work is produced from the application of a force through a distance. The technical definition is based on the realization that both force and displacement are vector quantities and it is the component of the force acting in the direction of the displacement that is used in the calculation of the amount of work, that is, work is the scalar product of the two vectors. In vector notation $w = -\mathbf{f} \cdot \mathbf{d} = -fd\cos\theta$, where θ is the angle between the force and the displacement. The negative sign is inserted to conform to the standard thermodynamic convention.

Heat is associated with a non-adiabatic process and is defined as the difference between the adiabatic work and the non-adiabatic work associated with the same change in state of the system. This is the formal (and best) definition of heat and is based on the definition of work. Work is a precisely defined mechanical concept. A less precise definition of heat is the statement that heat is the form of energy that is transferred between bodies in thermal contact with each other by virtue of a difference in temperature.

E2.2(a) In an expansion against constant pressure the internal pressure of the system is greater than the constant opposing pressure by a measurable non-infinitesimal amount. In a reversible expansion the internal pressure is only infinitesimally greater than the opposing pressure. Since the opposing pressure is greater in the reversible case, and in fact is at a maximum, the work generated in the reversible case is the maximum work.

E2.3(a) The difference results from the definition $H = U + PV$; hence $\Delta H = \Delta U + \Delta(PV)$. As $\Delta(PV)$ is not usually zero, except for isothermal processes in a perfect gas, the difference between ΔH and ΔU is a non-zero quantity. As shown in sections 2.4 and 2.5 of the text, ΔH can be interpreted as the heat associated with a process at constant pressure, and ΔU as the heat at constant volume.

Computer aided molecular modeling is now the method of choice for estimating standard reaction enthalpies, especially for large molecules with complex three-dimensional structures, but accurate numerical values are still difficult to obtain.

Numerical exercises

Assume that all gases are perfect unless stated otherwise. Note that $1\,\text{atm} = 1.01325\,\text{bar}$ exactly. Unless otherwise stated, thermochemical data are for 298.15 K.

E2.4(a) The physical definition of work is $dw = -F\,dz$ [2.6]

In a gravitational field the force is the weight of the object, which is $F = mg$

If g is constant over the distance the mass moves, dw may be intergrated to give the total work

$$w = -\int_{z_i}^{z_f} F\,dz = -\int_{z_i}^{z_f} mg\,dz = -mg(z_f - z_i) = -mgh \quad \text{where } h = (z_f - z_i)$$

(a) $w = (-1.0\,\text{kg}) \times (9.81\,\text{m s}^{-2}) \times (10\,\text{m}) = -98\,\text{J} = \boxed{98\,\text{J done}}$

(b) $w = (-1.0\,\text{kg}) \times (1.60\,\text{m s}^{-2}) \times (10\,\text{m}) = -16\,\text{J} = \boxed{16\,\text{J done}}$

E2.5(a) $w = -mgh$ [Exercise 2.4 **(a)**] $= (-65\,\text{kg}) \times (9.81\,\text{m s}^{-2}) \times (4.0\,\text{m})$
$= 2.6\,\text{kJ} = \boxed{2.6\,\text{kJ needed}}$

E2.6(a) This is an expansion against a constant external pressure; hence $w = -p_{ex}\Delta V$ [2.10]

$$p_{ex} = (1.0\,\text{atm}) \times (1.013 \times 10^5\,\text{Pa atm}^{-1}) = 1.0\bar{1} \times 10^5\,\text{Pa}$$

$$\Delta V = (100\,\text{cm}^2) \times (10\,\text{cm}) = 1.0 \times 10^3\,\text{cm}^3 = 1.0 \times 10^{-3}\,\text{m}^3$$

$$w = (-1.0\bar{1} \times 10^5\,\text{Pa}) \times (1.0 \times 10^{-3}\,\text{m}^3) = \boxed{-1.0 \times 10^2\,\text{J}}\;\text{as}\;1\,\text{Pa m}^3 = 1\,\text{J}$$

E2.7(a) For all cases $\Delta U = 0$, since the internal energy of a perfect gas depends only on temperature. (See *Molecular interpretation* 2.2 and Section 3.2 for a more complete discussion.) From the definition of enthalpy, $H = U + pV$, $\Delta H = \Delta U + \Delta(pV) = \Delta U + \Delta(nRT)$ (perfect gas). Hence, $\Delta H = 0$ as well, at constant temperature for all processes in a perfect gas.

(a) $\boxed{\Delta U = \Delta H = 0}$

$$w = -nRT\,\ln\left(\frac{V_f}{V_i}\right)\;[2.13]$$

$$= (-1.00\,\text{mol}) \times (8.314\,\text{J K}^{-1}\,\text{mol}^{-1}) \times (273\,\text{K}) \times \ln\left(\frac{44.8\,\text{L}}{22.4\,\text{L}}\right)$$

$$= -1.57 \times 10^3\,\text{J} = \boxed{-1.57\,\text{kJ}}$$

$$q = \Delta U - w\,(\text{First Law}) = 0 + 1.57\,\text{kJ} = \boxed{+1.57\,\text{kJ}}$$

(b) $\boxed{\Delta U = \Delta H = 0}$

$$w = -p_{ex}\Delta V\;[2.10] \quad \Delta V = (44.8 - 22.4)\,\text{L} = 22.4\,\text{L}$$

$$p_{ex} = p_f = \frac{nRT}{V_f} = \frac{(1.00\,\text{mol}) \times (0.08206\,\text{L atm K}^{-1}\,\text{mol}^{-1}) \times (273\,\text{K})}{44.8\,\text{L}} = 0.500\,\text{atm}$$

$$w = (-0.500\,\text{atm}) \times \left(\frac{1.013 \times 10^5\,\text{Pa}}{1\,\text{atm}}\right) \times (22.4\,\text{L}) \times \left(\frac{1\,\text{m}^3}{10^3\,\text{L}}\right)$$

$$= -1.13 \times 10^3\,\text{Pa m}^3 = -1.13 \times 10^3\,\text{J}$$

$$= \boxed{-1.13\,\text{kJ}}$$

$$q = \Delta U - w = 0 + 1.13\,\text{kJ} = \boxed{+1.13\,\text{kJ}}$$

(c) $\boxed{\Delta U = \Delta H = 0}$

$$\boxed{w = 0}\;[\text{free expansion}] \quad q = \Delta U - w = 0 - 0 = \boxed{0}$$

Comment. An isothermal free expansion of a perfect gas is also adiabatic.

E2.8(a) For a perfect gas at constant volume

$$\frac{p}{T} = \frac{nR}{V} = \text{constant}, \quad \text{hence,} \quad \frac{p_1}{T_1} = \frac{p_2}{T_2}$$

$$p_2 = \left(\frac{T_2}{T_1}\right) \times p_1 = \left(\frac{400\,\text{K}}{300\,\text{K}}\right) \times (1.00\,\text{atm}) = \boxed{1.33\,\text{atm}}$$

$$\Delta U = nC_{V,m}\Delta T [22b] = (n) \times \left(\tfrac{3}{2}R\right) \times (400\,\text{K} - 300\,\text{K})$$

$$= (1.00\,\text{mol}) \times \left(\tfrac{3}{2}\right) \times (8.314\,\text{J K}^{-1}\,\text{mol}^{-1}) \times (100\,\text{K})$$

$$= 1.25 \times 10^{3}\,\text{J} = \boxed{+1.25\,\text{kJ}}$$

$$\boxed{w = 0}\ [\text{constant volume}] \qquad q = \Delta U - w\ [\text{First Law}] = 1.25\,\text{kJ} - 0 = \boxed{+1.25\,\text{kJ}}$$

E2.9(a)   (a) $w = -p_{ex}\Delta V$ [2.10]

$$p_{ex} = (200\,\text{Torr}) \times (133.3\,\text{Pa Torr}^{-1}) = 2.66\overline{6} \times 10^{4}\,\text{Pa}$$

$$\Delta V = 3.3\,\text{L} = 3.3 \times 10^{-3}\,\text{m}^{3}$$

Therefore, $w = (-2.66\overline{6} \times 10^{4}\,\text{Pa}) \times (3.3 \times 10^{-3}\,\text{m}^{3}) = \boxed{-88\,\text{J}}$

(b) $w = -nRT\,\ln\dfrac{V_f}{V_i}$ [2.13]

$$n = \frac{4.50\,\text{g}}{16.04\,\text{g mol}^{-1}} = 0.280\overline{5}\,\text{mol}, \quad RT = 2.577\,\text{kJ mol}^{-1}, \quad V_i = 12.7\,\text{L}, \quad V_f = 16.0\,\text{L}$$

$$w = -(0.280\overline{5}\,\text{mol}) \times (2.577\,\text{kJ mol}^{-1}) \times \ln\left(\frac{16.0\,\text{L}}{12.7\,\text{L}}\right) = \boxed{-167\,\text{J}}$$

E2.10(a) $w = -nRT\,\ln\dfrac{V_f}{V_i}$ [2.13] $V_f = \dfrac{1}{3}V_i$

$$nRT = (5.20 \times 10^{-3}\,\text{mol}) \times (8.314\,\text{J K}^{-1}\,\text{mol}^{-1}) \times (260\,\text{K}) = 1.12\overline{4} \times 10^{2}\,\text{J}$$

$$w = -(1.12\overline{4} \times 10^{2}\,\text{J}) \times \ln\tfrac{1}{3} = \boxed{+123\,\text{J}}$$

E2.11(a) $\Delta H = \Delta H_{cond} = -\Delta H_{vap} = (-1\,\text{mol}) \times (40.656\,\text{kJ mol}^{-1}) = \boxed{-40.656\,\text{kJ}}$

Since the condensation is done isothermally and reversibly, the external pressure is constant at 1.00 atm. Hence,

$$q = q_p = \Delta H = \boxed{-40.656\,\text{kJ}}$$

$$w = -p_{ex}\Delta V\ [2.10] \quad \Delta V = V_{liq} - V_{vap} \approx -V_{vap}\ [V_{liq} \ll V_{vap}]$$

On the assumption that $H_2O(g)$ is a perfect gas, $V_{vap} = \dfrac{nRT}{p}$ and $p = p_{ex}$, since the condensation is done reversibly. Hence,

$$w = nRT = (1.00\,\text{mol}) \times (8.314\,\text{J K}^{-1}\,\text{mol}^{-1}) \times (373\,\text{K}) = +3.10 \times 10^{3}\,\text{J} = \boxed{+3.10\,\text{kJ}}$$

From eqn 26 $\Delta U = \Delta H - \Delta n_g RT \quad \Delta n_g = -1.00\,\text{mol}$

$$\Delta U = (-40.656\,\text{kJ}) + (1.00\,\text{mol}) \times (8.314\,\text{J K}^{-1}\,\text{mol}^{-1}) \times (373.15\,\text{K}) = \boxed{-37.55\,\text{kJ}}$$

E2.12(a)   The chemical reaction that occurs is

$$Mg(s) + 2HCl(aq) \rightarrow H_2(g) + MgCl_2(aq), \quad M(Mg) = 24.31\,\text{g mol}^{-1}$$

Work is done against the atmosphere by the expansion of the hydrogen gas produced in the reaction.

$$w = -p_{ex}\Delta V\ [2.10]$$

$$V_i = 0, \quad V_f = \frac{nRT}{p_f}, \quad p_f = p_{ex} \qquad w = -p_{ex}(V_f - V_i) = (-p_{ex}) \times \frac{nRT}{p_{ex}} = -nRT$$

$$n = \frac{15\,\text{g}}{24.31\,\text{g mol}^{-1}} = 0.61\overline{7}\,\text{mol}, \quad RT = 2.479\,\text{kJ mol}^{-1}$$

Hence, $w = (-0.61\overline{7}\,\text{mol}) \times (2.479\,\text{kJ mol}^{-1}) = \boxed{-1.5\,\text{kJ}}$

E2.13(a) $q = n\Delta H_{\text{fus}}^{\ominus} \quad \Delta H_{\text{fus}}^{\ominus} = 2.601\,\text{kJ mol}^{-1}$

$$n = \frac{750 \times 10^3\,\text{g}}{22.99\,\text{g mol}^{-1}} = 3.26\overline{2} \times 10^4\,\text{mol}$$

$$q = (3.26\overline{2} \times 10^4\,\text{mol}) \times (2.60\,\text{kJ mol}^{-1}) = \boxed{+8.48 \times 10^4\,\text{kJ}}$$

E2.14(a)   (a) $q = \Delta H$, since pressure is constant

$$\Delta H = \int_{T_i}^{T_f} dH, \quad dH = nC_{p,m}\,dT$$

$$d(H/\text{J}) = \{20.17 + 0.3665(T/\text{K})\}\,d(T/\text{K})$$

$$\Delta(H/\text{J}) = \int_{T_i}^{T_f} d(H/\text{J}) = \int_{298}^{473} \{20.17 + 0.3665(T/\text{K})\}\,d(T/\text{K})$$

$$= (20.17) \times (473 - 298) + \left(\frac{0.3665}{2}\right) \times \left(\frac{T}{\text{K}}\right)^2 \Big|_{298}^{473}$$

$$= (3.53\overline{0} \times 10^3) + (2.47\overline{25} \times 10^4)$$

$$q = \Delta H = \boxed{2.83 \times 10^4\,\text{J}} = \boxed{+28.3\,\text{kJ}}$$

$$w = -p_{ex}\Delta V \,[2.10], \quad p_{ex} = p$$

$$= -p\Delta V = -\Delta(pV)[\text{constant pressure}] = -\Delta(nRT)[\text{perfect gas}] = -nR\Delta T$$

$$= (-1.00\,\text{mol}) \times (8.314\,\text{J K}^{-1}\,\text{mol}^{-1}) \times (473\,\text{K} - 298\,\text{K}) = \boxed{-1.45 \times 10^3\,\text{J}}$$

$$= \boxed{-1.45\,\text{kJ}}$$

$$\Delta U = q + w = (28.3\,\text{kJ}) - (1.45\,\text{kJ}) = \boxed{+26.8\,\text{kJ}}$$

(b) The energy and enthalpy of a perfect gas depend on temperature alone (*Molecular interpretation 2.2* and *Exercise 2.7*); hence it does not matter whether the temperature change is brought about at constant volume or constant pressure; ΔH and ΔU are the same.

$$\Delta H = \boxed{+28.3\,\text{kJ}}, \quad \Delta U = \boxed{+26.8\,\text{kJ}}, \quad w = \boxed{0} \quad [\text{constant volume}]$$

$$q = \Delta U - w = \boxed{+26.8\,\text{kJ}}$$

E2.15(a)   For reversible adiabatic expansion

$$V_f T_f^c = V_i T_i^c \,[2.34] \quad \text{so} \quad T_f = T_i \left(\frac{V_i}{V_f}\right)^{1/c}$$

$$\text{where } c = \frac{C_{V,m}}{R} = \frac{C_{p,m} - R}{R} = \frac{(20.786 - 8.3145)\,\text{J K}^{-1}\,\text{mol}^{-1}}{8.3145\,\text{J K}^{-1}\,\text{mol}^{-1}} = 1.500$$

So the final temperature is

$$T_f = (273.15\,\text{K}) \times \left(\frac{1.0\,\text{L}}{3.0\,\text{L}}\right)^{1/1.500} = \boxed{13\bar{1}\,\text{K}}$$

E2.16(a)   Reversible adiabatic work is

$$w = C_V \Delta T = C_V \Delta T = n(C_{p,\text{m}} - R) \times (T_f - T_i)$$

where the temperatures are related by

$$T_f = T_i \left(\frac{V_i}{V_f}\right)^{1/c} \quad [2.35] \quad \text{where } c = \frac{C_{V,\text{m}}}{R} = \frac{C_{p,\text{m}} - R}{R} = 3.463$$

So $T_f = [(27.0 + 273.15)\,\text{K}] \times \left(\frac{500 \times 10^{-3}\,\text{L}}{3.00\,\text{L}}\right)^{1/3.463} = 179\,\text{K}$

and $w = \left(\frac{2.45\,\text{g}}{44.0\,\text{g mol}^{-1}}\right) \times [(37.11 - 8.3145)\,\text{J K}^{-1}\text{mol}^{-1}] \times (300 - 179)\,\text{K} = \boxed{194\,\text{J}}$

E2.17(a)   For reversible adiabatic expansion

$$p_f V_f^{\gamma} = p_i V_i^{\gamma} \quad [2.36] \quad \text{so} \quad p_f = p_i \left(\frac{V_i}{V_f}\right)^{\gamma} = (57.4\,\text{kPa}) \times \left(\frac{1.0\,\text{L}}{2.0\,\text{L}}\right)^{1.4} = \boxed{22\,\text{kPa}}$$

E2.18(a)   For reversible adiabatic expansion

$$p_f V_f^{\gamma} = p_i V_i^{\gamma} \quad [2.36] \quad \text{so} \quad p_f = p_i \left(\frac{V_i}{V_f}\right)^{\gamma}$$

We need p_i, which we can obtain from the perfect gas law

$$pV = nRT \quad \text{so} \quad p = \frac{nRT}{V}$$

$$p_i = \frac{\left(\frac{2.4\,\text{g}}{44\,\text{g mol}^{-1}}\right) \times (0.08206\,\text{L atm K}^{-1}\text{mol}^{-1}) \times (278\,\text{K})}{1.0\,\text{L}} = \boxed{1.2\,\text{atm}}$$

$$p_f = (1.2\,\text{atm}) \times \left(\frac{1.0\,\text{L}}{2.0\,\text{L}}\right)^{1.4} = \boxed{0.45\,\text{atm}}$$

E2.19(a)   The reaction for the combustion of butane is

$$C_4H_{10}(g) + \tfrac{13}{2}O_2(g) \rightarrow 4CO_2(g) + 5H_2O(l)$$

$$\Delta_c H^{\ominus}(C_4H_{10}, g) = (4) \times (\Delta_f H^{\ominus}(CO_2, g)) + (5) \times (\Delta_f H^{\ominus}(H_2O, g))$$
$$- (1) \times (\Delta_f H^{\ominus}(C_4H_{10}, g))$$

Solving for $\Delta_f H^{\ominus}(C_4H_{10}, g)$ and looking up the other data in Tables 2.5 and 2.6, we obtain

$$\Delta_f H^{\ominus}(C_4H_{10}, g) = (4) \times (-393.51\,\text{kJ mol}^{-1}) + (5) \times (-285.83\,\text{kJ mol}^{-1})$$

$$-(1) \times (-2878\,\text{kJ mol}^{-1})$$

$$= \boxed{-125\,\text{kJ mol}^{-1}}$$

Comment. This is very close to the value listed in Table 2.6. The small difference is undoubtedly the result of the error in the least precise value of the set of data, that for $\Delta_c H^{\ominus}$ (C_4H_{10}, g).

E2.20(a) $C_p = \dfrac{q_p}{\Delta T}$ [2.29] $= \dfrac{229\,J}{2.55\,K} = 89.8\,J\,K^{-1}$ $C_{p,m} = \dfrac{C_p}{n} = \dfrac{89.8\,J\,K^{-1}}{3.0\,mol} = \boxed{30\,J\,K^{-1}\,mol^{-1}}$

For a perfect gas

$$C_{p,m} - C_{V,m} = R \ [2.31]$$

$$C_{V,m} = C_{p,m} - R = (30 - 8.3)\,J\,K^{-1}\,mol^{-1} = \boxed{22\,J\,K^{-1}\,mol^{-1}}$$

E2.21(a) $q_p = \boxed{-1.2\,kJ}$ [energy left the sample] $\Delta H = q_p = \boxed{-1.2\,kJ}$

$C_p = \dfrac{q_p}{\Delta T} = \dfrac{-1.2\,kJ}{-15\,K} = \boxed{80\,J\,K^{-1}}$

E2.22(a) $q_p = C_p \Delta T$ [2.29] $= nC_{p,m}\Delta T = (3.0\,mol) \times (29.4\,J\,K^{-1}\,mol^{-1}) \times (25\,K) = \boxed{+2.2\,kJ}$

$\Delta H = q_p$ [2.24b] $= \boxed{+2.2\,kJ}$

$\Delta U = \Delta H - \Delta(pV)$(From $H \equiv U + pV$) $= \Delta H - \Delta(nRT)$[perfect gas] $= \Delta H - nR\Delta T$

$= (2.2\,kJ) - (3.0\,mol) \times (8.314\,J\,K^{-1}\,mol^{-1}) \times (25\,K) = (2.2\,kJ) - (0.62\,kJ)$

$= \boxed{+1.6\,kJ}$

E2.23(a) $q = \boxed{0}$ [adiabatic process]

$w = -p_{ex}\Delta V = (-600\,Torr) \times \left(\dfrac{1.013 \times 10^5\,Pa}{760\,Torr}\right) \times (40 \times 10^{-3}\,m^3) = \boxed{-3.2\,kJ}$

$\Delta U = w = \boxed{-3.2\,kJ}$ $[q = 0]$ or $C_V \Delta T = w$ $[C_V = nC_{V,m}; C_{V,m} = 21.1\,J\,K^{-1}\,mol^{-1}]$

or

$\Delta T = \dfrac{w}{C_V} = \dfrac{w}{nC_{V,m}} = \dfrac{-3.2 \times 10^3\,J}{(4.0\,mol) \times (21.1\,J\,K^{-1}\,mol^{-1})} = \boxed{-38\,K}$

$\Delta H = \Delta U + \Delta(pV) = \Delta U + nR\Delta T$

$= (-3.2\,kJ) + (4.0\,mol) \times (8.314\,J\,K^{-1}\,mol^{-1}) \times (-38\,K) = \boxed{-4.5\,kJ}$

Question. Calculate the final pressure of the gas.

E2.24(a) $q = \boxed{0}$ [adiabatic process]

$\Delta U = nC_{V,m}\Delta T$ [perfect gas] $= (3.0\,mol) \times (27.5\,J\,K^{-1}\,mol^{-1}) \times (50\,K) = \boxed{+4.1\,kJ}$

$w = \Delta U - q = 4.1\,kJ - 0 = \boxed{+4.1\,kJ}$

$\Delta H = \Delta U + nR\Delta T$ $[\Delta(pV) = \Delta(nRT) = nR\Delta T]$

$= (4.1\,kJ) + (3.0\,mol) \times (8.314\,J\,K^{-1}\,mol^{-1}) \times (50\,K) = \boxed{+5.4\,kJ}$

$V_i = \dfrac{nRT_i}{p_i} = \dfrac{(3.0\,mol) \times (8.206 \times 10^{-2}\,L\,atm\,K^{-1}\,mol^{-1}) \times (200\,K)}{(2.0\,atm)} = 24.6\,L$

$V_f = V_i \left(\dfrac{T_i}{T_f}\right)^c$ [2.34], $c = \dfrac{C_V}{R} = \dfrac{27.5\,J\,K^{-1}\,mol^{-1}}{8.314\,J\,K^{-1}\,mol^{-1}} = 3.31$

$$V_f = (24.6\,\text{L}) \times \left(\frac{200\,\text{K}}{250\,\text{K}}\right)^{3.31} = \boxed{11.8\,\text{L}}$$

$$p_f = \frac{nRT_f}{V_f} = \frac{(3.0\,\text{mol}) \times (8.206 \times 10^{-2}\,\text{L atm K}^{-1}\,\text{mol}^{-1}) \times (250\,\text{K})}{11.8\,\text{L}} = \boxed{5.2\,\text{atm}}$$

E2.25(a)
$$V_i = \frac{nRT_i}{p_i} = \frac{(1.0\,\text{mol}) \times (8.206 \times 10^{-2}\,\text{L atm K}^{-1}\,\text{mol}^{-1}) \times (310\,\text{K})}{3.25\,\text{atm}} = \boxed{7.8\overline{3}\,\text{L}}$$

$$\gamma = \frac{C_p}{C_V} = \frac{C_V + R}{C_V} = \frac{(20.8 + 8.31)\text{J K}^{-1}\,\text{mol}^{-1}}{20.8\,\text{J K}^{-1}\,\text{mol}^{-1}} = 1.40 \qquad \frac{1}{\gamma} = 0.714$$

$$V_f = V_i \left(\frac{p_i}{p_f}\right)^{1/\gamma} = (7.8\overline{3}\,\text{L}) \times \left(\frac{3.25\,\text{atm}}{2.50\,\text{atm}}\right)^{0.714} = \boxed{9.4\overline{4}\,\text{L}}$$

$$T_f = \frac{p_f V_f}{nR} = \frac{(2.50\,\text{atm}) \times (9.4\overline{4}\,\text{L})}{(1.0\,\text{mol}) \times (8.206 \times 10^{-2}\,\text{L atm K}^{-1}\,\text{mol}^{-1})} = \boxed{28\overline{8}\,\text{K}}$$

$$w = C_V(T_f - T_i)\,[2.33] = (20.8\,\text{J K}^{-1}\,\text{mol}^{-1}) \times (1.0\,\text{mol}) \times (288\,\text{K} - 310\,\text{K})$$
$$= \boxed{-0.46\,\text{kJ}}$$

E2.26(a) For this small temperature range α may be assumed to be constant; hence

$$dV = \left(\frac{\partial V}{\partial T}\right)_p dT\,[\text{pressure constant}] = \alpha V\,dT$$

$\Delta V \approx \alpha V \Delta T$ [the change in V is small; hence $V \approx$ constant]

Mercury, $\alpha = 1.82 \times 10^{-4}\,\text{K}^{-1}$,

$$\Delta V \approx (1.82 \times 10^{-4}\,\text{K}^{-1}) \times (1.0\,\text{cm}^3) \times (5\,\text{K}) \approx 9.\overline{1} \times 10^{-4}\,\text{cm}^3 = \boxed{+0.9\,\text{mm}^3}$$

E2.27(a) In an adiabatic process, $q = \boxed{0}$. Work against a constant external pressure is

$$w = -p_{ex}\Delta V = -(1.0\,\text{atm}) \times (1.01 \times 10^5\,\text{Pa atm}^{-1}) \times \frac{(20\,\text{cm}) \times (10\,\text{cm}^2)}{(100\,\text{cm m}^{-1})^3} = \boxed{-20\,\text{J}}$$

$$\Delta U = q + w = \boxed{-20\,\text{J}}$$

$$w = C_V \Delta T = n(C_{p,m} - R)\Delta T \quad \text{so} \quad \Delta T = \frac{w}{n(C_{p,m} - R)},$$

$$\Delta T = \frac{-20\,\text{J}}{(2.0\,\text{mol}) \times (37.11 - 8.3145\,\text{J K}^{-1}\,\text{mol}^{-1})} = \boxed{-0.35\,\text{K}}$$

$$\Delta H = \Delta U + \Delta(pV) = \Delta U + nR\Delta T$$
$$= -20\,\text{J} + (2.0\,\text{mol}) \times (8.3145\,\text{J K}^{-1}\,\text{mol}^{-1}) \times (-0.35\,\text{K}) = \boxed{-26\,\text{J}}$$

E2.28(a) The amount of Xe in the sample is

$$n = \frac{65.0\,\text{g}}{131.3\,\text{g mol}^{-1}} = 0.495\,\text{mol}$$

(a) For reversible adiabatic expansion

$$p_f V_f^{\gamma} = p_i V_i^{\gamma} \quad \text{so} \quad V_f = V_i \left(\frac{p_i}{p_f}\right)^{1/\gamma},$$

where $\quad \gamma = \dfrac{C_{p,m}}{C_{V,m}} \quad$ where $\quad C_{V,m} = (20.79 - 8.3145)\,\text{J K}^{-1}\text{mol}^{-1} = 12.48\,\text{J K}^{-1}\,\text{mol}^{-1}$, so $\gamma = \dfrac{20.79\,\text{J K}^{-1}\,\text{mol}^{-1}}{12.48\,\text{J K}^{-1}\,\text{mol}^{-1}} = 1.666,$

and $V_i = \dfrac{nRT_i}{p_i} = \dfrac{(0.495\,\text{mol}) \times (0.08206\,\text{L atm K}^{-1}\text{mol}^{-1}) \times (298\,\text{K})}{2.00\,\text{atm}} = 6.05\,\text{L},$

so $V_f = V_i \left(\dfrac{p_i}{p_f}\right)^{1/\gamma} = (6.05\,\text{L}) \times \left(\dfrac{2.00\,\text{atm}}{1.00\,\text{atm}}\right)^{(1/1.666)} = 9.17\,\text{L}$

$T_f = \dfrac{p_f V_f}{nR} = \dfrac{(1.00\,\text{atm}) \times (9.17\,\text{L})}{(0.495\,\text{mol}) \times (0.08206\,\text{L atm K}^{-1}\,\text{mol}^{-1})} = \boxed{226\,\text{K}}$

(b) For adiabatic expansion against a constant external pressure

$$w = -p_{\text{ex}}\Delta V = C_V \Delta T \quad \text{so} \quad -p_{\text{ex}}(V_f - V_i) = C_V(T_f - T_i)$$

In addition, the perfect gas law holds

$$p_f V_f = nRT_f$$

Solve the latter for T_f in terms of V_f, and insert into the previous relationship to solve for V_f

$$T_f = \dfrac{p_f V_f}{nR} \quad \text{so} \quad -p_{\text{ex}}(V_f - V_i) = C_V\left(\dfrac{p_f V_f}{nR} - T_i\right)$$

Collecting terms gives

$$C_V T_i + p_{\text{ex}} V_i = V_f\left(p_{\text{ex}} + \dfrac{C_V p_f}{nR}\right) \quad \text{so} \quad V_f = \dfrac{C_V T_i + p_{\text{ex}} V_i}{p_{\text{ex}} + \dfrac{C_{V,m} p_f}{R}},$$

$$V_f = \dfrac{(12.48\,\text{J K}^{-1}\,\text{mol}^{-1}) \times (0.495\,\text{mol}) \times (298\,\text{K}) + (1.00\,\text{atm}) \times (1.01 \times 10^5\,\text{Pa atm}^{-1}) \times \left(\dfrac{6.05\,\text{L}}{1000\,\text{L m}^{-3}}\right)}{\left(1.00\,\text{atm} + \dfrac{(12.48\,\text{J K}^{-1}\,\text{mol}^{-1}) \times (1.00\,\text{atm})}{8.3145\,\text{J K}^{-1}\,\text{mol}^{-1}}\right) \times (1.01 \times 10^5\,\text{Pa atm}^{-1})}$$

$$V_f = 9.71 \times 10^{-3}\,\text{m}^3$$

Finally, the temperature is

$$T_f = \dfrac{p_f V_f}{nR} = \dfrac{(1.00\,\text{atm}) \times (1.01 \times 10^5\,\text{Pa atm}^{-1}) \times (9.71 \times 10^{-3}\,\text{m}^3)}{(0.495\,\text{mol}) \times (8.3145\,\text{J K}^{-1}\,\text{mol}^{-1})} = \boxed{238\,\text{K}}$$

E2.29(a) $q_p = n\Delta_{\text{vap}}H^{\ominus}$ [constant pressure] $= (0.50\,\text{mol}) \times (26.0\,\text{kJ mol}^{-1}) = \boxed{+13\,\text{kJ}}$

$w = -p_{\text{ex}}\Delta V\ [10] \approx -p_{\text{ex}}V(\text{g})[V(\text{g}) \gg V(\text{l})] \approx -(p_{\text{ex}}) \times \left(\dfrac{nRT}{p_{\text{ex}}}\right) = -nRT$

Therefore, $w \approx (-0.50\,\text{mol}) \times (8.314\,\text{J K}^{-1}\,\text{mol}^{-1}) \times (250\,\text{K}) = \boxed{-1.0\,\text{kJ}}$

$\Delta H = q_p\ [2.24b] = \boxed{+13\,\text{kJ}} \qquad \Delta U = q + w = (13\,\text{kJ}) - (1.0\,\text{kJ}) = \boxed{+12\,\text{kJ}}$

E2.30(a) $C_6H_5C_2H_5(\text{l}) + \frac{21}{2}O_2(\text{g}) \rightarrow 8CO_2(\text{g}) + 5H_2O(\text{l})$

$\Delta_c H^{\ominus} = 8\Delta_f H^{\ominus}(CO_2, \text{g}) + 5\Delta_f H^{\ominus}(H_2O, \text{l}) - \Delta_f H^{\ominus}(C_6H_5C_2H_5, \text{l})$

$= \{(8) \times (-393.51) + (5) \times (-285.83) - (-12.5)\}\,\text{kJ mol}^{-1}$

$= \boxed{-4564.7\,\text{kJ mol}^{-1}}$

E2.31(a) The reaction is $C_6H_{12}(l) + H_2(g) \rightarrow C_6H_{14}(l) \quad \Delta_r H^\ominus =?$

From Table 2.5 and the information in the exercise

$$C_6H_{12}(l) + 9O_2(g) \rightarrow 6CO_2(g) + 6H_2O(l) \quad \Delta_c H^\ominus = -4003 \,\text{kJ mol}^{-1}$$

$$C_6H_{14}(l) + \tfrac{19}{2}O_2(g) \rightarrow 6CO_2(g) + 7H_2O(l) \quad \Delta_c H^\ominus = -4163 \,\text{kJ mol}^{-1}$$

The difference of these reactions is

$$C_6H_{12}(l) + H_2(l) \rightarrow C_6H_{14}(l) + \tfrac{1}{2}O_2(g) \quad \Delta_r H^\ominus = +160 \,\text{kJ mol}^{-1}$$

This reaction may be converted to the desired reaction by subtracting from it

$$H_2O(l) \rightarrow H_2(g) + \tfrac{1}{2}O_2(g) \quad \Delta_r H^\ominus = -\Delta_f H^\ominus(H_2O, l) = 285.83 \,\text{kJ mol}^{-1}$$

This gives $C_6H_{12}(l) + H_2(g) \rightarrow C_6H_{14}(l) \quad \Delta_r H^\ominus = \boxed{-126 \,\text{kJ mol}^{-1}}$

E2.32(a) First $\Delta_f H^\ominus[(CH_2)_3, g]$ is calculated, and then that result is used to calculate $\Delta_r H^\ominus$ for the isomerization.

$$(CH_2)_3(g) + \tfrac{9}{2}O_2(g) \rightarrow 3CO_2(g) + 3H_2O(l) \quad \Delta_c H^\ominus = -2091 \,\text{kJ mol}^{-1}$$

$$\Delta_f H^\ominus\{(CH_2)_3, g\} = -\Delta_c H^\ominus + 3\Delta_f H^\ominus(CO_2, g) + 3\Delta_f H^\ominus(H_2O, g)$$

$$= \{+2091 + (3) \times (-393.51) + (3) \times (-285.83)\} \,\text{kJ mol}^{-1}$$

$$= \boxed{+53 \,\text{kJ mol}^{-1}}$$

$$(CH_2)_3(g) \rightarrow C_3H_6(g) \quad \Delta_r H^\ominus =?$$

$$\Delta_r H^\ominus = \Delta_f H^\ominus(C_3H_6, g) - \Delta_f H^\ominus\{(CH_2)_3, g\}$$

$$= (20.42 - 53) \,\text{kJ mol}^{-1} = \boxed{-33 \,\text{kJ mol}^{-1}}$$

E2.33(a) The formation reaction of liquid methylacetate is

$$3C(s) + 3H_2(g) + O_2(g) \rightarrow CH_3COOOCH_3(l) \quad \Delta_f H^\ominus = -442 \,\text{kJ mol}^{-1}$$

$$\Delta U = \Delta H - \Delta n_g RT \,[2.26], \quad \Delta n_g = -4 \,\text{mol},$$

$$\Delta n_g RT = (-4 \,\text{mol}) \times (2.479 \,\text{kJ mol}^{-1}) = -9.916 \,\text{kJ}$$

Therefore $\Delta_f U^\ominus = (-442 \,\text{kJ mol}^{-1}) + (9.9 \,\text{kJ mol}^{-1}) = \boxed{-432 \,\text{kJ mol}^{-1}}$

E2.34(a) $C = \dfrac{q}{\Delta T} \,[2.17] \quad \text{and} \quad q = IVt \,[2.18]$

Thus $C = \dfrac{IVt}{\Delta T} = \dfrac{(3.20 \,\text{A}) \times (12.0 \,\text{V}) \times (27.0 \,\text{s})}{1.617 \,\text{K}} = \boxed{641 \,\text{J K}^{-1}} \quad (1 \,\text{J} = 1 \,\text{A V s})$

E2.35(a) For naphthalene the reaction is $C_{10}H_8(s) + 12O_2(g) \rightarrow 10CO_2(g) + 4H_2O(l)$

A bomb calorimeter gives $q_V = n\Delta_c U^\ominus$ rather than $q_p = n\Delta_c H^\ominus$; thus we need

$$\Delta_c U^\ominus = \Delta_c H^\ominus - \Delta n_g RT \,[2.26], \quad \Delta n_g = -2 \,\text{mol}$$

$$\Delta_c H^\ominus = -5157 \,\text{kJ mol}^{-1} \,[\text{Table 2.5}] \quad \text{assume } T \approx 298 \,\text{K}$$

$$\Delta_c U^\ominus = (-5157 \,\text{kJ mol}^{-1}) - (-2) \times (8.3 \times 10^{-3} \,\text{kJ K}^{-1} \,\text{mol}^{-1}) \times (298 \,\text{K})$$

$$= -5152 \,\text{kJ mol}^{-1}$$

$$|q| = |q_V| = |n\Delta_c U^{\ominus}| = \left(\frac{120 \times 10^{-3}\,\mathrm{g}}{128.18\,\mathrm{g\,mol^{-1}}}\right) \times (5152\,\mathrm{kJ\,mol^{-1}}) = 4.82\overline{3}\,\mathrm{kJ}$$

$$C = \frac{|q|}{\Delta T} = \frac{4.82\overline{3}\,\mathrm{kJ}}{3.05\,\mathrm{K}} = \boxed{1.58\,\mathrm{kJ\,K^{-1}}}$$

When phenol is used the reaction is

$$C_6H_5OH(s) + \tfrac{15}{2}O_2(g) \rightarrow 6CO_2(g) + 3H_2O(l)$$

$$\Delta_c H^{\ominus} = -3054\,\mathrm{kJ\,mol^{-1}} \text{ [Table 2.5]}$$

$$\Delta_c U^{\ominus} = \Delta_c H^{\ominus} - \Delta n_g RT, \quad \Delta n_g = -\tfrac{3}{2}$$

$$= (-3054\,\mathrm{kJ\,mol^{-1}}) + \left(\tfrac{3}{2}\right) \times (8.314 \times 10^{-3}\,\mathrm{kJ\,K^{-1}\,mol^{-1}}) \times (298\,\mathrm{K})$$

$$= -3050\,\mathrm{kJ\,mol^{-1}}$$

$$|q| = \left(\frac{100 \times 10^{-3}\,\mathrm{g}}{94.12\,\mathrm{g\,mol^{-1}}}\right) \times (3050\,\mathrm{kJ\,mol^{-1}}) = 3.24\overline{1}\,\mathrm{kJ}$$

$$\Delta T = \frac{|q|}{C} = \frac{3.24\overline{1}\,\mathrm{kJ}}{1.58\,\mathrm{kJ\,K^{-1}}} = \boxed{+2.05\,\mathrm{K}}$$

Comment. In this case $\Delta_c U^{\ominus}$ and $\Delta_c H^{\ominus}$ differed by ≈ 0.1 per cent. Thus, to within 3 significant figures, it would not have mattered if we had used $\Delta_c H^{\ominus}$ instead of $\Delta_c U^{\ominus}$, but for very precise work it would.

E2.36(a) $AgCl(s) \rightarrow Ag^+(aq) + Cl^-(aq)$

$$\Delta_{sol} H^{\ominus} = \Delta_f H^{\ominus}(Ag^+, aq) + \Delta_f H^{\ominus}(Cl^-, aq) - \Delta_f H^{\ominus}(AgCl, s)$$

$$= (105.58) + (167.16) - (-127.07)\,\mathrm{kJ\,mol^{-1}}$$

$$= \boxed{+65.49\,\mathrm{kJ\,mol^{-1}}}$$

E2.37(a) $NH_3(g) + SO_2(g) \rightarrow NH_3SO_2(s) \quad \Delta_r H^{\ominus} = -40\,\mathrm{kJ\,mol^{-1}}$

$$\Delta_r H^{\ominus} = \Delta_f H^{\ominus}(NH_3SO_2, s) - \Delta_f H^{\ominus}(NH_3, g) - \Delta_f H^{\ominus}(SO_2, g)$$

Solving for $\Delta_f H^{\ominus}(NH_3SO_2, s)$ yields

$$\Delta_f H^{\ominus}(NH_3SO_2, s) = \Delta_f H^{\ominus}(NH_3, g) + \Delta_f H^{\ominus}(SO_2, g) + \Delta_r H^{\ominus}$$

$$= (-46.11 - 296.83 - 40)\,\mathrm{kJ\,mol^{-1}} = \boxed{-383\,\mathrm{kJ\,mol^{-1}}}$$

E2.38(a)   **(a)** $\Delta_c H^{\ominus}(l) = \Delta_{vap} H^{\ominus} + \Delta_c H(g) = (15\,\mathrm{kJ\,mol^{-1}}) - (2220\,\mathrm{kJ\,mol^{-1}}) = \boxed{-2205\,\mathrm{kJ\,mol^{-1}}}$

   (b) $\Delta_c U^{\ominus}(l) = \Delta_c H^{\ominus}(l) - \Delta n_g RT, \quad \Delta n_g = -2$

$$= (-2205\,\mathrm{kJ\,mol^{-1}}) + (2) \times (2.479\,\mathrm{kJ\,mol^{-1}}) = \boxed{-2200\,\mathrm{kJ\,mol^{-1}}}$$

E2.39(a)   In each case $\Delta_r H^{\ominus} = \displaystyle\sum_{\text{Products}} \nu \Delta_f H^{\ominus} - \sum_{\text{Reactants}} \nu \Delta_f H^{\ominus}$ [2.42]

   (a) $\Delta_r H^{\ominus} = \Delta_f H^{\ominus}(N_2O_4, g) - 2\Delta_f H^{\ominus}(NO_2, g) = (9.16) - (2) \times (33.18)\,\mathrm{kJ\,mol^{-1}}$

$$= \boxed{-57.20\,\mathrm{kJ\,mol^{-1}}}$$

   (b) $\Delta_r H^{\ominus} = \Delta_f H^{\ominus}(NH_4Cl, s) - \Delta_f H^{\ominus}(NH_3, g) - \Delta_f H^{\ominus}(HCl, g)$

$$= \{(-314.43) - (-46.11) - (-92.31)\}\,\mathrm{kJ\,mol^{-1}} = \boxed{-176.01\,\mathrm{kJ\,mol^{-1}}}$$

E2.40(a) **(a)** reaction(3) $= (-2) \times$ reaction (1) + reaction (2) $\Delta n_g = -2$

The enthalpies of reactions are combined in the same manner as the equations (Hess's law).

$$\Delta_r H^\ominus(3) = (-2) \times \Delta_r H^\ominus(1) + \Delta_r H^\ominus(2)$$

$$= \{(-2) \times (-184.62) + (-483.64)\}\, \text{kJ mol}^{-1}$$

$$= \boxed{-114.40\,\text{kJ mol}^{-1}}$$

$$\Delta_r U^\ominus = \Delta_r H^\ominus - \Delta n_g RT\,[2.26] = (-114.40\,\text{kJ mol}^{-1}) - (-2) \times (2.48\,\text{kJ mol}^{-1})$$

$$= \boxed{-109.44\,\text{kJ mol}^{-1}}$$

(b) $\Delta_f H^\ominus$ refers to the formation of one mole of the compound, hence

$$\Delta_f H^\ominus(J) = \frac{\Delta_r H^\ominus(J)}{\nu_J}$$

$$\Delta_f H^\ominus(\text{HCl, g}) = \frac{-184.62}{2}\,\text{kJ mol}^{-1} = \boxed{-92.31\,\text{kJ mol}^{-1}}$$

$$\Delta_f H^\ominus(\text{H}_2\text{O, g}) = \frac{-483.64}{2}\,\text{kJ mol}^{-1} = \boxed{-241.82\,\text{kJ mol}^{-1}}$$

E2.41(a) $\Delta_r H^\ominus = \Delta_r U^\ominus + \Delta n_g RT\,[2.26];$ $\Delta n_g = +2$

$$= (-1373\,\text{kJ mol}^{-1}) + 2 \times (2.48\,\text{kJ mol}^{-1}) = \boxed{-1368\,\text{kJ mol}^{-1}}$$

Comment. As a number of these exercises have shown, the use of $\Delta_r H^\ominus$ as an approximation for $\Delta_r U^\ominus$ is often valid.

E2.42(a) In each case, the strategy is to combine reactions in such a way that the combination corresponds to the formation reaction desired. The enthalpies of the reactions are then combined in the same manner as the equations to yield the enthalpies of formation.

(a)

	$\Delta_r H^\ominus/(\text{kJ mol}^{-1})$
$\text{K(s)} + \frac{1}{2}\text{Cl}_2(\text{g}) \rightarrow \text{KCl(s)}$	-436.75
$\text{KCl(s)} + \frac{3}{2}\text{O}_2(\text{g}) \rightarrow \text{KClO}_3(\text{s})$	$\frac{1}{2} \times (89.4)$
$\text{K(s)} + \frac{1}{2}\text{Cl}_2(\text{g}) + \frac{3}{2}\text{O}_2(\text{g}) \rightarrow \text{KClO}_3(\text{s})$	-392.1

Hence, $\Delta_f H^\ominus(\text{KClO}_3, \text{s}) = \boxed{-392.1\,\text{kJ mol}^{-1}}$

(b)

	$\Delta_r H^\ominus/(\text{kJ mol}^{-1})$
$\text{Na(s)} + \frac{1}{2}\text{O}_2(\text{g}) + \frac{1}{2}\text{H}_2(\text{g}) \rightarrow \text{NaOH(s)}$	-425.61
$\text{NaOH(s)} + \text{CO}_2(\text{g}) \rightarrow \text{NaHCO}_3(\text{s})$	-127.5
$\text{C(s)} + \text{O}_2(\text{g}) \rightarrow \text{CO}_2(\text{g})$	-393.51
$\text{Na(s)} + \text{C(s)} + \frac{1}{2}\text{H}_2(\text{g}) + \frac{3}{2}\text{O}_2(\text{g}) \rightarrow \text{NaHCO}_3(\text{s})$	-946.6

Hence, $\Delta_f H^\ominus(\text{NaHCO}_3, \text{s}) = \boxed{-946.6\,\text{kJ mol}^{-1}}$

(c)

$$\Delta_r H^{\ominus}/(\text{kJ mol}^{-1})$$

$\frac{1}{2}N_2(g) + \frac{1}{2}O_2(g) \rightarrow NO(g)$	$+90.25$
$NO(g) + \frac{1}{2}Cl_2(g) \rightarrow NOCl(g)$	$-\frac{1}{2}(75.5)$

$$\frac{1}{2}N_2(g) + \frac{1}{2}O_2(g) + \frac{1}{2}Cl_2(g) \rightarrow NOCl(g) \qquad +52.5$$

Hence, $\Delta_f H^{\ominus}(\text{NOCl, g}) = \boxed{+52.5\,\text{kJ mol}^{-1}}$

E2.43(a) When the heat capacities of all substances participating in a chemical reaction are assumed to be constant over the range of temperatures involved Kirchoff's law [2.44] integrates to

$$\Delta_r H^{\ominus}(T_2) = \Delta_r H^{\ominus}(T_1) + \Delta_r C_p(T_2 - T_1) \quad \text{[Example 2.7]}$$

$$\Delta_r C_p = \sum_{\text{Products}} \nu C_{p,m}^{\ominus} - \sum_{\text{Reactants}} \nu C_{p,m}^{\ominus} \; [2.45]$$

$$\Delta_r C_p = C_p(N_2O_4, g) - 2\,C_p(NO_2, g) = (77.28) - (2) \times (37.20\,\text{J K}^{-1}\,\text{mol}^{-1})$$

$$= +2.88\,\text{J K}^{-1}\,\text{mol}^{-1}$$

$$\Delta_r H^{\ominus}(373\,\text{K}) = \Delta_r H^{\ominus}(298\,\text{K}) + \Delta_r C_p \Delta T$$

$$= (-57.20\,\text{kJ mol}^{-1}) + (2.88\,\text{J K}^{-1}) \times (75\,\text{K})$$

$$= \{(-57.20) + (0.22)\}\,\text{kJ mol}^{-1}$$

$$= \boxed{-56.98\,\text{kJ mol}^{-1}}$$

E2.44(a) **(a)** $\Delta_r H^{\ominus} = \sum_{\text{Products}} \nu \Delta_f H^{\ominus} - \sum_{\text{Reactants}} \nu \Delta_f H^{\ominus}$ [2.42]

$$\Delta_r H^{\ominus}(298\,\text{K}) = [(-110.53) - (-241.82)]\,\text{kJ mol}^{-1} = \boxed{+131.29\,\text{kJ mol}^{-1}}$$
$$\Delta_r U^{\ominus}(298\,\text{K}) = \Delta_r H^{\ominus}(298\,\text{K}) - \Delta n_g RT \; [2.26]$$
$$= (131.29\,\text{kJ mol}^{-1}) - (1) \times (2.48\,\text{kJ mol}^{-1})$$
$$= \boxed{+128.81\,\text{kJ mol}^{-1}}$$

(b) $\Delta_r H^{\ominus}(378\,\text{K}) = \Delta_r H^{\ominus}(298\,\text{K}) + \Delta_r C_p(T_2 - T_1)$ [Example 2.7]
$$\Delta_r C_p = C_{p,m}(CO, g) + C_{p,m}(H_2, g) - C_{p,m}(C, gr) - C_{p,m}(H_2O, g)$$
$$= (29.14 + 28.82 - 8.53 - 33.58) \times 10^{-3}\,\text{kJ K}^{-1}\,\text{mol}^{-1}$$
$$= 15.85 \times 10^{-3}\,\text{kJ K}^{-1}\,\text{mol}^{-1}$$
$$\Delta_r H^{\ominus}(378\,\text{K}) = (131.29\,\text{kJ mol}^{-1}) + (15.85 \times 10^{-3}\,\text{kJ K}^{-1}\,\text{mol}^{-1}) \times (80\,\text{K})$$
$$= (131.29 + 1.27)\,\text{kJ mol}^{-1} = \boxed{+132.56\,\text{kJ mol}^{-1}}$$
$$\Delta_r U^{\ominus}(378\,\text{K}) = \Delta_r H^{\ominus}(378\,\text{K}) - (1) \times (3.14\,\text{kJ mol}^{-1}) = (132.56 - 3.14)\,\text{kJ mol}^{-1}$$
$$= \boxed{+129.42\,\text{kJ mol}^{-1}}$$

Comment. The differences between both $\Delta_r H^{\ominus}$ and $\Delta_r U^{\ominus}$ at the two temperatures are small and justify the use of the approximation the $\Delta_r C_p$ is a constant.

E2.45(a) Since enthalpy is a state function, $\Delta_r H$ for the process (see Fig. 2.1)

$$Mg^{2+}(g) + 2Cl(g) + 2e^- \rightarrow MgCl_2(aq)$$

is independent of path; therefore the change in enthalpy for the path on the left is equal to the change in enthalpy for the path on the right. All numerical values are in kJ mol^{-1}.

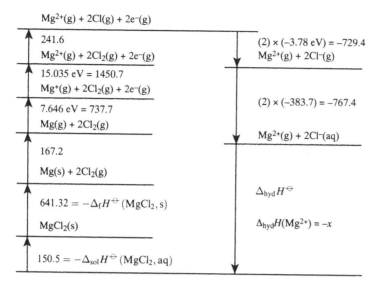

Figure 2.1

The cycle is the distance traversed upward along the left plus the distance traversed downward on the right. The sum of these distances is zero. Note that $E_{ea} = -\Delta_{eg}H^{\ominus}$. Therefore,

$$(150.5) + (641.32) + (167.2) + (737.7) + (1450.7)$$
$$+(241.6) + (-729.4) + (-767.4) + (-x) = 0$$

Solving to $x = 1892.2$, which yields

$$\Delta_{hyd}H^{\ominus}(Mg^{2+}) = \boxed{-1892.2\,kJ\,mol^{-1}}$$

E2.46(a) (a) Cyclohexane is composed of 6 $C(H)_2(C)_2$ groups, so

$$\Delta_f H^{\ominus} = 6 \times \Delta_f H^{\ominus}[C(H)_2(C)_2] = 6 \times (-20.7\,kJ\,mol^{-1})$$
$$= \boxed{-124.2\,kJ\,mol^{-1}}$$

(b) 2,4-dimethylhexane is composed of four $C(H)_3(C)$ groups, two $C(H)_2(C_2)$ groups, and two $C(H)(C)_3$ groups, so

$$\Delta_f H^{\ominus} = 4 \times (-42.17\,kJ\,mol^{-1}) + 2 \times (-20.7\,kJ\,mol^{-1}) + 2 \times (-6.19\,kJ\,mol^{-1})$$
$$= \boxed{-222.46\,kJ\,mol^{-1}}$$

Solutions to problems

Assume all gases are perfect unless stated otherwise. Unless otherwise stated, thermochemical data are for 298.15 K.

Solutions to numerical problems

P2.1 Since houses are not air-tight, some of the air originally in the house escapes to the outside and in the process does work against the atmosphere. Air may be assumed to have an effective molar mass of 29 g mol^{-1}.

mass of air $(20°C) = (1.21\,\text{kg m}^{-3}) \times 600\,\text{m}^3 = 726\,\text{kg}$

amount of air $(20°C) = n(20°C) = \dfrac{m}{M} = \dfrac{726\,\text{kg}}{0.029\,\text{kg mol}^{-1}} = 2.5\overline{03} \times 10^4\,\text{mol}$

The heating of a house is a constant-pressure process, hence for a constant volume

$$n \propto \dfrac{1}{T} \quad \text{and} \quad \dfrac{n(25°C)}{n(20°C)} = \dfrac{T(20°C)}{T(25°C)} = \dfrac{293\,\text{K}}{298\,\text{K}}$$

$$n(25°C) = \left(\dfrac{293\,\text{K}}{298\,\text{K}}\right) \times (2.5\overline{03} \times 10^4\,\text{mol}) = 2.4\overline{61} \times 10^4\,\text{mol}$$

Thus, $2.4\overline{61} \times 10^4$ mol of air has been heated from $20°C$ to $25°C$. In addition 4.2×10^2 mol of air which has escaped into the outside air at $20°C$ has done work against the atmosphere ($p_{ext} = 1.00$ atm).

$$V(\text{escaped air}) = \left(\dfrac{n(\text{escaped})}{n(\text{total})}\right) \times (600\,\text{m}^3) = \left(\dfrac{4.2 \times 10^2\,\text{mol}}{2.5 \times 10^4\,\text{mol}}\right) \times (600\,\text{m}^3) = 10\,\text{m}^3$$

$$q_p = \Delta H = nC_{p,\text{m}}\Delta T \ [2.28b]; \qquad \Delta U = nC_{V,\text{m}}\Delta T \ [2.22b]$$

Since the gas is assumed to be a perfect diatomic

$$C_{p,\text{m}} = \dfrac{7}{2}R, \qquad C_{V,\text{m}} = \dfrac{5}{2}R$$

Hence for the air which remained in the house

$$\Delta H(\text{internal}) = (2.4\overline{61} \times 10^4\,\text{mol}) \times \left(\tfrac{7}{2}\right) \times (8.314\,\text{J K}^{-1}\text{mol}^{-1}) \times (5\,\text{K}) = 3.6 \times 10^6\,\text{J}$$

$$= +3.6 \times 10^3\,\text{kJ}$$

$$\Delta U(\text{internal}) = (2.4\overline{61} \times 10^4\,\text{mol}) \times \left(\tfrac{5}{2}\right) \times (8.314\,\text{J K}^{-1}\text{mol}^{-1}) \times (5\,\text{K})$$

$$= +2.6 \times 10^6\,\text{J} = \boxed{+2.6 \times 10^3\,\text{kJ}}$$

The complete answer for $\Delta H(\text{total})$ must take into account the energy expended as work against the atmosphere by the expanding air

$$w(\text{by air}) = -p_{ext}\Delta V\,[2.10] = (-1.013 \times 10^5\,\text{Pa}) \times (10\,\text{m}^3) = -1.0 \times 10^6\,\text{J}$$

Thus the total heat which had to have been supplied to the air is

$$\Delta H(\text{total}) = \Delta H(\text{internal}) + w = (3.6 \times 10^3\,\text{kJ}) - (1.0 \times 10^3\,\text{kJ}) = \boxed{+2.6 \times 10^3\,\text{kJ}}$$

Hence, $\Delta H(\text{total}) = \Delta U(\text{internal})$

P2.2   (a) The work done on the gas in section B is

$$w_B = -nRT \ln\left(\dfrac{V_f}{V_i}\right)\,[2.13] = (-2.00\,\text{mol}) \times (8.314\,\text{J K}^{-1}\text{mol}^{-1}) \times (300\,\text{K})$$

$$\times \ln\left(\dfrac{1.00\,\text{L}}{2.00\,\text{L}}\right) = 3.46 \times 10^3\,\text{J}$$

Therefore, the work done by the gas in section A is $w_A = \boxed{-3.46 \times 10^3\,\text{J}}$

(b) $\boxed{\Delta U_B = 0}$ [constant temperature]

(c) $q_B = \Delta U_B - w_B = 0 - (3.46 \times 10^3 \, \text{J}) = \boxed{-3.46 \times 10^3 \, \text{J}}$

(d) Since the volume in section B is decreased by a factor of $\frac{1}{2}$, the pressure in B is doubled, and, since $p_A = p_B$, $p_{f,A} = 2p_{i,A}$. From the perfect gas law

$$\frac{T_{f,A}}{T_{i,A}} = \frac{p_{f,A} V_{f,A}}{p_{i,A} V_{i,A}} = \frac{(2p_{i,A}) \times (3.00 \, \text{L})}{(p_{i,A}) \times (2.00 \, \text{L})} = 3.00$$

Hence, $T_{f,A} = 3.00 \, T_{i,A} = (3.00) \times (300 \, \text{K}) = 900 \, \text{K}$

$$\Delta U_A = nC_{V,m}\Delta T = (2.00 \, \text{mol}) \times (20.0 \, \text{J K}^{-1}\text{mol}^{-1}) \times (600 \, \text{K}) = \boxed{+2.40 \times 10^4 \, \text{J}}$$

(e) $q_A = \Delta U_A - w_A = (2.40 \times 10^4 \, \text{J}) - (-3.46 \times 10^3 \, \text{J}) = \boxed{+2.75 \times 10^4 \, \text{J}}$

P2.3 The temperatures are readily obtained from the perfect gas equation, $T = \dfrac{pV}{nR}$

$$T_1 = \frac{(1.00 \, \text{atm}) \times (22.4 \, \text{L})}{(1.00 \, \text{mol}) \times (0.0821 \, \text{L atm mol}^{-1} \, \text{K}^{-1})} = \boxed{273 \, \text{K}}$$

Similarly, $T_2 = \boxed{546 \, \text{K}}$, $T_3 = \boxed{273 \, \text{K}}$

Step 1 → 2

$$w = -p_{ex}\Delta V = -p\Delta V = -nR\Delta T \quad (\Delta(pV) = \Delta(nRT))$$

$$w = -(1.00 \, \text{mol}) \times (8.314 \, \text{J K}^{-1} \, \text{mol}^{-1}) \times (546 - 273) \, \text{K} = \boxed{-2.27 \, \text{kJ}}$$

$$\Delta U = nC_{V,m}\Delta T = (1.00 \, \text{mol}) \times \tfrac{3}{2} \times (8.314 \, \text{J K}^{-1} \, \text{mol}^{-1}) \times (273 \, \text{K}) = \boxed{+3.40 \times 10^3 \, \text{J}}$$

$$q = \Delta U - w = (3.40 \times 10^3 + 2.27 \times 10^3) \, \text{J} = \boxed{+5.67 \times 10^3 \, \text{J}}$$

$$\Delta H = q = q_p = \boxed{+5.67 \times 10^3 \, \text{J}}$$

Step 2 → 3

$$w = \boxed{0} \quad \text{[constant volume]}$$

$$q = \Delta U = nC_{V,m}\Delta T = (1.00 \, \text{mol}) \times \left(\tfrac{3}{2}\right) \times (8.314 \, \text{J K}^{-1} \, \text{mol}^{-1}) \times (-273 \, \text{K})$$

$$= \boxed{-3.40 \, \text{kJ}}$$

From $H \equiv U + pV$

$$\Delta H = \Delta U + \Delta(pV) = \Delta U + \Delta(nRT) = \Delta U + nR\Delta T$$

$$= (-3.40 \times 10^3 \, \text{J}) + (1.00 \, \text{mol}) \times (8.314 \, \text{J K}^{-1} \, \text{mol}^{-1}) \times (-273 \, \text{K}) = \boxed{-5.67 \, \text{kJ}}$$

Step 3 → 1

ΔU and ΔH are $\boxed{\text{zero}}$ for an isothermal process in a perfect gas; hence

$$-q = w = -nRT \ln\frac{V_1}{V_3} = (-1.00 \, \text{mol}) \times (8.314 \, \text{J K}^{-1} \, \text{mol}^{-1}) \times (273 \, \text{K}) \times \ln\left(\frac{22.4 \, \text{L}}{44.8 \, \text{L}}\right)$$

$$= \boxed{+1.57 \times 10^3 \, \text{J}} \qquad q = \boxed{-1.57 \times 10^3 \, \text{J}}$$

Total cycle

State	p/atm	V/L	T/K
1	1.00	22.44	273
2	1.00	44.8	546
3	0.50	44.8	273

Step	Process	q/kJ	w/kJ	ΔU/kJ	ΔH/kJ
$1 \to 2$	p constant at p_{ex}	+5.67	−2.27	+3.40	+5.67
$2 \to 3$	V constant	−3.40	0	−3.40	−5.67
$3 \to 1$	Isothermal, reversible	−1.57	+1.57	0	0
Cycle		+0.70	−0.70	0	0

Comment. All values can be determined unambiguously. The net result of the overall process is that 700 J of heat has been converted to work.

P2.6   Since the volume is fixed, $\boxed{w = 0}$

since $\Delta U = q$ at constant volume, $\boxed{\Delta U = +2.35\,\text{kJ}}$

$\Delta H = \Delta U + \Delta(pV) = \Delta U + V\,\Delta p$ as $\Delta V = 0$. From the van der Waals equation [Table 1.6]

$$p = \frac{RT}{V_m - b} - \frac{a}{V_m^2} \qquad \Delta p = \frac{R\,\Delta T}{V_m - b} \qquad [\Delta V_m = 0 \text{ at constant volume}]$$

Therefore, $\Delta H = \Delta V + \dfrac{RV\,\Delta T}{V_m - b}$

From the data,

$$V_m = \frac{15.0\,\text{L}}{2.0\,\text{mol}} = 7.5\,\text{L mol}^{-1}, \; \Delta T = (341 - 300)\,\text{K} = 41\,\text{K}$$

$$V_m - b = (7.5 - 4.3 \times 10^{-2})\,\text{L mol}^{-1} = 7.4\overline{6}\,\text{L mol}^{-1}$$

$$\frac{RV\,\Delta T}{V_m - b} = \frac{(8.314\,\text{J K}^{-1}\,\text{mol}^{-1}) \times (15.0\,\text{L}) \times (41\,\text{K})}{7.4\overline{6}\,\text{L mol}^{-1}} = 0.68\,\text{kJ}$$

Therefore, $\Delta H = (2.35\,\text{kJ}) + (0.68\,\text{kJ}) = \boxed{+3.03\,\text{kJ}}$

P2.9   The heat supplied by the electric heater is

$$q_p = I\mathcal{V}t \; [2.18] = (0.232\,\text{A}) \times (12.0\,\text{V}) \times (650\,\text{s}) = 1.81 \times 10^3\,\text{J} = 1.81\,\text{kJ}$$

$$\Delta H = q_p = 1.81\,\text{kJ} \quad [\text{constant pressure}]$$

$$\Delta_{\text{vap}}H = \frac{\Delta H}{n} = \left(\frac{102\,\text{g mol}^{-1}}{1.871\,\text{g}}\right) \times (1.81\,\text{kJ}) = \boxed{+98.7\,\text{kJ mol}^{-1}}$$

$$\Delta_{\text{vap}}U = \Delta_{\text{vap}}H - \Delta n_g RT \; [2.26], \quad \Delta n_g = +1$$

$$= (98.7\,\text{kJ mol}^{-1}) - (8.314\,\text{J K}^{-1}\,\text{mol}^{-1}) \times (351\,\text{K}) = \boxed{+95.8\,\text{kJ mol}^{-1}}$$

P2.10   This cycle is represented in Figure 2.2

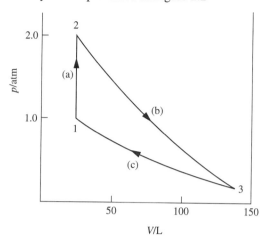

Figure 2.2

(a) We first calculate ΔU since ΔT is known ($\Delta T = 298$ K and then calculate q from the first law.)

$$\Delta U = nC_{V,m}\Delta T \ [2.21b]; \ C_{V,m} = C_{p,m} - R = \frac{7}{2}R - R = \frac{5}{2}R$$

$$\Delta U = (1.00\,\text{mol}) \times \left(\frac{5}{2}\right) \times (8.314\,\text{J K}^{-1}\,\text{mol}^{-1}) \times (298\,\text{K}) = 6.19 \times 10^3\,\text{J} = \boxed{+6.19\,\text{kJ}}$$

$$q = q_V = \Delta U - w = 6.19\,\text{kJ} - 0 \ \boxed{w = 0}, \text{constant volume}] = \boxed{+6.19\,\text{kJ}}$$

$$\Delta H = \Delta U + \Delta(pV) = \Delta U + \Delta(nRT) = \Delta U + nR\Delta T$$

$$= (6.19\,\text{kJ}) + (1.00\,\text{mol} \times (8.31 \times 10^{-3}\,\text{kJ mol}^{-1}) \times (298\,\text{K}) = \boxed{+8.67\,\text{kJ}}$$

(b) $\boxed{q = 0}$ (adiabatic)

$\Delta U(\text{b}) = -\Delta U(\text{a})$, since $\Delta T(\text{b}) = -\Delta T(\text{a})$ and the energy of a perfect gas depends on temperture alone.

$\Delta U = \boxed{-6.19\,\text{kJ}} = w$ [First law with $q = 0$]

$\Delta H(\text{b}) = -\Delta H(\text{a})$ since enthalpy of a perfect gas also depends only on temperature.

$\Delta H = \boxed{-8.67\,\text{kJ}}$

(c) $\Delta U = \Delta H = 0$ [isothermal process in perfect gas]

$q = -w$ [First law with $\Delta U = 0$]; $w = -nRT_1 \ln \dfrac{V_1}{V_3}$ [2.13]

$$V_2 = V_1 = \frac{nRT_1}{p_1} = \frac{(1.00\,\text{mol}) \times (0.08206\,\text{L atm K}^{-1}\,\text{mol}^{-1}) \times (298\,\text{K})}{1.00\,\text{atm}} = 24.4\bar{5}\,\text{L}$$

$$V_2 T_2^c = V_3 T_3^c; \ \text{hence} \quad V_3 = V_2 \left(\frac{T_2}{T_3}\right)^c \quad c = \frac{5}{2}$$

$$= (24.4\bar{5}\,\text{L}) \times \left(\frac{(2) \times (298\,\text{K})}{298\,\text{K}}\right)^{5/2} = 138.\bar{3}\,\text{L}$$

$$w = (-1.00\,\text{mol}) \times (8.314\,\text{J K}^{-1}\,\text{mol}^{-1}) \times (298\,\text{K}) \times \ln\left(\frac{22.4\bar{5}\,\text{L}}{138.3\,\text{L}}\right) = 4.29 \times 10^3\,\text{J} = \boxed{+4.29\,\text{kJ}}$$

$q = \boxed{-4.29\,\text{kJ}}$

For the entire cycle

$$\Delta U = \Delta H = \boxed{0}$$

$$q = (6.19\,\text{kJ}) + (0) - (4.29\,\text{kJ}) = \boxed{+1.90\,\text{kJ}}; \quad w = (0) - (6.19\,\text{kJ}) + (4.29\,\text{kJ}) = \boxed{-1.90\,\text{kJ}}$$

Comment. note that $q + w = 0$

P2.12   The formation reaction is

$$2C(s) + 3H_2(g) \rightarrow C_2H_6(g) \quad \Delta_f H^{\ominus}(T) = -84.68\,\text{kJ mol}^{-1}$$

In order to determine $\Delta_f H^{\ominus}(350\,\text{K})$ we employ Kirchhoff's law [2.44]; $T_2 = 350\,\text{K}$, $T_1 = 298\,\text{K}$

$$\Delta_f H^{\ominus}(T_2) = \Delta_f H^{\ominus}(T_1) + \int_{T_1}^{T_2} \Delta_r C_p \, dT$$

$$\Delta_r C_p = \sum_J \nu_J C_{p,m}(J) = C_{p,m}(C_2H_6) - 2C_{p,m}(C) - 3C_{p,m}(H_2)$$

From Table 2.2

$$C_{p,m}(C_2H_6)/(\text{J K}^{-1}\,\text{mol}^{-1}) = (14.73) + \left(\frac{0.1272}{\text{K}}\right)T$$

$$C_{p,m}(C, s)/(\text{J K}^{-1}\,\text{mol}^{-1}) = (16.86) + \left(\frac{4.77 \times 10^{-3}}{\text{K}}\right)T - \left(\frac{8.54 \times 10^5\,\text{K}^2}{T^2}\right)$$

$$C_{p,m}(H_2, g)/(\text{J K}^{-1}\,\text{mol}^{-1}) = (27.28) + \left(\frac{3.26 \times 10^{-3}}{\text{K}}\right)T + \left(\frac{0.50 \times 10^5\,\text{K}^2}{T^2}\right)$$

$$\Delta_r C_p/(\text{J K}^{-1}\,\text{mol}^{-1}) = (-100.83) + \left(\frac{0.1079\,T}{\text{K}}\right) + \left(\frac{1.56 \times 10^6\,\text{K}^2}{T^2}\right)$$

$$\int_{T_1}^{T_2} \frac{\Delta_r C_p \, dT}{\text{J K}^{-1}\,\text{mol}^{-1}} = (-100.83) \times (T_2 - T_1) + \left(\tfrac{1}{2}\right) \times (0.1079\,\text{K}^{-1}) \times (T_2{}^2 - T_1{}^2)$$

$$- (1.56 \times 10^6\,\text{K}^2) \times \left(\frac{1}{T_2} - \frac{1}{T_1}\right)$$

$$= (-100.83) \times (52\,\text{K}) + \left(\tfrac{1}{2}\right) \times (0.1079) \times (350^2 - 298^2)\,\text{K}$$

$$- (1.56 \times 10^6) \times \left(\frac{1}{350} - \frac{1}{298}\right)\text{K}$$

$$= -2.65 \times 10^3\,\text{K}$$

Multiplying by the units $\text{J K}^{-1}\,\text{mol}^{-1}$, we obtain

$$\int_{T_1}^{T_2} \Delta_r C_p \, dT = (-2.65 \times 10^3\,\text{K}) \times (\text{J K}^{-1}\,\text{mol}^{-1}) = -2.65 \times 10^3\,\text{J mol}^{-1}$$

$$= -2.65\,\text{kJ mol}^{-1}$$

Hence $\Delta_f H^{\ominus}(350\,\text{K}) = \Delta_f H^{\ominus}(298\,\text{K}) - 2.65\,\text{kJ mol}^{-1}$

$$= (-84.68\,\text{kJ mol}^{-1}) - (2.65\,\text{kJ mol}^{-1})$$

$$= \boxed{-87.33\,\text{kJ mol}^{-1}}$$

P2.13   The calorimeter is a constant-volume calorimeter as described in the text (Section 2.4); therefore

$$\Delta U = q_V$$

The calorimeter constant is determined from the data for the combustion of benzoic acid

$$\Delta U = \left(\frac{0.825\,\mathrm{g}}{122.12\,\mathrm{g\,mol^{-1}}}\right) \times (-3251\,\mathrm{kJ\,mol^{-1}}) = -21.9\overline{6}\,\mathrm{kJ}$$

Since $\Delta T = 1.940\,\mathrm{K}$, $C = \dfrac{|q|}{\Delta T} = \dfrac{21.9\overline{6}\,\mathrm{kJ}}{1.940\,\mathrm{K}} = 11.3\overline{2}\,\mathrm{kJ\,K^{-1}}$

For D-ribose, $\Delta U = -C\Delta T = (-11.3\overline{2}\,\mathrm{kJ\,K^{-1}}) \times (0.910\,\mathrm{K})$

Therefore, $\Delta_r U = \dfrac{\Delta U}{n} = (-11.3\overline{2}\,\mathrm{kJ\,K^{-1}}) \times (0.910\,\mathrm{K}) \times \left(\dfrac{150.13\,\mathrm{g\,mol^{-1}}}{0.727\,\mathrm{g}}\right) = -212\overline{7}\,\mathrm{kJ\,mol^{-1}}$

The combustion reaction for D-ribose is

$$C_5H_{10}O_5(s) + 5O_2(g) \rightarrow 5CO_2(g) + 5H_2O(l), \quad \Delta n_g = 0$$

$$\Delta_c H = \Delta_c U = \boxed{-2130\,\mathrm{kJ\,mol^{-1}}}$$

The enthalpy of formation is obtained from the sum

	$\Delta H/(\mathrm{kJ\,mol^{-1}})$
$5CO_2(g) + 5H_2O(l) \rightarrow C_5H_{10}O_5(s) + 5O_2(g)$	2130
$5C(s) + 5O_2(g) \rightarrow 5CO_2(g)$	$5 \times (-393.51)$
$5H_2(g) + \tfrac{5}{2}O_2(g) \rightarrow 5H_2O(l)$	$5 \times (-285.83)$
$5C(s) + 5H_2(g) + \tfrac{5}{2}O_2(g) \rightarrow C_5H_{10}O_5(s)$	-1267

Hence, $\Delta_f H = \boxed{-1267\,\mathrm{kJ\,mol^{-1}}}$

P2.15

$$H_3O^+(aq) + NaCH_3COO \cdot 3H_2O(s) \longrightarrow Na^+(aq) + CH_3COOH(aq) + 4H_2O(l)$$

$$n_{salt} = m_{salt}/M_{salt} = 1.3584\,\mathrm{g}/(136.08\,\mathrm{g\,mol^{-1}}) = 0.0099824\,\mathrm{mol}$$

Application of eqns 2.17 and 2.24b gives:

$$\Delta_r H_m = -\Delta_{calorimeter}H/n_{salt} = -C_{calorimeter+contents}\Delta T/n_{salt}$$
$$= (C_{calorimeter} + C_{solution})\Delta T/n_{salt}$$
$$= -(91.0\,\mathrm{J\,K^{-1}} + 4.144\,\mathrm{J\,K^{-1}mL^{-1}} \times 100\,\mathrm{mL}) \times (-0.397\,\mathrm{K})/0.0099824\,\mathrm{mol}$$
$$= 20.1\,\mathrm{kJ\,mol^{-1}}$$

Application of eqn 2.42 (the water coefficient is 3 not 4 because there is one water written on the left side of the reaction equation as part of the hydronium ion, H_3O^+) gives:

$$\Delta_r H_m^{\ominus} = 1 \times \Delta_f H_m^{\ominus}(Na^+(aq)) + 1 \times \Delta_f H_m^{\ominus}(CH_2COOH(aq)) + 3 \times \Delta_f H_m^{\ominus}(H_2O)$$
$$-1 \times \Delta_f H_m^{\ominus}(H^+(aq)) - 1 \times \Delta_f H_m^{\ominus}(NaCH_3COO \cdot 3H_2O)$$

Solving for $\Delta_f H_m^{\ominus} (Na^+(aq))$ and substituting $\Delta_f H_m^{\ominus}$ values found in both Table 2.6 and in the *Handbood of Chemistry and Physics*:

$$\Delta_f H_m^{\ominus} (Na^+(aq)) = \Delta_f H_m^{\ominus} - \Delta_f H_m^{\ominus} (CH_3COOH(aq)) - 3 \times \Delta_f H_m^{\ominus} (H_2O)$$
$$+ \Delta_f H_m^{\ominus} (H^+(aq)) + \Delta_f H_m^{\ominus} (NaCH_3COO \cdot 3H_2O)$$
$$\Delta_f H_m^{\ominus} (Na^+(aq)) = 20.1\,kJ\,mol^{-1} - (-485.76\,kJ\,mol^{-1}) - 3 \times (-285.83\,kJ\,mol^{-1})$$
$$+ (0\,kJ\,mol^{-1}) + (-1604\,kJ\,mol^{-1})$$
$$= \boxed{241\,kJ\,mol^{-1}}$$

P2.16 The reaction is

$$C_{60}(s) + 60O_2(g) \rightarrow 60CO_2(g)$$

Because the reaction does not change the number of moles of gas, $\Delta_c H = \Delta_c U$. Then

$$\Delta_c H^{\ominus} = (-36.0334\,kJ\,g^{-1}) \times (60 \times 12.011\,g\,mol^{-1}) = \boxed{-25\,968\,kJ\,mol^{-1}}$$
$$\Delta_c H^{\ominus} = 60\,\Delta_f H^{\ominus} (CO_2) - 60\,\Delta_f H^{\ominus} (O_2) - \Delta_f H^{\ominus} (C_{60}),$$

so

$$\Delta_f H^{\ominus} (C_{60}) = 60 \Delta_f H^{\ominus} (CO_2) - 60 \Delta_f H^{\ominus} (O_2) - \Delta_c H^{\ominus}$$
$$= [60(-393.51) - 60(0) - (-25968)]\,kJ\,mol^{-1} = \boxed{2357\,kJ\,mol^{-1}}$$

P2.18 **(a)** $\Delta_r H^{\ominus} = \Delta_f H^{\ominus} (SiH_2) + \Delta_f H^{\ominus} (H_2) - \Delta_f H^{\ominus} (SiH_4),$

$$= (274 + 0 - 34.3)\,kJ\,mol^{-1} = \boxed{240\,kJ\,mol^{-1}}$$

(b) $\Delta_r H^{\ominus} = \Delta_f H^{\ominus} (SiH_2) + \Delta_f H^{\ominus} (SiH_4) - \Delta_f H^{\ominus} (Si_2H_6),$

$$= (274 + 34.3 - 80.3)\,kJ\,mol^{-1} = \boxed{228\,kJ\,mol^{-1}}$$

P2.20 **(a)** (1) Heating at constant V, followed by cooling at constant p (Fig. 2.3(a)).
Isochoric step

$$pV^n = C \quad \text{or} \quad V = \left(\frac{C}{p}\right)^{1/n}$$

This says that $V = $ constant when $\boxed{n = \infty}$
Isobaric step

$$pV^n = C \quad \text{or} \quad p = \frac{C}{V^n}$$

This says that $p = $ constant when $\boxed{n = 0}$

(2) Adiabatic compression, followed by cooling at constant V (Fig. 2.3(b)).
Adiabat
$pV^{\gamma} = $ constant where $\gamma = C_p/C_V$ [2.37] so $pV^n = $ constant provided that $\boxed{n = \gamma}$
Isochore
(See part (1) above.) $\boxed{n = \infty}$

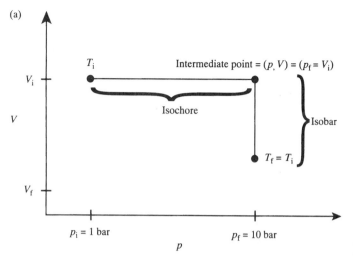

Figure 2.3(a)

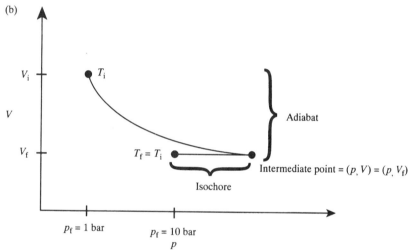

Figure 2.3(b)

(b) (1) *Initial state*

$$p_i = 1.00 \, \text{bar}, \qquad T_i = 298.15 \, \text{K}$$

$$V_i = \frac{nRT_i}{p_i} = \frac{(1 \, \text{mol}) \times (0.08314 \, \text{L bar K}^{-1} \, \text{mol}^{-1}) \times (298.15 \, \text{K})}{1.00 \, \text{bar}} = 24.79 \, \text{L}$$

Intermediate state

$$p = 10 \, \text{bar}, \qquad V = 24.79 \, \text{L}$$

$$T = \frac{pV}{nR} = \frac{(10.0 \, \text{bar}) \times (24.79 \, \text{L})}{(1 \, \text{mol}) \times (0.08314 \, \text{L bar K}^{-1} \, \text{mol}^{-1})} = 2981 \, \text{K}$$

Final state

$$p_f = 10.0 \, \text{bar}, \qquad T_f = 298.15 \, \text{K}$$

$$V_f = \frac{1}{10} V_i = 2.479 \, \text{L}$$

Isochoric step

$$w = -\int p\,dV = \boxed{0}$$

$$\Delta U = C_V \Delta T = (C_p - nR)\Delta T$$

$$= \tfrac{5}{2}nR\Delta T = \tfrac{5}{2}n(8.3145\,\mathrm{J\,K^{-1}\,mol^{-1}}) \times (2981 - 298)\,\mathrm{K}$$

$$\Delta U = \boxed{55.8\,\mathrm{kJ}}$$

$$q = \Delta U - w = \boxed{55.8\,\mathrm{kJ}}$$

$$\Delta H = \Delta U + \Delta(pV) = \Delta U + nR\Delta T$$

$$= 55.8\,\mathrm{kJ} + (1\,\mathrm{mol}) \times (8.3145\,\mathrm{J\,K^{-1}\,mol^{-1}}) \times (2981 - 298)\,\mathrm{K}$$

$$\Delta H = \boxed{78.1\,\mathrm{kJ}}$$

Isobaric step

$$w = -\int p\,dV = -p_{\mathrm{f}}\Delta V = -10\,\mathrm{bar} \times (2.479\,\mathrm{L} - 24.79\,\mathrm{L})$$

$$= 223\,\mathrm{L\,bar}\left(\frac{8.315\,\mathrm{J}}{0.08315\,\mathrm{L\,bar}}\right) = \boxed{22.3\,\mathrm{kJ}}$$

$$\Delta U = \tfrac{5}{2}nR\Delta T = \boxed{-55.8\,\mathrm{kJ}}$$

$$q = \Delta U - w = -55.8\,\mathrm{kJ} - 22.3\,\mathrm{kJ} = \boxed{-78.1\,\mathrm{kJ}}$$

$$\Delta H = \Delta U + \Delta(pV) = \Delta U + nR\Delta T = -78.1\,\mathrm{kJ}$$

Overall

$$w = w_{\mathrm{isochoric}} + w_{\mathrm{isobaric}} = \boxed{22.3\,\mathrm{kJ}}$$

$$\Delta U = \Delta U_{\mathrm{isochoric}} + \Delta U_{\mathrm{isobaric}} = \boxed{0}$$

$$q = q_{\mathrm{isochoric}} + q_{\mathrm{isobaric}} = 55.8\,\mathrm{kJ} - 78.1\,\mathrm{kJ}$$

$$q = \boxed{-22.3\,\mathrm{kJ}}$$

$$\Delta H = \Delta H_{\mathrm{isochoric}} + \Delta H_{\mathrm{isobaric}} = \boxed{0}$$

(2) *Intermediate state*

$$V = V_{\mathrm{f}}; \qquad \gamma = C_p/C_V = \left(\tfrac{7}{2}\right)/\left(\tfrac{5}{2}\right) = 7/5$$

$$p_i V_i^{\gamma} = p V_f^{\gamma}$$

$$p = \left(\frac{V_i}{V_f}\right)^{\gamma} p_i = \left(\frac{24.79\,\text{L}}{2.479\,\text{L}}\right)^{7/5} (1.00\,\text{bar}) = 25.12\,\text{bar}$$

$$T = \frac{pV}{nR} = \frac{(25.12\,\text{bar}) \times (2.479\,\text{L})}{(1\,\text{mol}) \times (0.08314\,\text{L bar K}^{-1}\,\text{mol}^{-1})} = 749\,\text{K}$$

Adiabatic step

$$q = \boxed{0}$$

$$\Delta U = C_V \Delta T = (1\,\text{mol}) \times \left(\tfrac{5}{2}\right) \times (8.3145\,\text{J K}^{-1}\,\text{mol}^{-1}) \times (749 - 298)\,\text{K}$$

$$\Delta U = \boxed{9.37\,\text{kJ}}$$

$$w = \Delta U - q = \boxed{9.37\,\text{kJ}}$$

$$\Delta H = \Delta U + \Delta(pV) = \Delta U + nR\Delta T$$

$$= 9.37\,\text{kJ} + (1\,\text{mol}) \times (8.3145\,\text{J K}^{-1}\,\text{mol}^{-1}) \times (749 - 298)\,\text{K}$$

$$\Delta H = \boxed{13.1\,\text{kJ}}$$

Isochoric step

$$w = -\int p\,dV = \boxed{0}$$

$$\Delta U = C_V \Delta T = \tfrac{5}{2} nR\Delta T$$

$$= \tfrac{5}{2}(1\,\text{mol}) \times (8.3145\,\text{J K}^{-1}\,\text{mol}^{-1}) \times (298 - 749)\text{K}$$

$$\Delta U = \boxed{-9.37\,\text{kJ}}$$

$$q = \Delta U - w = \boxed{-9.37\,\text{kJ}}$$

$$\Delta H = \Delta U + nR\Delta T$$

$$= -9.37\,\text{kJ} + (1\,\text{mol}) \times (8.3145\,\text{J K}^{-1}\,\text{mol}^{-1}) \times (298 - 749)\text{K}$$

$$\Delta H = \boxed{-13.1\,\text{kJ}}$$

Overall

$$w = w_{\text{adiabat}} + w_{\text{isochore}} = \boxed{9.37\,\text{kJ}}$$

$$q = q_{\text{adiabat}} + q_{\text{isochore}} = \boxed{-9.37\,\text{kJ}}$$

$$\Delta U = \Delta U_{\text{adiabat}} + \Delta U_{\text{isochore}} = \boxed{0}$$

$$\Delta H = \Delta H_{\text{adiabat}} + \Delta H_{\text{isochore}} = \boxed{0}$$

P2.22   (a) The process is polytropic provided that $pV^n = \text{constant} = C$. Taking the logarithm, $\ln p = \ln C - n \ln V$. Thus, if the plot of $\ln p$ against $\ln V$ is linear, the process is polytropic and the plot slope equals $-n$. The plot is shown in Fig. 2.4. A linear regression analysis of the plot provides the following

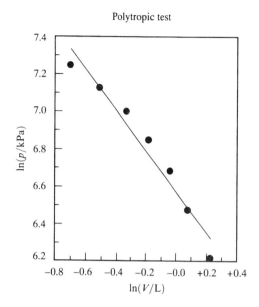

Polytropic test

Figure 2.4

$\ln C = 6.575,$ standard deviation $= 0.040;$ $C = e^{6.575}$ $C = \boxed{716.9}$

$n = \boxed{1.098},$ standard deviation $= 0.109$

$R = 0.9761$

Thus 97.61 per cent of the variation is apparently explained by the regression. However, the plot appears to vary noticeably from the linear in a systematic manner, so we conclude that the process is *not polytropic*. An R value of 0.9761, although apparently close to 1.000, does not satisfy the criterion of a good fit to a linear plot; an R value > 0.99 would be desirable.

The polytropic test plot suggests that a polynomial fit of the form

$$\ln(p/\text{kPa}) = a + b\ln(V/\text{L}) + C[\ln(V/\text{L})]^2$$

would provide a good empirical fit of the data.

Polynomial fit

$a = 6.6071,$ standard deviation $= 0.0083$

$b = -1.5217,$ standard deviation $= 0.0432$

$c = -0.08976,$ standard deviation $= 0.0795$

$R = 0.99928$

99.93 per cent of the variation is explained by the regression.

(b) Since the data have proven not to be polytropic, $p(V)$ is not a simple function that may be used in the work integral. In the absence of the simple integrand that can yield an analytic function for w, a numerical integration is appropriate.

$$w = - \int p_{ext} \, dV = - \int_{V_i=1.25\,L}^{V_f=0.50\,L} p \, dV \quad \text{[assuming mechanical equilibrium]}$$

$$= -(-685) \, \text{kPa L} \left(\frac{10^{-3} \, \text{m}^3}{L} \right) = \boxed{685 \, \text{J}}$$

(c) We know the values of p, T, and V for the initial state. This data can be used to calculate n. p and T can be used in the cubic polynomial form of the van der Waals equation. The root is V_m

$$V_m^3 - \left(b + \frac{RT}{p} \right) V_m^2 + \left(\frac{a}{p} \right) V_m - \frac{ab}{p} = 0 \quad \text{[Example 1.6]}$$

$$p = 5.00 \times 10^5 \, \text{Pa} \left(\frac{1 \, \text{atm}}{1.01325 \times 10^5 \, \text{Pa}} \right) = 4.93 \, \text{atm}$$

$$T = (273.15 + 30) \, \text{K} = 303.15 \, \text{K}$$

$$a = 4.225 \, \text{L}^2 \, \text{atm} \, \text{mol}^{-2}$$

$$b = 3.707 \times 10^{-2} \, \text{L} \, \text{mol}^{-1}$$

$$b + \frac{RT}{p} = 3.707 \times 10^{-2} \, \text{L} \, \text{mol}^{-1} + \frac{(0.0820578 \, \text{L} \, \text{atm} \, \text{K}^{-1} \, \text{mol}^{-1}) \times (303.15 \, \text{K})}{4.93 \, \text{atm}}$$

$$= 5.08 \, \text{L} \, \text{mol}^{-1}$$

$$\frac{a}{p} = \frac{4.225 \, \text{L}^2 \, \text{atm} \, \text{mol}^{-2}}{4.93 \, \text{atm}} = 0.857 \, \text{L}^2 \, \text{mol}^{-2}$$

$$\frac{ab}{p} = (0.857 \, \text{L}^2 \, \text{mol}^{-2}) \times (3.707 \times 10^{-2} \, \text{L} \, \text{mol}^{-1}) = 0.0318 \, \text{L}^3 \, \text{mol}^{-3}$$

Therefore, $V_m^3 - (5.08 \, \text{L} \, \text{mol}^{-1}) V_m^2 + (0.857 \, \text{L}^2 \, \text{mol}^{-2}) V_m - 0.0318 \, \text{L}^3 \, \text{mol}^{-3} = 0$

The real root of this equation is

$$V_m = 4.91 \, \text{L} \, \text{mol}^{-1}$$

$$n = V/V_m = 1.25 \, \text{L}/(4.91 \, \text{L} \, \text{mol}^{-1})$$

$$n = \boxed{0.255 \, \text{mol}}$$

Now we can use the van der Waals equation and the values of n, p, and V for the final state to calculate the final temperature. For the final state

$$V_m = \frac{0.50 \, \text{L}}{0.255 \, \text{mol}} = 1.96 \, \text{L} \, \text{mol}^{-1}$$

$$p = 1400 \, \text{kPa} \left(\frac{1 \, \text{atm}}{101.325 \, \text{kPa}} \right) = 13.82 \, \text{atm}$$

$$T = \frac{(V_m - b) \times \left(p + \frac{a}{V_m^2}\right)}{R} \quad [1.25]$$

$$= \frac{\left(1.96 \, \text{L mol}^{-1} - 0.03707 \, \text{L mol}^{-1}\right) \times \left(13.82 \, \text{atm} + \frac{4.225 \, \text{L}^2 \, \text{atm mol}^{-2}}{(1.96 \, \text{L mol}^{-1})^2}\right)}{0.082058 \, \text{L atm K}^{-1} \, \text{mol}^{-1}}$$

$$T = \boxed{350 \, \text{K or } 76.5^\circ \text{C}}$$

Solutions to theoretical problems

P2.24

$$w = -\int_{v_1}^{v_2} p \, dV \quad [2.12]$$

Inserting $V_m = \dfrac{V}{n}$ into the virial equation for p [1.22] we obtain

$$p = nRT\left(\frac{1}{V} + \frac{nB}{V^2} + \frac{n^2C}{V^3} + \cdots\right) \quad [V = nV_m]$$

Therefore, $w = -nRT \displaystyle\int_{V_1}^{V_2} \left(\frac{1}{V} + \frac{nB}{V^2} + \frac{n^2C}{V^3} + \cdots\right) dV$

$$= -nRT \ln\frac{V_2}{V_1} + n^2RTB\left(\frac{1}{V_2} - \frac{1}{V_1}\right) + \frac{1}{2}n^3RTC\left(\frac{1}{V_2^2} - \frac{1}{V_1^2}\right)$$

For $n = 1$ mol: $nRT = (1.0 \, \text{mol}) \times (8.314 \, \text{J K}^{-1} \, \text{mol}^{-1}) \times (273 \, \text{K}) = 2.2\overline{7} \, \text{kJ}$

From Table 1.3, $B = -21.7 \text{cm}^3\text{mol}^{-1}$, $C = 1200 \, \text{cm}^6 \, \text{mol}^{-2}$, so

$$n^2BRT = (1.0 \, \text{mol}) \times (-21.7 \, \text{cm}^3 \, \text{mol}^{-1}) \times (2.2\overline{7} \, \text{kJ}) = -49.\overline{3} \, \text{kJ cm}^3$$

$$\tfrac{1}{2}n^3CRT = \tfrac{1}{2}(1.0 \, \text{mol})^2 \times (1200 \, \text{cm}^6 \, \text{mol}^{-2}) \times (2.2\overline{7} \, \text{kJ}) = +136\overline{2} \, \text{kJ cm}^6$$

Therefore,

(a) $w = -2.2\overline{7} \, \text{kJ} \ln 2 - (49.\overline{3} \, \text{kJ}) \times \left(\dfrac{1}{1000} - \dfrac{1}{500}\right) + (136\overline{2} \, \text{kJ}) \times \left(\dfrac{1}{1000^2} - \dfrac{1}{500^2}\right)$

$= (-1.5\overline{7}) + (0.049) - (4.1 \times 10^{-3}) \, \text{kJ} = -1.5\overline{2} \, \text{kJ} = \boxed{-1.5 \, \text{kJ}}$

(b) A perfect gas corresponds to the first term of the expansion of p, so $w = -1.5\overline{7} \, \text{kJ} = \boxed{-1.6 \, \text{kJ}}$

P2.26

$$w = -\int_{V_1}^{V_2} p \, dV \quad \text{with } p = \frac{nRT}{V - nb} - \frac{n^2a}{V^2} \quad [\text{Table 1.6}]$$

Therefore, $w = -nRT \displaystyle\int_{V_1}^{V_2} \frac{dV}{V - nb} + n^2a \int_{V_1}^{V_2} \frac{dV}{V^2} = \boxed{-nRT \ln\left(\frac{V_2 - nb}{V_1 - nb}\right) - n^2a\left(\frac{1}{V_2} - \frac{1}{V_1}\right)}$

This expression can be interpreted more readily if we assume $V \gg nb$, which is certainly valid at all but the highest pressure. Then using the first term of the Taylor series expansion,

$$\ln(1 - x) = -x - \frac{x^2}{2} \ldots \text{ for } |x| \ll 1$$

$$\ln(V - nb) = \ln V + \ln\left(1 - \frac{nb}{V}\right) \approx \ln V - \frac{nb}{V}$$

and, after substitution

$$w \approx -nRT \ln\left(\frac{V_2}{V_1}\right) + n^2 bRT\left(\frac{1}{V_2} - \frac{1}{V_1}\right) - n^2 a\left(\frac{1}{V_2} - \frac{1}{V_1}\right)$$

$$\approx -nRT \ln\left(\frac{V_2}{V_1}\right) - n^2(a - bRT)\left(\frac{1}{V_2} - \frac{1}{V_1}\right)$$

$$\approx +w^0 - n^2(a - bRT)\left(\frac{1}{V_2} - \frac{1}{V_1}\right) = \text{perfect gas value plus van der Waals correction.}$$

w_0, the perfect gas value, is negative in expansion and positive in compression. Considering the correction term, in expansion $V_2 > V_1$, so $\left(\frac{1}{V_2} - \frac{1}{V_1}\right) < 0$. If attractive forces predominate, $a > bRT$ and the work done by the van der Waals gas is less in magnitude (less negative) than the prefect gas – the gas cannot easily expand. If repulsive forces predominate, $bRT > a$ and the work done by the van der Waals gas is greater in magnitude than the perfect gas – the gas easily expands.

(a) $w = -nRT \ln\left(\frac{V_f}{V_i}\right) = (-1.0\,\text{mol}^{-1}) \times (8.314\,\text{J K}^{-1}\,\text{mol}) \times (298\,\text{K}) \times \ln\left(\frac{2.0\,\text{L}}{1.0\,\text{L}}\right)$

$\quad = -1.7\bar{2} \times 10^3\,\text{J}$

$\quad = \boxed{-1.7\,\text{kJ}}$

(b) $w = w^0 - (1.0\,\text{mol})^2 \times (0 - (5.11 \times 10^{-2}\,\text{L mol}^{-1}) \times (8.314\,\text{J K}^{-1}\text{mol}^{-1}) \times (298\,\text{K}))$

$\quad \times \left(\frac{1}{2.0\,\text{L}} - \frac{1}{1.0\,\text{L}}\right)$

$\quad = (-1.7\bar{2} \times 10^3\,\text{J}) - (63\,\text{J}) = -1.7\bar{8} \times 10^3\,\text{J} = \boxed{-1.8\,\text{kJ}}$

(c) $w = w^0 - (1.0\,\text{mol})^2 \times (4.2\,\text{L}^2\,\text{atm mol}^{-2}) \times \left(\frac{1}{2.0\,\text{L}} - \frac{1}{1.0\,\text{L}}\right) = w^0 + 2.1\,\text{L atm}$

$\quad = (-1.7\bar{2} \times 10^3\,\text{J}) + (2.1\,\text{L}) \times \left(\frac{10^{-3}\,\text{m}^3}{1\,\text{L}}\right) \times (\text{atm}) \times \left(\frac{1.01 \times 10^5\,\text{Pa}}{1\,\text{atm}}\right)$

$\quad = (-1.7\bar{2} \times 10^3\,\text{J}) + (0.21 \times 10^3\,\text{J}) = \boxed{-1.5\text{kJ}}$

Schematically, the indicator diagrams for the cases (a), (b), and (c) would appear as in Figure 2.5. For case (b) the pressure is always greater than the perfect gas pressure and for case (c) always less. Therefore,

$$\int_{V_i}^{V_f} p\,dV\,(c) < \int_{V_i}^{V_f} p\,dV\,(a) < \int_{V_i}^{V_f} p\,dV\,(b)$$

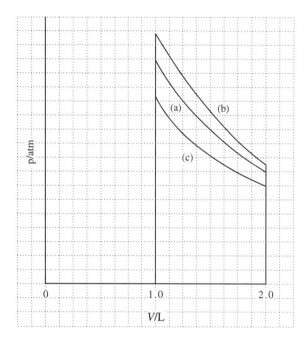

Figure 2.5

P2.28 The enthalpy of a perfect gas depends only on temperature; hence

$$\Delta H = nC_{p,\mathrm{m}}(T_\mathrm{f} - T_\mathrm{i})$$

This applies for any temperature change in a perfect gas including a reversible adiabatic one. The strategy is then to show that

$$\int_\mathrm{i}^\mathrm{f} \mathrm{d}H = \int_\mathrm{i}^\mathrm{f} V\,\mathrm{d}p = nC_{p,\mathrm{m}}(T_\mathrm{f} - T_\mathrm{i})$$

For a reversible, adiabatic change, $pV^\gamma = \mathrm{const}(A')$, so $V = \dfrac{A}{p^{1/\gamma}}$

$$\Delta H = A\int_\mathrm{i}^\mathrm{f} \frac{\mathrm{d}p}{p^{1/\gamma}} = \left\{\frac{A}{1 - \frac{1}{\gamma}}\right\} \times \left(\frac{1}{p^{1/\gamma - 1}}\right)\Bigg|_{p_\mathrm{i}}^{p_\mathrm{f}} = \left(\frac{\gamma A}{\gamma - 1}\right) \times \left(\frac{1}{p_\mathrm{f}^{1/\gamma - 1}} - \frac{1}{p_\mathrm{i}^{1/\gamma - 1}}\right)$$

$$= \left(\frac{\gamma A}{\gamma - 1}\right) \times \left(\frac{p_\mathrm{f}}{p_\mathrm{f}^{1/\gamma}} - \frac{p_\mathrm{i}}{p_\mathrm{i}^{1/\gamma}}\right)$$

$$= \left(\frac{\gamma}{\gamma - 1}\right) \times (p_\mathrm{f}V_\mathrm{f} - p_\mathrm{i}V_\mathrm{i}) = \left(\frac{nR\gamma}{\gamma - 1}\right) \times (T_\mathrm{f} - T_\mathrm{i})$$

$$\frac{\gamma}{\gamma - 1} = \frac{1}{\left(1 - \frac{1}{\gamma}\right)} = \frac{1}{\left(1 - \frac{C_{V,\mathrm{m}}}{C_{p,\mathrm{m}}}\right)} = \frac{C_{p,\mathrm{m}}}{C_{p,\mathrm{m}} - C_{V,\mathrm{m}}} = \frac{C_{p,\mathrm{m}}}{R}$$

Hence, $\boxed{\Delta H = nC_{p,\mathrm{m}}(T_\mathrm{f} - T_\mathrm{i})}$, and the supposition is proven.

Solutions to applications

P2.31   **(a)** $q_V = n\Delta_c U^{\ominus}$; hence

$$|\Delta_c U^{\ominus}| = \frac{q_V}{n} = \frac{C\Delta T}{n}[q_V = C \times \Delta T] = \frac{MC\Delta T}{m} \quad \left[n = \frac{m}{M}, \; m = \text{mass}\right]$$

Therefore, since $M = 180.16\,\text{mol}^{-1}$,

$$|\Delta_c U^{\ominus}| = \frac{(180.16\,\text{g mol}^{-1} \times (641\,\text{J K}^{-1}) \times (7.793\,\text{K})}{0.3212\,\text{g}} = \boxed{280\bar{2}\,\text{kJ mol}^{-1}}$$

(b) The complete aerobic oxidation is

$$C_{12}H_{22}O_{11} + 12O_2 \rightarrow 12CO_2 + 11H_2 \quad \Delta_c H^{\ominus} = -5645\,\text{kJ mol}^{-1}$$

The anaerobic hydrolysis to lactic acid is

$$C_{12}H_{22}O_{11} + H_2O \rightarrow 4CH_3CH(OH)COOH$$

$$\Delta H^{\ominus} = 4\Delta_f H^{\ominus}(\text{lactic acid}) - \Delta_f H^{\ominus}(\text{sucrose}) - \Delta_f H^{\ominus}(H_2O, l)$$

$$= \{(4) \times (-694.0) - (-2222) - (-285.8)\}\,\text{kJ mol}^{-1}$$

$$= -268\,\text{kJ mol}^{-1}$$

Therefore, $\Delta_c H^{\ominus}$ is $\boxed{\text{more exothermic by } 5376\,\text{kJ mol}^{-1}}$ than the hydrolysis reaction.

P2.32   From the definition of $H[H \equiv U + pV]$

$$\Delta_{trs}H - \Delta_{trs}U = \Delta(pV_m) = p\Delta V_m \quad [\text{constant pressure}]$$

$V_m = \dfrac{M}{\rho}$ where ρ is the density; therefore:

$$\Delta_{trs}H - \Delta_{trs}U = pM\Delta\frac{1}{\rho} = pM\left(\frac{1}{\rho(\text{d})} - \frac{1}{\rho(\text{gr})}\right)$$

$$= (500 \times 10^3 \times 10^5\,\text{Pa}) \times (12.01\,\text{g mol}^{-1}) \times \left(\frac{1}{3.52\,\text{g cm}^{-3}} - \frac{1}{2.27\,\text{g cm}^{-3}}\right)$$

$$= -9.39 \times 10^{10}\,\text{Pa cm}^3\,\text{mol}^{-1} = -9.39 \times 10^4\,\text{Pa m}^3\,\text{mol}^{-1}$$

$$= -9.39 \times 10^4\,\text{J mol}^{-1}$$

$$= \boxed{-93.9\,\text{kJ mol}^{-1}}$$

P2.33   **(a)** $q_p = \Delta_c H^{\ominus}$

Therefore, the heat outputs per mole are:

	butane	pentane	octane
$|\Delta_c H^{\ominus}/(\text{kJ mol}^{-1})|$	2878	3537	5471

(b) The heat outputs per gram are $|\Delta_c H^{\ominus}|/M$, and are:

$M/(\text{g mol}^{-1})$	58.13	72.15	114.23
$(\Delta_c H^{\ominus}/M)(\text{kJ g}^{-1})$	49.51	49.02	47.89

Comment. There is a little difference in the specific enthalpies of hydrocarbon fuels, so the decision to use one fuel as opposed to another is determined by other factors.

P2.34   Data: methane–octane normal alkanes

$\Delta_c H/(\text{kJ mol}^{-1})$	−890	−1560	−2220	−2878	−3537	−4163	−5471
$M/(\text{g mol}^{-1})$	16.04	30.07	44.10	58.13	72.15	86.18	114.23

Suppose that $\Delta_c H = kM^n$. There are two methods by which a regression analysis can be used to determine the values of k and n. If you have a software package that can perform a "power fit" of the type $Y = aX^b$, the analysis is direct using $Y = \Delta_c H$ and $X = M$. Then, $k = a$ and $n = b$, alternatively, taking the logarithm yields another equation—one of linear form

$$\ln |\Delta_c H| = \ln |k| + n \ln M \quad \text{where } k < 0$$

This equation suggests a linear regression fit of $\ln(\Delta_c H)$ against $\ln M$ (Fig. 2.6). The intercept is $\ln k$ and the slope is n. Linear regression fit

$$\ln |k| = 4.2112, \quad \text{standard deviation} = 0.0480; \quad k = -e^{4.2112} = \boxed{-67.44}$$

$$\boxed{n = 0.9253}, \quad \text{standard deviation} = 0.0121$$

$$R = 1.000$$

This is a good regression fit; essentially all of the variation is explained by the regression.

For decane the experimental value of $\Delta_c H$ equals $-6772.5 \text{ kJ mol}^{-1}$ (*CRC Handbook of Chemistry and Physics*). The predicted value is

$$\Delta_c H = kM^n = -67.44(142.28)^{(0.9253)} \text{ kJ mol}^{-1}$$

$$\boxed{\Delta_c H = -6625.5 \text{ kJ mol}^{-1}}$$

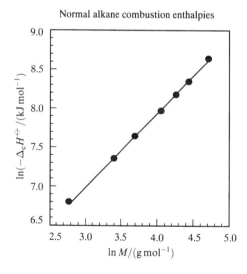

Normal alkane combustion enthalpies

Figure 2.6

$$\text{Per cent error of prediction} = \left[\left| \frac{-6772.5 - (-6625.5)}{-6625.5} \right| \right] \times 100$$

$$\text{Per cent error of prediction} = \boxed{2.17 \text{ per cent}}$$

P2.36   The three possible fates of the radical are

    (a) $tert\text{-}C_4H_9 \rightarrow sec\text{-}C_4H_9$,

    (b) $tert\text{-}C_4H_9 \rightarrow C_3H_6 + CH_3$,

    (c) $tert\text{-}C_4H_9 \rightarrow C_2H_4 + C_2H_5$.

The three corresponding enthalpy changes are

    (a)

$$\Delta_r H^{\ominus} = \Delta_f H^{\ominus}(sec\text{-}C_4H_9) - \Delta_f H^{\ominus}(tert\text{-}C_4H_9) = (67.5 - 51.3)\,\text{kJ mol}^{-1}$$

$$= \boxed{16.2\,\text{kJ mol}^{-1}}$$

    (b)

$$\Delta_r H^{\ominus} = \Delta_f H^{\ominus}(C_3H_6) + \Delta_f H^{\ominus}(CH_3) - \Delta_f H^{\ominus}(tert\text{-}C_4H_9)$$

$$= (20.42 + 145.49 - 51.3)\,\text{kJ mol}^{-1} = \boxed{114.6\,\text{kJ mol}^{-1}}$$

    (c)

$$\Delta_r H^{\ominus} = \Delta_f H^{\ominus}(C_2H_4) + \Delta_f H^{\ominus}(C_2H_5) - \Delta_f H^{\ominus}(tert\text{-}C_4H_9)$$

$$= (52.26 + 121.0 - 51.3)\,\text{kJ mol}^{-1} = \boxed{122.0\,\text{kJ mol}^{-1}}$$

3 The First Law: the machinery

Solutions to exercises

Discussion questions

E3.1(a) The change in a state function is independent of the path taken between the initial and final states; hence for the calculation of the change in that function, any convenient path may be chosen. This may greatly simplify the computation involved, and illustrates the power of thermodynamics.

E3.2(a) In a perfect gas there is no energy of interaction between the molecules of the gas, so changing the distance between the molecules by changing the volume of the gas cannot affect the internal energy of the gas. U is independent of V; hence π_T is zero.

E3.3(a) The Joule experiment showed that the internal energy of a gas at low pressures (a perfect gas) is zero. Hence in the calculation of energy changes for processes in a perfect gas one can ignore any effect due to a change in volume. This greatly simplifies the calculations involved because one can drop the first term of eqn 3.3 and need work only with $dU = C_V\,dT$. In a more sensitive apparatus, Joule would have observed a small temperature change upon expansion of the "real" gas. Joule's result holds exactly only in the limit of zero pressure where all gases can be considered perfect.

Numerical exercises

Assume that all gases are perfect and that all data refer to 298.15 K unless stated otherwise.

E3.4(a) **(a)** $\quad \dfrac{\partial^2 f}{\partial y \partial x} = \dfrac{\partial}{\partial y}(2xy) = 2x \qquad \dfrac{\partial^2 f}{\partial x \partial y} = \dfrac{\partial}{\partial x}(x^2 + 6y) = 2x$

(b) $\quad \dfrac{\partial^2 f}{\partial y \partial x} = \dfrac{\partial}{\partial y}(\cos xy - xy \sin xy)$

$$= -x \sin xy - x \sin xy - x^2 y \cos xy = -2x \sin xy - x^2 y \cos xy$$

$$\dfrac{\partial^2 f}{\partial x \partial y} = \dfrac{\partial}{\partial x}(-x^2 \sin xy) = -2x \sin xy - x^2 y \cos xy$$

E3.5(a) $\quad dz = \left(\dfrac{\partial z}{\partial x}\right)_y dx + \left(\dfrac{\partial z}{\partial y}\right)_x dy$ *[Further information 1.7]*

$$\left(\dfrac{\partial z}{\partial x}\right)_y = \left(\dfrac{\partial (ax^2 y^3)}{\partial x}\right)_y = 2axy^3 \qquad \left(\dfrac{\partial z}{\partial y}\right)_x = \left(\dfrac{\partial (ax^2 y^3)}{\partial y}\right)_x = 3ax^2 y^2$$

$$dz = \boxed{2axy^3\,dx + 3ax^2 y^2\,dy}$$

E3.6(a) **(a)** $\quad dz = \left(\dfrac{\partial z}{\partial x}\right)_y dx + \left(\dfrac{\partial z}{\partial y}\right)_x dy$ *[Further information 1.7]*

$$\left(\dfrac{\partial z}{\partial x}\right)_y = (2x - 2y + 2) \qquad \left(\dfrac{\partial z}{\partial y}\right)_x = (4y - 2x - 4)$$

$$dz = \boxed{(2x - 2y + 2)\,dx + (4y - 2x - 4)\,dy}$$

(b) $\quad \dfrac{\partial^2 z}{\partial y \partial x} = \dfrac{\partial}{\partial y}(2x - 2y + 2) = -2 \qquad \dfrac{\partial^2 z}{\partial x \partial y} = \dfrac{\partial}{\partial x}(4y - 2x - 4) = -2$

Comment. The total differential of a function is necessarily an exact differential.

E3.7(a) $\quad dz = \left(\dfrac{\partial z}{\partial x}\right)_y dx + \left(\dfrac{\partial z}{\partial y}\right)_x dy \quad$ [*Further information* 1.7]

$$\left(\dfrac{\partial z}{\partial x}\right)_y = \left(y + \dfrac{1}{x}\right) \qquad \left(\dfrac{\partial z}{\partial y}\right)_x = (x - 1)$$

$$dz = \boxed{\left(y + \dfrac{1}{x}\right) dx + (x - 1)\, dy}$$

A differential is exact if it satisfies the condition

$$\dfrac{\partial^2 z}{\partial y \partial x} = \dfrac{\partial^2 z}{\partial x \partial y} \quad [\textit{Further information } 1.7]$$

$$\left(\dfrac{\partial}{\partial y}\right)\left(\dfrac{\partial z}{\partial x}\right)_y = \left(\dfrac{\partial}{\partial y}\right)\left(y + \dfrac{1}{x}\right) = 1 \qquad \left(\dfrac{\partial}{\partial x}\right)\left(\dfrac{\partial z}{\partial y}\right)_x = \left(\dfrac{\partial}{\partial x}\right)(x - 1) = 1$$

Comment. The total differential of a function is necessarily exact as is demonstrated here by the reciprocity test.

E3.8(a) $\quad C_V = \left(\dfrac{\partial U}{\partial T}\right)_V$

$$\boxed{\left(\dfrac{\partial C_V}{\partial V}\right)_T = \left(\dfrac{\partial}{\partial V}\left(\dfrac{\partial U}{\partial T}\right)_V\right)_T = \left(\dfrac{\partial}{\partial T}\left(\dfrac{\partial U}{\partial V}\right)_T\right)_V} \qquad \text{[derivatives may be taken in any order]}$$

$$\left(\dfrac{\partial U}{\partial V}\right)_T = 0 \text{ for a perfect gas [Section 3.3]}$$

Hence, $\boxed{\left(\dfrac{\partial C_V}{\partial V}\right)_T = 0}$

E3.9(a) $\quad H = U + pV$

$$\left(\dfrac{\partial H}{\partial U}\right)_p = \boxed{1 + p\left(\dfrac{\partial V}{\partial U}\right)_p}$$

E3.10(a) $\quad V = V(p, T); \quad \text{hence, } dV = \boxed{\left(\dfrac{\partial V}{\partial p}\right)_T dp + \left(\dfrac{\partial V}{\partial T}\right)_p dT}$

We use $\alpha = \left(\dfrac{1}{V}\right)\left(\dfrac{\partial V}{\partial T}\right)_p$ [3.8] and $\kappa_T = -\left(\dfrac{1}{V}\right)\left(\dfrac{\partial V}{\partial p}\right)_T$ [3.14] and obtain

$$d\ln V = \dfrac{1}{V} dV = \left(\dfrac{1}{V}\right)\left(\dfrac{\partial V}{\partial p}\right)_T dp + \left(\dfrac{1}{V}\right)\left(\dfrac{\partial V}{\partial T}\right)_p dT = \boxed{-\kappa_T\, dp + \alpha\, dT}$$

E3.11(a) $\quad \left(\dfrac{\partial U}{\partial V}\right)_T = \left(\dfrac{\partial}{\partial V}\left(\dfrac{3}{2}nRT\right)\right)_T = \boxed{0}$

$$H = U + pV = U + nRT \quad [pV = nRT]$$

$$\left(\dfrac{\partial H}{\partial V}\right)_T = \left(\dfrac{\partial U}{\partial V}\right)_T + \left(\dfrac{\partial nRT}{\partial V}\right)_T = 0 + 0 = \boxed{0}$$

E3.12(a) $V = V(T, p)$;   hence, $dV = \left(\dfrac{\partial V}{\partial T}\right)_p dT + \left(\dfrac{\partial V}{\partial p}\right)_T dp$

Dividing each term by $(dT)_V$ we obtain

$$\left(\frac{\partial V}{\partial T}\right)_V = 0 = \left(\frac{\partial V}{\partial T}\right)_p \left(\frac{\partial T}{\partial T}\right)_V + \left(\frac{\partial V}{\partial p}\right)_T \left(\frac{\partial p}{\partial T}\right)_V$$

or $0 = \left(\dfrac{\partial V}{\partial T}\right)_p + \left(\dfrac{\partial V}{\partial p}\right)_T \left(\dfrac{\partial p}{\partial T}\right)_V$

or $\left(\dfrac{\partial p}{\partial T}\right)_V = -\dfrac{\left(\frac{\partial V}{\partial T}\right)_p}{\left(\frac{\partial V}{\partial P}\right)_T} = \dfrac{\left(\frac{1}{V}\right)\left(\frac{\partial V}{\partial T}\right)_p}{-\left(\frac{1}{V}\right)\left(\frac{\partial V}{\partial p}\right)_T} = \dfrac{\alpha}{\kappa_T}$

E3.13(a) $\mu = \left(\dfrac{\partial T}{\partial p}\right)_H$ [3.15] $= \lim\limits_{\Delta p \to 0} \left(\dfrac{\Delta T}{\Delta p}\right)_H \approx \dfrac{\Delta T}{\Delta p}$ [for μ constant over this temperature range]

$$\mu = \frac{-22\,\text{K}}{-31\,\text{atm}} = \boxed{0.71\,\text{K atm}^{-1}}$$

E3.14(a) $U_m = U_m(T, V_m)$; $dU_m = \left(\dfrac{\partial U_m}{\partial T}\right)_{V_m} dT + \left(\dfrac{\partial U_m}{\partial V_m}\right)_T dV_m{}_T$ $[\pi_T = (\partial U_m/\partial V)_T]$

For an isothermal expansion $dT = 0$; hence

$$dU_m = \left(\frac{\partial U_m}{\partial V_m}\right)_T dV_m = \frac{a}{V_m^2} dV_m$$

$$\Delta U_m = \int_{V_{m,1}}^{V_{m,2}} dU_m = \int_{V_{m,1}}^{V_{m,2}} \frac{a}{V_m^2} dV_m = a \int_{1.00\,\text{L mol}^{-1}}^{24.8\,\text{L mol}^{-1}} \frac{dV_m}{V_m^2} = -\frac{a}{V_m} \Big|_{1.00\,\text{L mol}^{-1}}^{24.8\,\text{L mol}^{-1}}$$

$$= -\frac{a}{24.8\,\text{L mol}^{-1}} + \frac{a}{1.00\,\text{L mol}^{-1}} = \frac{23.8a}{24.8\,\text{L mol}^{-1}} = 0.959\overline{7}\,\text{mol L}^{-1}a;$$

$a = 1.352\,\text{L}^2\,\text{atm mol}^{-2}$ [Table 1.6]

$\Delta U_m = (0.959\overline{7}\,\text{mol L}^{-1}) \times (1.352\,\text{L}^2\,\text{atm mol}^{-2})$

$$= (1.30\,\text{L atm mol}^{-1}) \times \left(\frac{10^{-3}\,\text{m}^3}{\text{L}}\right) \times \left(\frac{1.013 \times 10^5\,\text{Pa}}{\text{atm}}\right) = \boxed{+131\,\text{J mol}^{-1}}$$

$$w = -\int p\,dV_m$$

For a van der Waals gas

$$p = \frac{RT}{V_m - b} - \frac{a}{V_m^2}$$

Hence,

$$w = -\int \left(\frac{RT}{V_m - b}\right) dV_m + \int \frac{a}{V_m^2} dV_m = -q + \Delta U_m$$

Therefore,

$$q = \int_{1.0\,\text{L mol}^{-1}}^{24.8\,\text{L mol}^{-1}} \left(\frac{RT}{V_m - b}\right) dV_m = -RT \ln(V_m - b)\Big|_{1.00\,\text{L mol}^{-1}}^{24.8\,\text{L mol}^{-1}}$$

$$= RT \ln\left(\frac{(24.8) - (3.9 \times 10^{-2})}{(1.00) - (3.9 \times 10^{-2})}\right)$$

$$= (8.314\,\mathrm{J\,K^{-1}\,mol^{-1}}) \times (298\,\mathrm{K}) \times (3.25) = \boxed{+8.05 \times 10^3\,\mathrm{J\,mol^{-1}}}$$

$$w = -q + \Delta U_\mathrm{m} = -(8.05 \times 10^3\,\mathrm{J\,mol^{-1}}) + (131\,\mathrm{J\,mol^{-1}}) = \boxed{-7.92 \times 10^3\,\mathrm{J\,mol^{-1}}}$$

E3.15(a) $\alpha = \left(\dfrac{1}{V}\right)\left(\dfrac{\partial V}{\partial T}\right)_p$ [3.8]; $\alpha_{320} = \left(\dfrac{1}{V_{320}}\right)\left(\dfrac{\partial V}{\partial T}\right)_{p,320}$

$$V_{320} = V_{300}\{(0.75) + (3.9 \times 10^{-4}) \times (320) + (1.48 \times 10^{-6}) \times (320)^2\} = (V_{300}) \times (1.026)$$

$$\frac{1}{V_{320}} = \left(\frac{1}{1.02\overline{6}}\right) \times \left(\frac{1}{V_{300}}\right) = \frac{0.974}{V_{300}}$$

$$\left(\frac{\partial V}{\partial T}\right)_p = V_{300}(3.9 \times 10^{-4}/\mathrm{K} + 2.96 \times 10^{-6}\,T/\mathrm{K}^2)$$

$$\left(\frac{\partial V}{\partial T}\right)_{p,320} = V_{300}(3.9 \times 10^{-4}/\mathrm{K} + 2.96 \times 10^{-6} \times 320/\mathrm{K}) = 1.34 \times 10^{-3}\,\mathrm{K^{-1}}\,V_{300}$$

$$\alpha_{320} = \left(\frac{1}{V_{320}}\right)\left(\frac{\partial V}{\partial T}\right)_{p,320} = \left(\frac{0.974}{V_{300}}\right) \times (1.3 \times 10^{-3}\,\mathrm{K^{-1}}\,V_{300})$$

$$= (0.974) \times (1.34 \times 10^{-3}\,\mathrm{K^{-1}}) = \boxed{1.31 \times 10^{-3}\,\mathrm{K^{-1}}}$$

Comment. Knowledge of the density at 300 K is not required to solve this exercise, though it would be required to obtain V_{300} and V_{320} in absolute rather than relative form.

E3.16(a) $\kappa_T = -\left(\dfrac{1}{V}\right)\left(\dfrac{\partial V}{\partial p}\right)_T$ [3.14];   thus $\left(\dfrac{\partial V}{\partial p}\right)_T = -\kappa_T V$

$$dV = \left(\frac{\partial V}{\partial p}\right)_T dp \text{ [at constant } T\text{]};   \text{then } dV = -\kappa_T V\,dp \text{ or } \frac{dV}{V} = -\kappa_T\,dp$$

Substituting $V = \dfrac{m}{\rho}$ and $dV = -\dfrac{m}{\rho^2}\,d\rho$; $\dfrac{dV}{V} = -\dfrac{d\rho}{\rho} = -\kappa_T\,dp$

Therefore, $\dfrac{\delta\rho}{\rho} \approx \kappa_T\,\delta p$

For $\dfrac{\delta\rho}{\rho} = 0.08 \times 10^{-2} = 8 \times 10^{-4}$, $\delta p \approx \dfrac{8 \times 10^{-4}}{\kappa} = \dfrac{8 \times 10^{-4}}{7.35 \times 10^{-7}\,\mathrm{atm^{-1}}} = \boxed{1.\overline{1} \times 10^3\,\mathrm{atm}}$

E3.17(a) $\left(\dfrac{\partial H_\mathrm{m}}{\partial p}\right)_T = -\mu C_{p,\mathrm{m}} = (-0.25\,\mathrm{K\,atm^{-1}}) \times (29\,\mathrm{J\,K^{-1}\,mol^{-1}}) = \boxed{-7.2\,\mathrm{J\,atm^{-1}\,mol^{-1}}}$

$$dH = n\left(\frac{\partial H_\mathrm{m}}{\partial p}\right)_T dp = -n\mu C_{p,\mathrm{m}}\,dp$$

$$\Delta H = \int_{p_1}^{p_2} -n\mu C_{p,\mathrm{m}}\,dp = -n\mu C_{p,\mathrm{m}}(p_2 - p_1) \text{ [}\mu \text{ and } C_p \text{ are constants]}$$

$$\Delta H = -n\mu C_{p,\mathrm{m}}(-75\,\mathrm{atm}) = (-15\,\mathrm{mol}) \times (+7.2\,\mathrm{J\,atm^{-1}\,mol^{-1}}) \times (-75\,\mathrm{atm}) = +8.1\,\mathrm{kJ}$$

$$q(\text{supplied}) = +\Delta H = \boxed{+8.1\,\mathrm{kJ}}$$

E3.18(a) $\mu = \left(\dfrac{\partial T}{\partial p}\right)_H = \lim\limits_{\Delta p \to 0} \left(\dfrac{\Delta T}{\Delta p}\right)$

If Δp is not so large as to produce a ΔT which is a large fraction of T we may write approximately

$$\mu \approx \dfrac{\Delta T}{\Delta p} \quad \text{or} \quad \Delta p \approx \dfrac{\Delta T}{\mu}$$

For $\Delta T = -5.0\,\text{K}$,

$$\Delta p \approx \dfrac{-5.0\,\text{K}}{1.2\,\text{K atm}^{-1}} = \boxed{-4.2\,\text{atm}}$$

Solutions to problems

Assume that all gases are perfect and that all data refer to 298.15 K unless stated otherwise.

Solutions to numerical problems

P3.2 **(a)** $\mathrm{d}U_\mathrm{m} = \left(\dfrac{\partial U_\mathrm{m}}{\partial T}\right)_p \mathrm{d}T + \left(\dfrac{\partial U_\mathrm{m}}{\partial p}\right)_T \mathrm{d}p$

If we assume that $\left(\dfrac{\partial U_\mathrm{m}}{\partial p}\right)_T$ is small, and if the change in temperature is not large (10 K probably qualifies as small), then we may write

$$\Delta U_\mathrm{m} \approx \left(\dfrac{\partial U_\mathrm{m}}{\partial T}\right)_p \Delta T \qquad \left(\dfrac{\partial U_\mathrm{m}}{\partial T}\right)_p = C_{V,\mathrm{m}} + \alpha V_\mathrm{m} \left(\dfrac{\partial U}{\partial V}\right)_T = C_{V,\mathrm{m}} + \alpha \pi_T V_\mathrm{m} \quad [3.9]$$

Since $C_{p,\mathrm{m}} - C_{V,\mathrm{m}} = \alpha V_\mathrm{m}(p + \pi_T)$ [*Justification* 3.3]

$$\pi_T = \dfrac{C_{p,\mathrm{m}} - C_{V,\mathrm{m}}}{\alpha V_\mathrm{m}} - p$$

and hence

$$\left(\dfrac{\partial U_\mathrm{m}}{\partial T}\right)_p = C_{V,\mathrm{m}} + \alpha V_\mathrm{m}\left(\dfrac{C_{p,\mathrm{m}} - C_{V,\mathrm{m}}}{\alpha V} - p\right) = C_{p,\mathrm{m}} - \alpha p V_\mathrm{m}$$

$C_{p,\mathrm{m}} = 75.29\,\text{J K}^{-1}\,\text{mol}^{-1}$ [Table 2.6], $\quad \alpha = 2.1 \times 10^{-4}\,\text{K}^{-1}$ [Table 3.1]

$V_\mathrm{m} = 18.02\,\text{g mol}^{-1}/\rho$ [$\rho = 0.997\,\text{g cm}^{-3}$ at 25°C] $= 18.07 \times \text{cm}^3\,\text{mol}^{-1}$

$\quad = 18.07 \times 10^{-6}\,\text{m}^3\,\text{mol}^{-1}$

Therefore, $\left(\dfrac{\partial U_\mathrm{m}}{\partial T}\right)_p = (75.29\,\text{J K}^{-1}\,\text{mol}^{-1}) - (2.1 \times 10^{-4}\,\text{K}^{-1}) \times (1.013 \times 10^5\,\text{Pa})$

$\quad\quad\quad \times (18.07 \times 10^{-6}\,\text{m}^3\,\text{mol}^{-1})$

$\quad\quad\quad = (75.29\,\text{J K}^{-1}\,\text{mol}^{-1}) - (3.8 \times 10^{-4}\,\text{J K}^{-1}\,\text{mol}^{-1}) = 75.29\,\text{J K}^{-1}\,\text{mol}^{-1}$

Therefore, $\Delta U_\mathrm{m} \approx (75.29\,\text{J K}^{-1}\,\text{mol}^{-1}) \times (10\,\text{K}) = \boxed{+0.75\,\text{kJ mol}^{-1}}$

(b) $\mathrm{d}H_\mathrm{m} = \left(\dfrac{\partial H_\mathrm{m}}{\partial T}\right)_p \mathrm{d}T + \left(\dfrac{\partial H_\mathrm{m}}{\partial p}\right)_T \mathrm{d}p$

Assuming that $\left(\dfrac{\partial H_\mathrm{m}}{\partial p}\right)_T$ is small and that ΔT is not large, we may write

$$\Delta H_m \approx \left(\frac{\partial H_m}{\partial T}\right)_p \Delta T = C_{p,m}\Delta T = (75.29\,\mathrm{J\,K^{-1}\,mol^{-1}}) \times (10\,\mathrm{K})$$

$$= 7.5 \times 10^2\,\mathrm{J\,mol^{-1}} = \boxed{+0.75\,\mathrm{kJ\,mol^{-1}}}$$

The difference is

$$\Delta H_m - \Delta U_m = \alpha p V_m \Delta T = +3.8\,\mathrm{mJ\,mol^{-1}}$$

which is the change in energy as a result of doing expansion work.

P3.3 $\quad T_f = \left(\dfrac{p_f}{p_i}\right)^{1/c\gamma} \times (T_i)$ [Exercise 2.28 (a)]; hence, $\quad c\gamma \ln \dfrac{T_f}{T_i} = \ln \dfrac{p_f}{p_i}$

Since $c\gamma = \left(\dfrac{C_{V,m}}{R}\right) \times \left(\dfrac{C_{p,m}}{C_{V,m}}\right) = \dfrac{C_{p,m}}{R}$

$$C_{p,m} = R\frac{\ln\left(\frac{p_f}{p_i}\right)}{\ln\left(\frac{T_f}{T_i}\right)} = (8.314\,\mathrm{J\,K^{-1}\,mol^{-1}}) \times \left(\frac{\ln\left(\frac{613.85}{1522.2}\right)}{\ln\left(\frac{248.44}{298.15}\right)}\right) = \boxed{41.40\,\mathrm{J\,K^{-1}\,mol^{-1}}}$$

P3.4 $\quad dH = \left(\dfrac{\partial H}{\partial T}\right)_p dT + \left(\dfrac{\partial H}{\partial p}\right)_T dp \quad$ or $\quad dH = \left(\dfrac{\partial H}{\partial p}\right)_T dp \quad$ [constant temperature]

$$\left(\frac{\partial H_m}{\partial p}\right)_T = -\mu C_{p,m} = -\left(\frac{2a}{RT} - b\right)$$

$$= -\left(\frac{(2) \times (3.60\,\mathrm{L^2\,atm\,mol^{-2}})}{(0.0821\,\mathrm{L\,atm\,K^{-1}\,mol^{-1}}) \times (300\,\mathrm{K})} - (0.044\,\mathrm{L\,mol^{-1}})\right)$$

$$= -0.248\overline{3}\,\mathrm{L\,mol^{-1}}$$

$$\Delta H = \int_{p_i}^{p_f} dH = \int_{p_i}^{p_f} (-0.248\overline{3}\,\mathrm{L\,mol^{-1})}\,dp = -0.248\overline{3}(p_f - p_i)\,\mathrm{L\,mol^{-1}}$$

$$p = \frac{RT}{V_m - b} - \frac{a}{V_m^2}\quad [1.25b]$$

$$p_i = \left(\frac{(0.0821\,\mathrm{L\,atm\,K^{-1}\,mol^{-1}}) \times (300\,\mathrm{K})}{(20.0\,\mathrm{L\,mol^{-1}}) - (0.044\,\mathrm{L\,mol^{-1}})}\right) - \left(\frac{3.60\,\mathrm{L^2\,atm\,mol^{-2}}}{(20.0\,\mathrm{L\,mol^{-1}})^2}\right) = 1.22\overline{5}\,\mathrm{atm}$$

$$p_f = \left(\frac{(0.0821\,\mathrm{L\,atm\,K^{-1}\,mol^{-1}}) \times (300\,\mathrm{K})}{(10.0\,\mathrm{L\,mol^{-1}}) - (0.044\,\mathrm{L\,mol^{-1}})}\right) - \left(\frac{3.60\,\mathrm{L^2\,atm\,mol^{-2}}}{(10.0\,\mathrm{L\,mol^{-1}})^2}\right) = 2.43\overline{8}\,\mathrm{atm}$$

$$\Delta H = (-0.248\overline{3}\,\mathrm{L\,mol^{-1}}) \times (2.43\overline{8}\,\mathrm{atm} - 1.225\,\mathrm{atm})$$

$$= (-0.301\,\mathrm{L\,atm\,mol^{-1}}) \times \left(\frac{10^{-3}\,\mathrm{m^3}}{\mathrm{L}}\right) \times \left(\frac{1.013 \times 10^5\,\mathrm{Pa}}{\mathrm{atm}}\right) = \boxed{-30.5\,\mathrm{J\,mol^{-1}}}$$

P3.6 The system is shown in Fig. 3.1.

Initial equilibrium state

$m = 1.00\,\mathrm{mol}$ diatomic gas in each section

$p_i = 1.00\,\mathrm{bar}$

$T_i = 298\,\mathrm{K}$

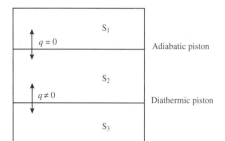

Figure 3.1

For each section $V_i = \dfrac{nRT_i}{p_i} = \dfrac{(1\,\text{mol}) \times (0.083145\,\text{L bar K}^{-1}\,\text{mol}^{-1}) \times (298\,\text{K})}{1.00\,\text{bar}} = 24.8\,\text{L}$

$V_{\text{total}} = 3V_i = 74.3\,\text{L} = \text{constant}$

Final equilibrium state

S_2 and S_3 have experienced an adiabatic process that has changed the temperature of S_3 to $T_3 = 348\,\text{K}$. Since S_2 and S_3 are separated by a diathermic wall, we can conclude that the final temperature of S_2 is identical to that of S_3 so that $T_2 = T_3$. For the same reason $V_2 = V_3$ and at equilibrium $p_1 = p_2 = p_3$ (so-called mechanical equilibrium).

$$T_3 = T_i \left(\frac{V_i}{V_3}\right)^{1/c} \quad [2.35] \quad \text{where } c = \frac{C_{V,\text{m}}}{R} = \frac{5}{2}$$

$$\left(\frac{T_3}{T_i}\right)^c = \frac{V_i}{V_3} \quad \text{or} \quad V_3 = V_i \left(\frac{T_i}{T_3}\right)^c$$

$$V_3 = 24.8\,\text{L} \left(\frac{298\,\text{K}}{348\,\text{K}}\right)^{5/2} = \boxed{16.8\,\text{L}} = V_2$$

$$V_1 = V - 2V_2 = 74.3\,\text{L} - 2(16.8\,\text{L}) = \boxed{40.7\,\text{L}}$$

$$p_3 = \frac{nRT_3}{V_3} = \frac{(1\,\text{mol}) \times (0.08315\,\text{L bar K}^{-1}\text{mol}^{-1}) \times (348\,\text{K})}{16.8\,\text{L}}$$

$$p_3 = \boxed{1.72\,\text{bar}} = p_1 = p_2$$

$$T_1 = \frac{p_1 V_1}{nR} = \frac{(1.72\,\text{bar}) \times (40.7\,\text{L})}{(1\,\text{mol}) \times (0.08315\,\text{L bar K}^{-1}\,\text{mol}^{-1})}$$

$$T_1 = \boxed{842\,\text{K}} \quad T_2 = T_3 = \boxed{348\,\text{K}}$$

$$\Delta U_1 = n_1 C_V \Delta T_1 = \frac{5}{2}(8.315\,\text{J K}^{-1}\,\text{mol}^{-1}) \times (1\,\text{mol}) \times (842\,\text{K} - 298\,\text{K})$$

$$\Delta U_1 = \boxed{11.3\,\text{kJ}}$$

$$\Delta U_3 = n_3 C_V \Delta T_3 = \frac{5}{2}(8.315\,\text{J K}^{-1}\,\text{mol}^{-1}) \times (1\,\text{mol}) \times (348\,\text{K} - 298\,\text{K})$$

$$\Delta U_3 = \boxed{1.04\,\text{kJ}} = \Delta U_2$$

$$\Delta U_{\text{total}} = \Delta U_1 + \Delta U_2 + \Delta U_3 = 11.3\,\text{kJ} + 2(1.04\,\text{kJ}) = \boxed{13.4\,\text{kJ}}$$

Notice that it does not matter whether the piston between chambers 2 and 3 is diathermic or adiabatic. The result is identical in these two cases because chamber 2 is receiving no heat from chamber 1 that can be distributed to chamber 3. That is, even with a diathermic piston between chambers 2 and 3 the heat flow from chamber 2 to chamber 3, q_{23}, equals zero.

The proof is as follows.

$$\Delta U_2 = w_2 + q_2 = C_V \Delta T_2$$

$$= -\int p\,dV_2 + \underbrace{q_{12}}_{0} + \underbrace{q_{32}}_{-q_{23}} = C_V \Delta T_2$$

Therefore

$$\Delta U_2 = -\int p\,dV_2 - q_{23} = C_V \Delta T_2 \qquad (1)$$

Likewise,

$$\Delta U_3 = -\int p\,dV_3 + q_{23} = C_V \Delta T_3 \qquad (2)$$

Subtracting (2) from (1):

$$-\int p\,d\underbrace{(V_2 - V_3)}_{0} - 2q_{23} = C_V \underbrace{(T_2 - T_3)}_{0}$$

Therefore, $q_{23} = 0$ for the presence of the diathermic wall with the particular heat flow of this problem.

P3.8

$$\mu = \left(\frac{\partial T}{\partial p}\right)_H = -\frac{1}{C_p}\left(\frac{\partial H}{\partial p}\right)_T$$

$$\mu = \frac{1}{C_p}\left\{T\left(\frac{\partial V}{\partial T}\right)_p - V\right\} \qquad \text{[Problem 3.24]}$$

But $V = \dfrac{RT}{p} + b$ or $\left(\dfrac{\partial V}{\partial T}\right)_p = \dfrac{R}{p}$

Therefore,

$$\mu = \frac{1}{C_p}\left\{\frac{RT}{p} - V\right\} = \frac{1}{C_p}\left\{\frac{RT}{p} - \frac{RT}{p} - b\right\} = \frac{-b}{C_p}$$

Since $b > 0$ and $C_p > 0$, we conclude that for this gas $\mu < 0$ or $\left(\dfrac{\partial T}{\partial p}\right)_H < 0$. This says that when the pressure drops during a Joule–Thomson expansion the temperature must $\boxed{\text{increase}}$

Solutions to theoretical problems

P3.9

$$\oint dz = \int_{(0,0)}^{(1,1)} dz + \int_{(1,1)}^{(0,0)} dz \quad \text{along } y = x$$

$$\int_{(0,0)}^{(1,1)} dz = \int_{(0,0)}^{(1,1)} (xy\,dx + xy\,dy) = \int_{(0,0)}^{(1,1)} (x^2\,dx + y^2\,dy) = \frac{1}{3} + \frac{1}{3} = \frac{2}{3}$$

along $y = x^2$

$$\int_{(1,1)}^{(0,0)} dz = \int_{(1,1)}^{(0,0)} (x^3\,dx + y^{3/2}\,dy) = -\frac{1}{4} - \frac{2}{5} = -\frac{13}{20}$$

The sum is $\left(\dfrac{2}{3} - \dfrac{13}{20}\right) = \dfrac{1}{60} \neq 0$. Therefore, this differential is $\boxed{\text{not exact}}$, since $\oint dz = 0$ if $dz =$ exact.

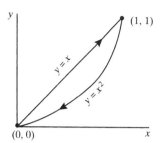

Figure 3.2

P3.10   A differential $df = g\,dx + h\,dy$ is exact if

$$\left(\frac{\partial g}{\partial y}\right)_x = \left(\frac{\partial h}{\partial x}\right)_y \quad \text{[\textit{Further information} 1.7]}$$

$$\left[\frac{\partial\left(\frac{RT}{p}\right)}{\partial T}\right]_p = \frac{R}{p} \quad \left[\frac{\partial(-R)}{\partial p}\right]_T = 0$$

Since $\dfrac{R}{p} \neq 0$, this differential is $\boxed{\text{not exact}}$, and hence cannot be the differential of a property of the system.

$$\frac{1}{T}\,dq = \frac{R}{p}\,dp - \frac{R}{T}\,dT \quad \left[\frac{\partial\left(\frac{R}{p}\right)}{\partial T}\right]_p = 0 \quad \left[\frac{\partial\left(-\frac{R}{T}\right)}{\partial p}\right]_T = 0$$

Therefore, $\dfrac{dq}{T}$ $\boxed{\text{is an exact differential}}$ and is the differential of a property of the system. This property will be identified with the entropy in Chapter 4.

P3.14   Using the permuter (Relation 3, *Further information* 1.7)

$$\left(\frac{\partial p}{\partial T}\right)_V = -\left(\frac{\partial p}{\partial V}\right)_T \left(\frac{\partial V}{\partial T}\right)_p$$

Substituting into the given expression for $C_p - C_V$

$$C_p - C_V = -T\left(\frac{\partial p}{\partial V}\right)_T \left(\frac{\partial V}{\partial T}\right)_p^2$$

Using the inverter [Relation 2, *Further information* 1.7]

$$C_p - C_V = -\frac{T\left(\frac{\partial V}{\partial T}\right)_p^2}{\left(\frac{\partial V}{\partial p}\right)_T}$$

With $pV = nRT$

$$\left(\frac{\partial V}{\partial T}\right)_p^2 = \left(\frac{nR}{p}\right)^2; \qquad \left(\frac{\partial V}{\partial p}\right)_T = -\frac{nRT}{p^2}$$

$$C_p - C_V = \frac{-T\left(\frac{nR}{p}\right)^2}{-\frac{nRT}{p^2}} = \boxed{nR}$$

P3.15 $\qquad$ $U = U(T, V)$

Hence, $dU = \left(\dfrac{\partial U}{\partial T}\right)_V dT + \left(\dfrac{\partial U}{\partial V}\right)_T dV = C_V\, dT + \pi_T\, dV$

Thus, if $\pi_T = 0$

$$\Delta U = \int_i^f dU = \int_{T_i}^{T_f} C_V\, dT$$

Therefore, if C_V and $\Delta T = T_f - T_i$ are known, ΔU may be calculated.

In the Joule experiment, a gas is expanded freely in a water bath (Figure 3.8 of text). Hence, $w = 0$. Heat transferred to the water bath may be determined from the change in temperature, ΔT, of the bath. Those gases for which $\Delta T = 0$ are defined as perfect gases. Since $\Delta T = 0$, $q = 0$, and $\Delta U = 0$.

$$\pi_T = \left(\frac{\partial U}{\partial V}\right)_T = \lim_{\Delta V \to 0} \left(\frac{\Delta U}{\Delta V}\right)_T = 0$$

Therefore, whether or not a process in a perfect gas is one of constant volume, $\Delta U = \displaystyle\int_{T_i}^{T_f} C_V\, dT$ applies.

P3.17 $\qquad$ $\mu \equiv \left(\dfrac{\partial T}{\partial p}\right)_H$

Use of the permuter (Relation 3, *Further information* 1.7) yields

$$\mu = -\frac{\left(\frac{\partial H}{\partial p}\right)_T}{C_{p,m}}$$

$$\left(\frac{\partial H}{\partial p}\right)_T = \left(\frac{\partial U}{\partial p}\right)_T + \left[\frac{\partial(pV_m)}{\partial p}\right]_T = \left(\frac{\partial U}{\partial V_m}\right)_T \left(\frac{\partial V_m}{\partial p}\right)_T + \left[\frac{\partial(pV_m)}{\partial p}\right]_T$$

Use the virial expansion of the van der Waals equation in terms of p. (See the solution to Problem 1.12 and Section 1.3 of the text.)

$$pV_m = RT\left[1 + \frac{1}{RT}\left(b - \frac{a}{RT}\right)p + \cdots\right]$$

$$\left[\frac{\partial(pV_m)}{\partial p}\right]_T \approx b - \frac{a}{RT}, \quad \left(\frac{\partial V_m}{\partial p}\right)_T \approx -\frac{RT}{p^2}$$

Substituting $\left(\dfrac{\partial H}{\partial p}\right)_T \approx \left(\dfrac{a}{V_m^2}\right) \times \left(-\dfrac{RT}{p^2}\right) + \left(b - \dfrac{a}{RT}\right) \approx \dfrac{-aRT}{(pV_m)^2} + \left(b - \dfrac{a}{RT}\right)$

Since $\left(\dfrac{\partial H}{\partial p}\right)_T$ is in a sense a correction term, that is, it approaches zero for a perfect gas, little error will be introduced by the approximation, $(pV_m)^2 = (RT)^2$.

Thus $\left(\dfrac{\partial H}{\partial p}\right)_T \approx \left(-\dfrac{a}{RT}\right) + \left(b - \dfrac{a}{RT}\right) = \left(b - \dfrac{2a}{RT}\right)$ and $\mu = \dfrac{\left(\frac{2a}{RT} - b\right)}{C_{p,m}}$

P3.19

$$H_m = H_m(T, p)$$

$$dH_m = \left(\frac{\partial H_m}{\partial T}\right)_p dT + \left(\frac{\partial H_m}{\partial p}\right)_T dp$$

Since $dT = 0$

$$dH_m = \left(\frac{\partial H_m}{\partial p}\right)_T dp \qquad \left(\frac{\partial H_m}{\partial p}\right)_T = -\mu C_{p,m} \text{ [15]} = -\left(\frac{2a}{RT} - b\right)$$

$$\Delta H_m = \int_{p_i}^{p_f} dH_m = -\int_{p_i}^{p_f} \left(\frac{2a}{RT} - b\right) dp = -\left(\frac{2a}{RT} - b\right)(p_f - p_i)$$

$$= -\left(\frac{(2) \times (1.352\,L^2\,atm\,mol^{-2})}{(0.08206\,L\,atm\,K^{-1}\,mol^{-1}) \times (300\,K)} - (0.0387\,L\,mol^{-1})\right)$$

$$\times (1.00\,atm - 500\,atm)$$

$$= (35.5\,atm) \times \left(\frac{10^{-3}\,m^3}{1\,L}\right) \times \left(\frac{1.013 \times 10^5\,Pa}{1\,atm}\right) = 3.60 \times 10^3\,J = \boxed{+3.60\,kJ}$$

Comment. Note that it is not necessary to know the value of $C_{p,m}$.

P3.22

$$\alpha = \frac{1}{V}\left(\frac{\partial V}{\partial T}\right)_p = \frac{1}{V\left(\frac{\partial T}{\partial V}\right)_p} \text{ [Relation 2, \textit{Further information} 1.7]}$$

$$= \frac{1}{V} \times \frac{1}{\left(\frac{T}{V-nb}\right) - \left(\frac{2na}{RV^3}\right) \times (V - nb)} \text{ [Problem 3.21]}$$

$$= \frac{(RV^2) \times (V - nb)}{(RTV^3) - (2na) \times (V - nb)^2}$$

$$\kappa_T = -\frac{1}{V}\left(\frac{\partial V}{\partial p}\right)_T = \frac{-1}{V\left(\frac{\partial p}{\partial V}\right)_T} \text{ [Relation 2]}$$

$$= -\frac{1}{V} \times \frac{1}{\left(\frac{-nRT}{(V-nb)^2}\right) + \left(\frac{2n^2a}{V^3}\right)} \text{ [Problem 3.20]}$$

$$= \boxed{\frac{V^2(V - nb)^2}{nRTV^3 - 2n^2a(V - nb)^2}}$$

Then $\dfrac{\kappa_T}{\alpha} = \dfrac{V - nb}{nR}$, implying that $\kappa_T R = \alpha(V_m - b)$

From the definitions of α and κ_T above

$$\frac{\kappa_T}{\alpha} = \frac{-\left(\frac{\partial V}{\partial p}\right)_T}{\left(\frac{\partial V}{\partial T}\right)_p} = \frac{-1}{\left(\frac{\partial p}{\partial V}\right)_T\left(\frac{\partial V}{\partial T}\right)_p} \text{ [Relation 2]}$$

$$= \left(\frac{\partial T}{\partial p}\right)_V \text{ [Chain relation, \textit{Further information} 1.7]}$$

$$= \frac{V - nb}{nR} \text{ [Problem 3.21]},$$

$$\kappa_T R = \frac{\alpha(V - nb)}{n}$$

Hence, $\kappa_T R = \alpha(V_m - b)$

P3.24 Work with the left-hand side of the relation to be proved and show that after manipulation using the general relations between partial derivatives and the given equation for $\left(\dfrac{\partial U}{\partial V}\right)_T$, the right-hand side is produced.

$$\left(\frac{\partial H}{\partial p}\right)_T = \left(\frac{\partial H}{\partial V}\right)_T \left(\frac{\partial V}{\partial p}\right)_T \quad \text{[change of variable]}$$

$$= \left(\frac{\partial(U + pV)}{\partial V}\right)_T \left(\frac{\partial V}{\partial p}\right)_T \quad \text{[definition of } H]$$

$$= \left(\frac{\partial U}{\partial V}\right)_T \left(\frac{\partial V}{\partial p}\right)_T + \left(\frac{\partial pV}{\partial V}\right)_T \left(\frac{\partial V}{\partial p}\right)_T$$

$$= \left\{T\left(\frac{\partial p}{\partial T}\right)_V - p\right\}\left(\frac{\partial V}{\partial p}\right)_T + \left(\frac{\partial pV}{\partial p}\right)_T \quad \left[\text{equation for } \left(\frac{\partial U}{\partial V}\right)_T\right]$$

$$= T\left(\frac{\partial p}{\partial T}\right)_V \left(\frac{\partial V}{\partial p}\right)_T - p\left(\frac{\partial V}{\partial p}\right)_T + V + p\left(\frac{\partial V}{\partial p}\right)_T$$

$$= T\left(\frac{\partial p}{\partial T}\right)_V \left(\frac{\partial V}{\partial p}\right)_T + V = \frac{-T}{\left(\frac{\partial T}{\partial V}\right)_p} + V \quad \text{[chain relation]}$$

$$= \boxed{-T\left(\frac{\partial V}{\partial T}\right)_p + V} \quad \text{[Relation 2, \textit{Further information} 1.7]}$$

P3.26 $\quad c = \left(\dfrac{RT\gamma}{M}\right)^{1/2}$, $\quad p = \rho\dfrac{RT}{M}$, $\quad$ so $\quad \dfrac{RT}{M} = \dfrac{p}{\rho}$; $\quad$ hence $\quad \boxed{c = \left(\dfrac{\gamma p}{\rho}\right)^{1/2}}$

For argon, $\gamma = \dfrac{5}{3}$, so $c = \left(\dfrac{(8.314\,\text{J K}^{-1}\,\text{mol}^{-1}) \times (298\,\text{K}) \times \frac{5}{3}}{39.95 \times 10^{-3}\,\text{kg mol}^{-1}}\right)^{1/2} = \boxed{322\,\text{m s}^{-1}}$

Solutions to applications

P3.28 The coefficient of thermal expansion is

$$\alpha = \frac{1}{V}\left(\frac{\partial V}{\partial T}\right)_p \approx \frac{\Delta V}{V\Delta T} \quad \text{so} \quad \Delta V = \alpha V \Delta T$$

This change in volume is equal to the change in height (sea level rise, Δh) times the area of the ocean (assuming that area remains constant). We will use α of pure water, although the oceans are complex solutions. For a $2°C$ rise

$$\Delta V = (2.1 \times 10^{-4}\,\text{K}^{-1}) \times (1.37 \times 10^9\,\text{km}^3) \times (2.0\,\text{K}) = 5.8 \times 10^5\,\text{km}^3$$

so $\Delta h = \dfrac{\Delta V}{A} = 1.6 \times 10^{-3}\,\text{km} = \boxed{1.6\,\text{m}}$

Since the rise in sea level is directly proportional to the rise in temperature, $\Delta T = 1°C$ would lead to $\Delta h = \boxed{0.80\,\text{m}}$ and $\Delta T = 3.5°C$ would lead to $\Delta h = \boxed{2.8\,\text{m}}$

Comment. More detailed models of climate change predict somewhat smaller rises, but the same order of magnitude.

P3.30   We compute μ from

$$\mu = -\frac{1}{C_p}\left(\frac{\partial H}{\partial p}\right)_T$$

and we estimate $\left(\dfrac{\partial H}{\partial p}\right)_T$ from the enthalpy and pressure data. We are given both enthalpy and heat capacity data on a mass basis rather than a molar basis; however, the masses will cancel, so we need not convert to a molar basis.

(a) At 300 K

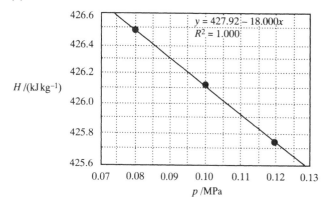

Figure 3.3(a)

The regression analysis gives the slope as $-18.0\,\text{J}\,\text{g}^{-1}\,\text{MPa}^{-1} \approx \left(\dfrac{\partial H}{\partial p}\right)_T$,

so $\mu = -\dfrac{-18.0\,\text{kJ}\,\text{kg}^{-1}\,\text{MPa}^{-1}}{0.7649\,\text{kJ}\,\text{kg}^{-1}\,\text{K}^{-1}} = \boxed{23.5\,\text{K}\,\text{MPa}^{-1}}$

(b) At 350 K

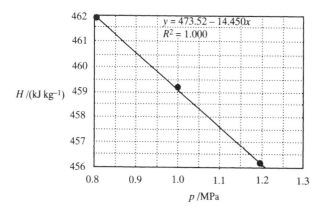

Figure 3.3(b)

The regression analysis gives the slope as $-14.5\,\text{J}\,\text{g}^{-1}\,\text{MPa}^{-1} \approx \left(\dfrac{\partial H}{\partial p}\right)_T$,

so $\mu = -\dfrac{-14.5\,\text{kJ}\,\text{kg}^{-1}\,\text{MPa}^{-1}}{1.0392\,\text{kJ}\,\text{kg}^{-1}\,\text{K}^{-1}} = \boxed{14.0\,\text{K}\,\text{MPa}^{-1}}$

4 The Second Law: the concepts

Solutions to exercises

Discussion questions

E4.1(a) We must remember that the second law of thermodynamics states only that the total entropy of both the system (here, the molecules organizing themselves into cells) and the surroundings (here, the medium) must increase in a naturally occurring process. It does not state that entropy must increase in a portion of the universe that interacts with its surroundings. In this case, the cells grow by using chemical energy from their surroundings (the medium) and in the process the increase in the entropy of the medium outweighs the decrease in entropy of the system. Hence, the second law is not violated.

E4.2(a) Refrigerators and heat pumps are thermodynamically similar and both can be thought of as heat engines operating in reverse. These devices are distinguished more by their applications than by the thermodynamic principles involved (see Section 4.2 and Box 4.1). Heat engines convert heat into work for the purpose of accomplishing a task, such as pulling a train. Refrigerators consume work for the purpose of extracting heat from a small space (the interior of the refrigerator) and in the process discharge heat to a large space. Heat pumps consume work for the purpose of extracting heat from a large space (the outside of a building) and in the process discharge heat to a small space (the inside of a building).

E4.3(a) All of these expressions are obtained from the fundamental relation of thermodynamics which is a combination of the first and second laws. (See the first equation in Justification 4.2.) It may be written as

$$-dU - p_{ex}dV + dw_{add} + TdS \geq 0$$

where we have divided the work into pressure-volume work and additional work. Under conditions of constant energy and volume and no additional work, that is, an isolated system, this relation reduces to

$$dS \geq 0$$

which is equivalent to $\Delta S_{tot} = \Delta S_{universe} \geq 0$, since the universe is an isolated system.

Under conditions of constant entropy and volume and no additional work, the fundamental relation reduces to

$$dU \leq 0.$$

Under conditions of constant temperature and volume, with no additional work, the relation reduces to

$$dA \leq 0,$$

where A is defined as $U - TS$.

Under conditions of constant temperature and pressure, with no additional work, the relation reduces to

$$dG \leq 0,$$

where G is defined as $U + pV - TS = H - TS$.

In all of the these relations, choosing the inequality provides the criteria for spontaneous change. Choosing the equal sign gives us the criteria for equilibrium under the conditions specified. See Exercise 4.3 (b).

Numerical exercises

Assume that all gases are perfect and that data refer to 298.15 K unless otherwise stated.

E4.4(a) Assume that the block is so large that its temperature does not change significantly as a result of the heat transfer. Then

$$\Delta S = \int_i^f \frac{dq_{rev}}{T} \text{ [4.3]} = \frac{1}{T} \int_i^f dq_{rev} \text{ [constant } T] = \frac{q_{rev}}{T}$$

(a) $\Delta S = \dfrac{25 \times 10^3 \text{ J}}{273.15 \text{ K}} = \boxed{92 \text{ J K}^{-1}}$ (b) $\Delta S = \dfrac{25 \times 10^3 \text{ J}}{373.15 \text{ K}} = \boxed{67 \text{ J K}^{-1}}$

E4.5(a) $S_m(T_f) = S_m(T_i) + \int_{T_i}^{T_f} \dfrac{C_{V,m}}{T} dT$ [4.20, with $C_{V,m}$ in place of $C_{p,m}$]

If we assume that neon is a perfect gas then $C_{V,m}$ may be taken to be constant and given by

$$C_{V,m} = C_{p,m} - R; \qquad C_{p,m} = 20.786 \text{ J K}^{-1} \text{ mol}^{-1} \text{ [Table 2.6]}$$
$$= (20.786 - 8.314) \text{ J K}^{-1} \text{ mol}^{-1}$$
$$= 12.472 \text{ J K}^{-1} \text{ mol}^{-1}$$

Integrating, we obtain

$$S_m(500 \text{ K}) = S_m(298 \text{ K}) + C_{V,m} \ln \frac{T_f}{T_i}$$
$$= (146.22 \text{ J K}^{-1} \text{ mol}^{-1}) + (12.472 \text{ J K}^{-1} \text{ mol}^{-1}) \ln \left(\frac{500 \text{ K}}{298 \text{ K}} \right)$$
$$= (146.22 + 6.45) \text{ J K}^{-1} \text{ mol}^{-1} = \boxed{152.67 \text{ J K}^{-1} \text{ mol}^{-1}}$$

E4.6(a) $\Delta S(\text{system}) = nC_{p,m} \ln \left(\dfrac{T_f}{T_i} \right)$ [4.20]

$$C_{p,m} = C_{V,m} + R[2.31] = \tfrac{3}{2}R + R = \tfrac{5}{2}R$$
$$\Delta S = (1.00 \text{ mol}) \times \left(\frac{5}{2} \right) \times (8.314 \text{ J K}^{-1} \text{ mol}^{-1}) \times \ln \left(\frac{573 \text{ K}}{373 \text{ K}} \right) = \boxed{8.92 \text{ J K}^{-1}}$$

E4.7(a) Since entropy is a state function, ΔS may be calculated from the most convenient path, which in this case corresponds to constant-pressure heating followed by constant-temperature compression.

$$\Delta S = nC_{p,m} \ln \left(\frac{T_f}{T_i} \right) \text{ [4.20, at } p_i] + nR \ln \left(\frac{V_f}{V_i} \right) \text{ [4.17, at } T_f]$$

Since pressure and volume are inversely related (Boyle's law), $\dfrac{V_f}{V_i} = \dfrac{p_i}{p_f}$. Hence,

$$\Delta S = nC_{p,m} \ln \left(\frac{T_f}{T_i} \right) - nR \ln \left(\frac{p_f}{p_i} \right) = (3.00 \text{ mol}) \times \left(\frac{5}{2} \right) \times (8.314 \text{ J K}^{-1} \text{ mol}^{-1}) \times \ln \left(\frac{398 \text{ K}}{298 \text{ K}} \right)$$
$$-(3.00 \text{ mol}) \times (8.314 \text{ J K}^{-1} \text{ mol}^{-1}) \times \ln \left(\frac{5.00 \text{ atm}}{1.00 \text{ atm}} \right)$$
$$= (18.0\bar{4} - 40.1\bar{4}) \text{ J K}^{-1} = \boxed{-22.1 \text{ J K}^{-1}}$$

Though $\Delta S(\text{system})$ is negative, the process can still occur spontaneously if $\Delta S(\text{total})$ is positive.

E4.8(a) $q = q_{\text{rev}} = \boxed{0}$ [adiabatic reversible process]

$$\Delta S = \int_i^f \frac{dq_{\text{rev}}}{T} = \boxed{0}$$

$\Delta U = nC_{V,\text{m}}\Delta T \ [2.22b] = (3.00\,\text{mol}) \times (27.5\,\text{J K}^{-1}\,\text{mol}^{-1}) \times (50\,\text{K}) = 4.1 \times 10^3\,\text{J}$

$\qquad = \boxed{+4.1\,\text{kJ}}$

$w = \Delta U$ [First Law with $q = 0$]

$\Delta H = nC_{p,\text{m}}\Delta T \ [2.28b]$

$C_{p,\text{m}} = C_{V,\text{m}} + R[2.31] = (27.5 + 8.3)\,\text{J K}^{-1}\,\text{mol}^{-1} = 35.8\,\text{J K}^{-1}\,\text{mol}^{-1}$

$\Delta H = (3.00\,\text{mol}) \times (35.8\,\text{J K}^{-1}\,\text{mol}^{-1}) \times (50\,\text{K}) = 5.4 \times 10^3\,\text{J} = \boxed{+5.4\,\text{kJ}}$

Comment. Neither initial nor final pressures and volumes are needed for the solution to this exercise.

E4.9(a) $\Delta S = nC_{V,\text{m}}\ln\left(\dfrac{T_f}{T_i}\right) + nR\ln\left(\dfrac{V_f}{V_i}\right)$ [Example 4.3]

$C_{V,\text{m}} = C_{p,\text{m}} - R = \frac{5}{2}R - R = \frac{3}{2}R$

$\Delta S = (1.00\,\text{mol}) \times \left(\dfrac{3}{2}\right) \times (8.314\,\text{J K}^{-1}\,\text{mol}^{-1})\ln\left(\dfrac{600\,\text{K}}{300\,\text{K}}\right)$

$\qquad + (1.00\,\text{mol}) \times (8.314\,\text{J K}^{-1}\,\text{mol}^{-1}) \times \ln\left(\dfrac{50.0\,\text{L}}{30.0\,\text{L}}\right) = \boxed{+12.9\,\text{J K}^{-1}}$

E4.10(a) $\Delta S = \dfrac{q_{\text{rev}}}{T}$ [constant temperature]

If reversible $q = q_{\text{rev}}$.

$q_{\text{rev}} = T\Delta S = (500\,\text{K}) \times (2.41\,\text{J K}^{-1}) = 1.21\,\text{kJ}$

$1.21\,\text{kJ} \neq 1.00\,\text{kJ} = q$

Therefore, the process is $\boxed{\text{not reversible.}}$

E4.11(a) **(a)** $\Delta H = \displaystyle\int_{T_1}^{T_2} nC_{p,\text{m}}\,dT$ [constant pressure]

$C_{p,\text{m}}/(\text{J K}^{-1}\,\text{mol}^{-1}) = (a + bT)$ [Table 2.2], $\quad a = 20.67,\ b = 12.38 \times 10^{-3}\,\text{K}^{-1}$

$\Delta H = \displaystyle\int_{T_1}^{T_2} n(a + bT)\,\text{J K}^{-1}\,\text{mol}^{-1}\,dT$

$\qquad = na(T_2 - T_1)\,\text{J K}^{-1}\,\text{mol}^{-1} + \frac{1}{2}nb(T_2^2 - T_1^2)\,\text{J K}^{-1}\,\text{mol}^{-1}$

$n = \dfrac{1.75 \times 10^3\,\text{g}}{26.98\,\text{g mol}^{-1}} = 64.8\bar{6}\,\text{mol}$

$\Delta H = \big[(64.8\bar{6}\,\text{mol}) \times (20.67) \times (265 - 300)\,\text{K}$

$\qquad + \left(\frac{1}{2}\right) \times (64.8\bar{6}\,\text{mol}) \times (12.38 \times 10^{-3}\,\text{K}^{-1}) \times (265^2 - 300^2)\,\text{K}^2\big]\text{J K}^{-1}\,\text{mol}^{-1}$

$\Delta H = -54.9 \times 10^3\,\text{J} = \boxed{-54.9\,\text{kJ}}$

(b) $\Delta S = \int_{T_1}^{T_2} \frac{nC_{p,\mathrm{m}}}{T}\, dT\,[4.19] = \int_{T_1}^{T_2} \frac{n(a + bT)}{T} \mathrm{J\,K^{-1}\,mol^{-1}}\, dT$

$= na\ln\left(\frac{T_2}{T_1}\right)\mathrm{J\,K^{-1}\,mol^{-1}} + nb(T_2 - T_1)\,\mathrm{J\,K^{-1}\,mol^{-1}}$

$= (64.86\,\mathrm{mol}) \times (20.67) \times \ln\left(\frac{265}{300}\right)\mathrm{J\,K^{-1}\,mol^{-1}}$

$+ (64.86\,\mathrm{mol}) \times (12.38 \times 10^{-3}\,\mathrm{K}) \times (265\,\mathrm{K} - 300\,\mathrm{K})\,\mathrm{J\,K^{-1}\,mol^{-1}}$

$\Delta S = \boxed{-195\,\mathrm{J\,K^{-1}}}$

E4.12(a) $\Delta S = nR\ln\left(\frac{V_f}{V_i}\right)\,[4.17]; \qquad \frac{p_i}{p_f} = \frac{V_f}{V_i}$ [Boyle's law]

$\Delta S = nR\ln\left(\frac{p_i}{p_f}\right) = \left(\frac{25\,\mathrm{g}}{16.04\,\mathrm{g\,mol^{-1}}}\right) \times (8.314\,\mathrm{J\,K^{-1}\,mol^{-1}}) \times \ln\left(\frac{18.5\,\mathrm{atm}}{2.5\,\mathrm{atm}}\right)$

$= \boxed{+26\,\mathrm{J\,K^{-1}}}$

E4.13(a) $\Delta S = nR\ln\left(\frac{V_f}{V_i}\right)\,[4.17]$

The number of moles (or nR) and then V_f need to be determined

$nR = \frac{p_i V_i}{T_i} = \frac{(1.00\,\mathrm{atm}) \times (15.0\,\mathrm{L})}{250\,\mathrm{K}} = \frac{(1.01\bar{3} \times 10^5\,\mathrm{Pa}) \times (15.0 \times 10^{-3}\,\mathrm{m^3})}{250\,\mathrm{K}} = 6.08\,\mathrm{J\,K^{-1}}$

$\ln\frac{V_f}{V_i} = \frac{\Delta S}{nR} = \frac{-5.0\,\mathrm{J\,K^{-1}}}{6.08\,\mathrm{J\,K^{-1}}} = -0.82\bar{3}$

Hence, $V_f = V_i e^{-0.82\bar{3}} = (15.0\,\mathrm{L}) \times (0.43\bar{9}) = \boxed{6.6\,\mathrm{L}}$

E4.14(a)   Find the common final temperature T_f by noting that the heat lost by the hot sample is gained by the cold sample

$-n_1 C_{p,\mathrm{m}}(T_f - T_{i1}) = n_2 C_{p,\mathrm{m}}(T_f - T_{i2})$

Hence, $T_f = \frac{n_1 T_{i1} + n_2 T_{i2}}{n_1 + n_2}$

Since $\frac{n_1}{n_2} = \frac{1}{2}$, $T_f = \frac{1}{3}(353\,\mathrm{K} + 2 \times 283\,\mathrm{K}) = 306\,\mathrm{K}$

The total change in entropy is that of the 50 g sample (ΔS_1) plus that of the 100 g sample (ΔS_2).

$\Delta S = \Delta S_1 + \Delta S_2 = n_1 C_{p,\mathrm{m}}\ln\frac{T_f}{T_{i1}} + n_2 C_{p,\mathrm{m}}\ln\frac{T_f}{T_{i2}}$ [constant pressure, 4.20]

$= \left(\frac{50\,\mathrm{g}}{18.02\,\mathrm{g\,mol^{-1}}}\right) \times (75.5\,\mathrm{J\,K^{-1}\,mol^{-1}}) \times \left(\ln\frac{306}{353} + 2\ln\frac{306}{283}\right) = \boxed{+2.8\,\mathrm{J\,K^{-1}}}$

E4.15(a)   Since the container is isolated, the heat flow is zero and therefore $\boxed{\Delta H = 0}$; since the masses of the bricks are equal, the final temperature must be their mean temperature, $50°\mathrm{C}$.

Specific heat capacities are heat capacities per gram and are related to the molar heat capacities by

$$C_s = \frac{C_m}{M} \quad [C_{p,m} \approx C_{V,m} = C_m]$$

So $nC_m = mC_s \ (nM = m)$

$$\Delta H(\text{individual}) = mC_s\Delta T = 1.00 \times 10^4 \, g \times 0.385 \, J\,K^{-1}\,g^{-1} \times (\pm 50\,K)$$
$$= \boxed{\pm 1.9 \times 10^2 \, kJ}$$

$$\Delta S = mC_s \ln\left(\frac{T_f}{T_i}\right) \quad [4.20]$$

$$\Delta S_1 = (10.0 \times 10^3 \, g) \times (0.385 \, J\,K^{-1}\,g^{-1}) \times \ln\left(\frac{323\,K}{273\,K}\right) = -5.541 \times 10^2 \, J\,K^{-1}$$

$$\Delta S_2 = (10.0 \times 10^3 \, g) \times (0.385 \, J\,K^{-1}\,g^{-1}) \times \ln\left(\frac{323}{273}\right) = 6.475 \times 10^2 \, J\,K^{-1}$$

$$\Delta S_{tot} = \Delta S_1 + \Delta S_2 = \boxed{+93.4 \, J\,K^{-1}}$$

Comment. The positive value of ΔS_{tot} corresponds to a spontaneous process.

E4.16(a) **(a)** $q = \boxed{0}$ [adiabatic]

(b) $w = -p_{ex}\Delta V [2.10] = -(1.01 \times 10^5 \, Pa) \times (20\,cm) \times (10\,cm^2) \times \left(\frac{10^{-6}\,m^3}{cm^3}\right) = \boxed{-20\,J}$

(c) $\Delta U = q + w = 0 - 20\,J = \boxed{-20\,J}$

(d) $\Delta U = nC_{V,m}\Delta T \ [2.21b]$

$$\Delta T = \frac{-20\,J}{(2.0\,mol) \times (28.8\,J\,K^{-1}\,mol^{-1})} = \boxed{-0.34\overline{7}\,K}$$

(e) $\Delta S = nC_{V,m} \ln\left(\frac{T_f}{T_i}\right) + nR \ln\left(\frac{V_f}{V_i}\right)$ [Example 4.3 and Exercise 4.9(a)]

$T_f = T_i - 0.34\overline{7}\,K = (298.15\,K) - (0.34\overline{7}\,K) = 297.80\overline{3}\,K$

$V_i = \frac{nRT}{p_i} = \frac{(2.0\,mol) \times (0.08206\,L\,atm\,K^{-1}\,mol^{-1}) \times (298.15\,K)}{10\,atm} = 4.89\overline{3}\,L$

$V_f = V_i + \Delta V = 4.89\overline{3} + 0.20\,L = 5.09\overline{3}\,L$

Substituting these values into the expression for ΔS above gives

$$\Delta S = (2.0\,mol) \times (28.8\,J\,K^{-1}\,mol^{-1}) \times \ln\left(\frac{297.80\overline{3}\,K}{298.15\,K}\right)$$

$$+ (2.0\,mol) \times (8.314\,J\,K^{-1}\,mol^{-1}) \ln\left(\frac{5.09\overline{3}\,L}{4.893}\right)$$

$$= (-0.067\overline{1} + 0.66\overline{6})\,J\,K^{-1} = \boxed{+0.60\,J\,K^{-1}}$$

E4.17(a) **(a)** $\Delta_{vap}S = \dfrac{\Delta_{vap}H}{T_b} = \dfrac{29.4 \times 10^3\,\text{J mol}^{-1}}{334.88\,\text{K}} = \boxed{+87.8\,\text{J K}^{-1}\,\text{mol}^{-1}}$

(b) If the vaporization occurs reversibly, $\Delta S_{tot} = 0$, so $\Delta S_{surr} = \boxed{-87.8\,\text{J K}^{-1}\,\text{mol}^{-1}}$

E4.18(a) In each case

$$\Delta_r S^{\ominus} = \sum_{\text{Products}} \nu S_m^{\ominus} - \sum_{\text{Reactants}} \nu S_m^{\ominus}\ [4.22]$$

with $S_m^{\ominus}$ values obtained from Tables 2.5 and 2.6.

(a) $\Delta_r S^{\ominus} = 2S_m^{\ominus}(CH_3COOH, l) - 2S_m^{\ominus}(CH_3CHO, g) - S_m^{\ominus}(O_2, g)$

$= [(2 \times 159.8) - (2 \times 250.3) - 205.14]\,\text{J K}^{-1}\,\text{mol}^{-1} = \boxed{-386.1\,\text{J K}^{-1}\,\text{mol}^{-1}}$

(b) $\Delta_r S^{\ominus} = 2S_m^{\ominus}(AgBr, s) + S_m^{\ominus}(Cl_2, g) - 2S_m^{\ominus}(AgCl, s) - S_m^{\ominus}(Br_2, l)$

$= [(2 \times 107.1) + (223.07) - (2 \times 96.2) - (152.23)]\,\text{J K}^{-1}\,\text{mol}^{-1}$

$= \boxed{+92.6\,\text{J K}^{-1}\,\text{mol}^{-1}}$

(c) $\Delta_r S^{\ominus} = S_m^{\ominus}(HgCl_2, s) - S_m^{\ominus}(Hg, l) - S_m^{\ominus}(Cl_2, g)$

$= [146.0 - 76.02 - 223.07]\,\text{J K}^{-1}\,\text{mol}^{-1} = \boxed{-153.1\,\text{J K}^{-1}\,\text{mol}^{-1}}$

E4.19(a) In each case we use

$$\Delta_r G^{\ominus} = \Delta_r H^{\ominus} - T\Delta_r S^{\ominus}\ [4.39]$$

along with

$$\Delta_r H^{\ominus} = \sum_{\text{Products}} \nu\Delta_f H^{\ominus} - \sum_{\text{Reactants}} \nu\Delta_f H^{\ominus}\ [2.42]$$

(a) $\Delta_r H^{\ominus} = 2\Delta_f H^{\ominus}(CH_3COOH, l) - 2\Delta_f H^{\ominus}(CH_3CHO, g)$

$= [2 \times (-484.5) - 2 \times (-166.19)]\,\text{kJ mol}^{-1} = -636.6\bar{2}\,\text{kJ mol}^{-1}$

$\Delta_r G^{\ominus} = -636.6\bar{2}\,\text{kJ mol}^{-1} - (298.15\,\text{K}) \times (-386.1\,\text{J K}^{-1}\,\text{mol}^{-1}) = \boxed{-521.5\,\text{kJ mol}^{-1}}$

(b) $\Delta_r H^{\ominus} = 2\Delta_f H^{\ominus}(AgBr, s) - 2\Delta_f H^{\ominus}(AgCl, s)$

$= [2 \times (-100.37) - 2 \times (-127.07)]\,\text{kJ mol}^{-1} = +53.40\,\text{kJ mol}^{-1}$

$\Delta_r G^{\ominus} = +53.40\,\text{kJ mol}^{-1} - (298.15\,\text{K}) \times (+92.6)\,\text{J K}^{-1}\,\text{mol}^{-1} = \boxed{+25.8\,\text{kJ mol}^{-1}}$

(c) $\Delta_r H^{\ominus} = \Delta_f H^{\ominus}(HgCl_2, s) = -224.3\,\text{kJ mol}^{-1}$

$\Delta_r G^{\ominus} = -224.3\,\text{kJ mol}^{-1} - (298.15\,\text{K}) \times (-153.1\,\text{J K}^{-1}\,\text{mol}^{-1}) = \boxed{-178.7\,\text{kJ mol}^{-1}}$

E4.20(a) In each case $\Delta_r G^{\ominus} = \displaystyle\sum_{\text{Products}} \nu\Delta_f G^{\ominus} - \sum_{\text{Reactants}} \nu\Delta_f G^{\ominus}\ [4.40]$

with $\Delta_f G^{\ominus}$ (J) values from Table 2.6.

(a) $\Delta_r G^{\ominus} = 2\Delta_f G^{\ominus}(CH_3COOH, l) - 2\Delta_f G^{\ominus}(CH_3CHO, g)$

$= [2 \times (-389.9) - 2 \times (-128.86)]\,\text{kJ mol}^{-1}$

$= \boxed{-522.1\,\text{kJ mol}^{-1}}$

(b) $\Delta_r G^{\ominus} = 2\Delta_f G^{\ominus}(AgBr, s) - 2\Delta_f G^{\ominus}(AgCl, s) = [2 \times (-96.90) - 2 \times (-109.79)]\,\text{kJ mol}^{-1}$

$= \boxed{+25.78\,\text{kJ mol}^{-1}}$

(c) $\Delta_r G^\ominus = \Delta_f G^\ominus(HgCl_2, s)$

$\qquad = \boxed{-178.6\,kJ\,mol^{-1}}$

Comment. In each case these values of $\Delta_r G^\ominus$ agree closely with the calculated values in Exercise **4.16(a)**.

E4.21(a) $\Delta_r G^\ominus = \Delta_r H^\ominus - T\Delta_r S\ [4.39] \quad \Delta_r H^\ominus = \sum_{Products} \nu\Delta_f H^\ominus - \sum_{Reactants} \nu\Delta_f H^\ominus\ [2.42]$

$\Delta_r S^\ominus = \sum_{Products} \nu S_m^\ominus - \sum_{Reactants} \nu S_m^\ominus\ [4.22]$

$\Delta_r H^\ominus = 2\Delta_f H^\ominus(H_2O, l) - 4\Delta_f H^\ominus(HCl, g) = \{2 \times (-285.83) - 4 \times (-92.31)\}\,kJ\,mol^{-1}$

$\qquad = -202.42\,kJ\,mol^{-1}$

$\Delta_r S^\ominus = 2S_m^\ominus(Cl_2, g) + 2S_m^\ominus(H_2O, l) - 4S_m^\ominus(HCl, g) - S_m^\ominus(O_2, g)$

$\qquad = [(2 \times 69.91) + (2 \times 223.07) - (4 \times 186.91) - (205.14)]\,J\,K^{-1}\,mol^{-1}$

$\qquad = -366.82\,J\,K^{-1}\,mol^{-1} = -0.36682\,kJ\,K^{-1}\,mol^{-1}$

$\Delta_r G^\ominus = -202.42\,kJ\,mol^{-1} - (298.15\,K) \times (-0.36682\,kJ\,K^{-1}\,mol^{-1}) = \boxed{-93.05\,kJ\,mol^{-1}}$

Question. Repeat the calculation based on $\Delta_f G^\ominus$ data of Table 2.6. What difference, if any, is there from the value above?

E4.22(a)   The formation reaction for phenol is

$6C(s) + 3H_2(g) + \tfrac{1}{2}O_2(g) \rightarrow C_6H_5OH(s)$

$\Delta_f G^\ominus = \Delta_f H^\ominus - T\Delta_f S^\ominus\ [4.39]$

$\Delta_f H^\ominus$ is to be obtained from $\Delta_c H^\ominus$ for phenol and data from Tables 2.5 and 2.6. Thus

$C_6H_5OH(s) + 7O_2(g) \rightarrow 6CO_2(g) + 3H_2O(l)$

$\Delta_c H^\ominus = 6\Delta_f H^\ominus(CO_2, g) + 3\Delta_f H^\ominus(H_2O, l) - \Delta_f H^\ominus(C_6H_5OH, s)$

Hence $\Delta_f H^\ominus(C_6H_5OH, s) = 6\Delta_f H^\ominus(CO_2, g) + 3\Delta_f H^\ominus(H_2O, l) - \Delta_c H^\ominus$

$\qquad\qquad = [6 \times (-393.51) + 3 \times (-285.83) - (-3054)]\,kJ\,mol^{-1}$

$\qquad\qquad = -164.\overline{55}\,kJ\,mol^{-1}$

$\Delta_f S^\ominus = \sum_{Products} \nu S_m^\ominus - \sum_{Reactants} \nu S_m^\ominus\ [4.22]$

$\Delta_f S^\ominus = S_m^\ominus(C_6H_5OH, s) - 6S_m^\ominus(C, s) - 3S_m^\ominus(H_2, g) - \tfrac{1}{2}S_m^\ominus(O_2, g)$

$\qquad = \left[144.0 - (6 \times 5.740) - (3 \times 130.68) - \left(\tfrac{1}{2} \times 205.14\right)\right]\,J\,K^{-1}\,mol^{-1}$

$\qquad = -385.0\overline{5}\,J\,K^{-1}\,mol^{-1}$

Hence $\Delta_f G^\ominus = -164.\overline{55}\,kJ\,mol^{-1} - (298.15\,K) \times (-385.0\overline{5}\,J\,K^{-1}\,mol^{-1}) = \boxed{-50\,kJ\,mol^{-1}}$

E4.23(a)   **(a)** $\Delta S(gas) = nR\ln\dfrac{V_f}{V_i}\ [4.17] = \left(\dfrac{14\,g}{28.02\,g\,mol^{-1}}\right) \times (8.314\,J\,K^{-1}\,mol^{-1}) \times (\ln 2)$

$\qquad = \boxed{+2.9\,J\,K^{-1}}$

$\Delta S(surroundings) = \boxed{-2.9\,J\,K^{-1}}$ [overall zero entropy production]

$\Delta S(total) = \boxed{0}$ [reversible process]

(b) $\Delta S(\text{gas}) = \boxed{+2.9\,\text{J K}^{-1}}$ [S a state function]

$\Delta S(\text{surroundings}) = \boxed{0}$ [the surroundings do not change]

$\Delta S(\text{total}) = \boxed{+2.9\,\text{J K}^{-1}}$

(c) $\Delta S(\text{gas}) = \boxed{0}$ [$q_{\text{rev}} = 0$]

$\Delta S(\text{surroundings}) = \boxed{0}$ [no heat is transferred to the surroundings]

$\Delta S(\text{total}) = \boxed{0}$

E4.24   The same final state is attained if the change takes place in two stages, one being isothermal compression

$$\Delta S_1 = nR \ln \frac{V_f}{V_i}[4.17] = nR \ln \frac{1}{2} = -nR \ln 2$$

and the second, heating at constant volume

$$\Delta S_2 = nC_{V,m} \ln \frac{T_f}{T_i}\ [4.20] = nC_{V,m} \ln 2$$

The overall entropy change is therefore

$$\Delta S = -nR \ln 2 + nC_{V,m} \ln 2 = \boxed{n(C_{V,m} - R) \ln 2}$$

E4.25(a) $CH_4(g) + 2O_2(g) \rightarrow CO_2(g) + 2H_2O(l)$

$$\Delta_r G^\ominus = \sum_{\text{Products}} v\Delta_f G^\ominus - \sum_{\text{Rectants}} v\Delta_f G^\ominus\ [4.40]$$

$$\Delta_r G^\ominus = \Delta_f G^\ominus(CO_2, g) + 2\Delta_f G^\ominus(H_2O, l) - \Delta_f G^\ominus(CH_4, g)$$

$$= \{-394.36 + (2 \times -237.13) - (-50.72)\}\,\text{kJ mol}^{-1} = -817.90\,\text{kJ mol}^{-1}$$

Therefore, the maximum non-expansion work is $\boxed{817.90\,\text{kJ mol}^{-1}}$ [since $|w_e| = |\Delta G|$].

E4.26(a) $\varepsilon_{\text{rev}} = 1 - \dfrac{T_c}{T_h}$ [4.11]

(a) $\varepsilon = 1 - \dfrac{333\,\text{K}}{373\,\text{K}} = \boxed{0.11}$ [11 per cent efficiency]

(b) $\varepsilon = 1 - \dfrac{353\,\text{K}}{573\,\text{K}} = \boxed{0.38}$ [38 per cent efficiency]

Solutions to problems

Assume that all gases are perfect and that data refer to 298.15 K unless otherwise stated.

Solutions to numerical problems

P4.3   **(a)** $q(\text{total}) = q(H_2O) + q(Cu) = 0$, hence $-q(H_2O) = q(Cu)$

$q(H_2O) = n(-\Delta_{\text{vap}}H) + nC_{p,m}(H_2O, l)\Delta T(H_2O)$

$q(Cu) = mC_s\Delta T(Cu), \quad C_s = 0.385\,\text{J K}^{-1}\text{g}^{-1}$

$(1.00\,\text{mol}) \times (40.656 \times 10^3\,\text{J mol}^{-1}) - (1.00\,\text{mol}) \times (75.3\,\text{J K}^{-1}\,\text{mol}^{-1}) \times (\theta - 100^\circ\text{C})$

$= (2.00 \times 10^3\,\text{g}) \times (0.385\,\text{J K}^{-1}\,\text{g}^{-1}) \times \theta$

Solving for θ, $\theta = 57.0°C = \boxed{330.2\,K}$

$q(Cu) = (2.00 \times 10^3\,g) \times (0.385\,J\,K^{-1}\,g^{-1}) \times (57.0\,K) = 4.39 \times 10^4\,J = \boxed{43.9\,kJ}$

$q(H_2O) = \boxed{-43.9\,kJ}$

$\Delta S(total) = \Delta S(H_2O) + \Delta S(Cu)$

$$\Delta S(H_2O) = \frac{-n\Delta_{vap}H}{T_b} + nC_{p,m}\ln\left(\frac{T_f}{T_i}\right) \quad [4.20]$$

$$= -\frac{(1.00\,mol) \times (40.656 \times 10^3\,J\,mol^{-1})}{373.2\,K}$$

$$+ (1.00\,mol) \times (75.3\,J\,K^{-1}\,mol^{-1}) \times \ln\left(\frac{330.2\,K}{373.2\,K}\right)$$

$$= -108.9\,J\,K^{-1} - 9.22\,J\,K^{-1} = \boxed{-118.\bar{1}\,J\,K^{-1}}$$

$$\Delta S(Cu) = mC_s \ln\frac{T_f}{T_i} = (2.00 \times 10^3\,g) \times (0.385\,J\,K^{-1}\,g^{-1}) \times \ln\left(\frac{330.2\,K}{273.2\,K}\right)$$

$$= \boxed{145.\bar{9}\,J\,K^{-1}}$$

$$\Delta S(total) = -118.\bar{1}\,J\,K^{-1} + 145.\bar{9}\,J\,K^{-1} = \boxed{28\,J\,K^{-1}}$$

This process is spontaneous since ΔS (surroundings) is zero and, hence, ΔS(universe) = ΔS(total) = positive.

(b) The volume of the container may be calculated from the perfect gas law.

$$V = \frac{nRT}{p} = \frac{(1.00\,mol) \times (0.08206\,L\,atm\,K^{-1}\,mol^{-1}) \times (373.2\,K)}{1.00\,atm} = 30.6\,L$$

At 57.0°C the vapor pressure of water is 130 Torr (HCP). The amount of water vapor present at equilibrium is then

$$n = \frac{pV}{RT} = \frac{(130\,Torr) \times \left(\frac{1\,atm}{760\,Torr}\right) \times (30.6\,L)}{(0.08206\,L\,atm\,K^{-1}\,mol^{-1}) \times (330.2\,K)} = 0.193\,mol$$

This is a substantial fraction of the original amount of water and cannot be ignored. Consequently the calculation needs to be redone taking into account the fact that only a part, n_1, of the vapor condenses into a liquid while the remainder $(1.00\,mol - n_1)$ remains gaseous. The heat flow involving water, then, becomes

$$q(H_2O) = -n_1\Delta_{vap}H + n_1C_{p,m}(H_2O, l)\Delta T(H_2O)$$
$$+ (1.00\,mol - n_1)C_{p,m}(H_2O, g)\Delta T(H_2O)$$

Because n_1 depends on the equilibrium temperature through

$$n_1 = 1.00\,mol - \frac{pV}{RT}$$

where p is the vapor pressure of water, we will have two unknowns (p and T) in the equation $-q(H_2O) = q(Cu)$. There are two ways out of this dilemma: (1) p may be expressed as a function of T by use of the Clapeyron equation (Chapter 6), or (2) by use of successive approximations. Redoing the calculation with

$$n_1 = (1.00\,\text{mol}) - (0.193\,\text{mol}) = 0.80\overline{7}\,\text{mol}$$

(noting that $C_{p,m}(\text{H}_2\text{O}, \text{g}) = (75.3 - 41.9)\,\text{J}\,\text{mol}^{-1}\,\text{K}^{-1}$ (Problem 4.1)) yields a final temperature of 47.2°C. At this temperature, the vapor pressure of water is 80.41 Torr, corresponding to

$$n_1 = (1.00\,\text{mol}) - (0.123\,\text{mol}) = 0.87\overline{7}\,\text{mol}$$

The recalculated final temperature is 50.8°C. The successive approximations eventually converge to yield a value of $\boxed{49.9^\circ\text{C} = 323.2\,\text{K}}$ for the final temperature. Using this value of the final temperature, the heat transferred and the various entropies are calculated as in part (**a**).

P4.4   This problem concerns the same system and the same changes of state as Problem 2.3. The final temperature of section A was there calculated to be 900 K.

(**a**) $\Delta S_A = n C_{V,m} \ln\left(\dfrac{T_{A,f}}{V_{A,i}}\right) + nR \ln\left(\dfrac{V_{A,f}}{V_{A,i}}\right)$ [Example 4.3]

$$= (2.0\,\text{mol}) \times (20\,\text{J}\,\text{K}^{-1}\,\text{mol}^{-1}) \times \ln\left(\frac{900\,\text{K}}{300\,\text{K}}\right)$$

$$+ (2.00\,\text{mol}) \times (8.314\,\text{J}\,\text{K}^{-1}\,\text{mol}^{-1}) \times \ln\left(\frac{3.00\,\text{L}}{2.00\,\text{L}}\right)$$

$$= \boxed{50.7\,\text{J}\,\text{K}^{-1}}$$

$$\Delta S_B = nR \ln\left(\frac{V_{B,f}}{V_{B,i}}\right) = (2.00\,\text{mol}) \times (8.314\,\text{J}\,\text{K}^{-1}\,\text{mol}^{-1}) \times \ln\left(\frac{1.00\,\text{L}}{2.00\,\text{L}}\right)$$

$$= \boxed{-11.5\,\text{J}\,\text{K}^{-1}}$$

(**b**)   In the solution to Problem 2.3 the reversible work in sections A and B was calculated:

$$w_A = -3.46 \times 10^3\,\text{J}, \qquad w_B = 3.46 \times 10^3\,\text{J}, \qquad w_{\text{max}} = w_{\text{rev}} = \Delta A \,[4.35]$$

But this relation holds only at constant temperature; hence

$$\Delta A_B = w_B = +3.46 \times 10^3\,\text{J} = \boxed{+3.46\,\text{kJ}} \,[\text{constant temperature}]$$

$$\Delta A_A \neq w_A \,[\text{temperature not constant}]$$

We might expect that ΔA_A is negative, since w_A is negative; but based on the information provided we can only state that it is $\boxed{\text{indeterminate.}}$

(**c**)   Under constant-temperature conditions

$$\Delta G = \Delta H - T\Delta S$$

In section B, $\Delta H_B = 0$ [constant temperature, perfect gas]

$$\Delta S_B = -11.5\,\text{J}\,\text{K}^{-1}$$

$$\Delta G_B = -T_B \Delta S_B = -(300\,\text{K}) \times (-11.5\,\text{J}\,\text{K}^{-1}) = \boxed{3.46 \times 10^3\,\text{J}}$$

ΔG_A is $\boxed{\text{indeterminate}}$ in both magnitude and sign. A resolution of this problem is only possible based on additional relations developed in Chapters 5 and 19.

(d) $\Delta S(\text{total system}) = \Delta S_A + \Delta S_B = (50.7 - 11.5)\,\text{J K}^{-1} = \boxed{+39.2\,\text{J K}^{-1}}$

If the process has been carried out reversibly as assumed in the statement of the problem we can say

$$\Delta S(\text{system}) + \Delta S(\text{surroundings}) = 0$$

Hence, $\Delta S(\text{surroundings}) = \boxed{-39.2\,\text{J K}^{-1}}$

Question. Can you design this process such that heat is added to section A reversibly?

P4.5

	Step 1	Step 2	Step 3	Step 4	Cycle
q	+11.5 kJ	0	−5.74 kJ	0	5.8 kJ
w	−11.5 kJ	−3.74 kJ	+5.74 kJ	3.74 kJ	−5.8 kJ
ΔU	0	−3.74 kJ	0	+3.74 kJ	0
ΔH	0	−6.23 kJ	0	+6.23 kJ	0
ΔS	+19.1 J K^{-1}	0	−19.1 J K^{-1}	0	0
ΔS_{sur}	−19.1 J K^{-1}	0	+19.1 J K^{-1}	0	0
ΔS_{tot}	0	0	0	0	0

Step 1

$$\Delta U = \Delta H = \boxed{0}\ [\text{isothermal}]$$

$$w = -nRT \ln\left(\frac{V_f}{V_i}\right) = nRT \ln\left(\frac{p_f}{p_i}\right)\ [\text{2.13, and Boyle's law}]$$

$$= (1.00\,\text{mol}) \times (8.314\,\text{J K}^{-1}\,\text{mol}^{-1}) \times (600\,\text{K}) \times \ln\left(\frac{1.00\,\text{atm}}{10.0\,\text{atm}}\right) = \boxed{-11.5\,\text{kJ}}$$

$$q = -w = \boxed{11.5\,\text{kJ}}$$

$$\Delta S = nR \ln\left(\frac{V_f}{V_i}\right)\ [\text{4.17}] = -nR \ln\left(\frac{p_f}{p_i}\right)\ [\text{Boyle's law}]$$

$$= -(1.00\,\text{mol}) \times (8.314\,\text{J K}^{-1}\,\text{mol}^{-1}) \times \ln\left(\frac{1.00\,\text{atm}}{10.0\,\text{atm}}\right) = \boxed{+19.1\,\text{J K}^{-1}}$$

$$\Delta S(\text{sur}) = -\Delta S(\text{system})\ [\text{reversible process}] = \boxed{-19.1\,\text{J K}^{-1}}$$

$$\Delta S_{\text{tot}} = \Delta S(\text{system}) + \Delta S(\text{sur}) = \boxed{0}$$

Step 2

$$q = \boxed{0}\ [\text{adiabatic}]$$

$$\Delta U = nC_{V,\text{m}}\Delta T\ [\text{2.21b}]$$

$$= (1.00\,\text{mol}) \times \left(\tfrac{3}{2}\right) \times (8.314\,\text{J K}^{-1}\,\text{mol}^{-1}) \times (300\,\text{K} - 600\,\text{K}) = \boxed{-3.74\,\text{kJ}}$$

$$w = \Delta U = \boxed{-3.74\,\text{kJ}}$$

$$\Delta H = \Delta U + \Delta(pV) = \Delta U + nR\Delta T$$

$$= (-3.74\,\text{kJ}) + (1.00\,\text{mol}) \times (8.314\,\text{J K}^{-1}\,\text{mol}^{-1}) \times (-300\,\text{K})$$

$$= \boxed{-6.23\,\text{kJ}}$$

$\Delta S = \Delta S(\text{sur}) = \boxed{0}$ [reversible adiabatic process]

$\Delta S_{\text{tot}} = \boxed{0}$

Step 3

These quantities may be calculated in the same manner as for *Step 1* or more easily as follows

$\Delta U = \Delta H = \boxed{0}$ [isothermal]

$\varepsilon_{\text{rev}} = 1 - \dfrac{T_c}{T_h}$ [4.11] $= 1 - \dfrac{300\,\text{K}}{600\,\text{K}} = 0.500 = 1 + \dfrac{q_c}{q_h}$ [4.10]

$q_c = -0.500\,q_h = -(0.500) \times (11.5\,\text{kJ}) = -5.74\,\text{kJ}$

$q_c = \boxed{-5.74\,\text{kJ}} \qquad w = -q_c = \boxed{5.74\,\text{kJ}}$

$\Delta S = -\Delta S$ (*Step 1*) [initial and final temperature reversed] $= \boxed{-19.1\,\text{J K}^{-1}}$

$\Delta S(\text{sur}) = -\Delta S(\text{system}) = \boxed{+19.1\,\text{J K}^{-1}}$

$\Delta S_{\text{tot}} = \boxed{0}$

Step 4

ΔU and ΔH are the negative of their values in *Step 2*. (Initial and final temperatures reversed.)

$\Delta U = \boxed{+3.74\,\text{kJ}}, \qquad \Delta H = \boxed{+6.23\,\text{kJ}}, \qquad q = \boxed{0}$ [adiabatic]

$w = \Delta U = \boxed{+3.74\,\text{kJ}}$

$\Delta S = \Delta S(\text{sur}) = \boxed{0}$ [reversible adiabatic process]

$\Delta S_{\text{tot}} = \boxed{0}$

Cycle

$\Delta U = \Delta H = \Delta S = \boxed{0}$ [$\Delta(\text{state function}) = 0$ for any cycle]

$\Delta S(\text{sur}) = 0$ [all reversible processes]

$\Delta S_{\text{tot}} = \boxed{0}$

$q(\text{cycle}) = (11.5 - 5.74)\,\text{kJ} = \boxed{5.8\,\text{kJ}} \qquad w(\text{cycle}) = -q(\text{cycle}) = \boxed{-5.8\,\text{kJ}}$

P4.6

	q	w	$\Delta U = \Delta H$	ΔS	ΔS_{sur}	ΔS_{tot}
Path (a)	2.74 kJ	−2.74 kJ	0	9.13 J K^{-1}	−9.13 J K^{-1}	0
Path (b)	1.66 kJ	−1.66 kJ	0	9.13 J K^{-1}	−5.53 J K^{-1}	3.60 J K^{-1}

Path (a)

$w = -nRT \ln\left(\dfrac{V_f}{V_i}\right) = -nRT \ln\left(\dfrac{p_i}{p_f}\right)$ [Boyle's law]

$= -(1.00\,\text{mol}) \times (8.314\,\text{J K}^{-1}\,\text{mol}^{-1}) \times (300\,\text{K}) \times \ln\left(\dfrac{3.00\,\text{atm}}{1.00\,\text{atm}}\right) = -2.74 \times 10^3\,\text{J}$

$= \boxed{-2.74\,\text{kJ}}$

$$\Delta H = \Delta U = \boxed{0} \text{ [isothermal process in perfect gas]}$$

$$q = \Delta U - w = 0 - (-2.74\,\text{kJ}) = \boxed{+2.74\,\text{kJ}}$$

$$\Delta S = \frac{q_{\text{rev}}}{T} \text{ [Example 4.1]} = \frac{2.74 \times 10^3\,\text{J}}{300\,\text{K}} = \boxed{+9.13\,\text{J K}^{-1}}$$

$$\Delta S_{\text{tot}} = \boxed{0} \text{ [reversible process]}$$

$$\Delta S_{\text{sur}} = \Delta S_{\text{tot}} - \Delta S = 0 - 9.13\,\text{J K}^{-1} = \boxed{-9.13\,\text{J K}^{-1}}$$

Path (b)

$$w = -p_{\text{ex}}(V_{\text{f}} - V_{\text{i}}) = -p_{\text{ex}}\left(\frac{nRT}{p_{\text{f}}} - \frac{nRT}{p_{\text{i}}}\right) \text{ [perfect gas]} = -nRT\left(\frac{p_{\text{ex}}}{p_{\text{f}}} - \frac{p_{\text{ex}}}{p_{\text{i}}}\right)$$

$$= -(1.00\,\text{mol}) \times (8.314\,\text{J K}^{-1}\,\text{mol}^{-1}) \times (300\,\text{K}) \times \left(\frac{1.00\,\text{atm}}{1.00\,\text{atm}} - \frac{1.00\,\text{atm}}{3.00\,\text{atm}}\right)$$

$$= -1.66 \times 10^3\,\text{J} = \boxed{-1.66\,\text{kJ}}$$

$$\Delta H = \Delta U = \boxed{0} \text{ [isothermal process in perfect gas]}$$

$$q = \Delta U - w = 0 - (-1.66\,\text{kJ}) = \boxed{+1.66\,\text{kJ}}$$

$$\Delta S = \frac{q_{\text{rev}}}{T} = \boxed{+9.13\,\text{J K}^{-1}} \quad [\Delta S \text{ is independent of path}]$$

$$\Delta S_{\text{sur}} = \frac{q_{\text{sur}}}{T_{\text{sur}}} = \frac{-q}{T_{\text{sur}}} = \frac{-1.66 \times 10^3\,\text{J}}{300\,\text{K}} = \boxed{-5.53\,\text{J K}^{-1}}$$

$$\Delta S_{\text{tot}} = \Delta S + \Delta S_{\text{sur}} = (9.13 - 5.53)\,\text{J K}^{-1} = \boxed{+3.60\,\text{J K}^{-1}}$$

P4.7

	q	$w = \Delta U$	ΔH	ΔS	ΔS_{sur}	ΔS_{tot}
Path (a)	0	$-9.1 \times 10^2\,\text{J}$	$-1.5 \times 10^3\,\text{J}$	0	0	0
Path (b)	0	$-7.5 \times 10^2\,\text{J}$	$-1.2 \times 10^3\,\text{J}$	$+1.12\,\text{J K}^{-1}$	0	$+1.12\,\text{J K}^{-1}$

$$C_{p,\text{m}} = C_{V,\text{m}} + R = \frac{3}{2}R + R = \frac{5}{2}R, \qquad \gamma = \frac{C_{p,\text{m}}}{C_{V,\text{m}}} = \frac{5}{3}, \qquad c = \frac{C_{V,\text{m}}}{R} = \frac{\frac{3}{2}R}{R} = \frac{3}{2}$$

(a) $\quad T_{\text{f}} = \left(\dfrac{V_{\text{i}}}{V_{\text{f}}}\right)^{1/c} T_{\text{i}} \text{ [2.35]} = \left(\dfrac{V_{\text{i}}}{V_{\text{f}}}\right)^{\gamma-1} T_{\text{i}} \quad \left[\dfrac{1}{c} = \dfrac{R}{C_{V,\text{m}}} = \dfrac{C_{p,\text{m}} - C_{V,\text{m}}}{C_{V,\text{m}}} = \gamma - 1\right]$

$$p_{\text{i}}V_{\text{i}}^{\gamma} = p_{\text{f}}V_{\text{f}}^{\gamma} \text{ [2.36]} \quad \text{or} \quad \frac{V_{\text{i}}}{V_{\text{f}}} = \left(\frac{p_{\text{f}}}{p_{\text{i}}}\right)^{1/\gamma}$$

Substituting into the expression for T_{f} above

$$T_{\text{f}} = \left(\frac{p_{\text{f}}}{p_{\text{i}}}\right)^{(\gamma-1)/\gamma} T_{\text{i}} = \left(\frac{p_{\text{i}}}{p_{\text{f}}}\right)^{(1-\gamma)/\gamma} T_{\text{i}}$$

$$= \left(\frac{1.00\,\text{atm}}{0.50\,\text{atm}}\right)^{[1-(5/3)]/(5/3)} \times (300\,\text{K}) = \boxed{227\,\text{K}}$$

$$w = \Delta U = nC_{V,\mathrm{m}}\Delta T = (1.00\,\mathrm{mol}) \times \left(\tfrac{3}{2}\right) \times (8.314\,\mathrm{J\,K^{-1}\,mol^{-1}}) \times (227.\bar{4} - 300\,\mathrm{K})$$

$$= \boxed{-9.1 \times 10^2\,\mathrm{J}}$$

$$\Delta H = nC_{p,\mathrm{m}}\Delta T = (1.00\,\mathrm{mol}) \times \left(\tfrac{5}{2}\right) \times (8.314\,\mathrm{J\,K^{-1}\,mol^{-1}}) \times (-72.\bar{6}\,\mathrm{K})$$

$$= \boxed{-1.5 \times 10^3\,\mathrm{J}}$$

$$\Delta S_{\mathrm{tot}} = \boxed{0}\ \text{[reversible process]} = \Delta S + \Delta S_{\mathrm{sur}}$$

$$\Delta S_{\mathrm{sur}} = \boxed{0}\ \text{[adiabatic process]};\quad \text{hence,}\quad \Delta S = \boxed{0}$$

(b) $\Delta U = w$ [adiabatic process]

$$\Delta U = nC_{V,\mathrm{m}}(T_{\mathrm{f}} - T_{\mathrm{i}})$$

$$w = -p_{\mathrm{ex}}(V_{\mathrm{f}} - V_{\mathrm{i}}) = -p_{\mathrm{ex}}\left(\frac{nRT_{\mathrm{f}}}{p_{\mathrm{f}}} - \frac{nRT_{\mathrm{i}}}{p_{\mathrm{i}}}\right)$$

Solving for T_{f}, with $p_{\mathrm{ex}} = p_{\mathrm{f}} = 0.50\,\mathrm{atm}$, $p_{\mathrm{i}} = 1.00\,\mathrm{atm}$

$$T_{\mathrm{f}} = T_{\mathrm{i}} \times \left\{\frac{C_{V,\mathrm{m}} + \left(\frac{p_{\mathrm{ex}}R}{p_{\mathrm{i}}}\right)}{C_{V,\mathrm{m}} + \left(\frac{p_{\mathrm{ex}}R}{p_{\mathrm{f}}}\right)}\right\} = (300\,\mathrm{K}) \times \left(\frac{\tfrac{3}{2}R + \tfrac{1}{2}R}{\tfrac{3}{2}R + R}\right) = (300\,\mathrm{K}) \times \frac{4}{5} = \boxed{240\,\mathrm{K}}$$

$$w = \Delta U = (1.00\,\mathrm{mol}) \times \left(\tfrac{3}{2}\right) \times (8.314\,\mathrm{J\,K^{-1}\,mol^{-1}}) \times (240\,\mathrm{K} - 300\,\mathrm{K})$$

$$= \boxed{-7.5 \times 10^2\,\mathrm{J}}$$

$$\Delta H = nC_{p,\mathrm{m}}\Delta T = (1.00\,\mathrm{mol}) \times \left(\tfrac{5}{2}\right) \times (8.314\,\mathrm{J\,K^{-1}\,mol^{-1}}) \times (-60\,\mathrm{K})$$

$$= \boxed{-1.2 \times 10^3\,\mathrm{J}}$$

$$\Delta S = nC_{p,\mathrm{m}}\ln\left(\frac{T_{\mathrm{f}}}{T_{\mathrm{i}}}\right) - nR\ln\left(\frac{p_{\mathrm{f}}}{p_{\mathrm{i}}}\right)\ \text{[Exercise 4.4]}$$

$$= (1.00\,\mathrm{mol}) \times \left(\frac{5}{2}\right) \times (8.314\,\mathrm{J\,K^{-1}\,mol^{-1}}) \times \ln\left(\frac{240\,\mathrm{K}}{300\,\mathrm{K}}\right)$$

$$- (1.00\,\mathrm{mol}) \times (8.314\,\mathrm{J\,K^{-1}\,mol^{-1}}) \times \ln\left(\frac{0.50\,\mathrm{atm}}{1.00\,\mathrm{atm}}\right) = \boxed{+1.12\,\mathrm{J\,K^{-1}}}$$

$$\Delta S_{\mathrm{sur}} = \boxed{0}\ \text{[adiabatic process]}$$

$$\Delta S_{\mathrm{tot}} = \Delta S + \Delta S_{\mathrm{sur}} = 1.12\,\mathrm{J\,K^{-1}} + 0 = \boxed{+1.12\,\mathrm{J\,K^{-1}}}$$

P4.11 $\quad \mathrm{C(s)} + \tfrac{1}{2}\mathrm{O_2(g)} + 2\mathrm{H_2(g)} \rightarrow \mathrm{CH_3OH(l)}, \quad \Delta n_{\mathrm{g}} = -2.5\,\mathrm{mol}$

$\Delta G = \Delta H - T\Delta S$ [constant temperature] $\Delta H = \Delta U + \Delta(pV)$

Therefore, $\Delta G = \Delta U - T\Delta S + \Delta(pV) = \Delta A + \Delta(pV)$ and

$$\Delta_{\mathrm{f}}A^{\ominus} = \Delta_{\mathrm{f}}G^{\ominus} - \Delta(pV) = \Delta_{\mathrm{f}}G^{\ominus} - \Delta n_{\mathrm{g}}(RT)\ \text{[perfect gases]} = \Delta_{\mathrm{f}}G^{\ominus} + 2.5RT$$

$$= [(-166.27) + (2.5) \times (2.479)]\,\mathrm{kJ\,mol^{-1}} = \boxed{-160.07\,\mathrm{kJ\,mol^{-1}}}$$

P4.13 **(a)** Under constant-temperature conditions

$$\Delta A = w_{\mathrm{max}}\ \text{[4.35]}$$

Since $\Delta A = \Delta G - \Delta(pV)$, it is convenient to first work part **(b)**.

(b) Under constant-temperature and pressure conditions

$$\Delta G = w_{add,max} = \Delta G \text{ [4.38]}$$

Using the same cycle as in Problem 4.1, with

$$\Delta C_{p,m} \equiv C_{p,m}(\text{liq}) - C_{p,m}(\text{gas}) \cdots$$
$$\Delta G(T) = \Delta H(T) - T\Delta S(T)$$

$$= \Delta H(T_f) - \Delta C_{p,m}(T - T_f) - T\left(\Delta S(T_f) - \Delta C_{p,m} \ln \frac{T}{T_f}\right)$$

$$= \Delta H(T_f) - \frac{T}{T_f}\Delta H(T_f) - \Delta C_{p,m}\left(T - T_f - T \ln \frac{T}{T_f}\right);$$

$$\Delta H(T_f) = -\Delta_{fus}H(T_f)$$

$$= \left(\frac{T}{T_f} - 1\right)\Delta_{fus}H(T_f) - \Delta C_{p,m}\left(T - T_f - T \ln \frac{T}{T_f}\right)$$

$$T = 268 \text{ K}, \quad T_f = 273 \text{ K}, \quad \Delta_{fus}H = 6.01 \text{ kJ mol}^{-1}, \quad \Delta C_{p,m} = +37.3 \text{ J K}^{-1} \text{ mol}^{-1}:$$

$$\Delta G(268 \text{ K}) = \left(\frac{268}{273} - 1\right) \times (6.01 \text{ kJ mol}^{-1}) - (37.3 \text{ J mol}^{-1})$$

$$\times \left(268 - 273 - 268 \ln \frac{268}{273}\right) = \boxed{-0.11 \text{ kJ mol}^{-1}}$$

Returning to part **(a)** we use

$$\Delta A = \Delta G - \Delta(pV) = \Delta G - p\Delta V \text{ [constant pressure]} = \Delta G - pM\Delta\left(\frac{1}{\rho}\right)$$

$$= (-0.11 \text{ kJ mol}^{-1}) - (1.013 \times 10^5 \text{ Pa}) \times (18.02 \times 10^{-3} \text{ kg mol}^{-1})$$

$$\times \left(\frac{1}{917 \text{ kg m}^{-3}} - \frac{1}{999 \text{ kg m}^{-3}}\right)$$

$$= (-0.11 \text{ kJ mol}^{-1}) - (1.6 \times 10^{-4} \text{ kJ mol}^{-1}) = -0.11 \text{ kJ mol}^{-1}$$

Therefore

(a) Maximum work is $\boxed{0.11 \text{ kJ mol}^{-1}}$

(b) Maximum non-expansion work is also $\boxed{0.11 \text{ kJ mol}^{-1}}$

However, there is a slight difference of $1.6 \times 10^{-4} \text{ kJ mol}^{-1}$ between the two values.

P4.14
$$S_m(T) = S_m(0) + \int_0^T \frac{C_{p,m} dT}{T} \text{ [4.19]}$$

From the data, draw up the following table

T/K	10	15	20	25	30	50
$\frac{C_{p,m}}{T}/(\text{J K}^{-2} \text{ mol}^{-1})$	0.28	0.47	0.540	0.564	0.550	0.428

T/K		70	100	150	200	250	298
$\dfrac{C_{p,m}}{T}/(\mathrm{J\,K^{-2}\,mol^{-1}})$		0.333	0.245	0.169	0.129	0.105	0.089

Plot $C_{p,m}/T$ against T (Fig. 4.1). This has been done on two scales. The region 0 to 10 K has been constructed using $C_{p,m} = aT^3$, fitted to the point at $T = 10$ K, at which $C_{p,m} = 2.8\,\mathrm{J\,K^{-1}\,mol^{-1}}$, so $a = 2.8 \times 10^{-3}\,\mathrm{J\,K^{-4}\,mol^{-1}}$. The area can be determined (primitively) by counting squares, which gives area A $= 38.28\,\mathrm{J\,K^{-1}\,mol^{-1}}$, area B (up to 0°C) $= 25.60\,\mathrm{J\,K^{-1}\,mol^{-1}}$, area B (up to 25°C) $= 27.80\,\mathrm{J\,K^{-1}\,mol^{-1}}$. Hence

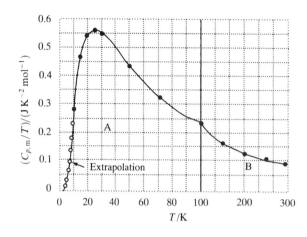

Figure 4.1

(a) $S_m(273\,\mathrm{K}) = S_m(0) + \boxed{63.88\,\mathrm{J\,K^{-1}\,mol^{-1}}}$

(b) $S_m(298\,\mathrm{K}) = S_m(0) + \boxed{66.08\,\mathrm{J\,K^{-1}\,mol^{-1}}}$

P4.16

$$\Delta_r G^{\ominus} = \Delta_r H^{\ominus} - T\Delta_r S^{\ominus} = 26.120\,\mathrm{kJ\,mol^{-1}}$$
$$\Delta_r H^{\ominus} = +55.000\,\mathrm{kJ\,mol^{-1}}$$

It is convenient to first work part **(b)**.

(b) Hence $\Delta_r S^{\ominus} = \dfrac{(55.000 - 26.120)\,\mathrm{kJ\,mol^{-1}}}{298.15\,\mathrm{K}} = \boxed{+96.864\,\mathrm{J\,K^{-1}\,mol^{-1}}}$

$$\Delta_r S^{\ominus} = 4S_m^{\ominus}(\mathrm{K^+, aq}) + S_m^{\ominus}([\mathrm{Fe(CN)_6}]^{4-}, \mathrm{aq}) + 3S_m^{\ominus}(\mathrm{H_2O, l})$$
$$- S_m^{\ominus}(\mathrm{K_4[Fe(CN)_6] \cdot 3H_2O, s})$$

(a) Therefore,

$$S_m^{\ominus}([\mathrm{Fe(CN)_6}]^{4-}, \mathrm{aq}) = \Delta_r S^{\ominus} - 4S_m^{\ominus}(\mathrm{K^+, aq}) - 3S_m^{\ominus}(\mathrm{H_2O, l})$$
$$+ S_m^{\ominus}(\mathrm{K_4[Fe(CN)_6] \cdot 3\,H_2O, s})$$
$$= [96.864 - (4 \times 102.5) - (3 \times 69.9) + (599.7)]\,\mathrm{J\,K^{-1}\,mol^{-1}}$$
$$= \boxed{+76.9\,\mathrm{J\,K^{-1}\,mol^{-1}}}$$

P4.18 Draw up the following table and proceed as in Problem 4.14.

T/K	14.14	16.33	20.03	31.15	44.08	64.81
$(C_{p,m}/T)/(\mathrm{J\,K^{-2}\,mol^{-1}})$	0.671	0.778	0.908	1.045	1.063	1.024

T/K	100.90	140.86	183.59	225.10	262.99	298.06
$(C_{p,m}/T)/(\mathrm{J\,K^{-2}\,mol^{-1}})$	0.942	0.861	0.787	0.727	0.685	0.659

Plot $C_{p,m}$ against T (Fig. 4.2(a)) and $C_{p,m}/T$ against T (Fig. 4.2(b)), extrapolating to $T = 0$ with $C_{p,m} = aT^3$ fitted at T=14.14 K, which gives $a = 3.36\,\mathrm{mJ\,K^{-4}\,mol^{-1}}$. Integration by determining the area under the curve then gives

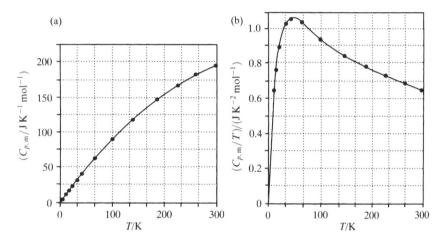

Figure 4.2

$$\int_0^{298\,\mathrm{K}} C_{p,m}\,dT = 34.4\,\mathrm{kJ\,mol^{-1}}, \quad \text{so} \quad H_m(298\,\mathrm{K}) = H_m(0) + \boxed{34.4\,\mathrm{kJ\,mol^{-1}}}$$

$$\int_0^{298\,\mathrm{K}} \frac{C_{p,m}\,dT}{T} = 243\,\mathrm{J\,K^{-1}\,mol^{-1}}, \quad \text{so} \quad S_m(298\,\mathrm{K}) = S_m(0) + \boxed{243\,\mathrm{J\,K^{-1}\,mol^{-1}}}$$

P4.19 The entropy at 200 K is calculated from

$$S_{m,200}^{\ominus} = S_{m,100}^{\ominus} + \int_{100\,\mathrm{K}}^{200\,\mathrm{K}} \frac{C_{p,m}\,dT}{T}$$

The integrand may be evaluated for each of the data points, and the numerical integration carried out by a standard procedure such as the trapezoid rule (taking the integral within any interval as the mean value of the integrand times the length of the interval). Programs for performing this integration are readily available for personal computers. Many graphing calculators will also perform this numerical integration.

The transformed data appear below.

T/K	100	120	140	150	160	180	200
$C_{p,m}/(\mathrm{J\,K^{-1}\,mol^{-1}})$	23.00	23.74	24.25	24.44	24.61	24.89	25.11
$\frac{C_{p,m}}{T}/(\mathrm{J\,K^{-2}\,mol^{-1}})$	0.230	0.1978	0.1732	0.1629	0.1538	0.1383	0.1256

Integration by the trapezoid rule yields

$$S^{\ominus}_{m,200} = (29.79 + 16.81)\,\mathrm{J\,K^{-1}\,mol^{-1}} = \boxed{46.60\,\mathrm{J\,K^{-1}\,mol^{-1}}}$$

Taking $C_{p,m}$ constant yields

$$S^{\ominus}_{m,200} = S^{\ominus}_{m,100} + C_{p,m}\ln(200\,\mathrm{K}/100\,\mathrm{K})$$

$$= [29.79 + 24.44\ln(200\,\mathrm{K}/100\,\mathrm{K})]\,\mathrm{J\,K^{-1}\,mol^{-1}} = \boxed{46.73\,\mathrm{J\,K^{-1}\,mol^{-1}}}$$

The difference is slight.

Solutions to theoretical problems

P4.21   Paths A and B in Fig. 4.3 are the reversible adiabatic paths which are assumed to cross at state 1. Path C (dashed) is an isothermal path which connects the adiabatic paths at states 2 and 3. Now go round the cycle ($1 \to 2$, step 1; $2 \to 3$, step 2; $3 \to 1$, step 3).

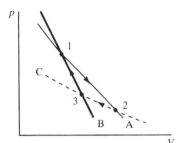

V **Figure 4.3**

Step 1 $\Delta U_1 = q_1 + w_1 = w_1$ $[q_1 = 0,\ \text{adiabatic}]$

Step 2 $\Delta U_2 = q_2 + w_2 = 0$ [isothermal step, energy depends on temperature only]

Step 3 $\Delta U_3 = q_3 + w_3 = w_3$ $[q_3 = 0,\ \text{adiabatic}]$

For the cycle $\Delta U = 0 = w_1 + q_2 + w_2 + w_3$ or $w(\text{net}) = w_1 + w_2 + w_3 = -q_2$

But, $\Delta U_1 = -\Delta U_3 [\Delta T_1 = -\Delta T_2]$; hence $w_1 = -w_3$, and $w(\text{net}) = w_2 = -q_2$, or $-w(\text{net}) = q_2$.

Thus, a net amount of work has been done by the system from heat obtained from a heat reservoir at the temperature of step 2, without at the same time transferring heat from a hot to a cold reservoir. This violates the Kelvin statement of the Second Law. Therefore, the assumption that the two adiabatic reversible paths may intersect is disproven.

Question. May any adiabatic paths intersect, reversible or not?

P4.24 $T = T(p, H)$

$$\mathrm{d}T = \left(\frac{\partial T}{\partial p}\right)_H \mathrm{d}p + \left(\frac{\partial T}{\partial H}\right)_p \mathrm{d}H$$

The Joule–Thomson expansion is a constant-enthalpy process (Section 3.4). Hence,

$$\mathrm{d}T = \left(\frac{\partial T}{\partial p}\right)_H \mathrm{d}p = \mu\,\mathrm{d}p$$

$$\Delta T = \int_{p_i}^{p_f} \mu \, dp = \mu \, \Delta p \quad [\mu \text{ is constant}]$$

$$= (0.21 \, \text{K atm}^{-1}) \times (1.00 \, \text{atm} - 100 \, \text{atm}) = \boxed{-21 \, \text{K}}$$

$T_f = T_i + \Delta T = (373 - 21) \, \text{K} = 352 \, \text{K} \, [\text{Mean } T = 363 \, \text{K}]$

$S = S(T, p)$

Therefore, $\quad dS = \left(\dfrac{\partial S}{\partial T}\right)_p dT + \left(\dfrac{\partial S}{\partial p}\right)_T dp$

$$\left(\frac{\partial S}{\partial T}\right)_p = \frac{C_p}{T} \qquad \left(\frac{\partial S}{\partial p}\right)_T = -\left(\frac{\partial V}{\partial T}\right)_p \qquad [\text{Table 5.1}]$$

For $V_m = \dfrac{RT}{p}(1 + Bp)$

$$\left(\frac{\partial V_m}{\partial T}\right)_p = \frac{R}{p}(1 + Bp)$$

Then

$$dS_m = \frac{C_{p,m}}{T} dT - \frac{R}{p}(1 + Bp) \, dp$$

or

$$dS_m = \frac{C_{p,m}}{T} dT - \frac{R}{p} dp - RB \, dp$$

Upon integration

$$\Delta S_m = \int_{T_1, p_1}^{T_2, p_2} dS_m = C_{p,m} \ln\left(\frac{T_2}{T_1}\right) - R \ln\left(\frac{p_2}{p_1}\right) - RB(p_2 - p_1)$$

$$= \frac{5}{2} R \ln\left(\frac{352}{373}\right) - R \ln\left(\frac{1}{100}\right) - R\left(-\frac{0.525 \, \text{atm}^{-1}}{363}\right) \times (-99 \, \text{atm})$$

$$= \boxed{+35.9 \, \text{J K}^{-1} \, \text{mol}^{-1}}$$

P4.25 The efficiency of any reversible engine in which the working substance is a perfect gas is given by

$$\varepsilon_{rev} = \frac{|w|}{q_h} \, [4.9] = 1 - \frac{T_c}{T_h} = 1 + \frac{q_{c,min}}{q_h} \, [4.8, \text{with } q_c = q_{c,min}]$$

Therefore, for a perfect gas   [*Justification* 4.1]

$$-\frac{q_{c,min}}{q_h} = \frac{T_c}{T_h} \quad \text{or} \quad -\frac{q_{T,min}}{q_h} = \frac{T}{T_h} \quad \text{and} \quad T = -\frac{q_{T,min}}{q_h} \times T_h$$

But for a reversible engine employing any working substance (including a perfect gas)

$$-\frac{q_{c,min}}{q_h} = \frac{T_c^a}{T_h^a} \quad [4.8, \text{Section 4.2(b)}],$$

where the symbol T^a is used to indicate the absolute temperature based on the Second Law. Thus,

$$T^a = -\frac{q_{T,min}}{q_h} \times T_h^a$$

Since $q_{T,\mathrm{min}}$ and q_h are experimentally measured heats, and are the same no matter what temperature scale is employed

$$\frac{T}{T_\mathrm{h}} = \frac{T^\mathrm{a}}{T_\mathrm{h}^\mathrm{a}}$$

Thus T and T^a differ from each other by at most a constant numerical factor which becomes 1 if T_h and T_h^a are both assigned the same value, say 273.16 at the triple point of water.

P4.28   Polytropic process: $pV^n = C =$ constant.

If $n = 0$, then $pV^n = pV^0 = p =$ constant and the process is isobaric.

If $n = 1$, then $pV^n = pV =$ constant. But the perfect gas equation of state says that $pV = nRT$ so we note that both of these equations are correct provided that $T =$ constant. The process is isothermal (Fig. 4.4(a)).

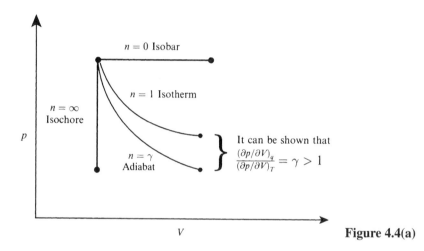

Figure 4.4(a)

If $n = \gamma$, then $pV^n = pV^\gamma = C$ and the process is adiabatic.

In the limit as $n \to \infty$ we need to examine the process equation in the form $V = \left(\dfrac{C}{p}\right)^{1/n}$. Then,

$$\lim_{n \to \infty} V = \left(\frac{C}{p}\right)^0 = \text{constant and we find that the process is isochoric (Fig. 4.4(a)).}$$

Consider the functional behaviour of $\partial V/\partial p$

$$\frac{\partial V}{\partial p} = \frac{\partial}{\partial p}\left(\frac{C}{p}\right)^{1/n} = -\frac{C^{1/n} p^{-\left(\frac{1}{n}+1\right)}}{n}$$

In all cases $p > 0$, $V > 0$, $C > 0$ and $p^{-\left(\frac{1}{n}+1\right)} > 0$.

Therefore we may write

$$\frac{\partial V}{\partial p} = -\frac{1}{n}\left|C^{1/n} p^{-\left(\frac{1}{n}+1\right)}\right|$$

and, in the special case for which $n < 0$, this becomes

$$\frac{\partial V}{\partial p} = \left|\frac{1}{n}C^{1/n} p^{-\left(\frac{1}{n}+1\right)}\right| \quad n < 0$$

This equation *cannot* be a description of a real substance because it says that, as pressure increases, volume increases.

Nevertheless, we can construct plots of p against V and T against S for these physically unrealizable cases (Fig. 4.4(b)).

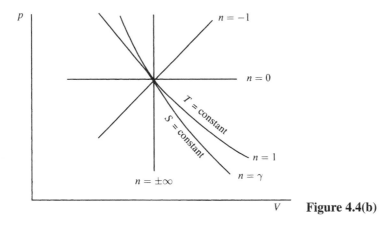

V **Figure 4.4(b)**

To determine the qualitative behaviour of $S(T)$ for each process, consider the following observations.

Adiabatic, reversible process [4.2]

$$dS = dq/T = 0 \quad \text{or} \quad S = \text{constant}$$

Isobaric process [4.19]

$$\left(\frac{\partial S}{\partial T}\right)_p = \frac{C_p}{T} > 0 \quad \text{so} \quad \left(\frac{\partial S}{\partial T}\right)_p \sim \frac{1}{T}$$

Isochoric process

$$dq_V = C_V \, dT$$
$$dS = \frac{dq_V}{T} = \frac{C_V \, dT}{T}$$

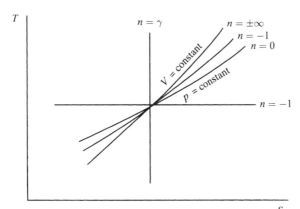

S **Figure 4.4(c)**

$$\left(\frac{\partial S}{\partial T}\right)_V = \frac{C_V}{T} > 0 \quad \text{so} \quad \left(\frac{\partial S}{\partial T}\right)_V \sim \frac{1}{T}$$

Because $C_p > C_V$ [3.21], $\left(\dfrac{\partial S}{\partial T}\right)_p > \left(\dfrac{\partial S}{\partial T}\right)_V$

Plots of T against S for all values of n, even the physically unrealizable cases are shown in Fig. 4.4(c).

Solutions to applications

P4.29   The groups in C_2H_5 are $(C-C)(H)_2$ and $C-(C)(H)_3$, so

$$S_{\text{int}}^{\ominus} = (135.9 + 126.8)\,\text{J}\,\text{K}^{-1}\,\text{mol}^{-1} = 262.7\,\text{J}\,\text{K}^{-1}\,\text{mol}^{-1},$$

and

$$S^{\ominus} = S_{\text{int}}^{\ominus} - R\ln\sigma = (262.7 - 8.3145\ln 6)\,\text{J}\,\text{K}^{-1}\,\text{mol}^{-1}$$
$$= \boxed{247.8\,\text{J}\,\text{K}^{-1}\,\text{mol}^{-1}}$$

The groups in sec-C_4H_9 are $C-(C)(H)_3$, $C-(C)(C)(H)_2$, $C-(C)_2(H)$ and $C-(C)(H)_3$, so

$$S_{\text{int}}^{\ominus} = (126.8 + 42.0 + 59.3 + 126.8)\,\text{J}\,\text{K}^{-1}\,\text{mol}^{-1} = 354.9\,\text{J}\,\text{K}^{-1}\,\text{mol}^{-1},$$

and

$$S^{\ominus} = S_{\text{int}}^{\ominus} - R\ln\sigma = (354.9 - 8.3145\ln 9)\,\text{J}\,\text{K}^{-1}\,\text{mol}^{-1}$$
$$= \boxed{336.6\,\text{J}\,\text{K}^{-1}\,\text{mol}^{-1}}$$

The groups in $tert$-C_4H_9 are $(C)-(C)_3$ and $3C-(C)(H)_3$, so

$$S_{\text{int}}^{\ominus} = [-29.2 + 3(126.8)]\,\text{J}\,\text{K}^{-1}\,\text{mol}^{-1} = 351.2\,\text{J}\,\text{K}^{-1}\,\text{mol}^{-1},$$

and

$$S^{\ominus} = S_{\text{int}}^{\ominus} - R\ln\sigma = (351.2 - 8.3145\ln 81)\,\text{J}\,\text{K}^{-1}\,\text{mol}^{-1}$$
$$= \boxed{314.7\,\text{J}\,\text{K}^{-1}\,\text{mol}^{-1}}$$

P4.32   The Otto cycle is represented in Fig. 4.5. Assume one mole of air.

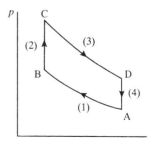

V **Figure 4.5**

$$\varepsilon = \frac{|w|_{\text{cycle}}}{|q_2|} \ [4.9]$$

$$w_{\text{cycle}} = w_1 + w_3 = \Delta U_1 + \Delta U_3 \ [q_1 = q_3 = 0] = C_V(T_B - T_A) + C_V(T_D - T_C) \ [2.32]$$

$$q_2 = \Delta U_2 = C_V(T_C - T_B)$$

$$\varepsilon = \frac{|T_B - T_A + T_D - T_C|}{|T_C - T_B|} = 1 - \left(\frac{T_D - T_A}{T_C - T_B}\right)$$

We know that

$$\frac{T_A}{T_B} = \left(\frac{V_B}{V_A}\right)^{1/c} \quad \text{and} \quad \frac{T_D}{T_C} = \left(\frac{V_C}{V_D}\right)^{1/c} \ [2.34]$$

and since $V_B = V_C$ and $V_A = V_D$, $\dfrac{T_A}{T_B} = \dfrac{T_D}{T_C}$, or $T_D = \dfrac{T_A T_C}{T_B}$

Then $\varepsilon = 1 - \dfrac{\frac{T_A T_C}{T_B} - T_A}{T_C - T_B} = 1 - \dfrac{T_A}{T_B}$ or $\boxed{\varepsilon = 1 - \left(\dfrac{V_B}{V_A}\right)^{1/c}}$

Assume $C_{V,m}(\text{air}) = \dfrac{5}{2}R$, then $c = \dfrac{2}{5}$.

For $\dfrac{V_A}{V_B} = 10$, $\varepsilon = 1 - \left(\dfrac{1}{10}\right)^{2/5} = \boxed{0.47}$

$$\Delta S_1 = \Delta S_3 = \Delta S_{\text{sur},1} = \Delta S_{\text{sur},3} = \boxed{0} \quad \text{[adiabatic reversible steps]}$$

$$\Delta S_2 = C_{V,m} \ln\left(\frac{T_C}{T_B}\right)$$

At constant volume $\left(\dfrac{T_C}{T_B}\right) = \left(\dfrac{p_C}{p_B}\right) = 5.0$

$$\Delta S_2 = \left(\tfrac{5}{2}\right) \times (8.314 \, \text{J K}^{-1} \, \text{mol}^{-1}) \times (\ln 5.0) = \boxed{+33 \, \text{J K}^{-1}}$$

$$\Delta S_{\text{sur},2} = -\Delta S_2 = \boxed{-33 \, \text{J K}^{-1}}$$

$$\Delta S_4 = -\Delta S_2 \left[\frac{T_C}{T_D} = \frac{T_B}{T_A}\right] = \boxed{-33 \, \text{J K}^{-1}}$$

$$\Delta S_{\text{sur},4} = -\Delta S_4 = \boxed{+33 \, \text{J K}^{-1}}$$

5 The Second Law: the machinery

Solutions to exercises

Discussion questions

E5.1(a) The Maxwell relations are relations between partial derivatives all of which are expressed in terms of functions of state (properties of the system). Partial derivatives can be thought of as a kind of shorthand for an experiment. Therefore, the partial derivative $\left(\frac{\partial S}{\partial V}\right)_T$ tells us how the entropy of the system changes when we change its volume under constant temperature conditions. But as entropy is not a property that can be measured directly (there are no entropy meters), it is important that the derivative (and hence the experiment) be transformed into a form that involves directly measurable properties. That is what the following Maxwell relation does for us.

$$\left(\frac{\partial S}{\partial V}\right)_T = \left(\frac{\partial p}{\partial T}\right)_V$$

Pressure, temperature, and volume are easily measured properties.

E5.2(a) The relation $(\partial G/\partial p)_T = V$ shows that the Gibbs function of a system increases with p at constant T in proportion to the magnitude of its volume. This makes good sense when one considers the definition of G, which is $G = U + pV - TS$. Hence, G is expected to increase with p in proportion to V when T is constant.

E5.3(a) The expression for the molar Gibbs energy for a perfect gas has a particularly simple and useful form, which is $G_m = G_m^\ominus + RT \ln(p/p^\ominus)$. It is of value to retain this simple form, where the second term on the right hand side of the expression shows how the Gibbs energy varies from its standard value for real systems as the system changes its state from the standard state. This can be accomplished by defining a new quantity called the fugacity, f, which has the same units as pressure and is defind so that the $G_m = G_m^\ominus + RT \ln(f/p^\ominus)$ holds true. This allows us to derive thermodynamically exact expressions from the Gibbs energy.

Numerical exercises

Assume all gases are perfect and that the temperature is 298.15 K unless stated otherwise.

E5.4(a) $\alpha = \left(\frac{1}{V}\right)\left(\frac{\partial V}{\partial T}\right)_p$ [3.8]; $\quad \kappa_T = -\left(\frac{1}{V}\right) \times \left(\frac{\partial V}{\partial p}\right)_T$ [3.14]

$$\left(\frac{\partial S}{\partial V}\right)_T = \left(\frac{\partial p}{\partial T}\right)_V = -\left(\frac{\partial V}{\partial T}\right)_p\left(\frac{\partial p}{\partial V}\right)_T \quad \text{[Relation no. 3, \textit{Further information} 1.7]}$$

$$= -\frac{\left(\frac{\partial V}{\partial T}\right)_p}{\left(\frac{\partial V}{\partial p}\right)_T} \quad \text{[Relation no. 2, \textit{Further information} 1.7]}$$

$$= -\frac{\left(\frac{1}{V}\right)\left(\frac{\partial V}{\partial T}\right)_p}{\left(\frac{1}{V}\right)\left(\frac{\partial V}{\partial p}\right)_T} = \boxed{+\frac{\alpha}{\kappa_T}}$$

E5.5(a) $\Delta G = nRT \ln\left(\frac{p_f}{p_i}\right)$ [Example 5.2] $= nRT \ln\left(\frac{V_i}{V_f}\right)$ [Boyle's law]

$$= (3.0 \times 10^{-3}\,\text{mol}) \times (8.314\,\text{J K}^{-1}\,\text{mol}^{-1}) \times (300\,\text{K}) \times \ln\left(\frac{36}{60}\right) = \boxed{-3.8\,\text{J}}$$

E5.6(a) $\Delta G = G_f - G_i$

$$\left(\frac{\partial G}{\partial T}\right)_p = -S \text{ [5.10]}; \quad \text{hence} \quad \left(\frac{\partial G_f}{\partial T}\right)_p = -S_f, \quad \text{and} \quad \left(\frac{\partial G_i}{\partial T}\right)_p = -S_i$$

$$\Delta S = S_f - S_i = -\left(\frac{\partial G_f}{\partial T}\right)_p + \left(\frac{\partial G_i}{\partial T}\right)_p = -\left(\frac{\partial (G_f - G_i)}{\partial T}\right)_p$$

$$= -\left(\frac{\partial \Delta G}{\partial T}\right)_p = -\frac{\partial}{\partial T}\left(-85.40\,\text{J} + 36.5\,\text{J} \times \frac{T}{\text{K}}\right)$$

$$= \boxed{-36.5\,\text{J\,K}^{-1}}$$

E5.7(a) $dG = -S\,dT + V\,dp$ [5.9] at constant T, $dG = V\,dp$; therefore

$$\Delta G = \int_{p_i}^{p_f} V\,dp$$

$$V = V_1(1 - \kappa_T p) \quad V_1 = V(1\,\text{atm}) = \frac{m}{\rho} \quad \kappa_T = 76.8 \times 10^{-6}\,\text{atm}^{-1} \quad \text{[Table 3.1]}$$

Then,

$$V = \frac{m}{p}(1 - 7.68 \times 10^{-5}\,\text{atm}^{-1}p)$$

$$\Delta G = \int_{1\,\text{atm}}^{3000\,\text{atm}} \frac{m}{\rho}(1 - 7.68 \times 10^{-5}\,\text{atm}^{-1}p)\,dp$$

$$= \int_{1\,\text{atm}}^{3000\,\text{atm}} \frac{m}{\rho}\,dp - \int_{1\,\text{atm}}^{3000\,\text{atm}} \frac{m}{\rho} \times (7.68 \times 10^{-5}\,\text{atm}^{-1})p\,dp$$

$$= \frac{m}{\rho} \times 2999\,\text{atm} - (7.68 \times 10^{-5}\,\text{atm}^{-1}) \times \frac{m}{\rho} \times (9.00 \times 10^6\,\text{atm}^3)$$

$$= \frac{m}{\rho}(2999\,\text{atm} - 691\,\text{atm}) = \frac{35\,\text{g}}{0.789\,\text{g\,cm}^{-3}} \times 2308\,\text{atm}$$

$$= 44.\overline{4}\,\text{cm}^3 \times \frac{10^{-6}\,\text{m}^3}{\text{cm}^3} \times 2308\,\text{atm} \times (1.013 \times 10^{-5}\,\text{Pa\,atm}^{-1})$$

$$= 10.\overline{4}\,\text{kJ} = \boxed{10\,\text{kJ}}$$

E5.8(a)   **(a)** $\Delta S = nR \ln\left(\dfrac{V_f}{V_i}\right)$ [4.17] $= nR \ln\left(\dfrac{p_i}{p_f}\right)$ [Boyle's law]

Taking inverse logarithms

$$p_f = p_i e^{-\Delta S/nR} = (3.50\,\text{atm}) \times e^{-(25.0\,\text{J\,K}^{-1})/(2.00 \times 8.314\,\text{J\,K}^{-1}\,\text{mol}^{-1})}$$

$$= (3.50\,\text{atm}) \times e^{1.50} = \boxed{15.7\,\text{atm}}$$

(b) $\Delta G = nRT \ln\left(\dfrac{p_f}{p_i}\right)$ [Example 5.2]

$$= -T\Delta S\ [\Delta H = 0, \text{ constant temperature, perfect gas}]$$

$$= (-330\,\text{K}) \times (-25.0\,\text{J\,K}^{-1}) = \boxed{+8.25\,\text{kJ}}$$

E5.9(a) $\Delta G_m = G_{m,f} - G_{m,i} = RT \ln\left(\dfrac{p_f}{p_i}\right)$ [5.16] $= (8.314\,\mathrm{J\,K^{-1}\,mol^{-1}}) \times (313\,\mathrm{K}) \times \ln\left(\dfrac{29.5}{1.8}\right)$

$$= \boxed{+7.3\,\mathrm{kJ\,mol^{-1}}}$$

E5.10(a) $G_m^0 = G_m^{\ominus} + RT \ln\left(\dfrac{p}{p^{\ominus}}\right)$ [5.16 with $G_m = G_m^0$]

$G_m = G_m^{\ominus} + RT \ln\left(\dfrac{f}{p^{\ominus}}\right)$ [5.18]

$G_m - G_m^0 = RT \ln \dfrac{f}{p}$ [5.18 minus 5.17]; $\dfrac{f}{p} = \phi$

$$= RT \ln \phi = (8.314\,\mathrm{J\,K^{-1}\,mol^{-1}}) \times (200\,\mathrm{K}) \times (\ln 0.72) = \boxed{-0.55\,\mathrm{kJ\,mol^{-1}}}$$

E5.11(a) $B' = \dfrac{B}{RT}$ [Problem 1.21] $= \dfrac{-81.7 \times 10^{-6}\,\mathrm{m^3\,mol^{-1}}}{(8.314\,\mathrm{J\,K^{-1}\,mol^{-1}}) \times (373\,\mathrm{K})} = \boxed{-2.63 \times 10^{-8}\,\mathrm{Pa^{-1}}}$

$\ln \phi = \displaystyle\int_0^p \dfrac{Z-1}{p}\,\mathrm{d}p$ [5.20]

$Z = 1 + B'p + C'p^2 + \cdots$

$\ln \phi = \displaystyle\int_0^p B'\,\mathrm{d}p = B'p$

$\phi = e^{B'p + \cdots}$ [truncating series after term in B']

$\quad = e^{(-2.63 \times 10^{-8}\,\mathrm{Pa^{-1}}) \times (50) \times (1.013 \times 10^5\,\mathrm{Pa})}$

$\quad = e^{-0.13\bar{3}} = \boxed{0.88}$

E5.12(a) $\Delta G = nV_m\Delta p$ [5.15] $= V\Delta p = (1.0 \times 10^{-3}\,\mathrm{m^3}) \times (99) \times (1.013 \times 10^5\,\mathrm{Pa})$

$$= 10\,\mathrm{kPa\,m^3} = \boxed{+10\,\mathrm{kJ}}$$

E5.13(a) $\Delta G_m = RT \ln \dfrac{p_f}{p_i}$ [5.16] $= (8.314\,\mathrm{J\,K^{-1}\,mol^{-1}}) \times (298\,\mathrm{K}) \times \ln\left(\dfrac{100.0}{1.0}\right) = \boxed{+11\,\mathrm{kJ\,mol^{-1}}}$

E5.14(a)   An equation of state is a functional relationship between the state properties, p, V_m, and T. From the definition

$$A \equiv U - TS \qquad \mathrm{d}A = \mathrm{d}U - T\,\mathrm{d}S - S\,\mathrm{d}T$$

Using [5.2] in eqn 4.31a, $\mathrm{d}A = -S\,\mathrm{d}T - p\,\mathrm{d}V_m$; hence

$$p = -\left(\dfrac{\partial A}{\partial V_m}\right)_T = -\dfrac{a}{V_m^2} + RT \times \left(\dfrac{1}{V_m - b}\right) = \boxed{\dfrac{RT}{V_m - b} - \dfrac{a}{V_m^2}}$$

which is the van der Waals equation.

E5.15(a) $\left(\dfrac{\partial S}{\partial V}\right)_T = \left(\dfrac{\partial p}{\partial T}\right)_V$ [Table 5.1]

For a van der Waals gas

$$p = \dfrac{nRT}{V - nb} - \dfrac{n^2 a}{V^2}$$

Hence, $\left(\dfrac{\partial S}{\partial V}\right)_T = \left(\dfrac{\partial p}{\partial T}\right)_V = \boxed{\dfrac{nR}{V - nb}}$

$$\mathrm{d}S = \left(\dfrac{\partial S}{\partial V}\right)_T \mathrm{d}V\,[\text{constant temperature}] = \left(\dfrac{\partial p}{\partial T}\right)_V \mathrm{d}V = \dfrac{nR}{V - nb}\,\mathrm{d}V$$

$$\Delta S = \int_{V_i}^{V_f} dS = \int_{V_i}^{V_f} \frac{nR}{V - nb} dV = \boxed{nR \ln\left(\frac{V_f - nb}{V_i - nb}\right)}$$

For a perfect gas $\Delta S = nR \ln\left(\dfrac{V_f}{V_i}\right)$

$$\frac{V_f - nb}{V_i - nb} > \frac{V_f}{V_i},$$

Therefore, ΔS will be greater for a van der Waals gas.

Solutions to problems

Solutions to numerical problems

P5.1   The Gibbs–Helmholtz equation [5.11] may be recast into an analogous equation involving ΔG and ΔH, since

$$\left(\frac{\partial \Delta G}{\partial T}\right)_p = \left(\frac{\partial G_f}{\partial T}\right)_p - \left(\frac{\partial G_i}{\partial T}\right)_p$$

and $\Delta H = H_f - H_i$

Thus, $\left(\dfrac{\partial}{\partial T} \dfrac{\Delta_r G^{\ominus}}{T}\right)_p = -\dfrac{\Delta_r H^{\ominus}}{T^2}$

$$d\left(\frac{\Delta_r G^{\ominus}}{T}\right) = \left(\frac{\partial}{\partial T} \frac{\Delta_r G^{\ominus}}{T}\right)_p dT\,[\text{constant pressure}] = -\frac{\Delta_r H^{\ominus}}{T^2} dT$$

$$\Delta\left(\frac{\Delta_r G^{\ominus}}{T}\right) = -\int_{T_c}^{T} \frac{\Delta_r H^{\ominus}\, dT}{T^2}$$

$$\approx -\Delta_r H^{\ominus} \int_{T_c}^{T} \frac{dT}{T^2} = \Delta_r H^{\ominus}\left(\frac{1}{T} - \frac{1}{T_c}\right) \quad [\Delta_r H^{\ominus} \text{ assumed constant}]$$

Therefore, $\dfrac{\Delta_r G^{\ominus}(T)}{T} - \dfrac{\Delta_r G^{\ominus}(T_c)}{T_c} \approx \Delta_r H^{\ominus}\left(\dfrac{1}{T} - \dfrac{1}{T_c}\right)$

and so

$$\Delta_r G^{\ominus}(T) = \frac{T}{T_c}\Delta_r G^{\ominus}(T_c) + \left(1 - \frac{T}{T_c}\right)\Delta_r H^{\ominus}(T_c)$$

$$= \tau \Delta_r G^{\ominus}(T_c) + (1 - \tau)\Delta_r H^{\ominus}(T_c) \quad \tau = \frac{T}{T_c}$$

For the reaction

$$2CO(g) + O_2(g) \rightarrow 2CO_2(g)$$

$$\Delta_r G^{\ominus}(T_c) = 2\Delta_f G^{\ominus}(CO_2, g) - 2\Delta_f G^{\ominus}(CO, g)$$

$$= [2 \times (-394.36) - 2 \times (-137.17)]\,\text{kJ mol}^{-1} = -514.38\,\text{kJ mol}^{-1}$$

$$\Delta_r H^{\ominus}(T_c) = 2\Delta_f H^{\ominus}(CO_2, g) - 2\Delta_f H^{\ominus}(CO, g)$$

$$= [2 \times (-393.51) - 2 \times (-110.53)]\,\text{kJ mol}^{-1} = -565.96\,\text{kJ mol}^{-1}$$

Therefore, since $\tau = \dfrac{375}{298.15} = 1.25\bar{8}$

$$\Delta_r G^{\ominus}(375\,\text{K}) = \{(1.25\overline{8}) \times (-514.38) + (1 - 1.25\overline{8}) \times (-565.96)\}\,\text{kJ mol}^{-1}$$

$$= \boxed{-501\,\text{kJ mol}^{-1}}$$

P5.3 A graphical integration of $\ln \phi = \int_0^p \left(\dfrac{Z-1}{p}\right)\,dp$ [5.20] is performed. We draw up the following table

p/atm	1	4	7	10	40	70	100
$10^3 \left(\dfrac{Z-1}{p}\right) \Big/ \text{atm}^{-1}$	−2.9	−3.01	−3.03	−3.04	−3.17	−3.19	−3.13

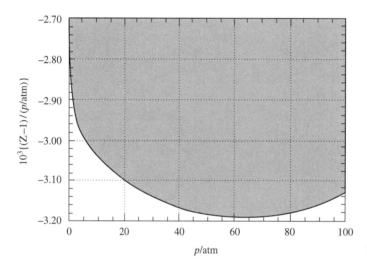

Figure 5.1

The points are plotted in Fig. 5.1. The integral is the shaded area, which has the value -0.313, so at 100 atm

$$\phi = e^{-0.313} = 0.73$$

and the fugacity of oxygen is 100 atm $\times 0.73 = \boxed{73\,\text{atm}}$

P5.4 The logarithm of the fugacity coefficient is given by

$$\ln \phi = \int_0^p \left(\frac{Z-1}{p}\right)\,dp$$

The integrand may be evaluated for each of the experimental points, and the numerical integration carried out by a standard procedure such as the trapezoid rule (taking the integral within any interval as the mean value of the intergrand times the length of the interval). The transformed data as computed using the trapezoid rule appear below, along with a plot of the integrand (Fig. 5.2).

p/bar	Z	$[(Z-1)/p]/\text{bar}^{-1}$	$\ln \phi$	ϕ
0.500	0.99412	−0.0118	−0.00294	0.997
1.013	0.98896	−0.0109	−0.00875	0.991
2.00	0.97942	−0.0103	−0.0192	0.981
3.00	0.96995	−0.0100	−0.0294	0.971
5.00	0.95133	−0.00973	−0.0491	0.952
10.00	0.90569	−0.00943	−0.0970	0.908
20.0	0.81227	−0.00939	−0.191	0.826
30.0	0.70177	−0.00994	−0.288	0.750
42.4	0.47198	−0.0125	−0.427	0.653
50.0	0.22376	−0.0155	−0.533	0.587
70.0	0.26520	−0.0105	−0.793	0.452
100.0	0.34920	−0.00651	−1.05	0.351
200	0.62362	−0.00188	−1.47	0.230
300	0.88288	−0.000390	−1.58	0.206
500	1.37109	0.000742	−1.55	0.213
1000	2.48836	0.00149	−0.989	0.372

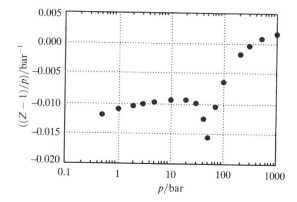

Figure 5.2

Solutions to theoretical problems

P5.6

$H \equiv U + pV$

$\mathrm{d}H = \mathrm{d}U + p\,\mathrm{d}V + V\,\mathrm{d}p = T\,\mathrm{d}S - p\,\mathrm{d}V\ [5.2] + p\,\mathrm{d}V + V\,\mathrm{d}p = T\,\mathrm{d}S + V\,\mathrm{d}p$

Since H is a state function, $\mathrm{d}H$ is exact, and it follows that

$$\left(\frac{\partial H}{\partial S}\right)_p = T \quad \text{and} \quad \boxed{\left(\frac{\partial V}{\partial S}\right)_p = \left(\frac{\partial T}{\partial p}\right)_S}$$

Similarly, $A \equiv U - TS$

$\mathrm{d}A = \mathrm{d}U - T\,\mathrm{d}S - S\mathrm{d}T = T\,\mathrm{d}S - p\,\mathrm{d}V\ [5.2] - T\,\mathrm{d}S - S\mathrm{d}T = -p\,\mathrm{d}V - S\,\mathrm{d}T$

Since $\mathrm{d}A$ is exact,

$$\boxed{\left(\frac{\partial S}{\partial V}\right)_T = \left(\frac{\partial p}{\partial T}\right)_V}$$

P5.7

$$\left(\frac{\partial S}{\partial V}\right)_T = \left(\frac{\partial p}{\partial T}\right)_V = \boxed{\frac{\alpha}{\kappa_T}} \quad \text{[Exercise 5.4(a)]}$$

$$\left(\frac{\partial V}{\partial S}\right)_p = \left(\frac{\partial T}{\partial p}\right)_S$$

$$\left(\frac{\partial T}{\partial p}\right)_S = -\left(\frac{\partial T}{\partial S}\right)_p \left(\frac{\partial S}{\partial p}\right)_T \text{ [permuter]} = -\frac{\left(\frac{\partial S}{\partial p}\right)_T}{\left(\frac{\partial S}{\partial T}\right)_p} \text{ [inversion]}$$

$$\left(\frac{\partial S}{\partial p}\right)_T = -\left(\frac{\partial V}{\partial T}\right)_p \quad \text{[Maxwell relation]}$$

$$= -\alpha V$$

at constant p

$$dS = \left(\frac{\partial S}{\partial T}\right)_p dT$$

and

$$dS = \frac{dq_{\text{rev}}}{T} = \frac{dH}{T} = \frac{C_p\, dT}{T} \quad [dq_p = dH]$$

Therefore, $\left(\dfrac{\partial S}{\partial T}\right)_p = \dfrac{C_p}{T}$

and $\left(\dfrac{\partial V}{\partial S}\right)_p = \boxed{\dfrac{\alpha T V}{C_p}}$

P5.9

$$\left(\frac{\partial S}{\partial V}\right)_T = \left(\frac{\partial p}{\partial T}\right)_V \text{ [Maxwell relation]}; \qquad \left(\frac{\partial p}{\partial T}\right)_V = \left(\frac{\partial}{\partial T}\left(\frac{nRT}{V}\right)\right)_V = \frac{nR}{V}$$

$$dS = \left(\frac{\partial S}{\partial V}\right)_T dV \text{ [constant temperature]} = nR\frac{dV}{V} = nR\, d\ln V$$

$$S = \int dS = \int nR\, d\ln V$$

$$S = nR\ln V + \text{constant} \quad \text{or} \quad S \propto R\ln V$$

P5.11   Start from the relation

$$dH = T\, dS + V\, dp \text{ [Problem 5.6]}$$

Divide by dV at constant T, which gives

$$\left(\frac{\partial H}{\partial V}\right)_T = T\left(\frac{\partial S}{\partial V}\right)_T + V\left(\frac{\partial p}{\partial V}\right)_T = T\left(\frac{\partial p}{\partial T}\right)_V \text{ [Maxwell relation]} + V\left(\frac{\partial p}{\partial V}\right)_T$$

Inserting $\left(\dfrac{\partial p}{\partial V}\right)_T = -\dfrac{\left(\frac{\partial T}{\partial V}\right)_p}{\left(\frac{\partial T}{\partial p}\right)_V}$ [permuter followed by inversion]

yields $\left(\dfrac{\partial H}{\partial V}\right)_V = \left[T - V\left(\dfrac{\partial T}{\partial V}\right)_p\right]\left(\dfrac{\partial p}{\partial T}\right)_V$

Now note that $-(V^2)\left(\dfrac{\partial p}{\partial T}\right)_V \left(\dfrac{\partial(T/V)}{\partial V}\right)_p = -(V^2)\left(\dfrac{\partial p}{\partial T}\right)_V$

$$\times \left[\left(\dfrac{1}{V}\right) \times \left(\dfrac{\partial T}{\partial V}\right)_p - \left(\dfrac{T}{V^2}\right)\right]$$

$$= \left[T - V\left(\dfrac{\partial T}{\partial V}\right)_p\right]\left(\dfrac{\partial p}{\partial T}\right)_V = \left(\dfrac{\partial H}{\partial V}\right)_T$$

which is the relation to be proved.

P5.14 $\quad \mu_J = \left(\dfrac{\partial T}{\partial V}\right)_U \qquad C_V = \left(\dfrac{\partial U}{\partial T}\right)_V$

$$\mu_J C_V = \left(\dfrac{\partial T}{\partial V}\right)_U \left(\dfrac{\partial U}{\partial T}\right)_V = \dfrac{-1}{\left(\dfrac{\partial V}{\partial U}\right)_T} \quad \text{[chain relation]}$$

$$= -\left(\dfrac{\partial U}{\partial V}\right)_T \quad \text{[inversion]} = p - T\left(\dfrac{\partial p}{\partial T}\right)_V \quad \text{[5.8]}$$

$$\left(\dfrac{\partial p}{\partial T}\right)_V = \dfrac{-1}{\left(\dfrac{\partial T}{\partial V}\right)_p \left(\dfrac{\partial V}{\partial p}\right)_T} \quad \text{[chain relation]} = \dfrac{-\left(\dfrac{\partial V}{\partial T}\right)_p}{\left(\dfrac{\partial V}{\partial p}\right)_T} = \dfrac{\alpha}{\kappa_T}$$

Therefore, $\boxed{\mu_J C_V = p - \dfrac{\alpha T}{\kappa_T}}$

P5.16 (a) Keeping only the first order terms of eqns. 1.21 and 1.22, $p = \dfrac{RT}{V_m}\left(1 + \dfrac{B}{V_m}\right)$ where

$B(T) = RTB'(T)$

Eqn 5.8 gives

$$\pi_T = T\left(\dfrac{\partial p}{\partial T}\right)_V - p$$

$$= T\left[\dfrac{R}{V_m}\left(1 + \dfrac{B}{V_m}\right) + \dfrac{RT}{V_m^2}\left(\dfrac{\partial B}{\partial T}\right)_V\right] - p$$

$$= \dfrac{RT^2}{V_m^2}\left(\dfrac{\partial B}{\partial T}\right)_V = \dfrac{RT^2}{V_m^2}\left(\dfrac{\partial(RTB')}{\partial T}\right)_V$$

$$= \left(\dfrac{RT}{V_m}\right)^2\left[B' + T\left(\dfrac{\partial B'}{\partial T}\right)_V\right]$$

$$= \left(\dfrac{RT}{V_m}\right)^2\left[B' + T\left(\dfrac{\partial}{\partial T}\right)_V (a + be^{-c/T^2})\right]$$

$$= \left(\dfrac{RT}{V_m}\right)^2\left[B' + T\left(\dfrac{2bc}{T^3}\right)e^{-c/T^2}\right]$$

$$= \left(\dfrac{RT}{V_m}\right)^2\left[B' + \dfrac{2c}{T^2}(B' - a)\right]$$

$$\pi_T(p, T) = \left(\frac{RT}{V_m(p, T)}\right)^2\left[\left(1 + \frac{2c}{T^2}\right)B'(T) - \frac{2ac}{T^2}\right]$$

$$B'(298.15\,\text{K}) = -0.1993\,\text{bar}^{-1} + 0.2002\,\text{bar}^{-1}e^{\frac{-1131\,K^2}{(298.15\,K)^2}} = -1.631 \times 10^{-3}\,\text{bar}^{-1}$$

Eqn 1.21 gives:

$$V_m(100\text{bar},\ 298.15\,\text{K}) = \left[\frac{(0.0831451\,\text{L bar K}^{-1}\text{mol}^{-1}) \times (298.15\,K)}{100\,\text{bar}}\right]$$

$$\times\left[1 - (1.631 \times 10^{-3}\,\text{bar}^{-1}) \times (100\,\text{bar})\right]$$

$$V_m(100\,\text{bar},\ 298.15\,\text{K}) = 0.2075\,\text{L mol}^{-1}$$

$$\pi_T(100\,\text{bar},\ 298.15\,\text{K}) = \left[\frac{(8.31451\,\text{J}\,\cancel{\text{K}^{-1}\,\text{mol}^{-1}}) \times (298.15\,\cancel{K})}{0.2075\,\cancel{\text{L mol}^{-1}}}\right]$$

$$\times\left[\left(1 + \frac{2(1131K^2)}{(298.15\,\text{K})^2} \times \left(-1.631 \times 10^{-3}\text{bar}^{-1}\right)\right.\right.$$

$$\left.\left. -\frac{2\left(-0.1993\,\text{bar}^{-1}\right) \times (1131\,\cancel{K^2})}{(298.15\,\cancel{K})^2}\right] \times \left(\frac{1\,\text{bar}}{10^5\text{Nm}^{-2}}\right) \times \left(\frac{1\text{L}}{10^{-3}\text{m}^{-3}}\right)\right)$$

$$\pi_T(100\,\text{bar}, 298.15\,\text{K}) = \left(1.427 \times 10^8\,\text{J}^2\,\text{L}^{-2}\right) \times (0.003399) \times \left(\frac{1\text{L}}{100\,\text{J}}\right)$$

$$= \boxed{4.851\,\text{J}\,\text{L}^{-1}}$$

(b) $\Delta G_m(T) = \displaystyle\int_{p_i}^{p_f} V_m(p, T)\,\mathrm{d}p$ (integral form of eqn 5.10 along an isotherm)

$$= RT\int_{p_i}^{p_f}\frac{1 + B'(T)p}{p}\,\mathrm{d}p \quad\text{(eqn. 1.21 along isotherm)}$$

$$= RT\left[\ln\left(\frac{p_f}{p_i}\right) + B'(T) \times (p_f - p_i)\right]$$

$$\Delta S_m(T) = -\int_{p_i}^{p_f}\left(\frac{\partial V_m}{\partial T}\right)_p\,\mathrm{d}p \quad\text{(Integral form of last eqn of Table 5.1 along an isotherm)}$$

$$= -\int_{pi}^{p_f}\frac{R}{p}\left[1 + B'(T)p + p\left[\left(\frac{2c}{T^2}\right) \times (B'(T) - a)\right]\right]\mathrm{d}p$$

$$= -R\left\{\ln\left(\frac{p_f}{p_i}\right) + \left[B'(T) + \left(\frac{2c}{T^2}\right) \times (B'(T) - a)\right](p_f - p_i)\right\}$$

$$\Delta G_m = \Delta H_m - T\Delta S_m \quad\text{or}\quad \Delta H_m(T) = \Delta G_m(T) + T\Delta S_m(T)$$

For $p_i = 1$ bar and $p_f = 100$ bar these equations yield

$$\Delta G_m(298.15\,\text{K}) = 11.02\,\text{kJ mol}^{-1}$$

$$\Delta S_m(298.15\,\text{K}) = -41.1\,\text{J K}^{-1}\,\text{mol}^{-1}$$

$$\Delta H_m(298.15\,\text{K}) = -1.23\,\text{kJ mol}^{-1}$$

P5.18
$$\left(\frac{\partial}{\partial T}\left(\frac{\Delta_r G}{T}\right)\right)_p = \frac{-\Delta_r H}{T^2} \text{ [5.11 and Problem 5.1]}$$

(a)
$$\int d\left(\frac{\Delta_r G}{T}\right) = -\int \frac{\Delta_r H \, dT}{T^2} \approx -\Delta_r H \int \frac{dT}{T^2} \text{ [}\Delta_r H \text{ constant]}$$

$$\frac{\Delta_r G'}{T'} - \frac{\Delta_r G}{T} = \Delta_r H \left(\frac{1}{T'} - \frac{1}{T}\right)$$

$$\Delta_r G' = \frac{T'}{T}\Delta_r G + \left(1 - \frac{T'}{T}\right)\Delta_r H$$

$$= \boxed{\tau \Delta_r G + (1 - \tau)\Delta_r H} \quad \text{with} \quad \tau = \frac{T'}{T} \text{ [Problem 5.1]}$$

(b) $\Delta_r H(T'') = \Delta_r H(T) + (T'' - T)\Delta_r C_p$ [given, T'' is the variable]

$$\frac{\Delta_r G'}{T'} - \frac{\Delta_r G}{T} = -\Delta_r H \int_T^{T'} \frac{dT''}{T''^2} - \Delta_r C_p \int_T^{T'} \frac{(T'' - T)\,dT''}{T''^2}$$

$$= \left(\frac{1}{T'} - \frac{1}{T}\right)\Delta_r H - \Delta_r C_p \ln\frac{T'}{T} - T\Delta_r C_p \left(\frac{1}{T'} - \frac{1}{T}\right)$$

Therefore, with $\tau = \dfrac{T'}{T}$

$$\Delta_r G' = \tau \Delta_r G + (1 - \tau)\Delta_r H - T'\Delta_r C_p \ln \tau - T\Delta_r C_p(1 - \tau)$$

$$= \boxed{\tau \Delta_r G + (1 - \tau)(\Delta_r H - T\Delta_r C_p) - T'\Delta_r C_p \ln \tau}$$

P5.20 $S = S(T, V)$

$$dS = \left(\frac{\partial S}{\partial T}\right)_V dT + \left(\frac{\partial S}{\partial V}\right)_T dV$$

$$T\,dS = T\left(\frac{\partial S}{\partial T}\right)_V dT + T\left(\frac{\partial S}{\partial V}\right)_T dV$$

Now, $\left(\dfrac{\partial S}{\partial T}\right)_V = \left(\dfrac{\partial S}{\partial U}\right)_V \left(\dfrac{\partial U}{\partial T}\right)_V = \dfrac{1}{T} \times C_V$ [5.4]

$$\left(\frac{\partial S}{\partial V}\right)_T = \left(\frac{\partial p}{\partial T}\right)_V \text{ [Maxwell relation]}$$

Hence, $\boxed{T\,dS = C_V\,dT + T\left(\dfrac{\partial p}{\partial T}\right)_V dV}$

For a reversible, isothermal expansion, $T\,dS = dq_{rev}$; therefore

$$dq_{rev} = T\left(\frac{\partial p}{\partial T}\right)_V dV = \frac{nRT}{V - nb}\,dV$$

$$q_{rev} = nRT \int_{V_i}^{V_f} \frac{dV}{V - nb} = \boxed{nRT \ln\left(\frac{V_f - nb}{V_i - nb}\right)}$$

P5.22 $$\left(\frac{\partial S_m}{\partial p}\right)_T = -\left(\frac{\partial V_m}{\partial T}\right)_p \quad \text{[Maxwell relation; Table 5.1]}$$

Using $V_m = M/\rho$ where M is molar mass and ρ is density,

$$\left(\frac{\partial S_m}{\partial p}\right)_T = -\left(\frac{\partial (M/\rho)}{\partial T}\right)_p = \boxed{\frac{M}{\rho^2}\left(\frac{\partial \rho}{\partial T}\right)_p = \left(\frac{\partial S_m}{\partial p}\right)_T}$$

When $\theta < 4°C$, $(\partial \rho / \partial T)_p > 0$. So we conclude that for $0°C \leq \theta \leq 4°C$ entropy $\boxed{\text{increases}}$ with an increase in pressure at constant T (i.e. $(\partial S_m/\partial p)_T > 0$).

When $\theta = 4°C$, $(\partial \rho / \partial T)_p = 0$, because this is the temperature of maximum density. Consequently, $(\partial S_m/\partial p)_T = 0$ and entropy $\boxed{\text{remains constant}}$ with pressure.

When $\theta > 4°C$, $(\partial \rho / \partial T) < 0$ so $(\partial S_m/\partial p)_T < 0$ and entropy $\boxed{\text{decreases}}$ with increasing pressure at constant T.

P5.23   (A) $dp = \left[2nRT/(V-nb)^2\right]dV + \left[R(V-nb)/nb^2\right]dT$

$$\left(\frac{\partial p}{\partial V}\right)_T = \frac{2nRT}{V-nb)^2}; \qquad \left(\frac{\partial p}{\partial T}\right)_V = \frac{R(V-nb)}{nb^2}$$

(B) $dp = -[nRT/(V-nb)^2]dV + [nR/(V-nb)]dT$

$$\left(\frac{\partial p}{\partial V}\right)_T = -\frac{nRT}{(V-nb)^2}; \qquad \left(\frac{\partial p}{\partial T}\right)_V = \frac{nR}{V-nb}$$

It is expected that for a real substance $(\partial p/\partial V)_T < 0$. Option (B) has this property, but option (A) has the unreal $(\partial p/\partial V)_T > 0$ for $V > b$. Additionally, it is expected that a real substance will have the property that $(\partial p/\partial T)_V > 0$ and $\lim\limits_{V\to\infty}\left(\frac{\partial p}{\partial T}\right)_V = 0$. Both option (A) and option (B) have $(\partial p/\partial T) > 0$. Option (B) has the desired properties as $V \to \infty$. However, Option (A) has a $(\partial p/\partial T)_V$ that "explodes" to infinity as $V \to \infty$ and this is definitely not good—it is physically unreal.

We conclude that Option (B) is the description of choice. Taking the guess that

$$\boxed{p = \frac{nRT}{V-nb}} + \text{constant}$$

we see that this equation has the desired properties that $(\partial p/\partial V)_T = -nRT/(V-nb)^2$ and $(\partial p/\partial T)_V = nR/(V-nb)$. It is the equation of state.

P5.24   The starting point for the calculation is eqn 5.20. To evaluate the integral, we need an analytical expression for Z, which can be obtained from the equation of state. We saw in Section 1.4 that the van der Waals coefficient a represents the attractions between molecules, so it may be set equal to zero in this calculation.

When we neglect a in the van der Waals equation, that equation becomes

$$p = \frac{RT}{V_m - b}$$

and hence

$$Z = 1 + \frac{bp}{RT}$$

The integral in eqn 5.20 that we require is therefore

$$\int_0^p \left(\frac{Z-1}{p}\right) dp = \int_0^p \left(\frac{b}{RT}\right) dp = \frac{bp}{RT}$$

Consequently, from eqns 5.20 and 5.19, the fugacity at the pressure p is

$$f = p e^{bp/RT}$$

From Table 1.5, $b = 3.71 \times 10^{-2} \, \text{L mol}^{-1}$, so $pb/RT = 1.516 \times 10^{-2}$, giving

$$f = (10.00 \, \text{atm}) \times e^{0.01515} = 10.2 \, \text{atm}$$

Comment. The effect of the repulsive term (as represented by the coefficient b in the van der Waals equation) is to increase the fugacity above the pressure, and so the effective pressure of the gas—its "escaping tendency"—is greater than if it were perfect.

P5.26

$$\ln \phi = \int_0^p \left(\frac{Z-1}{p}\right) dp \quad [5.20]$$

$$Z = 1 + \frac{B}{V_m} + \frac{C}{V_m^2} = 1 + B'p + C'p^2 + \cdots$$

with $B' = \dfrac{B}{RT}$, $\quad C' = \dfrac{C - B^2}{R^2 T^2}$ [Problem 1.21]

$$\frac{Z-1}{p} = B' + C'p + \cdots$$

Therefore, $\quad \ln \phi \quad = \quad \displaystyle\int_0^p B' \, dp \; + \; \int_0^p C'p \, dp \; + \; \cdots \quad = \quad B'p \; + \; \frac{1}{2} C'p^2 \; + \; \cdots \quad =$

$$\boxed{\frac{Bp}{RT} + \frac{(C - B^2)p^2}{2R^2 T^2} + \cdots}$$

For argon, $\dfrac{Bp}{RT} = \dfrac{(-21.13 \times 10^{-3} \, \text{L mol}^{-1}) \times (1.00 \, \text{atm})}{(8.206 \times 10^{-2} \, \text{L atm K}^{-1} \, \text{mol}^{-1}) \times (273 \, \text{K})} = -9.43 \times 10^{-4}$

$$\frac{(C - B^2)p^2}{2R^2 T^2} = \frac{\{(1.054 \times 10^{-3} \, \text{L}^2 \, \text{mol}^{-2}) - (-21.13 \times 10^{-3} \, \text{L mol}^{-1})^2\} \times (1.00 \, \text{atm})^2}{(2) \times \{(8.206 \times 10^{-2} \, \text{L atm K}^{-1} \, \text{mol}^{-1}) \times (273 \, \text{K})\}^2}$$

$$= 6.05 \times 10^{-7}$$

Therefore, $\ln \phi = (-9.43 \times 10^{-4}) + (6.05 \times 10^{-7}) = -9.42 \times 10^{-4}$; $\phi = 0.9991$

Hence, $f = (1.00 \, \text{atm}) \times (0.9991) = \boxed{0.99\overline{91} \, \text{atm}}$

Solutions to applications

P5.29 The relative increase in water vapor in the atmosphere at constant relative humidity is the same as the relative increase in the equilibrium vapor pressure of water. Examination of the molar Gibbs function will help us estimate this increase. At equilibrium, the vapor and liquid have the same molar Gibbs function. So at the current temperature

$$G_{m,\text{liq}}(T_0) = G_{m,\text{vap}}(T_0) \quad \text{so} \quad G_{m,\text{liq}}^{\ominus}(T_0) = G_{m,\text{vap}}^{\ominus}(T_0) + RT_0 \ln p_0,$$

where the subscript 0 refers to the current equilibrium and p is the pressure divided by the standard pressure. The Gibbs function changes with temperature as follows

$$(\partial G/\partial T) = -S \quad \text{so} \quad G_{m,\text{liq}}^{\ominus}(T_1) = G_{m,\text{liq}}^{\ominus}(T_0) - (\Delta T)S_{\text{liq}}^{\ominus}$$

and similarly for the vapor. Thus, at the higher temperature

$$G_{m,\text{liq}}^{\ominus}(T_0) - (\Delta T)S_{\text{liq}}^{\ominus} = G_{m,\text{vap}}^{\ominus}(T_0) - (\Delta T)S_{\text{vap}}^{\ominus} + R(T_0 + \Delta T) \ln p$$

Solving both of these expressions for $G_{m,\text{liq}}^{\ominus}(T_0) - G_{m,\text{vap}}^{\ominus}(T_0)$ and equating them leads to

$$(\Delta T)(S_{\text{liq}}^{\ominus} - S_{\text{vap}}^{\ominus}) + R(T_0 + \Delta T) \ln p = RT_0 \ln p_0$$

Isolating p leads to

$$\ln p = \frac{(\Delta T)(S_{\text{vap}}^{\ominus} - S_{\text{liq}}^{\ominus})}{R(T_0 + \Delta T)} + \frac{T_0 \ln p_0}{T_0 + \Delta T}$$

$$p = \exp\left(\frac{(\Delta T)(S_{\text{vap}}^{\ominus} - S_{\text{liq}}^{\ominus})}{R(T_0 + \Delta T)}\right) p_0^{(T_0/(T_0+\Delta T))}$$

So $\quad p = \exp\left(\frac{(2.0\,\text{K}) \times (188.83 - 69.91)\,\text{J mol}^{-1}\,\text{K}^{-1}}{(8.3145\,\text{J mol}^{-1}\,\text{K}^{-1}) \times (290 + 2.0\,\text{K})}\right) \times (0.0189)^{(290\,\text{K}/(290+2.0)\text{K})},$

$p = 0.0214$ which represents a $\boxed{13\ \text{percent}}$ increase.

P5.30 (a) The strong ionic bonds of sodium flouride and sodium chloride products along with the strong double bonds of carbon dioxide are expected to cause the reaction to be exothermic. Furthermore, the net production of gas is expected to cause an entropy increase. Both effects contribute to make the reaction exergonic and spontaneous to the right because $\Delta_r G = \Delta_r H - T\Delta_r S$. The reaction mixture should not be heated much beyond $270°C$ because the sodium oxolate will be pyrolyzed to carbonate at high temperatures.

(b) At 298.15 K:

$$\Delta_r G_{298}^{\ominus} = 2\Delta_f G_{298}^{\ominus}(\text{NaF}) + 2\Delta_f G_{298}^{\ominus}(\text{NaCl}) + \Delta_f G_{298}^{\ominus}(\text{graphite}) + 4\Delta_f G_{298}^{\ominus}(\text{CO}_2)$$

$$-\Delta_f G_{298}^{\ominus}(\text{CF}_2\text{Cl}_2) - 2\Delta G_{298}^{\ominus}(\text{Na}_2(\text{CO}_2)_2)$$

$$= 2(-545.13\,\text{kJ mol}^{-1}) + 2(-384.07\,\text{kJ mol}^{-1}) + 0\,\text{kJ mol}^{-1}$$

$$-4(-394.36\,\text{kJ mol}^{-1}) - (-452.63\,\text{kJ mol}^{-1})$$

$$-2(-1.208 \times 10^3\,\text{kJ mol}^{-1})(\text{estimated})$$

$$= \boxed{-566.4\,\text{kJ mol}^{-1}}$$

Similarly, $\Delta_r G_{298}^{\ominus} = 122.8\,\text{J K}^{-1}\,\text{mol}^{-1}$

The accurate calculation of $\Delta_r G(T)$ requires that data about the constant heat capacity as a function of T be assembled for each reactant and product.

$$\Delta_r C_p(T) = \sum_{\text{Products}} \nu C_p(T) - \sum_{\text{Reactants}} \nu C_p(T) \quad [2.45]$$

numerical integrations can be used to acquire $\Delta_r G(T)$.

$$\Delta_r G(T) = \Delta_r H(T) - T \Delta_r S(T) \; [4.39]$$

$$= \Delta_r H_{298}^{\ominus} + \int_{298}^{T} \Delta_r C_p(T) \, dT - T \left[\Delta_r S_{298}^{\ominus} + \int_{298}^{T} \frac{\Delta_r C_p(T)}{T} \, dT \right] \; [2.43 \text{ and } 4.19]$$

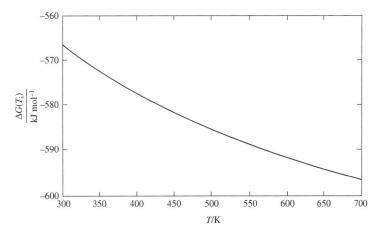

Figure 5.3(a)

Figure 5.3(b)

P5.33
$$G' = G + \int_0^p V \, dp = G + V_0 \int_0^p e^{-p/p^*} \, dp \; [V_0 \text{ is a constant}] = \boxed{G + p^* V_0 (1 - e^{-p/p^*})}$$

$$\Delta G = p^* V_0 (1 - e^{-p/p^*})$$

Since $e^{-p/p^*} < 1$ if $p > 0$, ΔG is positive. When the pressure is reduced to zero from a positive value, ΔG decreases to zero. Under constant-temperature conditions (p and V not constant) it is ΔA that determines the direction of natural change.

$$dA = -S \, dT - p \, dV \; [\text{Exercise } 5.14(\mathbf{a})] = -p \, dV \quad [\text{constant temperature}]$$

Since $dV = -\dfrac{V_0}{p^*} e^{-p/p^*} \, dp$, it is clear that $\boxed{V \text{ increases}}$ when pressure is relaxed [$dV > 0$ when $dp < 0$]. Substituting for dV, $dA = \dfrac{p}{p^*} V_0 e^{-p/p^*} \, dp$, so a decrease in p is spontaneous, as ΔA will be negative.

6 Physical transformations of pure substances

Solutions to exercises

Discussion questions

E6.1(a) Consider two phases of a system, labeled α and β. The phase with the lower chemical potential under the given set of conditions is the more stable phase. First consider the variation of μ with temperature at a fixed pressure. We have

$$\left(\frac{\partial \mu_\alpha}{\partial T}\right)_p = -S_\alpha \quad \text{and} \quad \left(\frac{\partial \mu_\beta}{\partial T}\right)_p = -S_\beta$$

Therefore, if S_β is larger in magnitude than S_α, the β phase will be favored over the α phase as temperature increases because its chemical potential decreases more rapidly with temperature than for the α phase. We also have

$$\left(\frac{\partial \mu_\alpha}{\partial P}\right)_T = V_{m,\alpha} \quad \text{and} \quad \left(\frac{\partial \mu_\beta}{\partial P}\right)_T = V_{m,\beta}$$

Therefore, if $V_{m,\alpha}$ is larger than $V_{m,\beta}$ the β phase will be favored over the α phase as pressure increases because the chemical potential of the β phase will not increase as rapidly with pressure as that of the α phase. See Example 6.1.

E6.2(a) At 400 K, the system is all vapor. As the system cools by the loss of heat energy, the temperature falls until the temperature reaches 373.15 K, when some liquid begins to form. As long as both the vapor and liquid phases are present, the temperature cannot drop below 373.15 K. At constant pressure, there are no degrees of freedom as long as two phases are present. As the system continues to lose heat energy, the proportion of liquid increases until eventually the entire system is in the liquid state at 373.15 K. With continued loss of heat, the temperature falls until 273.15 K is reached, at which point some solid begins to form. With continued cooling, eventually the entire system is converted to the solid form (ice). Continued loss of heat energy reduces the temperature further.

E6.3(a) The Clapeyron equation is exact and applies rigorously to all first-order phase transitions. It shows how pressure and temperature vary with respect to each other (temperature or pressure) along the phase boundary line, and in that sense, it defines the phase boundary line.

The Clausius–Clapeyron equation serves the same purpose, but it is not exact; its derivation involves approximations, in particular the assumptions that the perfect gas law holds and that the volume of condensed phases can be neglected in comparison to the volume of the gaseous phase. It applies only to phase transitions between the gaseous state and condensed phases.

Numerical exercises

E6.4(a) On the assumption that the vapor is a perfect gas and that $\Delta_{vap}H$ is independent of temperature, we may write

$$p = p^* e^{-\chi}, \quad \chi = \left(\frac{\Delta_{vap}H}{R}\right) \times \left(\frac{1}{T} - \frac{1}{T^*}\right) \text{ [6.12]}, \quad \ln\frac{p^*}{p} = \chi$$

$$\frac{1}{T} = \frac{1}{T^*} + \frac{R}{\Delta_{vap}H}\ln\frac{p^*}{p}$$

$$= \frac{1}{297.25\ \text{K}} + \left(\frac{8.314\ \text{J K}^{-1}\ \text{mol}^{-1}}{28.7 \times 10^3\ \text{J mol}^{-1}}\right)\ln\frac{400\ \text{Torr}}{500\ \text{Torr}} = 3.30\overline{0} \times 10^{-3}\ \text{K}^{-1}$$

Hence, $T = \boxed{303\ \text{K}} = \boxed{30^\circ\text{C}}$

E6.5(a)
$$\frac{dp}{dT} = \frac{\Delta_{trs}S}{\Delta_{trs}V} \quad [6.6]$$

$$\Delta_{fus}S = \Delta_{fus}V \times \left(\frac{dp}{dT}\right) \approx \Delta_{fus}V \times \frac{\Delta p}{\Delta T}$$

[$\Delta_{fus}S$ and $\Delta_{fus}V$ assumed independent of temperature.]

$$\Delta_{fus}S = [(163.3 - 161.0) \times 10^{-6}\,\text{m}^3\,\text{mol}^{-1}] \times \left(\frac{(100 - 1) \times (1.013 \times 10^5\,\text{Pa})}{(351.26 - 350.75)\,\text{K}}\right)$$

$$= \boxed{+45.2\overline{3}\,\text{J K}^{-1}\,\text{mol}^{-1}}$$

$$\Delta_{fus}H = T_f\Delta S = (350.75\,\text{K}) \times (45.23\,\text{J K}^{-1}\,\text{mol}^{-1}) = \boxed{+16\,\text{kJ mol}^{-1}}$$

E6.6(a) The expression for $\ln p$ is the indefinite integral of eqn 6.11

$$\int d\ln p = \int \frac{\Delta_{vap}H}{RT^2}\,dT; \quad \ln p = \text{constant} - \frac{\Delta_{vap}H}{RT}$$

Therefore, $\Delta_{vap}H = (2501.8\,\text{K}) \times R = (2501.8\,\text{K}) \times (8.314\,\text{J K}^{-1}\,\text{mol}^{-1}) = \boxed{+20.80\,\text{kJ mol}^{-1}}$

E6.7(a) (a) The indefinitely integrated form of eqn 6.11 is used as in Exercise 6.6.

$$\ln p = \text{constant} - \frac{\Delta_{vap}H}{RT}, \quad \text{or} \quad \log p = \text{constant} - \frac{\Delta_{vap}H}{2.303RT}$$

Therefore,

$$\Delta_{vap}H = (2.303) \times (1780\,\text{K}) \times R = (2.303) \times (1780\,\text{K}) \times (8.314\,\text{J K}^{-1}\,\text{mol}^{-1})$$

$$= \boxed{+34.08\,\text{kJ mol}^{-1}}$$

(b) The boiling point corresponds to $p = 1.000\,\text{atm} = 760\,\text{Torr}$.

$$\log 760 = 7.960 - \frac{1780\,\text{K}}{T_b}$$

$$T_b = \boxed{350.5\,\text{K}}$$

E6.8(a)
$$\Delta T \approx \frac{\Delta_{fus}V}{\Delta_{fus}S} \times \Delta p \quad [6.6, \text{ and Exercise 6.5}]$$

$$\approx \frac{T_f\Delta_{fus}V}{\Delta_{fus}H} \times \Delta p = \frac{T_f\Delta p M}{\Delta_{fus}H} = \Delta\left(\frac{1}{\rho}\right) \quad [V_m = M/\rho]$$

$$\approx \left(\frac{(278.6\,\text{K}) \times (999) \times (1.013 \times 10^5\,\text{Pa}) \times (78.12 \times 10^{-3}\,\text{kg mol}^{-1})}{10.59 \times 10^3\,\text{J mol}^{-1}}\right)$$

$$\times \left(\frac{1}{879\,\text{kg m}^{-3}} - \frac{1}{891\,\text{kg m}^{-3}}\right) \approx 3.18\,\text{K}$$

Therefore, at 1000 atm, $T_f \approx 278.6 + 3.18 = \boxed{281.8\,\text{K}}$ [8.7°C]

E6.9(a) The rate of loss of mass of water may be expressed as

$$\frac{dm}{dt} = \frac{dn}{dt} \times M_{H_2O}; \quad n = \frac{q}{\Delta_{vap}H}$$

$$\frac{dn}{dt} = \frac{\left(\frac{dq}{dt}\right)}{\Delta_{vap}H} = \frac{(1.2 \times 10^3 \, \text{W m}^{-2}) \times (50 \, \text{m}^2)}{44.0 \times 10^3 \, \text{J mol}^{-1}} = 1.4 \, \text{mol s}^{-1}$$

$$\frac{dm}{dt} = (1.4 \, \text{mol s}^{-1}) \times (18.02 \, \text{g mol}^{-1}) = \boxed{25 \, \text{g s}^{-1}}$$

E6.10(a) Assume perfect gas behavior.

$$n = \frac{pV}{RT} \qquad n = \frac{m}{M} \qquad V = 75 \, \text{m}^3$$

$$m = \frac{pVM}{RT}$$

(a) $\quad m = \dfrac{(24 \, \text{Torr}) \times (75 \times 10^3 \, \text{L}^3) \times (18.02 \, \text{g mol}^{-1})}{(62.364 \, \text{L Torr K}^{-1} \, \text{mol}^{-1}) \times (298.15 \, \text{K})} = \boxed{1.7 \, \text{kg}}$

(b) $\quad m = \dfrac{(98 \, \text{Torr}) \times (75 \times 10^3 \, \text{L}^3) \times (78.11 \, \text{g mol}^{-1})}{(62.364 \, \text{L Torr K}^{-1} \, \text{mol}^{-1}) \times (298.15 \, \text{K})} = \boxed{31 \, \text{kg}}$

(c) $\quad m = \dfrac{(1.7 \times 10^{-3} \, \text{Torr}) \times (75 \times 10^3 \, \text{L}^3) \times (200.59 \, \text{g mol}^{-1})}{(62.364 \, \text{L Torr K}^{-1} \, \text{mol}^{-1}) \times (298.15 \, \text{K})} = \boxed{1.4 \, \text{g}}$

Question. Assuming all the mercury vapor breathed remains in the body, how long would it take to accumulate 1.4 g? Make reasonable assumptions about the volume and frequency of a breath.

E6.11(a) The Clausius–Clapeyron equation [6.11] integrates to the form [6.12] which may be rewritten as

$$\ln\left(\frac{p_2}{p_1}\right) = \frac{\Delta_{vap}H}{R} \times \left(\frac{1}{T_1} - \frac{1}{T_2}\right)$$

(a) $\quad \ln\left(\dfrac{40 \, \text{Torr}}{10 \, \text{Torr}}\right) = \left(\dfrac{\Delta_{vap}H}{8.314 \, \text{J K}^{-1} \, \text{mol}^{-1}}\right) \times \left(\dfrac{1}{359.0 \, \text{K}} - \dfrac{1}{392.5 \, \text{K}}\right)$

$$1.38\overline{6} = \Delta_{vap}H \times (2.8\overline{6} \times 10^{-5} \, \text{J}^{-1} \, \text{mol})$$

$$\Delta_{vap}H = \boxed{48.\overline{5} \, \text{kJ mol}^{-1}}$$

(b) The normal boiling point corresponds to a vapor pressure of 760 Torr. Using the data at 119.3°C

$$\ln\left(\frac{760 \, \text{Torr}}{40 \, \text{Torr}}\right) = \left(\frac{48.\overline{5} \times 10^3 \, \text{J mol}^{-1}}{8.314 \, \text{J K}^{-1} \, \text{mol}^{-1}}\right) \times \left(\frac{1}{392.5 \, \text{K}} - \frac{1}{T_b}\right)$$

$$2.94\overline{4} = 14.\overline{86} - \frac{58\overline{31} \, \text{K}}{T_b}; \quad T_b = 48\overline{9} \, \text{K} = \boxed{21\overline{6}°\text{C}}$$

[The accepted value is 218°C.]

(c) $\quad \Delta_{vap}S(T_b) = \dfrac{\Delta_{vap}H(T_b)}{T_b} \approx \dfrac{48.5 \times 10^3 \, \text{J mol}^{-1}}{489 \, \text{K}} = \boxed{99 \, \text{J K}^{-1} \, \text{mol}^{-1}}$

E6.12(a) $\Delta T = T_f(50\,\text{bar}) - T_f(1\,\text{bar}) \approx \dfrac{T_f \Delta p M}{\Delta_{fus} H} \Delta\left(\dfrac{1}{\rho}\right)$ [Exercise 6.8]

$\Delta_{fus} H = 6.01\,\text{kJ mol}^{-1}$ [Table 2.3]

$\Delta T = \left(\dfrac{(273.15\,\text{K}) \times (49 \times 10^5\,\text{Pa}) \times (18 \times 10^{-3}\,\text{kg mol}^{-1})}{6.01 \times 10^3\,\text{J mol}^{-1}}\right)$

$\times \left(\dfrac{1}{1.00 \times 10^3\,\text{kg m}^{-3}} - \dfrac{1}{9.2 \times 10^2\,\text{kg m}^3}\right) = -0.35\,\text{K}$

$T_f(50\,\text{bar}) = (273.15\,\text{K}) - (0.35\,\text{K}) = \boxed{272.80\,\text{K}}$

E6.13(a) $\Delta_{vap} H = \Delta_{vap} U + \Delta_{vap}(pV) = 40.656\,\text{kJ mol}^{-1}$ [Table 2.3]

$\Delta_{vap}(pV) = p\Delta_{vap} V = p(V_{gas} - V_{liq}) \approx pV_{gas}$

$= RT$ [per mole of a perfect gas]

$= (8.314\,\text{J K}^{-1}\,\text{mol}^{-1}) \times (373.2\,\text{K}) = 3.102 \times 10^3\,\text{kJ mol}^{-1}$

$\text{Fraction} = \dfrac{\Delta_{vap}(pV)}{\Delta_{vap} H} = \dfrac{3.102 \times 10^3\,\text{kJ mol}^{-1}}{40.656\,\text{kJ mol}^{-1}} = \boxed{0.07630} \approx 7.6\ \text{percent}$

E6.14(a) $p = p^* e^{2\gamma V_m / rRT}$ [6.17]

$V_m = \dfrac{M}{\rho} = \dfrac{18.02\,\text{g mol}^{-1}}{0.9982\,\text{g cm}^{-3}} = 18.05\,\text{cm}^3\,\text{mol}^{-1} = 1.805 \times 10^{-5}\,\text{m}^3\,\text{mol}^{-1}$

$\dfrac{2\gamma V_m}{rRT} = \dfrac{(2) \times (7.275 \times 10^{-2}\,\text{N m}^{-1}) \times (1.805 \times 10^{-5}\,\text{m}^3\,\text{mol}^{-1})}{(1.0 \times 10^{-8}\,\text{m}) \times (8.314\,\text{J K}^{-1}\,\text{mol}^{-1}) \times (293\,\text{K})} = 0.10\overline{78}$

$p = (2.3\,\text{kPa}) \times e^{0.10\overline{78}} = \boxed{2.6\,\text{kPa}}$

E6.15(a) $\gamma = \tfrac{1}{2}\rho g h r$ [6.19] $= \left(\tfrac{1}{2}\right) \times (998.2\,\text{kg m}^{-3}) \times (9.807\,\text{m s}^{-2})$

$\times (4.96 \times 10^{-2}\,\text{m}) \times (3.00 \times 10^{-4}\,\text{m})$

$= 7.28 \times 10^{-2}\,\text{kg s}^{-2} = \boxed{7.28 \times 10^{-2}\,\text{N m}^{-1}}$

This value is in agreement with Table 6.1.

E6.16(a) $p_{in} - p_{out} = \dfrac{2\gamma}{r}$ [6.16] $= \dfrac{(2) \times (7.275 \times 10^{-2}\,\text{N m}^{-1})}{2.00 \times 10^{-7}\,\text{m}}$ [Table 6.1] $= \boxed{7.28 \times 10^5\,\text{Pa}}$

Comment. Pressure differentials for small droplets are quite large.

Solutions to problems
Solutions to numerical problems

P6.1 At the triple point, T_3, the vapor pressures of liquid and solid are equal, hence

$10.5916 - \dfrac{1871.2\,\text{K}}{T_3} = 8.3186 - \dfrac{1425.7\,\text{K}}{T_3}; \quad T_3 = \boxed{196.0\,\text{K}}$

$\log(p_3/\text{Torr}) = \dfrac{-1871.2\,\text{K}}{196.0\,\text{K}} + 10.5916 = 1.044\overline{7}; \quad p_3 = \boxed{11.1\,\text{Torr}}$

P6.2 Use the definite integral form of the Clausius–Clapeyron equation (Exercise 6.11(a)).

$\ln\left(\dfrac{p_2}{p_1}\right) = \dfrac{\Delta_{vap} H}{R} \times \left(\dfrac{1}{T_1} - \dfrac{1}{T_2}\right); \quad T_1 = \text{normal boiling point}; \quad p_1 = 1.000\,\text{atm}$

$$\ln(p_2/\text{atm}) = \left(\frac{20.25 \times 10^3 \,\text{J mol}^{-1}}{8.314 \,\text{J K}^{-1}\,\text{mol}^{-1}} \right) \times \left(\frac{1}{244.0 \,\text{K}} - \frac{1}{313.2 \,\text{K}} \right) = 2.206$$

$$p_2 = \boxed{9.07 \,\text{atm}} \approx 9 \,\text{atm}$$

Comment. Three significant figures are not really warranted in this answer because of the approximations employed.

P6.4   (a) $\left(\dfrac{\partial \mu(\text{l})}{\partial T} \right)_p - \left(\dfrac{\partial \mu(\text{s})}{\partial T} \right)_p = -S_m(\text{l}) + S_m(\text{s})$ [Section 6.7, eqn 13]

$$= -\Delta_{\text{fus}} S = \frac{-\Delta_{\text{fus}} H}{T_{\text{f}}}; \quad \Delta_{\text{fus}} H = 6.01 \,\text{kJ mol}^{-1} \text{ [Table 2.3]}$$

$$= \frac{-6.01 \,\text{kJ mol}^{-1}}{273.15 \,\text{K}} = \boxed{-22.0 \,\text{J K}^{-1}\,\text{mol}^{-1}}$$

(b) $\left(\dfrac{\partial \mu(\text{g})}{\partial T} \right)_p - \left(\dfrac{\partial \mu(\text{l})}{\partial T} \right)_p = -S_m(\text{g}) + S_m(\text{l}) = -\Delta_{\text{vap}} S$

$$= \frac{-\Delta_{\text{vap}} H}{T_{\text{b}}} = \frac{-40.6 \,\text{kJ mol}^{-1}}{373.15 \,\text{K}} = \boxed{-109.0 \,\text{J K}^{-1}\,\text{mol}^{-1}}$$

(c) $\Delta\mu \approx \left(\dfrac{\partial \mu}{\partial T} \right)_p \Delta T = -S_m \Delta T$ [6.1]

$$\Delta\mu(\text{l}) - \Delta\mu(\text{s}) = \mu(\text{l}, -5^{\circ}\text{C}) - \mu(\text{l}, 0^{\circ}\text{C}) - \mu(\text{s}, -5^{\circ}\text{C}) + \mu(\text{s}, 0^{\circ}\text{C})$$
$$= \mu(\text{l}, -5^{\circ}\text{C}) - \mu(\text{s}, -5^{\circ}\text{C})[\mu(\text{l}, 0^{\circ}\text{C}) = \mu(\text{s}, 0^{\circ}\text{C})]$$
$$\approx -\{S_m(\text{l}) - S_m(\text{s})\}\Delta T \approx -\Delta_{\text{fus}} S \Delta T$$
$$= -(5 \,\text{K}) \times (-22.0 \,\text{J K}^{-1}\,\text{mol}^{-1}) = \boxed{+11\bar{0} \,\text{J mol}^{-1}}$$

Since $\mu(\text{l}, -5^{\circ}\text{C}) > \mu(\text{s}, -5^{\circ}\text{C})$, there is a thermodynamic tendency to freeze.

P6.6 $\dfrac{\text{d}p}{\text{d}T} = \dfrac{\Delta_{\text{fus}} S}{\Delta_{\text{fus}} V}$ [6.6] $= \dfrac{\Delta_{\text{fus}} H}{T \Delta_{\text{fus}} V}$

$$\Delta T = \int_{T_{\text{m},1}}^{T_{\text{m},2}} \text{d}T = \int_{p_{\text{top}}}^{p_{\text{bot}}} \frac{T_m \Delta_{\text{fus}} V}{\Delta_{\text{fus}} H} \,\text{d}p$$

$$\Delta T \approx \frac{T_m \Delta_{\text{fus}} V}{\Delta_{\text{fus}} H} \times \Delta p \quad [T_m, \Delta_{\text{fus}} H, \text{ and } \Delta_{\text{fus}} V \text{ assumed constant}]$$

$$\Delta p = p_{\text{bot}} - p_{\text{top}} = \rho g h$$

Therefore

$$\Delta T = \frac{T_m \rho g h \Delta_{\text{fus}} V}{\Delta_{\text{fus}} H}$$

$$= \frac{(234.3 \,\text{K}) \times (13.6 \times 10^3 \,\text{kg m}^{-3}) \times (9.81 \,\text{m s}^{-2}) \times (10 \,\text{m}) \times (0.517 \times 10^{-6} \,\text{m}^3\,\text{mol}^{-1})}{2.292 \times 10^3 \,\text{J mol}^{-1}}$$

$$= 0.070 \,\text{K}$$

Therefore, the freezing point changes to $\boxed{234.4 \,\text{K}}$

P6.8　　$\dfrac{d \ln p}{dT} = \dfrac{\Delta_{vap}H}{RT^2}$ [6.11], yields upon indefinite integration

$$\ln p = \text{constant} - \frac{\Delta_{vap}H}{RT}$$

Therefore, plot $\ln p$ against $\dfrac{1}{T}$ and identify $\dfrac{-\Delta_{vap}H}{R}$ as its slope. Construct the following table

$\theta/°C$	0	20	40	50	70	80	90	100
T/K	273	293	313	323	343	353	363	373
1000 K/T	3.66	3.41	3.19	3.10	2.92	2.83	2.75	2.68
$\ln p$/Torr	2.67	3.87	4.89	5.34	6.15	6.51	6.84	7.16

The points are plotted in Fig. 6.1. The slope is -4569 K, so

$$\frac{-\Delta_{vap}H}{R} = -4569\,\text{K}, \quad \text{or} \quad \boxed{\Delta_{vap}H = +38.0\,\text{kJ mol}^{-1}}$$

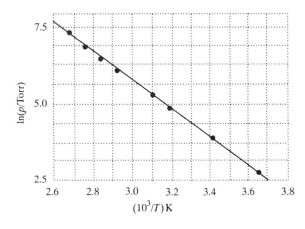

Figure 6.1

The normal boiling point occurs at $p = 760$ Torr, or at $\ln(p/\text{Torr}) = 6.633$, which from the figure corresponds to $1000\,\text{K}/T \approx 2.80$. Therefore, $\boxed{T_b = 357\,\text{K}\,(84°C)}$. The accepted value is 83°C.

P6.10　　The equations describing the coexistence curves for the three states are

(a) Solid–liquid boundary

$$p = p^* + \frac{\Delta_{fus}H}{\Delta_{fus}V}\ln\frac{T}{T^*} \quad [6.8]$$

(b) Liquid–vapor boundary

$$p = p^*e^{-\chi}, \quad \chi = \frac{\Delta_{vap}H}{R} \times \left(\frac{1}{T} - \frac{1}{T^*}\right) \quad [6.12]$$

(c) Solid–vapor boundary

$$p = p^* e^{-\chi}, \quad \chi = \frac{\Delta_{sub} H}{R} \times \left(\frac{1}{T} - \frac{1}{T^*} \right) \text{ [similar to 6.12]}$$

We need $\Delta_{sub} H = \Delta_{fus} H + \Delta_{vap} H = 41.4 \, \text{kJ mol}^{-1}$

$$\Delta_{fus} V = M \times \left(\frac{1}{\rho(l)} - \frac{1}{\rho(s)} \right) = \left(\frac{78.11 \, \text{g mol}^{-1}}{\text{g cm}^{-3}} \right) \times \left(\frac{1}{0.879} - \frac{1}{0.891} \right)$$

$$= +1.19\overline{7} \, \text{cm}^3 \, \text{mol}^{-1}$$

After insertion of these numerical values into the above equations, we obtain

(a) $$p = p^* + \left(\frac{10.6 \times 10^3 \, \text{J mol}^{-1}}{1.197 \times 10^{-6} \, \text{m}^3 \, \text{mol}^{-1}} \right) \ln \frac{T}{T^*}$$

$$= p^* + 8.85\overline{5} \times 10^9 \, \text{Pa} \ln \frac{T}{T^*} = p^* + (6.64 \times 10^7 \, \text{Torr}) \ln \frac{T}{T^*} \quad (1 \, \text{Torr} = 133.322 \, \text{Pa})$$

This line is plotted as a in Fig. 6.2, starting at $(p^*, T^*) = (36 \, \text{Torr}, 5.50°\text{C}(278.65 \, \text{K}))$.

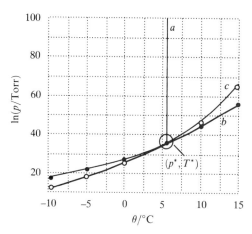

Figure 6.2

(b) $$\chi = \left(\frac{30.8 \times 10^3 \, \text{J mol}^{-1}}{8.314 \, \text{J K}^{-1} \, \text{mol}^{-1}} \right) \times \left(\frac{1}{T} - \frac{1}{T^*} \right) = (370\overline{5} \, \text{K}) \times \left(\frac{1}{T} - \frac{1}{T^*} \right)$$

$$p = p^* e^{-370\overline{5} \, \text{K} \times (1/T - 1/T^*)}$$

This equation is plotted as line b in Fig. 6.2, starting from $(p^*, T^*) = (36 \, \text{Torr}, 5.50°\text{C} (278.65 \, \text{K}))$.

(c) $$\chi = \left(\frac{41.4 \times 10^3 \, \text{J mol}^{-1}}{8.314 \, \text{J K}^{-1} \, \text{mol}^{-1}} \right) \times \left(\frac{1}{T} - \frac{1}{T^*} \right) = (498\overline{0} \, \text{K}) \times \left(\frac{1}{T} - \frac{1}{T^*} \right)$$

$$p = p^* e^{-498\overline{0} \, \text{K} \times (1/T - 1/T^*)}$$

These points are plotted as line c in Fig. 6.2, starting at $(36 \, \text{Torr}, 5.50°\text{C})$.

The lighter lines in Fig. 6.2 represent extensions of lines b and c into regions where the liquid and solid states respectively are not stable.

P6.13 Adapting eqn 6.12 to the sublimation curve and taking the natural logarithm of the equation gives:

$$\ln\left(\frac{p(T)}{p^*}\right) = -\frac{\Delta_{sub}H}{RT} + \frac{\Delta_{sub}H}{RT^*}$$

Using the reference temperature T^* at which the vapor pressure equals 1.00 Torr and recognizing that $\Delta_{sub}H/T^* = \Delta_{sub}S^*$ when $P^* = 1.00$ Torr gives:

$$\ln\left(\frac{p(T)}{1\,\text{Torr}}\right) = -\frac{\Delta_{sub}H}{RT} + \frac{\Delta_{sub}S^*}{R}$$

This equation says that a plot of measured $\ln\left(\frac{p(T)}{1\,\text{Torr}}\right)$ data against $1/T$ should be a linear plot. A linear regression fit of the plot (done with a scientific calculator or computer spreadsheet) gives the following values for the regression slope and intercept.

$$\text{slope} = -9.723 \times 10^3\,\text{K} = -\Delta_{sub}H/R$$
$$\text{intercept} = 24.184 = \Delta_{sub}S^*/R$$

Thus,

$$\Delta_{sub}H = 9.723 \times 10^3\,\text{K} \times R = 80.8\,\text{kJ mol}^{-1}$$
$$\Delta_{sub}S^* = 24.184 \times R = 201\,\text{J K}^{-1}\,\text{mol}^{-1}$$

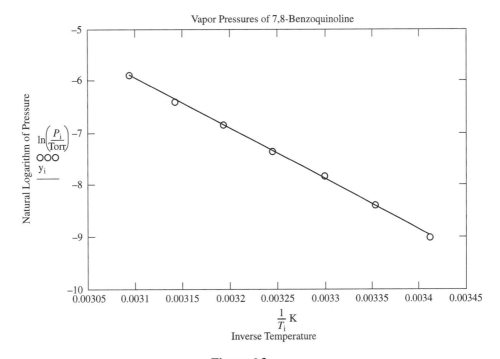

Figure 6.3

Solutions to theoretical problems

P6.15

$dH = C_p\,dT + V\,dp$, implying that $d\Delta H = \Delta C_p\,dT + \Delta V\,dp$

However, along a phase boundary dp and dT are related by

$$\frac{dp}{dT} = \frac{\Delta H}{T\Delta V} \quad \text{[Clapeyron equation, e.g. 6.6, 6.7, or 6.10]}$$

Therefore,

$$d\Delta H = \left(\Delta C_p + \Delta V \times \frac{\Delta H}{T\Delta V}\right) dT = \left(\Delta C_p + \frac{\Delta H}{T}\right) dT \quad \text{and} \quad \frac{dH}{dT} = \Delta C_p + \frac{\Delta H}{T}$$

Then, since

$$\frac{d}{dT}\left(\frac{\Delta H}{T}\right) = \frac{1}{T}\frac{d\Delta H}{dT} - \frac{\Delta H}{T^2} = \frac{1}{T}\left(\frac{d\Delta H}{dT} - \frac{\Delta H}{T}\right)$$

substituting the first result gives

$$\frac{d}{dT}\left(\frac{\Delta H}{T}\right) = \frac{\Delta C_p}{T}$$

Therefore, $d\left(\dfrac{\Delta H}{T}\right) = \dfrac{\Delta C_p\,dT}{T} = \boxed{\Delta C_p\,d\ln T}$

P6.18

In each phase the slopes are given by

$$\left(\frac{\partial \mu}{\partial T}\right)_p = -S_m \quad [6.1]$$

The curvatures of the graphs of μ against T are given by

$$\left(\frac{\partial^2 \mu}{\partial T^2}\right)_p = -\left(\frac{\partial S_m}{\partial T}\right)_p = \boxed{-\frac{1}{T} \times C_{p,m}} \quad \text{[Problem 5.7]}$$

Since $C_{p,m}$ is necessarily positive, the curvatures in all states of matter are necessarily negative. $C_{p,m}$ is often largest for the liquid state, though not always; but it is the ratio $C_{p,m}/T$ that determines the magnitude of the curvature, so no precise answer can be given for the state with greatest curvature. It depends upon the substance.

P6.19

(1) $V = V(T, p)$

$$dV = \left(\frac{\partial V}{\partial T}\right)_p dT + \left(\frac{\partial V}{\partial p}\right)_T dp$$

$$\left(\frac{\partial V}{\partial T}\right)_p = \alpha V, \qquad \left(\frac{\partial V}{\partial p}\right)_T = -\kappa_T V;$$

hence, $dV = \alpha V\,dT - \kappa_T V\,dp$

This equation applies to both phases 1 and 2, and since V is continuous through a second-order transition

$$\alpha_1\,dT - \kappa_{T,1}\,dp = \alpha_2\,dT - \kappa_{T,2}\,dp$$

Solving for $\dfrac{dp}{dT}$ yields $\boxed{\dfrac{dp}{dT} = \dfrac{\alpha_2 - \alpha_1}{\kappa_{T,2} - \kappa_{T,1}}}$

(2) $S_m = S_m(T, p)$

$$dS_m = \left(\frac{\partial S_m}{\partial T}\right)_p dT + \left(\frac{\partial S_m}{\partial p}\right)_T dp$$

$$\left(\frac{\partial S_m}{\partial T}\right)_p = \frac{C_{p,m}}{T} \text{ [Problem 5.7]} \qquad \left(\frac{\partial S_m}{\partial p}\right)_T = -\left(\frac{\partial V_m}{\partial T}\right)_p \text{ [Maxwell relation]}$$

$$= -\alpha V_m$$

Thus, $dS_m = \dfrac{C_{p,m}}{T} dT - \alpha V_m\, dp$.

This relation applies to both phases. For second-order transitions both S_m and V_m are continuous through the transition, $S_{m,1} = S_{m,2}$, $V_{m,1} = V_{m,2} = V_m$, so that

$$\frac{C_{p,m,1}}{T} dT - \alpha_1 V_m\, dp = \frac{C_{p,m,2}}{T} dT - \alpha_2 V_m\, dp$$

Solving for $\dfrac{dp}{dT}$ yields $\boxed{\dfrac{dp}{dT} = \dfrac{C_{p,m,2} - C_{p,m,1}}{T V_m(\alpha_2 - \alpha_1)}}$

The Clapeyron equation cannot apply because both ΔV and ΔS are zero through a second-order transition, resulting in an indeterminate form $\dfrac{0}{0}$.

Solutions to applications

P6.21   **(a)** The phase boundary is plotted in Fig. 6.4

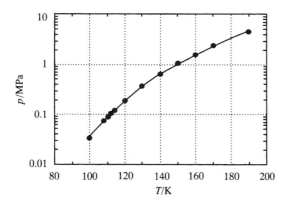

Figure 6.4

(b) The standard boiling point is the temperature at which the liquid is in equilibrium with the standard pressure of 1 bar (0.1 MPa). Interpolation of the plotted points gives $T_b = \boxed{112\,\text{K}}$

(c) The slope of the liquid–vapor coexistence curve is given by

$$\frac{dp}{dT} = \frac{\Delta_{vap}H}{T\Delta_{vap}V} \qquad \text{so} \quad \Delta_{vap}H = (T\Delta_{vap}V)\frac{dp}{dT}$$

The slope can be obtained graphically or by fitting the points nearest the boiling point. Then $dp/dT = 8.14 \times 10^{-3}\,\text{MPa K}^{-1}$, so

$$\Delta_{vap}H = (112\,\text{K}) \times \left(\frac{(8.89 - 0.0380)\,\text{L mol}^{-1}}{1000\,\text{L m}^3}\right) \times (8.14\,\text{kPa K}^{-1}) = \boxed{8.07\,\text{kJ mol}^{-1}}$$

7 Simple mixtures

Solutions to exercises

Discussion questions

E7.1(a)   At equilibrium, the chemical potentials of any component in both the liquid and vapor phases must be equal. This is justified by the requirement that for systems at equilibrium under constant temperature and pressure conditions, with no additional work, $\Delta G = 0$ [see Section 4.5(a) and the solution to Exercise 4.3(a)]. Here $\Delta G = \mu_i(v) - \mu_i(1)$, for all components, i, of the solution; hence their chemical potentials must be equal in the liquid and vapor phases.

E7.2(a)   A regular solution has an excess entropy of zero, but an excess enthalpy that is non-zero and dependent on composition, perhaps in the manner of eqn 7.30. We can think of a regular solution as one in which the different molecules of the solution are distributed randomly, as in an ideal solution, but have different energies of interaction with each other.

E7.3(a)   All of the colligative properties are a function of the concentration of the solute, which implies that the concentration can be determined by a measurement of these properties. See eqns 7.33, 7.34, 7.36, 7.37, and 7.40. Knowing the mass of the solute in solution then allows for a calculation of its molar mass. For example, the mole fraction of the solute is related to its mass as follows:

$$x_B = \frac{m_B/M_B}{m_B/M_B + m_A/M_A}$$

The only unknown in this expression is M_B which is easily solved for. See Example 7.5 for the details of how molar mass is determined from osmotic pressure.

Numerical exercises

E7.4(a)   Let A denote acetone and C chloroform. The total volume of the solution is

$$V = n_A V_A + n_C V_C$$

V_A and V_C are given; hence we need to determine n_A and n_C in 1.000 kg of the solution with the stated mole fraction. The total mass of the sample is $m = n_A M_A + n_C M_C$ (a). We also know that

$$x_A = \frac{n_A}{n_A + n_C}, \quad \text{implies that} \quad (x_A - 1)n_A + x_A n_C = 0$$

and hence that

$$-x_C n_A + x_A n_C = 0 \quad \text{(b)}$$

On solving (a) and (b), we find

$$n_A = \left(\frac{x_A}{x_C}\right) \times n_C, \quad n_C = \frac{m x_C}{x_A M_A + x_C M_C}$$

Since $x_C = 0.4693$, $x_A = 1 - x_C = 0.5307$,

$$n_C = \frac{(0.4693) \times (1000\,\text{g})}{[(0.5307) \times (58.08) + (0.4693) \times (119.37)]\,\text{g mol}^{-1}} = 5.404\,\text{mol}$$

$$n_A = \left(\frac{0.5307}{0.4693}\right) \times (5.404)\,\text{mol} = 6.111\,\text{mol}$$

The total volume, $V = n_A V_A + n_B V_B$, is therefore

$$V = (6.111\,\text{mol}) \times (74.166\,\text{cm}^3\,\text{mol}^{-1}) + (5.404\,\text{mol}) \times (80.235\,\text{cm}^3\,\text{mol}^{-1})$$

$$= \boxed{886.8\,\text{cm}^3}$$

E7.5(a)   Let A denote water and B ethanol. The total volume of the solution is

$$V = n_A V_A + n_B V_B$$

We are given V_A, we need to determine n_A and n_B in order to solve for V_B.

Assume we have $100\,\text{cm}^3$ of solution, then the mass of solution is

$$m = d \times V = (0.914\,\text{g cm}^{-3}) \times (100\,\text{cm}^3) = 91.4\,\text{g}$$

of which 45.7 g is water and 45.7 g ethanol.

$$100\,\text{cm}^3 = \left(\frac{45.7\,\text{g}}{18.02\,\text{g mol}^{-1}}\right) \times (17.4\,\text{cm}^3\,\text{mol}^{-1}) + \left(\frac{45.7\text{g}}{46.07\,\text{g mol}^{-1}}\right) \times V_B$$

$$= 44.1\overline{3}\,\text{cm}^3 + 0.992\overline{0}\,\text{mol} \times V_B$$

$$V_B = \frac{55.8\overline{7}\,\text{cm}^3}{0.9920\,\text{mol}} = \boxed{56.3\,\text{cm}^3\,\text{mol}^{-1}}$$

E7.6(a)   Check whether $\dfrac{p_B}{x_B}$ is equal to a constant (K_B)

x	0.005	0.012	0.019
p/x	6.4×10^3	6.4×10^3	$6.4 \times 10^3\,\text{kPa}$

Hence, $K_B \approx \boxed{6.4 \times 10^3\,\text{kPa}}$

E7.7(a)   In Exercise 7.6(a), the Henry's law constant was determined for concentrations expressed in mole fractions. Thus the concentration in molality must be converted to mole fraction.

$$m(\text{GeCl}_4) = 1000\,\text{g, corresponding to}$$

$$n(\text{GeCl}_4) = \frac{1000\,\text{g}}{214.39\,\text{g mol}^{-1}} = 4.664\,\text{mol}, \qquad n(\text{HCl}) = 0.10\,\text{mol}$$

Therefore, $x = \dfrac{0.10\,\text{mol}}{(0.10\,\text{mol}) + (4.664\,\text{mol})} = 0.021\overline{0}$

From $K_B = 6.4 \times 10^3\,\text{kPa}$ (Exercise 7.6(a)), $p = (0.021\overline{0} \times 6.4 \times 10^3\,\text{kPa}) = \boxed{1.3 \times 10^2\,\text{kPa}}$

E7.8(a)   Because the mole fraction of B is small,

$$x_B = \frac{n_B}{n_A + n_B} \approx \frac{n_B}{n_A}$$

The amount of solvent molecules in 1 kg of solvent of molar mass M is

$$n_A = \frac{1\,\text{kg}}{M}$$

Therefore,

$$x_B = \frac{n_B}{n_A} = n_B \times \frac{M}{1\,\text{kg}} = b_B \times M$$

where b_B is the molality of B. Hence, from eqn 6.34,

$$\Delta T = \left(\frac{RT^{*2}M}{\Delta_{vap}H}\right) b_B$$

and we can identify the ebullioscopic constant as

$$K_b = \frac{RT^{*2}M}{\Delta_{vap}H}$$

$$= \frac{(8.314\,\text{J K}^{-1}\,\text{mol}^{-1}) \times (349.9\,\text{K})^2 \times (153.81 \times 10^{-3}\,\text{kg mol}^{-1})}{30.0 \times 10^3\,\text{J mol}^{-1}}$$

$$= \boxed{5.22\,\text{K}/(\text{mol kg}^{-1})}$$

$$K_f = \frac{RT^{*2}M}{\Delta_{fus}H} \quad \text{[by analogy with the above]}$$

$$= \frac{(8.314\,\text{J K}^{-1}\,\text{mol}^{-1}) \times (250.3\,\text{K})^2 \times (153.81 \times 10^{-3}\,\text{kg mol}^{-1})}{2.47 \times 10^3\,\text{J mol}^{-1}}$$

$$= \boxed{32\,\text{K}/(\text{mol kg}^{-1})}$$

E7.9(a)   We assume that the solvent, benzene, is ideal and obeys Raoult's law.

Let B denote benzene and A the solute; then

$$p_B = x_B p_B^* \quad \text{and} \quad x_B = \frac{n_B}{n_A + n_B}$$

Hence, $p_B = \dfrac{n_B p_B^*}{n_A + n_B}$; which solves to

$$n_A = \frac{n_B(p_B^* - p_B)}{p_B}$$

Then, since $n_A = \dfrac{m_A}{M_A}$, where m_A is the mass of A present,

$$M_A = \frac{m_A p_B}{n_B(p_B^* - p_B)} = \frac{m_A M_B p_B}{m_B(p_B^* - p_B)}$$

From the data

$$M_A = \frac{(19.0\,\text{g}) \times (78.11\,\text{g mol}^{-1}) \times (386\,\text{Torr})}{(500\,\text{g}) \times (400 - 386)\,\text{Torr}} = \boxed{82\,\text{g mol}^{-1}}$$

E7.10(a) $M_B = \dfrac{\text{mass of B}}{n_B}$ [B = compound]

$n_B = \text{mass of CCl}_4 \times b_B$ [b_B = molality of B]

$$b_B = \frac{\Delta T}{K_f} \text{ [7.37]; thus}$$

$$M_B = \frac{\text{mass of B} \times K_f}{\text{mass of CCl}_4 \times \Delta T} \quad K_f = 30\,K/(mol\,kg^{-1})\text{ [Table 7.2]}$$

$$M_B = \frac{(100\,g) \times (30\,K\,kg\,mol^{-1})}{(0.750\,kg) \times (10.5\,K)} = \boxed{381\,g\,mol^{-1}}$$

E7.11(a) $\quad \Delta T = K_f b_B$ [7.37] $\quad b_B = \dfrac{n_B}{\text{mass of water}} \approx \dfrac{n_B}{V\rho}$ [dilute solution]

$$\rho \approx 10^3\,kg\,m^{-3} \text{ [density of solution} \approx \text{density of water]}$$

$$n_B \approx \frac{\Pi V}{RT} \text{ [7.40]} \qquad \Delta T \approx K_f \times \frac{\Pi}{RT\rho}$$

with $K_f = 1.86\,K/(mol\,kg^{-1})$ [Table 7.2]

$$\Delta T \approx \frac{(1.86\,K\,kg\,mol^{-1}) \times (120 \times 10^3\,Pa)}{(8.314\,J\,K^{-1}\,mol^{-1}) \times (300\,K) \times (10^3\,kg\,m^{-3})} = 0.089\,K$$

Therefore, the solution will freeze at about $\boxed{-0.09°C.}$

Comment. Osmotic pressures are inherently large. Even dilute solutions with small freezing point depressions have large osmotic pressures.

E7.12(a) $\quad \Delta_{mix}G = nRT\{x_A \ln x_A + x_B \ln x_B\}$ [7.18] $\quad x_A = x_B = 0.5, \quad n = \dfrac{pV}{RT}$

Therefore,

$$\Delta_{mix}G = (pV) \times \left(\frac{1}{2}\ln\frac{1}{2} + \frac{1}{2}\ln\frac{1}{2}\right) = -pV\ln 2$$

$$= (-1.0) \times (1.013 \times 10^5\,Pa) \times (5.0 \times 10^{-3}\,m^3) \times (\ln 2)$$

$$= -3.5 \times 10^2\,J = \boxed{-0.35\,kJ}$$

$$\Delta_{mix}S = -nR\{x_A \ln x_A + x_B \ln x_B\} = \frac{-\Delta_{mix}G}{T} \text{ [7.19]} = \frac{-0.35\,kJ}{298\,K} = \boxed{+1.2\,J\,K^{-1}}$$

E7.13(a) $\quad \Delta_{mix}S = -nR\sum_J x_J \ln x_J$ [7.19]

Therefore, for molar amounts,

$$\Delta_{mix}S = -R\sum_J x_J \ln x_J$$

$$= -R[(0.782\ln 0.782) + (0.209\ln 0.209) + (0.009\ln 0.009) + (0.0003\ln 0.0003)]$$

$$= 0.564R = \boxed{+4.7\,J\,K^{-1}\,mol^{-1}}$$

E7.14(a) Hexane and heptane form nearly ideal solutions, therefore eqn 7.19 applies.

$$\Delta_{mix}S = -nR(x_A \ln x_A + x_B \ln x_B) \text{ [7.19]}$$

We need to differentiate eqn 7.19 with respect to x_A and look for the value of x_A at which the derivative is zero. Since $x_B = 1 - x_A$, we need to differentiate

$$\Delta_{mix}S = -nR\{x_A \ln x_A + (1 - x_A)\ln(1 - x_A)\}$$

This gives $\left(\text{using } \dfrac{d \ln x}{dx} = \dfrac{1}{x}\right)$

$$\frac{d\Delta_{mix}S}{dx_A} = -nR\{\ln x_A + 1 - \ln(1 - x_A) - 1\} = -nR\ln\frac{x_A}{1 - x_A}$$

which is zero when $x_A = \frac{1}{2}$. Hence, the maximum entropy of mixing occurs for the preparation of a mixture that contains equal mole fractions of the two components.

(a) $\dfrac{n(\text{Hex})}{n(\text{Hep})} = 1 = \dfrac{\left(\frac{m(\text{Hex})}{M(\text{Hex})}\right)}{\left(\frac{m(\text{Hep})}{M(\text{Hep})}\right)}$

(b) $\dfrac{m(\text{Hex})}{m(\text{Hep})} = \dfrac{M(\text{Hex})}{M(\text{Hep})} = \dfrac{86.17\,\text{g mol}^{-1}}{100.20\,\text{g mol}^{-1}} = \boxed{0.8600}$

E7.15(a) $p = xK$ [7.26], $K = 0.167 \times 10^9\,\text{Pa}$ $x = \dfrac{n(\text{CO}_2)}{n(\text{CO}_2) + n(\text{H}_2\text{O})} \approx \dfrac{n(\text{CO}_2)}{n(\text{H}_2\text{O})}$

Therefore, with $1.00\,\text{kg}\,\text{H}_2\text{O}$

$$n(\text{CO}_2) \approx xn(\text{H}_2\text{O}) \quad \text{with} \quad n(\text{H}_2\text{O}) = \frac{1.00 \times 10^3\,\text{g}}{18.02\,\text{g mol}^{-1}} \quad \text{and} \quad x = \frac{p}{K}$$

Hence $n(\text{CO}_2) \approx \left(\dfrac{10^3\,\text{g}}{18.02\,\text{g mol}^{-1}}\right) \times \left(\dfrac{p}{0.167 \times 10^9\,\text{Pa}}\right) \approx (3.32 \times 10^{-7}\,\text{mol}) \times (p/\text{Pa})$

(a) $p = 0.10\,\text{atm} = 1.01 \times 10^4\,\text{Pa}$

Hence, $n(\text{CO}_2) = (3.32 \times 10^{-7}\,\text{mol}) \times (1.01 \times 10^4) = 3.4 \times 10^{-3}\,\text{mol}$. The solution is therefore $\boxed{3.4\,\text{mmol kg}^{-1}}$ in CO_2.

(b) $p = 1.0\,\text{atm}$; since $n \propto p$, the solution is $\boxed{34\,\text{mmol kg}^{-1}}$ in CO_2.

E7.16(a) Use the result established in Exercise 7.15(**a**) that the amount of CO_2 in 1 kg of water is given by

$$n(\text{CO}_2) = (3.32 \times 10^{-7}\,\text{mol}) \times (p/\text{Pa})$$

and substitute $p \approx (5.0) \times (1.01 \times 10^5\,\text{Pa}) = 5.05 \times 10^5\,\text{Pa}$, to give

$$n(\text{CO}_2) = (3.32 \times 10^{-7}\,\text{mol}) \times (5.05 \times 10^5) = 0.17\,\text{mol}$$

Hence, the molality of the solution is about $\boxed{0.17\,\text{mol kg}^{-1}}$ and, since molalities and molar concentrations for dilute aqueous solutions are approximately equal, the molar concentration is about $0.17\,\text{mol L}^{-1}$.

E7.17(a) $\Delta T = K_f b_B$ [7.37]; $K_f = 1.86\,\text{K kg mol}^{-1}$ [Table 7.2]

$$\Delta T = (1.86\,\text{K kg mol}^{-1}) \times \frac{\left(\frac{7.5\,\text{g}}{342.3\,\text{g mol}^{-1}}\right)}{0.25\,\text{kg}} = 0.16\,\text{K}$$

Hence, the freezing point will be approximately $\boxed{-0.16°\text{C}}$.

E7.18(a) The solubility in grams of anthracene per kg of benzene can be obtained from its mole fraction with use of the equation

$$\ln x_B = \frac{\Delta_{fus}H}{R} \times \left(\frac{1}{T^*} - \frac{1}{T} \right) \quad [7.39; \text{B, the solute, is anthracene}]$$

$$= \left(\frac{28.8 \times 10^3 \, \text{J mol}^{-1}}{8.314 \, \text{J K}^{-1} \, \text{mol}^{-1}} \right) \times \left(\frac{1}{490.15 \, \text{K}} - \frac{1}{298.15 \, \text{K}} \right) = -4.55$$

Therefore, $x_B = e^{-4.55} = 0.0106$

Since $x_B \ll 1$, $x(\text{anthracene}) \approx \dfrac{n(\text{anthracene})}{n(\text{benzene})}$

Therefore, in 1 kg of benzene,

$$n(\text{anthr.}) \approx x(\text{anthr.}) \times \left(\frac{1000 \, \text{g}}{78.11 \, \text{g mol}^{-1}} \right) \approx (0.0106) \times (12.80 \, \text{mol}) = 0.136 \, \text{mol}$$

The molality of the solution is therefore $0.136 \, \text{mol kg}^{-1}$. Since $M = 178 \, \text{g mol}^{-1}$, $0.136 \, \text{mol}$ corresponds to $\boxed{24 \, \text{g anthracene}}$ in 1 kg of benzene.

E7.19(a) The best value of the molar mass is obtained from values of the data extrapolated to zero concentration, since it is under this condition that eqn 7.40 applies.

$$\Pi V = n_B RT \ [7.40], \quad \text{so} \quad \Pi = \frac{mRT}{MV} = \frac{cRT}{M}, \quad c = \frac{m}{V}$$

$$\Pi = \rho g h \ [\text{hydrostatic pressure}], \quad \text{so} \quad h = \left(\frac{RT}{\rho g M} \right) c$$

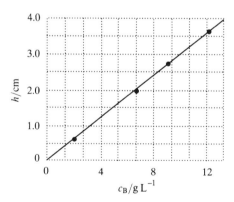

Figure 7.1

Hence, plot h against c and identify the slope as $\dfrac{RT}{\rho g M}$. Figure 7.1 shows the plot of the data.

The slope of the line is $0.29 \, \text{cm}/(\text{g L}^{-1})$, so

$$\frac{RT}{\rho g M} = \frac{0.29 \, \text{cm}}{\text{g L}^{-1}} = 0.29 \, \text{cm L g}^{-1} = 0.29 \times 10^{-2} \, \text{m}^4 \, \text{kg}^{-1}$$

Therefore,

$$M = \frac{RT}{(\rho g) \times (0.29 \times 10^{-2}\,\text{m}^4\,\text{kg}^{-1})}$$

$$= \frac{(8.314\,\text{J K}^{-1}\,\text{mol}^{-1}) \times (298.15\,\text{K})}{(1.004 \times 10^3\,\text{kg m}^{-3}) \times (9.81\,\text{m s}^{-2}) \times (0.29 \times 10^{-2}\,\text{m}^4\,\text{kg}^{-1})} = \boxed{87\,\text{kg mol}^{-1}}$$

E7.20(a) *For* A (Raoult's law basis; concentration in mole fraction)

$$a_A = \frac{p_A}{p_A^*}\ [7.45] = \frac{250\,\text{Torr}}{300\,\text{Torr}} = \boxed{0.833}; \qquad \gamma_A = \frac{a_A}{x_A} = \frac{0.833}{0.90} = \boxed{0.93}$$

For B (Henry's law basis; concentration in mole fraction)

$$a_B = \frac{p_B}{K_B}\ [7.52] = \frac{25\,\text{Torr}}{200\,\text{Torr}} = \boxed{0.125}; \qquad \gamma_B = \frac{a_B}{x_B} = \frac{0.125}{0.10} = \boxed{1.25}$$

For B (Henry's law basis; concentration in molality)

An equation analogous to eqn 7.52 is used $a_B = \dfrac{p_B}{K_B'}$ with a modified Henry's law constant K_B' which corresponds to the pressure of B in the limit of very low molalities.

$$p_B = \frac{b_B}{b^{\ominus}} \times K_B'$$

is analogous to $p_B = x_B K_B$. Since x_B and b_B are related as $b_B = \dfrac{x_B}{M_A x_A}$

K_B' and K_B are related as $K_B' = x_A M_A b^{\ominus} K_B$

We also need M_A

$$M_A = \frac{x_B}{x_A b_B} = \frac{0.10}{(0.90) \times (2.22\,\text{mol kg}^{-1})} = 0.050\,\text{kg mol}^{-1}$$

Then, $K_B' = (0.90) \times (0.050\,\text{kg mol}^{-1}) \times (1\,\text{mol kg}^{-1}) \times (200\,\text{Torr}) = 9.0\,\text{Torr}$

and $a_B = \dfrac{25\,\text{Torr}}{9.0\,\text{Torr}} = \boxed{2.8} \qquad \gamma_B = \dfrac{a_B}{\left(\frac{b_B}{b^{\ominus}}\right)} = \dfrac{2.8}{2.22} = \boxed{1.25}$

Comment. The two methods for the "solute" B give different values for the activities. This is reasonable since the chemical potentials in the reference states $\mu^{\dagger}$ and $\mu^{\ominus}$ are different.

Question. What are the activity and activity coefficient of B in the Raoult's law basis?

E7.21(a) In an ideal dilute solution the solvent (CCl_4) obeys Raoult's law and the solute (Br_2) obeys Henry's law; hence

$$p(CCl_4) = x(CCl_4)p^*(CCl_4)\ [23] = (0.950) \times (33.85\,\text{Torr}) = \boxed{32.2\,\text{Torr}}$$

$$p(Br_2) = x(Br_2)K(Br_2)\ [25] = (0.050) \times (122.36\,\text{Torr}) = \boxed{6.1\,\text{Torr}}$$

$$p(\text{Total}) = (32.2 + 6.1)\text{Torr} = \boxed{38.3\,\text{Torr}}$$

The composition of the vapor in equilibrium with the liquid is

$$y(CCl_4) = \frac{p(CCl_4)}{p(\text{Total})} = \frac{32.2\,\text{Torr}}{38.3\,\text{Torr}} = \boxed{0.841}$$

$$y(Br_2) = \frac{p(Br_2)}{p(\text{Total})} = \frac{6.1\,\text{Torr}}{38.3\,\text{Torr}} = \boxed{0.16}$$

E7.22(a)   Let A = acetone and M = methanol

$$y_A = \frac{p_A}{p_A + p_M} \text{ [Dalton's law]} = \frac{p_A}{760\,\text{Torr}} = 0.516$$

$$p_A = 392\,\text{Torr}, \quad p_M = 368\,\text{Torr}$$

$$a_A = \frac{p_A}{p_A^*}\,[42] = \frac{392\,\text{Torr}}{786\,\text{Torr}} = \boxed{0.499} \qquad a_M = \frac{p_M}{p_M^*} = \frac{368\,\text{Torr}}{551\,\text{Torr}} = \boxed{0.668}$$

$$\gamma_A = \frac{a_A}{x_A} = \frac{0.499}{0.400} = \boxed{1.25} \qquad \gamma_M = \frac{a_M}{x_M} = \frac{0.668}{0.600} = \boxed{1.11}$$

Solutions to problems

Solutions to numerical problems

P7.1 $p_A = y_A p$ and $p_B = y_B p$ (Dalton's law). Hence, draw up the following table

p_A/kPa	0	1.399	3.566	5.044	6.996	7.940	9.211	10.105	11.287	12.295
x_A	0	0.0898	0.2476	0.3577	0.5194	0.6036	0.7188	0.8019	0.9105	1
y_A	0	0.0410	0.1154	0.1762	0.2772	0.3393	0.4450	0.5435	0.7284	1

p_B/kPa	0	4.209	8.487	11.487	15.462	18.243	23.582	27.334	32.722	36.066
x_B	0	0.0895	0.1981	0.2812	0.3964	0.4806	0.6423	0.7524	0.9102	1
y_B	0	0.2716	0.4565	0.5550	0.6607	0.7228	0.8238	0.8846	0.9590	1

The data are plotted in Fig. 7.2.

We can assume, at the lowest concentrations of both A and B, that Henry's law will hold. The Henry's law constants are then given by

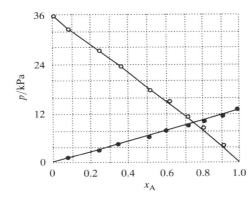

Figure 7.2

$$K_A = \frac{p_A}{x_A} = \boxed{15.58\,\text{kPa}} \text{ from the point at } x_A = 0.0898$$

$$K_B = \frac{p_B}{x_B} = \boxed{47.03\,\text{kPa}} \text{ from the point at } x_B = 0.0895$$

P7.2 $V_A = \left(\dfrac{\partial V}{\partial n_A}\right)_{n_B}$ [7.1, A $=$ NaCl(aq), B $=$ water] $= \left(\dfrac{\partial V}{\partial b}\right)_{n(H_2O)}$ mol^{-1} [with $b \equiv b/(\text{mol kg}^{-1})$]

$$= \left((16.62) + \tfrac{3}{2} \times (1.77) \times (b)^{1/2} + (2) \times (0.12b)\right) \text{cm}^3 \text{ mol}^{-1}$$

$$= \boxed{17.5 \text{ cm}^3 \text{ mol}^{-1}} \quad \text{when} \quad b = 0.100$$

For a solution consisting of 0.100 mol NaCl and 1.000 kg of water, corresponding to 55.49 mol H_2O, the total volume is given both by

$$V = [(1003) + (16.62) + (0.100) \times (1.77) \times (0.100)^{3/2} + (0.12) \times (0.100)^2] \text{cm}^3$$

$$= 1004.7 \text{ cm}^3$$

and by $V = n(\text{NaCl}) V_{\text{NaCl}} + n(H_2O) V_{H_2O}$ [7.3] $= (0.100 \text{ mol}) \times (17.5 \text{ cm}^3 \text{ mol}^{-1}) + (55.49 \text{ mol}) \times V_{H_2O}$.

Therefore, $V_{H_2O} = \dfrac{1004.7 \text{ cm}^3 - 1.75 \text{ cm}^3}{55.49 \text{ mol}} = \boxed{18.07 \text{ cm}^3 \text{ mol}^{-1}}$

Comment. Within four significant figures, this result is the same as the molar volume of pure water at 25°C.

Question. How does the partial molar volume of NaCl(aq) in this solution compare to molar volume of pure solid NaCl?

P7.3 Let $m(\text{CuSO}_4)$, which is the mass of $CuSO_4$ dissolved in 100 g of solution, be represented by

$$w = \frac{100 m_B}{m_A + m_B} = \text{mass percent of } CuSO_4$$

where m_B is the mass of $CuSO_4$ and m_A is the mass of water. Then using

$$\rho = \frac{m_A + m_B}{V} \qquad n_A = \frac{m_A}{M_A}$$

the procedure runs as follows

$$V_A = \left(\frac{\partial V}{\partial n_A}\right)_{n_B} = \left(\frac{\partial V}{\partial m_A}\right)_B M_A$$

$$= \frac{\partial}{\partial m_A}\left(\frac{m_A + m_B}{\rho}\right) \times M_A$$

$$= \frac{M_A}{\rho} + (m_A + m_B) M_A \frac{\partial}{\partial m_A} \frac{1}{\rho}$$

$$\frac{\partial}{\partial m_A}\frac{1}{\rho} = \left(\frac{\partial w}{\partial m_A}\right)\frac{\partial}{\partial w}\frac{1}{\rho} = \frac{-w}{m_A + m_B}\frac{\partial}{\partial w}\frac{1}{\rho}$$

Therefore,

$$V_A = \frac{M_A}{\rho} - w M_A \frac{\partial}{\partial w}\frac{1}{\rho}$$

and hence

$$\frac{1}{\rho} = \frac{V_A}{M_A} + w \frac{d}{dw}\left(\frac{1}{\rho}\right)$$

Therefore, plot $1/\rho$ against w and extrapolate the tangent to $w = 100$ to obtain V_B/M_B. For the actual procedure, draw up the following table

w	5	10	15	20
$\rho/(\text{g cm}^{-3})$	1.051	1.107	1.167	1.230
$1/(\rho/\text{g cm}^{-3})$	0.951	0.903	0.857	0.813

The values of $1/\rho$ are plotted against w in Fig. 7.3.

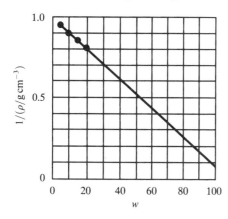

Figure 7.3

Four tangents are drawn to the curve at the four values of w. As the curve is a straight line to within the precision of the data, all four tangents are coincident and all four intercepts are equal at $0.075\,\text{g}^{-1}\,\text{cm}^3$. Thus

$$V(\text{CuSO}_4) = 0.075\,\text{g}^{-1}\,\text{cm}^3 \times 159.6\,\text{g mol}^{-1} = \boxed{12.0\,\text{cm}^3\,\text{mol}^{-1}}$$

P7.6

$$\Delta T = \frac{RT_f^{*2}x_B}{\Delta_{\text{fus}}H}\ [7.36], \quad x_B \approx \frac{n_B}{n(\text{CH}_3\text{COOH})} = \frac{n_B M(\text{CH}_3\text{COOH})}{1000\,\text{g}}$$

Hence, $\Delta T = \dfrac{n_B M R T_f^{*2}}{\Delta_{\text{fus}}H \times 1000\,\text{g}} = \dfrac{b_B M R T_f^{*2}}{\Delta_{\text{fus}}H}$ [b_B : molality of solution]

$$= b_B \times \left(\frac{(0.06005\,\text{kg mol}^{-1}) \times (8.314\,\text{J K}^{-1}\,\text{mol}^{-1}) \times (290\,\text{K})^2}{11.4 \times 10^3\,\text{J mol}^{-1}} \right)$$

$$= 3.68\,\text{K} \times b_B/(\text{mol kg}^{-1})$$

Giving for b_B, the apparent molality,

$$b_B = \nu b_B^0 = \frac{\Delta T}{3.68\,\text{K}}\,\text{mol kg}^{-1}$$

when b_B^0 is the actual molality and ν may be interpreted as the number of ions in solution per one formula unit of KCl. The apparent molar mass of KCl can be determined from the apparent molality by the relation

$$M_B(\text{apparent}) = \frac{b_B^0}{b_B} \times M_B^0 = \frac{1}{\nu} \times M_B^0 = \frac{1}{\nu} \times (74.56\,\text{g mol}^{-1})$$

where M_B^0 is the actual molar mass of KCl.

We can draw up the following table from the data.

$b_B^0/(\text{mol kg}^{-1})$	0.015	0.037	0.077	0.295	0.602
$\Delta T/K$	0.115	0.295	0.470	1.381	2.67
$b_B/(\text{mol kg}^{-1})$	0.0312	0.0802	0.128	0.375	0.726
$\nu = \dfrac{b_B}{b_B^0}$	2.1	2.2	1.7	1.3	1.2
$M_B(\text{app})/(\text{g mol}^{-1})$	26	34	44	57	62

A possible explanation is that the dissociation of KCl into ions is complete at the lower concentrations but incomplete at the higher concentrations. Values of ν greater than 2 are hard to explain, but could be a result of the approximations involved in obtaining equation 7.36.

See the original reference for further information about the interpretation of the data.

P7.8 (a) On a Raoult's law basis, $a = \dfrac{p}{p^*}$, $a = \gamma x$, and $\gamma = \dfrac{p}{xp^*}$. On a Henry's law basis, $a = \dfrac{p}{K}$, and

$\gamma = \dfrac{p}{xK}$. The vapor pressures of the pure components are given in the table of data and are:

$p_I^* = 353.4\,\text{Torr}$, $p_A^* = 280.4\,\text{Torr}$.

(b) The Henry's law constants are determined by plotting the data and extrapolating the low concentration data to $x = 1$. The data are plotted in Fig. 7.4. K_A and K_I are estimated as graphical tangents at $x_I = 1$ and $x_I = 0$, respectively. The values obtained are: $K_A = \boxed{450\,\text{Torr}}$ and $K_I = \boxed{465\,\text{Torr}}$

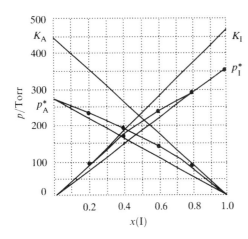

Figure 7.4

Then draw up the following table based on the values of the partial pressures obtained from the plots at the values of $x(I)$ given in the figure.

x_I	0	0.2	0.4	0.6	0.8	1.0
p_I/Torr	0	92	165	230	290	$353.4^{\ddagger}$
p_A/Torr	$280.4^{\dagger}$	230	185	135	80	0
$\gamma_I(R)$	—	1.30	1.17	1.09	1.03	$1.000\ [p_I/x_Ip_I^*]$
$\gamma_A(R)$	1.000	1.03	1.10	1.20	1.43	— $[p_A/x_Ap_A^*]$
$\gamma_I(H)$	1.000	0.990	0.887	0.824	0.780	$0.760\ [p_I/x_IK_I]$

†The value of p_A^*; ‡the value of p_I^*.

Question. In this problem both I and A were treated as solvents, but only I as a solute. Extend the table by including a row for γ_A (H).

P7.10 The partial molar volume of cyclohexane is

$$V_c = \left(\frac{\partial V}{\partial n_c} \right)_{p,T,n_2}$$

A similar expression holds for V_p. V_c can be evaluated graphically by plotting V against n_c and finding the slope at the desired point. In a similar manner, V_p can be evaluated by plotting V against n_p. To find V_c, V is needed at a variety of n_c while holding n_p constant, say at 1.0000 mol; likewise to find V_p, V is needed at a variety of n_p while holding n_c constant. The mole fraction in this system is

$$x_c = \frac{n_c}{n_c + n_p} \quad \text{so} \quad n_c = \frac{x_c n_p}{1 - x_c}$$

From n_c and n_p, the mass of the sample can be calculated, and the volume can be calculated from

$$V = \frac{m}{\rho} = \frac{n_c M_c + n_p M_p}{\rho}$$

The following table is drawn up

$n_c/\text{mol}(n_p = 1)$	V/cm^3	x_c	$\rho/\text{g cm}^{-3}$	$n_p/\text{mol}(n_c = 1)$	V/cm^3
2.295	529.4	0.6965	0.7661	0.4358	230.7
3.970	712.2	0.7988	0.7674	0.2519	179.4
9.040	1264	0.9004	0.7697	0.1106	139.9

These values are plotted in Fig. 7.5(a) and (b).

These plots show no curvature, so in this case, perhaps due to the limited number of data points, the molar volumes are independent of the mole numbers and are

$$V_c = \boxed{109.0 \text{ cm}^3 \text{ mol}^{-1}} \quad \text{and} \quad V_p = \boxed{279.3 \text{ cm}^3 \text{ mol}^{-1}}$$

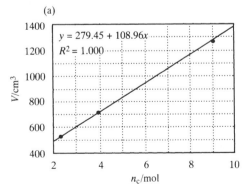
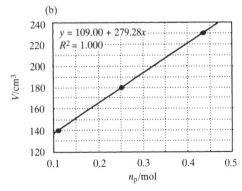

Figure 7.5

P7.12 The activity of a solvent is

$$a_A = \frac{p_A}{p_A^*} = x_A \gamma_A$$

so the activity coefficient is

$$\gamma_A = \frac{p_A}{x_A p_A^*} = \frac{y_A p}{x_A p_A^*}$$

where the last equality applies Dalton's law of partial pressures to the vapor phase.

Substituting the data, the following table of results is obtained.

p/kPa	x_T	y_T	γ_T	γ_E
23.40	0.000	0.000		
21.75	0.129	0.065	0.418	0.998
20.25	0.228	0.145	0.490	1.031
18.75	0.353	0.285	0.576	1.023
18.15	0.511	0.535	0.723	0.920
20.25	0.700	0.805	0.885	0.725
22.50	0.810	0.915	0.966	0.497
26.30	1.000	1.000		

P7.14 $S = S_0 e^{\tau/T}$ may be written in the form $\ln S = \ln S_0 + \dfrac{\tau}{T}$, which indicates that a plot of $\ln S$ against $1/T$ should be linear with slope τ and intercept $\ln S_0$. Linear regression analysis gives $\boxed{\tau = 165\,\text{K}}$, standard deviation $= 2\,\text{K}$

$$\ln(S_0/\text{mol L}^{-1}) = 2.990, \text{ standard deviation} = 0.007;\ S_0 = e^{2.990}\,\text{mol L}^{-1} = \boxed{19.89\,\text{mol L}^{-1}}$$

$$R = \boxed{0.99978}$$

The linear regression explains 99.98 per cent of the variation.

Equation 36 is

$$x_B = e^{-\left(\frac{\Delta_{\text{fus}}H}{R}\left(\frac{1}{T} - \frac{1}{T^*}\right)\right)} = e^{-\Delta_{\text{fus}}H/RT}e^{\Delta_{\text{fus}}H/RT^*}$$

Comparing to $S = S_0 e^{\tau/T}$, we see that

$$\boxed{S_0 = e^{-\Delta_{\text{fus}}H/RT^*}}$$

where T^* is the normal melting point of the solute and $\Delta_{\text{fus}}H$ is its heat of fusion $\boxed{\tau = \Delta_{\text{fus}}H/R}$

P7.15 (a) The following graph shows plots of H^E against x, where x is the mole fraction of 1-propanol in binary solutions with tripropyl amine. The symmetrical curve is that of a "regular" solution with $w = -1$. Not being symmetrical about $x = 0.5$, the plot for the alcohol–amine mixtures obviously demonstrates that the solutions are not regular. Mixing is highly exothermic at high alcohol concentrations and the exothermic maximum occurs when $x \cong 0.75$. Total hydrogen bond strength may be a maximum when 1 amine molecule is present for every 3 alcohol molecules.

(b) When χ is very small, mixtures have little proton donating alcohol and there is but a negligibly small enthalpy change that can be attributed to molecular reorganization of the amine bulk upon

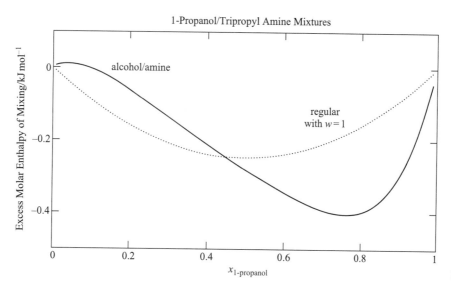

Figure 7.6

solution formation. In this case the excess enthalpy of mixing is composed of the endothermic breakage of donor-to-donor bonds (ΔH_{AH}), the endothermic breakage of amine acceptor-to-acceptor bonds (ΔH_B), and the exothermic formation of hydrogen bonds between donors and acceptors ($\Delta H_{\text{hydrogen bond}}$) The H^E value for very dilute (infinitely dilute) solutions is:

$$\lim_{x \to 0} H^E = \lim_{x \to 0} \sum_{r=0}^{3} A_r (1 - 2\chi)^r$$

$$= \sum_{r=0}^{3} A_r = 0.496 \, \text{kJ mol}^{-1}$$

$$\lim_{x \to 0} H^E = \sum_{r=0}^{3} A_r = \Delta H_{AH} + \Delta H_B + \Delta H_{\text{hydrogen bond}}$$

$$\Delta H_{\text{hydrogen bond}} = \sum_{r=0}^{3} A_r - \Delta H_{AH} - \Delta H_B$$

$$= 0.496 \, \text{kJ mol}^{-1} - (25.3 \, \text{kJ mol}^{-1}) - (0.121 \, \text{kJ mol}^{-1})$$

$$= -24.9 \, \text{kJ mol}^{-1}$$

T. M. Letcher and B. C. Bricknell, *J. Chem. Eng. Data*, **41**, 166–169 (1996).

Solutions to theoretical problems

P7.17

$$\mu_A = \left(\frac{\partial G}{\partial n_A} \right)_{n_B} \quad [7.4] = \mu_A^o + \left(\frac{\partial}{\partial n_A} (nG^E) \right)_{n_B} \quad [\mu_A^o \text{ is ideal value} = \mu_A^* + RT \ln x_A]$$

$$\left(\frac{\partial nG^E}{\partial n_A} \right)_{n_B} = G^E + n \left(\frac{\partial G^E}{\partial n_A} \right)_{n_B} = G^E + n \left(\frac{\partial x_A}{\partial n_A} \right)_B \left(\frac{\partial G^E}{\partial x_A} \right)_B$$

$$= G^E + n \times \frac{x_B}{n} \times \left(\frac{\partial G^E}{\partial x_A} \right)_B \quad [\partial x_A / \partial n_A = x_B / n]$$

$$= gRTx_A(1 - x_A) + (1 - x_A)gRT(1 - 2x_A)$$

$$= gRT(1 - x_A)^2 = gRTx_B^2$$

Therefore, $\mu_A = \boxed{\mu_A^* + RT \ln x_A + gRTx_B^2}$.

P7.19 $n_A dV_A + n_B dV_B = 0$ [Example 7.1]

Hence $\dfrac{n_A}{n_B} dV_A = -dV_B$

Therefore, by integration,

$$V_B(x_A) - V_B(0) = -\int_{V_A(0)}^{V_A(x_A)} \frac{n_A}{n_B} dV_A = -\int_{V_A(0)}^{V_A(x_A)} \frac{x_A dV_A}{1 - x_A} \quad [n_A = x_A n, \ n_B = x_B n]$$

Therefore, $V_B(x_A, x_B) = V_B(0, 1) - \displaystyle\int_{V_A(0)}^{V_A(x_A)} \frac{x_A \, dV_A}{1 - x_A}$

We should now plot $x_A/(1 - x_A)$ against V_A and estimate the integral. For the present purpose we integrate up to $V_A(0.5, \ 0.5) = 74.06 \text{ cm}^3 \text{ mol}^{-1}$ [Fig. 7.7], and use the data to construct the following table:

$V_A(\text{cm}^3 \text{ mol}^{-1})$	74.11	73.96	73.50	72.74
x_A	0.60	0.40	0.20	0
$x_A/(1 - x_A)$	1.50	0.67	0.25	0

The points are plotted in Fig. 7.7, and the area required is 0.30. Hence,

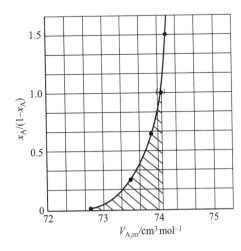

Figure 7.7

$$V(\text{CHCl}_3; \ 0.5, \ 0.5) = 80.66 \text{ cm}^3 \text{ mol}^{-1} - 0.30 \text{ cm}^3 \text{ mol}^{-1}$$

$$= \underline{80.36 \text{ cm}^3 \text{ mol}^{-1}}$$

P7.21 $\phi = -\dfrac{\ln a_A}{r}$

Therefore, $d\phi = -\dfrac{1}{r}d \ln a_A + \dfrac{1}{r^2}\ln a_A dr$

$d \ln a_A = \dfrac{1}{r}\ln a_A dr - r d\phi$

From the Gibbs–Duhem equation, $x_A\, d\mu_A = x_B\, d\mu_B = 0$, which implies that (since $\mu = \mu^{\ominus} + RT \ln a$, $d\mu_A = RT d \ln a_A$, $d\mu_B = RT d \ln a_B$)

$$d \ln a_B = -\dfrac{x_A}{x_B}\, d \ln a_A = -\dfrac{d \ln a_A}{r}$$

$$= -\dfrac{1}{r^2}\ln a_A\, dr = d\phi \text{ [from (b)]} = \dfrac{1}{r}\phi\, dr = dr = d\phi \text{ [from (a)]}$$

$$= \phi\, d \ln r + d\phi$$

Subtract $d \ln r$ from both sides, to obtain

$$d \ln \dfrac{a_B}{r} = (\phi - 1)\, d \ln r + d\phi = \dfrac{(\phi - 1)}{r}dr + d\phi$$

Then, by integration and noting that $\ln \left(\dfrac{a_B}{r}\right)_{r=0} = \ln \left(\dfrac{\gamma_B x_B}{r}\right)_{r=0} = \ln(\gamma_B)_{r=0} = \ln\ 1 = 0$

$$\ln \dfrac{a_B}{r} = \boxed{\phi - \phi(0) = \int_0^r \left(\dfrac{\phi - 1}{r}\right)dr}$$

Solutions to applications

P7.23   The sums of the equilibrium concentrations of Na^+ and Cl^- in each compartment are

$$[Na^+]_L + [Na^+]_R = [Cl^-]_L + [Cl^-]_R + \nu[P] = 2[Cl^-] + \nu[P]$$

Then use $[Cl^-] = 0.200\,\text{mol L}^{-1}$.
As $[P] = 9.1 \times 10^{-4}\,\text{mol L}^{-1}$, we find

$$[Na^+]_L - [Na^+]_R = \dfrac{6 \times (9.1 \times 10^{-4}\,\text{mol L}^{-1}) \times [Na^+]_L}{2 \times (0.200\,\text{mol L}^{-1}) + 6 \times (9.1 \times 10^{-4}\,\text{mol L}^{-1})}$$

and the sum above gives

$$\begin{aligned}[Na^+]_L + [Na^+]_R &= 2 \times (0.200\,\text{mol L}^{-1}) + 6 \times (9.1 \times 10^{-4}\,\text{mol L}^{-1}) \\ &= 0.405\,\text{mol L}^{-1}\end{aligned}$$

The solutions of these two equations are

$$[Na^+]_L = 0.204\,\text{mol L}^{-1} \quad [Na^+]_R = 0.201\,\text{mol L}^{-1}$$

Then

$$[Cl^-]_R = [Na^+]_R = 0.201\,\text{mol L}^{-1}$$
$$[Cl^-]_L = [Na^+]_L - 6[P] = 0.199\,\text{mol L}^{-1}$$

Comment. The Na^+ ions accumulate slightly in the compartment containing the macromolecule.

P7.25   We use eqn 7.41 in the form given in Example 7.5 with $\Pi = \rho g h$, then

$$\frac{\Pi}{c} = \frac{RT}{\overline{M}_n}\left(1 + \frac{B}{\overline{M}_n}c\right) = \frac{RT}{\overline{M}_n} + \frac{RTB}{\overline{M}_n^2}c$$

where c is the mass concentration of the polymer. Therefore plot Π/c against c. The intercept gives $RT/\overline{M}_n$ and the slope gives $RT/\overline{M}_n^2$.

The transformed data to plot are given in the table

$c/(\text{mg cm}^{-3})$	1.33	2.10	4.52	7.18	9.87
$(\Pi/c)/(\text{N m}^{-2}\ \text{mg}^{-1}\ \text{cm}^3)$	$22.5\overline{6}$	$24.2\overline{9}$	$29.2\overline{0}$	$34.2\overline{6}$	$39.5\overline{1}$

The plot is shown in Fig. 7.8. The intercept is $29.0\overline{9}\,\text{N m}^{-2}/(\text{mg cm}^{-3})$. The slope is $1.974\,\text{N m}^{-2}/(\text{mg cm}^{-3})^2$. Therefore

$$\overline{M}_n = \frac{RT}{29.0\overline{9}\,\text{N m}^{-2}/(\text{mg cm}^{-3})}$$

$$= \frac{8.3145\,\text{J K}^{-1}\,\text{mol}^{-1} \times 303.15\,\text{K}}{20.09\,\text{N m}^{-2}/(\text{mg cm}^{-3})} \times \left(\frac{1\text{g}}{10^3\,\text{mg}}\right) \times \left(\frac{10^6\,\text{cm}^3}{1\,\text{m}^3}\right)$$

$$= 1.25\overline{5} \times 10^5\,\text{g mol}^{-1} = \boxed{1.26 \times 10^5\,\text{g mol}^{-1}}$$

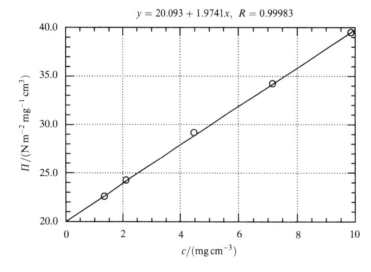

$$y = 20.093 + 1.9741x, \quad R = 0.99983$$

Figure 7.8

$$B = \frac{\overline{M}_n^2}{RT} \times 1.974\,\text{N m}^{-2}/(\text{mg cm}^{-3})^2$$

$$= \frac{\overline{M}_n}{\left(\frac{RT}{\overline{M}_n}\right)} \times 1.974\,\text{N m}^{-2}/(\text{mg cm}^{-3})^2$$

$$= \frac{1.25\overline{5} \times 10^5\,\text{g mol}^{-1} \times 1.974\,\text{N m}^{-2}/(\text{mg cm}^{-3})^2}{20.09\,\text{N m}^{-2}/(\text{mg cm}^{-3})}$$

$$= 1.23 \times 10^4\,\text{g mol}^{-1}/(\text{mg cm}^{-3})$$

$$= 1.23 \times 10^7\,\text{g mol}^{-1}/(\text{g cm}^{-3}) = \boxed{1.23 \times 10^4\,\text{L mol}^{-1}}$$

8   Phase diagrams

Solutions to exercises

Discussion questions

E8.1(a)   Phase: a state of matter that is uniform throughout, not only in chemical composition but also in physical state.

Constituent: any chemical species present in the system.

Component: a chemically independent constituent of the system. It is best understood in relation to the phrase "number of components" which is the minimum number of independent species necessary to define the composition of all the phases present in the system.

Degree of freedom (or variance): the number of intensive variables that can be changed without disturbing the number of phases in equilibrium.

E8.2(a)   See Figs 8.1(a) and (b).

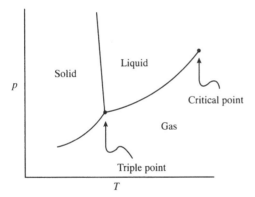

Figure 8.1(a)

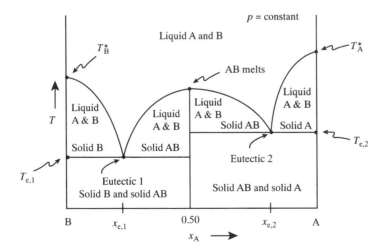

Figure 8.1(b)

E8.3(a)   See Fig 8.2.

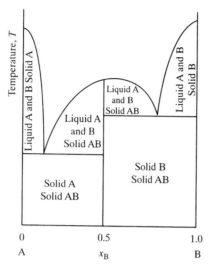

Figure 8.2

Numerical exercises

E8.4(a)   An expression for composition of the solution in terms of its vapour pressure is required. This is obtained from Dalton's law and Raoult's law as follows

$$p = p_A + p_B \text{ [Dalton's law]} = x_A p_A^* + (1 - x_A)p_B^*$$

Solving for x_A, $x_A = \dfrac{p - p_B^*}{p_A^* - p_B^*}$

For boiling under 0.50 atm (380 Torr) pressure, the combined vapour pressure, p, must be 380 Torr; hence $x_A = \dfrac{380 - 150}{400 - 150} = \boxed{0.920}$, $x_B = \boxed{0.080}$

The composition of the vapour is given by eqn 8.5

$$y_A = \frac{x_A p_A^*}{p_B^* + (p_A^* - p_B^*)x_A} = \frac{0.920 \times 400}{150 + (400 - 150) \times 0.920} = \boxed{0.968}$$

and $y_B = 1 - 0.968 = \boxed{0.032}$

E8.5(a)   The vapour pressures of components A and B may be expressed in terms of both their composition in the vapour and in the liquid. The pressures are the same whatever the expression; hence the expressions can be set equal to each other and solved for the composition.

$$p_A = y_A p = 0.350p = x_A p_A^* = x_A \times (575 \text{ Torr})$$
$$p_B = y_B p = (1 - y_A)p = 0.650p = x_B p_B^* = (1 - x_A) \times (390 \text{ Torr})$$

Therefore, $\dfrac{y_A p}{y_B p} = \dfrac{x_A p_A^*}{x_B p_B^*}$

Hence $\dfrac{0.350}{0.650} = \dfrac{575 x_A}{390(1 - x_A)}$

which solves to $x_A = \boxed{0.268}$, $x_B = 1 - x_A = \boxed{0.732}$

and, since $0.350p = x_A p_A^*$

$$p = \frac{x_A p_A^*}{0.350} = \frac{(0.268) \times (575 \, \text{Torr})}{0.350} = \boxed{440 \, \text{Torr}}$$

E8.6(a)   (a) Check to see if Raoult's law holds; if it does the solution is ideal.

$$p_A = x_A p_A^* = (0.6589) \times (957 \, \text{Torr}) = 630.6 \, \text{Torr}$$

$$p_B = x_A p_B^* = (0.3411) \times (379.5 \, \text{Torr}) = 129.\overline{4} \, \text{Torr}$$

$$p = p_A + p_B = 760 \, \text{Torr} = 1 \, \text{atm}$$

Since this is the pressure at which boiling occurs, Raoult's law holds and $\boxed{\text{the solution is ideal.}}$

   (b) $y_A = \dfrac{p_A}{p} \, [8.4] = \dfrac{630.6 \, \text{Torr}}{760 \, \text{Torr}} = \boxed{0.830}$ $y_B = 1 - y_A = 1.000 - 0.830 = \boxed{0.170}$

E8.7(a)   (a) $p(\text{total}) = p_{DE} + p_{DP} \, [\text{Dalton's law}] = x_{DE} p_{DE}^* + x_{DP} p_{DP}^* \, [\text{Raoult's law, 8.3}]$
   $x_{DE} = z_{DE}, \qquad x_{DP} = 1 - z_{DE} \, [\text{system all liquid}]$

$$p(\text{total}) = (0.60) \times (172 \, \text{Torr}) + (0.40) \times (128 \, \text{Torr}) = 10\overline{3} + 51 = \boxed{15\overline{4} \, \text{Torr}}$$

   (b) $y_{DE} = \dfrac{p_{DE}}{p} \, [8.4] = \dfrac{10\overline{3} \, \text{Torr}}{154 \, \text{Torr}} = \boxed{0.67}$ $y_{DP} = 1 - y_{DE} = \boxed{0.33}$

E8.8(a)   The data are plotted in Fig. 8.3. From the graph, the vapour in equilibrium with a liquid of composition
   (a) $x_M = 0.25$ is determined from the tie line labelled a in the figure extending from $x_M = 0.25$ to
   $\boxed{y_M = 0.36}$, (b) $x_0 = 0.25$ is determined from the tie line labelled b in the figure extending from
   $x_M = 0.75$ to $\boxed{y_M = 0.82.}$

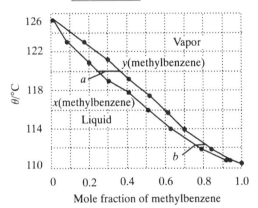

Figure 8.3

E8.9(a)   (a) Though there are three constituents, salt, water, and water vapor, there is an equilibrium condition
   between liquid water and its vapor. Hence, $\boxed{C = 2}$

   (b) Disregarding the water vapor for the reasons in (a) there are seven species: Na^+, H^+, $H_2PO_4^-$,
   HPO_4^{2-}, PO_4^{3-}, H_2O, OH^-. There are also three equilibria, namely

$$H_2PO_4^- \rightleftharpoons H^+ + HPO_4^{2-}$$

$$HPO_4^{2-} \rightleftharpoons H^+ + PO_4^{3-}$$

$$H^+ + OH^- \rightleftharpoons H_2O$$

(These could all be written as Brønsted equilibria without changing the conclusions.) There are also two conditions of electrical neutrality, namely

$$[Na^+] = [phosphates], \qquad [H^+] = [OH^-] + [phosphates]$$

where $[phosphates] = [H_2PO_4^-] + 2[HPO_4^{2-}] + 3[PO_4^{3-}]$. Hence, the number of independent components is

$$C = 7 - (3 + 2) = \boxed{2}$$

E8.10(a) $CuSO_4 \cdot 5H_2O(s) \rightleftharpoons CuSO_4(s) + 5H_2O(g)$

There are two solids, but one solid phase, as well as a gaseous phase; hence $\boxed{P = 2}$ Assuming all the water and $CuSO_4$ are formed by the dehydration, their amounts are then fixed by the equilibrium; hence $\boxed{C = 2}$

E8.11(a)   (a) The two components are Na_2SO_4 and H_2O (proton transfer equilibria to give HSO_4^- etc. do not change the number of independent components) so $\boxed{C = 2}$. There are three phases present (solid salt, liquid solution, vapour), so $\boxed{P = 3}$.

   (b) The variance is $F = C - P + 2 = 2 - 3 + 2 = \boxed{1}$

   Either pressure or temperature may be considered the independent variable, but not both as long as the equilibrium is maintained. If the pressure is changed, the temperature must be changed to maintain the equilibrium.

E8.12(a)   See Fig. 8.4.

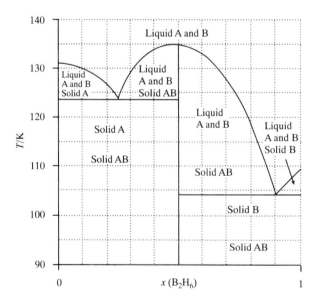

Figure 8.4

E8.13(a)   Refer to Fig. 8.26 of the text. At b_3 there are two phases with compositions $x_A = 0.18$ and $x_A = 0.70$; their abundances are in the ratio 0.13 (lever rule). Since $C = 2$ and $P = 2$ we have $F = 2$ (such as p and x). On heating, the phases merge, and the single-phase region is encountered. Then $F = 3$ (such as p, T and x). The liquid comes into equilibrium with its vapour when the isopleth cuts the phase line. At this temperature, and for all points up to b_1, $C = 2$ and $P = 2$, implying that $F = 2$ (for example p, x). The whole sample is a vapour above b_1.

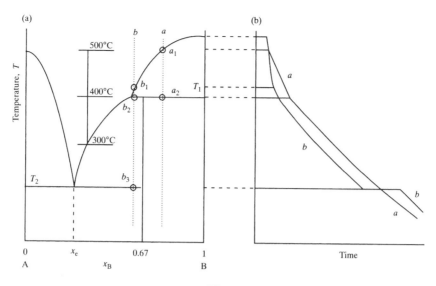

Figure 8.5

E8.14(a) The incongruent melting point (Section 8.6) is marked as $T_1 = 400°C$ in Fig. 8.5(a). The composition of the eutectic is marked as $x_e (\approx 0.30)$ in the figure. Its melting point is T_2 ($\approx 200°C$).

E8.15(a) The cooling curves are shown in Fig. 8.5(b). Note the breaks (abrupt change in slope) at temperatures corresponding to points a_1, a_2, b_1, b_2. Also note the eutectic halt at b_3.

E8.16(a) Refer to Fig. 8.6.

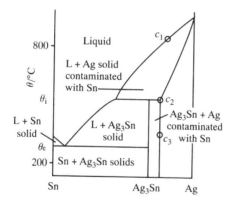

Figure 8.6

(a) The solubility of silver in tin at 800°C is determined by the point c_1 (at higher proportions of silver, the system separates into two phases). The point c_1 corresponds to $\boxed{80 \text{ per cent}}$ silver by mass.

(b) See point c_2. The compound Ag_3Sn decomposes at this temperature.

(c) The solubility of Ag_3Sn in silver is given point c_3 at 300°C.

E8.17(a) (a) See Figs 8.7(a) and (b).

(b) Follow line b in Fig. 8.7(a) down to the liquid line which intersects at point b_1. The vapor pressure at b_1 is $\approx \boxed{620 \text{ Torr}}$.

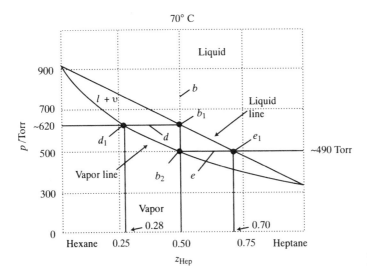

Figure 8.7(a)

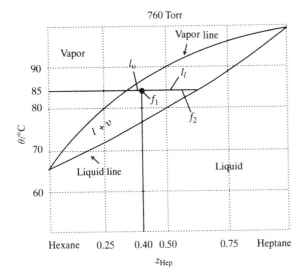

Figure 8.7(b)

(c) Follow line b in Fig. 8.7(a) down to the vapor line which intersects at point b_2. The vapor pressure at b_2 is $\approx$ $\boxed{490 \text{ Torr}}$. From points b_1 to b_2, the system changes from essentially all liquid to essentially all vapor.

(d) Consider tie line d; point b_1 gives the mole fractions of the liquid, which are

$$x(\text{Hep}) = 0.50 = 1 - x(\text{Hex}) \qquad x(\text{Hex}) = \boxed{0.50}$$

Point d_1 gives the mole fractions in the vapor which are

$$y(\text{Hep}) \approx 0.28 = 1 - y(\text{Hex}) \qquad y(\text{Hex}) \approx \boxed{0.72}$$

The initial vapor is richer in the more volatile component, hexane.

(e) Consider tie line e; point b_2 gives the mole fractions in the vapor, which are

$$y(\text{Hep}) = 0.50 = 1 - y(\text{Hex}) \qquad y(\text{Hex}) = \boxed{0.50}$$

Point e_1 gives the mole fractions in the liquid, which are

$$x(\text{Hep}) = 0.70 = 1 - x(\text{Hex}) \qquad x(\text{Hex}) = \boxed{0.30}$$

(f) Consider tie line f. The section, l_l, from point f_1 to the liquid line gives the relative amount of vapor; the section, l_v, from point f_1 to the liquid line gives the relative amount of liquid. That is

$$n_v l_v = n_l l_l \; [8.7] \quad \text{or} \quad \frac{n_v}{n_l} = \frac{l_l}{l_v} \approx \frac{6}{1}$$

Since the total amount is 2 mol, $n_v \approx \boxed{1.7}$ and $n_l \approx \boxed{0.3 \text{ mol}}$.

E8.18(a) The phase diagram is drawn in Fig. 8.8.

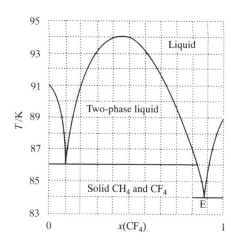

Figure 8.8

E8.19(a) The cooling curves are sketched in Fig. 8.9. Note the breaks and halts. The breaks correspond to changes in the rate of cooling due to the freezing out of a solid which releases its heat of fusion and thus slows down the cooling process. The halts correspond to the existence of three phases and hence no variance until one of the phases disappears.

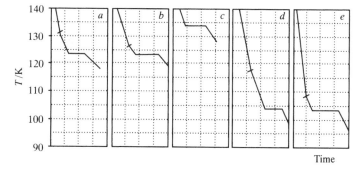

Figure 8.9

E8.20(a) The phase diagram is sketched in Fig. 8.10.

(a) The mixture has a single liquid phase at all compositions.

(b) When the composition reaches $x(C_6F_{14}) = 0.24$ the mixture separates into two liquid phases of compositions $x = 0.24$ and 0.48. The relative amounts of the two phases change until the

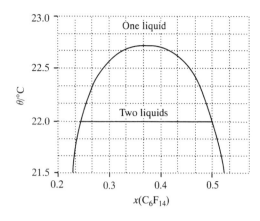

$\theta/°C$

$x(C_6F_{14})$

Figure 8.10

composition reaches $x = 0.48$. At all mole fractions greater than 0.48 in C_6F_{14} the mixture forms a single liquid phase.

Solutions to problems

Solutions to numerical problems

P8.1   (a)   The data, including that for pure chlorobenzene, are plotted in Fig. 8.11.

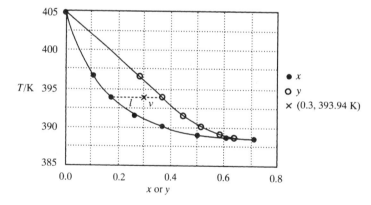

Figure 8.11

(b)   The smooth curve through the x, T data crosses $x = 0.300$ at 391.0 K, the boiling point of the mixture.

(c)   We need not interpolate data, for 393.94 K is a temperature for which we have experimental data. The mole fraction of 1-butanol in the liquid phase is 0.1700 and in the vapour phase 0.3691. According to the lever rule, the proportions of the two phases are in an inverse ratio of the distances their mole fractions are from the composition point in question. That is

$$\frac{n_{\text{liq}}}{n_{\text{vap}}} = \frac{v}{l} = \frac{0.3691 - 0.300}{0.300 - 0.1700} = \boxed{0.532}$$

P8.3 $p_A = a_A p_A^* = \gamma_A x_A p_A^*$ [7.45]

$$\gamma_A = \frac{p_A}{x_A p_A^*} = \frac{y_A p}{x_A p_A^*}$$

Sample calculation at 80 K

$$\gamma_{O_2}(80\,K) = \frac{0.11(100\,kPa)}{0.34(225\,Torr)}\left(\frac{760\,Torr}{101.325\,kPa}\right)$$

$$\gamma_{O_2}(80\,K) = 1.079$$

Summary

T/K	77.3	78	80	82	84	86	88	90.2
γ_{O_2}	—	0.877	1.079	1.039	0.995	0.993	0.990	0.987

To within the experimental uncertainties the solution appears to be ideal ($\gamma = 1$). The low value at 78 K may be caused by nonideality; however, the larger relative uncertainty in $y(O_2)$ is probably the origin of the low value.

A temperature–composition diagram is shown in Fig. 8.12(a). The near ideality of this solution is, however, best shown in the pressure–composition diagram of Fig. 8.12(b). The liquid line is essentially a straight line as predicted for an ideal solution.

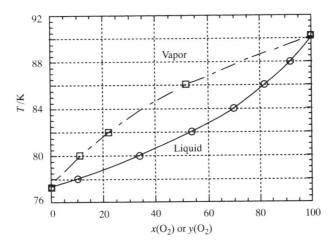

Figure 8.12(a)

P8.4   The phase diagram is shown in Fig. 8.13(a). The values of x_S corresponding to the three compounds are: (1) P_4S_3, 0.43; (2) P_4S_7, 0.64.(3) P_4S_{10}, 0.71.

The diagram has four eutectics labelled e_1, e_2, e_3, and e_4; eight two-phase liquid–solid regions, t_1 through t_8; and four two-phase solid regions, S_1, S_2, S_3, and S_4. The composition and physical state of the regions are as follows:

l: liquid S and P; S_1: solid P and solid P_4S_3; S_2: solid P_4S_3 and solid P_4S_7;
S_3: solid P_4S_7 and P_4S_{10}; S_4: solid P_4S_{10} and solid S
t_1: liquid P and S and solid P t_2: liquid P and S and solid P_4S_3
t_3: liquid P and S and solid P_4S_3 t_4: liquid P and S and solid P_4S_7
t_5: liquid P and S and solid P_4S_7 t_6: liquid P and S and solid P_4S_{10}
t_7: liquid P and S and solid P_4S_{10} t_8: liquid P and S and solid S

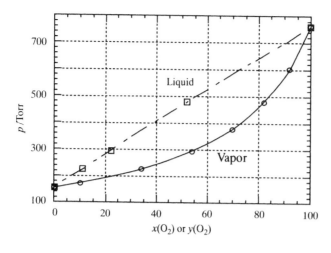

Figure 8.12(b)

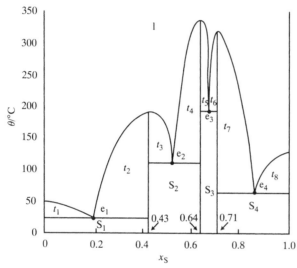

Figure 8.13(a)

A break in the cooling curve (Fig. 18.13(b)) occurs at point $b_1 \approx 125°C$ as a result of solid P_4S_3 forming; a eutectic halt occurs at point $e_1 \approx 20°C$.

P8.5   A compound with probable formula A_3B exists. It melts incongruently at 700 K, undergoing the peritectic reaction

$$A_3B(s) \rightarrow A(s) + (A + B, l)$$

The proportions of A and B in the product are dependent upon the overall composition and the temperature. A eutectic exists at 400 K and $x_B \approx 0.83$.

P8.7   The data are plotted in Fig. 8.15.

From the upper and lower extremes of the two-phase region we find $T_{uc} = \boxed{122°C}$ and $T_{lc} = \boxed{8°C}$ According to the phase diagram, miscibility is complete up to point a. Therefore, before that point is reached, $P = 1$, $C = 2$, implying that $F = 3(p, T, \text{and } x)$. Two phases occur at a corresponding to $w(MP) = 0.18$ and 0.84. At that point, $P = 2$, $C = 2$, and $F = 2(p, \text{and } x \text{ or } T)$. At the point a' there are two phases of composition $w = 0.18$ and 0.84. They are present in the ratio $\dfrac{a'' - a'}{a' - a} = 2$

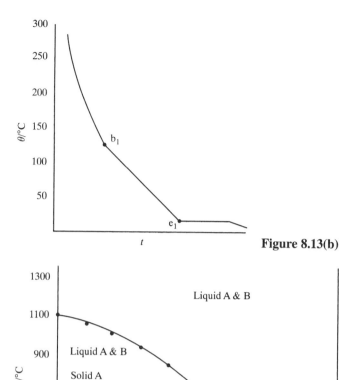

Figure 8.13(b)

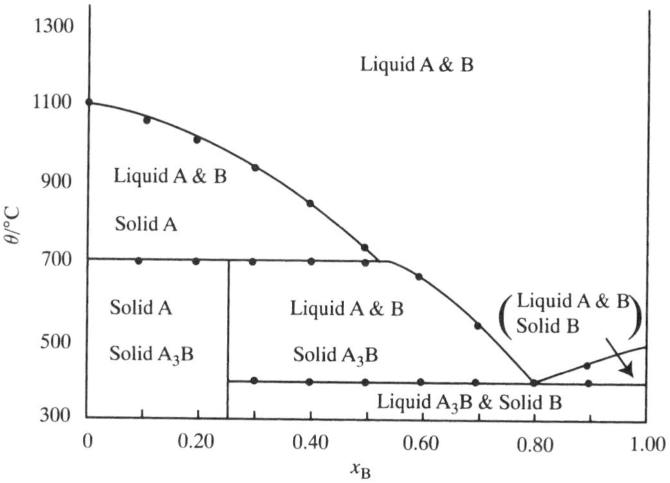

Figure 8.14

with the former dominant. At a'' there are still two phases with those compositions, but the former ($w = 0.18$) is present only as a trace. One more drop takes the system into the one-phase region.

P8.9

(a) The $\Delta_{mix}G(x_{Pb})$ curves show that at 1500 K lead and copper are totally miscible. They mix to form a homogeneous solution no matter what the relative amounts may be. However, the curve at 1300 K appears to have a small double minimum, which indicates two partially miscible phases (sections 7.4b and 8.5b) at temperatures lower than 1300 K (1100 K curve of the figure) there are two very distinct minimum and we expect two partially miscible phases. The upper critical temperature is about 1300 K at 1500 K,

$$F = C - P + 2 = 2 - 1 + 2 = \boxed{3} \quad \text{at } 1100\,\text{K,}$$
$$F = C - P + 2 = 2 - 2 + 2 = \boxed{2}$$

(b) When a homogeneous, equilibrium mixture with $x_{Pb} = 0.1$ is cooled from 1500 K to 1100 K, no phase separation occurs. The solution composition does not change.

If an $x_{Pb} = 0.7$ homogeneous, equilibrium mixture is cooled slowly, two partially miscible phases appear at about 1300 K. The separation occurs because the composition lies between two minimum on the $\Delta_{mix}G$ curve at 1300 K and phase separation lowers the total Gibbs energy.

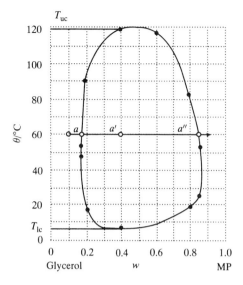

Figure 8.15

The composition of the two phases is determined by the equilibrium criterion $\mu_i(\alpha) = \mu_i(\beta)$ between the α and β phase. Since the chemical potential is the tangent of the $\Delta_{mix}G$ curve, we conclude that the straight line that is tangent to $\Delta_{mix}G(x)$ at two volumes of x (a double tangent) determine the composition of the two partially miscible phases. The 1100 K data is expanded (this can be done on a photocopy machine) so that the numerical values may be extracted more easily. The double tangent is drawn and the tangent points give the composition $\boxed{x_{Pb}(\alpha) = 0.19}$ and $\boxed{x_{Pb}(\beta) = 0.86}$. (Notice that the tangent points and the minimum do not normally coincide.) The relative amounts of the two phases is determined by the lever rule (eqn 8.7).

$$\frac{n_\alpha}{n_\beta} = \frac{l_\beta}{l_\alpha} = \frac{0.86 - 0.70}{0.70 - 0.19} = \boxed{0.36}$$

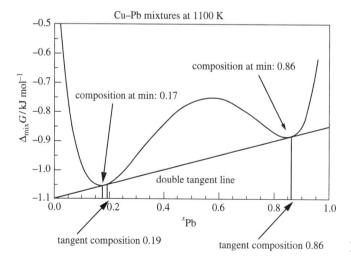

Figure 8.16

(c) Solubility at 1100 K is determined by the positions of the two minimum in the $\Delta_{mix}G$ curve. The maximum amount of lead that can be dissolved in copper yields a mixture that has $x_{Pb} = 0.17$, any more lead produces a second phase.

$$\text{solubility of Pb in Cu} = \left(\frac{0.17\,\text{mol Pb}}{0.83\,\text{mol Cu}}\right) \times \left(\frac{207.19\,\text{g Pb}}{1\,\text{mol Pb}}\right) \times \left(\frac{1\,\text{mol Cu}}{63.54\,\text{g Cu}}\right)$$

$$= \boxed{0.67\text{g Pb/gCu}}$$

The second minumum in the $\Delta_{mix}G$ curve at 1100 K is at $x_{Pb} = 0.86$.

$$\text{solubility of Cu in Pb} = \left(\frac{0.14\,\text{mol Cu}}{0.86\,\text{mol Pb}}\right) \times \left(\frac{63.54\,\text{g Cu}}{1\,\text{mol Cu}}\right) \times \left(\frac{1\,\text{mol Pb}}{207.19\,\text{g Pb}}\right)$$

$$= \boxed{0.050\,\text{g Cu/ g Pb}}$$

P8.11   The data are plotted in Fig. 8.17. At 360°C, $K_2FeCl_4(s)$ appears. The solution becomes richer in $FeCl_2$ until the temperature reaches 351°C, at which point $KFeCl_3(s)$ also appears. Below 351°C the system is a mixture of $K_2FeCl_4(s)$ and $KFeCl_3(s)$.

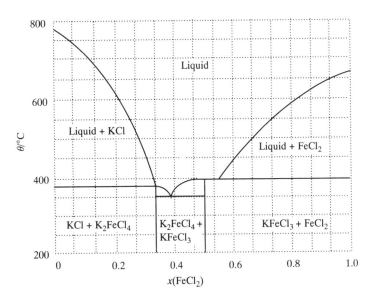

Figure 8.17

Solutions to theoretical problems

P8.13   The implication of this problem is that energy in the form of heat may be transferred between phases and that the volumes of the phases may also change. However, $U_\alpha + U_\beta = $ constant and $V_\alpha + V_\beta = $ constant. Hence,

$$dU_\beta = -d\,U_\alpha \text{ (b)} \quad \text{and} \quad dV_\beta = -dV_\alpha \text{ (c)}$$

The general condition of equilibrium in an isolated system is $dS = 0$; hence

$$dS = dS_\alpha + dS_\beta = 0 \text{ (a)}$$

$$S = S(U, V)$$

$$dS = \left(\frac{\partial S_\alpha}{\partial U_\alpha}\right)_{V_\alpha} dU_\alpha + \left(\frac{\partial S_\alpha}{\partial V_\alpha}\right)_{U_\alpha} dV_\alpha + \left(\frac{\partial S_\beta}{\partial U_\beta}\right)_{V_\beta} dU_\beta + \left(\frac{\partial S_\beta}{\partial V_\beta}\right)_{U_\beta} dV_\beta$$

Using conditions (b) and (c), and eqn 5.4

$$dS = \left(\frac{1}{T_\alpha} - \frac{1}{T_\beta}\right) dU_\alpha + \left(\frac{p_\alpha}{T_\alpha} - \frac{p_\beta}{T_\beta}\right) dV_\alpha = 0$$

The only way in which this expression may, in general, equal zero is for

$$\frac{1}{T_\alpha} - \frac{1}{T_\beta} = 0 \quad \text{and} \quad \frac{p_\alpha}{T_\alpha} - \frac{p_\beta}{T_\beta} = 0$$

Therefore, $\boxed{T_\alpha = T_\beta \text{ and } p_\alpha = p_\beta}$

Solution to applications

P8.15   The data are plotted in Fig. 8.18.

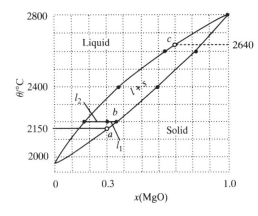

Figure 8.18

(a) As the solid composition $x(\text{MgO}) = 0.3$ is heated, liquid begins to form when the solid (lower) line is reached $\boxed{\text{at } 2150°\text{C.}}$

(b) From the tie line at 2200°C, the liquid composition is $y(\text{MgO}) = \boxed{0.18}$ and the solid $x(\text{MgO}) = \boxed{0.35.}$

The proportions of the two phases are given by the lever rule,

$$\frac{l_1}{l_2} = \frac{n(\text{liq})}{n(\text{sol})} = \frac{0.05}{0.12} = \boxed{0.4}$$

(c) Solidification begins at point c, corresponding to $\boxed{2640°\text{C}}$

P8.18   Define $GH(T) = G(T) - H_{\text{SER}} = a + bT$

To find the melting points of SiO_2 and Si as determined by the empirical $GH(T)$ functions, write the $\Delta_{\text{fus}}G(T)$ function in terms of $GH(T)$ functions.

$$SiO_2(s) \rightarrow SiO_2(l) \quad \Delta_{fus}G_{SiO_2}(T) = GH_{SiO_2(l)}(T) - GH_{SiO_2(s)}(T)$$

The melting point is the temperature for which $\Delta_{fus}G_{SiO_2}(T)$ equals zero. This temperature may be evaluated algebraically but it is more easily found with the root function of a scientific calculator or a computer software package. The empirical $GH(T)$ functions indicate that *silica melts at 1934.1 K* (The *CRC Handbook of Chemistry and Physics* reports a value of 1986 K.)

Likewise,

$$Si(s) \rightarrow Si(l) \quad \Delta_{fus}G_{Si}(T) = GH_{Si(l)}(T) - GH_{Si(s)}(T)$$

The root of $\Delta_{fus}G_{Si}(T)$ indicates that silicon melts at 1715.5 K. (The *CRC Handbook* reports a value of 1683 K.) The temperature dependence of the silica vapor pressure is evaluated with $\Delta_{sub}G(T)$ below the melting point and $\Delta_{vap}G(T)$ above the melting point.

$$SiO_2(s) \rightarrow SiO_2(g) \quad \Delta_{sub}G(T) = GH_{SiO_2(g)}(T) - GH_{SiO_2(s)}(T)$$
$$SiO_2(l) \rightarrow SiO_2(g) \quad \Delta_{vap}G(T) = GH_{SiO_2(g)}(T) - GH_{SiO_2(l)}(T)$$

The equilibrium constants for these phase equilibria equal the partial pressure of the gas, P_{SiO_2}, according to eqn 9.19

$$P_{SiO_2}(T) = 1 \text{ bar} \times e^{-\Delta_{sub}G(T)/RT} \quad \text{below 1934.1 K}$$
$$P_{SiO_2}(T) = 1 \text{ bar} \times e^{-\Delta_{vap}G(T)/RT} \quad \text{above 1934.1 K}$$

These functions are used to calculate silica vapor pressures in the 1800 K–2500 K range. Values are plotted against T in the following graph. Similar eqns are used to calculate the partial pressure of silicon, $P_{Si}(T)$. The plots show the vapor pressures to be very low in spite of the high temperatures. This indicates that binding forces are very strong in the condensed phases.

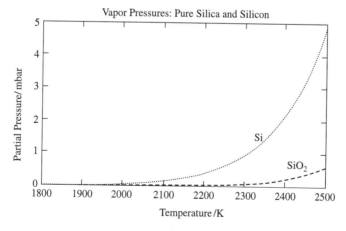

Figure 8.19

9 Chemical equilibrium

Solutions to exercises

Discussion questions

E9.1(a) The position of equilibrium is always determined by the condition that the reaction quotient, Q must equal the equilibrium constant, K. If the mixing in of an additional amount of reactant or product destroys that equality, then the reacting system will shift in such a way as to restore the equality. That implies that some of the added reactant or product must be removed by the reacting system and the amounts of other components will also be affected. These adjustments restore the concentrations to their (new) equilibrium values.

E9.2(a) (1) Response to change in pressure. The equilibrium constant is independent of pressure, but the individual partial pressures can change as the total pressure changes. This will happen when there is a difference, Δn_g, between the sums of the number of moles of gases on the product and reactant sides of the chemical equation. The requirement of an unchanged equilibrium constant implies that the side with the smaller number of moles of gas be favored as pressure increases.

(2) Response to change in temperature. Equation 9.26 (a) shows that K decreases with increasing temperature when the reaction is exothermic; thus the reaction shifts to the left, the opposite occurs in endothermic reactions. See Section 9.4 (a) for a more detailed discussion.

E9.3(a) The pH curve is show in Fig. 9.10 of the text.

E9.4(a) (1) It is assumed that activities of all species can be replaced with concentrations, namely, that the activity coefficients of all species involved in the reaction are 1.

(2) Contributions to $[H_3O^+]$ and $[OH^-]$ from the autoprotolysis of water are ignored.

(3) Contributions to $[A^-]$ from dissociation of remaining HA are ignored.

(4) Loss of $[HA]$ due to dissociation of remaining HA are ignored.

Numerical exercises

E9.5(a) $\Delta_r G^{\ominus} = -RT \ln K \ [9.8] = (-8.314 \, \text{J K}^{-1} \, \text{mol}^{-1}) \times (400 \, \text{K}) \times (\ln 2.07) = \boxed{-2.42 \, \text{kJ mol}^{-1}}$

E9.6(a) $\Delta_r G^{\ominus} = -RT \ln K \ [9.8]$

Taking inverse logarithms of both sides of this equation yields

$$K = e^{-\Delta_r G^{\ominus}/RT} = e^{+3.67 \times 10^3 \, \text{J mol}^{-1}/(8.314 \, \text{J K}^{-1} \, \text{mol}^{-1} \times 400 \, \text{K})} = \boxed{3.01}$$

E9.7(a) We draw up the following equilibrium table (Example 9.2). α is the equilibrium extent of dissociation.

	H_2O	H_2	O_2
Amount at equilibrium	$(1-\alpha)n$	αn	$\frac{1}{2}\alpha n$
Mole fraction	$\dfrac{1-\alpha}{1+\frac{1}{2}\alpha}$	$\dfrac{\alpha}{1+\frac{1}{2}\alpha}$	$\dfrac{\frac{1}{2}\alpha}{1+\frac{1}{2}\alpha}$
Partial pressure	$\dfrac{(1-\alpha)p}{1+\frac{1}{2}\alpha}$	$\dfrac{\alpha p}{1+\frac{1}{2}\alpha}$	$\dfrac{\frac{1}{2}\alpha p}{1+\frac{1}{2}\alpha}$

(a) $\quad K = \left(\prod_J a_J^{\nu_J}\right)_{\text{equilibrium}}$ [9.18]; $\quad a_J = \dfrac{p_J}{p^\ominus}$ [assume gases are perfect]

$$K = \frac{\left(\frac{p_{H_2}}{p^\ominus}\right)^2 \times \left(\frac{p_{O_2}}{p^\ominus}\right)}{\left(\frac{p_{H_2O}}{p^\ominus}\right)^2} \text{[9.18]} = \frac{\left(\frac{\alpha p}{(1+\frac{1}{2}\alpha)p^\ominus}\right)^2 \times \left(\frac{\frac{1}{2}\alpha p}{(1+\frac{1}{2}\alpha)p^\ominus}\right)}{\left(\frac{(1-\alpha)p}{(1+\frac{1}{2}\alpha)p^\ominus}\right)^2}$$

$$= \frac{\alpha^3 p}{2(1-\alpha)^2 \times \left(1+\frac{1}{2}\alpha\right)p^\ominus} = \frac{(0.0177)^3}{2(1-0.0177)^2 \times \left(1+\frac{1}{2} \times 0.0177\right)}$$

$$= 2.84\overline{8} \times 10^{-6} = \boxed{2.85 \times 10^{-6}}$$

(b) $\quad \Delta_r G^\ominus = -RT \ln K$ [9.8]

$$= -(8.314\,\text{J K}^{-1}\,\text{mol}^{-1}) \times (2257\,\text{K}) \times \ln(2.84\overline{8} \times 10^{-6}) = 2.40 \times 10^5\,\text{J mol}^{-1}$$

$$= \boxed{+240\,\text{kJ mol}^{-1}}$$

(c) $\quad \Delta_r G = \boxed{0}$ [the system is at equilibrium]

Comment. The equilibrium constant always applies to the reaction as written. If the reaction had been written as $H_2O(g) \rightleftharpoons H_2(g) + \frac{1}{2}O_2(g)$ as in Example 9.2 the value of K would have been 1.69×10^{-3}, which compares favourably to the approximate value 2.08×10^{-3} calculated there, at a slightly different temperature.

E9.8(a) We draw up the following equilibrium table

	N_2O_4	NO_2
Amount at equilibrium	$(1-\alpha)n$	$2\alpha n$
Mole fraction	$\dfrac{1-\alpha}{1+\alpha}$	$\dfrac{2\alpha}{1+\alpha}$
Partial pressure	$\dfrac{(1-\alpha)p}{1+\alpha}$	$\dfrac{2\alpha p}{1+\alpha}$

(a) Assuming the gases are perfect $a_J = \left(\dfrac{p_J}{p^\ominus}\right)$; hence

$$K = \frac{\left(\frac{p_{NO_2}}{p^\ominus}\right)^2}{\left(\frac{p_{N_2O_4}}{p^\ominus}\right)} \text{[9.18]} = \frac{4\alpha^2 p}{(1-\alpha^2)p^\ominus} = \frac{4\alpha^2}{(1-\alpha^2)} \quad [p = p^\ominus]$$

$$K = \frac{(4) \times (0.1846)^2}{1 - (0.1846)^2} = \boxed{0.1411}$$

(b) $\quad \Delta_r G^\ominus = -RT \ln K \text{[9.8]} = -(8.314\,\text{J K}^{-1}\,\text{mol}^{-1}) \times (298.2\,\text{K}) \times \ln(0.1411)$

$$= 4.855 \times 10^3\,\text{J mol}^{-1} = \boxed{+4.855\,\text{kJ mol}^{-1}}$$

(c) $\ln K(100^\circ\text{C}) = \ln K(25^\circ\text{C}) - \dfrac{\Delta_r H^{\ominus}}{R}\left(\dfrac{1}{373.2\,\text{K}} - \dfrac{1}{298.2\,\text{K}}\right)$ [9.28]

$\ln K(100^\circ\text{C}) = \ln(0.1411) - \left(\dfrac{57.2 \times 10^3\,\text{J mol}^{-1}}{8.314\,\text{J K}^{-1}\,\text{mol}^{-1}}\right) \times (-6.739 \times 10^{-4}\,\text{K}^{-1}) = 2.678$

$K(100^\circ\text{C}) = \boxed{14.556}$

Comment. In this case the increase in temperature results in a considerable shift in the equilibrium amounts of the substances in the reaction. The value of K changes from less than 1 to greater that 1. The value $\Delta_r G^{\ominus}$ calculated in (b) compares favorably to the value $4.73\,\text{kJ mol}^{-1}$ determined from the data of Table 2.6.

E9.9(a)   **(a)** $\Delta_r G^{\ominus} = \sum_J \nu_J \Delta_f G^{\ominus}\,(\text{J})$ [9.13]

$\nu(\text{Pb}) = 1, \qquad \nu(\text{CO}_2) = 1, \qquad \nu(\text{PbO}) = -1, \qquad \nu(\text{CO}) = -1$

The equation is

$0 = \text{Pb(s)} + \text{CO}_2(\text{g}) - \text{PbO(s)} - \text{CO(g)}$

$\Delta_r G^{\ominus} = \Delta_f G^{\ominus}(\text{Pb, s}) + \Delta_f G^{\ominus}(\text{CO}_2,\text{g}) - \Delta_f G^{\ominus}(\text{PbO, s, red}) - \Delta_f G^{\ominus}(\text{CO, g})$

$= (-394.36\,\text{kJ mol}^{-1}) - (-188.93\,\text{kJ mol}^{-1}) - (-137.17\,\text{kJ mol}^{-1})$

$= \boxed{-68.26\,\text{kJ mol}^{-1}}$

$\ln K = \dfrac{-\Delta_r G^{\ominus}}{RT}\,[9.8] = \dfrac{+68.26 \times 10^3\,\text{J mol}^{-1}}{(8.314\,\text{J K}^{-1}\,\text{mol}^{-1}) \times (298\,\text{K})} = 27.55; \quad K = \boxed{9.2 \times 10^{11}}$

(b) $\Delta_r H^{\ominus} = \Delta_f H^{\ominus}(\text{Pb, s}) + \Delta_f H^{\ominus}(\text{CO}_2,\text{g}) - \Delta_f H^{\ominus}(\text{PbO, s, red}) - \Delta_f H^{\ominus}(\text{CO, g})$
$= (-393.51\,\text{kJ mol}^{-1}) - (-218.99\,\text{kJ mol}^{-1}) - (-110.53\,\text{kJ mol}^{-1})$
$= -63.99\,\text{kJ mol}^{-1}$

$\ln K(400\,\text{K}) = \ln K(298) - \dfrac{\Delta_r H^{\ominus}}{R}\left(\dfrac{1}{400\,\text{K}} - \dfrac{1}{298\,\text{K}}\right)$ [9.28]

$= 27.55 - \left(\dfrac{-63.99 \times 10^3\,\text{J mol}^{-1}}{8.314\,\text{J K}^{-1}\,\text{mol}^{-1}}\right) \times (-8.557 \times 10^{-4}\,\text{K}^{-1}) = 20.9\overline{6}$

$K(400\,\text{K}) = \boxed{1.3 \times 10^9}$

$\Delta_r G^{\ominus}(400\,\text{K}) = -RT \ln K(400\,\text{K})[8] = -(8.314\,\text{J K}^{-1}\,\text{mol}^{-1}) \times (400\,\text{K}) \times (20.9\overline{6})$

$= -6.97 \times 10^4\,\text{J mol}^{-1} = \boxed{-69.7\,\text{kJ mol}^{-1}}$

Comment. $\Delta_r G^{\ominus}(400\,\text{K})$ could have been determined directly from its value at 298 K by using the integrated form of the Gibbs–Helmholtz equation, (third equation in *Justification* 9.3), rather than by first calculating $K(400\,\text{K})$ as in the solution above.

Question. What is the value of $\Delta_r G^{\ominus}(400\,\text{K})$ for this reaction obtained by the method suggested in the Comment?

E9.10(a) Draw up the following equilibrium table

	A	B	C	D	Total
Initial amounts/mol	1.00	2.00	0	1.00	4.00
Stated change/mol			+0.90		
Implied change/mol	−0.60	−0.30	+0.90	+0.60	
Equilibrium amounts/mol	0.40	1.70	0.90	1.60	4.60
Mole fractions	0.087	0.370	0.196	0.348	1.001

(a) The mole fractions are given in the table.

(b) $K_x = \prod_J x_J{}^{\nu_J}$ [analogous to eqn 9.18 and *Illustration* 9.4]

$$K_x = \frac{(0.196)^3 \times (0.348)^2}{(0.087)^2 \times (0.370)} = 0.32\bar{6} = \boxed{0.33}$$

(c) $p_J = x_J p$, $\qquad p = 1\,\text{bar}$, $\qquad p^{\ominus} = 1\,\text{bar}$

Assuming that the gases are perfect, $a_J = \dfrac{p_J}{p^{\ominus}}$, hence

$$K = \frac{(p_C/p^{\ominus})^3 \times (p_D/p^{\ominus})^2}{(p_A/p^{\ominus})^2 \times (p_B/p^{\ominus})}$$

$$= \frac{x_C^3 x_D^2}{x_A^2 x_B} \times \left(\frac{p}{p^{\ominus}}\right)^2 = K_x \quad \text{when } p = 1.00\,\text{bar} = \boxed{0.33}$$

(d) $\Delta_r G^{\ominus} = -RT \ln K = -(8.314\,\text{J K}^{-1}\,\text{mol}^{-1}) \times (298\,\text{K}) \times (\ln 0.32\bar{6})$

$$= \boxed{+2.8 \times 10^3\,\text{J mol}^{-1}}$$

E9.11(a) At 1280 K, $\Delta_r G^{\ominus} = +33 \times 10^3\,\text{J mol}^{-1}$; thus

$$\ln K_1(1280\,\text{K}) = -\frac{\Delta_r G^{\ominus}}{RT} = -\frac{33 \times 10^3\,\text{J mol}^{-1}}{(8.314\,\text{J K}^{-1}\,\text{mol}^{-1}) \times (1280\,\text{K})} = -3.1\bar{0}$$

$$K_1 = \boxed{0.045}$$

$$\ln K_2 = \ln K_1 - \frac{\Delta_r H^{\ominus}}{R}\left(\frac{1}{T_2} - \frac{1}{T_1}\right) \quad [9.28]$$

We look for the temperature T_2 that corresponds to $\ln K_2 = \ln(1) = 0$. This is the crossover temperature. Solving for T_2 from eqn 26 with $\ln K_2 = 0$, we obtain

$$\frac{1}{T_2} = \frac{R \ln K_1}{\Delta_r H^{\ominus}} + \frac{1}{T_1} = \left(\frac{(8.314\,\text{J K}^{-1}\,\text{mol}^{-1}) \times (-3.1\bar{0})}{224 \times 10^3\,\text{J mol}^{-1}}\right) + \left(\frac{1}{1280\,\text{K}}\right)$$

$$= 6.6\bar{6} \times 10^{-4}\,\text{K}^{-1}$$

$$T_2 = \boxed{1500\,\text{K}}$$

E9.12(a) Given $\ln K = -1.04 - \dfrac{1088\,\text{K}}{T} + \dfrac{1.51 \times 10^5\,\text{K}^2}{T^2}$

and since $\dfrac{\text{d}\ln K}{\text{d}(1/T)} = \dfrac{-\Delta_r H^\ominus}{R}$ [19.26b]

$$\dfrac{-\Delta_r H^\ominus}{R} = -1088\,\text{K} + \dfrac{(2) \times (1.51 \times 10^5\,\text{K}^2)}{T}$$

Then, at 400 K

$$\Delta_r H^\ominus = \left(1088\,\text{K} - \dfrac{3.02 \times 10^5\,\text{K}^2}{400\,\text{K}}\right) \times (8.314\,\text{J K}^{-1}\,\text{mol}^{-1}) = \boxed{+2.77\,\text{kJ mol}^{-1}}$$

$$\Delta_r G^\ominus = -RT \ln K\,[9.8] = RT \times \left(1.04 + \dfrac{1088\,\text{K}}{T} - \dfrac{1.51 \times 10^5\,\text{K}^2}{T^2}\right)$$

$$= RT \times \left(1.04 + \dfrac{1088\,\text{K}}{400\,\text{K}} - \dfrac{1.51 \times 10^5\,\text{K}^2}{(400\,\text{K})^2}\right) = +9.37\,\text{kJ mol}^{-1}$$

$$= \Delta_r H^\ominus - T\Delta_r S^\ominus\,[4.39]$$

Therefore, $\Delta_r S^\ominus = \dfrac{\Delta_r H^\ominus - \Delta_r G^\ominus}{T} = \dfrac{2.77\,\text{kJ mol}^{-1} - 9.37\,\text{kJ mol}^{-1}}{400\,\text{K}} = \boxed{-16.5\,\text{J K}^{-1}\,\text{mol}^{-1}}$

E9.13(a) Let B = borneol and I = isoborneol

$$\Delta_r G = \Delta G^\ominus + RT \ln Q\,[9.11], \quad Q = \dfrac{p_I}{p_B}\,[9.15]$$

$$p_B = x_B p = \dfrac{0.15\,\text{mol}}{0.15\,\text{mol} + 0.30\,\text{mol}} \times 600\,\text{Torr} = 200\,\text{Torr}; \quad p_I = p - p_B = 400\,\text{Torr}$$

$$Q = \dfrac{400\,\text{Torr}}{200\,\text{Torr}} = 2.00$$

$$\Delta_r G = (+9.4\,\text{kJ mol}^{-1}) + (8.314\,\text{J K}^{-1}\,\text{mol}^{-1}) \times (503\,\text{K}) \times (\ln 2.00) = \boxed{+12.3\,\text{kJ mol}^{-1}}$$

E9.14(a) $K_x = \displaystyle\prod_J x_J{}^{\nu_J}$ [analogous to eqn 9.18, *Justification* 9.2]

The relation of K_x to K is established in *Illustration* 9.4

$$K = \prod_J \left(\dfrac{p_J}{p^\ominus}\right)^{\nu_J} \left[9.18 \text{ with } a_J = \dfrac{p_J}{p^\ominus}\right]$$

$$= \prod_J x_J{}^{\nu_J} \times \left(\dfrac{p}{p^\ominus}\right)^{\sum_J \nu_J} [p_J = x_J p] = K_x \times \left(\dfrac{p}{p^\ominus}\right)^{\nu} \left[\nu \equiv \sum_J \nu_J\right]$$

Therefore, $K_x = K\left(\dfrac{p}{p^\ominus}\right)^{-\nu} \quad K_x \propto p^{-\nu}$ [K and $p^\ominus$ are constants]

$\nu = 1 + 1 - 1 = 1$, thus $K_x(2\,\text{bar}) = \frac{1}{2}K_x(1\,\text{bar})$; percentage change is 50 percent.

E9.15(a) Let B = borneol and I = isoborneol

$$K = K_x \times \left(\frac{p}{p^\ominus}\right)^\nu \text{ [Exercise 9.14]} = \frac{x_I}{x_B}[\nu = 1 - 1 = 0] = \frac{1 - x_B}{x_B}$$

Hence, $x_B = \dfrac{1}{1 + K} = \dfrac{1}{1 + 0.106} = 0.904$

$x_I = 0.096$

The initial amounts of the isomers are

$$n_B = \frac{7.50 \text{ g}}{M}, \qquad n_I = \frac{14.0 \text{ g}}{M}, \qquad n = \frac{21.5\overline{0} \text{ g}}{M}$$

The total amount remains the same, but at equilibrium

$$\frac{n_B}{n} = x_B = \boxed{0.904}, \qquad x_I = \boxed{0.096}, \qquad n_B = (0.904) \times \left(\frac{21.5\overline{0} \text{ g}}{M}\right)$$

The mass of borneol at equilibrium is therefore

$$m_B = n_B \times M = (0.904) \times (21.5\overline{0} \text{ g}) = 19.4 \text{ g}$$

and the mass of isoborneol is

$$m_I = n_I \times M = (0.096) \times (21.5\overline{0} \text{ g}) = 2.1 \text{ g}$$

E9.16(a) $\Delta_r G^\ominus = -RT \ln K$ [9.8]

Hence, a value of $\Delta_r G^\ominus < 0$ at 298 K corresponds to $K > 1$.

(a) $\Delta_r G^\ominus /(\text{kJ mol}^{-1}) = (-202.87) - (-95.30 - 16.45) = -91.12, \quad \boxed{K > 1}$

(b) $\Delta_r G^\ominus /(\text{kJ mol}^{-1}) = (3) \times (-856.64) - (2) \times (-1582.3) = +594.7, \quad \boxed{K < 1}$

(c) $\Delta_r G^\ominus /(\text{kJ mol}^{-1}) = (-100.4) - (-33.56) = -66.8, \quad \boxed{K > 1}$

E9.17(a) Le Chatelier's principle in the form of the rules in the first paragraph of Section 9.4 is employed. Thus we determine whether $\Delta_r H^\ominus$ is positive or negative by using the $\Delta_f H^\ominus$ values of Table 2.6.

(a) $\Delta_r H^\ominus /(\text{kJ mol}^{-1}) = (-314.43) - (-46.11 - 92.31) = -176.01$

(b) $\Delta_r H^\ominus /(\text{kJ mol}^{-1}) = (3) \times (-910.94) - (2) \times (-1675.7) = +618.6$

(c) $\Delta_r H^\ominus /(\text{kJ mol}^{-1}) = (-100.0) - (-20.63) = -79.4$

Since (a) and (c) are exothermic, an increase in temperature favors the reactants; $\boxed{\textbf{(b)}}$ is endothermic, and an increase in temperature favors the products.

E9.18(a) $\ln \dfrac{K'}{K} = \dfrac{\Delta_r H^\ominus}{R}\left(\dfrac{1}{T} - \dfrac{1}{T'}\right)$ [9.28]

Therefore, $\Delta_r H^\ominus = \dfrac{R \ln \frac{K'}{K}}{\left(\frac{1}{T} - \frac{1}{T'}\right)}$

$T' = 308 \, \text{K}$; hence, with $\dfrac{K'}{K} = \kappa$

$$\Delta_r H^\ominus = \frac{(8.314 \, \text{J K}^{-1} \, \text{mol}^{-1}) \times (\ln \kappa)}{\left(\frac{1}{298 \, \text{K}} - \frac{1}{308 \, \text{K}}\right)} = 76 \, \text{kJ mol}^{-1} \times \ln \kappa$$

Therefore

(a) $\kappa = 2$, $\quad \Delta_r H^\ominus = (76 \, \text{kJ mol}^{-1}) \times (\ln 2) = \boxed{+53 \, \text{kJ mol}^{-1}}$

(b) $\kappa = \frac{1}{2}$, $\quad \Delta_r H^\ominus = (76 \, \text{kJ mol}^{-1}) \times \left(\ln \frac{1}{2}\right) = \boxed{-53 \, \text{kJ mol}^{-1}}$

E9.19(a) $\quad \Delta_r G = \Delta G^\ominus + RT \ln Q$ [9.11]; $\quad Q = \displaystyle\prod_J a_J{}^{\nu_J}$ [9.15]

for $\frac{1}{2} N_2(g) + \frac{3}{2} H_2(g) \rightarrow NH_3(g)$

$$Q = \frac{\left(\dfrac{p(NH_3)}{p^\ominus}\right)}{\left(\dfrac{p(N_2)}{p^\ominus}\right)^{1/2} \left(\dfrac{p(H_2)}{p^\ominus}\right)^{3/2}} \qquad \left[a_J = \frac{p_J}{p^\ominus} \text{ for perfect gases}\right]$$

$$= \frac{p(NH_3) p^\ominus}{p(N_2)^{1/2} p(H_2)^{3/2}} = \frac{4.0}{(3.0)^{1/2} \times (1.0)^{3/2}} = \frac{4.0}{\sqrt{3.0}}$$

Therefore, $\Delta_r G = (-16.45 \, \text{kJ mol}^{-1}) + RT \ln \dfrac{4.0}{\sqrt{3.0}} = (-16.45 \, \text{kJ mol}^{-1}) + (2.07 \, \text{kJ mol}^{-1})$

$$= \boxed{-14.38 \, \text{kJ mol}^{-1}}.$$

Since $\Delta_r G < 0$, the spontaneous direction of reaction is $\boxed{\text{toward products}}$

E9.20(a) The reaction is

$$CaCO_3(s) \rightleftharpoons CaO(s) + CO_2(g)$$

For the purposes of this exercise we may assume that the required temperature is that temperature at which the $K = 1$ which corresponds to a pressure of 1 bar for the gaseous product. For $K = 1$, $\ln K = 0$ and $\Delta_r G^\ominus = 0$.

$$\Delta_r G^\ominus = \Delta_r H^\ominus - T \Delta_r S^\ominus = 0 \quad \text{when } \Delta_r H^\ominus = T \Delta_r S^\ominus$$

Therefore, the decomposition temperature (when $K = 1$) is

$$T = \frac{\Delta_r H^\ominus}{\Delta_r S^\ominus}$$

$$CaCO_3(s) \rightarrow CaO(s) + CO_2(g)$$

$$\Delta_r H^\ominus = (-635.09) - (393.51) - (-1206.9) \, \text{kJ mol}^{-1} = +178.3 \, \text{kJ mol}^{-1}$$

$$\Delta_r S^\ominus = (39.75) + (213.74) - (92.9) \, \text{J K}^{-1} \, \text{mol}^{-1} = +160.6 \, \text{J K}^{-1} \, \text{mol}^{-1}$$

$$T = \frac{178.3 \times 10^3 \, \text{J mol}^{-1}}{160.6 \, \text{J K}^{-1} \, \text{mol}^{-1}} = \boxed{1110 \, \text{K}} \; (840°C)$$

E9.21(a) (a) The half-way point corresponds to the condition

$$[Acid] = [Salt]$$

for which $pK_a = pH$ [9.42]

Hence, $\boxed{pK_a = 5.40}$ and $K_a = 10^{-5.40} = \boxed{4.0 \times 10^{-6}}$

(b) When the solution is $[Acid] = 0.015\,M$

$$pH = \tfrac{1}{2}pK_a - \tfrac{1}{2}\log[Acid]\,[40] = \tfrac{1}{2} \times (5.40) - \tfrac{1}{2} \times (-1.82) = \boxed{3.61}$$

E9.22(a) (a) NH_4Cl

In water, the NH_4^+ acts as an acid in the Brønsted equilibrium

$$NH_4^+(aq) + H_2O(l) \rightleftharpoons NH_3(aq) + H_3O^+(aq) \quad K_a = \frac{[H_3O^+][NH_3]}{[NH_4^+]}$$

$[NH_3] \approx [H_3O^+]$, because the water autoprotolysis can be ignored in the presence of a weak acid (NH_4^+); therefore,

$$K_a \approx \frac{[H_3O^+]^2}{[NH_4^+]} \approx \frac{[H_3O^+]^2}{S}$$

where S is the nominal concentration of the salt. Therefore,

$$[H_3O^+] \approx (SK_a)^{1/2}$$

and $pH \approx \tfrac{1}{2}pK_a - \tfrac{1}{2}\log S\,[9.45] \approx \tfrac{1}{2} \times (9.25) - \tfrac{1}{2} \times (\log 0.10) = \boxed{5.13}$

(b) $NaCH_3CO_2$

The $CH_3CO_2^-$ ion acts as a weak base

$$CH_3CO_2^-(aq) + H_2O(l) \rightleftharpoons CH_3COOH(aq) + OH^-(aq) \quad K_b = \frac{[CH_3COOH][OH^-]}{[CH_3CO_2^-]}$$

Then, since $[CH_3COOH] \approx [OH^-]$ and $[CH_3CO_2^-] \approx S$, the nominal concentration of the salt,

$$K_b \approx \frac{[OH^-]^2}{S}, \quad \text{implying that} \quad [OH^-] \approx (SK_b)^{1/2}$$

Therefore, $pOH = \tfrac{1}{2}pK_b - \tfrac{1}{2}\log S$

However, $pH + pOH = pK_w$, so $pH = pK_w - pOH$

$$pK_a + pK_b = pK_w, \quad \text{so} \quad pK_b = pK_w - pK_a$$

Therefore, $pH = pK_w - \tfrac{1}{2}(pK_w - pK_a) + \tfrac{1}{2}\log S = \tfrac{1}{2}pK_w + \tfrac{1}{2}pK_a + \tfrac{1}{2}\log S$

$$= \tfrac{1}{2} \times (14.00) + \tfrac{1}{2} \times (4.76) + \tfrac{1}{2} \times (\log 0.10) = \boxed{8.88}$$

(c) $CH_3COOH(aq) + H_2O(l) \rightleftharpoons H_3O^+(aq) + CH_3CO_2^-(aq)$ $K_a = \dfrac{[H_3O^+][CH_3CO_2^-]}{[CH_3COOH]}$

Since we can ignore the water autoprotolysis, $[H_3O^+] \approx [CH_3CO_2^-]$, so

$$K_a \approx \frac{[H_3O^+]^2}{A}$$

where $A = [CH_3COOH]$, the nominal acid concentration (the ionization is small). Therefore,

$$[H_3O^+] \approx (AK_a)^{1/2}, \quad \text{implying that} \quad pH \approx \tfrac{1}{2}pK_a - \tfrac{1}{2}\log A_0 \text{ [9.38]}$$

Hence, $pH \approx \tfrac{1}{2} \times (4.76) - \tfrac{1}{2} \times (\log 0.100) = \boxed{2.88}$

E9.23(a)   The pH of a solution in which the nominal salt concentration is S is

$$pH = \tfrac{1}{2}pK_w + \tfrac{1}{2}pK_a + \tfrac{1}{2}\log S \quad \text{[9.44, Exercise 9.22(b)]}$$

The volume of the solution at the stoichiometric point is

$$V = (25.00\,\text{mL}) + (25.00\,\text{mL}) \times \left(\frac{0.100\,\text{M}}{0.150\,\text{M}}\right) = 41.67\,\text{mL}$$

and the concentration of salt is

$$S = (0.100\,\text{M}) \times \left(\frac{25.00\,\text{mL}}{41.67\,\text{mL}}\right) = 0.0600\,\text{M}$$

Hence, with $pK_a = 3.86$,

$$pH = \tfrac{1}{2} \times (14.00) + \tfrac{1}{2} \times (3.86) + \tfrac{1}{2} \times (\log 0.0600) = \boxed{8.3}$$

E9.24(a)   One procedure is to plot eqn 9.40. An alternative procedure is to estimate some of the points using the expressions given in Fig. 9.10 of the text. Initially only the salt is present, and we use eqn 9.44 (as in Exercise 9.23)

$$pH = \tfrac{1}{2}pK_a + \tfrac{1}{2}pK_w + \tfrac{1}{2}\log S, \quad \log S = -1.00$$

$$= \tfrac{1}{2}(4.76 + 14.00 - 1.00) = 8.88 \tag{a}$$

When $A' \approx S$, use the Henderson–Hasselbalch equation [9.40]

$$pH = pK_a - \log\frac{A'}{S} = 4.76 - \log\frac{A'}{0.10} = 3.76 - \log A' \tag{b}$$

When so much acid has been added that $A' \gg S$, use the 'weak acid alone' formula [9.38]

$$pH = \tfrac{1}{2}pK_a - \tfrac{1}{2}\log A \tag{c}$$

We can draw up the following table

$A(\text{or}\,A')$	0	0.06	0.08	0.10	0.12	0.14		0.6	0.8	1.0
pH	8.88	4.98	4.86	4.76	4.68	4.61		2.49	2.43	2.38
Formula	(a)			(b)					(c)	

The results are plotted in Fig. 9.1.

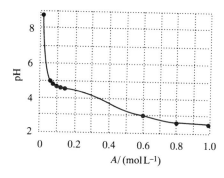

Figure 9.1

E9.25(a) According to the Henderson–Hasselbalch equation [9.41] the pH of a buffer varies about a central value given by pK_a. For the $\dfrac{[\text{acid}]}{[\text{salt}]}$ ratio to be neither very large nor very small we require $pK_a \approx pH$ (buffer).

(a) For pH $\approx$ 2.2 use $\boxed{Na_2HPO_4 + H_3PO_4}$ since

$$H_3PO_4 + H_2O \rightleftharpoons H_3O^+ + H_2PO_4^- \qquad pK_a = 2.16$$

(b) For pH $\approx$ 7 use $\boxed{NaH_2PO_4 + Na_2HPO_4}$ since

$$H_2PO_4^- + H_2O \rightleftharpoons H_3O^+ + HPO_4^{2-} \qquad pK_a = 7.2$$

Solutions to problems

Solutions to numerical problems

P9.1 (a) $\Delta_r G^\ominus = -RT \ln K = -(8.314\,\text{J\,K}^{-1}\,\text{mol}^{-1}) \times (298\,\text{K}) \times (\ln 0.164) = 4.48 \times 10^3\,\text{J\,mol}^{-1}$

$$= \boxed{+4.48\,\text{kJ\,mol}^{-1}}$$

(b) Draw up the following equilibrium table

	I_2	Br_2	IBr
Amounts	—	$(1-\alpha)n$	$2\alpha n$
Mole fractions	—	$\dfrac{(1-\alpha)}{(1+\alpha)}$	$\dfrac{2\alpha}{(1+\alpha)}$
Partial pressure	—	$\dfrac{(1-\alpha)p}{(1+\alpha)}$	$\dfrac{2\alpha p}{(1+\alpha)}$

$$K = \prod_J a_J^{\nu_J}\,[17] = \frac{\left(\dfrac{p_{IBr}}{p^\ominus}\right)^2}{\dfrac{p_{Br_2}}{p^\ominus}}\,[\text{perfect gases}] = \frac{\left\{(2\alpha)^2\,\dfrac{p}{p^\ominus}\right\}}{(1-\alpha)\times(1+\alpha)} = \frac{\left(4\alpha^2\,\dfrac{p}{p^\ominus}\right)}{1-\alpha^2} = 0.164$$

With $p = 0.164$ atm,

$$4\alpha^2 = 1 - \alpha^2 \qquad \alpha^2 = \tfrac{1}{5} \qquad \alpha = 0.447$$

$$p_{IBr} = \frac{2\alpha}{1+\alpha} \times p = \frac{(2)\times(0.447)}{1+0.447} \times (0.164\,\text{atm}) = \boxed{0.101\,\text{atm}}$$

(c) The equilibrium table needs to be modified as follows

$$p = p_{I_2} + p_{Br_2} + p_{IBr}$$

$$p_{Br_2} = x_{Br_2} p, \qquad p_{IBr} = x_{IBr} p, \qquad p_{I_2} = x_{I_2} p$$

with $x_{Br_2} = \dfrac{(1-\alpha)n}{(1+\alpha)n + n_{I_2}}$ [n = amount of Br_2 introduced into container]

and $x_{IBr} = \dfrac{2\alpha n}{(1+\alpha)n + n_{I_2}}$

K is constructed as above [9.18], but with these modified partial pressures. In order to complete the calculation additional data are required, namely, the amount of Br_2 introduced, n, and the equilibrium vapour pressure of $I_2(s)$. n_{I_2} can be calculated from a knowledge of the volume of the container at equilibrium which is most easily determined by successive approximations since p_{I_2} is small.

Question. What is the partial pressure of $IBr(g)$ if 0.0100 mol of $Br_2(g)$ is introduced into the container? The partial pressure of $I_2(s)$ at 25°C is 0.305 Torr.

P9.4 $CO_2(g) \rightleftharpoons CO(g) + \tfrac{1}{2}O_2(g)$

Draw up the following equilibrium table

	CO_2	CO	O_2
Amounts	$(1-\alpha)n$	αn	$\tfrac{1}{2}\alpha n$
Mole fractions	$\dfrac{(1-\alpha)}{\left(1+\frac{\alpha}{2}\right)}$	$\dfrac{\alpha}{\left(1+\frac{\alpha}{2}\right)}$	$\dfrac{\tfrac{1}{2}\alpha}{\left(1+\frac{\alpha}{2}\right)}$
Partial pressures	$\dfrac{(1-\alpha)p}{\left(1+\frac{\alpha}{2}\right)}$	$\dfrac{\alpha p}{\left(1+\frac{\alpha}{2}\right)}$	$\dfrac{\alpha p}{2\left(1+\frac{\alpha}{2}\right)}$

$$K = \left(\prod_J a_J{}^{\nu_J}\right)_{equilibrium} \quad [4.18] = \frac{\left(\frac{p_{CO}}{p^{\ominus}}\right) \times \left(\frac{p_{O_2}}{p^{\ominus}}\right)^{1/2}}{\left(\frac{p_{CO_2}}{p^{\ominus}}\right)} = \frac{\left(\frac{\alpha}{1+(\alpha/2)}\right) \times \left(\frac{\alpha/2}{1+(\alpha/2)}\right)^{1/2} \times \left(\frac{p}{p^{\ominus}}\right)^{1/2}}{\left(\frac{1-\alpha}{1+(\alpha/2)}\right)}$$

$$K \approx \frac{\alpha^{3/2}}{\sqrt{2}} \quad [\alpha \ll 1 \text{ at all the specified temperatures}]$$

$$\Delta_r G^{\ominus} = -RT \ln K \quad [9.8]$$

The calculated values of K and $\Delta_r G$ are given in the table below. From any two pairs of K and T, $\Delta_r H^{\ominus}$ may be calculated.

$$\ln K_2 = \ln K_1 - \frac{\Delta_r H^{\ominus}}{R}\left(\frac{1}{T_2} - \frac{1}{T_1}\right) \quad [9.28]$$

Solving for $\Delta_r H^{\ominus}$

$$\Delta_r H^{\ominus} = \frac{R \ln\left(\frac{K_2}{K_1}\right)}{\left(\frac{1}{T_1} - \frac{1}{T_2}\right)} \text{[Exercise 9.18]} = \frac{(8.314 \,\text{J K}^{-1}\,\text{mol}^{-1}) \times \ln\left(\frac{7.23\times10^{-6}}{1.22\times10^{-6}}\right)}{\left(\frac{1}{1395 \,\text{K}} - \frac{1}{1498 \,\text{K}}\right)}$$

$$= \boxed{3.00 \times 10^5 \,\text{J mol}^{-1}}$$

$$\Delta_r S^{\ominus} = \frac{\Delta_r H^{\ominus} - \Delta_r G^{\ominus}}{T}$$

The calculated values of $\Delta_r S^{\ominus}$ are also given in the table.

T/K	1395	1443	1498
$\alpha/10^{-4}$	1.44	2.50	4.71
$K/10^{-6}$	1.22	2.80	7.23
$\Delta_r G^{\ominus}/(kJ\,mol^{-1})$	158	153	147
$\Delta_r S^{\ominus}/(J\,K^{-1}\,mol^{-1})$	102	102	102

Comment. $\Delta_r S^{\ominus}$ is essentially constant over this temperature range but it is much different from its value at 25°C. $\Delta_r H^{\ominus}$, however, is only slightly different.

Question. What are the values of $\Delta_r H^{\ominus}$ and $\Delta_r S^{\ominus}$ at 25°C for this reaction?

P9.6 $\Delta_r G^{\ominus}(H_2CO, g) = \Delta_r G^{\ominus}(H_2CO, l) + \Delta_{vap} G^{\ominus}(H_2CO, l)$

For $H_2CO(l) \rightleftharpoons H_2CO(g)$, $K(vap) = \dfrac{p}{p^{\ominus}}$

$$\Delta_{vap} G^{\ominus} = -RT \ln K(vap) = -RT \ln \frac{p}{p^{\ominus}}$$

$$= -(8.314\,J\,K^{-1}\,mol^{-1}) \times (298\,K) \times \ln\left(\frac{1500\,Torr}{750\,Torr}\right) = -1.72\,kJ\,mol^{-1}$$

Therefore, for the reaction

$$CO(g) + H_2(g) \rightleftharpoons H_2CO(g),$$

$$\Delta_r G^{\ominus} = (+28.95) + (-1.72)\,kJ\,mol^{-1} = +27.23\,kJ\,mol^{-1}$$

Hence, $K = e^{(-27.23 \times 10^3\,J\,mol^{-1})/(8.314\,J\,K^{-1}\,mol^{-1}) \times (298\,K)} = e^{-10.99} = \boxed{1.69 \times 10^{-5}}$

P9.8   Draw up the following table using $H_2(g) + I_2 \rightleftharpoons 2HI(g)$

	H_2	I_2	HI	Total
Initial amounts/mol	0.300	0.400	0.200	0.900
Change/mol	$-x$	$-x$	$+2x$	
Equilibrium amounts/mol	$0.300 - x$	$0.400 - x$	$0.200 + 2x$	0.900
Mole fraction	$\dfrac{0.300 - x}{0.900}$	$\dfrac{0.400 - x}{0.900}$	$\dfrac{0.200 + 2x}{0.900}$	1

$$K = \frac{\left(\frac{p(HI)}{p^{\ominus}}\right)^2}{\left(\frac{p(H_2)}{p^{\ominus}}\right)\left(\frac{p(I_2)}{p^{\ominus}}\right)} = \frac{x(HI)^2}{x(H_2)x(I_2)}[p(J) = x_J p] = \frac{(0.200 + 2x)^2}{(0.300 - x)(0.400 - x)} = 870[given]$$

Therefore,

$$(0.0400) + (0.800x) + 4x^2 = (870) \times (0.120 - 0.700x + x^2) \quad or$$

$$866x^2 - 609.80x + 104.36 = 0$$

which solves to $x = 0.293[x = 0.411$ is exluded because x cannot exceed 0.300]. The final composition is therefore $\boxed{0.007\,mol\,H_2}$, $\boxed{0.107\,mol\,I_2}$, and $\boxed{0.786\,mol\,HI}$.

P9.11   Let a be the mole fraction of perylene, b that of benzo(e)pyrene, and c that of benzo(a)pyrene. The mole fractions add up to unity

$$a + b + c = 1$$

And the mole ratios of these isomers are equal to equilibrium constants

perylene $\rightleftharpoons$ benzo(e)pyrene $b/a = K_1$ (1)

benzo(e)pyrene $\rightleftharpoons$ benzo(a)pyrene $c/b = K_2$ (2)

(The third such equilibrium would not yield any additional information.) So we have three equations in three unknowns

$$b = aK_1 = c/K_2 \quad \text{so} \quad c = aK_1 K_2,$$

and $a + aK_1 + aK_1 K_2 = 1 = a(1 + K_1 + K_1 K_2) \quad \text{so} \quad a = (1 + K_1 + K_1 K_2)^{-1}$

We need the equilibrium constants, which are given by

$$K = \exp\left(\frac{-\Delta_r G^{\ominus}}{RT}\right) = \exp\left(\frac{-\Delta_r H^{\ominus}}{RT}\right) \exp\left(\frac{\Delta_r S^{\ominus}}{R}\right)$$

For reaction (1)

$$\Delta_r H^{\ominus} = (253.2 - 253.2)\,\text{kJ mol}^{-1} = 0.0$$

and $\Delta_r S^{\ominus} = (993.7 - 987.9)\,\text{J mol}^{-1}\,\text{K}^{-1} = 5.8\,\text{J mol}^{-1}\,\text{K}^{-1}$

so $K = \exp\left(\dfrac{5.8\,\text{J mol}^{-1}\,\text{K}^{-1}}{8.3145\,\text{J mol}^{-1}\,\text{K}^{-1}}\right) = 2.0$

For reaction (2)

$$\Delta_r H^{\ominus} = (262.4 - 253.2)\,\text{kJ mol}^{-1} = 9.2\,\text{kJ mol}^{-1}$$

and $\Delta_r S^{\ominus} = (999.4 - 993.7)\,\text{J mol}^{-1}\,\text{K}^{-1} = 5.7\,\text{J mol}^{-1}\,\text{K}^{-1}$

so $K = \exp\left(\dfrac{-9.2\,\text{kJ mol}^{-1}}{(8.3145\,\text{J mol}^{-1}\,\text{K}^{-1})(1000\,\text{K})}\right) \exp\left(\dfrac{5.7\,\text{J mol}^{-1}\,\text{K}^{-1}}{8.3145\,\text{J mol}^{-1}\,\text{K}^{-1}}\right) = 0.66$

Now we can go back and evaluate the terms

$$a = (1 + K_1 + K_1 K_2)^{-1} = (1 + 2.0 + 2.0 \times 0.66)^{-1} = \boxed{0.23}$$

$$b = aK_1 = (0.23) \times (2.0) = \boxed{0.46}$$

and $c = aK_1 K_2 = (0.23) \times (2.0) \times (0.66) = \boxed{0.30}$

P9.12   If we knew $\Delta_r H^{\ominus}$ for this reaction, we could calculate $\Delta_f H^{\ominus}(\text{HClO})$ from

$$\Delta_r H^{\ominus} = 2\Delta_f H^{\ominus}(\text{HClO}) - \Delta_f H^{\ominus}(\text{Cl}_2\text{O}) - \Delta_f H^{\ominus}(\text{H}_2\text{O})$$

We can find $\Delta_r H^{\ominus}$ if we know $\Delta_r G^{\ominus}$ and $\Delta_r S^{\ominus}$, since

$$\Delta_r G^{\ominus} = \Delta_r H^{\ominus} - T\Delta_r S^{\ominus}$$

And we can find $\Delta_r G^{\ominus}$ from the equilibrium constant.

$$K = \exp(-\Delta_r G^{\ominus}/RT) \quad \text{so} \quad \Delta_r G^{\ominus} = -RT \ln K,$$

$$\Delta_r G^\ominus = -(8.3145 \times 10^{-3} \text{ kJ K}^{-1} \text{ mol}^{-1}) \times (298 \text{ K}) \ln 8.2 \times 10^{-2}$$

$$= 6.2 \text{ kJ mol}^{-1}$$

$$\Delta_r H^\ominus = \Delta_r G^\ominus + T \Delta_r S^\ominus$$

$$= 6.2 \text{ kJ mol}^{-1} + (298 \text{ K}) \times (16.38 \times 10^{-3} \text{ kJ K}^{-1} \text{ mol}^{-1}),$$

$$\Delta_r H^\ominus = 11.1 \text{ kJ mol}^{-1}$$

Finally

$$\Delta_f H^\ominus (\text{HClO}) = 1/2[\Delta_r H^\ominus + \Delta_f H^\ominus (\text{Cl}_2\text{O}) + \Delta_f H^\ominus (\text{H}_2\text{O})],$$

$$\Delta_f H^\ominus (\text{HClO}) = 1/2[11.1 + 77.2 + (-241.82)] \text{ kJ mol}^{-1}$$

$$= \boxed{76.8 \text{ kJ mol}^{-1}}$$

Solutions to theoretical problems

P9.14 $K = K_\phi K_p$, but $\left(\dfrac{\partial K}{\partial p}\right)_T = 0$ [9.24]

Therefore, $\left(\dfrac{\partial K}{\partial p}\right)_T = K_\phi \left(\dfrac{\partial K_p}{\partial p}\right)_T + K_p \left(\dfrac{\partial K_\phi}{\partial p}\right)_T = 0$

which implies that $\left(\dfrac{\partial K_\phi}{\partial p}\right)_T = -\left(\dfrac{\partial K_p}{\partial p}\right)_T \left(\dfrac{K_\phi}{K_p}\right)$

and therefore if K_p increases with pressure, K_ϕ must decrease (because K_ϕ/K_p is positive).

P9.15 We draw up the following table using the stoichiometry $A + 3B \rightarrow 2C$ and $\Delta_{n_J} = \nu_J \xi$

	A	B	C	Total
Initial amount /mol	1	3	0	4
Change, Δn_J/mol	$-\xi$	-3ξ	$+2\xi$	
Equilibrium amount /mol	$1-\xi$	$3(1-\xi)$	2ξ	$2(2-\xi)$
Mole fraction	$\dfrac{1-\xi}{2(2-\xi)}$	$\dfrac{3(1-\xi)}{2(2-\xi)}$	$\dfrac{\xi}{2-\xi}$	1

$$K = \frac{\left(\frac{p_C}{p^\ominus}\right)^2}{\left(\frac{p_A}{p^\ominus}\right)\left(\frac{p_B}{p^\ominus}\right)^3} = \frac{x_C^2}{x_A x_B^3} \times \left(\frac{p^\ominus}{p}\right)^2 = \frac{\xi^2}{(2-\xi)^2} \times \frac{2(2-\xi)}{1-\xi} \times \frac{2^3(2-\xi)^3}{3^3(1-\xi)^3} \times \left(\frac{p^\ominus}{p}\right)^2$$

$$= \frac{16(2-\xi)^2 \xi^2}{27(1-\xi)^4} \times \left(\frac{p^\ominus}{p}\right)^2$$

Since K is independent of the pressure

$$\frac{(2-\xi)^2 \xi^2}{(1-\xi)^4} = a^2 \left(\frac{p}{p^\ominus}\right)^2 \qquad a^2 = \frac{27}{16} K, \text{ a constant}$$

Therefore $(2-\xi)\xi = a\left(\dfrac{p}{p^\ominus}\right) \times (1-\xi)^2$

$$\left(1 + \frac{ap}{p^{\ominus}}\right)\xi^2 - 2\left(1 + \frac{ap}{p^{\ominus}}\right)\xi + \frac{ap}{p^{\ominus}} = 0$$

which solves to $\boxed{\xi = 1 - \left(\dfrac{1}{1 + ap/p^{\ominus}}\right)^{1/2}}$

We choose the root with the negative sign because ξ lies between 0 and 1. The variation of ξ with p is shown in Fig. 9.2.

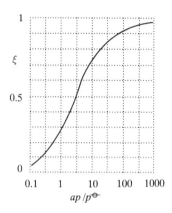

Figure 9.2

P9.17

$$\Delta_r G = \Delta_r H - T\Delta_r S$$

$$\Delta_r H' = \Delta_r H + \int_T^{T'} \Delta_r C_p \, dT \quad [2.44]$$

$$\Delta_r S' = \Delta_r S + \int_T^{T'} \frac{\Delta_r C_p}{T} \, dT \quad [4.19, \text{ with } \Delta_r S \text{ in place of } S]$$

$$\Delta_r G' = \Delta_r G + \int_T^{T'} \Delta_r C_p \, dT + (T - T')\Delta_r S - T'\int_T^{T'} \frac{\Delta_r C_p}{T} \, dT$$

$$= \Delta_r G + (T - T')\Delta_r S + \int_T^{T'} \left(1 - \frac{T'}{T}\right)\Delta_r C_p \, dT$$

$$\Delta_r C_p = \Delta a + T\Delta b + \frac{\Delta c}{T^2}$$

$$\left(1 - \frac{T'}{T}\right)\Delta_r C_p = \Delta a + T\Delta b + \frac{\Delta c}{T^2} - \frac{T'\Delta a}{T} - T'\Delta b - \frac{T'\Delta c}{T^3}$$

$$= \Delta a - T'\Delta b + T\Delta b - \frac{T'\Delta a}{T} + \frac{\Delta c}{T^2} - \frac{T'\Delta c}{T^3}$$

$$\int_T^{T'} \left(1 - \frac{T'}{T}\right)\Delta_r C_p \, dT = (\Delta a - T'\Delta b)(T' - T) + \frac{1}{2}(T'^2 - T^2)\Delta b - T'\Delta a \ln\frac{T'}{T}$$

$$+ \Delta c\left(\frac{1}{T} - \frac{1}{T'}\right) - \frac{1}{2}T'\Delta c\left(\frac{1}{T^2} - \frac{1}{T'^2}\right)$$

Therefore, $\boxed{\Delta_r G' = \Delta_r G + (T - T')\Delta_r S + \alpha\Delta a + \beta\Delta b + \gamma\Delta c}$

where $\alpha = T' - T - T' \ln \dfrac{T'}{T}$

$$\beta = \frac{1}{2}(T'^2 - T^2) - T'(T' - T)$$

$$\gamma = \frac{1}{T} - \frac{1}{T'} + \frac{1}{2}T'\left(\frac{1}{T'^2} - \frac{1}{T^2}\right)$$

For water,

$$H_2(g) + \tfrac{1}{2}O_2(g) \rightarrow H_2O(l) \quad \Delta_f G^{\ominus}(T) = -237.13\,\text{kJ mol}^{-1}$$

$$\Delta_r S^{\ominus}(T) = -163.34\,\text{J K}^{-1}\,\text{mol}^{-1}$$

$$\Delta a = a(H_2O) - a(H_2) - \tfrac{1}{2}a(O_2) = (75.29 - 27.88 - 14.98)\,\text{J K}^{-1}\,\text{mol}^{-1}$$

$$= +33.03\,\text{J K}^{-1}\,\text{mol}^{-1}$$

$$\Delta b = [(0) - (3.26 \times 10^{-3}) - (2.09 \times 10^{-3})]\,\text{J K}^{-2}\,\text{mol}^{-1} = -5.35 \times 10^{-3}\,\text{J K}^{-2}\,\text{mol}^{-1}$$

$$\Delta c = [(0) - (0.50 \times 10^5) + (0.83 \times 10^5)]\,\text{J K mol}^{-1} = +0.33 \times 10^5\,\text{J K mol}^{-1}$$

$$T = 298\,\text{K}, \qquad T' = 372\,\text{K, so}$$

$$\alpha = -8.5\,\text{K}, \qquad \beta = -2738\,\text{K}^2, \qquad \gamma = -8.288 \times 10^{-5}\,\text{K}^{-1}$$

and so

$$\begin{aligned}
\Delta_f G^{\ominus}(372\,\text{K}) &= (-237.13\,\text{kJ mol}^{-1}) + (-74\,\text{K}) \times (-163.34\,\text{J K}^{-1}\,\text{mol}^{-1}) \\
&\quad + (-8.5\,\text{K}) \times (33.03 \times 10^{-3}\,\text{kJ K}^{-1}\,\text{mol}^{-1}) \\
&\quad + (-2738\,\text{K}^2) \times (-5.35 \times 10^{-6}\,\text{kJ K}^{-2}\,\text{mol}^{-1}) \\
&\quad + (-8.288 \times 10^{-5}\,\text{K}^{-1}) \times (0.33 \times 10^2\,\text{kJ K mol}^{-1}) \\
&= [(-237.13) + (12.09) - (0.28) + (0.015) - (0.003)]\,\text{kJ mol}^{-1} \\
&= \boxed{-225.31\,\text{kJ mol}^{-1}}
\end{aligned}$$

Note that the β and γ terms are not significant (for this reaction and temperature range).

Solutions to applications

P9.19 The equilibrium to be considered is (A = gas)

$$A(g, 1\,\text{bar}) \rightleftharpoons A(\text{sol'n}) \quad K = \frac{(c/c^{\ominus})}{(p/p^{\ominus})} = \frac{s}{s^{\ominus}}$$

$$\Delta_r H^{\ominus} = -R \times \frac{d \ln K}{d\left(\frac{1}{T}\right)} \quad [9.26]$$

$$\ln K = \ln\left(\frac{s}{s^{\ominus}}\right) = 2.303 \log\left(\frac{s}{s^{\ominus}}\right)$$

$$\Delta_r H^{\ominus}(H_2) = -(2.303) \times (R) \times \frac{d}{d\left(\frac{1}{T}\right)}\left(-5.39 - \frac{768\,\text{K}}{T}\right)$$

$$= 2.303R \times 768\,\text{K} = \boxed{+14.7\,\text{kJ mol}^{-1}}$$

$$\Delta_r H^{\ominus}(CO) = -(2.303) \times (R) \times \frac{d}{d(\frac{1}{T})}\left(-5.98 - \frac{980\,K}{T}\right)$$

$$= 2.303 R \times 980\,K = \boxed{+18.8\,kJ\,mol^{-1}}$$

P9.21 The hydrolysis of M_gATP^{2-} is expected to be $\boxed{less}$ exergonic than the hydrolysis of ATP^{4-}. Electrostatic repulsions which tend to break the bonds between phosphate groups should be weaker in the Mg^{2+} complex than in ATP^{4-}, which has no charge compensation. Experimental data [R. A. Alberty and R. N. Goldberg, *Biochemistry 31*, 10610 (1992), and R. A. Alberty, *Biochem. Biophys. Acta 1207*, 1 (1994)] seem to support this conclusion, but the magnitude of the difference is highly dependent upon the concentration of Mg^{2+} and the pH. The relative complexing abilities of ATP^{4-} and ADP^{3-} with Mg^{2+} must also be taken into consideration.

P9.23 According to Henry's law

$$p_{CO_2} = x_{CO_2} K_H,$$

so $x_{CO_2} = \dfrac{p_{CO_2}}{K_H} = \dfrac{(3.6 \times 10^{-4}\,atm) \times (760\,Torr\,atm)}{1.25 \times 10^6\,Torr} = 2.2 \times 10^{-7}$

This mole fraction of aqueous CO_2 in natural rainwater corresponds to the following molality

$$b_{CO_2} = \frac{n_{CO_2}}{1\,kg\,H_2O} \approx \frac{n_{H_2O}x_{CO_2}}{1\,kg\,H_2O} = \frac{x_{CO_2}}{M_{H_2O}} = \frac{2.2 \times 10^{-7}}{18.015 \times 10^{-3}\,kg\,mol^{-1}}$$

$$= 1.2 \times 10^{-5}\,mol\,kg^{-1}$$

Assuming that aqueous CO_2 forms H_2CO_3 quantitatively and that H_2CO_3 is an ideally dilute solute, the reaction which determines the pH is

$$H_2CO_3 \rightleftharpoons H^+ + HCO_3^- \quad \text{with} \quad K = \frac{a_{H^+}a_{HCO_3^-}}{b_{H_2CO_3}} \approx \frac{a_{H^+}^2}{b_{H_2CO_3}}$$

The last approximation assumes that any other sources of H^+ or HCO_3^- are negligible. Assuming further that ionization is so small that $b_{H_2CO_3}$ is virtually unchanged by it, we have

$$a_{H^+} = \sqrt{K b_{H_2CO_3}} = \sqrt{(4.3 \times 10^{-7}) \times (1.2 \times 10^{-5})} = 2.3 \times 10^{-6}$$

So $pH = -\log a_{H^+} = \boxed{5.6\overline{4}}$

The pre-industrial atmosphere had $x_{CO_2} = 1.7 \times 10^{-7}$, $b_{CO_2} = 9.4 \times 10^{-6}\,mol\,kg^{-1}$, $a_{H^+} = 2.0 \times 10^{-6}$, and $pH = \boxed{5.7\overline{0}}$

P9.25 (a) Begin by writing equations for the equilibrium; use $\Delta_f G^{\ominus}$ tables to compute equilibrium constants

$$CO_2(aq) + H_2O(l) \rightleftharpoons H_2CO_3(aq) \tag{1}$$

$$\Delta_1 G^{\ominus} = 2.830\,kJ\,mol^{-1}; \qquad K_1 = 0.3193; \qquad \alpha_{H_2CO_3} = K_1\alpha_{CO_2}$$

$$H_2CO_3(aq) \rightleftharpoons H^+(aq) + HCO_3^-(aq) \tag{2}$$

$$\Delta_2 G^{\ominus} = 36.44\,kJ\,mol^{-1}; \qquad K_2 = 4.131 \times 10^{-7}; \qquad \alpha_{HCO_3^-} = K_2\alpha_{H_2CO_3}/[H^+]$$

$$HCO_3^-(aq) \rightleftharpoons H^+(aq) + CO_3^{2-}(aq) \tag{3}$$

$$\Delta_3 G^{\ominus} = 58.95\,\text{kJ mol}^{-1}; \quad K_3 = 4.704 \times 10^{-11}; \quad \alpha_{CO_3^{2-}} = K_3 \alpha_{HCO_3^-}/[H^+]$$

Conservation of carbon mass requires that

$$\alpha_{CO_2} + \alpha_{H_2CO_3} + \alpha_{HCO_3^-} + \alpha_{CO_3^{2-}} = 1$$

Substitution of the above α equations and solving for α_{CO_2} yields

$$\alpha_{CO_2} = (1 + K_1 + K_1 K_2 [H^+]^{-1} + K_1 K_2 K_3 [H^+]^{-2})^{-1}$$

The above equation is used to calculate α_{CO_2} at any pH$([H^+] = 10^{-pH}\,\text{mol L}^{-1})$. The α equations listed under reactions (1)–(3) can be used sequentially to calculate α_{HCO_3}, $\alpha_{HCO_3^-}$, and $\alpha_{CO_3^-}$ at the selected pH. At pH $= 7.4$, as seen in Fig. 9.3, $\alpha_{CO_2} = 0.216$, $\alpha_{H_2CO_3} = 0.069$, $\alpha_{HCO_3^-} = 0.715$, and $\alpha_{CO_3^{2-}} = 0.00084$. $\boxed{HCO_3^-, \; CO_2, \; \text{and } H_2CO_3}$ are the major species.

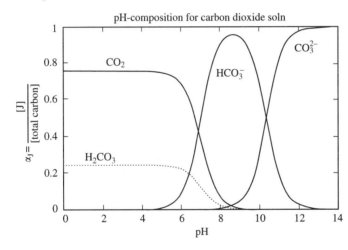

pH-composition for carbon dioxide soln

Figure 9.3

(b) The mole fraction of CO_2 in the oceans is estimated with the equation $x_{CO_2} = \dfrac{n_{CO_2}}{n_{H_2O}}$. Use T_c to represent the total number of moles of carbon containing species ($T_c = [CO_2] + [H_2CO_3] + [HCO_3^-] + [CO_3^{2-}]$, which approximately equals $[HCO_3^-]$ at pH $= 8$), we may write

$$x_{CO_2} = \frac{n_{CO_2}}{n_{H_2O}} = \frac{\alpha_{CO_2} T_c}{[H_2O] V_{oceans}} \quad \text{or} \quad T_c = \frac{x_{CO_2} [H_2O] V_{oceans}}{\alpha_{CO_2}}$$

where $[H_2O] = (1\,\text{g/cm}^3)(\text{mol}/18\text{g}) = 5.556 \times 10^4\,\text{mol/m}^3$ and $\alpha_{CO_2}(\text{pH} = 8) = 0.06863$. The mole fraction x_{CO_2} is evaluated with Henry's law (Table 7.1).

$$x_{CO_2} = \frac{P_{CO_2}}{K_{CO_2}} = \frac{(3.3 \times 10^{-4}\,\cancel{\text{atm}})}{1.25 \times 10^6\,\cancel{\text{Torr}}} \left(\frac{760\,\cancel{\text{Torr}}}{1\,\cancel{\text{atm}}}\right) = 2.006 \times 10^{-7}$$

Consequently,

$$T_c = \frac{(2.006 \times 10^{-7})(5.556 \times 10^4\,\text{mol m}^{-3})(1.37 \times 10^{18}\,\text{m}^3)}{0.06863}$$

$$= 2.225 \times 10^{17}\,\text{mol carbon}$$

This value of T_c is the maximum amount of carbon dissolved in the oceans under equilibrium conditions as it neglects the precipitation of carbonates in the presence of calcium cation – a very considerable sink. The model also assumes equilibrium of the ocean at all depths with atmospheric gases. The maximum mass of dissolved carbon is given by

$$T_c M_c = (2.225 \times 10^{17} \ \cancel{mol}) \left(\frac{0.012 \ \text{kg}}{\cancel{mol}} \right)$$

$$= 2.67 \times 10^{15} \ \text{kg} = \boxed{2.67 \times 10^{12} \ \text{tonne}}$$

P9.27 $\frac{1}{2} N_2 C_g + 3/2 \, H_2(g) \rightarrow NH_3(g)$ $\Delta \nu = -1/2$

First, calculate the standard reaction thermodynamic functions with formation thermodynamic properties found in the appendix (Table 2.6).

$$\Delta_r H^{\ominus}(298) = -46.11 \, \text{kJ} \quad \text{and} \quad \Delta_r S^{\ominus}(298) = -99.38 \, \text{JK}^{-1}$$

Use appendix information to define functions for the constant pressure heat capacity of reactants and products (Table 2.2). Define a function $\Delta_r C_p^{\ominus}(T)$ that makes it possible to calculate $\Delta_r C_p$ at 1 bar and any temperature (eqn 2.45). Define functions that make it possible to calculate the reaction enthalpy and entropy at 1 bar and any temperature (eqns 2.44 and 4.19).

$$\Delta_r H^{\ominus}(T) = \Delta_r H^{\ominus}(298) + \int_{298.15 \, \text{K}}^{T} \Delta_r C_p^{\ominus}(T) \, dT$$

$$\Delta_r S^{\ominus}(T) = \Delta_r S^{\ominus}(298) + \int_{298.15 \, \text{K}}^{T} \frac{\Delta_r C_p^{\ominus}(T)}{T} \, dT$$

For a prefect gas reaction mixture $\Delta_r H$ is independent of pressure at constant temperature. Consequently, $\Delta_r H(T, p) = \Delta_r H^{\ominus}(T)$. The pressure dependence of the reaction entropy may be evaluated with the expression:

$$\Delta_r S(T_p) = \Delta_r S^{\ominus}(T) + \sum_{\text{Products} - \text{Reactants}} \nu \int_{1 \, \text{bar}}^{p} \left(\frac{\partial S_m}{\partial P} \right)_T dP$$

$$= \Delta_r S^{\ominus}(T) - \sum_{\text{Products} - \text{Reactants}} \nu \int_{1 \, \text{bar}}^{p} \left(\frac{\partial V_m}{\partial T} \right)_p dp \quad \text{(Table 5.1)}$$

$$= \Delta_r S^{\ominus}(T) - \sum_{\text{Products} - \text{Reactants}} \nu \int_{1 \, \text{bar}}^{P} \frac{R}{p} \, dp$$

$$= \Delta_r S^{\ominus}(T) - \left[\sum_{\text{Products} - \text{Reactants}} \nu \right] R \ln \left(\frac{p}{1 \, \text{bar}} \right)$$

$$= \Delta_r S^{\ominus}(T) - 1/2 \, R \ln \left(\frac{p}{1 \, \text{bar}} \right)$$

The above two eqns make it possible to calculate $\Delta_r G(T, p)$.

$$\Delta_r G(T, p) = \Delta_r H(T, p) - T \, \Delta_r S(T, p)$$

Once the above functions have been defined on a scientific calculator or with mathematical software on a computer, the root function may be used to evaluate pressure where $\Delta_r G(T, p) = -500 \, \text{J}$ at a given temperature.

(i) (a) and (b) perfect gas mixture:

For $T = (450 + 273.15)$ K $= 723.15$ K, root$(\Delta_r G(723.15\,\text{K}, p) + 500\,\text{J}) = \boxed{156.5\ \text{bar}}$

For $T = (400 + 273.15)$ K $= 673.15$ K, root$(\Delta_r G(673.15\,\text{K}, p) + 500\,\text{J}) = \boxed{81.8\ \text{bar}}$

(ii) For a van der Waals gas mixture $\Delta_r H$ does depend upon pressure. The calculational equation is:

$$\Delta_r H(T, P) = \Delta_r H^{\ominus}(T) + \sum_{\text{Products}-\text{Reactants}} v \int_{1\,\text{bar}}^{p} \left(\frac{\partial H_m}{\partial p}\right)_T dp$$

$$= \Delta_r H^{\ominus}(T) + \sum_{\text{Products}-\text{Reactants}} v \int_{1\,\text{bar}}^{p} \left[V_m - T\left(\frac{\partial V_m}{\partial T}\right)_p\right] dp$$

(Theoretical Problem 5.10)

where $(\partial V_m/\partial T)_p = R(V_m - b)^{-1}(RT(V_m - b)^{-2} - 2aV_m^{-3})^{-1}$

and $V_m(T, p) = \text{root}\left(P - \dfrac{RT}{V_m - b} + \dfrac{a}{V_{m^2}}\right).$

The functional equation for $\Delta_r S$ calculations is:

$$\Delta_r S(T, P) = \Delta_r S^{\ominus}(T) - \sum_{\text{Products}-\text{Reactants}} v \int_{1\,\text{bar}}^{p} \left(\frac{\partial V_m}{\partial T}\right)_p dp$$

where $(\partial V_m/\partial T)_p$ and $V_m(T, p)$ are calculated as described above. As usual, $\Delta_r G(T, p) = \Delta_r H(T, p) - T\Delta_r S(T, p)$.

(a) and (b) van der Waals gas mixture:

For $T = 723.15$K, root$(\Delta_r G(723.15\,\text{K}, p) + 500\,\text{J}) = \boxed{132.5\ \text{bar}}$

For $T = 673.15$K, root$(\Delta_r G(673.15\,\text{K}, p) + 500\,\text{J}) = \boxed{73.7\ \text{bar}}$

(c) $\Delta_r G(T, p)$ isotherms $\boxed{\text{confirm}}$ Le Chatelier's principle. Along an isotherm, $\Delta_r G$ decreases as pressure increases. This corresponds to a shift to the right in the reaction equation and reduces the stress by shifting to the side that has fewer total moles of gas. Additionally the reaction is exothermic, so Chatelier's principle predicts a shift to the left with an increase in temperature. The isotherms confirm this as an increase in $\Delta_r G$ as temperature is increased at constant pressure. See Figure 9.4.

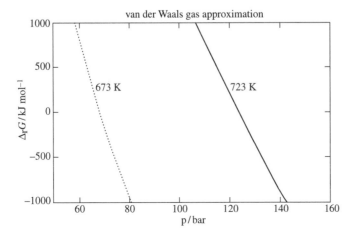

Figure 9.4

10 Equilibrium electrochemistry

Solutions to exercises

Discussion questions

E10.1(a) Let us only consider ions in aqueous solution. Values for the Gibbs energy of formation of ions in water are assigned relative to $H^+(aq)$, which implies $\Delta_f G^\ominus (H^+, aq) = 0$. Therefore, let us consider the Gibbs energy of formation of ions X^{z-} which can be obtained by dissociation of compounds H_z-X. Then values for other ions M^{z+} can be obtained from compounds $M_{z-}X_{z+}$. An important point to note is that $\Delta_f G^\ominus$ of ion X is not determined by the properties of X alone, but includes contributions from the dissociation, ionization, and hydration of hydrogen. Values are obtained by constructing a suitable cycle in which all the quantities in the cycle are known experimentally, or can be calculated theoretically. See Fig. 10.1 of the text. For example

$$\Delta_f G^\ominus (Cl^-, aq) = 1/2\Delta H(H\!-\!H) + I(H) + 1/2\Delta H(Cl\!-\!Cl) + E_{ea}(Cl)$$
$$+ \Delta_{hyd} G^\ominus (H)^+ + \Delta_{hyd} G^\ominus (Cl^-)$$

where the terms on the right side are, respectively, half the bond dissociation enthalpy of H_2, the ionization energy of atomic H, half the bond dissociation enthalpy of Cl_2, the electron affinity of Cl, and Gibbs energies of hydration of H^+ and Cl^-. These latter are the Gibbs energies of the processes leading to the formation of $Cl^-(aq)$ and $H^+(aq)$. Note that we have used the enthalpies rather than the Gibbs energies for bond dissociation, ionization, and electron affinity. See footnote 1 on p. 254 of the text for a justification of these substitutions.

E10.2(a) A galvanic cell uses a spontaneous chemical reaction to generate a potential difference and deliver an electric current to an external device. An electrolytic cell uses an external potential difference to drive a chemical reaction in the cell that is by itself non-spontaneous. In their essential features, these two kinds of cells can be considered opposites of each other, in the sense that an electrolytic cell can be thought of as a galvanic cell operating in the reverse direction. For some electrochemical cells, this is easy to accomplish. We say they are rechargeable. The most common example is the lead-acid battery used in automobiles. For many other cells, however, this kind of reversibility cannot be achieved. In Chapter 10, we study the potentials of cells operating under zero current or equilibrium, hence reversible, conditions. Working electrochemical cells are taken up in Chapter 29.

E10.3(a) The starting point is the Nernst equation for the cell (eqn 10.34),

$$E = E^\ominus - \frac{RT}{\nu F} \ln Q.$$

Measurement of E, with knowledge of the reaction quotient, Q, would seem to provide a straightforward determination of $E^\ominus$. But a problem arises because the calculation of Q requires not only knowledge of the concentrations of the species involved in the cell reaction but also of their activity coefficients. These coefficients are not usually available, so the calculation cannot be directly completed. However, at very low concentrations, the Debye–Huckel limiting law for the coefficients holds. The procedure then is to substitute the Debye–Huckel law for the activity coefficients into the specific form of the Nernst equation for the cell under investigation and then to take measurements of E as a function of concentration. From an extrapolation of the data to zero concentration where the law holds exactly, the standard potential of the cell can be obtained. See Section 10.5(b) for the details of this procedure and examples.

Numerical exercises

E10.4(a) $CuSO_4(aq)$ and $ZnSO_4(aq)$ are strong electrolytes; therefore the net ionic equation is

$$Zn(s) + Cu^{2+}(aq) \rightarrow Zn^{2+}(aq) + Cu(s)$$

$$\Delta_r H^{\ominus} = \Delta_f H^{\ominus}(Zn^{2+}, aq) - \Delta_f H^{\ominus}(Cu^{2+}, aq)$$

$$= (-153.89 \, kJ \, mol^{-1}) - (64.77 \, kJ \, mol^{-1}) = \boxed{-218.66 \, kJ \, mol^{-1}}$$

Comment. $SO_4^{2-}(aq)$ is a spectator ion and was ignored in the determination of $\Delta_r H^{\ominus}$ above. This is justifiable because $\Delta_r H^{\ominus}$ refers to the standard state of all species participating in the reaction.

E10.5(a) $HgCl_2(s) \rightleftharpoons Hg^{2+}(aq) + 2Cl^-(aq)$

$$K = \prod_J a_J^{\nu_J}$$

Since the solubility is expected to be low, we may (initially) ignore activity coefficients. Hence

$$K = \frac{b(Hg^{2+})}{b^{\ominus}} \times \frac{b^2(Cl^-)}{(b^{\ominus})^2} \qquad b(Cl^-) = 2b(Hg^{2+}) = 2s$$

$$K = \frac{s(2s)^2}{(b^{\ominus})^3} = \frac{4s^3}{(b^{\ominus})^3} \qquad s = b(Hg^{2+}) = \left(\frac{1}{4}K\right)^{1/3} b^{\ominus}$$

K may be determined from

$$\ln K = \boxed{\frac{-\Delta_r G^{\ominus}}{RT}} \quad \text{[Chapter 9]}$$

$$\Delta_r G^{\ominus} = \Delta_r G^{\ominus}(Hg^{2+}, aq) + 2\Delta_f G^{\ominus}(Cl^-, aq) - \Delta_r G^{\ominus}(HgCl_2, s)$$

$$= [(+164.40) + (2) \times (-131.23) - (-178.6)] \, kJ \, mol^{-1} = +80.54 \, kJ \, mol^{-1}$$

$$\ln K = \frac{-80.54 \times 10^3 \, J \, mol^{-1}}{(8.314 \, J \, K^{-1} \, mol^{-1}) \times (298.15 \, K)} = -32.49$$

Hence $K = 7.75 \times 10^{-15}$ and $s = \boxed{1.25 \times 10^{-5} \, mol \, L^{-1}}$

E10.6(a) A procedure similar to that outlined in Section 10.1 and Fig. 10.1 of the text is followed.

		$\Delta G^{\ominus}/(kJ \, mol^{-1})$		
		Cl^-	F^-	
Dissociation of H_2	$\frac{1}{2}H_2 \rightarrow H$	$+218$	$+218$	(Table 14.3)
Ionization of H	$H \rightarrow H^+ + e^-$	$+1312$	$+1312$	(Table 13.4)
Hydration of H^+	$H^+(g) \rightarrow H^+(aq)$	x	x	
Dissociation of X_2	$\frac{1}{2}X_2 \rightarrow X$	$+121$	78	(Table 14.3)
Electron gain by X	$X + e^- \rightarrow X^-$	-348.7	-322	(Table 13.5)
Hydration of X^-	$X^-(g) \rightarrow X^-(aq)$	y	y'	
Overall		$\Delta_f G^{\ominus}(Cl^-)$	$\Delta_f G^{\ominus}(F^-)$	

Hence, $\Delta_f G^{\ominus}(Cl^-) = x + y + 1302 \, kJ \, mol^{-1}$

$$\Delta_f G^{\ominus}(F^-) = x + y' + 1286 \, kJ \, mol^{-1}$$

and $\Delta_f G^{\ominus}(Cl^-) - \Delta_f G^{\ominus}(F^-) = y - y' + 16 \, kJ \, mol^{-1}$

The ratio of hydration Gibbs energies is

$$\frac{\Delta_{\text{solv}} G^{\ominus}(\text{F}^-)}{\Delta_{\text{solv}} G^{\ominus}(\text{Cl}^-)} = \frac{r(\text{Cl}^-)}{r(\text{F}^-)}[10.3] = \frac{181\,\text{pm}}{131\,\text{pm}}\,[\text{Table 21.3}] = 1.38$$

$$\Delta_{\text{solv}} G^{\ominus} = -\frac{z_i^2}{r_i/\text{pm}} \times (6.86 \times 10^4\,\text{kJ mol}^{-1})$$

$$z^2 = 1, \qquad r(\text{Cl}^-) = 181\,\text{pm} \;[\text{Table 21.3}]$$

$$\Delta_{\text{solv}} G^{\ominus}(\text{Cl}^-) = -\frac{6.86 \times 10^4\,\text{kJ mol}^{-1}}{181} = -379\,\text{kJ mol}^{-1}$$

$$\Delta_{\text{solv}} G^{\ominus}(\text{F}^-) = (1.38) \times (-379\,\text{kJ mol}^{-1}) = -523\,\text{kJ mol}^{-1}$$

and $\Delta_{\text{f}} G^{\ominus}(\text{Cl}^-) - \Delta_{\text{f}} G^{\ominus}(\text{F}^-) = [(-379) - (-523) + (16)]\,\text{kJ mol}^{-1} = +160\,\text{kJ mol}^{-1}$

Hence (Table 2.6)

$$\Delta_{\text{f}} G^{\ominus}(\text{F}^-) = [(-131.23) - (160)]\,\text{kJ mol}^{-1} = \boxed{-291\,\text{kJ mol}^{-1}}$$

(The "experimental" value, Table 2.6, is $-278.79\,\text{kJ mol}^{-1}$.)

E10.7(a) $\quad I = \frac{1}{2} \sum_i (b_i/b^{\ominus}) z_i^2 \;[10.18]$

and for an $M_p X_q$ salt, $(b_+/b^{\ominus}) = p(b/b^{\ominus})$, $(b_-/b^{\ominus}) = q(b/b^{\ominus})$, so

$$I = \frac{1}{2}(pz_+^2 + qz_-^2)\left(\frac{b}{b^{\ominus}}\right)$$

(a) $\quad I(\text{KCl}) = \frac{1}{2}(1 \times 1 + 1 \times 1)\left(\frac{b}{b^{\ominus}}\right) = \left(\frac{b}{b^{\ominus}}\right)$

(b) $\quad I(\text{FeCl}_3) = \frac{1}{2}(1 \times 3^2 + 3 \times 1)\left(\frac{b}{b^{\ominus}}\right) = 6\left(\frac{b}{b^{\ominus}}\right)$

(c) $\quad I(\text{CuSO}_4) = \frac{1}{2}(1 \times 2^2 + 1 \times 2^2)\left(\frac{b}{b^{\ominus}}\right) = 4\left(\frac{b}{b^{\ominus}}\right)$

E10.8(a) $\quad I = I(\text{KCl}) + I(\text{CuSO}_4) = \left(\frac{b}{b^{\ominus}}\right)(\text{KCl}) + 4\left(\frac{b}{b^{\ominus}}\right)(\text{CuSO}_4)\;[\text{Exercise 10.7(a)}]$

$$= (0.10) + (4) \times (0.20) = \boxed{0.90}$$

Comment. Note that the ionic strength of a solution of more than one electrolyte may be calculated by summing the ionic strengths of each electrolyte considered as a separate solution, as in the solution to this exercise, or by summing the product $\frac{1}{2}\left(\frac{b_i}{b^{\ominus}}\right)z_i^2$ for each individual ion, as in the definition of I [10.18].

E10.9(a) $\quad I = I(\text{KNO}_3) = \left(\frac{b}{b^{\ominus}}\right)(\text{KNO}_3) = 0.150$

Therefore, the ionic strengths of the added salts must be 0.100.

(a) $\quad I(\text{Ca(NO}_3)_2) = \frac{1}{2}(2^2 + 2)\left(\frac{b}{b^{\ominus}}\right) = 3\left(\frac{b}{b^{\ominus}}\right)$

Therefore, the solution should be made $\frac{1}{3} \times 0.100 \,\text{mol kg}^{-1} = 0.0333 \,\text{mol kg}^{-1}$ in $\text{Ca(NO}_3)_2$. The mass that should be added to 500 g of the solution is therefore

$$(0.500 \,\text{kg}) \times (0.0333 \,\text{mol kg}^{-1}) \times (164 \,\text{g mol}^{-1}) = \boxed{2.73 \,\text{g}}$$

(b) $\quad I(\text{NaCl}) = \left(\frac{b}{b^{\ominus}}\right)$; therefore, with $b = 0.100 \,\text{mol kg}^{-1}$

$$(0.500 \,\text{kg}) \times (0.100 \,\text{mol kg}^{-1}) \times (58.4 \,\text{g mol}^{-1}) = \boxed{2.92 \,\text{g}}$$

(We are neglecting the fact that the mass of solution is slightly different from the mass of solvent.)

E10.10(a) $\quad I(\text{KCl}) = (b/b^{\ominus}), \qquad I(\text{CuSO}_4) = 4(b/b^{\ominus})$ [Exercise 10.7(a)]

For $I(\text{KCl}) = I(\text{CuSO}_4)$, $(b/b^{\ominus})(\text{KCl}) = 4(b/b^{\ominus})(\text{CuSO}_4)$

Therefore, if $b(\text{KCl}) = 1.00 \,\text{mol kg}^{-1}$, we require $b(\text{CuSO}_4) = \boxed{0.25 \,\text{mol kg}^{-1}}$

E10.11(a) $\quad \gamma_{\pm} = (\gamma_+^p \gamma_-^q)^{1/s} \quad s = p + q$ [10.14]

For CaCl_2, $p = 1$, $q = 2$, $s = 3$, $\boxed{\gamma_{\pm} = (\gamma_+ \gamma_-^2)^{1/3}}$

E10.12(a) These concentrations are sufficiently dilute for the Debye–Hückel limiting law to give a good approximate value for the mean ionic activity coefficient. Hence

$$\log \gamma_{\pm} = -|z_+ z_-| A I^{1/2} \text{ [10.17]}$$

$$I = \frac{1}{2} \sum_i z_i^2 \left(\frac{b_i}{b^{\ominus}}\right) \text{ [10.18]} = \frac{1}{2}[(4 \times 0.010) + (1 \times 0.020) + (1 \times 0.030) + (1 \times 0.030)]$$

$$= \boxed{0.060}$$

$$\log \gamma_{\pm} = -2 \times 1 \times 0.509 \times (0.060)^{1/2} = -0.24\overline{94}; \qquad \gamma_{\pm} = 0.56\overline{3} = \boxed{0.56}$$

E10.13(a) $\quad I(\text{LaCl}_3) = \frac{1}{2}(3^2 + 3)\left(\frac{b}{b^{\ominus}}\right) = 6\left(\frac{b}{b^{\ominus}}\right) = 3.000$

From the limiting law [10.17]

$$\log \gamma_{\pm} = -0.509|z_+ z_-| I^{1/2} = (-0.509) \times (3) \times (3.000)^{1/2} = -2.64\overline{5}$$

Hence $\gamma_{\pm} = 2.3 \times 10^{-3}$

and the error is $\boxed{1 \times 10^4 \text{ per cent}}$

Comment. It is not surprising that the limiting law provides such a poor prediction of $\gamma_{\pm}$ for this $(3, 1)$ electrolyte at this high concentration.

E10.14(a) $\quad \log \gamma_{\pm} = -\dfrac{A|z_+ z_-| I^{1/2}}{1 + B I^{1/2}}$ [10.28]

Solving for B,

$$B = -\left(\frac{1}{I^{1/2}} + \frac{A|z_+ z_-|}{\log \gamma_{\pm}}\right)$$

For HBr, $I = \left(\dfrac{b}{b^{\ominus}}\right)$ and $|z_+z_-| = 1$; so

$$B = -\left(\dfrac{1}{(b/b^{\ominus})^{1/2}} + \dfrac{0.509}{\log \gamma_{\pm}}\right)$$

Hence, draw up the following table

$(b/b^{\ominus})$	5.0×10^{-3}	10.0×10^{-3}	20.0×10^{-3}
$\gamma_{\pm}$	0.930	0.907	0.879
B	2.01	2.01	2.02

The constancy of B indicates that the mean ionic activity coefficient of HBr obeys the extended Debye–Hückel law very well.

E10.15(a) $CaF_2(s) \rightleftharpoons Ca^{2+}(aq) + 2F^-(aq)$ $K_s = 3.9 \times 10^{-11}$

$\Delta_r G^{\ominus} = -RT \ln K_s$

$\qquad = -(8.314\,J\,K^{-1}\,mol^{-1}) \times (298.15\,K) \times (\ln 3.9 \times 10^{-11}) = +59.4\,kJ\,mol^{-1}$

$\qquad = \Delta_f G^{\ominus}(CaF_2, aq) - \Delta_f G^{\ominus}(CaF_2, s)$

Hence, $\Delta_f G^{\ominus}(CaF_2, aq) = \Delta G^{\ominus} + \Delta_f G^{\ominus}(CaF_2, s)$

$$= [59.4 - 1167]\,kJ\,mol^{-1} = \boxed{-1108\,kJ\,mol^{-1}}$$

E10.16(a) The Nernst equation may be applied to individual reduction potentials as well as to overall cell potentials (Section 10.5). Hence

$$E(H^+/H_2) = \dfrac{RT}{F} \ln \dfrac{a(H^+)}{(f_{H_2}/p^{\ominus})^{1/2}} \text{ [Equation prior to eqn 10.42]}$$

and $\Delta E = E_1 - E_2 = \dfrac{RT}{F} \ln \dfrac{a_1(H^+)}{a_2(H^+)} \ [f_{H_2} \text{ is constant}] = \dfrac{RT}{F} \ln \dfrac{\gamma_{\pm}b_1}{\gamma_{\pm}b_2} \quad [\gamma_+ \approx \gamma_{\pm}]$

$$= (25.7\,mV) \times \ln\left(\dfrac{(20.0) \times (0.879)}{(5.0) \times (0.930)}\right) = \boxed{34.2\,mV}$$

Comment. Strictly $a(H^+) = \gamma(H^+)b(H^+)$, but $\gamma(H^+)$ cannot be determined from the data provided. However, since the solution is dilute, it is a valid approximation to replace $\gamma(H^+)$ with $\gamma_{\pm}$.

E10.17(a) We begin by choosing, based on an educated guess, the right and left electrodes.

R: $Cl_2(g) + 2e^- \rightarrow 2Cl^-(aq)$ $E_R^{\ominus} = +1.36\,V$ [Table 10.7]

L: $Mn^{2+}(aq) + 2e^- \rightarrow Mn(s)$ $E_L^{\ominus} = ?$

The cell corresponding to these half-reaction is

$Mn|MnCl_2(aq)|Cl_2(g)|Pt$; $E_{cell}^{\ominus} = E_R^{\ominus} - E_L^{\ominus} = 1.36\,V - E^{\ominus}(Mn, Mn^{2+})$

Hence, $E^{\ominus}(Mn, Mn^{2+}) = 1.36\,V - 2.54\,V = \boxed{-1.18\,V}$

Comment. With this choice of the right and left electrodes $E_{cell}^{\ominus} > 0$; the opposite choice would have resulted in $E_{cell}^{\ominus} < 0$ and could not have corresponded to the thermodynamically spontaneous reaction given.

E10.18(a) The cell notation specifies the right and left electrodes. Note that for proper cancellation we must equalize the number of electrons in half-reactions being combined.

$$E^{\ominus}$$

(a) R: $2Ag^+(aq) + 2e^- \rightarrow 2Ag(s)$ $+0.80\,V$

L: $Zn^+(aq) + 2e^- \rightarrow Zn(s)$ $-0.76\,V$

Overall (R − L): $2Ag^+(aq) + Zn(s) \rightarrow 2Ag(s) + Zn^{2+}(aq)$ $+1.56\,V$

(b) R: $2H^+(aq) + 2e^- \rightarrow H_2(g)$ 0

L: $Cd^{2+}(aq) + 2e^- \rightarrow Cd(s)$ $-0.40\,V$

Overall (R − L): $Cd(s) + 2H^+(aq) \rightarrow Cd^{2+}(aq) + H_2(g)$ $+0.40\,V$

(c) R: $Cr^{3+}(aq) + 3e^- \rightarrow Cr(s)$ $-0.74\,V$

L: $3[Fe(CN)_6]^{3-}(aq) + 3e^- \rightarrow 3[Fe(CN)_6]^{4-}(aq)$ $+0.36\,V$

Overall (R − L): $Cr^{3+}(aq) + 3[Fe(CN)_6]^{4-}(aq) \rightarrow Cr(s)$
$$+ 3[Fe(CN)_6]^{3-}(aq) \quad -1.10\,V$$

Comment. Those cells for which $E^{\ominus} > 0$ may operate as spontaneous galvanic cells under standard conditions. Those for which $E^{\ominus} < 0$ may operate as non-spontaneous electrolytic cells. Recall that $E^{\ominus}$ informs us of the spontaneity of a cell under standard conditions only. For other conditions we require E.

E10.19(a) The conditions (concentrations, etc.) under which these reactions occur are not given. For the purposes of this exercise we assume standard conditions. The specification of the right and left electrodes is determined by the direction of the reaction as written. As always, in combining half-reactions to form an overall cell reaction we must write the half-reactions with equal number of electrons to ensure proper cancellation. We first identify the half-reactions, and then set up the corresponding cell.

$$E^{\ominus}$$

(a) R: $Cu^{2+}(aq) + 2e^- \rightarrow Cu(s)$ $+0.34\,V$

L: $Zn^{2+}(aq) + 2e^- \rightarrow Zn(s)$ $-0.76\,V$

Hence the cell is

$Zn(s)|ZnSO_4(aq)||CuSO_4(aq)|Cu(s)$ $+1.10\,V$

(b) R: $AgCl(s) + e^- \rightarrow Ag(s) + Cl^-(aq)$ $+0.22\,V$

L: $H^+(aq) + e^- \rightarrow \frac{1}{2}H_2(g)$ 0

and the cell is

$Pt|H_2(g)|H^+(aq)|AgCl(s)|Ag(s)$

or $Pt|H_2(g)|HCl(aq)|AgCl(s)|Ag(s)$ $+0.22\,V$

(c) R: $O_2(g) + 4H^+(aq) + 4e^- \rightarrow 2H_2O(l)$ $+1.23\,V$

L: $4H^+(aq) + 4e^- \rightarrow 2H_2(g)$ 0

and the cell is

$Pt|H_2(g)|H^+(aq), H_2O(l)|O_2(g)|Pt$ $+1.23\,V$

Comment. All of these cells have $E^{\ominus} > 0$, corresponding to a spontaneous cell reaction under standard conditions. If $E^{\ominus}$ had turned out to be negative, the spontaneous reaction would have been the reverse of the one given, with the right and left electrodes of the cell also reversed.

E10.20(a) See the solutions to Exercise 10.18(a), where we have used $E^\ominus = E_R^\ominus - E_L^\ominus$, with standard electrode potentials from Table 10.7.

E10.21(a) See the solutions to Exercise 10.19(a), where we have used $E^\ominus = E_R^\ominus - E_L^\ominus$.

E10.22(a) In each case find $E^\ominus = E_R^\ominus - E_L^\ominus$ from the data in Table 10.7, then use

$$\Delta_r G^\ominus = -vFE^\ominus \quad [10.32]$$

(a) $2Na(s) + 2H_2O(l) \rightarrow 2NaOH(aq) + H_2(g)$ $\qquad E^\ominus = +1.88\,V$ [Exercise 10.16(b)(a)]

Therefore, with $v = 2$

$$\Delta_r G^\ominus = (-2) \times (96.485\,kC\,mol^{-1}) \times (1.88\,V) = \boxed{-363\,kJ\,mol^{-1}}$$

(b) $2K(s) + 2H_2O(l) \rightarrow 2KOH(aq) + H_2(g)$

$E^\ominus = E^\ominus(H_2O, OH^-, H_2) - E^\ominus(K, K^+)$

$\qquad = -0.83\,V - (-2.93\,V) = +2.10\,V$ with $v = 2$

Therefore,

$$\Delta_r G^\ominus = (-2) \times (96.485\,kC\,mol^{-1}) \times (2.10\,V) = \boxed{-405\,kJ\,mol^{-1}}$$

E10.23(a) **(a)** $E^\ominus = \dfrac{-\Delta G^\ominus}{vF}[10.33] = \dfrac{+62.5\,kJ\,mol^{-1}}{(2) \times (96.485\,kC\,mol^{-1})} = \boxed{+0.324\,V}$

(b) $E^\ominus = E_R^\ominus - E_L^\ominus = E^\ominus(Fe^{3+}, Fe^{2+}) - E^\ominus(Ag, Ag_2CrO_4, CrO_4^{2-})$

Therefore, $E^\ominus(Ag, Ag_2CrO_4, CrO_4^{2-}) = E^\ominus(Fe^{3+}, Fe^{2+}) - E^\ominus$

$$= [+0.77 - 0.324]\,V = \boxed{+0.45\,V}$$

E10.24(a) When combining two half-reactions to correspond to an overall cell reaction which is spontaneous, the combination must be such that the electrons in the half-reactions cancel and that $E_{cell} > 0$. Thus

R: $O_2(g) + 4H^+(aq) + 4e^- \rightarrow 2H_2O(l)$ $\qquad E^\ominus = +1.23\,V$

L: $2Ag_2S(s) + 4e^- \rightarrow 4Ag(s) + 2S^{2-}(aq)$ $\quad E^\ominus = -0.69\,V$

R − L: $4Ag(s) + 2S^{2-}(aq) + O_2(g) + 4H^+(aq) \rightarrow 2Ag_2S(s) + 2H_2O(l)$

$E^\ominus = E_R^\ominus - E_L^\ominus = (1.23\,V) - (-0.69\,V) = \boxed{+1.92\,V}$

Comment. Under standard conditions $E = E^\ominus > 0$ and the reaction is spontaneous. Because of the large positive $E^\ominus$, the reaction is likely to remain spontaneous unless the conditions are changed drastically to make $E < 0$.

Question. Can you devise conditions such that $E < 0$?

E10.25(a)

$\qquad\qquad\qquad\qquad\qquad\qquad\qquad\qquad\qquad\qquad\qquad E^\ominus$

R: $Cd^{2+}(aq) + 2e^- \rightarrow Cd(s)$ $\qquad\qquad\qquad\qquad -0.40\,V$

L: $2AgBr(s) + 2e^- \rightarrow 2Ag(s) + 2Br^-(aq)$ $\qquad\qquad +0.07\,V$

Hence, overall (R − L)

$Cd^{2+}(aq) + 2Ag(s) + 2Br^-(aq) \rightarrow Cd(s) + 2AgBr(s)$ $\quad -0.47\,V$

$$Q = \frac{1}{a(Cd^{2+})a^2(Br^-)} \qquad E = E^\ominus + \frac{RT}{2F}\ln a(Cd^{2+})a^2(Br^-)$$

$$a(Cd^{2+}) = \gamma_+ b_+; \qquad a(Br^-) = \gamma_- b_- \qquad \left[b \equiv \frac{b}{b^\ominus} \right]$$

$$b_+ = 0.010 \, mol \, kg^{-1}, \qquad b_- = 0.050 \, mol \, kg^{-1}$$

We assume that $\gamma_+(Cd^{2+}) \approx \gamma_\pm\{Cd(NO_3)_2\}$ and that $\gamma_-(Br^-) \approx \gamma_\pm(KBr)$; hence

$$E = E^\ominus + \frac{RT}{2F} \ln b(Cd^{2+})b^2(Br^-) + \frac{2.303RT}{2F} \log \gamma_\pm\{Cd(NO_3)_2\}\gamma_\pm^2(KBr)$$

$$\log \gamma_\pm\{Cd(NO_3)_2\} \approx -A|z_+ z_-| \times I^{1/2}, \quad I = 3b = 0.030 \, mol \, kg^{-1}$$

$$\approx -(0.509) \times (2) \times (0.030)^{1/2} = -0.18$$

$$\log \gamma_\pm(KBr) \approx -A|z_+ z_-| \times I^{1/2}, \quad I = b$$

$$\approx -(0.509) \times (1) \times (0.050)^{1/2} = -0.11$$

Hence, $E = (-0.47 \, V) + \left(\dfrac{25.693 \, mV}{2} \right) \times \ln(0.010 \times 0.050^2)$

$$+ \left(\frac{(2.303) \times (25.693 \, mV)}{2} \right) \times (-0.18 + 2 \times (-0.11)) = \boxed{-0.62 \, V}$$

E10.26(a) In each case $\ln K = \dfrac{\nu F E^\ominus}{RT}$ [10.36]

(a) $Sn(s) + Sn^{4+}(aq) \rightleftharpoons 2Sn^{2+}(aq)$

R: $Sn^{4+} + 2e^- \rightarrow Sn^{2+}(aq)$ $\qquad$ $+0.15 \, V$ $\left.\right\}$ $E^\ominus = +0.29 \, V$

L: $Sn^{2+}(aq) + 2e^- \rightarrow Sn(s)$ $\qquad$ $-0.14 \, V$

$$\ln K = \frac{(2) \times (0.29 \, V)}{25.693 \, mV} = 22.\overline{6}, \qquad K = \boxed{6.5 \times 10^9}$$

(b) $Sn(s) + 2AgCl(s) \rightleftharpoons SnCl_2(aq) + 2Ag(s)$

R: $2AgCl(s) + 2e^- \rightarrow 2Ag(s) + 2Cl^-(aq)$ $\qquad$ $+0.22 \, V$ $\left.\right\}$ $+0.36 \, V$

L: $Sn^{2+}(aq) + 2e^- \rightarrow Sn(s)$ $\qquad$ $-0.14 \, V$

$$\ln K = \frac{(2) \times (0.36 \, V)}{25.693 \, mV} = +28.\overline{0}, \qquad K = \boxed{1.5 \times 10^{12}}$$

E10.27(a) We need to obtain $E^\ominus$ for the couple

(3) $Au^{3+}(aq) + 2e^- \rightarrow Au^+(aq)$

from the values of $E^\ominus$ for the couples

(1) $Au^+(aq) + e^- \rightarrow Au(s)$ $\qquad$ $E_1^\ominus = 1.69 \, V$

(2) $Au^{3+}(aq) + 3e^- \rightarrow Au(s)$ $\qquad$ $E_2^\ominus = 1.40 \, V$

We see that (3) = (2) − (1), therefore

$$\Delta_r G_3^\ominus = \Delta_r G_1^\ominus - \Delta_r G_2^\ominus$$

$$-\nu_3 F E_3^\ominus = -\nu_1 F E_1^\ominus - \nu_2 F E_2^\ominus$$

Solving for $E_3^{\ominus}$ we obtain

$$E_3^{\ominus} = \frac{\nu_2 E_2^{\ominus} - \nu_1 E_1^{\ominus}}{\nu_3} = \frac{(3) \times (1.40\,\text{V}) - (1) \times (1.69\,\text{V})}{2} = 1.26\,\text{V}$$

Then,

R: $Au^{3+}(aq) + 2e^- \rightarrow Au^+(aq)$ $E_R^{\ominus} = 1.26\,\text{V}$

L: $2Fe^{3+}(aq) + 2e^- \rightarrow 2Fe^{2+}(aq)$ $E_L^{\ominus} = 0.77\,\text{V}$

R − L: $2Fe^{2+}(aq) + Au^{3+}(aq) \rightarrow 2Fe^{3+}(aq) + Au^+(aq)$

$E^{\ominus} = E_R^{\ominus} - E_L^{\ominus} = (1.26\,\text{V}) - (0.77\,\text{V}) = \boxed{+0.49\,\text{V}}$

$\ln K = \dfrac{\nu F E^{\ominus}}{RT}[10.36] = \dfrac{(2) \times (0.49\,\text{V})}{25.7 \times 10^{-3}\,\text{V}} = 38.\overline{1}, \qquad K = \boxed{4 \times 10^{16}}$

E10.28(a)   First assume all activity coefficients are 1 and calculate K_s°, the ideal solubility product constant.

(1) $AgCl(s) \rightleftharpoons Ag^+(aq) + Cl^-(aq)$

Since all stoichiometric coefficients are 1

$$S(AgCl) = b(Ag^+) = b(Cl^-)$$

Hence, $K_s^{\circ} = \dfrac{b(Ag^+)b(Cl^-)}{b^{\ominus 2}} = \dfrac{b^2(Ag^+)}{b^{\ominus 2}} = \dfrac{S^2}{b^{\ominus 2}} = (1.34 \times 10^{-5})^2 = \boxed{1.80 \times 10^{-10}}$

(2) $BaSO_4(s) \rightleftharpoons Ba^{2+}(aq) + SO_4^{2-}(aq)$

$S(BaSO_4) = b(Ba^{2+}) = b(SO_4^{2-})$

As above, $K_s^{\circ} = \dfrac{S^2}{b^{\ominus 2}} = (9.51 \times 10^{-4})^2 = \boxed{9.04 \times 10^{-7}}$

Now redo the calculation taking into account the deviation of the activity coefficients from 1 in order to obtain K_s, the true thermodynamic solubility product constant. We assume that the activity coefficients can be estimated from the Debye–Hückel limiting law since the concentrations of the ions are low.

For both $AgCl(s)$ and $BaSO_4(s)$

$$a_+ = \frac{\gamma_+ b_+}{b^{\ominus}}, \qquad a_- = \frac{\gamma_- b_-}{b^{\ominus}}$$

$$K_s = a_+ a_- = \gamma_+ \gamma_- \left(\frac{b_+}{b^{\ominus}}\right) \times \left(\frac{b_-}{b^{\ominus}}\right) = \gamma_+ \gamma_- K_s^{\circ}, \qquad \gamma_+ \gamma_- = \gamma_{\pm}^2$$

Thus, $K_s = \gamma_{\pm}^2 K_s^{\circ}$

$\log \gamma_{\pm} = -|z_+ z_-| A I^{1/2}, \quad A = 0.509$

For AgCl, $I = S$, $|z_+ z_-| = 1$, and so

$\log \gamma_{\pm} = -(0.509) \times (1.34 \times 10^{-5})^{1/2} = -1.86 \times 10^{-3}, \qquad \gamma_{\pm} \approx 0.9957$

Hence, $K_s = \gamma_{\pm}^2 \times K_s^{\circ} \approx \boxed{0.991\,K_s^{\circ}}$

For $BaSO_4$, $I = 4S$, $|z_+z_-| = 4$, and so

$$\log \gamma_\pm = -(0.509) \times (4) \times \left[(4) \times (9.51 \times 10^{-4})\right]^{1/2} = -0.126, \qquad \gamma_\pm \approx 0.75$$

Hence, $K_s = \gamma_\pm^2 K_s^\circ \approx (0.75)^2 K_s^\circ \approx \boxed{0.56\, K_s^\circ}$

Thus, the neglect of activity coefficients is significant for $BaSO_4$.

E10.29(a) A Nernst equation can be written for a half-reaction as well as for a whole-cell reaction.

The half-reaction is (Table 10.7)

$$Cr_2O_7^{2-}(aq) + 14\,H^+(aq) + 6e^- \rightarrow 2Cr^{3+}(aq) + 7H_2O(l)$$

The reaction quotient is

$$Q = \frac{a^2(Cr^{3+})}{a(Cr_2O_7^{2-})a^{14}(H^+)} \qquad \nu = 6$$

Hence, $\boxed{E = E^\ominus - \dfrac{RT}{6F} \ln \dfrac{a^2(Cr^{3+})}{a(Cr_2O_7^{2-})a^{14}(H^+)}}$

E10.30(a) R: $2AgCl(s) + 2e^- \rightarrow 2Ag(s) + 2Cl^-(aq)$ $+0.22\,V$
L: $2H^+(aq) + 2e^- \rightarrow H_2(g)$ 0

Overall, R − L: $2AgCl(s) + H_2(g) \rightarrow 2Ag(s) + 2H^+(aq) + 2Cl^-(aq)$

$$Q = a^2(H^+)a^2(Cl^-)[\nu = 2] = a^4(H^+) \quad [\text{Assume } a(H^+) \approx a(Cl^-)]$$

Therefore, from the Nernst equation [10.34],

$$E = E^\ominus - \frac{RT}{2F} \ln a^4(H^+)$$

$$= E^\ominus - \frac{2RT}{F} \ln a(H^+) = E^\ominus + (2) \times (2.303) \times \left(\frac{RT}{F}\right) \times pH$$

Hence,

$$pH = \left(\frac{F}{(2) \times (2.303RT)}\right) \times (E - E^\ominus)$$

$$= \frac{E - 0.22\,V}{0.1183\,V} = \frac{(0.322\,V) - (0.22\,V)}{0.1183\,V} = \boxed{0.86}$$

Comment. This value of the pH corresponds roughly to a concentration of $H^+(aq)$ of about $0.1\,mol\,kg^{-1}$. At this rather high concentration the assumption that the activities of $H^+(aq)$ and $Cl^-(aq)$ are equal may not be justified.

E10.31(a) The left electrode contains no $AgBr(s)$; hence the electrode reactions are

R: $AgBr(s) + e^- \rightarrow Ag(s) + Br^-(aq)$

L: $Ag^+(aq) + e^- \rightarrow Ag(s)$

Overall: $AgBr(s) \rightarrow Ag^+(aq) + Br^-(aq)$

Therefore, since the cell reaction is the solubility equilibrium, for a saturated solution there is no further tendency to dissolve and so $\boxed{E = 0}$

E10.32(a)   R: $Ag^+(aq) + e^- \rightarrow Ag(s)$ +0.80 V $\left.\right\}$ $E^\ominus = E_R^\ominus - E_L^\ominus = 0.95\,V$
          L: $AgI(s) + e^- \rightarrow Ag(s) + I^-(aq)$ $-0.15\,V$

Overall (R − L): $Ag^+(aq) + I^-(aq) \rightarrow AgI(s)$ $\nu = 1$

$$\ln K = \frac{\nu F E^\ominus}{RT}[10.36] = \frac{0.95\,V}{25.693 \times 10^{-3}\,V} = 36.9\overline{75}$$

$$K = \boxed{\overline{1} \times 10^{16}}$$

However, $K_s = K^{-1}$ since the solubility equilibrium is written as the reverse of the cell reaction. Therefore, (b) $K_s = \boxed{\overline{1} \times 10^{-16}}$. The solubility is obtained from $b(Ag^+) \approx b(I^-)$ and $S = b(Ag^+)$, so $K_s \approx b^2(Ag^+)$ implying that (a) $S = (K_s)^{1/2} = \boxed{\overline{1} \times 10^{-8}\,mol\,kg^{-1}}$.

Solutions to problems

Solutions to numerical problems

P10.2   **(a)** $I = \frac{1}{2}\left\{\left(\frac{b}{b^\ominus}\right)_+ z_+^2 + \left(\frac{b}{b^\ominus}\right)_- z_-^2\right\}[10.19] = 4\left(\frac{b}{b^\ominus}\right)$

For $CuSO_4$, $I = (4) \times (1.0 \times 10^{-3}) = \boxed{4.0 \times 10^{-3}}$

For $ZnSO_4$, $I = (4) \times (3.0 \times 10^{-3}) = \boxed{1.2 \times 10^{-2}}$

(b) $\log \gamma_\pm = -|z_+ z_-| A I^{1/2}$

$\log \gamma_\pm(CuSO_4) = -(4) \times (0.509) \times (4.0 \times 10^{-3})^{1/2} = -0.12\overline{88}$

$\gamma_\pm(CuSO_4) = \boxed{0.74}$

$\log \gamma_\pm(ZnSO_4) = -(4) \times (0.509) \times (1.2 \times 10^{-2})^{1/2} = -0.22\overline{30}$

$\gamma_\pm(ZnSO_4) = \boxed{0.60}$

(c) The reaction in the Daniell cell is

$$Cu^{2+}(aq) + SO_4^{2-}(aq) + Zn(s) \rightarrow Cu(s) + Zn^{2+}(aq) + SO_4^{2-}(aq)$$

Hence, $Q = \dfrac{a(Zn^{2+})a(SO_4^{2-}, R)}{a(Cu^{2+})a(SO_4^{2-}, L)}$

$$= \frac{\gamma_+ b_+(Zn^{2+})\gamma_- b_-(SO_4^{2-}, R)}{\gamma_+ b_+(Cu^{2+})\gamma_- b_-(SO_4^{2-}, L)} \quad \left[b \equiv \frac{b}{b^\ominus} \text{ here and below}\right]$$

where the designations R and L refer to the right and left sides of the equation for the cell reaction and all b are assumed to be unitless, that is, $\dfrac{b}{b^\ominus}$.

$b_+(Zn^{2+}) = b_-(SO_4^{2-}, R) = b(ZnSO_4)$

$b_+(Cu^{2+}) = b_-(SO_4^{2-}, L) = b(CuSO_4)$

Therefore,

$$Q = \frac{\gamma_\pm^2(ZnSO_4)b^2(ZnSO_4)}{\gamma_\pm^2(CuSO_4)b^2(CuSO_4)} = \frac{(0.60)^2 \times (3.0 \times 10^{-3})^2}{(0.74)^2 \times (1.0 \times 10^{-3})^2} = 5.9\overline{2} = \boxed{5.9}$$

(d) $\quad E^\ominus = -\frac{\Delta_r G^\ominus}{\nu F} [10.33] = \frac{-(-212.7 \times 10^3 \, J \, mol^{-1})}{(2) \times (9.6485 \times 10^4 \, C \, mol^{-1})} = \boxed{+1.102 \, V}$

(e) $\quad E = E^\ominus = -\frac{25.693 \times 10^{-3} \, V}{\nu} \ln Q = (1.102 \, V) - \left(\frac{25.693 \times 10^{-3} \, V}{2}\right) \ln(5.9\overline{2})$

$$= (1.102 \, V) - (0.023 \, V) = \boxed{+1.079 \, V}$$

P10.3 The electrode half-reactions and their potentials are

$$E^\ominus$$

R: $\quad Q(aq) + 2H^+(aq) + 2e^- \rightarrow QH_2(aq) \qquad\qquad 0.6994 \, V$

L: $\quad Hg_2Cl_2(s) + 2e^- \rightarrow 2Hg(l) + 2Cl^-(aq) \qquad\quad 0.2676 \, V$

Overall $(R - L)$: $\quad Q(aq) + 2H^+(aq) \rightarrow QH_2(aq) + Hg_2Cl_2(s) \quad 0.4318 \, V$

$$Q\text{(reaction quotient)} = \frac{a(QH_2)}{a(Q)a^2(H^+)a^2(Cl^-)}$$

Since quinhydrone is an equimolecular complex of Q and QH_2, $m(Q) = m(QH_2)$, and since their activity coefficients are assumed to be 1 or to be equal, we have $a(QH_2) \approx a(Q)$. Thus

$$Q = \frac{1}{a^2(H^+)a^2(Cl^-)} \qquad E = E^\ominus - \frac{25.7 \, mV}{\nu} \ln Q \quad [\textit{Illustration p. 260}]$$

$$\ln Q = \frac{\nu(E^\ominus - E)}{25.7 \, mV} = \frac{(2) \times (0.4318 - 0.190) \, V}{25.7 \times 10^{-3} \, V} = 18.8\overline{2} \qquad Q = 1.\overline{49} \times 10^8$$

$$a^2(H^+) = (\gamma_+ b_+)^2; \qquad a^2(Cl^-) = (\gamma_- b_-)^2 \quad \left[b \equiv \frac{b}{b^\ominus}\right]$$

For $HCl(aq)$, $b_+ = b_- = b$, and if the activity coefficients are assumed equal, $a^2(H^+) = a^2(Cl^-)$; hence

$$Q = \frac{1}{a^2(H^+)a^2(Cl^-)} = \frac{1}{a^4(H^+)}$$

Thus, $a(H^+) = \left(\frac{1}{Q}\right)^{1/4} = \left(\frac{1}{1.49 \times 10^8}\right)^{1/4} = 9 \times 10^{-3}$

$pH = -\log a(H^+) = \boxed{2.0}$

P10.4 **(a)** $\quad E = E^\ominus - \frac{25.693 \, mV}{\nu} \ln Q \quad [10.34, 25°C]$

$$Q = a(Zn^{2+})a^2(Cl^-)$$

$$= \gamma_+ \left(\frac{b}{b^\ominus}\right)(Zn^{2+})\gamma_-^2 \left(\frac{b}{b^\ominus}\right)^2 (Cl^-) \qquad b(Zn^{2+}) = b, \ b(Cl^-) = 2b, \ \gamma_+ \gamma_-^2 = \gamma_\pm^3$$

Therefore, $Q = \gamma_\pm^3 \times 4b^3 \quad \left[b \equiv \frac{b}{b^\ominus} \text{ here and below}\right]$

and $E = E^{\ominus} - \dfrac{25.693\,\text{mV}}{2}\ln(4b^3\gamma_{\pm}^3) = E^{\ominus} - \left(\dfrac{3}{2}\right) \times (25.693\,\text{mV}) \times \ln(4^{1/3}b\gamma_{\pm})$

$$= \boxed{E^{\ominus} - (38.54\,\text{mV}) \times \ln(4^{1/3}b) - (38.54\,\text{mV})\ln(\gamma_{\pm})}$$

(b) $E^{\ominus}(\text{Cell}) = E_{R}^{\ominus} - E_{L}^{\ominus} = E^{\ominus}(\text{Hg}_2\text{Cl}_2, \text{Hg}) - E^{\ominus}(\text{Zn}^{2+}, \text{Zn})$

$$= (0.2676\,\text{V}) - (-0.7628\,\text{V}) = \boxed{+1.0304\,\text{V}}$$

(c) $\Delta_r G = -\nu F E = -(2) \times (9.6485 \times 10^4\,\text{C}\,\text{mol}^{-1}) \times (1.2272\,\text{V}) = \boxed{-236.81\,\text{kJ}\,\text{mol}^{-1}}$

$\Delta_r G^{\ominus} = -\nu F E^{\ominus} = -(2) \times (9.6485 \times 10^4\,\text{C}\,\text{mol}^{-1}) \times (1.0304\,\text{V}) = \boxed{-198.84\,\text{kJ}\,\text{mol}^{-1}}$

$\ln K = -\dfrac{\Delta_r G^{\ominus}}{RT} = \dfrac{1.9884 \times 10^5\,\text{J}\,\text{mol}^{-1}}{(8.3145\,\text{J}\,\text{K}^{-1}\,\text{mol}^{-1}) \times (298.15\,\text{K})} = 80.211 \quad K = \boxed{6.84 \times 10^{34}}$

(d) From part **(a)**

$$1.2272\,\text{V} = 1.0304\,\text{V} - (38.54\,\text{mV}) \times \ln(4^{1/3} \times 0.0050) - (38.54\,\text{mV}) \times \ln\gamma_{\pm}$$

$\ln\gamma_{\pm} = -\dfrac{(1.2272\,\text{V}) - (1.0304\,\text{V}) - (0.186\overline{4}\,\text{V})}{0.03854\,\text{V}} = -0.269\overline{8}; \qquad \gamma_{\pm} = \boxed{0.763}$

(e) $\log\gamma_{\pm} = -|z_-z_+|AI^{1/2}$ [10.17]

$$I = \frac{1}{2}\sum_i z_i^2\left(\frac{b_i}{b^{\ominus}}\right)\ \text{[10.18]}$$

$b(\text{Zn}^{2+}) = b = 0.0050\,\text{mol}\,\text{kg}^{-1} \qquad b(\text{Cl}^-) = 2b = 0.010\,\text{mol}\,\text{kg}^{-1}$

$I = \frac{1}{2}[(4) \times (0.0050) + (0.010)] = 0.015$

$\log\gamma_{\pm} = -(2) \times (0.509) \times (0.015)^{1/2} = -0.12\overline{5}; \qquad \gamma_{\pm} = \boxed{0.75}$

This compares remarkably well to the value obtained from experimetal data in part **(d)**.

(f) $\Delta_r S = -\left(\dfrac{\partial\Delta_r G}{\partial T}\right)_p$

$$= \nu F\left(\frac{\partial E}{\partial T}\right)_p \text{[10.45]} = (2) \times (9.6485 \times 10^4\,\text{C}\,\text{mol}^{-1}) \times (-4.52 \times 10^{-4}\,\text{V}\,\text{K}^{-1})$$

$$= \boxed{-87.2\,\text{J}\,\text{K}^{-1}\,\text{mol}^{-1}}$$

$\Delta_r H = \Delta_r G + T\Delta_r S = (-236.81\,\text{kJ}\,\text{mol}^{-1}) + (298.15\,\text{K}) \times (-87.2\,\text{J}\,\text{K}^{-1}\,\text{mol}^{-1})$

$$= \boxed{-262.4\,\text{kJ}\,\text{mol}^{-1}}$$

P10.5 $\text{H}_2(\text{g})|\text{HCl}(\text{aq})|\text{Hg}_2\text{Cl}_2(\text{s})|\text{Hg}(\text{l})$

$E = E^{\ominus} - \dfrac{RT}{F}\ln a(\text{H}^+)a(\text{Cl}^-)$ [10.34]

$a(\text{H}^+) = \gamma_+ b_+ = \gamma_+ b; \qquad a(\text{Cl}^-) = \gamma_- b_- = \gamma_- b \qquad \left[b = \dfrac{b}{b^{\ominus}}\text{here and below}\right]$

$a(\text{H}^+)a(\text{Cl}^-) = \gamma_+\gamma_- b^2 = \gamma_{\pm}^2 b^2$

$E = E^{\ominus} - \dfrac{2RT}{F}\ln b - \dfrac{2RT}{F}\ln\gamma_{\pm}$ (a)

Converting from natural logarithms to common logarithms (base 10) in order to introduce the Debye–Hückel expression, we obtain

$$E = E^{\ominus} - \frac{(2.303) \times 2RT}{F} \log b - \frac{(2.303) \times 2RT}{F} \log \gamma_{\pm}$$

$$= E^{\ominus} - (0.1183\,\text{V}) \log b - (0.1183\,\text{V}) \log \gamma_{\pm}$$

$$= E^{\ominus} - (0.1183\,\text{V}) \log b - (0.1183\,\text{V}) \left[-|z_{+}z_{-}|AI^{1/2} \right]$$

$$= E^{\ominus} - (0.1183\,\text{V}) \log b + (0.1183\,\text{V}) \times A \times b^{1/2} \quad [I = b]$$

Rearranging,

$$E + (0.1183\,\text{V}) \log b = E^{\ominus} + \text{constant} \times b^{1/2}$$

Therefore, plot $E + (0.1183\,\text{V}) \log b$ against $b^{1/2}$, and the intercept at $b = 0$ is $E^{\ominus}/\text{V}$. Draw up the following table

$b/(\text{mmol kg}^{-1})$	1.6077	3.0769	5.0403	7.6938	10.9474
$\left(\dfrac{b}{b^{\ominus}}\right)^{1/2}$	0.04010	0.05547	0.07100	0.08771	0.1046
$E/\text{V} + (0.1183) \log b$	0.27029	0.27109	0.27186	0.27260	0.27337

The points are plotted in Fig. 10.1. The intercept is at 0.26840, so $E^{\ominus} = +0.26840\,\text{V}$. A least–squares best fit gives $E^{\ominus} = \boxed{+0.26843\,\text{V}}$ and a coefficient of determination equal to 0.99895.

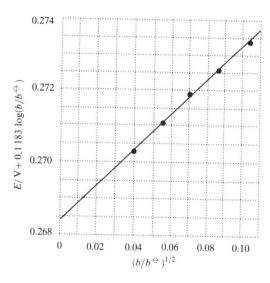

Figure 10.1

For the activity coefficients we obtain from equation (a)

$$\ln \gamma_{\pm} = \frac{E^{\ominus} - E}{2RT/F} - \ln \frac{b}{b^{\ominus}} = \frac{0.26843 - E/\text{V}}{0.05139} - \ln \frac{b}{b^{\ominus}}$$

and we draw up the following table

$b/(\text{mmol kg}^{-1})$	1.6077	3.0769	5.0403	7.6938	10.9474
$\ln \gamma_{\pm}$	−0.03465	−0.05038	−0.06542	−0.07993	−0.09500
$\gamma_{\pm}$	0.9659	0.9509	0.9367	0.9232	0.9094

P10.8 $H_2(g)|HCl(b)|AgCl(s)|Ag(s)$

$\frac{1}{2}H_2(g) + AgCl(s) \rightarrow HCl(aq) + Ag(s)$

$E = E^{\ominus} - \frac{RT}{F}\ln a(H^+)a(Cl^-) = E^{\ominus} - \frac{2RT}{F}\ln b - \frac{2RT}{F}\ln\gamma_{\pm}$

$[b \equiv \dfrac{b}{b^{\ominus}}$, here and below]

$= E^{\ominus} - \frac{2RT}{F}\ln b - (2)\times(2.303)\frac{RT}{F}\log\gamma_{\pm}$

$= E^{\ominus} - \frac{2RT}{F}\ln b - (2)\times(2.303)\frac{RT}{F}\left[-0.509b^{1/2} + kb\right]\quad [I = b]$

Therefore, with $\dfrac{2RT}{F}\times 2.303 = 0.1183$ V,

$E/V + 0.1183\log b - 0.0602b^{1/2} = E^{\ominus}/V - 0.1183kb$

hence, with $y = E/V + 0.1183\log b - 0.0602b^{1/2}$,

$$\boxed{y = E^{\ominus}/V - 0.1183kb}$$

We now draw up the following table

$b/(\text{mmol kg}^{-1})$	123.8	25.63	9.138	5.619	3.215
y	0.2135	0.2204	0.2216	0.2218	0.2221

(a) The data are plotted in Fig. 10.2, and extrapolate to 0.2223 V; hence $E^{\ominus} = \boxed{+0.2223\,\text{V}}$

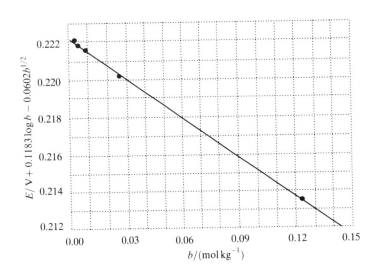

Figure 10.2

(b) $E = E^{\ominus} - \dfrac{2RT}{F}\ln b - \dfrac{2RT}{F}\ln\gamma_{\pm}$

and so $\ln\gamma_{\pm} = \dfrac{E^{\ominus} - E - 0.0514\,\text{V}\ln b}{0.0514\,\text{V}} = \dfrac{(0.2223) - (0.3524) - (0.0514)\ln(0.100)}{0.0514}$

$= -0.228\overline{5},\quad$ implying that $\gamma_{\pm} = \boxed{0.796}$

Since $a(H^+) = \dfrac{\gamma_\pm b}{b^\ominus}$, $a(H^+) = (0.796) \times (0.100) = 0.0796$, and hence

$$pH = -\log a(H^+) = -\log(0.0796) = \boxed{1.10}$$

P10.9   According to the Debye–Hückel limiting law

$$\log \gamma_\pm = -0.509|z_+ z_-|I^{1/2} = -0.509 \left(\frac{b}{b^\ominus}\right)^{1/2} \quad [10.17]$$

We draw up the following table

$b/(\text{mmol kg}^{-1})$	1.0	2.0	5.0	10.0	20.0
$I^{1/2}$	0.032	0.045	0.071	0.100	0.141
$\gamma_\pm(\text{calc})$	0.964	0.949	0.920	0.889	0.847
$\gamma_\pm(\text{exp})$	0.9649	0.9519	0.9275	0.9024	0.8712
$\log \gamma_\pm(\text{calc})$	−0.0161	−0.0228	−0.0360	−0.0509	−0.0720
$\log \gamma_\pm(\text{exp})$	−0.0155	−0.0214	−0.0327	−0.0446	−0.0599

The points are plotted against $I^{1/2}$ in Fig. 10.3. Note that the limiting slopes of the calculated and experimental curves coincide. A sufficiently good value of B in the extended Debye–Hückel law may be obtained by assuming that the constant A in the extended law is the same as A in the limiting law. Using the data at $20.0\,\text{mmol kg}^{-1}$ we may solve for B.

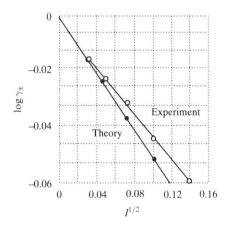

Figure 10.3

$$B = -\frac{A}{\log \gamma_\pm} - \frac{1}{I^{1/2}} = -\frac{0.509}{(-0.0599)} - \frac{1}{0.141} = 1.40\overline{5}$$

Thus,

$$\log \gamma_\pm = -\frac{0.509 I^{1/2}}{1 + 1.40\overline{5} I^{1/2}}$$

In order to determine whether or not the fit is improved, we use the data at $10.0\,\text{mmol kg}^{-1}$

$$\log \gamma_\pm = \frac{-(0.509) \times (0.100)}{(1) + (1.405) \times (0.100)} = -0.0446$$

which fits the data almost exactly. The fits to the other data points will also be almost exact.

P10.11   **(a)** We seek a redox couple with a reduction potential that is more negative than that of Eu^{3+}/Eu but not as negative as that of Yb^{3+}/Yb, so that the standard potential of the reduction of Eu^{3+} is positive while that of Yb^{3+} (and all the other lanthanides) is negative. The couple, then, must have a reduction potential $\boxed{\text{between} -1.991 \text{ and} -2.19 \text{ V}}$. In addition, there must not be any more favourable reaction for the reducing agent (e.g. further oxidation with a potential of greater than 2.19 V). $\boxed{\text{Scandium}}$ fits these criteria; the Sc^{3+}/Sc couple has $E^{\ominus} = -2.09$ V.

(b) For Eu to deposit spontaneously, we must have $E > 0$ for the reaction

$$Eu^{3+} + Sc \rightarrow Eu + Sc^{3+}$$

The cell potential is given by the Nernst equation

$$E_{Eu} = E_{Eu}^{\ominus} - \frac{RT}{3F} \ln Q = E_{Eu}^{\ominus} - \frac{RT}{3F} \ln \frac{a(Sc^{3+})}{a(Eu^{3+})} > 0$$

$$\frac{a(Sc^{3+})}{a(Eu^{3+})} < \exp\left(\frac{3FE_{Eu}^{\ominus}}{RT}\right)$$

where $E_{Eu}^{\ominus} = -1.991$ V $- (-2.09$ V$) = 0.10$ V

We must have $E < 0$ for the reaction

$Yb^{3+} + Sc \rightarrow Yb + Sc^{3+}$, so

$$E_{Yb} = E_{Yb}^{\ominus} - \frac{RT}{3F} \ln Q = E_{Yb}^{\ominus} - \frac{RT}{3F} \ln \frac{a(Sc^{3+})}{a(Yb^{3+})} < 0$$

$$\frac{a(Sc^{3+})}{a(Yb^{3+})} > \exp\left(\frac{3FE_{Yb}^{\ominus}}{RT}\right) \quad \text{or} \quad \frac{a(Yb^{3+})}{a(Sc^{3+})} > \exp\left(\frac{3FE_{Yb}^{\ominus}}{RT}\right)$$

where $E_{Yb}^{\ominus} = -2.19$ V $- (-2.09$ V $= -0.10$ V

Clearly both criteria require the proper amount of scandium to be used, so there are two criteria which must be satisfied separately. The question, however, asked about the implications for the ratio of lanthanides. So, bearing in mind the requirement that the proper amount of scandium be present, the constraint on the lanthanides is

$$\frac{a(Yb^{3+})}{a(Eu^{3+})} < \exp\left(\frac{3(96485 \,\text{C mol}^{-1}) \times (0.20 \,\text{V})}{(8.3145 \,\text{J mol}^{-1} \,\text{K}^{-1}) \times (298 \,\text{K})}\right) = \boxed{1.4 \times 10^{10}}$$

Thus, the separation could work over a large range of concentrations.

P10.13 $Hg_2Cl_2(s) + 2e^- \rightleftharpoons 2Hg(s) + 2Cl^-(aq)$ $\qquad\qquad E_{cathode}^{\ominus} = 0.27$ V

$\qquad\qquad\qquad H_2(g) \rightleftharpoons H^+(aq) + 2e^-$ $\qquad\qquad\qquad E_{anode}^{\ominus} = 0.00$ V

$Hg_2Cl_2(s) + H_2(g) \rightleftharpoons 2H^+(aq) + 2Hg(s) + 2Cl^-(aq)$

$E^{\ominus} = E_{cathode}^{\ominus} - E_{anode}^{\ominus} = 0.27$ V

$\nu = 2$

(a) $\quad E = E^{\ominus} - \frac{RT}{\nu F} \ln Q = E^{\ominus} - \frac{RT}{\nu F} \ln \left[\frac{a_{Cl^-}^2 \, a_{H^+}^2}{f_{H_2}/p^{\ominus}}\right]$

assuming that the ion activity coefficients are not affected by pressure and that H_2 behaves as a perfect gas

$$\Delta E = E(p_2) - E(p_1) = \frac{RT}{\nu F} \ln \frac{f_2}{f_1} = \frac{RT}{\nu F} \ln \frac{p_2}{p_1}$$

Alternatively, with $p_{ref} = 1.00$ atm this can be written as

$$\Delta E(p) = \frac{RT}{vF} \ln \frac{p}{p_{ref}} \tag{1}$$

$$\left(\frac{\partial \Delta E}{\partial p} \right)_T = \frac{RT}{vFp} \tag{2}$$

Equation 1 indicates that a plot of ΔE against $\ln p$ should be linear if the perfect gas assumption is valid. The accompanying plot is indeed linear below about 100 atm but is considerably nonlinear at higher pressures.

(b) An empirical equation for $\Delta E(p)$ is suggested by the substitution of $f = \phi p$ [5.19] into the equation for ΔE,

$$\Delta E = \frac{RT}{vF} \ln \frac{f_2}{f_1} = \frac{RT}{vF} \left\{ \ln \frac{p_2}{p_1} + \ln \frac{\phi_2}{\phi_1} \right\}$$

Equation 5.20 relates ϕ to p and the virial coefficients

$$\ln \phi = B'p + \tfrac{1}{2}C'p^2 + \cdots$$

which suggests the empirical form

$$\Delta E = \frac{RT}{vF} \{ \ln p + c_1 + c_2 p + c_3 p^2 + \cdots \}$$

where the values of c_1, c_2 and c_3 are regression parameters. Regression fit with $\dfrac{RT}{vF} = \dfrac{25.693 \, mV}{2} = 12.847 \, mV$ yields

$c_1 = -0.01685,$ standard deviation $= 0.02471$

$c_2 = 6.288 \times 10^{-4} \, atm^{-1},$ standard deviation $= 1.362 \times 10^{-4} atm^{-1}$

$c_3 = 7.663 \times 10^{-8} atm^{-1},$ standard deviation $= 1.355 \times 10^{-7} atm^{-1}$

$R = 0.999\,760$

The standard deviation of c_1 allows for the conclusion that $c_1 = 0$. The large standard deviation of c_3 also indicates that its inclusion is superfluous. Redoing the regression analysis with the simpler form

$$\boxed{\Delta E = \frac{RT}{vF} \{ \ln p + Cp \}} \qquad \boxed{\left(\frac{\partial E}{\partial p} \right) = \frac{RT}{vF} \left\{ \frac{1}{p} + C \right\}}$$

We find that

$$\boxed{C = 6.665 \times 10^{-4} atm^{-1}}, \quad \text{standard deviation} = 2.4 \times 10^{-5} atm^{-1}$$

$R = 0.999\,400$

The correlation coefficient is modestly smaller for the simpler regression fit, the standard deviation of the regression coefficient is much smaller. The simpler equation is the form of choice to describe this data (Fig. 10.4).

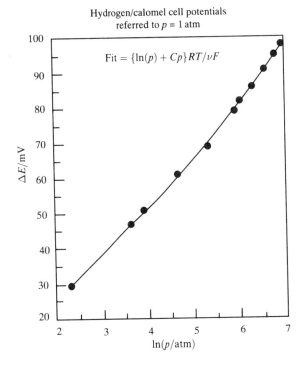

Figure 10.4

(c)
$$\Delta E = \frac{RT}{\nu F}\left\{\ln\frac{p}{p_{ref}} + \ln\frac{\phi}{\phi_{ref}}\right\} \tag{3}$$

where $\ln\phi = \int_0^p \left(\frac{Z-1}{p}\right)dp$

van der Waals equation of state

$$p = \left[\frac{RT}{V_m - b}\right] - \frac{a}{V_m^2}; \qquad dp = \frac{-RT\,dV_m}{(V_m - b)^2} + \frac{2a\,dV_m}{V_m^3} \quad \text{at constant } T$$

$$\ln\phi = \int_0^p \left(\frac{\frac{pV_m}{RT} - 1}{p}\right)dp = \int_\infty^{V_m} \left(\frac{V_m}{RT} - \frac{1}{p}\right) \times \left\{\frac{-RT}{(V_m - b)^2} + \frac{2a}{V_m^3}\right\}dV_m$$

The integral may be numerically performed and substituted into eqn (1) for the evaluation of ΔE.
Empirical virial equation

$$Z = 1 + Ap + Bp^2 \quad \text{where} \quad A = 5.37 \times 10^{-4}\,\text{atm}^{-1} \quad \text{and} \quad B = 3.5 \times 10^{-8}\,\text{atm}^{-2}$$

$$\ln\phi = \int_0^p \left[\frac{(1 + Ap + Bp^2) - 1}{p}\right]dp = \int_0^p [A + Bp]\,dp$$

$$\ln\phi = Ap + \frac{Bp^2}{2}; \qquad \ln\phi_{ref} = A\,\text{atm} + B\,\text{atm}^2/2$$

$$\boxed{\Delta E = \frac{RT}{\nu F}\left\{\ln\left(\frac{p}{\text{atm}}\right) + A(p - 1\,\text{atm}) + \frac{B}{2}(p - 1\,\text{atm})^2\right\}}$$

(d) From eqn 1

$$\phi/\phi_{\text{ref}} = \left(\frac{p_{\text{ref}}}{p}\right) e^{\nu F \Delta E/RT} = \left(\frac{1\,\text{atm}}{p}\right) e^{\nu F \Delta E/RT}$$

p/atm	10	38	51	108	210	380	430	560	720	900	1020
ΔE/mV	29.5	47	51	61	69	79	82	86	91	95	98
ϕ/ϕ_{ref}		0.9937	1.0211	1.0387	1.0683	1.0241	1.2326	1.3758	1.4423	1.6555	1.8082 2.0151

As pressure increases, the values of ϕ are greater than 1 and increase. This means the repulsive forces dominate.

P10.14
$$\text{HA(aq)} \rightarrow \text{H}^+(\text{aq}) + \text{A}^-(\text{aq})$$

Molalities $\quad (1-\alpha)b \qquad \alpha b \qquad \alpha b$

$$K_a = \frac{a(\text{H}^+)a(\text{A}^-)}{a(\text{HA})} = \frac{\gamma_{\pm}^2 b(\text{H}^+)b(\text{A}^-)}{b(\text{HA})} = \gamma_{\pm}^2 K_a' \quad \left[b \equiv \frac{b}{b^{\ominus}}\right]$$

$$K_a' = \frac{b(\text{H}^+)b(\text{A}^-)}{b(\text{HA})} = \frac{\alpha^2 b}{1-\alpha}$$

Hence,

$$\log K_a' = \log K_a - 2\log \gamma_{\pm} = \log K_a + 2AI^{1/2} \text{ [Debye–Hückel limiting law]}$$

$$= \log K_a + 2A(\alpha b)^{1/2} \quad [I = \alpha b]$$

We therefore construct the following table

$\dfrac{1000b}{b^{\ominus}}$	0.0280	0.1114	0.2184	1.0283	2.414	5.9115
$1000\left(\dfrac{\alpha b}{b^{\ominus}}\right)^{1/2}$	3.89	6.04	7.36	11.3	14.1	17.9
$10^5 \times K_a'$	1.768	1.779	1.781	1.799	1.809	1.822
$\log K_a'$	−4.753	−4.750	−4.749	−4.745	−4.743	−4.739

$\log K_a'$ is plotted against $\left(\dfrac{\alpha b}{b^{\ominus}}\right)^{1/2}$ in Fig. 10.5, and we see that a good straight line is obtained.

P10.16 Electrochemical Cell Equation:

$$\tfrac{1}{2}\,\text{H}_2\,(\text{g, 1 bar}) + \text{AgCl(s)} \rightleftharpoons \text{H}^+(\text{aq}) + \text{Cl}^-(\text{aq}) + \text{Ag(s)}$$

where $f(\text{H}_2) = 1\,\text{bar} = p^{\ominus} \quad a_{\text{cl}^-} = \gamma_{\text{cl}^-} b$

Weak acid Equlibrium:

$$\text{BH}^+ \rightleftharpoons \text{B} + \text{H}^+$$

where $b_{\text{BH}} = b_{\text{B}} = b$

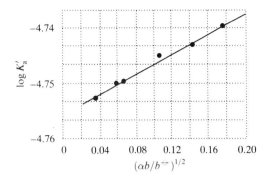

Figure 10.5

and $K_a = a_B\, a_{H^+}/a_{BH} = \gamma_B\, b\, a_{H^+}/\gamma_{BH}\, b = \gamma_B\, a_{H^+}/\gamma_{BH}$

or $a_H = \gamma_{BH}\, K_a/\gamma_B$

Ionic strength (neglect b_{H^+} because $b_H \ll b$):

$I = \frac{1}{2}\{z_{BH}^2\, b_{BH} + z_{cl^-}^2\, b_{cl^-}\} = b$

according to the Nerst equation (eqn 10.34):

$$E = E^\circ - \frac{RT}{F}\ln\left(\frac{a_H + a_{Cl^-}}{f(H_2/p^{\ominus})}\right) = E^{\ominus} - \frac{RT\ln(10)}{F}\log(a_H + a_{Cl^-})$$

$$\frac{F}{RT\ln(10)}\left(E - E^{\ominus}\right) = -\log(a_{H^+}\gamma_{Cl^-}\, b) = -\log\left(\frac{K_a\, \gamma_{BH}\, \gamma_{Cl^-}\, b}{\gamma_B}\right)$$

$$= pK_a - \log(b) - 2\log(\gamma_{\pm})$$

$$\frac{F}{RT\ln(10)}\left(E - E^{\ominus}\right) = pK_a - \log(b) + \frac{2A\sqrt{b}}{1 + B\sqrt{b}} - 2\,kb$$

where $A = 0.5091$.

The expression to the left of the above equality is experimental data that is a function of b. The parameters pK_a, B, and k on the right side are systematically varied with a mathematical regression software until the right side fits the left side in a least squares sense.

$$\boxed{\begin{aligned} pK_a &= 6.736,\ B = 1.997\,\text{kg}^{0.5}\,\text{mol}^{-0.5} \\ k &= -0.121\,\text{kg}\,\text{mol}^{-1} \end{aligned}}$$

$$\gamma_{\pm} = 10^{\left(\frac{-AI^{1/2}}{1 + BI^{1/2}} + kb\right)}$$

P10.17   (a) $BH^+ \rightleftharpoons B + H^+$ K_a

Using mathematical software to determine the parameters a_0, a_1, and a_2 with a regression fit of the data, it is found that

$$_pK_a = a_0 + a_1/(T/K) + a_2\ln(T/K)$$

where $a_0 = -10.057$, $a_1 = 2282$, and $a_2 = 1.604$.

The thermodynamic properties of the reaction can be evaluated with the $pK_a(T)$ function and eqns 9.19, 9.26, and 4.39. See Fig. 10.7

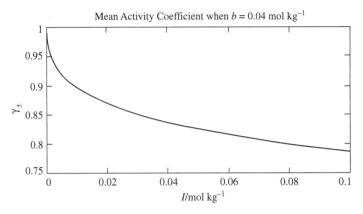

Figure 10.6

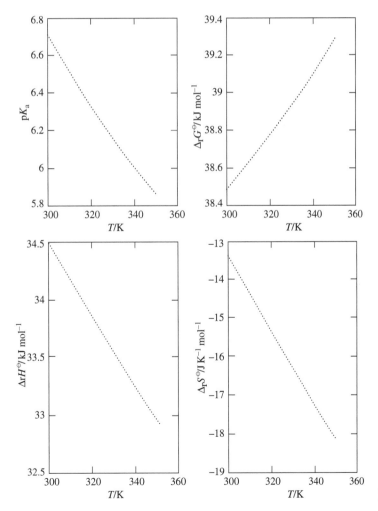

Figure 10.7

$$\Delta_r G^{\ominus}(T) = RT \ln(10)\, pK_a(T)$$

$$\Delta_r H^{\ominus}(T) = R \ln(10)(a_1 - a_2 T) \quad (a_1 \text{ in Kelvin})$$

$$\Delta_r s^{\ominus}(T) = \frac{\Delta_r H^{\ominus}(T) - \Delta_r G^{\ominus}(T)}{T}$$

Solutions to theoretical problems

P10.20 $MX(s) \rightleftharpoons M^+(aq) + X^-(aq)$

$$K_s = a(M^+)a(X^-) = b(M^+)b(X^-)\gamma_{\pm}^2 \quad \left[b \equiv \frac{b}{b^{\ominus}}\right]$$

$$b(M^+) = b(X^-) = S = I$$

$$\ln \gamma_{\pm} = 2.303 \log \gamma_{\pm} = (-2.303) \times (0.509) \times S^{1/2} = -1.172 S^{1/2}$$

$$\gamma_{\pm} = e^{-(1.172 S^{1/2})},$$

Hence, $\dfrac{K_s}{\gamma_{\pm}^2} = S^2$ implying that

$$S = \frac{K_s^{1/2}}{\gamma_{\pm}} = \boxed{K_s e^{(1.172 S^{1/2})}}$$

P10.23 $A(s) \rightleftharpoons A(l)$

$$\mu_A^*(s) = \mu_A^*(l) + RT \ln a_A$$

and $\Delta_{fus} G = \mu_A^*(l) - \mu_A^*(s) = -RT \ln a_A$

Hence, $\ln a_A = \dfrac{-\Delta_{fus} G}{RT}$

$$\frac{d \ln a_A}{dT} = -\frac{1}{R} \frac{d}{dT}\left(\frac{\Delta_{fus} G}{T}\right) = \frac{\Delta_{fus} H}{RT^2} \quad \text{[Gibbs–Helmholtz]}$$

For $\Delta T = T_f^* - T$, $d\Delta T = -dT$ and

$$\frac{d \ln a_A}{d\Delta T} = \frac{-\Delta_{fus} H}{RT^2} \approx \frac{-\Delta_{fus} H}{RT_f^2}$$

But $K_f = \dfrac{RT_f^2 M_A}{\Delta_{fus} H}$ [Chapter 7]

Therefore,

$$\frac{d \ln a_A}{d\Delta T} = \frac{-M_A}{K_f} \quad \text{and} \quad d \ln a_A = \frac{-M_A\, d\Delta T}{K_f}$$

According to the Gibbs–Duhem equation (Chapter 7)

$$n_A\, d\mu_A + n_B\, d\mu_B = 0$$

which implies that

$$n_A\, d \ln a_A + n_B\, d \ln a_B = 0 \quad [\mu = \mu^{\ominus} + RT \ln a]$$

and hence that $d \ln a_A = -\dfrac{n_B}{n_A} d \ln a_B$

Hence, $\dfrac{d \ln a_B}{d \Delta T} = \dfrac{n_A M_A}{n_B K_f} = \dfrac{1}{b_B K_f}$ [for $n_A M_A = 1$ kg]

We know from the Gibbs–Duhem equation that

$$x_A \, d \ln a_A + x_B \, d \ln a_B = 0$$

and hence that $\displaystyle\int d \ln a_A = - \int \dfrac{x_B}{x_A} \, d \ln a_B$

Therefore $\ln a_A = - \displaystyle\int \dfrac{x_B}{x_A} \, d \ln a_B$

The osmotic coefficient was defined in Problem 7.21 as

$$\phi = -\dfrac{1}{r} \ln a_A = -\dfrac{x_A}{x_B} \ln a_A$$

Therefore,

$$\phi = \dfrac{x_A}{x_B} \int \dfrac{x_B}{x_A} \, d \ln a_B = \dfrac{1}{b} \int_0^b b \, d \ln a_B = \dfrac{1}{b} \int_0^b b \, d \ln \gamma b = \dfrac{1}{b} \int_0^b b \, d \ln b + \dfrac{1}{b} \int_0^b b \, d \ln \gamma$$

$$= 1 + \dfrac{1}{b} \int_0^b b \, d \ln \gamma$$

From the Debye–Hückel limiting law,

$$\ln \gamma = -A' b^{1/2}   [A' = 2.303 A]$$

Hence, $d \ln \gamma = -\tfrac{1}{2} A' b^{-1/2} \, db$ and so

$$\phi = 1 + \dfrac{1}{b} \left(-\dfrac{1}{2} A' \right) \int_0^b b^{1/2} \, db = 1 - \dfrac{1}{2} \left(\dfrac{A'}{b} \right) \times \dfrac{2}{3} b^{3/2} = \boxed{1 - \dfrac{1}{3} A' b^{1/2}}$$

Comment. For the depression of the freezing point in a 1,1–electrolyte

$$\ln a_A = \dfrac{-\Delta_{fus} G}{RT} + \dfrac{\Delta_{fus} G}{RT^*}$$

and hence $-r\phi = \dfrac{-\Delta_{fus} H}{R} \left(\dfrac{1}{T} - \dfrac{1}{T^*} \right)$

Therefore, $\phi = \dfrac{\Delta_{fus} H x_A}{R x_B} \left(\dfrac{1}{T} - \dfrac{1}{T^*} \right) = \dfrac{\Delta_{fus} H x_A}{R x_B} \left(\dfrac{T^* - T}{TT^*} \right) \approx \dfrac{\Delta_{fus} H x_A \Delta T}{R x_B T^{*2}}$

$$\approx \dfrac{\Delta_{fus} H \Delta T}{\nu R b_B T^{*2} M_A}$$

where $\nu = 2$. Therefore, since $K_f = \dfrac{M R T^{*2}}{\Delta_{fus} H}$

$$\boxed{\phi = \dfrac{\Delta T}{2 b_B K_f}}$$

P10.24 $PX_\nu(s) \rightleftharpoons p^{\nu+}(aq) + \nu X^-(aq)$

This process is a solubility equilibrium described by a solubility constant K_s

$$K_s = a_{p^{\nu+}} a_{x^-}^\nu$$

Introducing activity coefficients and concentrations, b, we obtain

$$K_s = b_{p^{\nu+}} b_{X^-}^\nu \gamma_\pm^{\nu+1}$$

at low to moderate ionic strengths we can use the Debye–Hückel limiting law as a good approximation for $\gamma_\pm$,

$$\log \gamma_\pm = -|z_+ z_-|A I^{\frac{1}{2}} = -|z_+|A I^{\frac{1}{2}}$$

addition of a salt, such as $(NH_4)_2SO_4$ causes I to increase and $\log \gamma_\pm$ to become more negative and $\gamma_\pm$ will decrease. However, K_s is a true equilibrium constant and remains unchanged. Therefore, the concentration of $P^{\nu+}$ increases and the protein solubility increases proportionately.

We may also explain this effect with the use of Le Chatelier's principle. As the ionic strength increases by the addition of an inert electrolyte such as $(NH_4)_2SO_4$, the ions of the protein that are in solution attract one another less strongly, so that the equilibrium is shifted in the direction of increased solubility.

The explanation of the salting out effect is somewhat more complicated and can be related to the failure the Debye–Hückel limiting law at higher ionic strengths. At high ionic strengths we may write

$$\log \gamma_\pm = -|z_+|A I^{1/2} + KI$$

where K is the salting out constant. At low concentrations of inert salt, $I^{1/2} > I$, and salting in occurs, but at high concentrations, $I > I^{1/2}$. And salting out occurs. The Le Chatelier's principle explanation is that the water molecules are tied up by ion–dipole interactions and become unavailable for solvating the protein, thereby leading to decreased solubility.

P10.26   **(a)** The Nernst equation appropriate to the fluoride selective electrode is

$$E = E_{ap} + \beta \frac{RT}{F} \ln(a_{F^-} + k_{F^-,OH^-} a_{OH^-})$$

at 298 K, this may be written, after setting $\beta \approx 1$,

$$E = E_{ap} + 0.05916 \, V \log(a_{F^-} + k_{F^-,OH^-} a_{OH^-})$$

(b) At high pH, a_{OH^-} is large, and the second term inside the parentheses may be a significant fraction of a_{F^-}. At low pH, F^- is converted to HF, to which the electrode is insensitive. The activities of the species involved are related to each other through K_a.

$$K_a = \frac{a_{H^+} a_{F^-}}{a_{HF}}, \quad a_{F^-} = \frac{K_a a_{HF}}{a_{H^+}} = \frac{3.5 \times 10^{-4} a_{HF}}{a_{H^+}}$$

a_{H^+} and a_{OH^-} are related through $K_w = a_{H^+} a_{OH^-}$.

$$E = E_{ap} + 0.05916 \, V \log \left[a_{F^-} + k_{F^-,OH^-} \left(\frac{K_w}{a_{H^+}} \right) \right]$$

In the following analysis, let us set all activity coefficients equal to 1.

Let us draw up the following table for $E - E_{ap}$

pH [F⁻]	4	5	6	7	8	9
10^{-7}	−0.414	−0.414	−0.414	−0.412	−0.396	−0.353
10^{-6}	−0.355	−0.355	−0.355	−0.355	−0.353	−0.337
10^{-5}	−0.296	−0.296	−0.296	−0.296	−0.296	−0.293
10^{-4}	−0.237	−0.237	−0.237	−0.237	−0.237	−0.236
10^{-3}	−0.177	−0.177	−0.177	−0.177	−0.177	−0.177
10^{-2}	−0.118	−0.118	−0.118	−0.118	−0.118	−0.118
10^{-1}	−0.059	−0.059	−0.059	−0.059	−0.059	−0.059
1	0	0	0	0	0	0

We see that at pH ≤ 8 the emf responds linearly to $\log a_{F^-}$. At pH $= 5$ and below, the ratio

$$\frac{a_{HF}}{a_{F^-}} = \frac{a_{H^+}}{K_a} = \frac{a_{H^+}}{3.5 \times 10^{-4}} = \frac{10^{-5}}{3.5 \times 10^{-4}} = 0.029$$

indicates that a significant fraction (>0.03) of F^- has been removed from the test solution. Therefore, the acceptable pH range for the use of this electrode is $\boxed{5 < \text{pH} < 8}$.

P10.28 **(a)** The cell reaction is

$$H_2(g) + \frac{1}{2}O_2(g) \rightarrow H_2O(l)$$

$$\Delta_r G^\ominus = \Delta_f G^\ominus (H_2O, 1) = -237.13 \, \text{kJ mol}^{-1} \, [\text{Table 2.6}]$$

$$E^\ominus = -\frac{\Delta_r G^\ominus}{\nu F} [10.33] = \frac{+237.13 \, \text{kJ mol}^{-1}}{(2) \times (96.485 \, \text{kC mol}^{-1})} = \boxed{+1.23 \, \text{V}}$$

(b) $C_4H_{10}(g) + \frac{13}{2}O_2(g) \rightarrow 4CO_2(g) + 5H_2O(l)$

$$\begin{aligned}
\Delta_f G^\ominus &= 4\Delta_f G^\ominus (CO_2, g) + 5\Delta_f G^\ominus (H_2O, 1) - \Delta_f G^\ominus (C_4H_{10}, g) \\
&= E(4) \times (-394.36) + (5) \times (-237.13) - (-17.03)] \, \text{kJ mol}^{-1} \\
&\quad [\text{Tables 2.5 and 2.6}] \\
&= -2746.06 \, \text{kJ mol}^{-1}
\end{aligned}$$

In this reaction the number of electrons transferred, ν is not immediately apparent as in part **(a)**. To find ν we break the cell reaction down into half-reactions as follows

R: $\frac{13}{2}O_2(g) + 26e^- + 26H^+(aq) \rightarrow 13H_2O(l)$

L: $4CO_2(g) + 26e^- + 26H^+(aq) \rightarrow C_4H_{10}(g) + 8H_2O(l)$

R − L: $C_4H_{10}(g) + \frac{13}{2}O_2(g) \rightarrow 4CO_2(g) + 8H_2O(l)$

Hence, $\nu = 26$.

Therefore, $E = \dfrac{-\Delta G^\ominus}{\nu F} = \dfrac{+2746.06 \, \text{kJ mol}^{-1}}{(26) \times (96.485 \, \text{kC mol}^{-1})} = \boxed{+1.09 \, \text{V}}$

Part 2: Structure

11 Quantum theory: introduction and principles

Solutions to exercises

Discussion questions

E11.1(a) At the end of the nineteenth century and the beginning of the twentieth, there were many experimental results on the properties of matter and radiation that could not be explained on the basis of established physical principles and theories. Here we list only some of the most significant.

(1) The energy density distribution of blackbody radiation as a function of wavelength.
(2) The heat capacities of monatomic solids such as copper metal.
(3) The absorption and emission spectra of atoms and molecules, especially the line spectra of atoms.
(4) The frequency dependence of the kinetic energy of emitted electrons in the photoelectric effect.
(5) The diffraction of electrons by crystals in a manner similar to that observed for X-rays.

E11.2(a) The heat capacities of monatomic solids are primarily a result of the energy acquired by vibrations of the atoms about their equilibrium positions. If this energy can be acquired continuously, we expect that the equipartition of energy principle should apply. This principle states that for each direction of motion and for each kind of energy (potential and kinetic) the associated energy should be $1/2\,kT$. For three directions and both kinds of motion, a total of $3\,kT$, which gives a heat capacity of $3\,k$ per atom, or $3\,R$ per mole, independent of temperature. But the experiments show a temperature dependence. The heat capacity falls steeply below $3\,R$ at low temperatures. Einstein showed that by allowing the energy of the atomic oscillators to be quantized according to Planck's formula, rather than continuous, this temperature dependence could be explained. The physical reason is that at low temperatures only a few atomic oscillators have enough energy to populate the higher quantized levels, at higher temperatures more of them can acquire the energy to become active.

E11.3(a) If the wavefunction describing the linear momentum of a particle is precisely known, the particle has a definite state of linear momentum; but then according to the uncertainty principle, the position of the particle is completely unknown as demonstrated in the derivation leading to eqn 11.24. Conversely, if the position of a particle is precisely known, its linear momentum cannot be described by a single wavefunction, but rather by a superposition of many wavefunctions, each corresponding to a different value for the linear momentum. Thus all knowledge of the linear momentum of the particle is lost. In the limit of an infinite number of superposed wavefunctions, the wavepacket illustrated in Fig. 11.27 turns into the sharply spiked packet shown in Fig. 11.26. But the requirement of the superposition of an infinite number of momentum wavefunctions in order to locate the particle means a complete lack of knowledge of the momentum.

Numerical exercises

E11.4(a) The power radiated divided by the area is the excitance, M

$$M = \sigma T^4 \ [11.2b]$$

Hence, the power, P is

$$P = \sigma T^4 \times A = (5.67 \times 10^{-8}\,\text{W m}^{-2}\,\text{K}^{-4}) \times (1500\,\text{K})^4 \times (6.0\,\text{m}^2) = \boxed{1.7\,\text{MW}}$$

E11.5(a) The Wien displacement law [11.1] is used to obtain the wavelength corresponding to the maximum (greatest intensity) in Fig. 11.1 of the text.

$$\lambda_{max} = \frac{c_2}{5T}[c_2 = 1.44\,\text{cm K}] = \frac{1.44\,\text{cm K}}{(5) \times (11000\,\text{K})} = 2.62 \times 10^{-5}\,\text{cm}$$

$$= 2.62 \times 10^{-7}\,\text{m} = \boxed{262\,\text{nm}}$$

Comment. This wavelength is in the ultraviolet region of the electromagnetic spectrum. Compare to the Sun which is the subject of Exercise 11.15.

E11.6(a) $\lambda = \dfrac{h}{p}[11.13] = \dfrac{h}{mv}$

Hence, $v = \dfrac{h}{m\lambda} = \dfrac{6.63 \times 10^{-34}\,\text{J s}}{(9.11 \times 10^{-31}\,\text{kg}) \times (0.030\,\text{m})} = \boxed{0.0242\,\text{m s}^{-1}}$ very slow!

E11.7(a) $\lambda = \dfrac{h}{p}[11.13] = \dfrac{h}{mv} = \dfrac{6.626 \times 10^{-34}\,\text{J s}}{(9.109 \times 10^{-31}\,\text{kg}) \times \left(\frac{1}{137}\right) \times (2.998 \times 10^8\,\text{m s}^{-1})}$

$$= 3.32\overline{4} \times 10^{-10}\,\text{m} = \boxed{332\,\text{pm}}$$

Comment. One wavelength of the matter wave of an electron with this velocity just fits in the first Bohr orbit. The velocity of the electron in the first Bohr classical orbit is thus $\frac{1}{137}c$.

Question. What is the wavelength of an electron with velocity approaching the speed of light? Such velocities can be achieved with particle accelerators.

E11.8(a)   If we assume that photons obey a relation analogous to the deBroglie relation we may write

$$p = \frac{h}{\lambda} = \frac{6.626 \times 10^{-34}\,\text{J s}}{750 \times 10^{-9}\,\text{m}} = \boxed{8.83 \times 10^{-28}\,\text{kg m s}^{-1}}$$

For an electron with the same momentum

$$v = \frac{p}{m} = \frac{8.83 \times 10^{-28}\,\text{kg m s}^{-1}}{9.11 \times 10^{-31}\,\text{kg}} = \boxed{9.69 \times 10^2\,\text{m s}^{-1}}$$

E11.9(a)   This is essentially the photoelectric effect with the work function Φ being the ionization energy I. Hence,

$$\frac{1}{2}m_e v^2 = h\nu - I = \frac{hc}{\lambda} - I$$

Solving for λ

$$\lambda = \frac{hc}{I + \frac{1}{2}mv^2} = \frac{(6.626 \times 10^{-34}\,\text{J s}) \times (2.998 \times 10^8\,\text{m s}^{-1})}{(3.44 \times 10^{-18}\,\text{J}) + \left(\frac{1}{2}\right) \times (9.109 \times 10^{-31}\,\text{kg}) \times (1.03 \times 10^6\,\text{m s}^{-1})^2}$$

$$= 5.06 \times 10^{-8}\,\text{m} = \boxed{50.6\,\text{nm}}$$

Question. What is the energy of the photon?

E11.10(a) $\Delta p \approx 0.0100\,\text{per cent of } p_0 = p_0 \times (1.00 \times 10^{-4}) = m_p v \times (1.00 \times 10^{-4}) \quad (p_0 = m_p v)$

$$\Delta q \approx \frac{\hbar}{2\Delta p}[11.41] \approx \frac{1.055 \times 10^{-34}\,\text{J s}}{(2) \times (1.673 \times 10^{-27}\,\text{kg}) \times (4.5 \times 10^5\,\text{m s}^{-1}) \times (1.00 \times 10^{-4})}$$

$$\approx 7.0\overline{1} \times 10^{-10}\,\text{m}, \quad \text{or} \quad \boxed{0.70\,\text{nm}}$$

E11.11(a) $\quad E = h\nu = \dfrac{hc}{\lambda}, \qquad E(\text{per mole}) = N_A E = \dfrac{N_A hc}{\lambda}$

$hc = (6.62608 \times 10^{-34}\,\text{J s}) \times (2.99792 \times 10^{8}\,\text{m s}^{-1}) = 1.986 \times 10^{-25}\,\text{J m}$

$N_A hc = (6.02214 \times 10^{23}\,\text{mol}^{-1}) \times (1.986 \times 10^{-25}\,\text{J m}) = 0.1196\,\text{J m mol}^{-1}$

Thus, $E = \dfrac{1.986 \times 10^{-25}\,\text{J m}}{\lambda};\qquad E(\text{per mole}) = \dfrac{0.1196\,\text{J m mol}^{-1}}{\lambda}$

We can therefore draw up the following table

λ/nm	E/J	$E/(\text{kJ mol}^{-1})$
(a) 600	3.31×10^{-19}	199
(b) 550	3.61×10^{-19}	218
(c) 400	4.97×10^{-19}	299

E11.12(a) Assuming that the H atom is free and stationary, if a photon is absorbed, the atom acquires its momentum p. It therefore reaches a speed v such that $p = mv$. Thus,

$$v = \frac{p}{m_H} = \frac{p}{1.674 \times 10^{-27}\,\text{kg}}$$

$$[m_H = 1.008\,\text{u} = (1.008) \times (1.6605 \times 10^{-27}\,\text{kg}) = 1.674 \times 10^{-27}\,\text{kg}]$$

We draw up the following table using the information in the table above and $p = \dfrac{h}{\lambda}$.

λ/nm	$p/(\text{kg m s}^{-1})$	$v/(\text{m s}^{-1})$
600	1.10×10^{-27}	0.66
550	1.20×10^{-27}	0.72
400	1.66×10^{-27}	0.99

E11.13(a) The total energy emitted in a period τ is $P\tau$. The energy of a photon of 650 nm light is $E = \dfrac{hc}{\lambda}$ with $\lambda = 650$ nm. The total number of photons emitted in an interval τ is then the total energy divided by the energy per photon.

$$N = \frac{P\tau}{E} = \frac{P\tau\lambda}{hc}$$

DeBroglie's relation applies to each photon and thus the total momentum imparted to the glow-worm is

$$p = \frac{Nh}{\lambda} = \frac{P\tau\lambda}{hc} \times \frac{h}{\lambda} = \frac{P\tau}{c}$$

$$P = 0.10\,\text{W} = 0.10\,\text{J s}^{-1}, \qquad \tau = 10\,\text{y}, \qquad p = mv$$

Hence the final speed is

$$v = \frac{P\tau}{cm} = \frac{(0.10\,\text{J s}^{-1}) \times (3.16 \times 10^{8}\,\text{s})}{(2.998 \times 10^{8}\,\text{m s}^{-1}) \times (5.0 \times 10^{-3}\,\text{kg})} = \boxed{21\,\text{m s}^{-1}}$$

Comment. Note that the answer is independent of the wavelength of the radiation emitted: the greater the wavelength the smaller the photon momentum, but the greater the number of photons emitted.

Question. If this glow-worm eventually turns into a firefly which glows for 1 s intervals while flying with a speed of $0.1 \, \text{m s}^{-1}$, what additional speed does the 1 s glowing impart to the firefly? Ignore any frictional effects of air.

E11.14(a) Power is energy per unit time; hence

$$N = \frac{P}{h\nu}[P = \text{power in J s}^{-1}] = \frac{P\lambda}{hc}$$

$$= \frac{P\lambda}{(6.626 \times 10^{-34} \, \text{J s}) \times (2.998 \times 10^8 \, \text{m s}^{-1})} = \frac{(P/\text{W}) \times (\lambda/\text{nm}) \, \text{s}^{-1}}{1.99 \times 10^{-16}}$$

$$= 5.03 \times 10^{15} (P/\text{W}) \times (\lambda/\text{nm}) \, \text{s}^{-1}$$

(a) $N = (5.03 \times 10^{15}) \times (1.0) \times (550 \, \text{s}^{-1}) = \boxed{2.8 \times 10^{18} \, \text{s}^{-1}}$

(b) $N = (5.03 \times 10^{15}) \times (100) \times (550 \, \text{s}^{-1}) = \boxed{2.8 \times 10^{20} \, \text{s}^{-1}}$

E11.15(a) From Wien's law,

$$T\lambda_{\text{max}} = \frac{1}{5}c_2, \quad c_2 = 1.44 \, \text{cm K [11.1]}$$

Therefore, $T = \dfrac{1.44 \, \text{cm K}}{(5) \times (480 \times 10^{-7} \, \text{cm})} = \boxed{6000 \, \text{K}}$

E11.16(a) $E_K = \dfrac{1}{2}mv^2 = h\nu - \Phi = \dfrac{hc}{\lambda} - \Phi$ [11.12]

$\Phi = 2.14 \, \text{eV} = (2.14) \times (1.602 \times 10^{-19} \, \text{J}) = 3.43 \times 10^{-19} \, \text{J}$

(a) $\dfrac{hc}{\lambda} = \dfrac{(6.626 \times 10^{-34} \, \text{J s}) \times (2.998 \times 10^8 \, \text{m s}^{-1})}{700 \times 10^{-9} \, \text{m}} = 2.84 \times 10^{-19} \, \text{J} < \Phi,$

so $\boxed{\text{no ejection}}$ occurs.

(b) $\dfrac{hc}{\lambda} = 6.62 \times 10^{-19} \, \text{J}$

$E_K = \frac{1}{2}mv^2 = (6.62 - 3.43) \times 10^{-19} \, \text{J} = \boxed{3.19 \times 10^{-19} \, \text{J}}$

$v = \left(\dfrac{2E_K}{m}\right)^{1/2} = \left(\dfrac{(2) \times (3.19 \times 10^{-19} \, \text{J})}{9.109 \times 10^{-31} \, \text{kg}}\right)^{1/2} = \boxed{837 \, \text{km s}^{-1}}$

E11.17(a) $\Delta E = \hbar\omega = h\nu = \dfrac{h}{T} \quad \left[T = \text{period} = \dfrac{1}{\nu} = \dfrac{2\pi}{\omega}\right]$

(a) $\Delta E = \dfrac{6.626 \times 10^{-34} \, \text{J s}}{10^{-15} \, \text{s}} = \boxed{7 \times 10^{-19} \, \text{J}},$

corresponding to $N_A \times (7 \times 10^{-19} \, \text{J}) = \boxed{400 \, \text{kJ mol}^{-1}}$

(b) $\Delta E = \dfrac{6.626 \times 10^{-34} \, \text{J s}}{10^{-14} \, \text{s}} = \boxed{7 \times 10^{-20} \, \text{J}}, \quad \boxed{40 \, \text{kJ mol}^{-1}}$

(c) $\Delta E = \dfrac{6.626 \times 10^{-34} \, \text{J s}}{1 \, \text{s}} = \boxed{7 \times 10^{-34} \, \text{J}}, \quad \boxed{4 \times 10^{-13} \, \text{kJ mol}^{-1}}$

E11.18(a) $\lambda = \dfrac{h}{p} = \dfrac{h}{mv}$ [11.13]

(a) $\lambda = \dfrac{6.626 \times 10^{-34}\,\text{J s}}{(1.0 \times 10^{-3}\,\text{kg}) \times (1.0 \times 10^{-2}\,\text{m s}^{-1})} = \boxed{6.6 \times 10^{-29}\,\text{m}}$

(b) $\lambda = \dfrac{6.626 \times 10^{-34}\,\text{J s}}{(1.0 \times 10^{-3}\,\text{kg}) \times (1.00 \times 10^{5}\,\text{m s}^{-1})} = \boxed{6.6 \times 10^{-36}\,\text{m}}$

(c) $\lambda = \dfrac{6.626 \times 10^{-34}\,\text{J s}}{(4.003) \times (1.6605 \times 10^{-27}\,\text{kg}) \times (1000\,\text{m s}^{-1})} = \boxed{99.7\,\text{pm}}$

Comment. The wavelengths in **(a)** and **(b)** are smaller than the dimensions of any known particle, whereas that in **(c)** is comparable to atomic dimensions.

Question. For stationary particles, $v = 0$, corresponding to an infinite wavelength. What meaning can be ascribed to this result?

E11.19(a)   The minimum uncertainty in position and momentum is given by the uncertainly principle in the form

$\Delta p \Delta q \geq \tfrac{1}{2}\hbar$ [11.41], with the choice of the equality

$\Delta p = m\Delta v$

$\Delta v_{min} = \dfrac{\hbar}{2m\Delta q} = \dfrac{1.055 \times 10^{-34}\,\text{J s}}{(2) \times (0.500\,\text{kg}) \times (1.0 \times 10^{-6}\,\text{m})} = \boxed{1.1 \times 10^{-28}\,\text{m s}^{-1}}$

$\Delta q_{min} = \dfrac{\hbar}{2m\Delta v} = \dfrac{1.055 \times 10^{-34}\,\text{J s}}{(2) \times (5.0 \times 10^{-3}\,\text{kg}) \times (1 \times 10^{-5}\,\text{m s}^{-1})} = \boxed{1 \times 10^{-27}\,\text{m}}$

Comment. These uncertainties are extremely small; thus, the ball and bullet are effectively classical particles.

Question. If the ball were stationary (no uncertainty in position) the uncertainty in speed would be infinite. Thus, the ball could have a very high speed, contradicting the fact that it is stationary. What is the resolution of this apparent paradox?

E11.20(a)   In this case the work function is the ionization energy of the electron

$\dfrac{1}{2}mv^2 = h\nu - I$ [11.12], $\nu = \dfrac{c}{\lambda}$

$I = \dfrac{hc}{\lambda} - \dfrac{1}{2}mv^2$

$= \dfrac{(6.626 \times 10^{-34}\,\text{J s}) \times (2.998 \times 10^{8}\,\text{m s}^{-1})}{150 \times 10^{-12}\,\text{m}} - \left(\dfrac{1}{2}\right) \times (9.109 \times 10^{-31}\,\text{kg})$

$\times (2.14 \times 10^{7}\,\text{m s}^{-1})^2$

$= \boxed{1.12 \times 10^{-15}\,\text{J}}$

Solutions to problems

Solutions to numerical problems

P11.1   A cavity approximates an ideal black body; hence the Planck distribution applies

$$\rho = \frac{8\pi hc}{\lambda^5} \left(\frac{1}{e^{hc/\lambda kT} - 1} \right) \quad [11.5]$$

Since the wavelength range is small (5 nm) we may write as a good approximation

$$\Delta E = \rho \Delta \lambda, \quad \lambda \approx 652.5 \, \text{nm}$$

$$\frac{hc}{\lambda k} = \frac{(6.626 \times 10^{-34} \, \text{J s}) \times (2.998 \times 10^8 \, \text{m s}^{-1})}{(6.525 \times 10^{-7} \, \text{m}) \times (1.381 \times 10^{-23} \, \text{J K}^{-1})} = 2.205 \times 10^4 \, \text{K}$$

$$\frac{8\pi hc}{\lambda^5} = \frac{(8\pi) \times (6.626 \times 10^{-34} \, \text{J s}) \times (2.998 \times 10^8 \, \text{m s}^{-1})}{(652.5 \times 10^{-9} \, \text{m})^5} = 4.221 \times 10^7 \, \text{J m}^{-4}$$

$$\Delta E = (4.221 \times 10^7 \, \text{J m}^{-4}) \times \left(\frac{1}{e^{(2.205 \times 10^4 \, \text{K})/T} - 1} \right) \times (5 \times 10^{-9} \, \text{m})$$

(a) $T = 298 \, \text{K}, \quad \Delta E = \dfrac{0.211 \, \text{J m}^{-3}}{e^{(2.205 \times 10^4)/298} - 1} = \boxed{1.6 \times 10^{-33} \, \text{J m}^{-3}}$

(b) $T = 3273 \, \text{K}, \quad \Delta E = \dfrac{0.211 \, \text{J m}^{-3}}{e^{(2.205 \times 10^4)/3273} - 1} = \boxed{2.5 \times 10^{-4} \, \text{J m}^{-3}}$

Comment. The energy density in the cavity does not depend on the volume of the cavity, but the total energy in any given wavelength range does, as well as the total energy over all wavelength ranges.

Question. What is the total energy in this cavity within the range 650–655 nm at the stated temperatures?

P11.2 $\lambda_{\text{max}} T = \dfrac{hc}{5k}, \quad \left[11.1, \text{ and } c_2 = \dfrac{hc}{k} \right]$

Therefore, $\lambda_{\text{max}} = \dfrac{hc}{5k} \times \dfrac{1}{T}$ and if we plot λ_{max} against $\dfrac{1}{T}$ we can obtain h from the slope. We draw up the following table

$\theta/^\circ\text{C}$	1000	1500	2000	2500	3000	3500
T/K	1273	1773	2273	2773	3273	3773
$10^4/(T/\text{K})$	7.86	5.64	4.40	3.61	3.06	2.65
$\lambda_{\text{max}}/\text{nm}$	2181	1600	1240	1035	878	763

The points are plotted in Fig. 11.1. From the graph, the slope is $2.73 \times 10^6 \, \text{nm}/(1/\text{K})$, that is,

$$\frac{hc}{5k} = 2.73 \times 10^6 \, \frac{\text{nm}}{1/\text{K}} = 2.73 \times 10^{-3} \, \text{m K}$$

and $h = \dfrac{(5) \times (1.38066 \times 10^{-23} \, \text{J K}^{-1}) \times (2.73 \times 10^{-3} \, \text{m K})}{2.99792 \times 10^8 \, \text{m s}^{-1}} = \boxed{6.29 \times 10^{-34} \, \text{J s}}$

Comment. Planck's estimate of the constant h in his first paper of 1900 on black body radiation was $6.55 \times 10^{-27} \, \text{erg sec} (1 \, \text{erg} = 10^{-7} \, \text{J})$ which is remarkably close to the current value of

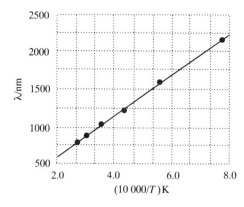

Figure 11.1

6.626×10^{-34} J s. Also from his analysis of the experimental data he obtained values of k (the Boltzmann constant), N_A (the Avogadro constant), and e (the fundamental charge). His values of these constants remained the most accurate for almost 20 years.

P11.4   The full solution of the Schrödinger equation for the problem of a particle in a one-dimensional box is given in Chapter 12. Here we need only the wavefunction which is provided. It is the square of the wavefunction that is related to the probability. Here $\psi^2 = \dfrac{2}{L} \sin^2 \dfrac{\pi x}{L}$ and the probability that the particle will be found between a and b is

$$P(a, b) = \int_a^b \psi^2 \, dx \text{ [Section 11.4]}$$

$$= \frac{2}{L} \int_a^b \sin^2 \frac{\pi x}{L} \, dx = \left(\frac{x}{L} - \frac{1}{2\pi} \sin \frac{2\pi x}{L} \right) \Big|_a^b$$

$$= \frac{b - a}{L} - \frac{1}{2\pi} \left(\sin \frac{2\pi b}{L} - \sin \frac{2\pi a}{L} \right)$$

$L = 10.0$ nm

(a) $P(4.95, 5.05) = \dfrac{0.10}{10.0} - \dfrac{1}{2\pi} \left(\sin \dfrac{(2\pi) \times (5.05)}{10.0} - \sin \dfrac{(2\pi) \times (4.95)}{10.0} \right)$

$$= 0.010 + 0.010 = \boxed{0.020}$$

(b) $P(1.95, 2.05) = \dfrac{0.10}{10.0} - \dfrac{1}{2\pi} \left(\sin \dfrac{(2\pi) \times (2.05)}{10.0} - \sin \dfrac{(2\pi) \times (1.95)}{10.0} \right)$

$$= 0.010 - 0.0031 = \boxed{0.007}$$

(c) $P(9.90, 10.0) = \dfrac{0.10}{10.0} - \dfrac{1}{2\pi} \left(\sin \dfrac{(2\pi) \times (10.0)}{10.0} - \sin \dfrac{(2\pi) \times (9.90)}{10.0} \right)$

$$= 0.010 - 0.009993 = \boxed{7 \times 10^{-6}}$$

(d) $P(5.0, 10.0) = \boxed{0.5}$ [by symmetry]

(e) $P\left(\dfrac{1}{3}L, \dfrac{2}{3}L \right) = \dfrac{1}{3} - \dfrac{1}{2\pi} \left(\sin \dfrac{4\pi}{3} - \sin \dfrac{2\pi}{3} \right) = \boxed{0.61}$

P11.6   The average position (angle) is given by:

$$\langle \phi \rangle = \int \psi^* \phi \psi \, d\tau = \int_0^{2\pi} \frac{e^{im\phi}}{(2\pi)^{1/2}} \phi \frac{e^{-im\phi}}{(2\pi)^{1/2}} \, d\phi = \frac{1}{2\pi} \int_0^{2\pi} \phi \, d\phi = \frac{1}{2\pi} \frac{\phi^2}{2} \Big|_0^{2\pi} = \boxed{\pi}.$$

Note: this result applies to all values of the quantum number m, for it drops out of the calculation.

P11.8   The expectation value of the commutator is:

$$\langle [\hat{x}, \hat{p}] \rangle = \int \psi^* [\hat{x}, \hat{p}] \psi \, d\tau.$$

First evaluate the commutator acting on the wavefunction. The commutator of the position and momentum operators is defined as

$$[\hat{x}, \hat{p}] = \hat{x}\hat{p} - \hat{p}\hat{x} = x \times \frac{\hbar}{i} \frac{d}{dx} - \frac{\hbar}{i} \frac{d}{dx} x,$$

so the commutator acting on the wavefunction is

$$[\hat{x}, \hat{p}]\psi = x \times \frac{\hbar}{i} \frac{d\psi}{dx} - \frac{\hbar}{i} \frac{d}{dx}(x\psi),$$

where $\Psi = (2a)^{1/2} e^{-ax}$.

Evaluating this expression yields

$$[\hat{x}, \hat{p}]\psi = \frac{x\hbar}{i}(2a)^{1/2} a \, e^{-ax} - \frac{\hbar}{i}[(2a)^{1/2} e^{-ax} + xa(2a)^{1/2} e^{-ax}],$$

$$[\hat{x}, \hat{p}]\psi = \frac{\hbar(2a)^{1/2} e^{-ax}}{i}(xa - 1 - xa) = i\hbar(2a)^{1/2} e^{-ax},$$

which is just $i\hbar$ times the original wavefunction. Putting this result into the expectation value yields:

$$\langle [\hat{x}, \hat{p}] \rangle = \int_0^\infty (2a)^{1/2} e^{-ax} (i\hbar)(2a)^{1/2} e^{-ax} \, dx = 2ia\hbar \int_0^\infty e^{-2ax} \, dx$$

$$\langle [\hat{x}, \hat{p}] \rangle = 2ia\hbar \times \frac{e^{-2ax}}{-2a} \Big|_0^\infty = \boxed{i\hbar}.$$

Note: Although the commutator is a well defined and useful operator in quantum mechanics, it does not correspond to an observable quantity. Thus one need not be concerned about obtaining an imaginary expectation value.

Solutions to theoretical problems

P11.11   **(a)** With a little manipulation, a small-wavelength approximation of the Planck distribution can be derived that has the same form as Wien's formula. First examine the Planck distribution,

$$\rho_{\text{Planck}} = \frac{8\pi hc}{\lambda^5 (e^{hc/\lambda kT} - 1)},$$

for small-wavelength behavior. The factor λ^{-5} gets large as λ itself gets small, but the other factor, namely $\frac{1}{e^{hc/\lambda kT} - 1}$ gets small even faster. Focus on that factor, and try to express it in terms of a single decaying exponential (as in Wien's formula), at least in the small-λ limit.

Multiplying it by one in the form of $\dfrac{e^{-hc/\lambda kT}}{e^{-hc/\lambda kT}}$, yields $\dfrac{e^{-hc/\lambda kT}}{1 - e^{-hc/\lambda kT}}$, where $e^{-hc/\lambda kT}$ is small, so let us call it ε. The factor, then, becomes $\dfrac{\varepsilon}{1 - \varepsilon}$, which can be expressed as a power series in ε as $\varepsilon(1 + \varepsilon \cdots)$. For sufficiently small wavelengths, then, the Planck distribution may be approximated as:

$$\rho_{\text{Planck}} \approx \frac{8\pi hc\varepsilon}{\lambda^5} = \frac{8\pi hce^{-hc/\lambda kT}}{\lambda^5}.$$

This has the same form as Wien's formula:

$$\rho_{\text{Wien}} = \frac{a}{\lambda^5} e^{-b/\lambda kT}.$$

Comparing the two formulas gives the values of the Wien constants:

$$a = \boxed{8\pi hc} \qquad \text{and} \qquad b = \boxed{hc}.$$

(b) The wavelength at which the Wien distribution is a maximum is found by setting the derivative of the distribution function to zero:

$$\frac{d\rho_{\text{Wein}}}{d\lambda} = 0 = \frac{a}{\lambda^5} e^{-b/\lambda kT} \left(\frac{b}{\lambda^2 kT} \right) - \frac{5a}{\lambda^6} e^{-b/\lambda kT} = \frac{a}{\lambda^6} e^{-b/\lambda kT} \left(\frac{b}{\lambda kT} - 5 \right),$$

so $\dfrac{b}{\lambda kT} - 5 = 0$ and $\lambda_{\text{max}} = \dfrac{b}{5kT} = \dfrac{hc}{5kT}$.

Putting this in the same form as the Wien displacement law, we get:

$$T\lambda_{\text{max}} = \frac{1}{5} c_2, \quad \text{where } c_2 = \frac{hc}{k},$$

as was demonstrated in Problem 11.9.

The Stefan–Boltzmann law gives the energy density as a function of temperature. The energy density is related to the distribution function by:

$$dE = \rho \, d\lambda \quad \text{so} \quad E = \int_0^\infty \rho \, d\lambda.$$

The energy density implied by the Wien distribution is:

$$E = \int_0^\infty \frac{a}{\lambda^5} e^{-b/\lambda kT} \, d\lambda.$$

Integration by parts several times yields:

$$E = e^{-b/\lambda kT} \left(\left(\frac{b}{k\lambda} \right)^3 + 3 \left(\frac{b}{k\lambda} \right)^2 T + \frac{6bT^2}{k\lambda} + 6T^3 \right) \frac{aTk^4}{b^4} \bigg|_0^\infty = \frac{6aT^4 k^4}{b^4},$$

$$E = \frac{48\pi k^4 T^4}{h^3 c^3},$$

in other words, a constant times T^4, consistent with the Stefan–Boltzmann law.

P11.13   We require $\int \psi^* \psi \, d\tau = 1$, and so write $\psi = Nf$ and find N for the given f.

(a) $N^2 \int_0^L \sin^2 \dfrac{n\pi x}{L} \, dx = \dfrac{1}{2} N^2 \int_0^L \left(1 - \cos \dfrac{2n\pi x}{L} \right) dx$ [trigonometric identity]

$$= \dfrac{1}{2} N^2 \left(x - \dfrac{L}{2n\pi} \sin \dfrac{2n\pi x}{L} \right) \Big|_0^L$$

$$= \dfrac{L}{2} N^2 = 1 \quad \text{if} \boxed{N = \left(\dfrac{2}{L} \right)^{1/2}}$$

(b) $N^2 \int_{-L}^L c^2 \, dx = 2N^2 c^2 L = 1 \quad \text{if} \boxed{N = \dfrac{1}{c(2L)^{1/2}}}$

(c) $N^2 \int_0^\infty e^{-2r/a} r^2 \, dr \int_0^\pi \sin\theta \, d\theta \int_0^{2\pi} d\phi \qquad [d\tau = r^2 \sin\theta \, dr \, d\theta \, d\phi]$

$$= N^2 \left(\dfrac{a^3}{4} \right) \times (2) \times (2\pi) = 1 \quad \text{if} \boxed{N = \dfrac{1}{(\pi a^3)^{1/2}}}$$

(d) $N^2 \int_0^\infty r^2 \times r^2 e^{-r/a} \, dr \int_0^\pi \sin^3\theta \, d\theta \int_0^{2\pi} \cos^2\phi \, d\phi \qquad [x = r\cos\phi \sin\theta]$

$$= N^2 4! a^5 \times \dfrac{4}{3} \times \pi = 32\pi a^5 N^2 = 1 \quad \text{if} \boxed{N = \dfrac{1}{(32\pi a^5)^{1/2}}}$$

We have used $\int \sin^3\theta \, d\theta = -\dfrac{1}{3}(\cos\theta)(\sin^2\theta + 2)$, as found in tables of integrals and

$$\int_0^{2\pi} \cos^2\phi \, d\phi = \int_0^{2\pi} \sin^2\phi \, d\phi$$

by symmetry with $\int_0^{2\pi} (\cos^2\phi + \sin^2\phi) \, d\phi = \int_0^{2\pi} d\phi = 2\pi$

P11.15   In each case form $\hat{\Omega} f$. If the result is ωf where ω is a constant, then f is an eigenfunction of the operator $\hat{\Omega}$ and ω is the eigenvalue [11.30].

(a) $\dfrac{d}{dx} e^{ikx} = ik \, e^{ikx}; \quad \boxed{\text{yes; eigenvalue} = ik}$

(b) $\dfrac{d}{dx} \cos kx = -k \sin kx; \quad$ no.

(c) $\dfrac{d}{dx} k = 0; \quad \boxed{\text{yes; eigenvalue} = 0}$

(d) $\dfrac{d}{dx} kx = k = \dfrac{1}{x} kx; \quad$ no [$1/x$ is not a constant].

(e) $\dfrac{d}{dx} e^{-\alpha x^2} = -2\alpha x \, e^{-\alpha x^2}; \quad$ no [$-2\alpha x$ is not a constant].

P11.17 Follow the procedure of Problem 11.15

(a) $\dfrac{d^2}{dx^2} e^{ikx} = -k^2 e^{ikx}$; yes; eigenvalue = $\boxed{-k^2}$

(b) $\dfrac{d^2}{dx^2} \cos kx = -k^2 \cos kx$; yes; eigenvalue = $\boxed{-k^2}$

(c) $\dfrac{d^2}{dx^2} k = 0$; yes; eigenvalue = $\boxed{0}$

(d) $\dfrac{d^2}{dx^2} kx = 0$; yes; eigenvalue = $\boxed{0}$

(e) $\dfrac{d^2}{dx^2} e^{-\alpha x^2} = (-2\alpha + 4\alpha^2 x^2)e^{-\alpha x^2}$; no.

$\boxed{\text{Hence, } \text{(a, b, c, d) are eigenfunctions of } \dfrac{d^2}{dx^2}; \text{ (b, d) are eigenfunctions of } \dfrac{d^2}{dx^2}, \text{ but not of } \dfrac{d}{dx}}$

P11.18 $\psi = (\cos\chi)e^{ikx} + (\sin\chi)\,e^{-ikx} = c_1 e^{ikx} + c_2\,e^{-ikx}$. The linear momentum operator is

$$\hat{p}_x = \frac{\hbar}{i}\frac{d}{dx} \quad [11.32]$$

As demonstrated in the text (*Justification* 11.2), e^{ikx} is an eigenfunction of $\hat{p}_x$ with eigenvalue $+k\hbar$; likewise e^{-ikx} is an eigenfunction of $\hat{p}_x$ with eigenvalue $-k\hbar$. Therefore, by the principle of linear superposition (Section 11.5(d), *Justification* 11.3),

(a) $P = c_1^2 = \boxed{\cos^2\chi}$

(b) $P = c_2^2 = \boxed{\sin^2\chi}$

(c) $c_1^2 = 0.90 = \cos^2\chi$, so $\cos\chi = 0.95$

$c_2^2 = 0.10 = \sin^2\chi$, so $\sin\chi = \pm 0.32$; hence

$\boxed{\psi = 0.95 e^{ikx} \pm 0.32 e^{-ikx}}$

P11.21 $\langle r \rangle = N^2 \displaystyle\int \psi^* r \psi\, d\tau, \qquad \langle r^2 \rangle = N^2 \displaystyle\int \psi^* r^2 \psi\, d\tau$

(a) $\psi = \left(2 - \dfrac{r}{a_0}\right) e^{-r/2a_0}, \qquad N = \left(\dfrac{1}{32\pi a_0^3}\right)^{1/2}$ [Problem 11.14]

$$\langle r \rangle = \frac{1}{32\pi a_0^3} \int_0^\infty r\left(2 - \frac{r}{a_0}\right)^2 r^2 e^{-r/a_0}\, dr \times 4\pi \quad \left[\int_0^\pi \sin\theta\, d\theta \int_0^{2\pi} d\phi = 4\pi\right]$$

$$= \frac{1}{8a_0^3} \int_0^\infty \left(4r^3 - \frac{4r^4}{a_0} + \frac{r^5}{a_0^2}\right) e^{-r/a_0}\, dr$$

$$= \frac{1}{8a_0^3}(4 \times 3! a_0^4 - 4 \times 4! a_0^4 + 5! a_0^4) = \boxed{6a_0} \quad \left[\int_0^\infty x^n e^{-ax}\, dx = \frac{n!}{a^{n+1}}\right]$$

$$\langle r^2 \rangle = \frac{1}{8a_0^3} \int_0^\infty \left(4r^4 - \frac{4r^5}{a_0} + \frac{r^6}{a_0^2} \right) e^{-r/a_0} \, dr = \frac{1}{8a_0^3} (4 \times 4! - 4 \times 5! + 6!) a_0^5$$

$$= \boxed{42a_0^2}$$

(b) $\quad \psi = Nr \sin\theta \cos\phi \, e^{-r/2a_0}, \qquad N = \left(\frac{1}{32\pi a_0^5} \right)^{1/2} \quad$ [Problem 11.14]

$$\langle r \rangle = \frac{1}{32\pi a_0^5} \int_0^\infty r^5 e^{-r/a_0} \, dr \times \frac{4\pi}{3} = \frac{1}{24a_0^5} \times 5! a_0^6 = \boxed{5a_0}$$

$$\langle r^2 \rangle = \frac{1}{24a_0^5} \int_0^\infty r^6 e^{-r/a_0} \, dr = \frac{1}{24a_0^5} \times 6! a_0^7 = \boxed{30a_0^2}$$

P11.22 $\quad \psi = \left(\frac{1}{\pi a_0^3} \right)^{1/2} e^{-r/a_0}$ [Example 11.4]

(a) $\quad \langle V \rangle = \int \psi^* \hat{V} \psi \, d\tau \quad \left[V = -\frac{e^2}{4\pi\varepsilon_0 r}, \text{ Section 13.1} \right]$

$$\langle V \rangle = \int \psi^* \left(\frac{-e^2}{4\pi\varepsilon_0} \cdot \frac{1}{r} \right) \psi \, d\tau = \frac{1}{\pi a_0^3} \left(\frac{-e^2}{4\pi\varepsilon_0} \right) \int_0^\infty r \, e^{-2r/a_0} \, dr \times 4\pi$$

$$= \frac{1}{\pi a_0^3} \left(\frac{-e^2}{4\pi\varepsilon_0} \right) \times \left(\frac{a_0}{2} \right)^2 \times 4\pi = \boxed{\frac{-e^2}{4\pi\varepsilon_0 a_0}}$$

(b) For three-dimensional systems such as the hydrogen atom the kinetic energy operator is

$$\hat{T} = -\frac{\hbar^2}{2m_e} \nabla^2 \text{ [Table 11.1, } m_e \approx \mu \text{ for the hydrogen atom]}$$

$$\nabla^2 = \frac{\partial^2}{\partial r^2} + \frac{2}{r} \frac{\partial}{\partial r} + \frac{1}{r^2} \Lambda^2 = \left(\frac{1}{r} \right) \times \left(\frac{\partial^2}{\partial r^2} \right) r + \frac{1}{r^2} \Lambda^2$$

$$\Lambda^2 \psi = 0 \quad [\psi \text{ has no angular coordinates}]$$

$$\nabla^2 \psi = \left(\frac{1}{\pi a_0^3} \right)^{1/2} \times \left(\frac{1}{r} \right) \times \left(\frac{d^2}{dr^2} \right) r \, e^{-r/a_0}$$

$$= \left(\frac{1}{\pi a_0^3} \right)^{1/2} \times \left[-\left(\frac{2}{a_0 r} \right) + \frac{1}{a_0^2} \right] e^{-r/a_0}$$

Then, $\langle T \rangle = -\left(\dfrac{\hbar^2}{2m_e} \right) \times \left(\dfrac{1}{\pi a_0^3} \right) \displaystyle\int_0^{2\pi} d\phi \int_0^\pi \sin\theta \, d\theta \int_0^\infty \left[-\left(\frac{2}{a_0 r} \right) + \left(\frac{1}{a_0^2} \right) \right] e^{-2r/a_0} r^2 \, dr$

$$= -\left(\frac{2\hbar^2}{m_e a_0^3} \right) \int_0^\infty \left[-\left(\frac{2r}{a_0} \right) + \left(\frac{r^2}{a_0^2} \right) \right] e^{-2r/a_0} \, dr$$

$$= -\left(\frac{2\hbar^2}{m_e a_0^3} \right) \times \left(-\frac{a_0}{4} \right) \left[\int_0^\infty x^n e^{-ax} \, dx = \boxed{\frac{n!}{a^{n+1}}} \right] = \boxed{\frac{\hbar^2}{2m_e a_0^2}}$$

Inserting $a_0 = \dfrac{4\pi\varepsilon_0\hbar^2}{m_e e^2}$ [Chapter 13]

$$\langle T \rangle = \frac{e^2}{8\pi\varepsilon_0 a_0} = -\frac{1}{2}\langle V \rangle$$

P11.24   The quantity $\hat{\Omega}_1\hat{\Omega}_2 - \hat{\Omega}_2\hat{\Omega}_1$ [*Illustration* 11.2] is referred to as the commutator of the operators $\hat{\Omega}_1$ and $\hat{\Omega}_2$. In obtaining the commutator it is necessary to realize that the operators operate on functions; thus, we form

$$\hat{\Omega}_1\hat{\Omega}_2 f(x) - \hat{\Omega}_2\hat{\Omega}_1 f(x)$$

(a) $\dfrac{\mathrm{d}}{\mathrm{d}x}\hat{x} f(x) = \hat{x}\dfrac{\mathrm{d}f(x)}{\mathrm{d}x} + f(x)$

$\hat{x}\dfrac{\mathrm{d}}{\mathrm{d}x} f(x) = x\dfrac{\mathrm{d}f(x)}{\mathrm{d}x}$

$\left(\dfrac{\mathrm{d}}{\mathrm{d}x}\hat{x} - \hat{x}\dfrac{\mathrm{d}}{\mathrm{d}x}\right) f(x) = f(x)$

Thus, $\left(\dfrac{\mathrm{d}}{\mathrm{d}x}\hat{x} - \hat{x}\dfrac{\mathrm{d}}{\mathrm{d}x}\right) = \boxed{1}$

(b) $\dfrac{\mathrm{d}}{\mathrm{d}x}\hat{x}^2 f(x) = x^2 f'(x) + 2xf(x)$

$\hat{x}^2\dfrac{\mathrm{d}}{\mathrm{d}x} f(x) = x^2 f'(x)$

$\left(\dfrac{\mathrm{d}}{\mathrm{d}x}\hat{x}^2 - \hat{x}^2\dfrac{\mathrm{d}}{\mathrm{d}x}\right) f(x) = 2xf(x)$

Thus, $\left(\dfrac{\mathrm{d}}{\mathrm{d}x}\hat{x}^2 - \hat{x}^2\dfrac{\mathrm{d}}{\mathrm{d}x}\right) = \boxed{2x}$

(c) $p_x = \dfrac{\hbar}{i}\dfrac{\mathrm{d}}{\mathrm{d}x}$

Therefore $a = \left(\hat{x} + \hbar\dfrac{\mathrm{d}}{\mathrm{d}x}\right)$ and $a^\dagger = \left(\hat{x} - \hbar\dfrac{\mathrm{d}}{\mathrm{d}x}\right)$

Then $aa^\dagger f(x) = \dfrac{1}{2}\left(\hat{x} + \hbar\dfrac{\mathrm{d}}{\mathrm{d}x}\right) \times \left(\hat{x} - \hbar\dfrac{\mathrm{d}}{\mathrm{d}x}\right) f(x)$

and $a^\dagger a f(x) = \dfrac{1}{2}\left(\hat{x} - \hbar\dfrac{\mathrm{d}}{\mathrm{d}x}\right) \times \left(\hat{x} + \hbar\dfrac{\mathrm{d}}{\mathrm{d}x}\right) f(x)$

The terms in $\hat{x}^2$ and $\left(\dfrac{\mathrm{d}}{\mathrm{d}x}\right)^2$ obviously drop out when the difference is taken and are ignored in what follows; thus

$$aa^\dagger f(x) = \frac{1}{2}\left(-\hat{x}\hbar\frac{\mathrm{d}}{\mathrm{d}x} + \hbar\frac{\mathrm{d}}{\mathrm{d}x}x\right) f(x)$$

$$a^\dagger a f(x) = \frac{1}{2}\left(x\hbar\frac{\mathrm{d}}{\mathrm{d}x} - \hbar\frac{\mathrm{d}}{\mathrm{d}x}x\right) f(x)$$

These expressions are the negative of each other, therefore

$$(aa^\dagger - a^\dagger a)f(x) = \hbar\frac{\mathrm{d}}{\mathrm{d}x}\hat{x}f(x) - \hbar\hat{x}\frac{\mathrm{d}}{\mathrm{d}x}f(x)$$

$$= \hbar\left(\frac{\mathrm{d}}{\mathrm{d}x}\hat{x} - \hat{x}\frac{\mathrm{d}}{\mathrm{d}x}\right)f(x) = \hbar f(x) \text{ [from (a)]}$$

Therefore, $(aa^\dagger - a^\dagger a) = \boxed{\hbar}$

Solutions to applications

P11.25 $\lambda_{max} = \dfrac{1.44\,\mathrm{cm\,K}}{5T}$ [11.1]

$$= \frac{1.44\,\mathrm{cm\,K}}{5(5800\,\mathrm{K})} = 5.0 \times 10^{-5}\,\mathrm{cm}\left(\frac{10^9\,\mathrm{nm}}{10^2\,\mathrm{cm}}\right)$$

$\lambda_{max} = \boxed{500\,\mathrm{nm, blue\text{–}green}}$ [see Fig. 16.1 in the text]

P11.26 $I = aI + M = aI + \sigma T^4$ so $T = \left(\dfrac{I(1-a)}{\sigma}\right)^{1/4} = \left(\dfrac{(343\,\mathrm{W\,m^{-2}}) \times (1 - 0.30)}{5.67 \times 10^{-8}\,\mathrm{W\,m^{-2}\,K^{-4}}}\right)^{1/4}$

$$= \boxed{255\,\mathrm{K}}$$

where I is the incoming energy flux, a the albedo (fraction of incoming radiation absorbed), M the excitance and σ the Stefan–Boltzmann constant. Wien's displacement law relates the temperature to the wavelength of the most intense radiation

$$T\lambda_{max} = c_2/5 \quad \text{so} \quad \lambda_{max} = \frac{c_2}{5T} = \frac{1.44\,\mathrm{cm\,K}}{5(255\,\mathrm{K})}$$

$$= 1.13 \times 10^{-3}\,\mathrm{cm} = \boxed{11.3\,\mu\mathrm{m}} \text{ in the infrared.}$$

P11.28   (a) $CH_4(g) \rightarrow C(\text{graphite}) + 2H_2(g)$

$$\Delta_r G^\ominus = -\Delta_f G^\ominus(CH_4) = -(-50.72\,\mathrm{kJ\,mol^{-1}}) = 50.72\,\mathrm{kJ\,mol^{-1}} \quad \text{at } T$$

$$\Delta_r H^\ominus = -\Delta_f H^\ominus(CH_4) = -(-74.81\,\mathrm{kJ\,mol^{-1}}) = 74.81\,\mathrm{kJ\,mol^{-1}} \quad \text{at } T$$

We want to find the temperature at which $\Delta_r G^\ominus(T) = 0$. Below this temperature methane is stable with respect to decomposition into the elements. Above this temperature it is unstable. Assuming that the heat capacities are basically independent of temperature

$$\Delta_r C_p^\ominus(T) \approx \Delta_r C_p^\ominus(T) = [8.527 + 2(28.824) - 35.31]\,\mathrm{J\,K^{-1}\,mol^{-1}}$$

$$\approx 30.865\,\mathrm{J\,K^{-1}\,mol^{-1}}$$

$$\Delta_r H^\ominus(T) = \Delta_r H^\ominus(T) + \int_T^T \Delta_r C_p^\ominus(T)\,\mathrm{d}T \text{ [2.44]}$$

$$= \Delta_r H^\ominus(T) + \Delta_r C_p^\ominus \times (T - T)$$

$$\left(\frac{\partial}{\partial T}\left(\frac{\Delta_r G^\ominus}{T}\right)\right)_p = -\frac{\Delta_r H^\ominus}{T^2} \quad [5.13]$$

At constant pressure (the standard pressure)

$$\int_{\boldsymbol{\mathfrak{T}}}^{T} d(\Delta_r G^\ominus / T) = -\int_{\boldsymbol{\mathfrak{T}}}^{T} \frac{\Delta_r H^\ominus}{T^2}\,dT$$

$$\frac{\Delta_r G^\ominus (T)}{T} = \frac{\Delta_r G^\ominus (\boldsymbol{\mathfrak{T}})}{\boldsymbol{\mathfrak{T}}} - \int_{\boldsymbol{\mathfrak{T}}}^{T} \frac{\Delta_r H^\ominus (\boldsymbol{\mathfrak{T}}) + \Delta_r C_p^\ominus \times (T - \boldsymbol{\mathfrak{T}})}{T^2}\,dT$$

$$= \frac{\Delta_r G^\ominus (\boldsymbol{\mathfrak{T}})}{\boldsymbol{\mathfrak{T}}} - [\Delta_r H^\ominus (\boldsymbol{\mathfrak{T}}) - \Delta_r C_p^\ominus \times \boldsymbol{\mathfrak{T}}] \int_{\boldsymbol{\mathfrak{T}}}^{T} \frac{1}{T^2}\,dT - \Delta_r C_p^\ominus \int_{\boldsymbol{\mathfrak{T}}}^{T} \frac{1}{T}\,dT$$

$$= \frac{\Delta_r G^\ominus (\boldsymbol{\mathfrak{T}})}{\boldsymbol{\mathfrak{T}}} + [\Delta_r H^\ominus (\boldsymbol{\mathfrak{T}}) - \Delta_r C_p^\ominus \times \boldsymbol{\mathfrak{T}}] \times \left[\frac{1}{T} - \frac{1}{\boldsymbol{\mathfrak{T}}}\right] - \Delta_r C_p^\ominus \ln\left(\frac{T}{\boldsymbol{\mathfrak{T}}}\right)$$

The value of T for which $\Delta_r G^\ominus (T) = 0$ can be determined by examination of a plot (Fig. 11.2) of $\dfrac{\Delta_r G^\ominus (T)}{T}$ against T.

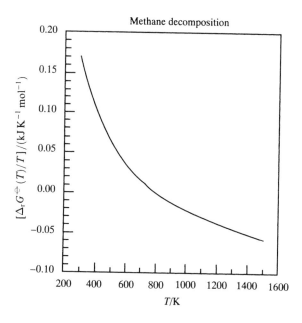

Figure 11.2

$$\frac{\Delta_r G^\ominus (\boldsymbol{\mathfrak{T}})}{\boldsymbol{\mathfrak{T}}} = 50.72\,\text{kJ mol}^{-1}/298.15\,\text{K} = 0.1701\,\text{kJ K}^{-1}\,\text{mol}^{-1}$$

$$\Delta_r H^\ominus (\boldsymbol{\mathfrak{T}}) - \Delta_r C_p^\ominus \times \boldsymbol{\mathfrak{T}} = 74.81\,\text{kJ mol}^{-1} - (30.865\,\text{J K}^{-1}\,\text{mol}^{-1}) \times (298\,\text{K})$$

$$\times \left(\frac{10^{-3}\,\text{kJ}}{\text{J}}\right)$$

$$= 65.61\,\text{kJ mol}^{-1}$$

$$\Delta_r C_p^{\ominus} = (30.865 \, \text{J K}^{-1} \, \text{mol}^{-1}) \times \left(\frac{10^{-3} \, \text{kJ}}{\text{J}}\right) = 0.030\,865 \, \text{kJ K}^{-1} \, \text{mol}^{-1}$$

With the estimate of constant $\Delta_r C_p^{\ominus}$, methane is unstable above 825 K.

(b) $\lambda_{\text{max}} = \dfrac{\frac{1}{5}(1.44 \, \text{cm K})}{T}$ [11.1]

$$\lambda_{\text{max}} = \frac{\frac{1}{5}(1.44 \, \text{cm K})}{1000 \, \text{K}} = 2.88 \times 10^{-4} \, \text{cm} \left(\frac{10^9 \, \text{nm}}{10^2 \, \text{cm}}\right)$$

$$\boxed{\lambda_{\text{max}}(1000 \, \text{K}) = 2880 \, \text{nm}}$$

(c)   Excitance ratio $= \dfrac{M(\text{brown dwarf})}{M(\text{Sun})} = \dfrac{\sigma T_{\text{brown dwarf}}^4}{\sigma T_{\text{Sun}}^4}$ [11.2b]

$$= \frac{(1000 \, \text{K})^4}{(6000 \, \text{K})^4} = \boxed{7.7 \times 10^{-4}}$$

Energy density ratio $= \dfrac{\rho(\text{brown dwarf})}{\rho(\text{Sun})}$

$$= \frac{\frac{8\pi hc}{\lambda^5}\left(\frac{1}{e^{(hc/\lambda kT_{\text{brown dwarf}})}-1}\right)}{\frac{8\pi hc}{\lambda^5}\left(\frac{1}{e^{(hc/\lambda kT_{\text{Sun}})}-1}\right)} \quad [11.5]$$

$$= \frac{e^{(hc/\lambda kT_{\text{Sun}})} - 1}{e^{(hc/\lambda kT_{\text{brown dwarf}})} - 1}$$

The energy density ratio is a function of λ so we will calculate the ratio at λ_{max} of the brown dwarf.

$$\frac{hc}{\lambda_{\text{brown dwarf}} k} = \frac{(6.62 \times 10^{-34} \, \text{J s}) \times (3.00 \times 10^8 \, \text{m s}^{-1})}{(2880 \times 10^{-9} \, \text{m}) \times (1.381 \times 10^{-23} \, \text{J K}^{-1})}$$

$$= 4998 \, \text{K}$$

Energy density ratio $= \dfrac{e^{\frac{4998 \, \text{K}}{T_{\text{Sun}}}} - 1}{e^{\frac{4998}{T_{\text{brown dwarf}}}} - 1}$

$$= \frac{e^{\frac{4998}{6000}} - 1}{e^{\frac{4998}{1000}} - 1} = \frac{1.300}{147}$$

$$= \boxed{8.8 \times 10^{-3}}$$

(d) The wavelength of visible radiation is between about 700 nm (red) and 420 nm (violet). (See text Fig. 16.1.)

$$\text{Fraction of visible energy density} = \frac{1}{aT^4} \left| \int_{700 \, \text{nm}}^{420 \, \text{nm}} \rho(\lambda) \, d\lambda \right| \quad [11.2a, 11.5]$$

$$= \frac{c}{4\sigma T^4} \left| \int_{700 \, \text{nm}}^{420 \, \text{nm}} \rho(\lambda) \, d\lambda \right|$$

As an estimate, let us suppose that $\rho(\lambda)$ doesn't vary too drastically in the visible at 1000 K. Then,

$$\left| \int_{700\,\text{nm}}^{420\,\text{nm}} \rho(\lambda)\,d\lambda \right| \sim \rho(560\,\text{nm}) \times (700\,\text{nm} - 420\,\text{nm})$$

$$\sim \left(\frac{8\pi hc}{(560 \times 10^{-9}\,\text{m})^5} \right) \times \left(\frac{1}{e^{\left(\left(\frac{4998\,\text{K}}{1000\,\text{K}} \right) \times \left(\frac{2880\,\text{nm}}{560\,\text{nm}} \right) \right)} - 1} \right)$$

$$\times \left(280 \times 10^{-9}\,\text{m} \right)$$

$$= \frac{8\pi (6.626 \times 10^{-34}\,\text{J s}) \times (3.00 \times 10^8\,\text{m s}^{-1})}{1.97 \times 10^{-25}\,\text{m}^4} \left(\frac{1}{e^{25.70} - 1} \right)$$

$$= 1.75 \times 10^{-10}\,\text{J m}^{-3}$$

$$\text{fraction of visible energy density} \sim \frac{(3.00 \times 10^8\,\text{m s}^{-1}) \times (1.75 \times 10^{-10}\,\text{J m}^{-3})}{4(5.67 \times 10^{-8}\,\text{W m}^{-2}\,\text{K}^{-4}) \times (1000\,\text{K})^4}$$

$$\sim \boxed{2.31 \times 10^{-7}}$$

Very little of the brown dwarf's radiation is in the visible. It doesn't shine brightly.

12 Quantum theory: techniques and applications

Solutions to exercises

Discussion questions

E12.1(a) In quantum mechanics, particles are said to have wave characteristics. The fact of the existence of the particle then requires that the wavelengths of the waves representing them be such that the wave does not experience destructive interference upon reflection by a barrier or in its motion around a closed loop. This requirement restricts the wavelength to values $\lambda = 2/n \times L$, where L is the length of the path and n is a positive integer. Then using the relations $\lambda = h/p$ and $E = p^2/2m$, the energy is quantized at $E = n^2 h^2/8mL^2$. This derivation applies specifically to the particle in a box, the derivation is similar for the particle on a ring; the same principles apply (see Section 12.6).

E12.2(a) The lowest energy level possible for a confined quantum mechanical system is the zero-point energy, and zero-point energy is not zero energy. The system must have at least that minimum amount of energy even at absolute zero. The physical reason is that if the particle is confined, its position is not completely uncertain, and therefore its momentum, and hence its kinetic energy, cannot be exactly zero. The particle in a box, the harmonic oscillator, the particle on a ring or on a sphere, the hydrogen atom, and many other systems we will encounter, all have zero-point energy.

E12.3(a) Fermions are particles with half-integral spin, 1/2, 3/2, 5/2, ..., whereas bosons have spin, 0, 1, 2, ... All fundamental particles that make up matter have spin 1/2 and are fermions, but composite particles can be either fermions or bosons.

Fermions: electrons, protons, neutrons, ^{3}He, ...

Bosons: photons, deuterons (^{2}H), ^{12}C, ...

Numerical exercises

E12.4(a)
$$E = \frac{n^2 h^2}{8m_e L^2} \quad [12.7]$$

$$\frac{h^2}{8m_e L^2} = \frac{(6.626 \times 10^{-34}\,\text{J s})^2}{(8) \times (9.109 \times 10^{-31}\,\text{kg}) \times (1.0 \times 10^{-9}\,\text{m})^2} = 6.02 \times 10^{-20}\,\text{J}$$

The conversion factors required are

$$E/(\text{kJ mol}^{-1}) = \frac{N_A}{10^3} E/\text{J}$$

$$1\,\text{eV} = 1.602 \times 10^{-19}\,\text{J}; \qquad 1\,\text{cm}^{-1} = 1.986 \times 10^{-23}\,\text{J}$$

(a)
$$E_2 - E_1 = (4-1)\frac{h^2}{8m_e L^2} = (3) \times (6.02 \times 10^{-20}\,\text{J})$$
$$= 18.06 \times 10^{-20}\,\text{J}$$
$$= \boxed{1.81 \times 10^{-19}\,\text{J}}, \boxed{110\,\text{kJ mol}^{-1}}, \boxed{1.1\,\text{eV}}, \boxed{9100\,\text{cm}^{-1}}$$

(b)
$$E_6 - E_5 = (36-25)\frac{h^2}{8m_e L^2} = \frac{11h^2}{8m_e L^2}$$
$$= (11) \times (6.02 \times 10^{-20}\,\text{J})$$
$$= \boxed{6.6 \times 10^{-19}\,\text{J}}, \boxed{400\,\text{kJ mol}^{-1}}, \boxed{4.1\,\text{eV}}, \boxed{33\,000\,\text{cm}^{-1}}$$

Comment. The energy level separations increase as n increases.

Question. For what value of n is $E_{n+1} - E_n$ for the system of this exercise equal to the ionization energy of the H-atom which is $13.6\,\text{eV}$?

E12.5(a)   The wavefunctions are

$$\psi_n = \left(\frac{2}{L}\right)^{1/2} \sin\left(\frac{n\pi x}{L}\right) \quad [12.7]$$

The required probability is

$$P = \int_{0.49L}^{0.51L} \psi_n^2 \, dx \approx \psi_n^2 \Delta x$$

(a) $\psi_1^2 = \left(\frac{2}{L}\right) \sin^2\left(\frac{\pi x}{L}\right) = \left(\frac{2}{L}\right) \sin^2\left(\frac{\pi}{2}\right) [x \approx 0.50L] = \left(\frac{2}{L}\right) \quad \left[\sin\frac{\pi}{2} = 1\right]$

$$P = \left(\frac{2}{L}\right) \times 0.02L = \boxed{0.04}$$

(b) $\psi_2^2 = \left(\frac{2}{L}\right) \sin^2\left(\frac{2\pi x}{L}\right) = \left(\frac{2}{L}\right) \sin^2 \pi = \boxed{0}$

E12.6(a)   The wavefunction for a particle in the state $n = 1$ in a square-well potential is

$$\psi_1 = \left(\frac{2}{L}\right)^{1/2} \sin\left(\frac{\pi x}{L}\right)$$

and

$$\hat{p} = \frac{\hbar}{i} \frac{d}{dx}$$

$$\langle p \rangle = \int_0^L \psi_1^* \hat{p} \psi_1 \, dx = \frac{2\hbar}{iL} \int_0^L \sin\left(\frac{\pi x}{L}\right) \frac{d}{dx} \sin\left(\frac{\pi x}{L}\right) dx$$

$$= \frac{2\pi\hbar}{iL^2} \int_0^L \sin\left(\frac{\pi x}{L}\right) \cos\left(\frac{\pi x}{L}\right) dx = \boxed{0}$$

$$\hat{p}^2 = -\hbar^2 \frac{d^2}{dx^2}$$

$$\langle p^2 \rangle = -\frac{2\hbar^2}{L} \int_0^L \sin\left(\frac{\pi x}{L}\right) \frac{d^2}{dx^2} \sin\left(\frac{\pi x}{L}\right) dx = \left(\frac{2\hbar^2}{L}\right) \times \left(\frac{\pi}{L}\right)^2 \int_0^L \sin^2 ax \, dx \quad \left[a = \frac{\pi}{L}\right]$$

$$= \left(\frac{2\hbar^2}{L}\right) \times \left(\frac{\pi}{L}\right)^2 \left(\frac{1}{2}x - \frac{1}{4a}\sin 2ax\right)\Big|_0^L = \left(\frac{2\hbar^2}{L}\right) \times \left(\frac{\pi}{L}\right)^2 \times \left(\frac{L}{2}\right) = \boxed{\frac{h^2}{4L^2}}$$

Comment. The expectation value of $\hat{p}$ is zero because on average the particle moves to the left as often as the right.

E12.7(a) $\psi_3 = \left(\dfrac{2}{L}\right)^{1/2} \sin\left(\dfrac{3\pi x}{L}\right)$

$P(x) \propto \psi_3^2 \propto \sin^2\left(\dfrac{3\pi x}{L}\right)$

The maxima and minima in $P(x)$ correspond to $\dfrac{dP(x)}{dx} = 0$.

$$\frac{dP(x)}{dx} \propto \sin\left(\frac{3\pi x}{L}\right)\cos\left(\frac{3\pi x}{L}\right) \propto \sin\left(\frac{6\pi x}{L}\right) \qquad [2\sin\alpha\cos\alpha = \sin 2\alpha]$$

$\sin\theta = 0$ when $\theta = \left(\dfrac{6\pi x}{L}\right) = n'\pi, n' = 0, 1, 2, \ldots$ which corresponds to $x = \dfrac{n'L}{6}, n' \leq 6$.

$n' = 0, 2, 4$, and 6 correspond to minima in ψ_3, leaving $n' = 1, 3$, and 5 for the maxima, that is

$$\boxed{x = \frac{L}{6}, \frac{L}{2} \text{ and } \frac{5L}{6}}$$

Comment. Maxima in ψ^2 correspond to maxima *and* minima in ψ itself, so one can also solve this exercise by finding all points where $\dfrac{d\psi}{dx} = 0$

E12.8(a) $E = (n_1^2 + n_2^2 + n_3^2) \times \left(\dfrac{h^2}{8mL^2}\right)$ [3-dimensional analogue of eqn 12.19]

$$E_{111} = \frac{3h^2}{8mL^2}, \qquad 3E_{111} = \frac{9h^2}{8mL^2}$$

Hence, we require the values of n_1, n_2, and n_3 that make

$$n_1^2 + n_2^2 + n_3^2 = 9$$

Therefore, $(n_1, n_2, n_3) = (1, 2, 2), (2, 1, 2)$, and $(2, 2, 1)$ and the degeneracy is $\boxed{3}$.

Question. What is the smallest multiple of the lowest energy, E_{111} for which $E_{n_1 n_2 n_3}$ does not exist?

E12.9(a) $E = (n_1^2 + n_2^2 + n_3^2) \times \left(\dfrac{h^2}{8mL^2}\right) = \dfrac{K}{L^2}, \quad K = (n_1^2 + n_2^2 + n_3^2) \times \left(\dfrac{h^2}{8m}\right)$

$$\frac{\Delta E}{E} = \frac{\dfrac{K}{(0.9L)^2} - \dfrac{K}{L^2}}{\dfrac{K}{L^2}} = \frac{1}{0.81} - 1 = \boxed{0.23}, \quad \text{or} \quad \boxed{23 \text{ per cent}}$$

E12.10(a) $E = \left(v + \dfrac{1}{2}\right)\hbar\omega, \quad \omega = \left(\dfrac{k}{m}\right)^{1/2}$ [12.32]

The zero-point energy corresponds to $v = 0$; hence

$$E_0 = \frac{1}{2}\hbar\omega = \frac{1}{2}\hbar\left(\frac{k}{m}\right)^{1/2} = \left(\frac{1}{2}\right) \times (1.055 \times 10^{-34}\,\text{J s}) \times \left(\frac{155\,\text{N m}^{-1}}{2.33 \times 10^{-26}\,\text{kg}}\right)^{1/2}$$

$$= \boxed{4.30 \times 10^{-21}\,\text{J}}$$

E12.11(a) $\Delta E = E_{v+1} - E_v = \left(v + 1 + \dfrac{1}{2}\right)\hbar\omega - \left(v + \dfrac{1}{2}\right)\hbar\omega = \hbar\omega = \hbar\left(\dfrac{k}{m}\right)^{1/2}$ [12.32]

Hence $k = m\left(\dfrac{\Delta E}{\hbar}\right)^2 = (1.33 \times 10^{-25}\,\text{kg}) \times \left(\dfrac{4.82 \times 10^{-21}\,\text{J}}{1.055 \times 10^{-34}\,\text{J s}}\right)^2 = \boxed{278\,\text{N m}^{-1}}$

$[1\,\text{J} = 1\,\text{N m}]$

E12.12(a) The requirement for a transition to occur is that $\Delta E(\text{system}) = E(\text{photon})$.

$\Delta E(\text{system}) = \hbar\omega$ [Exercise 12.11(**a**)]

$E(\text{photon}) = h\nu = \dfrac{hc}{\lambda}$

Therefore, $\dfrac{hc}{\lambda} = \dfrac{h\omega}{2\pi} = \left(\dfrac{h}{2\pi}\right) \times \left(\dfrac{k}{m}\right)^{1/2}$

$\lambda = 2\pi c\left(\dfrac{m}{k}\right)^{1/2} = (2\pi) \times (2.998 \times 10^8\,\text{m s}^{-1}) \times \left(\dfrac{1.673 \times 10^{-27}\,\text{kg}}{855\,\text{N m}^{-1}}\right)^{1/2}$

$= 2.63 \times 10^{-6}\,\text{m} = \boxed{2.63\,\mu\text{m}}$

E12.13(a) Since $\lambda \propto m^{1/2}$, $\lambda_{\text{new}} = 2^{1/2}$, $\lambda_{\text{old}} = (2^{1/2}) \times (2.63\,\mu\text{m}) = \boxed{3.72\,\mu\text{m}}$

The change in wavelength is $\lambda_{\text{new}} - \lambda_{\text{old}} = 1.09\,\mu\text{m}$.

E12.14(a) **(a)** $\omega = \left(\dfrac{g}{l}\right)^{1/2}$ [elementary physics]

$\Delta E = \hbar\omega$ [harmonic oscillator level separations, Exercise 12.11(**a**)]

$= (1.055 \times 10^{-34}\,\text{J s}) \times \left(\dfrac{9.81\,\text{m s}^{-2}}{1\,\text{m}}\right)^{1/2} = \boxed{3.3 \times 10^{-34}\,\text{J}}$

(b) $\Delta E = h\nu = (6.626 \times 10^{-34}\,\text{J Hz}^{-1}) \times (5\,\text{Hz}) = \boxed{3.3 \times 10^{-33}\,\text{J}}$

E12.15(a) The Schrödinger equation for the linear harmonic oscillator is

$$-\dfrac{\hbar^2}{2m}\dfrac{d^2\psi}{dx^2} + \dfrac{1}{2}kx^2\psi = E\psi \text{ [12.31]}$$

The ground-state wavefunction is

$$\psi_0 = N_0 e^{-y^2/2}, \quad \text{where } y = \left(\dfrac{mk}{\hbar^2}\right)^{1/4} x \text{ [12.35]} = \dfrac{x}{\alpha}$$

with $\alpha = \left(\dfrac{\hbar^2}{mk}\right)^{1/4} = \left(\dfrac{\hbar^2}{m^2\omega^2}\right)^{1/4}$; $\quad k = \dfrac{\hbar^2}{m\alpha^4}$ (a)

Thus, $\psi_0 = N_0 e^{-x^2/2\alpha^2}$.

Performing the operations

$$\dfrac{d\psi_0}{dx} = \left(-\dfrac{1}{\alpha^2}x\right)\psi_0$$

$$\frac{d^2\psi_0}{dx^2} = \left(-\frac{1}{\alpha^2}x\right) \times \left(-\frac{1}{\alpha^2}x\right) \times \psi_0 - \frac{1}{\alpha^2}\psi_0 = \frac{x^2}{\alpha^4}\psi_0 - \frac{1}{\alpha^2}\psi_0 = \left(\frac{x^2}{\alpha^4} - \frac{1}{\alpha^2}\right)\psi_0$$

Thus,

$$-\frac{\hbar^2}{2m}\left(\frac{x^2}{\alpha^4} - \frac{1}{\alpha^2}\right)\psi_0 + \frac{1}{2}kx^2\psi_0 = E_0\psi_0$$

which implies

$$E_0 = \frac{-\hbar^2}{2m}\left(\frac{x^2}{\alpha^4} - \frac{1}{\alpha^2}\right) + \frac{1}{2}kx^2 \tag{b}$$

But E_0 is a constant, independent of x; therefore the terms which contain x must drop out, which is possible only if

$$-\frac{\hbar^2}{2m\alpha^4} + \frac{1}{2}k = 0$$

which is consistent with $k = \dfrac{\hbar^2}{m\alpha^4}$ as in (a). What is left in (b) is

$$E_0 = \frac{\hbar^2}{2m\alpha^2} = \hbar\omega \quad \left[\text{using } \omega = \left(\frac{k}{m}\right)^{1/2} \text{ and } k = \frac{\hbar^2}{m\alpha^4}\right]$$

Therefore, ψ_0 is a solution of the Schrödinger equation with energy $\frac{1}{2}\hbar\omega$.

E12.16(a) As described in Exercise 12.14(**b**), for the vibrations of a diatomic molecule $\mu = \dfrac{m}{2}$ must be substituted for m. Thus

$$E_0 = \frac{1}{2}\hbar\omega = \frac{1}{2}\hbar\left(\frac{k}{\mu}\right)^{1/2} = \frac{1}{2}\hbar\left(\frac{2k}{m}\right)^{1/2}$$

$$m(^{35}Cl) = 34.96888\,u = (34.9688\,u) \times (1.66054 \times 10^{-27}\,kg/u) = 5.807 \times 10^{-26}\,kg$$

$$E_0 = \left(\frac{1.05457 \times 10^{-34}\,J\,s}{2}\right) \times \left(\frac{(2) \times (329\,N\,m^{-1})}{5.807 \times 10^{-26}\,kg}\right)^{1/2} = \boxed{5.61 \times 10^{-21}\,J}$$

E12.17(a) We require

$$\int \psi^*\psi \, d\tau = 1$$

that is

$$\int_0^{2\pi} N^2 e^{-im_l\phi} e^{im_l\phi} \, d\phi = \int_0^{2\pi} N^2 \, d\phi = 2\pi N^2 = 1$$

$$N^2 = \frac{1}{2\pi} \qquad N = \boxed{\left(\frac{1}{2\pi}\right)^{1/2}}$$

E12.18(a) Magnitude of angular momentum $= \{l(l+1)\}^{1/2}\hbar$ [12.66a]

Projection on arbitrary axis $= m_l\hbar$ [12.66b]

Thus,

$$\text{Magnitude} = (2^{1/2}) \times \hbar = \boxed{1.49 \times 10^{-34}\,\text{J s}}$$

$$\text{Possible projections} = \boxed{0}, \quad \pm\hbar = \boxed{0, \pm1.05 \times 10^{-34}\,\text{J s}}$$

E12.19(a) The diagrams are drawn by forming a vector of length $\{j(j+1)\}^{1/2}$, with $j = s$ or l as appropriate, and with a projection m_j on the z-axis (see Fig. 12.1). Each vector represents the edge of a cone around the z-axis (that for $m_j = 0$ represents the side view of a disk perpendicular to z).

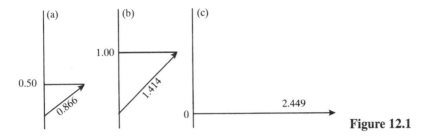

Figure 12.1

Solutions to problems

Solutions to numerical problems

P12.1
$$E = \frac{n^2h^2}{8mL^2}, \quad E_2 - E_1 = \frac{3h^2}{8mL^2}$$

We take $m(\text{O}_2) = (32.000) \times (1.6605 \times 10^{-27}\,\text{kg})$, and find

$$E_2 - E_1 = \frac{(3) \times (6.626 \times 10^{-34}\,\text{J s})^2}{(8) \times (32.00) \times (1.6605 \times 10^{-27}\,\text{kg}) \times (5.0 \times 10^{-2}\,\text{m})^2} = \boxed{1.24 \times 10^{-39}\,\text{J}}$$

We set $E = \dfrac{n^2h^2}{8mL^2} = \dfrac{1}{2}kT$ and solve for n.

From above $\dfrac{h^2}{8mL^2} = \dfrac{E_2 - E_1}{3} = 4.13 \times 10^{-40}\,\text{J}$; then

$$n^2 \times (4.13 \times 10^{-40}\,\text{J}) = \left(\tfrac{1}{2}\right) \times (1.381 \times 10^{-23}\,\text{J K}^{-1}) \times (300\,\text{K}) = 2.07 \times 10^{-21}\,\text{J}$$

We find $n = \left(\dfrac{2.07 \times 10^{-21}\,\text{J}}{4.13 \times 10^{-40}\,\text{J}}\right)^{1/2} = \boxed{2.2 \times 10^9}$

At this level,

$$E_n - E_{n-1} = \{n^2 - (n-1)^2\} \times \frac{h^2}{8mL^2} = (2n-1) \times \frac{h^2}{8mL^2} \approx (2n) \times \frac{h^2}{8mL^2}$$

$$= (4.4 \times 10^9) \times (4.13 \times 10^{-40}\,\text{J}) \approx \boxed{1.8 \times 10^{-30}\,\text{J}} \quad [\text{or } 1.1\,\text{J mol}^{-1}]$$

P12.2
$$\omega = \left(\frac{k}{\mu}\right)^{1/2} \quad [12.32, \text{ with } \mu \text{ in place of } m]$$

Also, $\omega = 2\pi\nu = \dfrac{2\pi c}{\lambda} = 2\pi c\tilde{\nu}$

Therefore $k = \omega^2 \mu = 4\pi^2 c^2 \tilde{\nu}^2 \mu = \dfrac{4\pi^2 c^2 \tilde{\nu}^2 m_1 m_2}{m_1 + m_2}$.

We draw up the following table using the information inside the back cover

	$^1H^{35}Cl$	$^1H^{81}Br$	$^1H^{127}I$	$^{12}C^{16}O$	$^{14}N^{16}O$
$\tilde{\nu}/m^{-1}$	299000	265000	231000	217000	190400
$10^{27} m_1/kg$	1.6735	1.6735	1.6735	19.926	23.253
$10^{27} m_2/kg$	58.066	134.36	210.72	26.560	26.560
$k/(N\,m^{-1})$	516	412	314	1902	1595

Therefore, the order of stiffness, is $CO > NO > HCl > HBr > HI$.

P12.3

$E = \dfrac{m_l^2 \hbar^2}{2I}$ [12.49] $= \dfrac{m_l^2 \hbar^2}{2mr^2}$ $[I = mr^2]$

$E_0 = 0$ $[m_l = 0]$

$E_1 = \dfrac{\hbar^2}{2mr^2} = \dfrac{(1.055 \times 10^{-34}\,J\,s)^2}{(2) \times (1.008) \times (1.6605 \times 10^{-27}\,kg) \times (160 \times 10^{-12}\,m)^2} = \boxed{1.30 \times 10^{-22}\,J}$

$[1.96 \times 10^{11}\,Hz]$

The minimum angular momentum is $\boxed{\pm\hbar}$

P12.5

(a) Treat the small step in the potential energy function as a perturbation in the energy operator:

$$H^{(1)} = \begin{cases} 0 & \text{for } 0 \le x \le 1/2(L-a) \text{ and } 1/2(L+a) \le x \le L \\ \varepsilon & \text{for } 1/2(L-a) \le x \le 1/2(L+a) \end{cases}$$

The first-order correction to the ground-state energy, E_1, is:

$$E_1^{(1)} = \int_0^L \psi_1^{(0)*} H^{(1)} \psi_1^{(0)} dx = \int_{1/2(L-a)}^{1/2(L+a)} \left(\frac{2}{L}\right)^{1/2} \sin\left(\frac{\pi x}{L}\right) \varepsilon \left(\frac{2}{L}\right)^{1/2} \sin\left(\frac{\pi x}{L}\right) dx,$$

$$E_1^{(1)} = \frac{2\varepsilon}{L} \int_{1/2(L-a)}^{1/2(L+a)} \sin^2\left(\frac{\pi x}{L}\right) dx = \frac{\varepsilon}{L\pi}\left(\pi x - L\cos\left(\frac{\pi x}{L}\right)\sin\left(\frac{\pi x}{L}\right)\right)\Big|_{1/2(L-a)}^{1/2(L+a)},$$

$$E_1^{(1)} = \frac{\varepsilon a}{L} - \frac{\varepsilon}{\pi}\cos\left(\frac{\pi(L+a)}{2L}\right)\sin\left(\frac{\pi(L+a)}{2L}\right) + \frac{\varepsilon}{\pi}\cos\left(\frac{\pi(L-a)}{2L}\right)\sin\left(\frac{\pi(L-a)}{2L}\right).$$

This expression can be simplified considerably with a few trigonometric identities. The product of sine and cosine is related to the sine of twice the angle:

$$\cos\left(\frac{\pi(L\pm a)}{2L}\right)\sin\left(\frac{\pi(L\pm a)}{2L}\right) = 1/2\sin\left(\frac{\pi(L\pm a)}{L}\right) = 1/2\sin\left(\pi \pm \frac{\pi a}{L}\right),$$

and the sine of a sum can be written in a particularly simple form since one of the terms in the sum is π:

$$\sin\left(\pi \pm \frac{\pi a}{L}\right) = \sin\pi\cos\left(\frac{\pi a}{L}\right) \pm \cos\pi\sin\left(\frac{\pi a}{L}\right) = \mp\sin\left(\frac{\pi a}{L}\right).$$

Thus $E_1^{(1)} = \boxed{\dfrac{\varepsilon a}{L} + \dfrac{\varepsilon}{\pi}\sin\left(\dfrac{\pi a}{L}\right)}$.

(b) If $a = L/10$, the first-order correction to the ground-state energy is:

$$E_1^{(1)} = \boxed{\varepsilon/10 + \frac{\varepsilon}{\pi}\sin(\pi/10)} = \boxed{0.1984\varepsilon}$$

P12.7 The second-order correction to the ground-state energy, E_1, is:

$$E_1^{(2)} = \sum_{n=2}^{\infty} \frac{\left| \int_0^L \psi_n^{(0)*} H^{(1)} \psi_1^{(0)} \, dx \right|^2}{E_1^{(0)} - E_n^{(0)}},$$

where $H^{(1)} = mgx$, $\psi_n^{(0)} = \sin\dfrac{n\pi x}{L}$, and $E_n = \dfrac{n^2 h^2}{8mL^2}$.

The denominator in the sum is:

$$E_1^{(0)} - E_n^{(0)} = \frac{h^2}{8mL^2} - \frac{n^2 h^2}{8mL^2} = \frac{(1-n^2)h^2}{8mL^2}.$$

The integral in the sum is:

$$\int_0^L \psi_n^{(0)*} H^{(1)} \psi_1^{(0)} \, dx = \int_0^L \left(\frac{2}{L}\right)^{1/2} \sin\left(\frac{n\pi x}{L}\right) mgx \left(\frac{2}{L}\right)^{1/2} \sin\left(\frac{\pi x}{L}\right) dx,$$

$$\int_0^L \psi_n^{(0)*} H^{(1)} \psi_1^{(0)} \, dx = \frac{2mg}{L} \int_0^L x \sin ax \sin bx \, dx,$$

where $a = n\pi/L$ and $b = \pi/L$

The integral formulas given with the problem allow this integral to be expressed as:

$$-\frac{2mg}{L}\frac{d}{da}\int_0^L \cos ax \sin bx \, dx = -mg\frac{d}{da}\left(\frac{\cos(a-b)x}{2(a-b)} - \frac{\cos(a+b)x}{2(a+b)}\right)\Bigg|_0^L$$

$$= -\frac{2mg}{L}\left(\frac{-x\sin(a-b)x}{2(a-b)} - \frac{\cos(a-b)x}{2(a-b)^2} + \frac{x\sin(a+b)x}{2(a+b)} + \frac{\cos(a+b)x}{2(a+b)^2}\right)\Bigg|_0^L.$$

The arguments of the trigonometric functions at the upper limit are:

$$(a-b)L = (n-1)\pi \quad \text{and} \quad (a+b)L = (n+1)\pi.$$

Therefore, the sine terms vanish. Similarly, the cosines are ± 1 depending on whether the argument is an even or odd multiple of π; they simplify to $(-1)^{n+1}$. At the lower limit, the sines are still zero, and the cosines are all 1. The integral, evaluated at its limits with π/L factors pulled out from the as and bs in its denominator, becomes:

$$\frac{mgL}{\pi^2}\left(\frac{(-1)^{n+1}-1}{(n-1)^2} - \frac{(-1)^{n+1}-1}{(n+1)^2}\right) = \frac{mgL[(-1)^{n+1}-1]}{\pi^2}\left(\frac{(n+1)^2 - (n-1)^2}{(n-1)^2(n+1)^2}\right),$$

$$= \frac{4mgL[(-1)^{n+1}-1]n}{\pi^2(n^2-1)^2}.$$

The second-order correction, then, is:

$$E_1^{(2)} = \sum_{n=2}^{\infty} \frac{\left(\frac{4mgL[(-1)^{n+1}-1]n}{\pi^2(n^2-1)^2}\right)^2}{\frac{(1-n^2)h^2}{8mL^2}} = -\frac{128m^3g^2L^4}{\pi^4h^2}\sum_{n=2}^{\infty}\frac{[(-1)^n+1]^2n^2}{(n^2-1)^5}.$$

Note that the terms with odd n vanish. Therefore, the sum can be rewritten, changing n to $2k$, as:

$$E_1^{(2)} = -\frac{2048m^3g^2L^4}{\pi^4h^2}\sum_{k=1}^{\infty}\frac{k^2}{(4k^2-1)^5}.$$

The sum converges rapidly to 4.121×10^{-3}, as can easily be verified numerically; in fact, to three significant figures, terms after the first do not affect the sum. So the second-order correction is:

$$\boxed{E_1^{(2)} = -\frac{0.08664m^3g^2L^4}{h^2}.}$$

The first-order correction to the ground-state wavefunction is also a sum:

$$\psi_0^{(1)} = \sum_n c_n \psi_n^{(0)},$$

where $c_n = -\dfrac{\int_0^L \psi_n^{(0)*}H^1\psi_1^{(0)}\,dx}{E_n^{(0)}-E_1^{(0)}} = -\dfrac{\frac{4mgL[(-1)^{n+1}-1]n}{\pi^2(n^2-1)^2}}{\frac{(n^2-1)h^2}{8mL^2}} = \boxed{\dfrac{32m^2gL^3[(-1)^n+1]^n}{\pi^2h^2(n^2-1)^3}}.$ Once again, the odd n terms vanish.

How does the first-order correction alter the wavefunction? Recall that the perturbation raises the potential energy near the top of the box (near L) much more than near the bottom (near $x = 0$); therefore, we expect the probability of finding the particle near the bottom to be enhanced compared with that of finding it near the top. Because the zero-order ground-state wavefunction is positive throughout the interior of the box, we thus expect the wavefunction itself to be raised near the bottom of the box and lowered near the top. In fact, the correction terms do just this. First, note that the basis wavefunctions with odd n are symmetric with respect to the center of the box; therefore, they would have the same effect near the top of the box as near the bottom. The coefficients of these terms are zero: they do not contribute to the correction. The even-n basis functions all start positive near $x = 0$ and end negative near $x = L$; therefore, such terms must be multiplied by positive coefficients (as the result provides) to enhance the wavefunction near the bottom and diminish it near the top.

Solutions to theoretical problems

P12.9   The text defines the transmission probability and expresses it as the ratio of $|A'|^2/|A|^2$, where the coefficients A and A' are introduced in eqns 12.22 and 12.25. Eqns 12.26 and 12.27 list four equations for the six unknown coefficients of the full wavefunction. Once we realize that we can set B' to zero, these equations in five unknowns are:

(a) $A + B = C + D$,

(b) $Ce^{\kappa L} + De^{-\kappa L} = A'e^{ikL}$,

(c) $ikA - ikB = \kappa C - \kappa D$,

(d) $\kappa Ce^{\kappa L} - \kappa De^{-\kappa L} = ikA'e^{ikL}$.

We need A' in terms of A alone, which means we must eliminate B, C, and D. Notice that B appears only in eqns (a) and (c). Solving these equations for B and setting the results equal to each other yields:

$$B = C + D - A = A - \frac{\kappa C}{ik} + \frac{\kappa D}{ik}.$$

Solve this equation for C:

$$C = \frac{2A + D\left(\frac{\kappa}{ik} - 1\right)}{\frac{\kappa}{ik} + 1} = \frac{2Aik + D(\kappa - ik)}{\kappa + ik}.$$

Now note that the desired A' appears only in (b) and (d). Solve these for A' and set them equal:

$$A' = e^{-ikL}(Ce^{\kappa L} + De^{\kappa L}) = \frac{\kappa e^{-ikL}}{ik}(Ce^{\kappa L} - De^{-\kappa L}).$$

Solve the resulting equation for C, and set it equal to the previously obtained expression for C:

$$C = \frac{\left(\frac{\kappa}{ik} + 1\right)De^{-2\kappa L}}{\frac{\kappa}{ik} - 1} = \frac{(\kappa + ik)De^{-2\kappa L}}{\kappa - ik} = \frac{2Aik + D(\kappa - ik)}{\kappa + ik}.$$

Solve this resulting equation for D in terms of A:

$$\frac{(\kappa + ik)^2 e^{-2\kappa L} - (\kappa - ik)^2}{(\kappa - ik)(\kappa + ik)}D = \frac{2Aik}{\kappa + ik},$$

so $D = \dfrac{2Aik(\kappa - ik)}{(\kappa + ik)^2 e^{-2\kappa L} - (\kappa - ik)^2}.$

Substituting this expression back into an expression for C yields:

$$C = \frac{2Aik(\kappa + ik)e^{-2\kappa L}}{(\kappa + ik)^2 e^{-2\kappa L} - (\kappa - ik)^2}.$$

Substituting for C and D in the expression for A' yields:

$$A' = e^{-ikL}(Ce^{\kappa L} + De^{-\kappa L}) = \frac{2Aike^{-ikL}}{(\kappa + ik)^2 e^{-2\kappa L} - (\kappa - ik)^2}[(\kappa + ik)e^{-\kappa L} + (\kappa - ik)e^{-\kappa L}],$$

$$A' = \frac{4Aik\kappa e^{-\kappa L}e^{-ikL}}{(\kappa + ik)^2 e^{-2\kappa L} - (\kappa - ik)^2} = \frac{4Aik\kappa e^{-ikL}}{(\kappa + ik)^2 e^{-\kappa L} - (\kappa - ik)^2 e^{\kappa L}}.$$

The transmission coefficient is:

$$T = \frac{|A'|^2}{|A|^2} = \left(\frac{4Aik\kappa e^{-ikL}}{(\kappa + ik)^2 e^{-\kappa L} - (\kappa - ik)^2 e^{\kappa L}}\right)\left(\frac{-4Ak\kappa e^{ikL}}{(\kappa - ik)^2 e^{-\kappa L} - (\kappa + ik)^2 e^{\kappa L}}\right).$$

The denominator is worth expanding separately in several steps. It is:

$$(\kappa + ik)^2(\kappa - ik)^2 e^{-2\kappa L} - (\kappa - ik)^4 - (\kappa + ik)^4 + (\kappa - ik)^2(\kappa + ik)^2 e^{2\kappa L}$$
$$= (\kappa^2 + k^2)^2(e^{2\kappa L} + e^{-2\kappa L}) - (\kappa^2 - 2i\kappa k - k^2)^2 - (\kappa^2 + 2i\kappa k - k^2)^2$$
$$= (\kappa^4 + 2\kappa^2 k^2 + k^4)(e^{2\kappa L} + e^{-2\kappa L}) - (2\kappa^4 - 12\kappa^2 k^2 + 2k^2).$$

If the $12\kappa^2 k^2$ term were $-4\kappa^2 k^2$ instead, we could collect terms still further (completing the square), but of course we must also account for the difference between those quantities, making the denominator:

$$(\kappa^4 + 2\kappa^2 k^2 + k^4)(e^{2\kappa L} - 2 + e^{-2\kappa L}) + 16\kappa^2 k^2 = (\kappa^2 + k^2)^2(e^{2\kappa L} - e^{-2\kappa L})^2 + 16\kappa^2 k^2$$

So the coefficient is:

$$T = \frac{16 k^2 \kappa^2}{(\kappa^2 + k^2)^2(e^{2\kappa L} - e^{-2\kappa L})^2 + 16\kappa^2 k^2}.$$

We are almost there. To get to eqn 12.28a, we invert the expression:

$$T = \left(\frac{(\kappa^2 + k^2)^2(e^{2\kappa L} - e^{-2\kappa L})^2 + 16\kappa^2 k^2}{16 k^2 \kappa^2}\right)^{-1} = \left(\frac{(\kappa^2 + k^2)^2(e^{2\kappa L} - e^{-2\kappa L})^2}{16 k^2 \kappa^2} + 1\right)^{-1}.$$

Finally, we try to express $\dfrac{(\kappa^2 + k^2)}{k^2 \kappa^2}$ in terms of a ratio of energies, $\varepsilon = E/V$. Eqns 12.22 and 12.24 define k and κ. The factors involving, 2, $\hbar$, and the mass cancel leaving $\kappa \propto (V - E)^{1/2}$ and $k \propto E^{1/2}$, so:

$$\frac{(\kappa^2 + k^2)^2}{k^2 \kappa^2} = \frac{[E + (V - E)]^2}{E(V - E)} = \frac{V^2}{E(V - E)} = \frac{1}{\varepsilon(1 - \varepsilon)},$$

which makes the transmission coefficient:

$$T = \left(\frac{(e^{2\kappa L} - e^{-2\kappa L})^2}{16\varepsilon(1 - \varepsilon)} + 1\right)^{-1}.$$

P12.11   We assume that the barrier begins at $x = 0$ and that the barrier extends in the positive x direction.

(a) $P = \displaystyle\int_{\text{Barrier}} \psi^2 \, d\tau = \int_0^\infty N^2 e^{-2\kappa x} \, dx = \boxed{\dfrac{N^2}{2\kappa}}$

(b) $\langle x \rangle = \displaystyle\int_0^\infty x \psi^2 \, dx = N^2 \int_0^\infty x e^{-2\kappa x} \, dx = \dfrac{N^2}{(2\kappa)^2} = \boxed{\dfrac{N^2}{4\kappa^2}}$

Question. Is N a normalization constant?

P12.13 $\langle E_K \rangle \equiv \langle T \rangle = \displaystyle\int_{-\infty}^{+\infty} \psi^* \hat{T} \psi \, dx$ with $\hat{T} \equiv \dfrac{\hat{p}^2}{2m}$ and $\hat{p} = \dfrac{\hbar}{i}\dfrac{d}{dx}$

$\hat{T} = -\dfrac{\hbar^2}{2m}\dfrac{d^2}{dx^2} = -\dfrac{\hbar^2}{2m\alpha^2}\dfrac{d^2}{dy^2} = -\dfrac{1}{2}\hbar\omega\dfrac{d^2}{dy^2}$, $\left[x = \alpha y, \; \alpha^2 = \dfrac{\hbar}{m\omega}\right]$ which implies that

$\hat{T}\psi = -\dfrac{1}{2}\hbar\omega\left(\dfrac{d^2\psi}{dy^2}\right)$

We then use $\psi = N H e^{-y^2/2}$, and obtain

$$\dfrac{d^2\psi}{dy^2} = N\dfrac{d^2}{dy^2}(He^{-y^2/2}) = N\{H'' - 2yH' - H + y^2 H\}e^{-y^2/2}$$

From Table 12.1

$$H_v'' - 2yH_v' = -2vH_v$$

$$y^2 H_v = y\left(\tfrac{1}{2}H_{v+1} + vH_{v-1}\right) = \tfrac{1}{2}\left(\tfrac{1}{2}H_{v+2} + (v+1)H_v\right) + v\left(\tfrac{1}{2}H_v + (v-1)H_{v-2}\right)$$

$$= \tfrac{1}{4}H_{v+2} + v(v-1)H_{v-2} + \left(v+\tfrac{1}{2}\right)H_v$$

Hence, $\dfrac{d^2\psi}{dy^2} = N\left[\tfrac{1}{4}H_{v+2} + v(v-1)H_{v-2} - \left(v+\tfrac{1}{2}\right)H_v\right]e^{-y^2/2}$ Therefore,

$$\langle T\rangle = N^2\left(-\tfrac{1}{2}\hbar\omega\right)\int_{-\infty}^{+\infty} H_v\left[\tfrac{1}{4}H_{v+2} + v(v-1)H_{v-2} - \left(v+\tfrac{1}{2}\right)H_v\right]e^{-y^2}\,dx$$

$$[dx = \alpha\,dy]$$

$$= \alpha N^2\left(-\tfrac{1}{2}\hbar\omega\right)\left[0 + 0 - \left(v+\tfrac{1}{2}\right)\pi^{1/2}2^v v!\right]$$

$$\left[\int_{-\infty}^{+\infty} H_v H_{v'}e^{-y^2}\,dy = 0 \quad \text{if } v' \neq v, \text{ marginal note above Table 12.1}\right]$$

$$= \boxed{\tfrac{1}{2}\left(v+\tfrac{1}{2}\right)\hbar\omega} \quad \left[N_v^2 = \frac{1}{\alpha\pi^{1/2}2^v v!}, \text{ Example 12.2}\right]$$

P12.15 **(a)** $\langle x\rangle = \displaystyle\int_0^L \left(\tfrac{2}{L}\right)^{1/2}\sin\left(\tfrac{n\pi x}{L}\right)x\left(\tfrac{2}{L}\right)^{1/2}\sin\left(\tfrac{n\pi x}{L}\right)\,dx$

$$= \left(\tfrac{2}{L}\right)\int_0^L x\sin^2 ax\,dx \quad \left[a = \tfrac{n\pi}{L}\right]$$

$$= \left(\tfrac{2}{L}\right)\times\left(\frac{x^2}{4} - \frac{x\sin 2ax}{4a} - \frac{\cos 2ax}{8a^2}\right)\Bigg|_0^L = \left(\tfrac{2}{L}\right)\times\left(\frac{L^2}{4}\right)$$

$$= \frac{L}{2} \quad \text{[by symmetry also]}$$

$$\langle x^2\rangle = \frac{2}{L}\int_0^L x^2\sin^2 ax\,dx = \left(\tfrac{2}{L}\right)\times\left[\frac{x^3}{6} - \left(\frac{x^2}{4a} - \frac{1}{8a^3}\right)\sin 2ax - \frac{x\cos 2ax}{4a^2}\right]\Bigg|_0^L$$

$$= \left(\tfrac{2}{L}\right)\times\left(\frac{L^3}{6} - \frac{L^3}{4n^2\pi}\right) = \frac{L^2}{3}\left(1 - \frac{1}{6n^2\pi^2}\right)$$

$$\Delta x = \left[\frac{L^2}{3}\left(1 - \frac{1}{6n^2\pi^2}\right) - \frac{L^2}{4}\right]^{1/2} = \boxed{L\left(\frac{1}{12} - \frac{1}{2\pi^2 n^2}\right)^{1/2}}$$

$\langle p\rangle = 0$ [by symmetry, also see Exercise 12.5(a)]

$\langle p^2\rangle = \dfrac{n^2 h^2}{4L^2}$ [from $E = \dfrac{p^2}{2m}$, also Exercise 12.5(a)]

$$\Delta p = \left(\frac{n^2 h^2}{4L^2}\right)^{1/2} = \boxed{\frac{nh}{2L}}$$

$$\Delta p\,\Delta x = \frac{nh}{2L}\times L\left(\frac{1}{12} - \frac{1}{2\pi^2 n^2}\right)^{1/2} = \frac{nh}{2\sqrt{3}}\left(1 - \frac{1}{24\pi^2 n^2}\right)^{1/2} > \frac{\hbar}{2}$$

(b) $\langle x\rangle = \alpha^2\displaystyle\int_{-\infty}^{+\infty}\psi^2 y\,dy[x = \alpha y] = 0$ [by symmetry, y is an odd function]

$$\langle x^2\rangle = \frac{2}{k}\left\langle\tfrac{1}{2}kx^2\right\rangle = \frac{2}{k}\langle V\rangle$$

since $2\langle T \rangle = b\langle V \rangle [12.46, \langle T \rangle \equiv E_K] = 2\langle V \rangle$ $\left[V = ax^b = \dfrac{1}{2}kx^2, b = 2\right]$

or $\langle V \rangle = \langle T \rangle = \dfrac{1}{2}\left(v + \dfrac{1}{2}\right)\hbar\omega$ [Problem 12.13]

$$\langle x^2 \rangle = \left(v + \frac{1}{2}\right) \times \left(\frac{\hbar\omega}{k}\right) = \left(v + \frac{1}{2}\right) \times \left(\frac{\hbar}{\omega m}\right) = \left(v + \frac{1}{2}\right) \times \left(\frac{\hbar^2}{mk}\right)^{1/2} \quad [12.42]$$

$$\boxed{\Delta x = \left[\left(v + \frac{1}{2}\right)\frac{\hbar}{\omega m}\right]^{1/2}}$$

$\langle p \rangle = 0$ [by symmetry, or by noting that the integrand is always an odd function of x]

$\langle p^2 \rangle = 2m\langle T \rangle = (2m) \times \left(\dfrac{1}{2}\right) \times \left(v + \dfrac{1}{2}\right) \times \hbar\omega$ [Problem 12.13]

$$\boxed{\Delta p = \left[\left(v + \frac{1}{2}\right)\hbar\omega m\right]^{1/2}}$$

$$\Delta p \Delta x = \left(v + \frac{1}{2}\right)\hbar \geq \frac{\hbar}{2}$$

Comment. Both results show a consistency with the uncertainty principle in the form $\Delta p \Delta q \geq \dfrac{\hbar}{2}$ as given in Section 11.6, eqn 11.41.

P12.16 $\quad \mu \equiv \displaystyle\int \psi_{v'} x \psi_v \, dx = \alpha^2 \int \psi_{v'} y \psi_v \, dy \quad [x = \alpha y]$

$y\psi_v = N_v \left(\dfrac{1}{2}H_{v+1} + vH_{v-1}\right) e^{-y^2/2}$ [marginal note above Table 12.1]

Hence

$$\mu = \alpha^2 N_v N_{v'} \int \left(\frac{1}{2}H_{v'}H_{v+1} + vH_{v'}H_{v-1}\right) e^{-y^2} \, dy = 0 \quad \text{unless } v' = v \pm 1$$

[marginal note above Table 12.1]

For $v' = v + 1$

$$\mu = \frac{1}{2}\alpha^2 N_v N_{v+1} \int H_{v+1}^2 e^{-y^2} \, dy = \frac{1}{2}\alpha^2 N_v N_{v+1} \pi^{1/2} 2^{v+1}(v+1)! = \boxed{\alpha\left(\frac{v+1}{2}\right)^{1/2}}$$

For $v' = v - 1$

$$\mu = v\alpha^2 N_v N_{v-1} \int H_{v-1}^2 e^{-y^2} \, dy = v\alpha^2 N_v N_{v-1} \pi^{1/2} 2^{v-1}(v-1)! = \boxed{\alpha\left(\frac{v}{2}\right)^{1/2}}$$

No other values of v' result in a nonzero value for μ; hence, no other transitions are allowed.

P12.19 The Schrödinger equation is

$$-\frac{\hbar}{2m}\nabla^2\psi = E\psi \quad [12.60, \text{ with } V = 0]$$

$$\nabla^2\psi = \frac{1}{r}\frac{\partial^2(r\psi)}{\partial r^2} + \frac{1}{r^2}\Lambda^2\psi \text{ [Table 11.1]}$$

since $r = $ constant, the first term is eliminated and the Schrödinger equation may be rewritten

$$-\frac{\hbar^2}{2mr^2}\Lambda^2\psi = E\psi \quad \text{or} \quad -\frac{\hbar^2}{2I}\Lambda^2\psi = E\psi \quad [I = mr^2] \quad \text{or} \quad \Lambda^2\psi = -\frac{2IE\psi}{\hbar^2}$$

Now use $\psi = Y_{l,m_l}$ and see if they satisfy this equation.

(a) $\quad \Lambda^2 Y_{0,0} = \boxed{0}\,[l = 0,\, m_l = 0]$, implying that $\quad E = 0$

and angular momentum $= \boxed{0}$ [from $\{l(l+1)\}^{1/2}\hbar$]

(b) $\quad \Lambda^2 Y_{2,-1} = -2(2+1)Y_{2,-1} \quad [l = 2]$, and hence

$$-2(2+1)Y_{2,-1} = -\frac{2IE}{\hbar^2}Y_{2,-1}, \quad \text{implying that} \quad \boxed{E = \frac{3\hbar^2}{I}}$$

and the angular momentum is $\{2(2+1)\}^{1/2}\hbar = \boxed{6^{1/2}\hbar}$

(c) $\quad \Lambda^2 Y_{3,3} = -3(3+1)Y_{3,3} \quad [l = 3]$, and hence

$$-3(3+1)Y_{3,3} = -\frac{2IE}{\hbar^2}Y_{3,3}, \quad \text{implying that} \quad \boxed{E = \frac{6\hbar^2}{I}}$$

and the angular momentum is $\{3(3+1)\}^{1/2}\hbar = \boxed{2\sqrt{3}\hbar}$

P12.21   From the diagram in Fig. 12.2, $\cos\theta = \dfrac{m_l}{\{l(l+1)\}^{1/2}}$ and hence $\boxed{\theta = \arccos\dfrac{m_l}{\{l(l+1)\}^{1/2}}}$

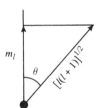

Figure 12.2

For an α electron, $m_s = +\frac{1}{2}$, $s = \frac{1}{2}$ and (with $m_l \to m_s,\, l \to s$)

$$\theta = \arccos\frac{\frac{1}{2}}{\left(\frac{3}{4}\right)^{1/2}} = \arccos\frac{1}{\sqrt{3}} = \boxed{54°44'}$$

The minimum angle occurs for $m_l = l$:

$$\lim_{l\to\infty}\theta_{\min} = \lim_{l\to\infty}\arccos\left(\frac{l}{\{l(l+1)\}^{1/2}}\right) = \lim_{l\to\infty}\arccos\frac{l}{l} = \arccos 1 = \boxed{0}$$

P12.23 $l = r \times p = \begin{vmatrix} i & j & k \\ \hat{x} & \hat{y} & \hat{z} \\ \hat{p}_x & \hat{p}_y & \hat{p}_z \end{vmatrix}$ [see any book treating the vector product of vectors]

$$= i(\hat{y}\hat{p}_z - \hat{z}\hat{p}_y) + j(\hat{z}\hat{p}_x - \hat{x}\hat{p}_z) + k(\hat{x}\hat{p}_y - \hat{y}\hat{p}_x)$$

Therefore,

$$\hat{l}_x = (\hat{y}\hat{p}_z - \hat{z}\hat{p}_y) = \boxed{\frac{\hbar}{i}\left(y\frac{\partial}{\partial z} - z\frac{\partial}{\partial y}\right)}$$

$$\hat{l}_y = (\hat{z}\hat{p}_x - \hat{x}\hat{p}_z) = \boxed{\frac{\hbar}{i}\left(z\frac{\partial}{\partial x} - x\frac{\partial}{\partial z}\right)}$$

$$\hat{l}_z = (\hat{x}\hat{p}_y - \hat{y}\hat{p}_x) = \boxed{\frac{\hbar}{i}\left(x\frac{\partial}{\partial y} - y\frac{\partial}{\partial x}\right)}$$

We have used $\hat{p}_x = \dfrac{\hbar}{i}\dfrac{\partial}{\partial x}$, etc. The commutator of $\hat{l}_x$ and $\hat{l}_y$ is $(\hat{l}_x\hat{l}_y - \hat{l}_y\hat{l}_x)$. We note that the operations always imply operation on a function. We form

$$\hat{l}_x\hat{l}_y f = -\hbar^2\left(y\frac{\partial}{\partial z} - z\frac{\partial}{\partial y}\right)\left(z\frac{\partial}{\partial x} - x\frac{\partial}{\partial z}\right)f$$

$$= -\hbar^2\left(yz\frac{\partial^2 f}{\partial z\partial x} + y\frac{\partial f}{\partial x} - yx\frac{\partial^2 f}{\partial z^2} - z^2\frac{\partial^2 f}{\partial y\partial x} + zx\frac{\partial^2 f}{\partial z\partial y}\right)$$

$$\hat{l}_y\hat{l}_x f = -\hbar^2\left(z\frac{\partial}{\partial x} - x\frac{\partial}{\partial z}\right)\left(y\frac{\partial}{\partial z} - z\frac{\partial}{\partial y}\right)f$$

$$= -\hbar^2\left(zy\frac{\partial^2 f}{\partial x\partial z} - z^2\frac{\partial^2 f}{\partial x\partial y} - xy\frac{\partial^2 f}{\partial z^2} + xz\frac{\partial^2 f}{\partial z\partial y} + x\frac{\partial f}{\partial y}\right)$$

Since multiplication and differentiation are each commutative, the results of the operation $\hat{l}_x\hat{l}_y$ and $\hat{l}_y\hat{l}_x$ differ only in one term. For $\hat{l}_y\hat{l}_x f$, $x\dfrac{\partial f}{\partial y}$ replaces $y\dfrac{\partial f}{\partial x}$. Hence, the commutator of the operations,

$(\hat{l}_x\hat{l}_y - \hat{l}_y\hat{l}_x)$, is $-\hbar^2\left(y\dfrac{\partial}{\partial x} - x\dfrac{\partial}{\partial y}\right)$ or $\boxed{-\dfrac{\hbar}{i}\hat{l}_z.}$

Comment. We also would find

$$(\hat{l}_y\hat{l}_z - \hat{l}_z\hat{l}_y) = -\frac{\hbar}{i}\hat{l}_x \quad \text{and} \quad (\hat{l}_z\hat{l}_x - \hat{l}_x\hat{l}_z) = -\frac{\hbar}{i}\hat{l}_y$$

P12.24 Upon making the operator substitutions

$$p_x = \frac{\hbar}{i}\frac{\partial}{\partial x} \quad \text{and} \quad p_y = \frac{\hbar}{i}\frac{\partial}{\partial y}$$

into $\hat{l}_z$ we find

$$\hat{l}_z = \frac{\hbar}{i}\left(x\frac{\partial}{\partial y} - y\frac{\partial}{\partial x}\right)$$

But $\dfrac{\partial}{\partial \phi} = \dfrac{\partial x}{\partial \phi}\dfrac{\partial}{\partial x} + \dfrac{\partial y}{\partial \phi}\dfrac{\partial}{\partial y} + \dfrac{\partial z}{\partial \phi}\dfrac{\partial}{\partial z}$ which is the chain rule of partial differentiation.

$$\frac{\partial x}{\partial \phi} = \frac{\partial}{\partial \phi}(r\sin\theta\cos\phi) = -r\sin\theta\sin\phi = -y$$

$$\frac{\partial y}{\partial \phi} = \frac{\partial}{\partial \phi}(r\sin\theta\sin\phi) = r\sin\theta\cos\phi = x$$

$$\frac{\partial z}{\partial \phi} = 0$$

Thus,

$$\frac{\partial}{\partial \phi} = -y\frac{\partial}{\partial x} + x\frac{\partial}{\partial y}$$

Upon substitution,

$$\hat{l}_z = \frac{\hbar}{i}\frac{\partial}{\partial \phi} = -i\hbar\frac{\partial}{\partial \phi}$$

Solutions to applications

P12.26 **(a)** $L = (21) \times (1.40 \times 10^{-10}\,\text{m}) = 2.94 \times 10^{-10}\,\text{m}$

$$\Delta E = E_{12} - E_{11} = (2n+1) \times \frac{h^2}{8mL^2}[12.9] = (2 \times 11 + 1) \times \frac{h^2}{8mL^2}$$

$$= \frac{(23) \times (6.626 \times 10^{-34}\,\text{J s})^2}{(8) \times (9.11 \times 10^{-31}\,\text{kg}) \times (2.94 \times 10^{-9})^2}$$

$$= 1.60\overline{3} \times 10^{-19}\,\text{J} = \boxed{1.60 \times 10^{-19}\,\text{J}}$$

(b) $\nu = \dfrac{\Delta E}{h} = \dfrac{1.60\overline{3} \times 10^{-19}\,\text{J}}{6.626 \times 10^{-34}\,\text{J s}} = 2.42 \times 10^{14}\,\text{s}^{-1} = \boxed{2.42 \times 10^{14}\,\text{Hz}}$

(c) The wavefunctions are

$$\psi_n = \left(\frac{2}{L}\right)^{1/2}\sin\left(\frac{n\pi x}{L}\right)$$

$$P_n = \int_{x=\frac{10}{21}L}^{x=\frac{11}{21}L}\left(\frac{2}{L}\right)\sin^2\left(\frac{n\pi x}{L}\right)dx = \left(\frac{2}{L}\right) \times \left[\frac{1}{2}x - \frac{L}{4n\pi}\sin\left(\frac{2n\pi x}{L}\right)\right]\Big|_{x=\frac{10}{21}L}^{x=\frac{11}{21}L}$$

$$= \left(\frac{2}{L}\right) \times \left[\frac{1}{42}L - \frac{L}{4n\pi}\left(\sin\frac{2n\pi \times 11}{21} - \sin\frac{2n\pi \times 10}{21}\right)\right]$$

We use

$$\sin\alpha - \sin\beta = 2\cos\frac{1}{2}(\alpha+\beta)\sin\frac{1}{2}(\alpha-\beta) = 2\cos n\pi\,\sin\left(\frac{n\pi}{21}\right)$$

Then

$$P_n = \frac{1}{21} - \frac{1}{n\pi}\left[\cos n\pi\,\sin\left(\frac{n\pi}{21}\right)\right]$$

We draw up the following table

n	$\cos n\pi$	$\sin\left(\dfrac{n\pi}{21}\right)$	$-\dfrac{1}{n\pi}\left[\cos n\pi\,\sin\left(\dfrac{n\pi}{21}\right)\right]$	P_n
1	−1	0.1490	0.04744	0.09506
2	+1	0.2948	−0.04691	0.00071
3	−1	0.4339	0.04604	0.09366
4	+1	0.5633	−0.04482	0.00279
5	−1	0.6801	−0.04330	0.09092
6	+1	0.7818	−0.04148	0.00614
7	−1	0.8660	+0.03938	0.08000
8	+1	0.9309	−0.03704	0.01058
9	−1	0.9749	+0.03448	0.08210
10	+1	0.9972	−0.03174	0.01588
11	−1	0.9972	+0.02886	0.07648

The sum of the P_n column is 0.56131. Since two electrons occupy each 'orbital' the number of electrons on average between C11 and C12 is

$$2 \times 0.56131 = \boxed{1.12262}$$

Comment. Note that the result in (c) is independent of the experimental value of L and thus is an exact number which can be written in as many figures as desired. The calculated result is consistent with ordinary chemical reasoning, namely, that each bond in a conjugated polyene has approximately one electron associated with it, but the result also indicates some slight bunching up toward the center.

P12.28   The rate of tunneling is proportional to the transmission probability, so a ratio of tunneling rates is equal to the corresponding ratio of transmission probabilities (given in eqn 12.28a). The desired factor is T_1/T_2, where the subscripts denote the tunneling distances in nanometers:

$$\frac{T_1}{T_2} = \frac{1 + \dfrac{(e^{\kappa L_2} - e^{-\kappa L_2})^2}{16\varepsilon(1-\varepsilon)}}{1 + \dfrac{(e^{\kappa L_1} - e^{-\kappa L_1})^2}{16\varepsilon(1-\varepsilon)}}.$$

If $\dfrac{(e^{\kappa L_2} - e^{-\kappa L_2})^2}{16\varepsilon(1-\varepsilon)} \gg 1,$

then $\dfrac{T_1}{T_2} \approx \dfrac{(e^{-\kappa L_2} - e^{-\kappa L_2})^2}{(e^{\kappa L_1} - e^{-\kappa L_1})^2} \approx e^{2\kappa(L_2-L_1)} = e^{2(7/nm)(2.0-1.0)nm} = \boxed{1.2 \times 10^6}.$

This is, the tunneling rate increases about a million-fold. Note: if the first approximation does not hold, we need more information, namely $\varepsilon = E/V$. If the first approximation is valid, then the second is also likely to be valid, namely that the negative exponential is negligible compared to the positive one.

P12.30   Assuming that one can identify the CO peak in the infrared spectrum of the CO-myoglobin complex, taking infrared spectra of each of the isotopic variants of CO-myoglobin complexes can show which atom binds to the haem group and determine the $C\equiv O$ force constant. Compare isotopic variants to $^{12}C^{16}O$ as the standard; when an isotope changes but the vibrational frequency does not, then the atom whose isotope was varied is the atom that binds to the haem. See table below, which includes predictions of the wavenumber of all isotopic variants compared to that of $\tilde{\nu}(^{12}C^{16}O)$. (As usual, the better the experimental results agree with the whole set of predictions, the more confidence one would have with the conclusion.)

Wavenumber	If O binds	If C binds
$\tilde{\nu}(^{12}C^{18}O)$	$\tilde{\nu}(^{12}C^{16}O)$†	$(16/18)^{1/2}\tilde{\nu}(^{12}C^{16}O)$
$\tilde{\nu}(^{13}C^{16}O)$	$(12/13)^{1/2}\tilde{\nu}(^{12}C^{16}O)$	$\tilde{\nu}(^{12}C^{16}O)$†
$\tilde{\nu}(^{13}C^{18}O)$	$(12/13)^{1/2}\tilde{\nu}(^{12}C^{16}O)$	$(16/18)\tilde{\nu}(^{12}C^{16}O)$

† That is, no change compared to the standard.

The wavenumber is related to the force constant as follows:

$$\omega = 2\pi c\tilde{\nu} = \left(\frac{k}{m}\right)^{1/2} \quad \text{so} \quad k = m(2\pi c\tilde{\nu})^2,$$

$$k = m(1.66 \times 10^{-27}\,\text{kg u}^{-1})[(2\pi)(2.998 \times 10^{10}\,\text{cm s}^{-1})\tilde{\nu}(^{12}C^{16}O)]^2,$$

and $\quad k/(\text{kg s}^{-1}) = (5.89 \times 10^{-5})(m/u)[\tilde{\nu}(^{12}C^{16}O)/\text{cm}^{-1}]^2.$

Here m is the mass of the atom that is not bound, *i.e.*, 12 u if O is bound and 16 u if C is bound. (Of course, one can compute k from any of the isotopic variants, and one take k to be a mean derived from all the relevant data.)

13 Atomic structure and atomic spectra

Solutions to exercises

Discussion questions

E13.1(a) The Schrodinger equation for the hydrogen atom is a six-dimensional partial differential equation, three dimensions for each particle in the atom. One cannot directly solve a multidimensional differential equation, it must be broken down into one-dimensional equations. This is the separation of variables procedure. The choice of coordinates is critical in this process. The separation of the Schrodinger equation can be accomplished in a set of coordinates that are natural to the system, but not in others. These natural coordinates are those directly related to the description of the motion of the atom. The atom as a whole (center of mass) can move from point to point in three-dimensional space. The natural coordinates for this kind of motion are the Cartesian coordinates of a point in space. The internal motion of the electron with respect to the proton is most naturally described with spherical polar coordinates. So the six-dimensional Schrodinger equation is first separated into two three-dimensional equations, one for the motion of the center of mass, the other for the internal motion. The separation of the center of mass equation and its solution is fully discussed in Section 12.2. The equation for the internal motion is separable into three one-dimensional equations, one in the angle ϕ, another in the angle θ, and a third in the distance r. The solutions of these three one-dimensional equations can be obtained by standard techniques and were already well known long before the advent of Quantum Mechanics. Another choice of coordinates would not have resulted in the separation of the Schrodinger equation just described. For the details of the separation procedure, see Sections 13.1 and 12.7.

E13.2(a) The selection rules are:

$$\Delta n = \pm 1, \pm 2, \ldots \qquad \Delta l = \pm 1 \qquad \Delta m_l = 0, \pm 1$$

In a spectroscopic transition the atom emits or absorbs a photon. Photons have a spin angular momentum of 1. Therefore, as a result of the transition, the angular momentum of the electromagnetic field has changed by $\pm 1\hbar$. The principle of the conservation of angular momentum then requires that the angular momentum of the atom has undergone an equal and opposite change in angular momentum. Hence, the selection rule on $\Delta l = \pm 1$. The principle quantum number n can change by any amount since n does not directly relate to angular momentum. The selection rule on Δm_l is harder to account for on basis of these simple considerations alone. One has to evaluate the transition dipole moment between the wavefunctions representing the initial and final states involved in the transition. See *Justification* 13.5 for an example of this procedure.

E13.3(a) See Section 13.4(d) of the text and any General Chemistry book, for example, Sections 1.11–1.14 of P. Atkins and L. Jones, *Chemical Principles*, W. H. Freeman, and Co., New York (1999).

E13.4(a) In the crudest form of the orbital approximation, the many-electron wavefunctions for atoms are represented as a simple product of one-electron wavefunctions. At a somewhat more sophisticated level, the many electron wavefunctions are written as linear combinations of such simple product functions that explicitly satisfy the Pauli exclusion principle. Relatively good one-electron functions are generated by the Hartree–Fock self-consistent field method described in Section 13.5. If we place no restrictions on the form of the one-electron functions, we reach the Hartree–Fock limit which gives us the best value of the calculated energy within the orbital approximation. The orbital approximation is based on the disregard of significant portions of the electron–electron interaction terms in the many-electron Hamiltonian, so we cannot expect that it will be quantitatively accurate. By abandoning the orbital approximation, we could in principle obtain essentially exact energies; however, there are significant conceptual advantages to retaining the orbital approximation. Increased accuracy can be obtained by reintroducing the neglected electron–electron interaction terms and including their effects on the energies of the atom by a form of perturbation theory similar to that described in Section 13.9. For a more complete discussion consult the references listed under *Further reading*.

Numerical exercises

E13.5(a) This is essentially the photoelectric effect [eqn 11.12 of Section 11.2] with the ionization energy of the ejected electron being the work function Φ.

$$h\nu = \tfrac{1}{2}m_e v^2 + I$$

$$I = h\nu - \frac{1}{2}m_e v^2 = (6.626 \times 10^{-34}\,\text{J Hz}^{-1}) \times \left(\frac{2.998 \times 10^8\,\text{m s}^{-1}}{58.4 \times 10^{-9}\,\text{m}}\right)$$

$$- \left(\tfrac{1}{2}\right) \times (9.109 \times 10^{-31}\,\text{kg}) \times (1.59 \times 10^6\,\text{m s}^{-1})^2$$

$$= 2.25 \times 10^{-18}\,\text{J, corresponding to } \boxed{14.0\,\text{eV}}$$

E13.6(a) $R_{2,0} \propto \left(2 - \dfrac{\rho}{2}\right) e^{-\rho/4}$ with $\rho = \dfrac{2r}{a_0}$ [Table 13.1]

$$\frac{dR}{dr} = \frac{2}{a_0}\frac{dR}{d\rho} = \frac{2}{a_0}\left(-\frac{1}{2} - \frac{1}{2} + \frac{1}{8}\rho\right) e^{-\rho/4} = 0 \quad \text{when } \rho = 8$$

Hence, the wavefunction has an extremum at $r = \boxed{4a_0}$. Since $2 - \dfrac{\rho}{2} < 0$, $\psi < 0$ and the extremum is a minimum (more formally: $\dfrac{d^2\psi}{dr^2} > 0$ at $\rho = 8$).

The second extremum is at $\boxed{r = 0}$. It is not a minimum and in fact is a physical maximum, though not one that can be obtained by differentiation. To see that it is maximum substitute $\rho = 0$ into $R_{2,0}$.

E13.7(a) The radial nodes correspond to $R_{3,0} = 0$. $R_{3,0} \propto 6 - 2\rho + \dfrac{1}{9}\rho^2$ (Table 13.1); the radial nodes occur at

$$6 - 2\rho + \tfrac{1}{9}\rho^2 = 0, \quad \text{or} \quad \rho = 3(3 \pm \sqrt{3}) = 1.27 \text{ and } 4.73.$$

Since $r = \dfrac{\rho a_0}{2}$, the radial nodes occur at $\boxed{101\,\text{pm and } 376\,\text{pm}}$

E13.8(a) $R_{1,0} = N e^{-r/a_0}$

$$\int_0^\infty R^2 r^2\, dr = 1 = \int_0^\infty N^2 r^2\, e^{-2r/a_0}\, dr = N^2 \times \frac{2!}{\left(\frac{2}{a_0}\right)^3} = 1 \quad \left[\int_0^\infty x^n e^{-ax}\, dx = \frac{n!}{a^{n+1}}\right]$$

$$N^2 = \frac{4}{a_0^3}, \qquad \boxed{N = \frac{2}{a_0^{3/2}}}$$

Thus,

$$R_{1,0} = 2\left(\frac{1}{a_0}\right)^{3/2} e^{-r/a_0},$$

which agrees with Table 13.1.

E13.9(a) This exercise has already been solved in Problem 12.17 by use of the virial theorem. Here we will solve it by straightforward integration.

$$\psi_{1,0,0} = R_{1,0}Y_{0,0} = \left(\frac{1}{\pi a_0^3}\right)^{1/2} e^{-r/a_0} \text{ [Tables 12.3 and 13.1]}$$

The potential energy operator is

$$V = -\frac{Ze^2}{4\pi\varepsilon_0} \times \left(\frac{1}{r}\right) = -k\left(\frac{1}{r}\right)$$

$$\langle V\rangle = -k\left\langle\frac{1}{r}\right\rangle \left[k = \frac{e^2}{4\pi\varepsilon_0}\right] = -k\int_0^\infty\int_0^\pi\int_0^{2\pi}\left(\frac{1}{\pi a_0^3}\right)e^{-r/a_0}\left(\frac{1}{r}\right)e^{-r/a_0}r^2\,dr\sin\theta\,d\theta\,d\phi$$

$$= -k \times (4\pi) \times \left(\frac{1}{\pi a_0^3}\right)\int_0^\infty re^{-2r/a_0}\,dr = -k \times \left(\frac{4}{a_0^3}\right) \times \left(\frac{a_0^2}{4}\right) = -k\left(\frac{1}{a_0}\right)$$

$$\left[\text{We have used } \int_0^\pi \sin\theta\,d\theta = 2, \int_0^{2\pi} d\phi = 2\pi, \text{ and } \int_0^\infty x^n e^{-ax}\,dx = \frac{n!}{a^{n+1}}\right]$$

Hence,

$$\langle V\rangle = -\frac{e^2}{4\pi\varepsilon_0 a_0} = \boxed{2E_{1s}}$$

The kinetic energy operator is $-\dfrac{\hbar^2}{2\mu}\nabla^2$ [13.7]; hence

$$\langle E_K\rangle \equiv \langle T\rangle = \int \psi_{1s}^* \left(-\frac{\hbar^2}{2\mu}\right)\nabla^2\psi_{1s}\,d\tau$$

$$\nabla^2\psi_{1s} = \frac{1}{r}\frac{\partial^2(r\psi_{1s})}{\partial r^2} + \frac{1}{r^2}\Lambda^2\psi_{1s} \text{ [Problem 12.19]}$$

$$= \left(\frac{1}{\pi a_0^3}\right)^{1/2} \times \left(\frac{1}{r}\right) \times \left(\frac{d^2}{dr^2}\right)re^{-r/a_0}$$

$$[\Lambda^2\psi_{1s} = 0, \psi_{1s} \text{ contains no angular variables}]$$

$$= \left(\frac{1}{\pi a_0^3}\right)^{1/2}\left[-\left(\frac{2}{a_0 r}\right) + \left(\frac{1}{a_0^2}\right)\right]e^{-r/a_0}$$

$$\langle T\rangle = -\left(\frac{\hbar^2}{2\mu}\right) \times \left(\frac{1}{\pi a_0^3}\right)\int_0^\infty\left[-\left(\frac{2}{a_0 r}\right) + \left(\frac{1}{a_0^2}\right)\right]e^{-2r/a_0}r^2\,dr \times \int_0^\pi \sin\theta\,d\theta\int_0^{2\pi} d\phi$$

$$= -\left(\frac{2\hbar^2}{\mu a_0^3}\right)\int_0^\infty\left[-\left(\frac{2r}{a_0}\right) \times \left(\frac{r^2}{a_0^2}\right)\right]e^{-2r/a_0}\,dr = -\left(\frac{2\hbar^2}{\mu a_0^3}\right) \times \left(-\frac{a_0}{4}\right) = \frac{\hbar^2}{2\mu a_0^2}$$

$$= \boxed{-E_{1s}}$$

Hence, $\langle T\rangle + \langle V\rangle = 2E_{1s} - E_{1s} = E_{1s}$

Comment. E_{1s} may also be written as

$$E_{1s} = -\frac{\mu e^4}{32\pi^2 \varepsilon_0^2 \hbar^2}$$

Question. Are the three different expressions for E_{1s} given in this exercise all equivalent?

E13.10(a) $P_{2s} = 4\pi r^2 \psi_{2s}^2$

$$\psi_{2s} = \frac{1}{2\sqrt{2}} \left(\frac{Z}{a_0}\right)^{3/2} \times \left(2 - \frac{\rho}{2}\right) e^{-\rho/4} \quad \left[\rho = \frac{2Zr}{a_0}\right]$$

$$P_{2s} = 4\pi \left(\frac{a_0 \rho}{2Z}\right)^2 \times \left(\frac{1}{8}\right) \times \left(\frac{Z}{a_0}\right)^3 \left(2 - \frac{\rho}{2}\right)^2 e^{-\rho/2}$$

$$P_{2s} = k\rho^2 \left(2 - \frac{\rho}{2}\right)^2 e^{-\rho/2} \quad \left[k = \frac{\pi Z}{8a_0} = \text{constant}\right]$$

The most probable value of r, or equivalently, ρ is where

$$\frac{d}{d\rho} \left\{\rho^2 \left(2 - \frac{\rho}{2}\right)^2 e^{-\rho/2}\right\} = 0$$

$$\propto \left\{2\rho \left(2 - \frac{\rho}{2}\right)^2 - 2\rho^2 \left(2 - \frac{\rho}{2}\right) - \rho^2 \left(2 - \frac{\rho}{2}\right)^2\right\} e^{-\rho/2} = 0$$

$$\propto \rho(\rho - 4)(\rho^2 - 12\rho + 16) = 0 \quad [e^{-\rho/2} \text{ is never zero, except as } \rho \to \infty]$$

Thus, $\rho^* = 0$, $\rho^* = 4$, $\rho^* = 6 \pm 2\sqrt{5}$

The principal (outermost) maximum is at $\rho^* = 6 + 2\sqrt{5}$

Hence, $r^* = (6 + 2\sqrt{5})\frac{a_0}{2Z} = \boxed{5.24\frac{a_0}{Z}}$

E13.11(a)   Indentify l and use angular momentum $= \{l(l + 1)\}^{1/2}\hbar$.

(a) $l = 0$, so angular momentum $= 0$ **(b)** $l = 0$, so angular momentum $= 0$

(c) $l = 2$, so angular momentum $= \sqrt{6}\hbar$

The total number of nodes is equal to $n - 1$ and the number of angular nodes is equal to l; hence the number of radial nodes is equal to $n - l - 1$. We can draw up the following table

	1s	3s	3d
n, l	1,0	3,0	3,2
Angular nodes	0	0	2
Radial nodes	0	2	0

E13.12(a)   We use the Clebsch–Gordan series [13.46] in the form

$$j = l + s, l + s - 1, \ldots, |l - s| \quad \text{[lower-case for a single electron]}$$

(a) $l = 2$, $\quad s = \frac{1}{2}$; so $j = \boxed{\frac{5}{2}, \frac{3}{2}}$ **(b)** $l = 3$, $\quad s = \frac{1}{2}$; so $j = \boxed{\frac{7}{2}, \frac{5}{2}}$

E13.13(a) The Clebsch–Gordan series for $\boxed{l = 1}$ and $s = \frac{1}{2}$ leads to $j = \frac{3}{2}$ and $\frac{1}{2}$.

E13.14(a) The energies are $E = -\dfrac{hc\mathcal{R}_H}{n^2}$ [Table 13.13], and the orbital degeneracy g of an energy level of principal quantum number n is

$$g = \sum_{l=0}^{n-1}(2l-1) = 1 + 3 + 5 + \cdots + 2n - 1 = \frac{(1 + 2n - 1)n}{2} = n^2$$

(a) $E = -hc\mathcal{R}_H$ implies that $n = 1$, so $\boxed{g = 1}$ [the $1s$ orbital].

(b) $E = -\dfrac{hc\mathcal{R}_H}{9}$ implies that $n = 3$, so $\boxed{g = 9}$ ($3s$ orbital, the three $3p$ orbitals, and the five $3d$ orbitals).

(c) $E = -\dfrac{hc\mathcal{R}_H}{25}$ implies that $n = 5$, so $\boxed{g = 25}$ (the $5s$ orbital, the three $5p$ orbitals, the five $5d$ orbitals, the seven $5f$ orbitals, the nine $5g$ orbitals).

E13.15(a) The letter D indicates that $L = 2$, the superscript 1 is the value of $2S + 1$, so $S = 0$ and the subscript 2 is the value of J. Hence, $\boxed{L = 2, \; S = 0, \; J = 2}$

E13.16(a) Here we use the probability density function ψ^2, rather than the radial distribution function P, since we are seeking the probability at a point, namely $\psi^2 \, d\tau$

The probability density varies as

$$\psi^2 = \frac{1}{\pi a_0^3} e^{-2r/a_0} \text{ [From 13.20]}$$

Therefore, the maximum value is at $r = 0$ and ψ^2 is 50 per cent of the maximum when

$$e^{-2r/a_0} = 0.50$$

implying that $r = -\frac{1}{2}a_0 \ln 0.50$, which is at $r = \boxed{0.35a_0}$ [18 pm]

E13.17(a) The selection rules for a many-electron atom are given by the set [13.47]. For a single-electron transition these amount to $\Delta n = $ any integer; $\Delta l = \pm 1$. Hence

(a) $2s \rightarrow 1s$; $\Delta l = 0$, $\boxed{\text{forbidden}}$ **(b)** $2p \rightarrow 1s$; $\Delta l = -1$, $\boxed{\text{allowed}}$

(c) $3d \rightarrow 2p$; $\Delta l = -1$, $\boxed{\text{allowed}}$

E13.18(a) For a given l there are $2l + 1$ values of m_l and hence $2l + 1$ orbitals. Each orbital may be occupied by two electrons. Hence the maximum occupancy is $2(2l + 1)$. Draw up the following table

	l	$2(2l+1)$		l	$2(2l+1)$
(a) $1s$	0	$\boxed{2}$	**(c)** $3d$	2	$\boxed{10}$
(b) $3p$	1	$\boxed{6}$	**(d)** $6g$	4	$\boxed{18}$

E13.19(a) **(a)** $\quad 1s^2 2s^2 2p^6 3s^2 3p^6 3d^8 = \boxed{[\text{Ar}]3d^8}$

(b) All subshells except $3d$ are filled and hence have no net spin. Applying Hund's rule to $3d^8$ shows that there are two unpaired spins. The paired spins do not contribute to the net spin, hence we consider only $s_1 = \frac{1}{2}$ and $s_2 = \frac{1}{2}$. The Clebsch–Gordan series [13.44, $l \to s$] produces

$$S = s_1 + s_2, \ldots, |s_1 - s_2|, \quad \text{hence} \quad \boxed{S = 1, 0}$$

$$M_S = -S, -S + 1, \ldots, S$$

For $S = 1$, $\boxed{M_S = -1, 0, +1}$

$$S = 0, \quad \boxed{M_S = 0}$$

E13.20(a) Use the Clebsch–Gordan series in the form

$$S' = s_1 + s_2, s_1 + s_2 - 1, \ldots, |s_1 - s_2|$$

and

$$S = S' + s_1, S' + s_1 - 1, \ldots, |S' - s_1|$$

in succession. The multiplicity is $2S + 1$.

(a) $S = \frac{1}{2} + \frac{1}{2}, \frac{1}{2} - \frac{1}{2} = \boxed{1, 0}$ with multiplicities $\boxed{3, 1}$ respectively

(b) $S' = 1, 0$; then $S = \boxed{\frac{3}{2}, \frac{1}{2}}$ [from 1], and $\boxed{\frac{1}{2}}$ [from 0], with multiplicities $\boxed{4, 2, 2}$

E13.21(a) These electrons are not equivalent (different subshells), hence all the terms that arise from the vector model and the Clebsch–Gordan series are allowed (Example 13.6).

$$L = l_1 + l_2, \ldots, |l - l_2| \, [13.44] = 2 \text{ only}$$
$$S = s_1 + s_2, \ldots, |s_1 - s_2| = 1, 0$$

The allowed terms are then ^{3}D and ^{1}D. The possible values of J are given by

$$J = L + S, \ldots, |L - S| \, [13.46] = 3, \, 2, \, 1 \text{ for } {}^3\text{D and } 2 \text{ for } {}^1\text{D}$$

The allowed complete term symbols are then

$$\boxed{{}^3\text{D}_3, \, {}^3\text{D}_2, \, {}^3\text{D}_1, \, {}^1\text{D}_2}$$

The $\boxed{{}^3\text{D} \text{ set of terms are the lower in energy}}$ [Hund's rule]

Comment. Hund's rule in the form given in the text does not allow the energies of the triplet terms to be distinguished. Experimental evidence indicates that ^{3}D$_1$ is lowest.

E13.22(a) Use the Clebsch–Gordan series in the form

$$J = L + S, L + S - 1, \ldots, |L - S|$$

The number of states (M_J values) is $2J + 1$ in each case.

(a) $L = 0$, $S = 0$; hence $\boxed{J = 0}$ and there is only $\boxed{1}$ state ($M_J = 0$).

(b) $L = 1$, $S = \frac{1}{2}$; hence $J = \boxed{\frac{3}{2}, \frac{1}{2}}$ ($^2P_{3/2}$, $^2P_{1/2}$) with 4, 2 states respectively.

(c) $L = 1$, $S = 1$; hence $J = \boxed{2, 1, 0}$ ($^3P_2$, $^3P_1$, $^3P_0$) with 5, 3, 1 states respectively.

E13.23(a) Closed shells and subshells do not contribute to either L or S and thus are ignored in what follows.

(a) Li[He]$2s^1$: $S = \frac{1}{2}$, $L = 0$; $J = \frac{1}{2}$, so the only term is $\boxed{^2S_{1/2}}$

(b) Na[He]$3p^1$: $S = \frac{1}{2}$, $L = 1$; $J = \frac{3}{2}, \frac{1}{2}$, so the terms are $\boxed{^2P_{3/2} \text{ and } ^2P_{1/2}}$

Solutions to problems
Solutions to numerical problems

P13.1 All lines in the hydrogen spectrum fit the Rydberg formula

$$\frac{1}{\lambda} = \mathcal{R}_H \left(\frac{1}{n_1^2} - \frac{1}{n_2^2} \right) \quad \left[13.1, \text{ with } \tilde{\nu} = \frac{1}{\lambda} \right] \quad \mathcal{R}_H = 109\,677\,\text{cm}^{-1}$$

Find n_1 from the value of λ_{max}, which arises from the transition $n_1 + 1 \to n_1$

$$\frac{1}{\lambda_{max}\mathcal{R}_H} = \frac{1}{n_1^2} - \frac{1}{(n_1 + 1)^2} = \frac{2n_1 + 1}{n_1^2(n_1 + 1)^2}$$

$$\lambda_{max}\mathcal{R}_H = \frac{n_1^2(n_1 + 1)^2}{2n_1 + 1} = (12\,368 \times 10^{-9}\,\text{m}) \times (109\,677 \times 10^2\,\text{m}^{-1}) = 135.65$$

Since $n_1 = 1, 2, 3$, and 4 have already been accounted for, try $n_1 = 5, 6, \ldots$. With $n_1 = 6$ we get $\frac{n_1^2(n_1 + 1)^2}{2n_1 + 1} = 136$. Hence, the Humphreys series is $\boxed{n_2 \to 6}$ and the transitions are given by

$$\frac{1}{\lambda} = (109\,677\,\text{cm}^{-1}) \times \left(\frac{1}{36} - \frac{1}{n_2^2} \right), \quad n_2 = 7, 8, \ldots$$

and occur at 12 372 nm, 7503 nm, 5908 nm, 5129 nm, ..., 3908 nm (at $n_2 = 15$), converging to 3282 nm as $n_2 \to \infty$, in agreement with the quoted experimental result.

P13.3 A Lyman series corresponds to $n_1 = 1$; hence

$$\tilde{\nu} = \mathcal{R}_{\text{Li}^{2+}} \left(1 - \frac{1}{n^2} \right), \quad n = 2, 3, \ldots \quad \left[\tilde{\nu} = \frac{1}{\lambda} \right]$$

Therefore, if the formula is appropriate, we expect to find that $\tilde{\nu} \left(1 - \frac{1}{n^2} \right)^{-1}$ is a constant ($\mathcal{R}_{\text{Li}^{2+}}$). We therefore draw up the following table.

n	2	3	4
$\tilde{\nu}/\text{cm}^{-1}$	740 747	877 924	925 933
$\tilde{\nu}\left(1 - \dfrac{1}{n^2}\right)^{-1} \Big/ \text{cm}^{-1}$	987 663	987 665	987 662

Hence, the formula does describe the transitions, and $\boxed{\mathcal{R}_{\text{Li}^{2+}} = 987\,663\,\text{cm}^{-1}}$. The Balmer transitions lie at

$$\tilde{\nu} = \mathcal{R}_{\text{Li}^{2+}} \left(\frac{1}{4} - \frac{1}{n^2}\right) \quad n = 3, 4, \dots$$

$$= (987\,663\,\text{cm}^{-1}) \times \left(\frac{1}{4} - \frac{1}{n^2}\right) = \boxed{137\,175\,\text{cm}^{-1}}, \boxed{185\,187\,\text{cm}^{-1}}, \dots$$

The ionization energy of the ground-state ion is given by

$$\tilde{\nu} = \mathcal{R}_{\text{Li}^{2+}} \left(1 - \frac{1}{n^2}\right), \quad n \to \infty$$

and hence corresponds to

$$\tilde{\nu} = 987\,663\,\text{cm}^{-1}, \quad \text{or} \quad \boxed{122.5\,\text{eV}}$$

P13.6 The ground term is $[\text{Ar}]4s^1\,{}^2S_{1/2}$ and the first excited is $[\text{Ar}]4p^1\,{}^2P$. The latter has two levels with $J = 1 + \frac{1}{2} = \frac{3}{2}$ and $J = 1 - \frac{1}{2} = \frac{1}{2}$ which are split by spin–orbit coupling (Section 13.8). Therefore, ascribe the transitions to $\boxed{{}^2P_{3/2} \to {}^2S_{1/2}}$ and $\boxed{{}^2P_{1/2} \to {}^2S_{1/2}}$ (since both are allowed). For these values of J, the splitting is equal to $\frac{3}{2}A$ (Example 13.5). Hence, since

$$(766.70 \times 10^{-7}\,\text{cm})^{-1} - (770.11 \times 10^{-7}\,\text{cm})^{-1} = 57.75\,\text{cm}^{-1}$$

we can conclude that $A = \boxed{38.50\,\text{cm}^{-1}}$

P13.8 The Rydberg constant for positronium ($\mathcal{R}_{\text{Ps}}$) is given by

$$\mathcal{R}_{\text{Ps}} = \frac{\mathcal{R}}{1 + \frac{m_e}{m_e}} = \frac{\mathcal{R}}{1 + 1} = \frac{1}{2}\mathcal{R} \quad [13.18; \text{ also Problem 13.7};\ m(\text{positron}) = m_e]$$

$$= 54\,869\,\text{cm}^{-1} \quad [\mathcal{R} = 109\,737\,\text{cm}^{-1}]$$

Hence

$$\tilde{\nu} = \frac{1}{\lambda} = (54\,869\,\text{cm}^{-1}) \times \left(\frac{1}{4} - \frac{1}{n^2}\right), \quad n = 3, 4, \dots$$

$$= \boxed{7621\,\text{cm}^{-1}}, \boxed{10\,288\,\text{cm}^{-1}}, \boxed{11\,522\,\text{cm}^{-1}}, \dots$$

The binding energy of Ps is

$$E = -hc\mathcal{R}_{\text{Ps}}, \quad \text{corresponding to } (-)54\,869\,\text{cm}^{-1}$$

The ionization energy is therefore $54\,869\,\text{cm}^{-1}$, or $\boxed{6.80\,\text{eV}}$

P13.9 **(a)** The splitting of adjacent energy levels is related to the difference in wavenumber of the spectral lines as follows:

$$hc\Delta\tilde{\nu} = \Delta E = \mu_B B, \quad \text{so} \quad \Delta\tilde{\nu} = \frac{\mu_B B}{hc} = \frac{(9.274 \times 10^{-24}\,\text{J T}^{-1})(2\,\text{T})}{(6.626 \times 10^{-34}\,\text{J s})(2.998 \times 10^{10}\,\text{cm s}^{-1})}$$

$$\Delta\tilde{\nu} = \boxed{0.9\,\text{cm}^{-1}}.$$

(b) Transitions induced by absorbing visible light have wavenumbers in the tens of thousands of reciprocal centimeters, so normal Zeeman splitting is small compared to the difference in energy of the states involved in the transition. Take a wavenumber from the middle of the visible spectrum as typical:

$$\tilde{\nu} = \frac{1}{\lambda} = \frac{1}{600\,\text{nm}}\left(\frac{10^9\,\text{nm m}^{-1}}{10^2\,\text{cm m}^{-1}}\right) = 1.7 \times 10^4\,\text{cm}^{-1}.$$

Or take the Balmer series as an example, as suggested in the problem; the Balmer wavenumbers are (eqn 13.1):

$$\tilde{\nu} = R_H\left(\frac{1}{2^2} - \frac{1}{n^3}\right).$$

The smallest Balmer wavenumber is

$$\tilde{\nu} = (109677\,\text{cm}^{-1}) \times (1/4 - 1/9) = 15233\,\text{cm}^{-1},$$

and the upper limit is

$$\tilde{\nu} = (109677\,\text{cm}^{-1}) \times (1/4 - 0) = 27419\,\text{cm}^{-1}.$$

Solutions to theoretical problems

P13.11   In each case we need to calculate $\langle r \rangle$. The radial wavefunctions (Table 13.1) rather than the radial distribution functions are appropriate for the purpose

$$
\begin{aligned}
\langle r \rangle_{2p} &= \int_0^\infty R_{21} r R_{21} r^2\,\mathrm{d}r \quad \left[\rho = \frac{2Zr}{a_0}\right] \\
&= \left(\frac{Z}{a_0}\right)^3 \times \left(\frac{1}{4\sqrt{6}}\right)^2 \int_0^\infty r^3 \rho^2 e^{-\rho/2}\,\mathrm{d}r \text{ [Table 13.1]} \\
&= \left(\frac{Z}{a_0}\right)^3 \times \left(\frac{1}{96}\right) \times \left(\frac{a_0}{2Z}\right)^4 \int_0^\infty \rho^5 e^{-\rho/2}\,\mathrm{d}\rho = \left(\frac{1}{96}\right) \times \left(\frac{a_0}{16Z}\right) \times (7680) = \frac{5a_0}{Z} \\
\langle r \rangle_{2s} &= \int_0^\infty R_{20} r R_{20} r^2\,\mathrm{d}r = \left(\frac{Z}{a_0}\right)^3 \times \left(\frac{1}{8}\right) \times \left(\frac{a_0}{2Z}\right)^4 \int_0^\infty \rho^3 \left(2 - \frac{1}{2}\rho\right)^2 e^{-\rho/2}\,\mathrm{d}\rho \\
&= \frac{a_0}{128Z} \int_0^\infty \left(4\rho^3 - 2\rho^4 + \frac{\rho^5}{4}\right) e^{-\rho/2}\,\mathrm{d}\rho = \frac{a_0}{128Z}(768) = \frac{6a_0}{Z}
\end{aligned}
$$

Comment. We conclude that the $2p$ orbital in hydrogen is on average closer to the nucleus. This is not necessarily true in heavier atoms where $E(2p) > E(2s)$.

P13.13   In each case we need to show that

$$\int_{\text{all space}} \psi_1^* \psi_2\,\mathrm{d}\tau = 0$$

(a)

$$\int_0^\infty \int_0^\pi \int_0^{2\pi} \psi_{1s} \psi_{2s} r^2\,\mathrm{d}r\,\sin\theta\,\mathrm{d}\theta\,\mathrm{d}\phi \overset{?}{=} 0$$

$$\left.\begin{aligned} \psi_{1s} &= R_{1,0} Y_{0,0} \\ \psi_{2s} &= R_{2,0} Y_{0,0} \end{aligned}\right\} \quad Y_{0,0} = \left(\frac{1}{4\pi}\right)^{\frac{1}{2}} \text{ [Table 12.3]}$$

Since $Y_{0,0}$ is a constant, the integral over the radial functions determines the orthogonality of the functions.

$$\int_0^\infty R_{1,0} R_{2,0} r^2 \, dr$$

$$R_{1,0} \propto e^{-\rho/2} = e^{-Zr/a_0} \left[\rho = \frac{2Zr}{a_0} \right]$$

$$R_{2,0} \propto (2 - \rho/2) e^{-\rho/4} = \left(\frac{2 - Zr}{a_0} \right) e^{-Zr/2a_0} \left[\rho = \frac{2Zr}{a_0} \right]$$

$$\int_0^\infty R_{1,0} R_{2,0} r^2 \, dr \propto \int_0^\infty e^{-Zr/a_0} \left(2 - \frac{Zr}{a_0} \right) r^2 \, dr$$

$$= \int_0^\infty 2 e^{-(3/2)Zr/a_0} r^2 \, dr - \int_0^\infty \frac{Z}{a_0} e^{-(3/2)Zr/a_0} r^3 \, dr$$

$$= \frac{2 \times 2!}{\left(\frac{3}{2} \frac{Z}{a_0} \right)^3} - \left(\frac{Z}{a_0} \right) \times \frac{3!}{\left(\frac{3}{2} \frac{Z}{a_0} \right)^4} = \boxed{0}$$

Hence, the functions are orthogonal.

(b) We use the p_x and p_y orbitals in the form given in Section 13.2(f), eqn 13.25

$$p_x \propto x, \quad p_y \propto y$$

Thus

$$\int_{\text{all space}} p_x p_y \, dx \, dy \, dz \propto \int_{-\infty}^{+\infty} \int_{-\infty}^{+\infty} \int_{-\infty}^{+\infty} xy \, dx \, dy \, dz$$

This is an integral of an odd function of x and y over the entire range of variable from $-\infty$ to $+\infty$, therefore, the $\boxed{\text{integral is zero}}$. More explicitly we may perform the integration using the orbitals in the form (Section 13.2(f), eqn 13.25)

$$p_x = f(r) \sin\theta \cos\phi \quad p_y = f(r) \sin\theta \sin\phi$$

$$\int_{\text{all space}} p_x p_y r^2 \, dr \, \sin\theta \, d\theta \, d\phi = \int_0^\infty f(r)^2 r^2 \, dr \int_0^\pi \sin^2\theta \, d\theta \int_0^{2\pi} \cos\phi \sin\phi \, d\phi$$

The first factor is nonzero since the radial functions are normalized. The second factor is $\dfrac{\pi}{2}$. The third factor is zero. Therefore, the product of the integrals is $\boxed{\text{zero}}$ and the functions are orthogonal.

P13.16 We use the p_x and p_y orbitals in the form (Section 13.2(f))

$$p_x = rf(r) \sin\theta \cos\phi \qquad p_y = rf(r) \sin\theta \sin\phi$$

and use $\cos\phi = \dfrac{1}{2}(e^{i\phi} + e^{-i\phi})$ and $\sin\phi = \dfrac{1}{2i}(e^{i\phi} - e^{-i\phi})$ then

$$p_x = \frac{1}{2} rf(r) \sin\theta (e^{i\phi} + e^{-i\phi}) \qquad p_y = \frac{1}{2i} rf(r) \sin\theta (e^{i\phi} - e^{-i\phi})$$

$$\hat{l}_z = \frac{\hbar}{i} \frac{\partial}{\partial \phi} \quad \text{[Problem 12.24 and Section 12.6 and eqn 12.57]}$$

$$\hat{l}_z p_x = \frac{\hbar}{2} r f(r) \sin \theta \, e^{i\phi} - \frac{\hbar}{2} r f(r) \sin \theta \, e^{-i\phi} = i\hbar p_y \neq \text{constant} \times p_x$$

$$\hat{l}_z p_y = \frac{\hbar}{2i} r f(r) \sin \theta \, e^{i\phi} + \frac{\hbar}{2i} r f(r) \sin \theta \, e^{-i\phi} = -i\hbar p_x \neq \text{constant} \times p_y$$

Therefore, neither p_x nor p_y are eigenfunctions of $\hat{l}_z$. However, $\boxed{p_x + i p_y \text{ and } p_x - i p_y}$ are eigenfunctions

$$p_x + i p_y = r f(r) \sin \theta \, e^{i\phi} \qquad p_x - i p_y = r f(r) \sin \theta \, e^{-i\phi}$$

since both $e^{i\phi}$ and $e^{-i\phi}$ are eigenfunctions of $\hat{l}_z$ with eigenvalues $+h$ and $-h$.

P13.17 The general rule to use in deciding commutation properties is that operators having no variable in common will commute with each other. We first consider the commutation of $\hat{l}_z$ with the Hamiltonian. This is most easily solved in spherical polar coordinates.

$$\hat{l}_z = \frac{\hbar}{i} \frac{\partial}{\partial \phi} \quad \text{[Problem 12.24 and Section 12.6 and eqn 12.57]}$$

$$H = -\frac{\hbar^2}{2\mu} \nabla^2 + V \quad [13.7] \quad V = -\frac{Ze^2}{4\pi \varepsilon_0 r}$$

Since V has no variable in common with $\hat{l}_z$, this part of the Hamiltonian and $\hat{l}$ commute.

$$\nabla^2 = \text{terms in } r \text{ only} + \text{terms in } \theta \text{ only} + \frac{1}{r^2 \sin^2 \theta} \frac{\partial^2}{\partial \phi^2} \quad \text{[Table 11.1]}$$

The terms in r only and θ only necessarily commute with $\hat{l}_z$ (ϕ only). The final term in ∇^2 contains $\frac{\partial^2}{\partial \phi^2}$ which commutes with $\frac{\partial}{\partial \phi}$, since an operator necessarily commutes with itself. By symmetry we can deduce that if H commutes with $\hat{l}_z$ it must also commute with $\hat{l}_x$ and $\hat{l}_y$ since they are related to each other by a simple transformation of coordinates. This proves useful in establishing the commutation of l^2 and H. We form

$$\hat{l}^2 = \hat{l} \cdot \hat{l} = (i\hat{l}_x + j\hat{l}_y + k\hat{l}_z) \cdot (i\hat{l}_x + j\hat{l}_y + k\hat{l}_z) = \hat{l}_x^2 + \hat{l}_y^2 + \hat{l}_z^2$$

If H commutes with each of $\hat{l}_x, \hat{l}_y$, and $\hat{l}_z$ it must commute with $\hat{l}_x^2, \hat{l}_y^2$, and $\hat{l}_z^2$. Therefore it also commutes with $\hat{l}^2$. Thus H commutes with both $\hat{l}^2$ and $\hat{l}_z$.

Comment. As described at the end of Section 11.6, the physical properties associated with non-commuting operators cannot be simultaneously known with precision. However, since $H, \hat{l}^2$, and $\hat{l}_z$ commute we may simultaneously have exact knowledge of the energy, the total orbital angular momentum, and the projection of the orbital angular momentum along an arbitrary axis.

P13.18
$$\psi_{1s} = \left(\frac{1}{\pi a_0^3}\right)^{1/2} e^{-r/a_0} \quad [13.20]$$

The probability of the electron being within a sphere of radius r' is

$$\int_0^{r'} \int_0^{\pi} \int_0^{2\pi} \psi_{1s}^2 r^2 \, dr \sin\theta \, d\theta \, d\phi$$

We set this equal to 0.90 and solve for r'. The integral over θ and ϕ gives a factor of 4π; thus

$$0.90 = \frac{4}{a_0^3} \int_0^{r'} r^2 e^{-2r/a_0} \, dr$$

$\int_0^{r'} r^2 e^{-2r/a_0} \, dr$ is integrated by parts to yield

$$-\frac{a_0 r^2 e^{-2r/a_0}}{2}\bigg|_0^{r'} + a_0 \left[-\frac{a_0 r e^{-2r/a_0}}{2}\bigg|_0^{r'} + \frac{a_0}{2}\left(-\frac{a_0 e^{-2r/a_0}}{2}\right)\bigg|_0^{r'}\right]$$

$$= -\frac{a_0 (r')^2 e^{-2r'/a_0}}{2} - \frac{a_0^2 r'}{2} e^{-2r'/a_0} - \frac{a_0^3}{4} e^{-2r'/a_0} + \frac{a_0^3}{4}$$

Multiplying by $\dfrac{4}{a_0^3}$ and factoring e^{-2r'/a_0}

$$0.90 = \left[-2\left(\frac{r'}{a_0}\right)^2 - 2\left(\frac{r'}{a_0}\right) - 1\right]e^{-2r'/a_0} + 1 \quad \text{or} \quad 2\left(\frac{r'}{a_0}\right)^2 + 2\left(\frac{r'}{a_0}\right) + 1 = 0.10 e^{2r'/a_0}$$

It is easiest to solve this numerically. It is seen that $\boxed{r' = 2.66\, a_0}$ satisfies the above equation.

P13.19
To determine the atomic diameters, we assume a body-centred cubic crystal array (Fig. 13.1) that has two atoms per unit cell. In this array there is an atom at each corner of the cube and one in the middle. The cubic diagonal is two atomic diameters (d) in length and equals $3^{1/2}$ times the length of the unit cell edge.

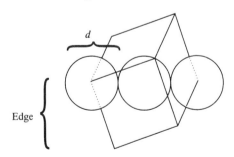

Figure 13.1

$$V_{\text{unit cell}} = \frac{(\text{edge})^3}{\text{unit cell}} = \left(\frac{V_{\text{m}}}{N_{\text{A}}}\right) \times \left(\frac{2\ \text{atoms}}{\text{unit cell}}\right) = \left(\frac{M}{\rho N_{\text{A}}}\right) \times \left(\frac{2\ \text{atoms}}{\text{unit cell}}\right)$$

$$\left(\frac{2d}{\sqrt{3}}\right)^3 \approx \frac{2M}{\rho N_{\text{A}}} \quad \text{or} \quad \boxed{d \simeq \frac{\sqrt{3}}{2}\left(\frac{2M}{\rho N_{\text{A}}}\right)^{1/3}} \ \text{bcc array}$$

$$d_{\text{H}} = \frac{\sqrt{3}}{2} \left[\frac{2(1.008 \,\text{g mol}^{-1}) \times \left(\frac{\text{m}^3}{10^6 \text{cm}^3}\right)}{(0.071 \,\text{g cm}^{-3}) \times (6.022 \times 10^{23} \,\text{mol}^{-1})} \right]^{1/3} = 3.14 \times 10^{-10} \,\text{m} = \boxed{0.314 \,\text{nm}}$$

$$d_{\text{U}} = \frac{\sqrt{3}}{2} \left[\frac{2(238.0 \,\text{g mol}^{-1}) \times \left(\frac{\text{m}^3}{10^6 \text{cm}^3}\right)}{(18.95 \,\text{g cm}^{-3}) \times (6.022 \times 10^{23} \,\text{mol}^{-1})} \right]^{1/3} = 3.01 \times 10^{-10} \,\text{m} = \boxed{0.301 \,\text{nm}}$$

Therefore,

$$d_{\text{H}} \approx d_{\text{U}} \approx 0.3 \,\text{nm}$$

It is reasonable to suppose that the radius of an atom can be approximated as equal to the average distance of a valence-shell electron from the nucleus. Substituting the effective nuclear charge (Z_{eff}) for the nuclear charge of a hydrogenic atom, the estimate becomes

$$\text{atomic radius} \approx \langle r \rangle_{n_{\text{valence shell}}} \approx \frac{(n_{\text{valence shell}})^2 a_0}{Z_{\text{eff}}}$$

This indicates that, when comparing atomic radii from left-to-right across a period of the periodic table, radii should decrease in proportion to (Z_{eff}). For a period of transition metals, the radii are approximately equal because the added electrons effectively screen the outermost electrons from the increased nuclear charge.

In a similar manner, the first ionization energy can be approximated with the proportionality

$$I \propto \frac{(Z_{\text{eff}})^2}{(n_{\text{valence}})^2}$$

This says that, in going across a period, I will vary more dramatically than does the atomic radius because of the $(Z_{\text{eff}})^2$ factor which provides greater variance than does $(Z_{\text{eff}})^{-1}$.

P13.22 $E = \text{kinetic energy} + \text{potential energy}$

$$E_{\text{classical}} = \frac{m_e v^2}{2} - \frac{Z e^2}{4\pi \varepsilon_0 a_0} \quad \text{where } v = \text{electron speed}$$

Using the quantum energy for $n = 1$, $Z = 1$, and $\mu \approx m_e$

$$E_1 = -\frac{m_e e^4}{2(4\pi \varepsilon_0)^2 \hbar^2} \quad [13.13]$$

$$= -\frac{e^2}{8\pi \varepsilon_0 a_0} \quad [13.15]$$

Therefore,

$$-\frac{e^2}{8\pi \varepsilon_0 a_0} = \frac{m_e v^2}{2} - \frac{e^2}{4\pi \varepsilon_0 a_0}$$

$$\text{or } v = \left(\frac{e^2}{4\pi \varepsilon_o m_e a_0} \right)^{1/2}$$

$$= \left(\frac{(1.602 \times 10^{-19} \,\text{C})^2}{(1.113 \times 10^{-10} \,\text{J}^{-1} \text{C}^2 \,\text{m}^{-1}) \times (9.109 \times 10^{-31} \,\text{kg}) \times (5.29 \times 10^{-11} \,\text{m})} \right)^{1/2}$$

$$= 2.19 \times 10^6 \, \text{m s}^{-1} \left(\frac{c}{3.00 \times 10^8 \, \text{m s}^{-1}} \right)$$

$v = \boxed{0.00729c}$ The classical speed is 0.73 per cent of the speed of light.

$$\text{Electric field strength } \mathcal{E} = \frac{e}{4\pi \varepsilon_0 a_0^2} = \frac{1.602 \times 10^{-19} \, \text{C}}{(1.113 \times 10^{-10} \, \text{J}^{-1} \, \text{C}^2 \, \text{m}^{-1}) \times (5.29 \times 10^{-11} \, \text{m})^2}$$

$$= \boxed{5.14 \times 10^{11} \, \text{V m}^{-1}}$$

We will also determine the magnetic field strength, $\mathcal{H}$, at the nucleus of the hydrogen atom. The magnetic field $\mathbf{B}$ at the nucleus is created by the current produced by the classical electron in the Bohr orbit for which $n = 1$ and $r = a_0$. Magnetic field strength at nucleus $\mathcal{H} = \dfrac{|\mathbf{B}|}{\mu_0} = \dfrac{\mathcal{B}}{\mu_0}$ where μ_0 is the vacuum permeability.

The electron in the classic Bohr orbit may be viewed as being electric current $I = e/\text{orbit period} = e/(2\pi a_0/v) = ev/(2\pi a_0)$ where v is the electron speed. The current creates the magnetic field and the Biot law relates the two (see Fig. 13.2)

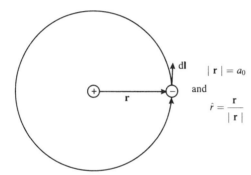

Figure 13.2

$$d\mathbf{B} = \frac{\mu_0}{4\pi} \frac{I \, d\mathbf{l} \times \hat{\mathbf{r}}}{r^2}$$

Since we are interested in the magnetic field at the nucleus, note that Biot's law says that $\mathbf{B}$ will be perpendicular to the plane of the orbit (the vector cross-product says this). In terms of magnitudes alone the magnetic field is given by

$$d\mathcal{B} = \frac{\mu_0}{4\pi} \frac{I \, dl \sin\theta}{r^2} = \frac{\mu_0}{4\pi} \frac{I \, dl}{r^2} \quad (\theta = 90°)$$

Since both I and r are constants for the orbit ($r = a_0$) this is easily integrated over one orbit.

$$\int_{\text{one orbit}} d\mathcal{B} = \frac{\mu_0 I}{4\pi a_0^2} \int_{\text{one orbit}} dl = \frac{\mu_0 I}{4\pi a_0^2} (2\pi a_0)$$

$$\mathcal{B} = \frac{\mu_0 I}{2a_0} = \frac{\mu_0 ev}{4\pi a_0^2} \quad \left[I = \frac{ev}{2\pi a_0} \right]$$

$$= \frac{(4\pi \times 10^{-7} \, \text{J A}^{-2} \, \text{m}^{-1}) \times (1.602 \times 10^{-19} \, \text{C}) \times (2.19 \times 10^6 \, \text{m s}^{-1})}{4\pi \times (5.29 \times 10^{-11} \, \text{m})^2} = \boxed{12.5 \, \text{T}}$$

$$[\text{T} = \text{J A}^{-1} \, \text{m}^{-2}]$$

$$\text{magnetic field strength at nucleus} = \mathcal{H}_{\text{nucleus}} = \frac{\mathcal{B}}{\mu_0} = \frac{ev}{4\pi a_0^2}$$

$$= \frac{(1.6022 \times 10^{-19}\,\text{C}) \times (2.19 \times 10^6\,\text{m s}^{-1})}{4\pi (5.29 \times 10^{-11}\,\text{m})^2}$$

$$\mathcal{H}_{\text{nucleus}} = \boxed{9.98 \times 10^6\,\text{A m}^{-1}}$$

The electric field strength of the atom is about 5000 times larger than the dielectric strength of a mica capacitor. The magnetic field strength is the same order of magnitude as that in a superconducting magnet used in NMR experiments. (See Chapter 18.)

P13.23   Refer to Problems 13.8 and 13.20 and their solutions.

$$\mu_{\text{H}} = \frac{m_e m_p}{m_e + m_p} \approx m_e \quad [m_p = \text{mass of proton}]$$

$$\mu_{\text{Ps}} = \frac{m_e m_{\text{pos}}}{m_e + m_{\text{pos}}} = \frac{m_e}{2} \quad [m_{\text{pos}} = \text{mass of positron} = m_e]$$

$$a_0 = r(n = 1) = \frac{4\pi \hbar^2 \varepsilon_0}{e^2 m_e} \quad [13.15 \text{ and Problem } 13.20]$$

To obtain a_{Ps}, the radius of the first Bohr orbit of positronium, we replace m_e with $\mu_{\text{Ps}} = \frac{m_e}{2}$; hence,

$$\boxed{a_{\text{Ps}} = 2a_0} = \frac{8\pi \hbar^2 \varepsilon_0}{e^2 m_e}$$

The energy of the first Bohr orbit of positronium is

$$E_{1,\text{Ps}} = -hc R_{\text{Ps}} = -\frac{hc}{2} R_\infty \quad [\text{Problem } 13.8]$$

Thus, $\boxed{E_{1,\text{Ps}} = \tfrac{1}{2} E_{1,\text{H}}}$

Question. What modifications are required in these relations when the finite mass of the hydrogen nucleus is recognized?

P13.24   **(a)** The emission line for the transitions $n + 1 \rightarrow n$ is the first line of the Lyman series when $n = 1$, or the first line of the Balmer series when $n = 2$, or the first line of the Paschen series when $n = 3$, etc. Let $\tilde{\nu}$ be the wavenumber of this line.

Then,

$$\tilde{\nu} = \frac{|\Delta E|}{hc} = \frac{E_{n+1} - E_n}{hc}$$

For hydrogenic atoms $E_n = -\dfrac{C\mu Z^2}{n^2}$ where $C = \dfrac{e^4}{32\pi^2 \varepsilon_0^2 \hbar^2}$ [13.13]

$$\tilde{\nu} = \frac{C\mu Z^2}{hc} \left\{ -\frac{1}{(n+1)^2} - \left(-\frac{1}{n^2} \right) \right\} = \frac{C\mu Z^2}{hc} \left\{ \frac{2n+1}{n^2(n+1)^2} \right\}$$

$$\tilde{\nu}_{He} = \frac{4C\mu_{He}}{hc}\left\{\frac{2n+1}{n^2(n+1)^2}\right\}; \qquad \tilde{\nu}_{H} = \frac{C\mu_{H}}{hc}\left\{\frac{2n+1}{n^2(n+1)^2}\right\}$$

Therefore,

$$\gamma = \frac{\frac{1}{4}\tilde{\nu}_{He} - \tilde{\nu}_{H}}{\tilde{\nu}_{H}} = \frac{\frac{1}{4}\left\{\frac{4C\mu_{He}}{hc}\right\}\left\{\frac{2n+1}{n^2(n+1)^2}\right\} - \frac{C\mu_{H}}{hc}\left\{\frac{2n+1}{n^2(n+1)^2}\right\}}{\frac{C\mu_{H}}{hc}\left\{\frac{2n+1}{n^2(n+1)^2}\right\}}$$

$$\boxed{\gamma = \frac{\mu_{He} - \mu_{H}}{\mu_{H}}}$$

(b) Using wavelength data

$$\tilde{\nu}_{H} = \frac{1}{\lambda_{H}} = \frac{1}{121.5664\,\text{nm}}\left(\frac{10^9\,\text{nm}}{10^2\,\text{cm}}\right) = 82\,259.57\,\text{cm}^{-1}$$

$$\tilde{\nu}_{He^+} = \frac{1}{\lambda_{He}} = \frac{1}{30.3779\,\text{nm}}\left(\frac{10^9\,\text{nm}}{10^2\,\text{cm}}\right) = 329\,186.67\,\text{cm}^{-1}$$

Therefore

$$\gamma_{2\rightarrow1} = \frac{\frac{1}{4}(329\,186.67\,\text{cm}^{-1}) - 82\,259.57\,\text{cm}^{-1}}{82\,259.57\,\text{cm}^{-1}} = 4.5098 \times 10^{-4}$$

$$\gamma = (\mu_{He}/\mu_{H}) - 1 \quad \text{or} \quad \frac{\mu_{H}}{\mu_{He}} = \frac{1}{1+\gamma} = 0.999\,549\,223$$

$$\frac{\mu_{H}}{\mu_{He}} = \left(\frac{m_p m_e}{m_p + m_e}\right)\left(\frac{4m_p + m_e}{4m_p m_e}\right) = \frac{4m_p + m_e}{4(m_p + m_e)}$$

But, $m_p \approx m_H - m_e$ so

$$\frac{\mu_{H}}{\mu_{He}} \approx \frac{4(m_H - m_e) + m_e}{4m_H} = \frac{4m_H - 3m_e}{4m_H} = 1 - \frac{3}{4}\left(\frac{m_e}{m_H}\right)$$

Therefore,

$$\frac{m_H}{m_e} = \frac{3/4}{\left(1 - \frac{\mu_H}{\mu_{He}}\right)} = \frac{3}{4(1 - 0.999\,549\,223)} = \boxed{1663}$$

Using Rydberg constants $\mathcal{R}_A$.
Since $\mathcal{R} \propto \mu$,

$$\frac{\mu_H}{\mu_{He}} = \frac{\mathcal{R}_H}{\mathcal{R}_{He}} = \frac{109\,677.7\,\text{cm}^{-1}}{109\,722.4\,\text{cm}^{-1}} = 0.999\,592\,608$$

Therefore,

$$\frac{m_H}{m_e} = \frac{3}{4\left(1 - \frac{\mu_H}{\mu_{He}}\right)} = \frac{3}{4(1 - 0.999\,592\,608)} = \boxed{1841}$$

The value for the ratio calculated from the Rydberg constants is very close to the best modern accepted value of 1836. The ratio m_H/m_e is very sensitive to errors in the (older) wavelength data and that may account for the discrepancy between the two values.

Solutions to applications

P13.27 $E_n = -\dfrac{hc\mathcal{R}_H}{n^2}$ where $\mathcal{R}_H = 109\,677\,\text{cm}^{-1}$ [13.13 with 13.17]

For $n = 100$

$$\Delta E = E_{n+1} - E_n = -hc\mathcal{R}_H \left(\frac{1}{101^2} - \frac{1}{100^2}\right) = 1.97 \times 10^{-6}\, hc\mathcal{R}$$

$$\tilde{\nu} = \frac{\Delta E}{hc} = 1.97 \times 10^{-6}\mathcal{R} = \boxed{0.216\,\text{cm}^{-1}}$$

$$\langle r \rangle_{n,l} = n^2 \left\{1 + \frac{1}{2}\left(1 - \frac{l(l+1)}{n^2}\right)\right\}\frac{a_0}{Z}\quad\text{[Example 13.2]}$$

$$\langle r \rangle_{100} \approx \frac{n^2 a_0}{Z} = 100^2 a_0 = 10^4 a_0 = \boxed{529\,\text{nm}}$$

$$I = E_\infty - E_n = -E_n = \frac{hc\mathcal{R}_H}{n^2}$$

$$I_{100} = 10^{-4}hc\mathcal{R}_H\quad\text{so}\quad \boxed{\frac{I_{100}}{hc} = 10.9677\,\text{cm}^{-1}}$$

At T

$$\frac{kT}{hc} = \frac{(1.38 \times 10^{-23}\,\text{J K}^{-1}) \times (298\,\text{K})\left(\frac{\text{m}}{10^2\,\text{cm}}\right)}{(6.63 \times 10^{-34}\,\text{J s}) \times (3.00 \times 10^8\,\text{m s}^{-1})} = 207\,\text{cm}^{-1}$$

so the thermal energy is readily available to ionize the state $n = 100$. Let $v_{\min}$ be the minimum speed required for collisional ionization. Then

$$\frac{1}{2}\frac{m_H v_{\min}^2}{hc} = \frac{I_{100}}{hc}$$

$$v_{\min} = \left[\frac{2hc}{m_H}\left(\frac{I_{100}}{hc}\right)\right]^{1/2}$$

$$= \sqrt{\frac{2(6.63 \times 10^{-34}\,\text{J s}) \times (3.00 \times 10^8\,\text{m s}^{-1}) \times (10.97\,\text{cm}^{-1})}{(1.008 \times 10^{-3}\,\text{kg mol}^{-1}) \times (6.022 \times 10^{23}\,\text{mol}^{-1})^{-1} \times \left(\frac{\text{m}}{10^2\,\text{cm}}\right)}}$$

$$\boxed{v_{\min} = 511\,\text{m s}^{-1}}\quad\text{[very slow for an H atom]}$$

The radius of a Bohr orbit is $a_n \approx n^2 a_0$; hence the geometric cross-section $\pi a_n^2 \approx n^4 \pi a_0^2$. For $n = 1$ this is $8.8 \times 10^{-21}\,\text{m}^2$; for $n = 100$, it is $\boxed{8.8 \times 10^{-13}\,\text{m}^2}$. Thus a neutral H atom in its ground state is likely to pass right by the $n = 100$ Rydberg atom, leaving it undisturbed, since it is largely empty space.

The radial wavefunction for $n = 100$ will have 99 radial nodes and an extremely small amplitude above $\dfrac{r}{a_0} \approx 20$. For large values of n we expect the radial wavefunction [13.16] to be governed largely by the product of ρ^{n-1} and $e^{-\rho/2n}$ and thus to approach a smoothly decreasing function of distance as the exponential will predominate over the power term.

P13.28 **(a)** Compute the ratios v_{star}/v for all three lines. We are given wavelength data, so we can use:

$$\frac{v_{star}}{v} = \frac{\lambda}{\lambda_{star}}$$

The ratios are:

$$\frac{438.392\,\text{nm}}{438.882\,\text{nm}} = 0.998884, \quad \frac{440.510\,\text{nm}}{441.000\,\text{nm}} = 0.998889, \quad \text{and} \quad \frac{441.510\,\text{nm}}{442.020\,\text{nm}} = 0.998846.$$

The frequencies of the stellar lines are all less than those of the stationary lines, so we infer that the star is $\boxed{\text{receding}}$ from earth. The Doppler effect follows:

$$v_{receding} = vf \quad \text{where} \quad f = \left(\frac{1 - s/c}{1 + s/c}\right)^{1/2}, \quad \text{so}$$

$$f^2(1 + s/c) = (1 - s/c), \quad (f^2 + 1)s/c = 1 - f^2, \quad s = \frac{1 - f^2}{1 + f^2}c$$

Our average value of f is 0.998873. (Note: the uncertainty is actually greater than the significant figures here imply, and a more careful analysis would treat uncertainty explicitly.) So the speed of recession with respect to the earth is:

$$s = \left(\frac{1 - 0.997747}{1 + 0.997747}\right)c = \boxed{1.128 \times 10^{-3}c} = \boxed{3.381 \times 10^5\,\text{m s}^{-1}}.$$

(b) One could compute the star's radial velocity with respect to the sun if one knew the earth's speed with respect to the sun along the sun–star vector at the time of the spectral observation. This could be estimated from quantities available through astronomical observation: the earth's orbital velocity times the cosine of the angle between that velocity vector and the earth–star vector at the time of the spectral observation. (The earth–star direction, which is observable by earth-based astronomers, is practically identical to the sun–star direction, which is technically the direction needed.) Alternatively, repeat the experiment half a year later. At that time, the earth's motion with respect to the sun is approximately equal in magnitude and opposite in direction compared to the original experiement. Averaging f values over the two experiments would yield f values in which the earth's motion is effectively averaged out.

14 Molecular structure

Solutions to exercises

Discussion questions

E14.1(a) Our comparison of the two theories will focus on the manner of construction of the trial wavefunctions for the hydrogen molecule in the simplest versions of both theories. In the valence bond method, the trial function is a linear combination of two simple product wavefunctions, in which one electron resides totally in an atomic orbital on atom A, and the other totally in an orbital on atom B. See eqns 14.1 and 14.2, as well as Fig. 14.2. There is no contribution to the wavefunction from products in which both electrons reside on either atom A or B. So the valence bond approach undervalues, by totally neglecting, any ionic contribution to the trial function. It is a totally covalent function. The molecular orbital function for the hydrogen molecule is a product of two functions of the form of eqn 14.7, one for each electron, that is

$$\psi = [A(1) \pm B(1)][A(2) \pm B(2)] = A(1)A(2) + B(1)B(2) + A(1)B(2) + B(1)A(2)$$

This function gives as much weight to the ionic forms as to the covalent forms. So the molecular orbital approach greatly overvalues the ionic contributions. At these crude levels of approximation, the valence bond method gives dissociation energies closer to the experimental values. However, more sophisticated versions of the molecular orbital approach is the method of choice for obtaining quantitative results on both diatomic and polyatomic molecules. See Sections 14.7–14.9.

E14.2(a) Both the Pauling and Mulliken methods for measuring the attracting power of atoms for electrons seem to make good chemical sense. If we look at eqn 14.20 (the Pauling scale), we see that if $D(A—B)$ were equal to $1/2[D(A—A) + D(B—B)]$ the calculated electronegativity difference would be zero, as expected for completely non-polar bonds. Hence, any increased strength of the A—B bond over the average of the A—B and B—B bonds, can reasonably be thought of as being due to the polarity of the A—B bond, which in turn is due to the difference in electronegativity of the atoms involved. Therefore, this difference in bond strengths can be used as a measure of electronegativity difference. To obtain numerical values for individual atoms, a reference state (atom) for electronegativity must be established. The value for fluorine is arbitrarily set at 4.0.

   The Mulliken scale may be more intuitive than the Pauling scale because we are used to thinking of ionization energies and electron affinities as measures of the electron attracting powers of atoms. The choice of factor $1/2$, however, is arbitrary, though reasonable, and no more arbitrary than the specific form of eqn 14.20 that defines the Pauling scale.

E14.3(a) Both methods parameterize, rather than calculate, the energy integrals that arise in molecular orbital theory. In the simple Hückel method, the overlap integral is also parameterized, whereas in the extended method, this integral is evaluated. In the simple method, the energy integrals, α and β, are always considered to be adjustable parameters; their numerical values emerge only at the end of the calculation by comparison to experimental energies. The extended method is less arbitrary, the energy integrals are related to ionization energies of the atoms (see eqn 14.47). The simple method has three other rather drastic approximations, listed in Section 14.7(a) of the text which eliminate many terms from the secular determinant and make it easier to solve. Ease of solution was important in the early days of Quantum Chemistry before the advent of computers, and without the use of these approximations, calculations on polyatomic molecules would have been difficult to accomplish.

   The simple Hückel method is usually applied only to the calculation of π-electron energies in conjugated organic systems, whereas the extended method is an all-valence electron calculation with a much greater range of applicability. The simple method is based on the assumption of the separability of the σ- and π-electron systems in the molecule. This is a very crude approximation and works best when the energy level pattern is determined largely by the symmetry of the molecule. (See Chapter 15.)

E14.4(a)   In general, for a molecular orbital expressed as a linear combination of atomic orbitals with coefficients c_i, the atomic populations are $\rho_i = c_i{}^2$, and represent the probability of finding the electron in atomic orbital i. The overlap population, ρ_{ij}, is the probability that the electron will be found in the overlap region between two bonded atoms i and j and is proportional to the overlap integral, S_{ij}, through the relation $\rho_{ij} = 2c_ic_jS_{ij}$. If we apportion one-half of the overlap population to each of the bonded atoms, we obtain the gross orbital population.

Numerical exercises

E14.5(a)   Refer to Fig. 14.30 of the text. Place two of the valence electrons in each orbital starting with the lowest energy orbital, until all valence electrons are used up. Apply Hund's rule to the filling of degenerate orbitals.

(a) Li_2 (2 electrons) $\boxed{1\sigma^2 \quad b = 1}$ **(b)** Be_2 (4 electrons) $\boxed{1\sigma^2 2\sigma^{*2} \quad b = 0}$

(c) C_2 (8 electrons) $\boxed{1\sigma^2 2\sigma^{*2} 1\pi^4 \quad b = 2}$

E14.6(a)   Note that CO and CN^- are isoelectronic with N_2 and that NO is isoelectronic with N_2^-, hence use Fig. 14.30 of the text.

(a)   CO (10 electrons) $\boxed{1\sigma^2 2\sigma^{*2} 1\pi^4 3\sigma^2} \quad b = 3$

(b)   NO (11 electrons) $\boxed{1\sigma^2 2\sigma^{*2} 1\pi^4 3\sigma^2 2\pi^{*1}} \quad b = 2.5$

(c) CN^- (10 electrons) $\boxed{1\sigma^2 2\sigma^{*2} 1\pi^4 3\sigma^2} \quad b = 3$

E14.7(a) B_2 (6 electrons): $1\sigma^2 2\sigma^{*2} 1\pi^2 \quad b = 1$
C_2 (8 electrons): $1\sigma^2 2\sigma^{*2} 1\pi^4 \quad b = 2$

The bond orders of B_2 and C_2 are respectively 1 and 2; so $\boxed{C_2}$ should have the greater bond dissociation enthalpy. The experimental values are approximately 4 eV and 6 eV respectively.

E14.8(a)   We can use a version of Fig. 14.28 of the text, but with the energy levels of F lower than those of Xe as in Fig. 14.1.

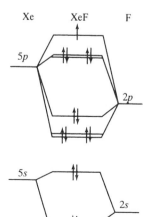

Figure 14.1

For XeF we insert 15 valence electrons. Since the bond order is increased when XeF^+ is formed from XeF (because an electron is removed from an antibonding orbital), XeF^+ will have a shorter bond length than XeF.

E14.9(a) The electron configurations are used to determine the bond orders. Larger bond order corresponds qualitatively to shorter bond length.

The bond orders of NO and N_2 are 2.5 and 3, respectively (Exercises 14.5(**b**) and 14.6(**a**)); hence N_2 should have the shorter bond length. The experimental values are 115 pm and 110 pm, respectively.

E14.10(a) We need to demonstrate that $\int \psi^2 \, d\tau = 1$, where $\psi = \dfrac{s + \sqrt{2}p}{\sqrt{3}}$.

$$\int \psi^2 \, d\tau = \frac{1}{3} \int (s + \sqrt{2}p)^2 \, d\tau = \frac{1}{3} \int (s^2 + 2p^2 + 2\sqrt{2}sp) \, d\tau = \frac{1}{3}(1 + 2 + 0) = 1$$

as $\int s^2 \, d\tau = 1$, $\int p^2 \, d\tau = 1$, and $\int sp \, d\tau = 0$ [orthogonality]

E14.11(a) We evaluate $\int (\psi_A + \psi_B)(\psi_A - \psi_B) \, d\tau$ and look at the result. If the integral is zero, then they are mutually orthogonal.

$$\int (\psi_A^2 - \psi_B^2) \, d\tau = 1 - 1 = \boxed{0}$$

Hence, they are orthogonal.

E14.12(a) The trial function obeys the boundary conditions of a particle in a box, so it is $\boxed{\text{appropriate}}$. The energy that corresponds to this wavefunction is eqn 14.23:

$$E_{\text{trial}} = \frac{\int \psi^* H \psi \, d\tau}{\int \psi^* \psi \, d\tau}.$$

Express ψ as $xL - x^2$. The Hamiltonion acting on the trial function is:

$$H\psi = \frac{-\hbar^2}{2m} \frac{d^2}{dx^2}(xL - x^2) = \frac{\hbar^2}{m}.$$

The numerator is:

$$\int \psi^* H \psi \, d\tau = \int_0^L (xL - x^2)\frac{\hbar^2}{m} dx = \frac{\hbar^2}{m}(x^2 L/2 - x^3/3)\Big|_0^L = \frac{\hbar^2 L^3}{6m}.$$

The denominator is:

$$\int \psi^* \psi \, d\tau = \int_0^L (xL - x^2)^2 \, dx = \int_0^L (x^2 L^2 - 2x^3 L + x^4) \, dx,$$

$$\int \psi^* \psi \, d\tau = (x^3 L^2/3 - x^4 L/2 + x^5/5)\Big|_0^L = L^5/30.$$

The energy is:

$$E_{\text{trial}} = \frac{\hbar^2 L^3/6m}{L^5/30} = \boxed{\frac{5\hbar^2}{mL^2}} = \frac{5h^2}{4\pi^2 mL^2}.$$

The exact ground-state energy is:

$$E = \frac{h^2}{8mL^2}.$$

Thus the trial energy is greater by a factor of:

$$\frac{5/4\pi^2}{1/8} = \boxed{1.013}.$$

For further elaboration, see W. Tandy Grubbs, "Variational Methods Applied to the Particle in the Box" (http://www.monmouth.edu/~tzielins/mathcad/TGrubbs/doc002.htm) in an on-line collection of Mathcad Documents for Physical Chemistry.

E14.13(a) The energy that corresponds to the trial function is eqn 14.23:

$$E_{\text{trial}} = \frac{\int \psi^* H \psi \, d\tau}{\int \psi^* \psi \, d\tau}.$$

Take this expression piece by piece. The Hamiltonian operator is:

$$H = -\frac{\hbar^2}{2\mu} \nabla^2 - \frac{e^2}{4\pi\varepsilon_0 r}.$$

Noting that the trial function is a function of distance only (not of angles), we can omit the angular pieces of the kinetic energy operator and write:

$$H\psi = -\frac{\hbar^2}{2\mu} \left(\frac{d^2\psi}{dr^2} + \frac{2}{r}\frac{d\psi}{dr} \right) - \frac{e^2\psi}{4\pi\varepsilon_0 r}.$$

We need the derivatives

$$\frac{d\psi}{dr} = \frac{dAe^{-ar^2}}{dr} = -2arAe^{-ar^2}, \text{ and}$$

$$\frac{d^2\psi}{dr^2} = -2aAe^{-ar^2} - 2arAe^{-ar^2}(-2ar) = -2aAe^{-ar^2}(1 - 2ar^2).$$

The Hamiltonian acting on the trial function is:

$$H\psi = -\frac{\hbar^2}{2\mu} \left(-2aAe^{-ar^2}(1 - 2ar^2) - \frac{2}{r}(2arAe^{-ar^2}) \right) - \frac{e^2Ae^{-ar^2}}{4\pi\varepsilon_0 r},$$

$$H\psi = \frac{\hbar^2 aAe^{-ar^2}}{\mu}(3 - 2ar^2) - \frac{e^2Ae^{-ar^2}}{4\pi\varepsilon_0 r}.$$

The integrals need be integrated only over r, for integrations over angles in numerator and denominator will contribute cancelling factors of 4π. The numerator is:

$$\int \psi^* H \psi \, d\tau = \int_0^\infty Ae^{-ar^2} \left(\frac{\hbar^2 aAe^{-ar^2}}{\mu}(3 - 2ar^2) - \frac{e^2Ae^{-ar^2}}{4\pi\varepsilon_0 r} \right) r^2 \, dr,$$

$$\int \psi^* H \psi \, d\tau = A^2 \int_0^\infty e^{-2ar^2} \left(\frac{\hbar^2 a}{\mu}(3r^2 - 2ar^4) - \frac{e^2 r}{4\pi \varepsilon_0} \right) dr.$$

We need the following definite integrals:

$$\int_0^\infty x^{2n} e^{-bx^2} \, dx = \frac{1 \cdot 3 \cdot 5 \cdots (2n-1)}{2^{n+1} b^n} \left(\frac{\pi}{b} \right)^{1/2} \quad \text{with } b = 2a \text{ and } n = 1 \text{ and } 2;$$

and $\displaystyle\int_0^\infty x^{2n+1} e^{-bx^2} \, dx = \frac{n!}{2b^{n+1}}$ with $b = 2a$ and $n = 0$.

So the numerator works out to be:

$$\int \psi^* H \psi \, d\tau = A^2 \left(\frac{\hbar^2 a}{\mu} \right) \left(\frac{3}{2^2(2a)} - \frac{2a \cdot 3}{2^3(2a)^2} \right) \left(\frac{\pi}{2a} \right)^{1/2} - A^2 \left(\frac{e^2}{4\pi \varepsilon_0} \right) \frac{1}{2(2a)},$$

$$\int \psi^* H \psi \, d\tau = \left(\frac{3A^2 \hbar^2}{16\mu} \right) \left(\frac{\pi}{2a} \right)^{1/2} - \frac{A^2 e^2}{16\pi \varepsilon_0 a}.$$

The denominator is:

$$\int \psi^* \psi \, d\tau = \int_0^\infty (Ae^{-ar^2}) r^2 dr = A^2 \int_0^\infty r^2 e^{-2ar^2} \, dr = \frac{A^2}{2^2(2a)} \left(\frac{\pi}{2a} \right)^{1/2} = \frac{A^2}{8a} \left(\frac{\pi}{2a} \right)^{1/2}.$$

Finally, the energy is:

$$E_{\text{trial}} = \frac{\left(\frac{3A^2 \hbar^2}{16\mu} \right) \left(\frac{\pi}{2a} \right)^{1/2} - \frac{A^2 e^2}{16\pi \varepsilon_0 a}}{\frac{A^2}{8a} \left(\frac{\pi}{2a} \right)^{1/2}} = \boxed{\frac{3a\hbar^2}{2\mu} - \frac{e^2}{\varepsilon_0} \left(\frac{a}{2\pi^3} \right)^{1/2}}.$$

E14.14(a) The molecular orbitals of the fragments and the molecular orbitals that they form are shown in Fig. 14.3.

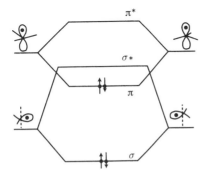

Figure 14.3

Comment. Note that the π-bonding orbital must be lower in energy than the σ-antibonding orbital for π-bonding to exist in ethene.

Question. Would the ethene molecule exist if the order of the energies of the π and σ^* orbitals were reversed?

E14.15(a) In setting up the secular determinant we use the approximation of Section 14.9

$$\textbf{(a)}\quad \begin{vmatrix} \alpha - E & \beta & 0 \\ \beta & \alpha - E & \beta \\ 0 & \beta & \alpha - E \end{vmatrix} = 0 \qquad \textbf{(b)}\quad \begin{vmatrix} \alpha - E & \beta & \beta \\ \beta & \alpha - E & \beta \\ \beta & \beta & \alpha - E \end{vmatrix} = 0$$

The atomic orbital basis is $1s_A$, $1s_B$, $1s_C$ in each case; in linear H_3 we ignore A, C overlap because A and C are not neighbouring atoms; in triangular H_3 we include it because they are.

E14.16(a) The structures are numbered to match the row and column numbers shown in the determinants:

anthracene   phenanthrene

(a) The secular determinant of anthracene in the Hückel approximation is:

	1	2	3	4	5	6	7	8	9	10	11	12	13	14
1	$\alpha-E$	β	0	0	0	0	0	0	0	0	0	0	0	β
2	β	$\alpha-E$	β	0	0	0	0	0	0	0	0	0	0	0
3	0	β	$\alpha-E$	β	0	0	0	0	0	0	0	β	0	0
4	0	0	β	$\alpha-E$	β	0	0	0	0	0	0	0	0	0
5	0	0	0	β	$\alpha-E$	β	0	0	0	β	0	0	0	0
6	0	0	0	0	β	$\alpha-E$	β	0	0	0	0	0	0	0
7	0	0	0	0	0	β	$\alpha-E$	β	0	0	0	0	0	0
8	0	0	0	0	0	0	β	$\alpha-E$	β	0	0	0	0	0
9	0	0	0	0	0	0	0	β	$\alpha-E$	β	0	0	0	0
10	0	0	0	0	β	0	0	0	β	$\alpha-E$	β	0	0	0
11	0	0	0	0	0	0	0	0	0	β	$\alpha-E$	β	0	0
12	0	0	β	0	0	0	0	0	0	0	β	$\alpha-E$	β	0
13	0	0	0	0	0	0	0	0	0	0	0	β	$\alpha-E$	β
14	β	0	0	0	0	0	0	0	0	0	0	0	β	$\alpha-E$

(b) The secular determinant of phenanthrene in the Hückel approximation is:

	1	2	3	4	5	6	7	8	9	10	11	12	13	14
1	$\alpha-E$	β	0	0	0	0	0	0	0	0	0	0	0	β
2	β	$\alpha-E$	β	0	0	0	0	0	0	0	0	0	0	0
3	0	β	$\alpha-E$	β	0	0	0	0	0	0	0	β	0	0
4	0	0	β	$\alpha-E$	β	0	0	0	0	0	0	0	0	0
5	0	0	0	β	$\alpha-E$	β	0	0	0	0	0	0	0	0
6	0	0	0	0	β	$\alpha-E$	β	0	0	0	β	0	0	0
7	0	0	0	0	0	β	$\alpha-E$	β	0	0	0	0	0	0
8	0	0	0	0	0	0	β	$\alpha-E$	β	0	0	0	0	0
9	0	0	0	0	0	0	0	β	$\alpha-E$	β	0	0	0	0
10	0	0	0	0	0	0	0	0	β	$\alpha-E$	β	0	0	0
11	0	0	0	0	0	β	0	0	0	β	$\alpha-E$	β	0	0
12	0	0	β	0	0	0	0	0	0	0	β	$\alpha-E$	β	0
13	0	0	0	0	0	0	0	0	0	0	0	β	$\alpha-E$	β
14	β	0	0	0	0	0	0	0	0	0	0	0	β	$\alpha-E$

Solutions to problems

Solutions to numerical problems

P14.2   Draw up the following table

R/a_0	0	1	2	3	4	5	6	7	8	9	10
S	1.000	0.858	0.586	0.349	0.189	0.097	0.047	0.022	0.010	0.005	0.002

The points are plotted in Fig. 14.4.

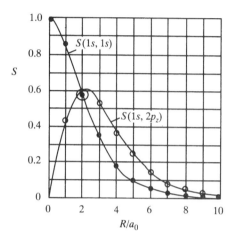

Figure 14.4

P14.3   The s orbital begins to spread into the region of negative amplitude of the p orbital. When their centres coincide, the region of positive overlap cancels the negative region. Draw up the following table

R/a_0	0	1	2	3	4	5	6	7	8	9	10
S	0	0.429	0.588	0.523	0.379	0.241	0.141	0.078	0.041	0.021	0.01

The points are plotted in Fig. 14.4. The maximum overlap occurs at $\boxed{R = 2.1a_0}$.

P14.4   Quantitatively correct values of the total amplitude require the properly normalized functions

$$\psi\pm = \left(\frac{1}{2(1 \pm S)}\right)^{1/2} \{\psi(A) \pm (B)\}\text{[Example 14.2]}$$

We first calculate the overlap integral at $R = 106\,\text{pm} = 2a_0$. (The expression for the overlap integral, S is given in Problem 14.2.)

$$S = \left(1 + 2 + \tfrac{1}{3}(2)^2\right)e^{-2} = 0.586$$

Then $N_+ = \left(\dfrac{1}{2(1 + S)}\right)^{1/2} = \left(\dfrac{1}{2(1 + 0.586)}\right)^{1/2} = 0.561$

$$N_- = \left(\frac{1}{2(1 - S)}\right)^{1/2} = \left(\frac{1}{2(1 - 0.586)}\right)^{1/2} = 1.09\bar{9}$$

We then calculate with $\psi = \left(\dfrac{1}{\pi a_0^3}\right)^{1/2} e^{-r_A/a_0}$, $\psi_{\pm} = N_{\pm}\left(\dfrac{1}{\pi a_0^3}\right)^{1/2}\{e^{-r_A/a_0} \pm e^{-r_B/a_0}\}$ with r_A and r_B both measured from nucleus A, that is

$$\psi_{\pm} = N_{\pm}\left(\frac{1}{\pi a_0^3}\right)^{1/2}\{e^{-|z|/a_0} \pm e^{-|z-R|/a_0}\}$$

with z measured from A along the axis toward B. We draw up the following table with $R = 106\,\text{pm}$ and $a_0 = 5.29\,\text{pm}$.

z/pm	-100	-80	-60	-40	-20	0	20	40
$\dfrac{\psi_+}{\left(\frac{1}{\pi a_0^3}\right)^{1/2}}$	0.096	0.14	0.20	0.30	0.44	0.64	0.49	0.42
$\dfrac{\psi_-}{\left(\frac{1}{\pi a_0^3}\right)^{1/2}}$	0.14	0.21	0.31	0.45	0.65	0.95	0.54	0.20

z/pm	60	80	100	120	140	160	180	200
$\dfrac{\psi_+}{\left(\frac{1}{\pi a_0^3}\right)^{1/2}}$	0.42	0.47	0.59	0.49	0.33	0.23	0.16	0.11
$\dfrac{\psi_-}{\left(\frac{1}{\pi a_0^3}\right)^{1/2}}$	-0.11	-0.43	-0.81	-0.73	-0.50	-0.34	-0.23	-0.16

The points are plotted in Fig. 14.5.

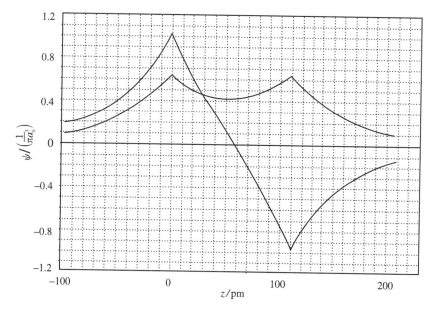

Figure 14.5

P14.8 $E_H = E_1 = -hc\mathcal{R}_H$ [Section 13.2(b)]

Draw up the following table using the data in question and using

$$\frac{e^2}{4\pi\varepsilon_0 R} = \frac{e^2}{4\pi\varepsilon_0 a_0} \times \frac{a_0}{R} = \frac{e^2}{4\pi\varepsilon_0 \times (4\pi\varepsilon_0\hbar^2/m_e e^2)} \times \frac{a_0}{R}$$

$$= \frac{m_e e^4}{16\pi^2\varepsilon_0^2\hbar^2} \times \frac{a_0}{R} = E_h \times \frac{a_0}{R} \quad \left[E_h \equiv \frac{m_e e^4}{16\pi^2\varepsilon_0^2\hbar^2} = 2hc\mathcal{R}_H \right]$$

so that $\dfrac{\left(\frac{e^2}{4\pi\varepsilon_0 R}\right)}{E_h} = \dfrac{a_0}{R}$.

R/a_0	0	1	2	3	4	∞
$\dfrac{\left(\frac{e^2}{4\pi\varepsilon_0 R}\right)}{E_h}$	∞	1	0.500	0.333	0.250	0
$(V_1 + V_2)/E_h$	2.000	1.465	0.843	0.529	0.342	0
$(E - E_H)/E_h$	∞	0.212	-0.031	-0.059	-0.038	0

The points are plotted in Fig. 14.6.

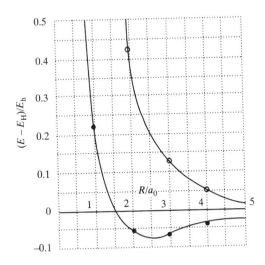

Figure 14.6

The minimum occurs at $R = 2.5a_0$, so $R = 130\,\text{pm}$. At the bond length

$$E - E_H = -0.07E_h = -1.91\,\text{eV}$$

Hence, the dissociation energy is predicted to be about $\boxed{1.9\,\text{eV}}$ and the equilibrium bond length about $\boxed{130\,\text{pm}}$.

P14.9   We proceed as in Problem 14.8 and draw up the following table

R/a_0	0	1	2	3	4	∞
$\dfrac{\left(\frac{e^2}{4\pi\varepsilon_0 R}\right)}{E_h}$	∞	1	0.500	0.333	0.250	0
$(V_1 - V_2)/E_h$	0	-0.007	0.031	0.131	0.158	0
$(E - E_H)/E_h$	∞	1.049	0.425	0.132	0.055	0

The points are also plotted in Fig. 14.6. The contribution V_2 decreases rapidly because it depends on the overlap of the two orbitals.

P14.11   The electron configuration of F_2 is $1\sigma_g^2 2\sigma_u^{*2} 3\sigma_g^2 1\pi_u^4 2\pi_g^{*4}$; that of F_2^- is $1\sigma_g^2 2\sigma_u^{*2} 3\sigma_g^2 1\pi_u^4 2\pi_g^{*4} 4\sigma_u^{*1}$.
So F_2^- has one more antibonding electron than does F_2, suggesting a lower bond order (1/2 versus 1) and therefore a weaker bond. By definition a weaker bond has a smaller dissociation energy (hence the difference in D_e). Weaker bonds tend to be longer (hence the difference in r_e) and less stiff (hence the difference in $\tilde{\nu}$, reflecting a difference in the force constant k) than stronger bonds between similar atoms.

P14.14
$$E_n = \frac{n^2 h^2}{8mL^2}, \quad n = 1, 2, \ldots \quad \text{and} \quad \psi_n = \left(\frac{2}{L}\right)^{1/2} \sin\left(\frac{n\pi x}{L}\right)$$

Two electrons occupy each level (by the Pauli principle), and so butadiene (in which there are four π electrons) has two electrons in ψ_1 and two electrons in ψ_2

$$\psi_1 = \left(\frac{2}{L}\right)^{1/2} \sin\left(\frac{\pi x}{L}\right) \qquad \psi_2 = \left(\frac{2}{L}\right)^{1/2} \sin\left(\frac{2\pi x}{L}\right)$$

These orbitals are sketched in Fig. 14.7(a).

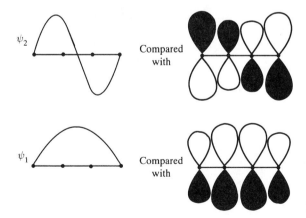

Figure 14.7(a)

The minimum excitation energy is

$$\Delta E = E_3 - E_2 = 5\left(\frac{h^2}{8m_e L^2}\right)$$

In $CH_2{=}CH{-}CH{=}CH{-}CH{=}CH{-}CH{=}CH_2$ there are eight π electrons to accommodate, so the HOMO will be ψ_4 and the LUMO ψ_5. From the particle-in-a-box solutions (Chapter 12)

$$\Delta E = E_5 - E_4 = (25 - 16)\frac{h^2}{8m_e L^2} = \frac{9h^2}{8m_e L^2}$$

$$= \frac{(9) \times (6.626 \times 10^{-34}\,\text{J s})^2}{(8) \times (9.109 \times 10^{-31}\,\text{kg}) \times (1.12 \times 10^{-9}\,\text{m})^2} = 4.3 \times 10^{-19}\,\text{J}$$

which corresponds to $\boxed{2.7\,\text{eV}}$. The HOMO and LUMO are

$$\psi_n = \left(\frac{2}{L}\right)^{1/2} \sin\left(\frac{n\pi x}{L}\right)$$

with $n = 4, 5$ respectively; the two wavefunctions are sketched in Fig. 14.7(b).

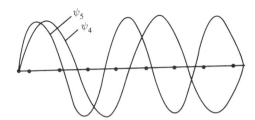

Figure 14.7(b)

Comment. It follows that

$$\lambda = \frac{hc}{\Delta E} = \frac{(6.626 \times 10^{-34}\,\text{J s}) \times (2.998 \times 10^8\,\text{m s}^{-1})}{4.3 \times 10^{-19}\,\text{J}} = 4.6 \times 10^{-7}\,\text{m}, \quad \text{or} \quad \boxed{460\,\text{nm.}}$$

The wavelength 460 nm corresponds to blue light; so the molecule is likely to appear $\boxed{\text{orange}}$ in white light (since blue is subtracted).

P14.15 (a) The secular determinant $|A_N|$ (eqn 14.26) for an open chain of N π-conjugated carbon atoms has the following form where $x = (\alpha - E)/\beta$.

$$|A_N| = \begin{vmatrix} x & 1 & 0 & 0 & \cdots \\ 1 & x & 1 & 0 & \cdots \\ 0 & 1 & x & 1 & \cdots \\ 0 & 0 & 1 & x & \cdots \\ \vdots & \vdots & \vdots & \vdots & \end{vmatrix} = 0$$

The secular matrix A has elements $a_{i,j}$ for $1 \le i, j \le N$. Each element $a_{i,j}$ has the minor $M_{i,j}$ that is formed by removing the ith row and the jth column and fusing the remaining elements into a new determinant. For example, the minor of $a_{1,1}(= x)$ is a determinant that looks exactly like $|A_N|$, but its order equals $N - 1$, so we label this minor as $|A_{N-1}|$. The minor of $a_{1,2}(= 1)$ has the order $N - 1$ and the form:

$$M_{1,2} = \begin{vmatrix} 1 & 1 & 0 & 0 & \cdots \\ 1 & x & 1 & 0 & \cdots \\ 0 & 1 & x & 1 & \cdots \\ 0 & 0 & 1 & x & \cdots \\ \vdots & \vdots & \vdots & \vdots & \end{vmatrix} = |A_{N-2}| - \begin{vmatrix} 0 & 1 & 0 & \cdots \\ 0 & x & 1 & \cdots \\ 0 & 1 & x & \cdots \\ 0 & 0 & 1 & \cdots \\ \vdots & \vdots & \vdots & \end{vmatrix} \begin{array}{l} = 0 \left(\begin{array}{l}\text{because of a}\\\text{column of zeroes}\end{array}\right) \end{array} = |A_{N-2}| = 0$$

We can now expand $|A_N|$ in terms of the first row elements. According to the rules of linear algebra,

$$|A_N| = \sum_{k=1}^{N} a_{1,k} \times (-1)^{1+k} \times M_{1,k}$$

$$= a_{1,1}M_{1,1} - a_{1,2}M_{1,2} + \sum_{k=3}^{N}(0) \times (-1)^{1+k} \times M_{1,k}$$

$$|A_N| = x|A_{N-1}| - |A_{N-2}|$$

Each determinant expands into a characteristic polynomial of x, P_N, so we may write the expression $\boxed{P_N(x) = xP_{N-1}(x) - P_{N-2}(x)}$ which is a reccurence equation.

By comparing the expansion of $|A_2|$ with the above expression, we find that $P_0 = 1$ and $P_1 = x$.

$$P_2 = |A_2| = \begin{vmatrix} x & 1 \\ 1 & x \end{vmatrix} = \underbrace{x \cdot x - 1}_{\text{direct expansion}} = \underbrace{xP_1 - P_0}_{\text{recurrence eqn}}$$

(b) The roots of the characteristic polynomials $P_N(k)$ give the allowed π energies for the molecules of the homologous series. The recurrence equation is used to generate the characteristic polynomials, the roots are found with mathematical software.

Since $x = (\alpha - E/\beta)$, the allowed energies are calculated with $E = \alpha - \beta x$ where x values are the polynomial roots. For example, the roots of the hexatriene ($N = 6$) secular determinant, and the resultant π-energy level diagram are found to be:

$$P_6(x) = x^6 - 5x^4 + 6x^2 - 1$$

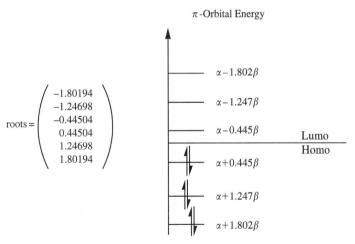

The ground electronic configuration of hexatriene has two electrons in each of the lowest energy π orbitals. The UV $\pi^* \leftarrow \pi$ absorption occurs between the $\alpha + 0.445\beta$ state and the $\alpha - 0.445\beta$ state, so simple Hückel MO theory predicts an absorption at -0.890β.

The U.V. absorptions of the homologous series should equal the energy of the LUMO minus the energy of the HOMO. For simple Hückel theory,

$$\tilde{\nu}_N = \tilde{E}_{LUMO} - \tilde{E}_{HOMO} = (\alpha - x_{LUMO}\beta) - (\alpha - x_{HOMO}\beta)$$

$$= (x_{HOMO} - x_{LUMO})\beta$$

The value of $x_{HOMO} - x_{LUMO}$ for each open chain conjugated hydrocarbon may be evaluated with the technique used for hexatriene. Alternatively, it can be proven that the kth root ($k = 1, 2, \ldots, N$) of the Nth characteristic polynomial is given by $x_k = -2\cos\left(\dfrac{k\pi}{N+1}\right)$.

The Pauli Exclusion Principle, makes $k_{HOMO} = N/2$ and $k_{LUMO} = k_{HOMO} + 1$. So,

$$\tilde{v}_N = 2\left\{\cos\left(\frac{(N+2)\pi}{2(N+1)}\right) - \cos\left(\frac{N\pi}{2(N+1)}\right)\right\}\beta$$

Substance	N	$x_{HOMO} - x_{LUMO}$	$\tilde{v}_{obs}/\text{cm}^{-1}$
Ethene	2	−2.0000	61,500
Butadiene	4	−1.2361	46,080
Hexatriene	6	−0.8901	39,750
Octatetraene	8	−0.6946	32,900

Simple Hückel theory predicts that the slope of the linear regression of $\tilde{v}_{obs}$ against $x_{HOMO} - x_{LUMO}$ equals β. Mathematical software gives the linear regression coefficients as:

$$\tilde{v}_{obs} = \text{intercept} + \text{slope} \times (x_{HOMO} - x_{LUMO})$$

where intercept $= 1.959 \times 10^4\ \text{cm}^{-1}$

slope $= \boxed{-2.114 \times 10^4\ \text{cm}^{-1} = \beta}$

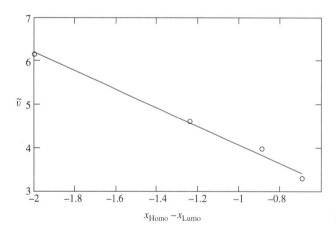

Figure 14.8

Simple Hückel theory predicts that the intercept of the above plot should equal zero and it is unable to explain the non-zero value.

(c) For octatetraene ($N = 8$)

$$E_{deloc} = E_\pi - 8(\alpha + \beta)$$

$$= 2\sum_{k=1}^{4}(\alpha - \beta x_k) - 8(\alpha + \beta)$$

$$= -2\beta\sum_{k=1}^{4} x_k - 8\beta$$

$$= -2\beta \sum_{k=1}^{4} \left\{ -2\cos\left(\frac{k\pi}{9}\right) \right\} - 8\beta$$

$$= 9.518\beta - 8\beta$$

$$\boxed{E_{\text{deloc}} = 1.518\beta = -3.208 \times 10^4 \text{cm}^{-1}}$$

The delocalizaton significantly lowers the π-system energy from that expected from the localized ethene model.

P14.16 **(a)** The table displays computed orbital energies and experimental $\pi^* \leftarrow \pi$ wavenumbers of ethene and the first few conjugated linear polyenes.

Species	$E_{\text{LUMO}}/\text{eV}^*$	$E_{\text{HOMO}}/\text{eV}^*$	$\Delta E/\text{eV}^*$	$\tilde{\nu}/\text{cm}^{-1}$
C_2H_4	1.2282	−10.6411	11.8693	61500
C_4H_6	0.2634	−9.4671	9.7305	46080
C_6H_8	−0.2494	−8.8993	8.6499	39750
C_8H_{10}	−0.5568	−8.5767	8.0199	32900
$C_{10}H_{12}$	−0.7556	−8.3755	7.6199	

*Semi-empirical, PM3 level, PC Spartan Pro™

(b) A plot of the computed energy difference vs. experimental wavenumbers appears below. The computed points fall on a rather good straight line. Of course a better fit can be obtained to a quadratic and a perfect fit to a cubic polynomial; however, the improvement would be slight and the justification even more slight. The linear least-squares best fit is:

$$\boxed{\Delta E/\text{eV} = 3.3534 + 1.3791 \times 10^{-4}\,\tilde{\nu}/\text{cm}^{-1}} \qquad (r^2 = 0.994)$$

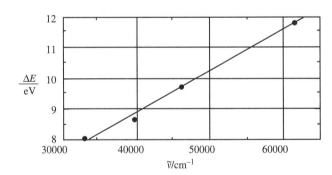

Figure 14.9

(c) The fitting procedure is necessary because the orbital energies are only approximate. Remember that an orbital wavefunction is itself an approximation. A semi-empirical computation is a further approximation. If the orbitals were exact, then we would expect the energy difference to be directly proportional to the spectroscopic wavenumbers with the following proportionality:

$$\Delta E = hc\tilde{\nu} = \frac{(6.626 \times 10^{-34}\,\text{J s})(2.998 \times 10^{10}\,\text{cm s}^{-1})\tilde{\nu}}{(1.602 \times 10^{-19}\,\text{J/eV})},$$

so $\Delta E/\text{eV} = 1.240 \times 10^{-4}\,\tilde{\nu}/\text{cm}^{-1}$.

Clearly this is different than the fit reported above. A further illustration of why the fitting procedure is necessary can be discerned by comparing the table from part **(a)** to a corresponding table based on a different computational model, namely Hartree–Fock computations with an STO-3G basis set:

Species	E_{LUMO}/eV^*	E_{HOMO}/eV^*	$\Delta E/eV^*$
C_2H_4	8.9335	−9.1288	18.0623
C_4H_6	6.9667	−7.5167	14.4834
C_6H_8	6.0041	−6.6783	12.6824
C_8H_{10}	5.4488	−6.1811	11.6299
$C_{10}H_{12}$	5.0975	−5.8621	10.9596

Ab initio, STO-3G,PC Spartan Pro™

Obviously these energy differences are not the same as the PM3 differences computed above. Nor are they energy differences that correspond to the experimental frequencies.

(d) Invert the fitted equation obtained in (b) above:

$$\tilde{\nu}/cm^{-1} = \frac{\Delta E/eV - 3.3534}{1.3791 \times 10^{-4}}.$$

So for $C_{10}H_{12}$, we expect a transition at:

$$\tilde{\nu}/cm^{-1} = \frac{7.6199 - 3.3534}{1.3791 \times 10^{-4}} = \boxed{30937\,cm^{-1}}.$$

Note: The STO-3G data fit a straight line equally well. That fit can also be used to estimate the transition in $C_{10}H_{12}$:

$$\tilde{\nu}/cm^{-1} = \frac{\Delta E/eV - 3.8311}{2.3045 \times 10^{-4}},$$

so for $C_{10}H_{12}$ we expect a transition at

$$\tilde{\nu}/cm^{-1} = \frac{10.9596 - 3.8311}{2.3045 \times 10^{-4}} = 30933\,cm^{-1}.$$

Even though the computations differed considerably in detail, with the calibration procedure they result in nearly identical predictions.

P14.18 (a) The standard enthalpy of formation ($\Delta_f H^{\ominus}/kJ\,mol^{-1}$) of ethene and the first few linear polyenes is listed below.

	Species computed*	Experimental[†]	% error
C_2H_4	69.580	52.46694	32.6
C_4H_6	129.834	108.8 ± 0.79	19.3
		111.9 ± 0.96	16.0
C_6H_8	188.523	168. ± 3	12.2
C_8H_{104}	246.848	295.9[‡]	16.6

*Semi-empirical, PM3 level, PC Sparaton Pro™
[†]hhttp://webbook.nist.gov/chemistry/
[‡]Pedley, Naylor, and Kirby, *Thermodynamic Data of Organic Compounds.*

(b) The % error, shown in the table, is defined by:

$$\% \text{ error} = \frac{\Delta_f H^{\ominus}(\text{calc}) - \Delta_f H^{\ominus}(\text{expt})}{\Delta_f H^{\ominus}(\text{expt})} \times 100\%.$$

(c) For all of the molecules, the computed enthalpies of formation exceed the experimental values by much more than the uncertainty in the experimental value. This observation serves to illustrate that molecular modeling software is not a substitute for experimentation when it comes to quantitative measures. It is also worth noting, however, that the experimental uncertainty can vary a great deal. The NIST database reports $\Delta_f H^{\ominus}$ for C_2H_4 to seven significant figures (with no explicit uncertainty). Even if the figure is not accurate to 1 part in 5000000, it is clearly a very precisely known quantity—as one should expect in such a familiar and well studied substance. The database lists two different determinations for $\Delta_f H^{\ominus}(C_4H_6)$, and the experimental values differ by more than the uncertainty claimed for each; a critical evaluation of the experimental data is called for. The uncertainty claimed for $\Delta_f H^{\ominus}(C_6H_8)$ is greater still (but still only about 2%). Finally, it should go without saying that not all of the figures reported by the molecular modeling software are physically significant.

Solutions to theoretical problems

P14.20   We need to determine if $E_- + E_+ > 2E_H$.

$$E_- + E_+ = -\frac{V_1 - V_2}{1 - S} + \frac{e^2}{4\pi\varepsilon_0 R} - \frac{V_1 + V_2}{1 + S} + \frac{e^2}{4\pi\varepsilon_0 R} + 2E_H$$

$$= -\frac{\{(V_1 - V_2) \times (1 + S) + (1 - S) \times (V_1 + V_2)\}}{(1 - S) \times (1 + S)} + \frac{2e^2}{4\pi\varepsilon_0 R} + 2E_H$$

$$= \frac{2(SV_2 - V_1)}{1 - S^2} + \frac{2e^2}{4\pi\varepsilon_0 R} + 2E_H$$

The nuclear repulsion term is always positive, and always tends to raise the mean energy of the orbitals above E_H. The contribution of the first term is difficult to assess. Where $S \approx 0$, $SV_2 \approx 0$ and $V_1 \approx 0$, and its contribution is dominated by the nuclear repulsion term. Where $S \approx 1$, $SV_2 \approx V_1$ and once again the nuclear repulsion term is dominant. At intermediate values of S, the first term is negative, but of smaller magnitude than the nuclear repulsion term. Thus in all cases $E_- + E_+ > 2E_H$.

P14.22   (a) $\psi = e^{-kr}$ $H = -\frac{\hbar^2}{2\mu}\nabla^2 - \frac{e^2}{4\pi\varepsilon_o r}$

$$\int \psi^2 \, d\tau = \int_0^\infty r^2 e^{-2kr} dr \int_0^\pi \sin\theta \, d\theta \int_0^{2\pi} d\phi = \frac{\pi}{k^3}$$

$$\int \psi \frac{1}{r} \psi \, d\tau = \int_0^\infty r e^{-2kr} dr \int_0^\pi \sin\theta \, d\theta \int_0^{2\pi} d\phi = \frac{\pi}{k^2}$$

$$\int \psi \nabla^2 \psi \, d\tau = \int \psi \frac{1}{r} \frac{d^2}{dr^2}(r e^{-kr}) d\tau = \int \psi \left(k^2 - \frac{2k}{r}\right) \psi \, d\tau$$

$$= \frac{\pi}{k} - \frac{2\pi}{k} = -\frac{\pi}{k}$$

Therefore

$$\int \psi H \psi \, d\tau = \frac{\hbar^2}{2\mu} \times \frac{\pi}{k} - \frac{e^2}{4\pi\varepsilon_0} \times \frac{\pi}{k^2}$$

and

$$E = \frac{\left(\frac{\hbar^2 \pi}{2\mu k}\right) - \left(\frac{e^2 \pi}{4\pi\varepsilon_0 k^2}\right)}{\pi/k^3} = \frac{\hbar^2 k^2}{2\mu} - \frac{e^2 k}{4\pi\varepsilon_0}$$

$$\frac{dE}{dk} = 2\left(\frac{\hbar^2}{2\mu}\right)k - \frac{e^2}{4\pi\varepsilon_0} = 0 \quad \text{when } k = \frac{e^2\mu}{4\pi\varepsilon_0\hbar^2}$$

The optimum energy is therefore

$$E = -\frac{e^4\mu}{32\pi^2\varepsilon_0^2\hbar^2} = \boxed{-hc\mathcal{R}_H} \text{ the exact value.}$$

(b) $\psi = e^{-kr^2}$, H as before.

$$\int \psi^2 \, d\tau = \int_0^\infty e^{-2kr^2}r^2 \, dr \int_0^\pi \sin\theta \, d\theta \int_0^{2\pi} d\phi = \frac{\pi}{2}\left(\frac{\pi}{2k^3}\right)^{1/2}$$

$$\int \psi\frac{1}{r}\psi \, d\tau = \int_0^\infty re^{-2kr^2} \, dr \int_0^\pi \sin\theta \, d\theta \int_0^{2\pi} d\phi = \frac{\pi}{k}$$

$$\int \psi\nabla^2\psi \, d\tau = -2\int \psi(3k - 2k^2r^2)\psi \, d\tau$$

$$= -2\int_0^\infty (3kr^2 - 2k^2r^4)e^{-2kr^2} \, dr \int_0^\pi \sin\theta \, d\theta \int_0^{2\pi} d\phi$$

$$= -8\pi\left\{\left(\frac{3k}{8}\right) \times \left(\frac{\pi}{2k^3}\right)^{1/2} - \frac{3k^2}{16}\left(\frac{\pi}{2k^5}\right)^{1/2}\right\}$$

Therefore

$$E = \frac{3\hbar^2k}{2\mu} - \frac{e^2k^{1/2}}{\varepsilon_0(2\pi)^{1/2}}$$

$$\frac{dE}{dk} = 0 \quad \text{when } k = \frac{e^4\mu^2}{18\pi^3\varepsilon_0^2\hbar^4}$$

and the optimum energy is therefore

$$E = -\frac{e^4\mu}{12\pi^3\varepsilon_0^2\hbar^2} = \boxed{-\frac{8}{3\pi} \times hc\mathcal{R}_H}$$

Since $8/3\pi < 1$, the energy in **(a)** is lower than in **(b)**, and so the exponential wavefunction is better than the Gaussian.

Solutions to applications

P14.24 **(a)**
$$\begin{vmatrix} \alpha - E & \beta & \beta \\ \beta & \alpha - E & \beta \\ \beta & \beta & \alpha - E \end{vmatrix} = 0$$

$$(\alpha - E)\begin{vmatrix} \alpha - E & \beta \\ \beta & \alpha - E \end{vmatrix} - \beta\begin{vmatrix} \beta & \beta \\ \beta & \alpha - E \end{vmatrix} + \beta\begin{vmatrix} \beta & \alpha - E \\ \beta & \beta \end{vmatrix} = 0$$

$$(\alpha - E) \times \{(\alpha - E)^2 - \beta^2\} - \beta\{\beta(\alpha - E) - \beta^2\} + \beta\{\beta^2 - (\alpha - E)\beta\} = 0$$

$$(\alpha - E) \times \{(\alpha - E)^2 - \beta^2\} - 2\beta^2\{\alpha - E - \beta\} = 0$$

$$(\alpha - E) \times (\alpha - E - \beta\} - \times(\alpha - E + \beta) - 2\beta^2(\alpha - E - \beta) = 0$$

$$(\alpha - E - \beta) \times \{(\alpha - E) \times (\alpha - E + \beta) - 2\beta^2\} = 0$$

$$(\alpha - E - \beta) \times \{(\alpha - E) \times (\alpha - E + 2\beta) - \beta(\alpha - E) - 2\beta^2\} = 0$$
$$(\alpha - E - \beta) \times \{(\alpha - E) \times (\alpha - E + 2\beta) - \beta(\alpha - E) + 2\beta\} = 0$$
$$(\alpha - E - \beta) \times (\alpha - E + 2\beta) \times (\alpha - E - \beta) = 0$$

Therefore, the desired roots are $E = \boxed{\alpha - \beta, \alpha - \beta, \text{ and } \alpha + 2\beta}$ The energy level diagram is shown in Fig. 14.10.

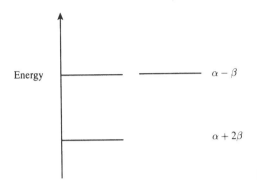

Energy

$\alpha - \beta$

$\alpha + 2\beta$

Figure 14.10

The binding energies are shown in the following table.

Species	Number of e^-	Binding energy
H_3^{2+}	1	$\alpha + 2\beta$
H_3^+	2	$2(\alpha + 2\beta) = 2\alpha + 4\beta$
H_3	3	$2(\alpha + 2\beta) + (\alpha - \beta) = 3\alpha + 3\beta$
H_3^-	4	$2(\alpha + 2\beta) + 2(\alpha - \beta) = 4\alpha + 2\beta$

(b)

$$H_3^+(g) \to 2H(g) + H^+(g) \quad \Delta H_1 = 849\,\text{kJ mol}^{-1}$$
$$\underline{H^+(g) + H_2(g) \to H_3^+(g) \qquad\qquad \Delta H_2 = ?}$$
$$H_2(g) \to 2H(g) \qquad\qquad \Delta H_3 = 2(217.97) - 0\,\text{kJ mol}^{-1}$$

$$\Delta H_2 = \Delta H_3 - \Delta H_1 = 2(217.97) - 849\,\text{kJmol}^{-1}$$

$$\Delta H_2 = \boxed{-413\,\text{kJmol}^{-1}}$$

This is only slightly less than the binding energy of H_2 ($435.94\,\text{kJ mol}^{-1}$)

(c) $\quad 2\alpha + 4\beta = -\Delta H_1 = -849\,\text{kJ mol}^{-1}$

$$\beta = \frac{-\Delta H_1 - 2\alpha}{4} \quad \text{where } \Delta H_1 = 849\,\text{kJ mol}^{-1}$$

Species	Binding energy
H_3^{2+}	$\alpha + 2\beta = -\dfrac{\Delta H_1}{2} = \boxed{-425\,\text{kJ mol}^{-1}}$
H_3^+	$2\alpha + 4\beta = -\Delta H_1 = \boxed{-849\,\text{kJ mol}^{-1}}$
H_3	$3\alpha + 3\beta = 3\left(\alpha - \dfrac{\Delta H_1 + 2\alpha}{4}\right) = 3\left(\dfrac{1}{2}\alpha - \dfrac{\Delta H_1}{4}\right) = \boxed{3(\alpha/2) - 212\,\text{kJ mol}^{-1}}$
H_3^-	$4\alpha + 2\beta = 4\alpha - \dfrac{\Delta H_1 + 2\alpha}{2} = 3\alpha - \dfrac{\Delta H_1}{2} = \boxed{3\alpha - 425\,\text{kJ mol}^{-1}}$

As α is a negative quantity, all four of these species are expected to be stable.

P14.26   This question refers to six 1,4-benzoquinones: the unsubstituted, four methyl-substituted, and a dimethyldimethoxy species. The table below defines the molecules and displays reduction potentials and computed LUMO energies.

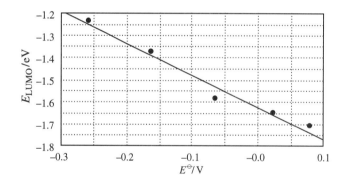

Species	R2	R3	R5	R6	$E^{\ominus}$/V	E_{LUMO}/eV*
1	H	H	H	H	0.078	−1.706
2	CH$_3$	H	H	H	0.023	−1.651
3	CH$_3$	H	CH$_3$	H	−0.067	−1.583
4	CH$_3$	CH$_3$	CH$_3$	H	−0.165	−1.371
5	CH$_3$	CH$_3$	CH$_3$	CH$_3$	−0.260	−1.233
6	CH$_3$	CH$_3$	CH$_3$O	CH$_3$O		−1.446

*Semi-empirical, PM3 level, PC Spartan Pro™

(a)   The calculations for the species 1–5 are plotted below in Fig. 14.11. The figure shows that a linear relationship exists between the reduction potential and the LUMO energy is consistent with these calculations.

Figure 14.11

(b)   The linear least-squares fit from the plot of E_{LUMO} vs. $E^{\ominus}$ is:

$$E_{LUMO}/eV = -1.621 - 1.435 \, E^{\ominus}/V \qquad (r^2 = 0.972)$$

Solving for $E^{\ominus}$ yields:

$$E^{\ominus}/V = -(E_{LUMO}/eV + 1.621)/1.435.$$

Substituting the computed LUMO energy for compound 6 (a model of ubiquinone) yields

$$E^{\ominus} = [-(-1.446 + 1.621)/1.435] = \boxed{-0.122 \, V}.$$

(c)   The model of plastoquinone defined in the problem is compound 4 in the table above. Its experimental reducing potential is known; however, a comparison to the ubiquinone analog based on

E_{LUMO} ought to use a computed reducing potential:

$$E^{\ominus} = [-(-1.371 + 1.621)/1.435] = \boxed{-0.174\,\text{V}}.$$

The better oxidizing agent is the one that is more easily reduced, the one with the less negative reduction potential. Thus we would expect compound 6 to be a better oxidizing agent than compound 4, and $\boxed{\text{ubiquinone a better oxidizing agent that plastoquinone}}$.

(d) Box 10.1 states that ubiquinone acts as an oxidizing agent (oxidizing NADH and $FADN_2$) in respiration (*i.e.*, in the overall oxidation of glucose by oxygent). Plastoquinone, on the other hand, acts as a reducing agent (reducing oxidized plastocyanin) in photosynthesis. In these processes, the better oxidant acts as an oxidant, and the better reductant as a reductant. (Note, however, that Box 10.1 describes other steps in which both species are recycled to their original forms: reduced ubiquinone is oxidized by iron(III) and oxidized plastoquinone is reduced by water.)

15 Molecular symmetry

Solutions to exercises

Discussion questions

E15.1(a) The point group to which a molecule belongs is determined by the symmetry elements it possesses. Therefore the first step is to examine a model (which can be a mental picture) of the molecule for all its symmetry elements. All possible symmetry elements are described in Section 15.1. We list all that apply to the molecule of interest and then follow the assignment procedure summarized by the flow diagram in Figure 15.14 of the text.

E15.2(a) The dipole moment is a fixed property of a molecule and as a result it must remain unchanged through any symmetry operation of the molecule. Recall that the dipole moment is a vector quantity and therefore both its magnitude and direction must be unaffected by the operation. That can only be the case if the dipole moment is coincident with *all* of the symmetry elements of the molecule. Hence molecules belonging to point groups containing symmetry elements that do not satisfy this criterion can be eliminated. Molecules with a center of symmetry cannot possess a dipole moment because any vector is changed through inversion. Molecules with more than one C_n axis cannot be polar since a vector cannot be coincident with more than one axis simultaneously. If the molecule has a plane of symmetry, the dipole moment must lie in the plane; if it has more than one plane of symmetry, the dipole moment must lie in the axis of intersection of these planes. A molecule can also be polar if it has one plane of symmetry and no C_n. Examination of the character tables at the end of the data section shows that the only point groups that satisfy these restrictions are C_s, C_n, and C_{nv}.

E15.3(a) A representative is a mathematical operator (usually a matrix) that represents the physical symmetry operation. The set of all these mathematical operators corresponding to all the operations of the group is called a representation. See Section 16.4(a) for examples.

E15.4(a) Selection rules tell us which transition probabilities between energy levels are non-zero, namely, which spectroscopic transitions will have a non-zero intensity. The intensities are given by the transition moment integral, eqn 15.9, which has the form of the integral of the product of three functions as described by eqn 15.8. Without actually having to perform the integrations involved, group theory can tell us which of these integrals will be non-zero, and hence tell us which are the allowed transitions. Such integrals will be non-zero only if the representation of the triple product in the point group of the molecule spans A_1 or contains a component that spans A_1. In practice, it is usually sufficient to work with the product of the characters of the representations, rather than the matrix representatives themselves. See Examples 15.6 and 15.7.

Numerical exercises

E15.5(a) The elements, other than the $\boxed{\text{identity } E}$, are a $\boxed{C_3 \text{ axis}}$ and $\boxed{\text{three vertical mirror planes } \sigma_v}$. The symmetry axis passes through the C–Cl nuclei (Fig. 15.1). The mirror planes are defined by the three ClCH planes.

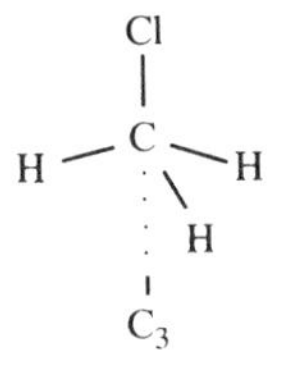

Figure 15.1

E15.6(a) Only molecules belonging to the groups C_n, C_{nv}, and C_s may be polar [Section 15.3(a)]; hence of the molecules listed only (a) pyridine and (b) nitroethane are polar.

E15.7(a) We refer to the character table for C_{4v} at the end of the *Data section*. We then use the procedure illustrated in Example 15.6, and draw up the following table of characters and their products

	E	$2C_4$	C_2	$2\sigma_v$	$2\sigma_d$	
$f_3 = p_z$	1	1	1	1	1	A_1
$f_2 = z$	1	1	1	1	1	A_1
$f_1 = p_x$	2	0	−2	0	0	E
$f_1 f_2 f_3$	2	0	−2	0	0	

The number of times that A_1 appears is 0 [since 2, 0, −2, 0, 0 are the characters of E itself], and so the integral is necessarily zero .

E15.8(a) We proceed as in Example 15.7, considering all three components of the electric dipole moment operator, μ.

Component of μ:	x			y			z		
A_1	1	1	1	1	1	1	1	1	1
$\Gamma(\mu)$	2	−1	0	2	−1	0	1	1	1
A_2	1	1	−1	1	1	−1	1	1	−1
$A_1 \Gamma(\mu) A_2$	2	−1	0	2	−1	0	1	1	−1
		E			E			A_2	

Since A_1 is not present in any product, the transition dipole moment must be zero.

E15.9(a) We first determine how x and y individually transform under the operations of the group. Using these results we determine how the products xy transforms. The transform of xy is the product of the transforms of x and y.

Under each operation the functions transform as follows.

	E	C_2	C_4	σ_v	σ_d
x	x	$-x$	y	x	$-y$
y	y	$-y$	$-x$	$-y$	$-x$
xy	xy	xy	$-xy$	$-xy$	xy
x	1	1	−1	−1	1

From the C_{4v} character table, we see that this set of characters belongs of B_2.

E15.10 In each molecule we must look for an improper rotation axis, perhaps in a disguised form ($S_1 = \sigma$, $S_2 = i$) (Section 15.3). If present the molecule cannot be chiral. D_{2h} contains i and C_{3h} contains σ_h ; therefore, molecules belonging to these point groups cannot be chiral. (Refer to Section 15.2.)

E15.11 In constructing the multiplication table it is convenient to consider the effects of the operation on an object or molecule belonging to that group. The molecule ion $[\text{Pt(NH}_2\text{C}_2\text{H}_4\text{NH}_2)_2]^{2+}$ belongs to D_2 (Fig. 15.2).

Alternatively, we may consider the effect of the operation on a point in space.

$$C_2(x, y, z) \rightarrow -x, -y, z$$
$$C_2'(x, y, z) \rightarrow x, -y, -z$$
$$C_2''(x, y, z) \rightarrow -x, y, -z$$

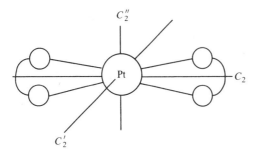

Figure 15.2

By inspection of the outsome of successive operations we can construct the following table

First operation	E	C_2	C_2'	C_2''
Second operation	E	E	C_2	C_2'	C_2''
	C_2	C_2	E	C_2''	C_2'
	C_2'	C_2'	C_2''	E	C_2
	C_2''	C_2''	C_2'	C_2	E

E15.12 List the symmetry elements of the objects (the principal ones, not necessarily all the implied ones), then use the remarks in Section 15.2, and Fig. 15.3 below. Also refer to Figs 15.14 and 15.15 of the text.

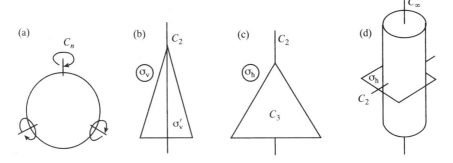

Figure 15.3

(a) Sphere: an infinite number of symmetry axes; therefore $\boxed{R_3}$

(b) Isosceles triangles: E, C_2, σ_v, and σ_v'; therefore $\boxed{C_{2v}}$

(c) Equilateral triangle: $\underbrace{E, C_3, C_2, \sigma_h}$

$$\underbrace{D_3}$$

$$\boxed{D_{3h}}$$

(d) Cylinder: $E, C_\infty, C_2, \sigma_h$; therefore $\boxed{D_{\infty h}}$

E15.13 **(a)** NO_2: $E, C_2, \sigma_v, \sigma_v'$; $\boxed{C_{2v}}$ **(b)** N_2O: $E, C_\infty, C_2, \sigma_v$; $\boxed{C_{\infty v}}$

(c) $CHCl_3$: $E, C_3, 3\sigma_v$; $\boxed{C_{3v}}$ **(d)** $CH_2{=}CH_2$: $E, C_2, 2C_2', \sigma_h$: $\boxed{D_{2h}}$

E15.14 **(a)** *cis*-CHCl=CHCl; $E, C_2, \sigma_v, \sigma_v'$; $\boxed{C_{2v}}$ **(b)** *trans*-CHCl=CHCl; E, C_2, σ_h, i; $\boxed{C_{2h}}$

E15.15 **(a)** Only molecules belonging to the point groups C_n, C_{nv}, and C_s may be polar (Section 15.3(a)); hence of the molecules listed $\boxed{NO_2}$ (C_{2v}), $\boxed{N_2O}$ $(C_{\infty v})$, $\boxed{CHCl_3}$ (C_{3v}) and $\boxed{cis\text{-}CHBr{=}CHBr}$ (C_{2v}) are polar.

(b) All the molecules listed possess an axis of improper rotation, S_n. (See the character tables at the end of the *Data section* which list the symmetry elements for the point groups involved. Note that a centre of inversion, i, is equivalent to S_2 and a mirror plane is equivalent to S_1.) Therefore, (Section 15.3(b)) $\boxed{\text{none}}$ of these molecules is chiral.

E15.16 Recall $p_x \propto x$, $p_y \propto y$, $p_z \propto z$, $d_{xy} \propto xy$, $d_{xz} \propto xz$, $d_{yz} \propto yz$, $d_{z^2} \propto z^2$, $d_{x^2-y^2} \propto x^2 - y^2$ (Section 13.2).

Now refer to the C_{2v} character table. The s orbital spans A_1 and the p orbitals of the central N atom span $A_1(p_z)$, $B_1(p_x)$, and $B_2(p_y)$. Therefore, $\boxed{\text{no orbitals}}$ span A_2, and hence $p_x(A) - p_x(B)$ is a nonbonding combination. If d orbitals are available, as they are in S of the SO_2 molecule, we could form a molecular orbital with $\boxed{d_{xy}}$, which is a basis for A_2.

E15.17 The electric dipole moment operator transforms as $x(B_1)$, $y(B_2)$, and $z(A_1)$ (C_{2v} character table). Transitions are allowed if $\int \psi_f^* \mu \psi_i \, d\tau$ is nonzero (Example 15.7), and hence are forbidden unless $\Gamma_f \times \Gamma(\mu) \times \Gamma_i$ contains A_1. Since $\Gamma_i = A_1$, this requires $\Gamma_f \times \Gamma(\mu) = A_1$. Since $B_1 \times B_1 = A_1$ and $B_2 \times B_2 = A_1$, and $A_1 \times A_1 = A_1$, x-polarized light may cause a transition to a B_1 term, y-polarized light to a B_2 term, and z-polarized light to an A_1 term.

E15.18 **(a)** The point group of benzene is D_{6h}. In D_{6h} μ spans $E_{1u}(x, y)$ and $A_{2u}(z)$, and the ground term is A_{1g}. Then, using $A_{2u} \times A_{1g} = A_{2u}$, $E_{1u} \times A_{1g} = E_{1u}$, $A_{2u} \times A_{2u} = A_{1g}$, and $E_{1u} \times E_{1u} = A_{1g} + A_{2g} + E_{2g}$, we conclude that the upper term is $\boxed{\text{either } E_{1u} \text{ or } A_{2u}}$.

(b) Naphthalene belongs to D_{2h}. In D_{2h} itself, the components span $B_{3u}(x)$, $B_{2u}(y)$, and $B_{1u}(z)$ and the ground term is A_g. Hence, since $A_g \times \Gamma = \Gamma$ in this group, the upper terms are $\boxed{B_{3u} \ (x\text{-polarized})}$, $\boxed{B_{2u} \ (y\text{-polarized})}$, and $\boxed{B_{1u} \ (z\text{-polarized})}$.

E15.19 We consider the integral

$$I = \int_{-a}^{a} f_1 f_2 \, d\theta = \int_{-a}^{a} \sin\theta \cos\theta \, d\theta$$

and hence draw up the following table for the effect of operations in the group C_s (see Fig. 15.4)

	E	σ_h
$f_1 = \sin\theta$	$\sin\theta$	$-\sin\theta$
$f_2 = \cos\theta$	$\cos\theta$	$\cos\theta$

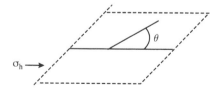

Figure 15.4

In terms of characters

	E	σ_h	
f_1	1	-1	A''
f_2	1	1	A'
$f_1 f_2$	1	-1	A''

Solutions to problems

P15.1   **(a)** Staggered CH_3CH_3: $E, C_3, C_2, 3\sigma_d$; $\boxed{D_{3d}}$ [see Fig. 15.6b of the text]

        (b) Chair C_6H_{12}: $E, C_3, C_2, 3\sigma_d$; $\boxed{D_{3d}}$

            Boat C_6H_{12}: $E, C_2, \sigma_v, \sigma_v'$; $\boxed{C_{2v}}$

        (c) B_2H_6: $E, C_2, 2C_2', \sigma_h$; $\boxed{D_{2h}}$

        (d) $[Co(en)_3]^{3+}$: $E, 2C_3, 3C_2$; $\boxed{D_3}$

        (e) Crown S_8: $E, C_4, C_2, 4C_2', 4\sigma_d, 2S_8$; $\boxed{D_{4d}}$

        Only boat C_6H_{12} may be polar, since all the others are D point groups. Only $[Co(en)_3]^{3+}$ belongs to a group without an improper rotation axis ($S_1 = \sigma$), and hence is chiral.

P15.2   The operations are illustrated in Fig. 15.5. Note that $R^2 = E$ for all the operations of the groups, that $ER = RE = R$ always, and that $RR' = R'R$ for this group. Since $C_2\sigma_h = i$, $\sigma_h i = C_2$, and $iC_2 = \sigma_h$ we can draw up the following group multiplication table

	E	C_2	σ_h	i
E	E	C_2	σ_h	i
C_2	C_2	E	i	σ_h
σ_h	σ_h	i	E	C_2
i	i	σ_h	C_2	E

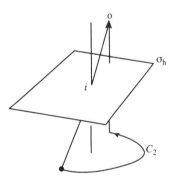

Figure 15.5

The $\boxed{\textit{trans}\text{-CHCl}=\text{CHCl}}$ molecule belongs to the group C_{2h}.

Comment. Note that the multiplication table for C_{2h} can be put into a one-to-one correspondence with the multiplication table of D_2 obtained in Exercise 15.7. We say that they both belong to the same abstract group and are isomorphous.

Question. Can you find another abstract group of order 4 and obtain its multiplication table? There is only one other.

P15.4   Refer to Fig. 15.3 of the text. Place orbitals h_1 and h_2 on the H atoms and s, p_x, p_y, and p_z on the O atom. The z-axis is the C_2 axis; x lies perpendicular to σ_v', y lies perpendicular to σ_v. Then draw up the following table of the effect of the operations on the basis

	E	C_2	σ_v	σ_v'
h_1	h_1	h_2	h_2	h_1
h_2	h_2	h_1	h_1	h_2
s	s	s	s	s
p_x	p_x	$-p_x$	p_x	$-p_x$
p_y	p_y	$-p_y$	$-p_y$	p_y
p_z	p_z	p_z	p_z	p_z

Express the columns headed by each operations R in the form

$$(\text{new}) = D(R)(\text{original})$$

where $D(R)$ is the 6×6 representative of the operation R. We use the rules of matrix multiplication set out in *Justification* 15.1

(i) $E\!:(h_1, h_2, s, p_x, p_y, p_z) \leftarrow (h_1, h_2, s, p_x, p_y, p_z)$ is reproduced by the 6×6 unit matrix

(ii) $C_2\!:(h_2, h_1, s, -p_x, -p_y, p_z) \leftarrow (h_1, h_2, s, p_x, p_y, p_z)$ is reproduced by

$$D(C_2) = \begin{bmatrix} 0 & 1 & 0 & 0 & 0 & 0 \\ 1 & 0 & 0 & 0 & 0 & 0 \\ 0 & 0 & 1 & 0 & 0 & 0 \\ 0 & 0 & 0 & -1 & 0 & 0 \\ 0 & 0 & 0 & 0 & -1 & 0 \\ 0 & 0 & 0 & 0 & 0 & 1 \end{bmatrix}$$

(iii) $\sigma_v\!:(h_2, h_1, s, p_x, -p_y, p_z) \leftarrow (h_1, h_2, s, p_x, p_y, p_z)$ is reproduced by

$$D(\sigma_v) = \begin{bmatrix} 0 & 1 & 0 & 0 & 0 & 0 \\ 1 & 0 & 0 & 0 & 0 & 0 \\ 0 & 0 & 1 & 0 & 0 & 0 \\ 0 & 0 & 0 & 1 & 0 & 0 \\ 0 & 0 & 0 & 0 & -1 & 0 \\ 0 & 0 & 0 & 0 & 0 & 1 \end{bmatrix}$$

(iv) $\sigma_v'\!:(h_1, h_2, s, -p_x, p_y, p_z) \leftarrow (h_1, h_2, s, p_x, p_y, p_z)$ is reproduced by

$$D(\sigma_v') = \begin{bmatrix} 1 & 0 & 0 & 0 & 0 & 0 \\ 0 & 1 & 0 & 0 & 0 & 0 \\ 0 & 0 & 1 & 0 & 0 & 0 \\ 0 & 0 & 0 & -1 & 0 & 0 \\ 0 & 0 & 0 & 0 & 1 & 0 \\ 0 & 0 & 0 & 0 & 0 & 1 \end{bmatrix}$$

(a) To confirm the correct representation of $C_2\sigma_v = \sigma_v'$ we write

$$D(C_2)D(\sigma_v) = \begin{bmatrix} 0 & 1 & 0 & 0 & 0 & 0 \\ 1 & 0 & 0 & 0 & 0 & 0 \\ 0 & 0 & 1 & 0 & 0 & 0 \\ 0 & 0 & 0 & -1 & 0 & 0 \\ 0 & 0 & 0 & 0 & -1 & 0 \\ 0 & 0 & 0 & 0 & 0 & 1 \end{bmatrix} \begin{bmatrix} 0 & 1 & 0 & 0 & 0 & 0 \\ 1 & 0 & 0 & 0 & 0 & 0 \\ 0 & 0 & 1 & 0 & 0 & 0 \\ 0 & 0 & 0 & 1 & 0 & 0 \\ 0 & 0 & 0 & 0 & -1 & 0 \\ 0 & 0 & 0 & 0 & 0 & 1 \end{bmatrix}$$

$$= \begin{bmatrix} 1 & 0 & 0 & 0 & 0 & 0 \\ 0 & 1 & 0 & 0 & 0 & 0 \\ 0 & 0 & 1 & 0 & 0 & 0 \\ 0 & 0 & 0 & -1 & 0 & 0 \\ 0 & 0 & 0 & 0 & 1 & 0 \\ 0 & 0 & 0 & 0 & 0 & 1 \end{bmatrix} = D(\sigma_v')$$

(b) Similarly, to confirm the correct representation of $\sigma_v\sigma_v' = C_2$, we write

$$\begin{bmatrix} 0 & 1 & 0 & 0 & 0 & 0 \\ 1 & 0 & 0 & 0 & 0 & 0 \\ 0 & 0 & 1 & 0 & 0 & 0 \\ 0 & 0 & 0 & 1 & 0 & 0 \\ 0 & 0 & 0 & 0 & -1 & 0 \\ 0 & 0 & 0 & 0 & 0 & 1 \end{bmatrix} \begin{bmatrix} 1 & 0 & 0 & 0 & 0 & 0 \\ 0 & 1 & 0 & 0 & 0 & 0 \\ 0 & 0 & 1 & 0 & 0 & 0 \\ 0 & 0 & 0 & -1 & 0 & 0 \\ 0 & 0 & 0 & 0 & 1 & 0 \\ 0 & 0 & 0 & 0 & 0 & 1 \end{bmatrix}$$

$$= \begin{bmatrix} 0 & 1 & 0 & 0 & 0 & 0 \\ 1 & 0 & 0 & 0 & 0 & 0 \\ 0 & 0 & 1 & 0 & 0 & 0 \\ 0 & 0 & 0 & -1 & 0 & 0 \\ 0 & 0 & 0 & 0 & -1 & 0 \\ 0 & 0 & 0 & 0 & 0 & 1 \end{bmatrix} = D(C_2)$$

(a) The characters of the representatives are the sums of their diagonal elements:

E	C_2	σ_v	σ_v'
6	0	2	4

(b) The characters are not those of any one irreducible representation, so the representation is reducible.

(c) The sum of the characters of the specified sum is

	E	C_2	σ_v	σ_v'
$3A_1$	3	3	3	3
B_1	1	−1	1	−1
$2B_2$	2	−2	−2	2
$3A_1 + B_1 + 2B_2$	6	0	2	4

which is the same as the original. Therefore the representation is $3A_1 + B_1 + 2B_2$.

P15.5   We examine how the operations of the C_{3v} group affect $l_z = xp_y - yp_x$ when applied to it. The transformation of x, y, and z, and by analogy p_x, p_y, and p_z, are as follows (see Fig. 15.6)

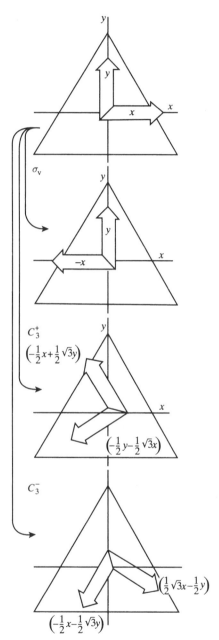

Figure 15.6

$$E(x, y, z) \rightarrow (x, y, z)$$

$$\sigma_v(x, y, z) \rightarrow (-x, y, z)$$

$$\sigma_v'(x, y, z) \rightarrow (x, -y, z)$$

$$\sigma_v''(x, y, z) \rightarrow (x, y, -z)$$

$$C_3^+(x, y, z) \rightarrow \left(-\tfrac{1}{2}x + \tfrac{1}{2}\sqrt{3}y, -\tfrac{1}{2}\sqrt{3}x - \tfrac{1}{2}y, z\right)$$

$$C_3^-(x, y, z) \rightarrow \left(-\tfrac{1}{2}x - \tfrac{1}{2}\sqrt{3}y, \tfrac{1}{2}\sqrt{3}x - \tfrac{1}{2}y, z\right)$$

The characters of all σ operations are the same, as are those of both C_3 operations (see the C_{3v} character table); hence we need consider only one operation in each class.

$$El_z = xp_y - yp_x = l_z$$

$$\sigma_v l_z = -xp_y + yp_x = -l_z \quad [(x, y, z) \rightarrow (-x, y, z)]$$

$$C_3^+ l_z = \left(-\tfrac{1}{2}x + \tfrac{1}{2}\sqrt{3}y\right) \times \left(-\tfrac{1}{2}\sqrt{3}p_x - \tfrac{1}{2}p_y\right) - \left(-\tfrac{1}{2}\sqrt{3}x - \tfrac{1}{2}y\right) \times \left(-\tfrac{1}{2}p_x + \tfrac{1}{2}\sqrt{3}p_y\right)$$

$$\left[(x, y, z) \rightarrow \left(-\tfrac{1}{2}x + \tfrac{1}{2}\sqrt{3}y, -\tfrac{1}{2}\sqrt{3}x - \tfrac{1}{2}y, z\right)\right]$$

$$= \tfrac{1}{4}(\sqrt{3}xp_x + xp_y - 3yp_x - \sqrt{3}yp_y - \sqrt{3}xp_x + 3xp_y - yp_x + \sqrt{3}yp_y)$$

$$= xp_y - yp_x = l_z$$

The representatives of E, σ_v, and C_3^+ are therefore all one-dimensional matrices with characters $1, -1, 1$ respectively. It follows that l_z is a basis for A_2 (see the C_{3v} character table).

P15.7   The multiplication table is

	1	σ_x	σ_y	σ_z
1	1	σ_x	σ_y	σ_z
σ_x	σ_x	1	$i\sigma_z$	$-i\sigma_y$
σ_y	σ_y	$-i\sigma_z$	1	$i\sigma_x$
σ_z	σ_z	$i\sigma_y$	$-i\sigma_x$	1

The matrices $\boxed{\text{do not form a group}}$ since the products $i\sigma_z, i\sigma_y, i\sigma_x$ and their negatives are not among the four given matrices.

P15.11   (a) We work through the flow diagram in the text (Fig. 15.14). We note that this complex with freely rotating CF_3 groups is not linear, it has no C_n axes with $n > 2$, but it does have C_2 axes; in fact it has two C_2 axes perpendicular to whichever C_2 we call principal, and it has a σ_h. Therefore, the point group is $\boxed{D_{2h}}$

(b) The plane shown in Fig. 15.7 is a mirror plane so long as the CF_3 groups each have a CF bond in the plane. (i) If the CF_3 groups are staggered, then the Ag–CN axis is still a C_2 axis; however, there are no other C_2 axes. The Ag–CF_3 axis is an S_2 axis, though, which means that the Ag atom is at an inversion centre. Continuing with the flow diagram, there is a σ_h (the plane shown in the figure). So the point group is $\boxed{C_{2h}}$ (ii) If the CF_3 groups are eclipsed, then the axis through the Ag and perpendicular to the plane of the Ag bonds is still a C_2 axis; however, neither of the Ag bond axes is a C_2 axis. There is no σ_h, but there are two σ_v planes (the plane shown and the plane perpendicular to it and the Ag bond plane). So the point group is $\boxed{C_{2v}}$

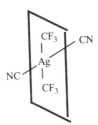

Figure 15.7

P15.13 (a) C_{2v}. The functions x^2, y^2, and z^2 are invariant under all operations of the group, and so $z(5z^2 - 3r^2)$ transforms as $z(A_1)$, $y(5y^2 - 3r^2)$ as $y(B_2)$, $x(5x^2 - 3r^2)$ as $x(B_1)$, and likewise for $z(x^2 - y^2)$, $y(x^2 - z^2)$, and $x(z^2 - y^2)$. The function xyz transforms as $B_1 \times B_2 \times A_1 = A_2$. Therefore, in group C_{2v}, $f \rightarrow \boxed{2A_1 + A_2 + 2B_1 + 2B_2}$

(b) C_{3v}. In C_{3v}, z transforms as A_1, and hence so does z^3. From the C_{3v} character table, $(x^2 - y^2, xy)$ is a basis for E, and so $(xyz, z(x^2 - y^2))$ is a basis for $A_1 \times E = E$. The linear combinations $y(5y^2 - 3r^2) + 5y(x^2 - z^2) \propto y$ and $x(5x^2 - 3r^2) + 5x(z^2 - y^2) \propto x$ are a basis for E. Likewise, the two linear combinations orthogonal to these are another basis for E. Hence, in the group C_{3v}, $f \rightarrow \boxed{A_1 + 3E}$

(c) T_d. Make the inspired guess that the f orbitals are a basis of dimension $3 + 3 + 1$, suggesting the decomposition $T + T + A$. Is the A representation A_1 or A_2? We see from the character table that the effect of S_4 discriminates between A_1 and A_2. Under S_4, $x \rightarrow y, y \rightarrow -x, z \rightarrow -z$, and so $xyz \rightarrow xyz$. The character is $\chi = 1$, and so xyz spans A_1. Likewise, $(x^3, y^3, z^3) \rightarrow (y^3, -x^3, -z^3)$ and $\chi = 0 + 0 - 1 = -1$. Hence, this trio spans T_2. Finally,

$$\{x(z^2 - y^2), y(z^2 - x^2), z(x^2 - y^2)\} \rightarrow \{y(z^2 - x^2), -x(z^2 - y^2), -z(y^2 - x^2)\}$$

resulting in $\chi = 1$, indicating T_1. Therefore, in T_d, $f \rightarrow \boxed{A_1 + T_1 + T_2}$

(d) O_h. Anticipate an $A + T + T$ decomposition as in the other cubic group. Since x, y, and z all have odd parity, all the irreducible representatives will be u. Under S_4, $xyz \rightarrow xyz$ (as in (c)), and so the representation is A_{2u} (see the character table). Under S_4, $(x^3, y^3, z^3) \rightarrow (y^3, -x^3, -z^3)$, as before, and $\chi = -1$, indicating T_{1u}. In the same way, the remaining three functions span T_{2u}. Hence, in O_h, $f \rightarrow \boxed{A_{2u} + T_{1u} + T_{2u}}$

(The shapes of the orbitals are shown in *Inorganic Chemistry*, D. F. Shriver, P. W. Atkins, and C. H. Langford, Oxford University Press and W. H. Freeman & Co (1998).)

The f orbitals will cluster into sets according to their irreducible representations. Thus **(a)** $f \rightarrow A_1 + T_1 + T_2$ in T_d symmetry, and there is one nondegenerate orbital and two sets of triply degenerate orbitals. **(b)** $f \rightarrow A_{2u} + T_{1u} + T_{2u}$, and the pattern of splitting (but not the order of energies) is the same.

P15.15 We begin by drawing up the following table

	N2s	N2p_x	N2p_y	N2p_z	O2p_x	O2p_y	O2p_z	O'2p_x	O'2p_y	O'2p_z	χ
E	N2s	N2p_x	N2p_y	N2p_z	O2p_x	O2p_y	O2p_z	O'2p_x	O'2p_y	O'2p_z	10
C_2	N2s	$-$N2p_x	$-$N2p_y	N2p_z	$-$O'2p_x	$-$O'2p_y	O'2p_z	$-$O2p_x	$-$O2p_y	O2p_z	0
σ_v	N2s	N2p_x	$-$N2p_y	N2p_z	O'2p_x	$-$O'2p_y	O'2p_z	O2p_x	$-$O2p_y	O2p_z	2
σ_v'	N2s	$-$N2p_x	N2p_y	N2p_z	$-$O2p_x	O2p_y	O2p_z	$-$O'2p_x	O'2p_y	O'2p_z	4

The character set (10, 0, 2, 4) decomposes into $\boxed{4A_1 + 2B_1 + 3B_2 + A_2}$. We then form symmetry-adapted linear combinations as described in Section 15.5

$\psi(A_1) = $N2$s$	(column 1)	$\psi(B_1) = $O2$p_x + $O'2$p_x$	(column 5)
$\psi(A_1) = $N2$p_z$	(column 4)	$\psi(B_2) = $N2$p_y$	(column 3)
$\psi(A_1) = $O2$p_z + $O'2$p_z$	(column 7)	$\psi(B_2) = $O2$p_y + $O'2$p_y$	(column 6)
$\psi(A_1) = $O2$p_y + $O'2$p_y$	(column 9)	$\psi(B_2) = $O2$p_z + $O'2$p_z$	(column 7)
$\psi(B_1) = $N2$p_x$	(column 2)	$\psi(A_2) = $O2$p_x + $O'2$p_x$	(column 5)

(The other columns yield the same combinations.)

P15.18 (a) For a photon to induce a spectroscopic transition, the transition moment $\langle \mu \rangle$ must be nonzero. The transition moment is the integral $\int \psi_f^* \mu \psi_i \, d\tau$, where the dipole moment operator has components proportional to the Cartesian coordinates. The integral vanishes unless the integrand, or at least some part of it, belongs to the totally symmetric representation of the molecule's point group. We can answer the first part of the question without reference to the character table, by considering the character of the integrand under inversion. Each component of μ has u character, but each state has g character; the integrand is g × g × u = u, so the integral vanishes and the

transition is not allowed .

(b) However, if a vibration breaks the inversion symmetry, a look at the I character table shows that the components of μ have T_1 character. To find the character of the integrand, we multiply together the characters of its factors. For the transition to T_1

	E	$12C_5$	$12C_5^2$	$20C_3$	$15C_2$
A_1	1	1	1	1	1
$\mu(T_1)$	3	$\frac{1}{2}(1+\sqrt{5})$	$\frac{1}{2}(1-\sqrt{5})$	0	-1
T_1	3	$\frac{1}{2}(1+\sqrt{5})$	$\frac{1}{2}(1-\sqrt{5})$	0	-1
Integrand	9	$\frac{1}{2}(3+\sqrt{5})$	$\frac{1}{2}(3-\sqrt{5})$	0	1

The decomposition of the characters of the integrand into those of the irreducible representations is difficult to do by inspection, but when accomplished it is seen to contain A_1. Therefore the transition to T_1 would become allowed. It is easier to use the formula below which is obtained from what is referred to as the 'little orthogonality theorem' of group theory. (See the *Justification* in Section 15.5 of the 5th edition of this text.) The coefficient of A_1 in the integrand is given as

$$c_{A_1} = \frac{1}{h}\sum_C g(C)\chi(C) = \{9 + 12[\tfrac{1}{2}(3+\sqrt{5})] + 12[\tfrac{1}{2}(3-\sqrt{5})] + 20(0) + 15(1)\}/60 = 1$$

So the integrand contains A_1, and the transition to T_1 would become allowed . For the transition to G

	E	$12C_5$	$12C_5^2$	$20C_3$	$15C_2$
A_1	1	1	1	1	1
$\mu(T_1)$	3	$\frac{1}{2}(1+\sqrt{5})$	$\frac{1}{2}(1-\sqrt{5})$	0	-1
G	4	-1	-1	1	0
Integrand	12	$-\frac{1}{2}(1+\sqrt{5})$	$-\frac{1}{2}(1-\sqrt{5})$	0	0

The little orthogonality theorem gives the coefficient of A_1 in the integrand as

$$c_{A_1} = \frac{1}{h}\sum_C g(C)\chi(C) = \{12 + 12[-\tfrac{1}{2}(1+\sqrt{5})]$$
$$+ 12[-\tfrac{1}{2}(1-\sqrt{5})] + 20(0) + 15(0)\}/60 = 0$$

So the integrand does not contain A_1, and the transition to G would still be forbidden .

Solutions to applications

P15.19 The shape of this molecule is shown in Fig. 15.8.

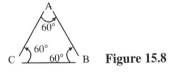

Figure 15.8

(a) Symmetry elements $\boxed{E, 2C_3, 3C_2, \sigma_h, 2S_3, 3\sigma_v}$

Point group $\boxed{D_{3h}}$

(b) $D(E) = \begin{pmatrix} 1 & 0 & 0 \\ 0 & 1 & 0 \\ 0 & 0 & 1 \end{pmatrix} = D(\sigma_h)$

$D(C_3) = \begin{pmatrix} 0 & 0 & 1 \\ 1 & 0 & 0 \\ 0 & 1 & 0 \end{pmatrix}, \qquad D(C_3') = D^2(C_3) = \begin{pmatrix} 0 & 1 & 0 \\ 0 & 0 & 1 \\ 1 & 0 & 0 \end{pmatrix}$

$D(S_3) = D(C_3), \qquad D(S_3') = D^2(S_3) = D(C_3')$

C_3' and S_3' are counter clockwise rotations.

σ_v is through A and perpendicular to B–C.

σ_v' is through B and perpendicular to A–C.

σ_v'' is through C and perpendicular to A–B.

$D(\sigma_v) = \begin{pmatrix} 1 & 0 & 0 \\ 0 & 0 & 1 \\ 0 & 1 & 0 \end{pmatrix}, \qquad D(\sigma_v') = \begin{pmatrix} 0 & 0 & 1 \\ 0 & 1 & 0 \\ 1 & 0 & 0 \end{pmatrix},$

$D(\sigma_v'') = \begin{pmatrix} 0 & 1 & 0 \\ 1 & 0 & 0 \\ 0 & 0 & 1 \end{pmatrix}$

$D(C_2) = D(\sigma_v), \qquad D(C_2') = D(\sigma_v'), \qquad D(C_2'') = D(\sigma_v''),$

(c) Example of elements of group multiplication table

$D(C_3)D(C_2) = \begin{pmatrix} 0 & 0 & 1 \\ 1 & 0 & 0 \\ 0 & 1 & 0 \end{pmatrix} \begin{pmatrix} 1 & 0 & 0 \\ 0 & 0 & 1 \\ 0 & 1 & 0 \end{pmatrix}$

$= \begin{pmatrix} 0 & 1 & 0 \\ 1 & 0 & 0 \\ 0 & 0 & 1 \end{pmatrix} = D(\sigma_v'')$

$D(\sigma_v')D(\sigma_v) = \begin{pmatrix} 0 & 0 & 1 \\ 0 & 1 & 0 \\ 1 & 0 & 0 \end{pmatrix} \begin{pmatrix} 1 & 0 & 0 \\ 0 & 0 & 1 \\ 0 & 1 & 0 \end{pmatrix}$

$= \begin{pmatrix} 0 & 1 & 0 \\ 0 & 0 & 1 \\ 1 & 0 & 0 \end{pmatrix} = D(C_3')$

D_{3h}	E	C_3	C_2	σ_v	σ_v'	σ_h	$\cdots$
E	E	C_3	C_2	σ_v	σ_v'	σ_h	$\cdots$
C_3	C_3	C_3'	σ_v''	σ_v''	σ_v	C_3	$\cdots$
C_2	C_2	σ_v'	E	E	C_3	C_2	$\cdots$
σ_v	σ_v	σ_v'	E	E	C_3	σ_v	$\cdots$
σ_v'	σ_v'	σ_v''	C_3'	C_3'	E	σ_v'	$\cdots$
σ_h	σ_h	C_3	C_2	σ_v	σ_v'	E	$\cdots$
$\vdots$	$\vdots$	$\vdots$	$\vdots$	$\vdots$	$\vdots$	$\vdots$	$\ddots$

(d) First, determine the number of s orbitals (the basis has three s orbitals) that have unchanged positions after application of each symmetry species of the D_{3h} point group.

D_{3h}	E	$2C_3$	$3C_2$	σ_h	$2S_3$	$3\sigma_v$
Unchanged basis members	3	0	1	3	0	1

This is not one of the irreducible representations reported in the D_{3h} character table but inspection shows that it is identical to $A_1' + E'$. This allows us to conclude that the three s orbitals span $\boxed{A_1' + E'}$

Comment. The multiplication table in part (c) is not strictly speaking *the* group multiplication; it is instead the multiplication table for the matrix representations of the group in the basis under consideration.

P15.20   The point group for the square H_4 molecule is D_{4h} with $h = 16$ symmetry species. To find the irreducible representations or symmetry species spanned by four s orbitals, we use the methodology of Section 15.5c.

D_{4h}	E	$2C_4$	C_2	$2C_2'$	$2C_2''$	i	$2S_4$	σ_h	$2\sigma_v$	$2\sigma_d$
Number of unchanged basis members	4	0	0	2	0	0	0	4	2	0

The basis representation is obviously a linear combination of the D_{4h} symmetry species; it reducible. Only the E, $2C_2'$, σ_h and $2\sigma_v$ symmetry elements contribute (The others have factors of zero) to the number of times symmetry species Γ contributes ($a(\Gamma)$) to the representation of the basis.

	E		$2C_2'$		σ_h		$2\sigma_v$
$a(A_{1g}) = \frac{1}{16}\{4\cdot1\cdot1$	$+$	$2\cdot2\cdot1$	$+$	$4\cdot1\cdot1$	$+$	$2\cdot2\cdot1\} = 1$
$a(A_{2g}) = \frac{1}{16}\{4\cdot1\cdot1$	$+$	$2\cdot2\cdot(-1)$	$+$	$4\cdot1\cdot1$	$+$	$2\cdot2\cdot(-1)\} = 0$
$a(B_{1g}) = \frac{1}{16}\{4\cdot1\cdot1$	$+$	$2\cdot2\cdot1$	$+$	$4\cdot1\cdot1$	$+$	$2\cdot2\cdot1\} = 1$
$a(B_{2g}) = \frac{1}{16}\{4\cdot1\cdot1$	$+$	$2\cdot2\cdot(-1)$	$+$	$4\cdot1\cdot1$	$+$	$2\cdot2\cdot(-1)\} = 0$
$a(E_g) = \frac{1}{16}\{4\cdot1\cdot2$	$+$	$2\cdot2\cdot0$	$+$	$4\cdot1\cdot(-2)$	$+$	$2\cdot2\cdot0\} = 0$
$a(A_{1u}) = \frac{1}{16}\{4\cdot1\cdot1$	$+$	$2\cdot2\cdot1$	$+$	$4\cdot1\cdot(-1)$	$+$	$2\cdot2\cdot(-1)\} = 0$
$a(A_{2u}) = \frac{1}{16}\{4\cdot1\cdot1$	$+$	$2\cdot2\cdot(-1)$	$+$	$4\cdot1\cdot(-1)$	$+$	$2\cdot2\cdot1\} = 0$
$a(B_{1u}) = \frac{1}{16}\{4\cdot1\cdot1$	$+$	$2\cdot2\cdot1$	$+$	$4\cdot1\cdot(-1)$	$+$	$2\cdot2\cdot(-1)\} = 0$
$a(B_{2u}) = \frac{1}{16}\{4\cdot1\cdot1$	$+$	$2\cdot2\cdot(-1)$	$+$	$4\cdot1\cdot(-1)$	$+$	$2\cdot2\cdot1\} = 0$
$a(E_u) = \frac{1}{16}\{4\cdot1\cdot2$	$+$	$2\cdot2\cdot0$	$+$	$4\cdot1\cdot2$	$+$	$2\cdot2\cdot0\} = 1$

The basis spans $\boxed{A_{1g} + B_{1g} + E_u}$

P15.22   Can the E_u excited state be reached by a dipole transition from the A_{1g} ground state? Only if the representation of the product $\psi_f^* \mu \psi_i$ includes the totally symmetric species A_{1g}. The z component of the dipole operator belongs to symmetry species A_{2u}, and the x and y components belong to E_u. So the products we must consider are $E_u A_{2u} A_{1g}$ and $E_u E_u A_{1g}$. For z-polarized transitions, the relevant characters are:

	E	$2C_4$	C_2	$2C_2'$	$2C_2''$	i	$2S_4$	σ_h	$2\sigma_v$	$2\sigma_d$
E_u	2	0	-2	0	0	-2	0	2	0	0
A_{2u}	1	1	1	-1	-1	-1	-1	-1	1	1
A_{1g}	1	1	1	1	1	1	1	1	1	1
$E_u A_{2u} A_{1g}$	2	0	-2	0	0	2	0	-2	0	0

To see whether $E_u A_{2u} A_{1g}$ contains A_{1g}, we would multiply the characters of the $E_u A_{2u} A_{1g}$ by the characters of A_{1g}, sum those products, and divide the sum by the order h of the group; since the characters of A_{1g} are all 1, we can simply sum the characters of $E_u A_{2u} A_{1g}$. Since they sum to zero, the product $E_u A_{2u} A_{1g}$ does not contain A_{1g}, and the $\boxed{z\text{-polarized transition is not allowed}}$

For x- or y-polarized transitions:

	E	$2C_4$	C_2	$2C_2'$	$2C_2''$	i	$2S_4$	σ_h	$2\sigma_v$	$2\sigma_d$
E_u	2	0	-2	0	0	-2	0	2	0	0
E_u	2	0	-2	0	0	-2	0	2	0	0
A_{1g}	1	1	1	1	1	1	1	1	1	1
$E_u E_u A_{1g}$	4	0	4	0	0	4	0	4	0	0

Summing the characters of $E_u E_u A_{1g}$, yields 16, the order of the group. Therefore the product $E_u E_u A_{1g}$ does contain A_{1g}, and the $\boxed{\text{transition is allowed}}$.

16 Spectroscopy 1: rotational and vibrational spectra

Solutions to exercises

Discussion questions

E16.1(a) A Fourier transform spectrometer is a device that collects data on the intensities of spectroscopic transitions as a function of one physical property and converts it into intensity data as a function of another property. The conversion is done through a mathematical procedure called Fourier transformation. A brief discussion of the mathematics involved is given in Sections 16.1(c) and 18.1(c). The pair of properties involved in FT spectrometers are complementary observables (conjugate variables) of each other, that is, the product of the uncertainties obeys the uncertainty relations (Section 11.6). As explained in Section 16.1(c), for IR spectroscopy, there are experimental advantages to collecting the intensity data as a function of the position of a movable mirror in the path of one part of a split beam of the radiation before the beam is recombined and enters the sample. The Fourier transformation then converts the intensity data to a function of wavenumber and that is how the spectrum is displayed. The conjugate pair of properties involved in NMR spectroscopy are time and frequency. There are experimental advantages to collecting the intensity data in the time domain and converting it to the frequency domain in which it is displayed.

E16.2(a) The gross selection rules tell us which are the allowed spectroscopic transitions. For both microwave and infrared spectroscopy, the allowed transitions depend on the existence of an oscillating dipole moment which can stir the electromagnetic field into oscillation (and vice versa for absorption). For microwave rotational spectroscopy, this implies that the molecule must have a permanent dipole moment, which is equivalent to an oscillating dipole when the molecule is rotating. See Figure 16.23. In the case of infrared vibrational spectroscopy, the physical basis of the gross selection rule is that the molecule have a structure that allows for the existence of an oscillating dipole moment when the molecule vibrates. Polar molecules necessarily satisfy this requirement, but non-polar molecules may also have a fluctuating dipole moment upon vibration. See Figure 16.35.

E16.3(a) The answer to this question depends precisely on what is meant by equilibrium bond length. See the solution to Problem 16.23 where it is demonstrated that the centrifugally distorted bond length r_c is given by the relation

$$r_c = \frac{r_e}{1 - m_{eff}\omega^2/k}.$$

The angular velocity depends upon the quantum number J through the relation

$$\omega^2 = J(J+1)\hbar^2/m_{eff}^2 r_e^4;$$

thus, the distortion is greater for higher rotational energy levels. But the equilibrium bond length r_e remains constant, if by that term one means the value of r corresponding to a vibrating non-rotating molecule with $J = 0$. However, if one describes the vibration of the molecule in a higher rotational state as having a new "equilibrium" distance r_c, the potential energy of vibration will also be different. It is lowered by the amount shown in eqn 16.40, that is, $-D_J J^2(J+1)^2$. A detailed analysis of the combined effects of rotation and vibration is quite complicated. The treatment in Section 16.12 ignores the effects of centrifugal distortion and anharmonicity. See the references under *Further Reading* for a more through discussion.

Numerical exercises

E16.4(a) The ratio of coefficients A/B is

$$\frac{A}{B} = \frac{8\pi h\nu^3}{c^3} \quad [16.18]$$

The frequency is

$$\nu = \frac{c}{\lambda} \quad \text{so} \quad \frac{A}{B} = \frac{8\pi h}{\lambda^3}$$

(a) $\dfrac{A}{B} = \dfrac{8\pi(6.626 \times 10^{-34}\,\text{J s})}{(70.8 \times 10^{-12}\,\text{m})^3} = \boxed{0.0469\,\text{J m}^{-3}\,\text{s}}$

(b) $\dfrac{A}{B} = \dfrac{8\pi h}{\lambda^3} = \dfrac{8\pi(6.626 \times 10^{-34}\,\text{J s})}{(500 \times 10^{-9}\,\text{m})^3} = \boxed{1.33 \times 10^{-13}\,\text{J m}^{-3}\,\text{s}}$

(c) $\dfrac{A}{B} = \dfrac{8\pi h}{\lambda^3} = 8\pi h\tilde{\nu}^3 = 8\pi\left[6.62 \times 10^{-34}\,\text{J s} \times 3000\,\text{cm}^{-1} \times (10^{-2}\,\text{m}^{-1}/1\,\text{cm}^{-1})\right]$

$$= \boxed{4.50 \times 10^{-16}\,\text{J m}^{-3}\,\text{s}}$$

Comment. Comparison of these ratios shows that the relative importance of spontaneous transitions decreases as the frequency decreases. The quotient $\dfrac{A}{B}$ has units. A unitless quotient is $\dfrac{A}{B\rho}$ with ρ given by eqn 14.

Question. What are the ratios $\dfrac{A}{B\rho}$ for the radiation of **(a)** through **(c)** and what additional conclusions can you draw from these results?

E16.5(a) A source approaching an observer appears to be emitting light of frequency

$$\nu_{\text{approaching}} = \frac{\nu}{1 - \frac{v}{c}} \quad [16.23]$$

Since $\nu \propto \dfrac{1}{\lambda}$, $\lambda_{\text{obs}} = \left(1 - \dfrac{v}{c}\right)\lambda$

$v = 80\,\text{km h}^{-1} = 22.\bar{2}\,\text{m s}^{-1}$. Hence,

$$\lambda_{\text{obs}} = \left(1 - \frac{22.\bar{2}\,\text{m s}^{-1}}{2.998 \times 10^8\,\text{m s}^{-1}}\right) \times (660\,\text{nm}) = \boxed{0.999\,999\,925 \times 660\,\text{nm}}$$

E16.6(a) $\delta\tilde{\nu} \approx \dfrac{5.31\,\text{cm}^{-1}}{\tau/\text{ps}}$ [16.26], implying that $\tau \approx \dfrac{5.31\,\text{ps}}{\delta\tilde{\nu}/\text{cm}^{-1}}$

(a) $\tau \approx \dfrac{5.31\,\text{ps}}{0.1} = \boxed{53\,\text{ps}}$ **(b)** $\tau \approx \dfrac{5.31\,\text{ps}}{1} = \boxed{5\,\text{ps}}$

E16.7(a) $\delta\tilde{\nu} \approx \dfrac{5.31\,\text{cm}^{-1}}{\tau/\text{ps}}$ [16.26]

(a) $\tau \approx 1.0 \times 10^{13}\,\text{s} = 0.10\,\text{ps}$, implying that $\delta\tilde{\nu} \approx \boxed{53\,\text{cm}^{-1}}$

(b) $\tau \approx (100) \times (1.0 \times 10^{-13}\,\text{s}) = 10\,\text{ps}$, implying that $\delta\tilde{\nu} \approx \boxed{0.53\,\text{cm}^{-1}}$

E16.8(a)   NO is a linear rotor and we assume there is little centrifugal distortion; hence

$$F(J) = BJ(J + 1) \, [16.38]$$

with $B = \dfrac{\hbar}{4\pi c I}$, $I = m_{\mathrm{eff}} R^2$ [Table 16.1], and

$$m_{\mathrm{eff}} = \dfrac{m_N m_O}{m_N + m_O} \quad \text{[nuclide masses from inside back cover of the text]}$$

$$= \left(\dfrac{(14.003 \, \mathrm{u}) \times (15.995 \, \mathrm{u})}{(14.003 \, \mathrm{u}) + (15.995 \, \mathrm{u})} \right) \times (1.6605 \times 10^{-27} \, \mathrm{kg \, u^{-1}}) = 1.240 \times 10^{-26} \, \mathrm{kg}$$

Then, $I = (1.240 \times 10^{-26} \, \mathrm{kg}) \times (1.15 \times 10^{-10} \, \mathrm{m})^2 = 1.64\bar{0} \times 10^{-46} \, \mathrm{kg \, m^2}$

and $B = \dfrac{1.0546 \times 10^{-34} \, \mathrm{J \, s}}{(4\pi) \times (2.998 \times 10^8 \, \mathrm{m \, s^{-1}}) \times (1.64\bar{0} \times 10^{-46} \, \mathrm{kg \, m^2})} = 170.\bar{7} \, \mathrm{m^{-1}} = 1.70\bar{7} \, \mathrm{cm^{-1}}$

The wavenumber of the $J = 4 \leftarrow 3$ transition is

$$\tilde{\nu} = 2B(J + 1)[43] = 8B[J = 3] = (8) \times (1.70\bar{7} \, \mathrm{cm^{-1}}) = 13.6 \, \mathrm{cm^{-1}}$$

The frequency is

$$\nu = \tilde{\nu} c = (13.6\bar{5} \, \mathrm{cm^{-1}}) \times \left(\dfrac{10^2 \, \mathrm{m^{-1}}}{1 \, \mathrm{cm^{-1}}} \right) \times (2.998 \times 10^8 \, \mathrm{m \, s^{-1}}) = \boxed{4.09 \times 10^{11} \, \mathrm{Hz}}$$

Question. What is the percentage change in these calculated values if centrifugal distortion is included?

E16.9(a)   **(a)** The wavenumber of the transition is related to the rotational constant by

$$hc\tilde{\nu} = \Delta E = hcB[J(J + 1) - (J - 1)J] = 2hcBJ \, [16.32, \, 16.34]$$

where J refers to the upper state ($J = 3$). The rotational constant is related to molecular structure by

$$B = \dfrac{\hbar}{4\pi c I} \, [16.31]$$

where I is moment of inertia. Putting these expressions together yields

$$\tilde{\nu} = 2BJ = \dfrac{\hbar J}{2\pi c I} \quad \text{so} \quad I = \dfrac{hJ}{c\tilde{\nu}} = \dfrac{(1.0546 \times 10^{-34} \, \mathrm{J \, s}) \times (3)}{2\pi (2.998 \times 10^{10} \, \mathrm{cm \, s^{-1}}) \times (63.56 \, \mathrm{cm^{-1}})}$$

$$= \boxed{2.642 \times 10^{-47} \, \mathrm{kg \, m^2}}$$

(b) The moment of inertia is related to the bond length by

$$I = m_{\mathrm{eff}} R^2 \quad \text{so} \quad R = \sqrt{\dfrac{I}{m_{\mathrm{eff}}}}$$

$$m_{\mathrm{eff}}^{-1} = m_H^{-1} + m_{Cl}^{-1} = \dfrac{(1.0078 \, \mathrm{u})^{-1} + (34.9688 \, \mathrm{u})^{-1}}{1.66054 \times 10^{-27} \, \mathrm{kg \, u^{-1}}} = 6.1477 \times 10^{26} \, \mathrm{kg^{-1}}$$

and $R = \sqrt{(6.1477 \times 10^{26} \, \mathrm{kg^{-1}}) \times (2.642 \times 10^{-47} \, \mathrm{kg \, m^2})} = 1.274 \times 10^{-10} \, \mathrm{m} = \boxed{127.4 \, \mathrm{pm}}$

E16.10(a) If the spacing of lines is constant, the effects of centrifugal distortion are negligible. Hence we may use for the wavenumbers of the transitions

$$F(J) - F(J - 1) = 2BJ \ [16.34]$$

Since $J = 1, 2, 3, \ldots$, the spacing of the lines is $2B$

$$12.604\,\mathrm{cm}^{-1} = 2B$$

$$B = 6.302\,\mathrm{cm}^{-1} = 6.302 \times 10^2\,\mathrm{m}^{-1}$$

$$I = \frac{\hbar}{4\pi c B} \ [\text{Problem 16.4}] = m_{\mathrm{eff}} R^2$$

$$\frac{\hbar}{4\pi c} = \frac{1.0546 \times 10^{-34}\,\mathrm{J\,s}}{(4\pi) \times (2.9979 \times 10^8\,\mathrm{m\,s}^{-1})} = 2.7993 \times 10^{-44}\,\mathrm{kg\,m}$$

$$I = \frac{2.7993 \times 10^{-44}\,\mathrm{kg\,m}}{6.302 \times 10^2\,\mathrm{m}^{-1}} = \boxed{4.442 \times 10^{-47}\,\mathrm{kg\,m}^2}$$

$$m_{\mathrm{eff}} = \frac{m_{\mathrm{Al}} m_{\mathrm{H}}}{m_{\mathrm{Al}} + m_{\mathrm{H}}}$$

$$= \left(\frac{(26.98) \times (1.008)}{(26.98) + (1.008)} \right) \mathrm{u} \times (1.6605 \times 10^{-27}\,\mathrm{kg\,u}^{-1}) = 1.613\bar{6} \times 10^{-27}\,\mathrm{kg}$$

$$R = \left(\frac{I}{m_{\mathrm{eff}}} \right)^{1/2} = \left(\frac{4.442 \times 10^{-47}\,\mathrm{kg\,m}^2}{1.6136 \times 10^{-27}\,\mathrm{kg}} \right)^{1/2} = 1.659 \times 10^{-10}\,\mathrm{m} = \boxed{165.9\,\mathrm{pm}}$$

E16.11(a) $\quad B = \dfrac{\hbar}{4\pi c I} \ [16.31], \quad$ implying that $\quad I = \dfrac{\hbar}{4\pi c B}$

Then, with $I = m_{\mathrm{eff}} R^2, \quad R = \left(\dfrac{\hbar}{4\pi m_{\mathrm{eff}} c B} \right)^{1/2}$

We use $m_{\mathrm{eff}} = \dfrac{m_1 m_2}{m_1 + m_2} = \dfrac{(126.904) \times (34.9688)}{(126.904) + (34.9688)}\,\mathrm{u} = 27.4146\,\mathrm{u}$

and hence obtain

$$R = \left(\frac{1.05457 \times 10^{-34}\,\mathrm{J\,s}}{(4\pi) \times (27.4146) \times (1.66054 \times 10^{-27}\,\mathrm{kg}) \times (2.99792 \times 10^{10}\,\mathrm{cm\,s}^{-1}) \times (0.1142\,\mathrm{cm}^{-1})} \right)^{1/2}$$

$$= \boxed{232.1\,\mathrm{pm}}$$

E16.12(a) The determination of two unknowns requires data from two independent experiments and the equation which relates the unknowns to the experimental data. In this exercise two independently determined values of B for two isotopically different HCN molecules are used to obtain the moments of inertia of the molecules and from these, by use of the equation for the moment of inertia of linear triatomic rotors (Table 16.1), the interatomic distances R_{HC} and R_{CN} are calculated.

Rotational constants which are usually expressed in wavenumbers (cm^{-1}) are sometimes expressed in frequency units (Hz). The conversion between the two is

$$B/\mathrm{Hz} = c \times B/\mathrm{cm}^{-1} \quad [c \text{ in } \mathrm{cm\,s}^{-1}]$$

Thus, $B(\text{in Hz}) = \dfrac{\hbar}{4\pi I}$ and $I = \dfrac{\hbar}{4\pi B}$

Let, $^1H = H$, $^2H = D$, $R_{HC} = R_{DC} = R$, $R_{CN} = R'$. Then

$$I(\text{HCN}) = \frac{1.05457 \times 10^{-34}\,\text{J s}}{(4\pi) \times (4.4316 \times 10^{10}\,\text{s}^{-1})} = 1.8937 \times 10^{-46}\,\text{kg m}^2$$

$$I(\text{DCN}) = \frac{1.05457 \times 10^{-34}\,\text{J s}}{(4\pi) \times (3.6208 \times 10^{10}\,\text{s}^{-1})} = 2.3178 \times 10^{-46}\,\text{kg m}^2$$

and from Table 16.1 with isotopic masses from the inside back cover

$$I(\text{HCN}) = m_H R^2 + m_N R'^2 - \frac{(m_H R - m_N R')^2}{m_H + m_C + m_N}$$

$$I(\text{HCN}) = \left[(1.0078 R^2) + (14.0031 R'^2) - \left(\frac{(1.0078 R - 14.0031 R')^2}{1.0078 + 12.0000 + 14.0031} \right) \right] \text{u}$$

Multiplying through by $m/\text{u} = (m_H + m_C + m_N)/\text{u} = 27.0109$

$$27.0109 \times I(\text{HCN}) = \{27.0109 \times (1.0078 R^2 + 14.0031 R'^2) - (1.0078 R - 14.0031 R')^2\}\,\text{u}$$

or $\left(\dfrac{27.0109}{1.66054 \times 10^{-27}\,\text{kg}} \right) \times (1.8937 \times 10^{-46}\,\text{kg m}^2) = 3.0804 \times 10^{-18}\,\text{m}^2$

$$= \{27.0109 \times (1.0078 R^2 + 14.0031 R'^2) - (1.0078 R - 14.0031 R')^2\} \tag{a}$$

In a similar manner we find for DCN

$$\left(\frac{28.0172}{1.66054 \times 10^{-27}\,\text{kg}} \right) \times (2.3178 \times 10^{-46}\,\text{kg m}^2) = 3.9107 \times 10^{-18}\,\text{m}^2$$

$$= \{28.0172 \times (2.0141 R^2 + 14.0031 R'^2) - (2.0141 R - 14.0031 R')^2\} \tag{b}$$

Thus there are two simultaneous quadratic equations (a) and (b) to solve for R and R'. These equations are most easily solved by readily available computer programs or by successive approximations. The results are

$$R = 1.065 \times 10^{-10}\,\text{m} = \boxed{106.5\,\text{pm}} \quad \text{and} \quad R' = 1.156 \times 10^{-10}\,\text{m} = \boxed{115.6\,\text{pm}}$$

These values are easily verified by direct substitution into the equations and agree well with the accepted values $R_{HC} = 1.064 \times 10^{-10}\,\text{m}$ and $R_{CN} = 1.156 \times 10^{-10}\,\text{m}$.

E16.13(a) The Stokes lines appear at

$$\tilde{\nu}(J + 2 \leftarrow J) = \tilde{\nu}_i - 2B(2J + 3) \text{ [16.49a]} \quad \text{with } J = 0, \ \tilde{\nu} = \tilde{\nu}_i - 6B$$

Since $B = 1.9987\,\text{cm}^{-1}$ (Table 16.2), the Stokes line appears at

$$\tilde{\nu} = (20487) - (6) \times (1.9987\,\text{cm}^{-1}) = \boxed{20\,475\,\text{cm}^{-1}}$$

E16.14(a) The separation of lines is $4B$ [Section 16.7, eqns 16.49a and 16.49b], so $B = 0.2438\,\text{cm}^{-1}$. Then we use

$$R = \left(\frac{\hbar}{4\pi m_{\text{eff}} cB} \right)^{1/2}$$

with $m_{\text{eff}} = \frac{1}{2} m(^{35}\text{Cl}) = \left(\frac{1}{2} \right) \times (34.9688\,\text{u}) = 17.4844\,\text{u}$

Therefore

$$R = \left(\frac{1.05457 \times 10^{-34} \, \text{J s}}{(4\pi) \times (17.4844) \times (1.6605 \times 10^{-27} \, \text{kg}) \times (2.9979 \times 10^{10} \, \text{cm s}^{-1}) \times (0.2438 \, \text{cm}^{-1})} \right)^{1/2}$$

$$= 1.989 \times 10^{-10} \, \text{m} = \boxed{198.9 \, \text{pm}}$$

E16.15(a) Polar molecules show a pure rotational absorption spectrum. Therefore, select the polar molecules based on their well-known structures. Alternatively, determine the point groups of the molecules and use the rule that only molecules belonging to C_n, C_{nv}, and C_s may be polar, and in the case of C_n and C_{nv}, that dipole must lie along the rotation axis. Hence the polar molecules are

(b) HCl   **(d)** CH_3Cl **(e)** CH_2Cl_2

Their point group symmetries are

(b) $C_{\infty v}$ **(d)** C_{3v} **(e)** $C_{2h} (trans)$, $C_{2v} (cis)$

Comment. Note that the *cis* form of CH_2Cl_2 is polar, but the *trans* form is not.

E16.16(a) We select those molecules with an anisotropic polarizability. A practical rule to apply is that spherical rotors do not have anisotropic polarizabilities. Therefore **(c)** CH_4 is inactive. All others are active.

E16.17(a) $\omega = 2\pi\nu = \left(\frac{k}{m} \right)^{1/2}$

$$k = 4\pi^2\nu^2 m = 4\pi^2 \times (2.0 \, \text{s}^{-1})^2 \times (1.0 \, \text{kg}) = 1.6 \times 10^2 \, \text{kg s}^{-2} = \boxed{1.6 \times 10^2 \, \text{N m}^{-1}}$$

E16.18(a) $\omega = \left(\frac{k}{m_{\text{eff}}} \right)^{1/2}$ [16.57]

The fractional difference is

$$\frac{\omega' - \omega}{\omega} = \frac{\left(\frac{k}{m'_{\text{eff}}} \right)^{1/2} - \left(\frac{k}{m_{\text{eff}}} \right)^{1/2}}{\left(\frac{k}{m_{\text{eff}}} \right)^{1/2}} = \frac{\left(\frac{1}{m'_{\text{eff}}} \right)^{1/2} - \left(\frac{1}{m_{\text{eff}}} \right)^{1/2}}{\left(\frac{1}{m_{\text{eff}}} \right)^{1/2}} = \left(\frac{m_{\text{eff}}}{m'_{\text{eff}}} \right)^{1/2} - 1$$

$$= \left(\frac{m(^{23}\text{Na})m(^{35}\text{Cl})\{m(^{23}\text{Na}) + m(^{37}\text{Cl})\}}{\{m(^{23}\text{Na}) + m(^{35}\text{Cl})\}m(^{23}\text{Na})m(^{37}\text{Cl})} \right)^{1/2} - 1$$

$$= \left(\frac{m(^{35}\text{Cl})}{m(^{37}\text{Cl})} \times \frac{m(^{23}\text{Na}) + m(^{37}\text{Cl})}{m(^{23}\text{Na}) + m(^{35}\text{Cl})} \right)^{1/2} - 1$$

$$= \left(\frac{34.9688}{36.9651} \times \frac{22.9898 + 36.9651}{22.9898 + 34.9688} \right)^{1/2} - 1 = -0.01089$$

Hence, the difference is $\boxed{1.089 \, \text{per cent}}$

E16.19(a) $\omega = \left(\frac{k}{m_{\text{eff}}} \right)^{1/2}$ [16.57]; $\omega = 2\pi\nu = 2\pi \left(\frac{c}{\lambda} \right) = 2\pi c \tilde{\nu}$

Therefore, $k = m_{\text{eff}} \omega^2 = 4\pi^2 m_{\text{eff}} c^2 \tilde{\nu}^2$, $m_{\text{eff}} = \frac{1}{2} m(^{35}\text{Cl})$

$$= (4\pi^2) \times \left(\frac{34.9688}{2} \right) \times (1.66054 \times 10^{-27} \, \text{kg}) \times [(2.997924 \times 10^{10} \, \text{cm s}^{-1}) \times (564.9 \, \text{cm}^{-1})]^2$$

$$= \boxed{328.7 \, \text{N m}^{-1}}$$

E16.20(a) We write, with $N' = N$ (upper state) and $N = N$ (lower state)

$$\frac{N'}{N} = e^{-h\nu/kT} \text{[from Boltzmann distribution]} = e^{-hc\tilde{\nu}/kT}$$

$$\frac{hc\tilde{\nu}}{k} = (1.4388 \text{ cm K}) \times (559.7 \text{ cm}^{-1}) \text{ [inside front cover]} = 805.3 \text{ K}$$

$$\frac{N(\text{upper})}{N(\text{lower})} = e^{-805.3 \text{ K}/T}$$

(a) $\dfrac{N(\text{upper})}{N(\text{lower})} = e^{-805.3/298} = \boxed{0.067}$ $(1:15)$ (b) $\dfrac{N(\text{upper})}{N(\text{lower})} = e^{-805.3/500} = \boxed{0.20}$ $(1:5)$

E16.21(a) $\omega = \left(\dfrac{k}{m_{\text{eff}}}\right)^{1/2}$ [16.57],   so $k = m_{\text{eff}}\omega^2 = 4\pi^2 m_{\text{eff}} c^2 \tilde{\nu}^2$

$$m_{\text{eff}} = \frac{m_1 m_2}{m_1 + m_2} [16.56]$$

$$m_{\text{eff}}(\text{H}^{19}\text{F}) = \frac{(1.0078) \times (18.9984)}{(1.0078) + (18.9984)} \text{ u} = 0.9570 \text{ u}$$

$$m_{\text{eff}}(\text{H}^{35}\text{Cl}) = \frac{(1.0078) \times (34.9688)}{(1.0078) + (34.9688)} \text{ u} = 0.9796 \text{ u}$$

$$m_{\text{eff}}(\text{H}^{81}\text{Br}) = \frac{(1.0078) \times (80.9163)}{(1.0078) + (80.9163)} \text{ u} = 0.9954 \text{ u}$$

$$m_{\text{eff}}(\text{H}^{127}\text{I}) = \frac{(1.0078) \times (126.9045)}{(1.0078) + (126.9045)} \text{ u} = 0.9999 \text{ u}$$

We draw up the following table

	HF	HCl	HBr	HI
$\tilde{\nu}/\text{cm}^{-1}$	4141.3	2988.9	2649.7	2309.5
m_{eff}/u	0.9570	0.9796	0.9954	0.9999
$k/(\text{N m}^{-1})$	$\boxed{967.0}$	$\boxed{515.6}$	$\boxed{411.8}$	$\boxed{314.2}$

Note the order of stiffness HF > HCl > HBr > HI.

Question. Which ratio, $\dfrac{k}{B(\text{A}-\text{B})}$ or $\dfrac{\tilde{\nu}}{B(\text{A}-\text{B})}$, where $B(\text{A}-\text{B})$ are the bond energies of Table 14.3, is the more nearly constant across the series of hydrogen halides? Why?

E16.22(a) Data on three transitions are provided. Only two are necessary to obtain the value of $\tilde{\nu}$ and x_e. The third datum can then be used to check the accuracy of the calculated values.

$$\Delta G(v = 1 \leftarrow 0) = \tilde{\nu} - 2\tilde{\nu}x_e = 1556.22 \text{ cm}^{-1} [16.64]$$

$$\Delta G(v = 2 \leftarrow 0) = 2\tilde{\nu} - 6\tilde{\nu}x_e = 3088.28 \text{ cm}^{-1} [16.65]$$

Multiply the first equation by 3, then subtract the second.

$$\tilde{\nu} = (3) \times (1556.22 \text{ cm}^{-1}) - (3088.28 \text{ cm}^{-1}) = \boxed{1580.38 \text{ cm}^{-1}}$$

Then from the first equation

$$x_e = \frac{\tilde{\nu} - 1556.22\,\text{cm}^{-1}}{2\tilde{\nu}} = \frac{(1580.38 - 1556.22)\,\text{cm}^{-1}}{(2) \times (1580.38\,\text{cm}^{-1})} = \boxed{7.644 \times 10^{-3}}$$

x_e data are usually reported as $x_e\tilde{\nu}$ which is

$$x_e\tilde{\nu} = 12.08\,\text{cm}^{-1}$$

$$\Delta G(v = 3 \leftarrow 0) = 3\tilde{\nu} - 12\tilde{\nu}x_e$$
$$= (3) \times (1580.38\,\text{cm}^{-1}) - (12) \times (12.08\,\text{cm}^{-1}) = 4596.18\,\text{cm}^{-1}$$

which is very close to the experimental value.

E16.23(a) $\Delta G_{v+1/2} = \tilde{\nu} - 2(v + 1)x_e\tilde{\nu}$ [16.64] where $\Delta G_{v+1/2} = G(v + 1) - G(v)$

Therefore, since

$$\Delta G_{v+1/2} = (1 - 2x_e)\tilde{\nu} - 2vx_e\tilde{\nu}$$

a plot of $\Delta G_{v+1/2}$ against v should give a straight line which gives $(1 - 2x_e)\tilde{\nu}$ from the intercept at $v = 0$ and $-2x_e\tilde{\nu}$ from the slope. We draw up the following table

v	0	1	2	3	4
$G(v)/\text{cm}^{-1}$	1481.86	4367.50	7149.04	9826.48	12399.8
$\Delta G_{v+1/2}/\text{cm}^{-1}$	2885.64	2781.54	2677.44	2573.34	

The points are plotted in Fig. 16.1. The intercept lies at 2885.6 and the slope is $\dfrac{-312.3}{3} = -104.1$; hence $x_e\tilde{\nu} = 52.1\,\text{cm}^{-1}$.

Since $\tilde{\nu} - 2x_e\tilde{\nu} = 2885.6\,\text{cm}^{-1}$, it follows that $\tilde{\nu} = 2989.8\,\text{cm}^{-1}$.

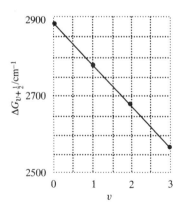

Figure 16.1

The dissociation energy may be obtained by assuming that the molecule is described by a Morse potential and that the constant D_e in the expression for the potential is an adequate first approximation for it. Then

$$D_e = \frac{\tilde{\nu}}{4x_e}\,[16.62] = \frac{\tilde{\nu}^2}{4x_e\tilde{\nu}} = \frac{(2989.8\,\text{cm}^{-1})^2}{(4) \times (52.1\,\text{cm}^{-1})} = 42.9 \times 10^3\,\text{cm}^{-1}, \qquad 5.32\,\text{eV}$$

However, the depth of the potential well D_e differs from D_0, the dissociation energy of the bond, by the zero-point energy; hence

$$D_0 = D_e - \tfrac{1}{2}\tilde{\nu} \text{ [Fig. 16.37, p. 515 of text]}$$

$$= (42.9 \times 10^3\,\text{cm}^{-1}) - \left(\tfrac{1}{2}\right) \times (2889.8\,\text{cm}^{-1})$$

$$= 41.5 \times 10^3\,\text{cm}^{-1} = \boxed{5.15\,\text{eV}}$$

E16.24(a) The R branch obeys the relation

$$\tilde{\nu}_R(J) = \tilde{\nu} + 2B(J+1) \text{ [16.69c]}$$

Hence, $\tilde{\nu}_R(2) = \tilde{\nu} + 6B = (2648.98) + (6) \times (8.465\,\text{cm}^{-1})$ [Table 16.2] $= \boxed{2699.77\,\text{cm}^{-1}}$

E16.25(a) See *Illustration* 16.5. Select those molecules in which a vibration gives rise to a change in dipole moment. It is helpful to write down the structural formulas of the compounds. The infrared active compounds are

(b) HCl   **(c)** CO_2 **(d)** H_2O

Comment. A more powerful method for determining infrared activity based on symmetry considerations is described in Section 16.17. Also see Exercises 16.27–16.28.

E16.26(a) The number of normal modes of vibration is given by (Section 16.17)

$$N_{\text{vib}} = \begin{cases} 3N - 5 \text{ for linear molecules} \\ 3N - 6 \text{ for nonlinear molecules} \end{cases}$$

where N is the number of atoms in the molecule. Hence, since none of these molecules are linear,

(a) 3   **(b)** 6   **(c)** 12

Comment. Even for moderately sized molecules the number of normal modes of vibration is large and they are usually difficult to visualize.

E16.27(a) See Figs 16.49 (H_2O, bent) and 16.48 (CO_2, linear) of the text as well as Example 16.7 and *Illustration* 16.7. Decide which modes correspond to (i) a changing electric dipole moment, (ii) a changing polarizability, and take note of the exclusion rule (Sections 16.16 and 16.17).

(a) Nonlinear: all modes both infrared and Raman active.

(b) Linear: the symmetric stretch is infrared inactive but Raman active.

The antisymmetric stretch is infrared active and (by the exclusion rule) Raman inactive. The two bending modes are infrared active and therefore Raman inactive.

E16.28(a) The uniform expansion is depicted in Fig. 16.2.

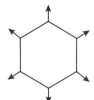

Figure 16.2

Benzene is centrosymmetric, and so the exclusion rule applies (Section 16.16). The mode is infrared inactive (symmetric breathing leaves the molecular dipole moment unchanged at zero), and therefore the mode may be $\boxed{\text{Raman active}}$ (and is). In group theoretical terms, the breathing mode has symmetry A_{1g} in D_{6h}, which is the point group for benzene, and the quadratic forms $x^2 + y^2$ and z^2 have this symmetry (see the character table for C_{6h}, a subgroup of D_{6h}). Hence, the mode is Raman active.

E16.29(a) Use the character table for the group C_{2v} (and see Example 16.6). The rotations span $A_2 + B_1 + B_2$. The translations span $A_1 + B_1 + B_2$. Hence the normal modes of vibration span the difference, $\boxed{4A_1 + A_2 + 2B_1 + 2B_2}$

Comment. $A_1, B_1,$ and B_2 are infrared active; all modes are Raman active.

Solutions to problems

Solutions to numerical problems

P16.2
$$\frac{\delta\lambda}{\lambda} = \frac{2}{c}\left(\frac{2kT\ln 2}{m}\right)^{1/2} \quad [16.24]$$

$$= \left(\frac{2}{2.998 \times 10^8 \text{ m s}^{-1}}\right) \times \left(\frac{(2) \times (1.381 \times 10^{-23}\,\text{J K}^{-1}) \times (298\,\text{K}) \times (\ln 2)}{(m/u) \times (1.6605 \times 10^{-27}\,\text{kg})}\right)^{1/2}$$

$$= \frac{1.237 \times 10^{-5}}{(m/u)^{1/2}}$$

(a) For $^1\text{H}^{35}\text{Cl}$, $m \approx 36\,\text{u}$, so $\dfrac{\delta\lambda}{\lambda} \approx \boxed{2.1 \times 10^{-6}}$

(b) For $^{127}\text{I}^{35}\text{Cl}$, $m \approx 162\,\text{u}$, so $\dfrac{\delta\lambda}{\lambda} \approx \boxed{9.7 \times 10^{-7}}$

For the second part of the problem, we also need

$$\frac{\delta\tilde{\nu}}{\tilde{\nu}} = \frac{\delta\nu}{\nu} = \frac{2}{c}\left(\frac{2kT\ln 2}{m}\right)^{1/2} \quad [16.24] = \frac{\delta\lambda}{\lambda} \quad \boxed{\left[\frac{\delta\lambda}{\lambda} \ll 1\right]}$$

(a) For HCl, $\nu(\text{rotation}) \approx 2Bc \approx (2) \times (10.6\,\text{cm}^{-1}) \times (2.998 \times 10^{10}\,\text{cm s}^{-1})$
$$\approx 6.4 \times 10^{11}\,\text{s}^{-1} \quad\text{or}\quad 6.4 \times 10^{11}\,\text{Hz}$$

Therefore, $\delta\nu(\text{rotation}) \approx (2.1 \times 10^{-6}) \times (6.4 \times 10^{11}\,\text{Hz}) = \boxed{1.3\,\text{MHz}}$

$\tilde{\nu}(\text{vibration}) \approx 2991\,\text{cm}^{-1}$ [Table 16.2]; therefore

$$\delta\tilde{\nu}(\text{vibration}) \approx (2.1 \times 10^{-6}) \times (2991\,\text{cm}^{-1}) = \boxed{0.0063\,\text{cm}^{-1}}$$

(b) For ICl, $\nu(\text{rotation}) \approx (2) \times (0.1142\,\text{cm}^{-1}) \times (2.998 \times 10^{10}\,\text{cm s}^{-1}) \approx 6.8 \times 10^9\,\text{Hz}$
$\delta\nu(\text{rotation}) \approx (9.7 \times 10^{-7}) \times (6.8 \times 10^9\,\text{Hz}) = \boxed{6.6\,\text{kHz}}$
$\tilde{\nu}(\text{vibration}) \approx 384\,\text{cm}^{-1}$
$\delta\tilde{\nu}(\text{vibration}) \approx (9.7 \times 10^{-7}) \times (384\,\text{cm}^{-1}) \approx \boxed{0.0004\,\text{cm}^{-1}}$

Comment. ICl is a solid which melts at 27.2°C and has a significant vapour pressure at 25°C.

P16.4 Rotational line separations are $2B$ (in wavenumber units), $2Bc$ (in frequency units), and $(2B)^{-1}$ in wavelength units. Hence the transitions are separated by $\boxed{596\,\text{GHz}}$, $\boxed{19.9\,\text{cm}^{-1}}$, and $\boxed{0.503\,\text{mm}}$.

Ammonia is a symmetric rotor (Section 16.4) and we know that

$$B = \frac{\hbar}{4\pi c I_\perp} \quad [16.37]$$

and from Table 16.1,

$$I_\perp = m_A R^2 (1 - \cos\theta) + \left(\frac{m_A m_B}{m}\right) R^2 (1 + 2\cos\theta)$$

$m_A = 1.6735 \times 10^{-27}$ kg, $m_B = 2.3252 \times 10^{-26}$ kg, and $m = 2.8273 \times 10^{-26}$ kg with $R = 101.4$ pm and $\theta = 106°47'$, which gives

$$\begin{aligned}
I_\perp &= (1.6735 \times 10^{-27}\,\text{kg}) \times (101.4 \times 10^{-12}\,\text{m})^2 \times (1 - \cos 106°47') \\
&\quad + \left(\frac{(1.6735 \times 10^{-27}) \times (2.3252 \times 10^{-26}\,\text{kg}^2)}{2.8273 \times 10^{-26}\,\text{kg}}\right) \\
&\quad \times (101.4 \times 10^{-12}\,\text{m})^2 \times (1 + 2\cos 106°47') \\
&= 2.815\bar{8} \times 10^{-47}\,\text{kg m}^2
\end{aligned}$$

Therefore,

$$B = \frac{1.05457 \times 10^{-34}\,\text{J s}}{(4\pi) \times (2.9979 \times 10^8\,\text{m s}^{-1}) \times (2.815\bar{8} \times 10^{-47}\,\text{kg m}^2)} = 994.1\,\text{m}^{-1} = \boxed{9.941\,\text{cm}^{-1}}$$

which is in accord with the data.

P16.6   Rotation about any axis perpendicular to the C_6 axis may be represented in its essentials by rotation of the pseudolinear molecule in Fig. 16.3(a) about the x-axis in the figure.

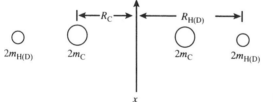

Figure 16.3(a)

The data allow for a determination of R_C and $R_{H(D)}$ which may be decomposed into R_{CC} and $R_{CH(D)}$.

$$I_H = 4m_H R_H^2 + 4m_C R_C^2 = 147.59 \times 10^{-47}\,\text{kg m}^2$$
$$I_D = 4m_D R_D^2 + 4m_C R_C^2 = 178.45 \times 10^{-47}\,\text{kg m}^2$$

Subtracting I_H from I_D (assume $R_H = R_D$) yields

$$4(m_D - m_H)R_H^2 = 30.86 \times 10^{-47}\,\text{kg m}^2$$
$$4(2.0141\,\text{u} - 1.0078\,\text{u}) \times (1.66054 \times 10^{-27}\,\text{kg u}^{-1}) \times (R_H^2) = 30.86 \times 10^{-47}\,\text{kg m}^2$$
$$R_H^2 = 4.616\bar{9} \times 10^{-20}\,\text{m}^2 \qquad R_H = 2.149 \times 10^{-10}\,\text{m}$$

$$R_C^2 = \frac{(147.59 \times 10^{-47}\,\text{kg m}^2) - (4m_H R_H^2)}{4m_C}$$

$$= \frac{(147.59 \times 10^{-47}\,\text{kg m}^2) - (4) \times (1.0078\,\text{u}) \times (1.66054 \times 10^{-27}\,\text{kg u}^{-1}) \times (4.616\bar{9} \times 10^{-20}\,\text{m}^2)}{(4) \times (12.011\,\text{u}) \times (1.66054 \times 10^{-27}\,\text{kg u}^{-1})}$$

$$= 1.4626 \times 10^{-20}\,\text{m}^2$$

$$R_C = 1.209 \times 10^{-10}\,\text{m}$$

Figure 16.3 (b) shows the relation between R_H, R_C, R_{CC}, and R_{CH}.

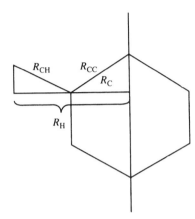

Figure 16.3(b)

$$R_{CC} = \frac{R_C}{\cos 30°} = \frac{1.209 \times 10^{-10}\,\text{m}}{0.8660} = 1.396 \times 10^{-10}\,\text{m} = \boxed{139.6\,\text{pm}}$$

$$R_{CH} = \frac{R_H - R_C}{\cos 30°} = \frac{0.940 \times 10^{-10}}{0.8660} = 1.08\bar{5} \times 10^{-10} = \boxed{108.\bar{5}\,\text{pm}}$$

$$R_{CD} = R_{CH}$$

Comment. These values are very close to the interatomic distances quoted by Herzberg in *Electronic spectra and electronic structure of polyatomic molecules*, p. 666 (*Further reading*, Chapter 17), which are 139.7 and 108.4 pm respectively.

P16.7 The separations between neighbouring lines are

$$20.81,\ 20.60,\ 20.64,\ 20.52,\ 20.34,\ 20.37,\ 20.26 \quad \text{mean: } 20.51\,\text{cm}^{-1}$$

Hence $B = \left(\frac{1}{2}\right) \times (20.51\,\text{cm}^{-1}) = 10.26\,\text{cm}^{-1}$ and

$$I = \frac{\hbar}{4\pi cB} = \frac{1.05457 \times 10^{-34}\,\text{J s}}{(4\pi) \times (2.99793 \times 10^{10}\,\text{cm s}^{-1}) \times (10.26\,\text{cm}^{-1})} = \boxed{2.728 \times 10^{-47}\,\text{kg m}^2}$$

$$R = \left(\frac{I}{m_{\text{eff}}}\right)^{1/2} \quad [\text{Table 16.1}] \quad \text{with } m_{\text{eff}} = 1.6266 \times 10^{-27}\,\text{kg [Exercise 16.3(a)]}$$

$$= \left(\frac{2.728 \times 10^{-47}\,\text{kg m}^2}{1.6266 \times 10^{-27}\,\text{kg}}\right)^{1/2} = \boxed{129.5\,\text{pm}}$$

Comment. A more accurate value would be obtained by ascribing the variation of the separations to centrifugal distortion, and not taking a simple average. Alternatively, the effect of centrifugal distortion could be minimized by plotting the observed separations against J, fitting them to a smooth curve, and extrapolating that curve to $J = 0$. Since $B \propto \dfrac{1}{I}$ and $I \propto m_{\text{eff}}$, $B \propto \dfrac{1}{m_{\text{eff}}}$. Hence, the corresponding lines in $^2H^{35}Cl$ will lie at a factor

$$\frac{m_{\text{eff}}(^1H^{35}Cl)}{m_{\text{eff}}(^2H^{35}Cl)} = \frac{1.6266}{3.1624} = 0.5144$$

to low frequency of $^1H^{35}Cl$ lines. Hence, we expect lines at $\boxed{10.56, 21.11, 31.67, \ldots \text{cm}^{-1}}$

P16.9 $R = \left(\dfrac{\hbar}{4\pi\mu cB}\right)^{1/2}$ and $\nu = 2cB(J + 1)$ [16.44, with $\nu = c\tilde{\nu}$]

We use $\mu(\text{CuBr}) \approx \dfrac{(63.55) \times (79.91)}{(63.55) + (79.91)}\text{u} = 35.40\,\text{u}$

and draw up the following table:

J	13	14	15	
ν/MHz	84421.34	90449.25	96476.72	
B/cm^{-1}	0.10057	0.10057	0.10057	$\left[B = \dfrac{\nu}{2c(J + 1)}\right]$

Hence, $R = \left(\dfrac{1.05457 \times 10^{-34}\,\text{J s}}{(4\pi) \times (35.40) \times (1.6605 \times 10^{-27}\,\text{kg}) \times (2.9979 \times 10^{10}\,\text{cm s}^{-1}) \times (0.10057\,\text{cm}^{-1})}\right)^{1/2}$

$= \boxed{218\,\text{pm}}$

P16.11   Plot frequency against J as in Fig. 16.4.

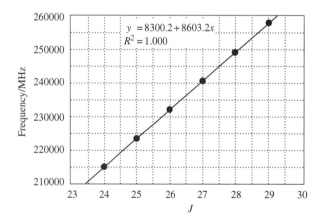

Figure 16.4

The rotational constant is related to the wavenumbers of observed transitions by

$$\tilde{\nu} = 2B(J + 1) = \frac{\nu}{c} \quad \text{so} \quad \nu = 2Bc(J + 1)$$

A plot of ν versus J, then, has a slope of $2Bc$. From Fig. 16.4, the slope is 8603 MHz, so

$$B = \frac{8603 \times 10^6 \, \text{s}^{-1}}{2(2.998 \times 10^8 \, \text{m s}^{-1})} = \boxed{14.35 \, \text{m}^{-1}}$$

The most highly populated energy level is roughly

$$J_{\text{max}} = \left(\frac{kT}{2hcB}\right)^{1/2} - \frac{1}{2}$$

so $J_{\text{max}} = \left(\frac{(1.381 \times 10^{-23} \, \text{J K}^{-1}) \times (298 \, \text{K})}{(6.626 \times 10^{-34} \, \text{J s}) \times (8603 \times 10^6 \, \text{s}^{-1})}\right)^{1/2} - \frac{1}{2} = \boxed{26}$ at 298 K

and $J_{\text{max}} = \left(\frac{(1.381 \times 10^{-23} \, \text{J K}^{-1}) \times (100 \, \text{K})}{(6.626 \times 10^{-34} \, \text{J s}) \times (8603 \times 10^6 \, \text{s}^{-1})}\right)^{1/2} - \frac{1}{2} = \boxed{15}$ at 100 K

P16.13 The Lewis structure is

$$[\ddot{\text{O}}{=}\text{N}{=}\ddot{\text{O}}]^+$$

VSEPR indicates that the ion is $\boxed{\text{linear}}$ and has a centre of symmetry. The activity of the modes is consistent with the rule of mutual exclusion; none is both infrared and Raman active. These transitions may be compared to those for CO_2 (Fig. 16.48 of the text) and are consistent with them. The Raman active mode at $1400 \, \text{cm}^{-1}$ is due to a symmetric stretch ($\tilde{\nu}_1$), that at $2360 \, \text{cm}^{-1}$ to the antisymmetric stretch ($\tilde{\nu}_3$) and that at $540 \, \text{cm}^{-1}$ to the two perpendicular bending modes ($\tilde{\nu}_2$). There is a combination band, $\tilde{\nu}_1 + \tilde{\nu}_3 = 3760 \, \text{cm}^{-1} \approx 3735 \, \text{cm}^{-1}$, which shows a weak intensity in the infrared.

P16.16 $V(R) = hcD_e\{1 - e^{-a(R-R_e)}\}^2$ [16.61]

$$\tilde{\nu} = \frac{\omega}{2\pi c} = 936.8 \, \text{cm}^{-1} \qquad x_e\tilde{\nu} = 14.15 \, \text{cm}^{-1}$$

$$a = \left(\frac{m_{\text{eff}}}{2hcD_e}\right)^{1/2}\omega \quad x_e = \frac{\hbar a^2}{2m_{\text{eff}}\omega} \quad D_e = \frac{\tilde{\nu}}{4x_e}$$

$$m_{\text{eff}}(\text{RbH}) \approx \frac{(1.008) \times (85.47)}{(1.008) + (85.47)}\text{u} = 1.654 \times 10^{-27} \, \text{kg}$$

$$D_e = \frac{\tilde{\nu}^2}{4x_e\tilde{\nu}} = \frac{(936.8 \, \text{cm}^{-1})^2}{(4) \times (14.15 \, \text{cm}^{-1})} = 15505 \, \text{cm}^{-1} \quad (1.92 \, \text{eV})$$

$$a = 2\pi\nu \left(\frac{m_{\text{eff}}}{2hcD_e}\right)^{1/2} [16.61] = 2\pi c\tilde{\nu} \left(\frac{m_{\text{eff}}}{2hcD_e}\right)^{1/2}$$

$$= (2\pi) \times (2.998 \times 10^{10} \, \text{cm s}^{-1}) \times (936.8 \, \text{cm}^{-1})$$

$$\times \left(\frac{1.654 \times 10^{-27} \, \text{kg}}{(2) \times (15505 \, \text{cm}^{-1}) \times (6.626 \times 10^{-34} \, \text{J s}) \times (2.998 \times 10^{10} \, \text{cm s}^{-1})}\right)^{1/2}$$

$$= 9.144 \times 10^9 \, \text{m}^{-1} = 9.144 \, \text{nm}^{-1} = \frac{1}{0.1094 \, \text{nm}}$$

Therefore, $\frac{V(R)}{hcD_e} = \{1 - e^{-(R-R_e)/(0.1094 \, \text{nm})}\}^2$

with $R_e = 236.7\,\text{pm}$. We draw up the following table

R/pm	50	100	200	300	400	500	600	700	800
$V/(hcD_e)$	20.4	6.20	0.159	0.193	0.601	0.828	0.929	0.971	0.988

These points are plotted in Fig. 16.5 as the line labelled $J = 0$

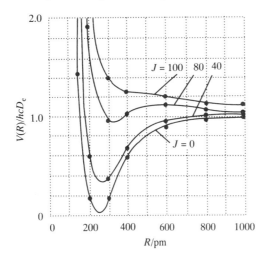

Figure 16.5

For the second part, we note that $B \propto \dfrac{1}{R^2}$ and write

$$V_J^* = V + hcB_e J(J+1) \times \left(\frac{R_e^2}{R^2}\right)$$

with B_e the equilibrium rotational constant, $B_e = 3.020\,\text{cm}^{-1}$.

We then draw up the following table using the values of V calculated above

R/pm	50	100	200	300	400	600	800	1000
$\dfrac{R_e}{R}$	4.73	2.37	1.18	0.79	0.59	0.39	0.30	0.24
$\dfrac{V}{hcD_e}$	20.4	6.20	0.159	0.193	0.601	0.929	0.988	1.000
$\dfrac{V_{40}^*}{hcD_e}$	27.5	7.99	0.606	0.392	0.713	0.979	1.016	1.016
$\dfrac{V_{80}^*}{hcD_e}$	48.7	13.3	1.93	0.979	1.043	1.13	1.099	1.069
$\dfrac{V_{100}^*}{hcD_e}$	64.5	17.2	2.91	1.42	1.29	1.24	1.16	1.11

These points are also plotted in Fig. 16.5.

P16.17　　(a) In the harmonic approximation

$$D_e = D_0 + \tfrac{1}{2}\tilde{\nu} \quad \text{so} \quad \tilde{\nu} = 2(D_e - D_0)$$

$$\tilde{\nu} = \frac{2(1.51 \times 10^{-23}\,\text{J} - 2 \times 10^{-26}\,\text{J})}{(6.626 \times 10^{-34}\,\text{J s}) \times (2.998 \times 10^8\,\text{m s}^{-1})} = \boxed{152\,\text{m}^{-1}}$$

The force constant is related to the vibrational frequency by

$$\omega = \left(\frac{k}{m_{eff}}\right)^{1/2} = 2\pi\nu = 2\pi c\tilde{\nu} \quad \text{so} \quad k = (2\pi c\tilde{\nu})^2 m_{eff}$$

The effective mass is

$$m_{eff} = \tfrac{1}{2}m = \tfrac{1}{2}(4.003\,\text{u}) \times (1.66 \times 10^{-27}\,\text{kg u}^{-1}) = 3.32 \times 10^{-27}\,\text{kg}$$

$$k = [2\pi(2.998 \times 10^8\,\text{m s}^{-1}) \times (152\,\text{m}^{-1})]^2 \times (3.32 \times 10^{-27}\,\text{kg})$$

$$= \boxed{2.72 \times 10^{-4}\,\text{kg s}^{-2}}$$

The moment of inertia is

$$I = m_{eff}R_e^2 = (3.32 \times 10^{-27}\,\text{kg}) \times (297 \times 10^{-12}\,\text{m})^2 = \boxed{2.93 \times 10^{-46}\,\text{kg m}^2}$$

The rotational constant is

$$B = \frac{\hbar}{4\pi cI} = \frac{1.0546 \times 10^{-34}\,\text{J s}}{4\pi(2.998 \times 10^8\,\text{m s}^{-1}) \times (2.93 \times 10^{-46}\,\text{kg m}^2)} = \boxed{95.5\,\text{m}^{-1}}$$

(b) In the Morse potential

$$x_e = \frac{\tilde{\nu}}{4D_e} \quad \text{and} \quad D_e = D_0 + \frac{1}{2}\left(1 - \frac{1}{2}x_e\right)\tilde{\nu} = D_0 + \frac{1}{2}\left(1 - \frac{\tilde{\nu}}{8D_e}\right)\tilde{\nu}$$

This rearranges to a quadratic equation in $\tilde{\nu}$

$$\frac{\tilde{\nu}^2}{16D_e} - \frac{1}{2}\tilde{\nu} + D_e - D_0 = 0 \quad \text{so} \quad \tilde{\nu} = \frac{\frac{1}{2} - \sqrt{\left(\frac{1}{2}\right)^2 - \frac{4(D_e - D_0)}{16D_e}}}{2(16D_e)^{-1}}$$

$$\tilde{\nu} = 4D_e\left(1 - \sqrt{\frac{D_0}{D_e}}\right)$$

$$= \frac{4(1.51 \times 10^{-23}\,\text{J})}{(6.626 \times 10^{-34}\,\text{J s}) \times (2.998 \times 10^8\,\text{m s}^{-1})}\left(1 - \sqrt{\frac{2 \times 10^{-26}\,\text{J}}{1.51 \times 10^{-23}\,\text{J}}}\right)$$

$$= \boxed{293\,\text{m}^{-1}}$$

and $\quad x_e = \dfrac{(293\,\text{m}^{-1}) \times (6.626 \times 10^{-34}\,\text{J s}) \times (2.998 \times 10^8\,\text{m s}^{-1})}{4(1.51 \times 10^{-23}\,\text{J})} = \boxed{0.96}$

P16.18 **(a)** The data table of each vibration–rotation band proceeds from the lowest energy transition to the highest. These may be indexed with $i = 0, 1, 2, \ldots, N-1$ where $N = 40$. For example, $\tilde{\nu}_0 = 2059.6\,\text{cm}^{-1}$ and $\tilde{\nu}_2 = 2068.8\,\text{cm}^{-1}$ in the fundamental band ($1 \leftarrow 0$). Overtone bands are indexed in the same way. The line spacing is about $4\,\text{cm}^{-1}$ except between $\tilde{\nu}_{19}$ and $\tilde{\nu}_{20}$ where it is about $8\,\text{cm}^{-1}$. This means that the Q branch is missing (Section 16.12) because the $\Delta J = 0$ transition is forbidden. We conclude that the data $\tilde{\nu}_0$ to $\tilde{\nu}_{19}$ are lines of the P branch ($\Delta J = -1$)

and the data $\tilde{\nu}_{20}$ to $\tilde{\nu}_{39}$ are lines of the R branch ($\Delta J = +1$). J and m assignment are:

	line	J	m

P branch $\left\{ \begin{array}{ccc} \vdots & \vdots & \vdots \\ \tilde{\nu}_{18} & 2 & -2 \\ \tilde{\nu}_{19} & 1 & -1 \end{array} \right\}$ $m = -J$

R branch $\left\{ \begin{array}{ccc} \tilde{\nu}_{20} & 0 & 1 \\ \tilde{\nu}_{21} & 1 & 2 \\ \vdots & \vdots & \vdots \end{array} \right\}$ $m = J + 1$

The vibration–rotation energy of the (v, J) state is based upon eqn 16.68 with the addition of terms for anharmonicity (eqn 16.62), centrifugal distortion (eqn 16.40), and vibration–rotation coupling.

$$S(v, J) = (v + 1/2)\tilde{\nu} - (v + 1/2)^2 x_e \tilde{\nu}$$
$$+ BJ(J + 1) - D_J J^2 (J + 1)^2 - a(v + 1/2)J(J + 1)$$

The equation for $S(v, J)$ is used to deduce transition lines for the P and R branches. Lines of the P branch $(m = -J)$ are:

$$\Delta S(v, J) = S(v, J - 1) - S(0, J)$$

$$\boxed{\Delta S(v, m) = A_0 + A_1 m + A_2 m^2 + A_3 m^3}$$

where
$$A_0(v, \tilde{\nu}, x_e \tilde{\nu}) = v\{\tilde{\nu} - (v + 1)x_e \tilde{\nu}\}$$
$$A_1(v, B, a) = 2B - a(v + 1)$$
$$A_2(v, a) = -av$$
$$A_3(D_J) = -4D_J$$

an identical equation describes the R branch $(m = J + 1)$.

(b) There are a number of ways for finding the values of the parameter set $(\tilde{\nu}, x_e \tilde{\nu}, B, D_J, a)$. All of them adjust the parameter values until the $\Delta S(v, m)$ equation fits the experimental spectra according to some criteria. One method selects values so that the sum of the squares of errors (SSE) is minimized.

$$\text{SSE}(\tilde{\nu}, x_e \tilde{\nu}, B, D_J, a) = \underbrace{\sum_{i=0}^{N-1} \{\tilde{\nu}_i - \Delta S(1, m_i)\}^2}_{\substack{\text{fundamental} \\ (v=1)}} + \underbrace{\sum_{i=0}^{N-1} \{\tilde{\nu}_i - \Delta S(2, m_i)\}^2}_{\substack{\text{1st overtone} \\ (v=2)}}$$

$$+ \underbrace{\sum_{i=0}^{N-1} \{\tilde{\nu}_i - \Delta S(3, m_i)\}^2}_{\substack{\text{2nd overtone} \\ (v=3)}}$$

Powerful mathematical software such as the Given/Minerr function of Mathcad is available for performing the minimization.

It is found that $\boxed{\tilde{\nu} = 2169.3 \, \text{cm}^{-1}}$, $\boxed{x_e \tilde{\nu} = 13.14 \, \text{cm}^{-1}}$,

$\boxed{B = 1.930 \, \text{cm}^{-1}}$, $\boxed{a = 0.01747 \, \text{cm}^{-1}}$, and $\boxed{D_J = 4.178 \times 10^{-6} \, \text{cm}^{-1}}$.

(c) $I_e = \dfrac{h}{8\pi^2 cB} = \boxed{1.450 \times 10^{-46}\,\text{kg m}^2}$ [16.31]

$$R_e = \sqrt{\dfrac{I}{\mu}} = \sqrt{\dfrac{1.450 \times 10^{-46}\,\text{kg m}^2}{1.139 \times 10^{-26}\,\text{kg}}} = \boxed{112.8\,\text{pm}}$$ [Table 16.1]

Centrifugal distortion increases the bond length and moment of inertia. Bond lengths and moments of inertia for each vibrational state may be calculated by subtracting the centrifugal distortion factor $a(v = 1/2)$ from the rotational constant.

$$I(v) = \dfrac{h}{8\pi^2 c\{B - a(v + 1/2)\}} \quad \text{and} \quad R(v) - \sqrt{\dfrac{I(v)}{\mu}}$$

v	$I(v)/(10^{-46}\,\text{kg m}^2)$	$R(v)/\text{pm}$
0	1.457	113.1
1	1.470	113.6
2	1.484	114.1
3	1.498	114.7

$$D_e = \dfrac{\tilde{v}^2}{4x_e\tilde{v}} = \boxed{89533\,\text{cm}^{-1}}$$ [16.62]

$D_0 = D_e - \text{zero point energy}$

$$= D_e - \left\{\tfrac{1}{2}\tilde{v} - \tfrac{1}{4}x_e\tilde{v}\right\} = \boxed{88452\,\text{cm}^{-1}}$$

P16.20   **(a)** Follow the flow chart in Fig. 15.14. CH_3Cl is not linear, it has a C_3 axis (only one), it does not have C_2 axes perpendicular to C_3, it has no σ_h, but does have 3 σ_v planes; so it belongs to $\boxed{C_{3v}}$.

(b) The number of normal modes of a non-linear molecule is $3N - 6$, where N is the number of atoms. So CH_3Cl has $\boxed{\text{nine}}$ normal modes.

(c) To determine the symmetry of the normal modes, consider how the cartesian axes of each atom are transformed under the symmetry operations of the C_{3v} group; the 15 cartesian displacements constitute the basis here. All 15 cartesian axes are left unchanged under the identity, so the character of this operation is 15. Under a C_3 operation, the H atoms are taken into each other, so they do not contribute to the character of C_3. The z axes of the C and Cl atoms, are unchanged, so they contribute 2 to the character of C_3; for these two atoms

$$x \to -\dfrac{x}{2} + \dfrac{3^{1/2}y}{2} \quad \text{and} \quad y \to -\dfrac{y}{2} + \dfrac{3^{1/2}x}{2},$$

so there is a contribution of $-1/2$ to the character from each of these coordinates in each of these atoms. In total, then $\chi = 0$ for C_3. To find the character of σ_v, call one of the σ_v planes the yz plane; it contains C, Cl, and one H atom. The y and z coordinates of these three atoms are unchanged, but the x coordinates are taken into their negatives, contributing $6 - 3 = 3$ to the character for this operation; the other two atoms are interchanged, so they contribute nothing to the character. To find the irreproducible representations that this basis spans, we multiply its characters by the characters of the irreproducible representations, sum those products, and divide the sum by the order h of the group (as in section 15.5(a)). The table below illustrates the procedure.

	E	$2C_2$	$3\sigma_v$		E	$2C_2$	$3\sigma_v$	sum/h
basis	15	0	3					
A_1	1	1	1	basis × A_1	15	0	3	3
A_2	1	1	-1	basis × A_2	15	0	-3	2
E	2	-1	0	basis × E	30	0	0	5

Of these 15 modes of motion, 3 are translations (an A_1 and an E) and 3 rotations (an A_2 and an E); we subtract these to leave the vibrations, which span

$2A_1 + A_2 + 3E$ (two A_1 modes, one A_2 mode, and 3 doubly-degenerate E modes).

(d) Any mode whose symmetry species is the same as that of x, y, or z is infrared active. Thus all but the A_2 mode are infrared active.

(e) Only modes whose symmetry species is the same as a quadratic form may be Raman active. Thus all but the A_2 mode may be Raman active.

Solutions to theoretical problems

P16.24   Refer to the flow chart in Fig. 15.14. Yes at the first question (linear?) leads to linear point groups and therefore linear rotors. If the molecule is not linear, then yes at the next question (two or more C_n with $n > 2$?) leads to cubic and icosahedral groups and therefore spherical rotors. If the molecule is not a spherical rotor, yes at the next question leads to symmetric rotors if the highest C_n has $n > 2$; if not, the molecule is an asymmetric rotor.

(a) CH_4: not linear, but more than two $C_n (n > 2)$, so spherical rotor.

(b) CH_3CN: not linear, C_3 (only one of them), so symmetric rotor.

(c) CO_2: linear, so linear rotor.

(d) CH_3OH: not linear, no C_n, so asymmetric rotor.

(e) benzene: not linear, C_6, but only one high-order axis, so symmetric rotor.

(f) pyridine: not linear, C_2 is highest rotational axis, so a symmetric rotor.

P16.25 $N \propto g e^{-E/kT}$ [Boltzmann distribution, Chapters 2 and 19]

$N_J \propto g_J e^{-Eg/kT} \propto (2J + 1)e^{-hcBJ(J+1)/kT}$ [$g_J = 2J + 1$ for a diatomic rotor]

The maximum population occurs when

$$\frac{d}{dJ}N_J \propto \left\{ 2 - (2J + 1)^2 \times \left(\frac{hcB}{kT} \right) \right\} e^{-hcBJ(J+1)/kT} = 0$$

and, since the exponential can never be zero at a finite temperature, when

$$(2J + 1)^2 \times \left(\frac{hcB}{kT} \right) = 2$$

or when $J_{max} = \left(\frac{kT}{2hcB} \right)^{1/2} - \frac{1}{2}$

For ICl, with $\dfrac{kT}{hc} = 207.22 \text{ cm}^{-1}$ (inside front cover)

$$J_{max} = \left(\frac{207.22 \text{ cm}^{-1}}{0.2284 \text{ cm}^{-1}} \right)^{1/2} - \frac{1}{2} = \boxed{30}$$

For a spherical rotor, $N_J \propto (2J + 1)^2 e^{-hcBJ(J+1)/kT}$ [$g_J = (2J + 1)^2$]

and the greatest population occurs when

$$\frac{dN_J}{dJ} \propto \left(8J + 4 - \frac{hcB(2J+1)^3}{kT}\right) e^{-hcBJ(J+1)/kT} = 0$$

which occurs when

$$4(2J+1) = \frac{hcB(2J+1)^3}{kT}$$

or at $J_{max} = \boxed{\left(\frac{kT}{hcB}\right)^{1/2} - \frac{1}{2}}$

For CH$_4$, $J_{max} = \left(\frac{207.22\,\text{cm}^{-1}}{5.24\,\text{cm}^{-1}}\right)^{1/2} - \frac{1}{2} = \boxed{6}$

P16.27   The $1000\,\text{cm}^{-1}$ infrared absorbance is due to the vibration state transition $1, \text{g} \leftarrow 0, \text{u}$. This is a mode in which the nitrogen atom moves toward the plane of the hydrogens while the plane moves toward the nitrogen. Because of the finite height of the barrier, the wavefunctions for this vibrational mode have small but non-zero values within the barrier. In this situation it is possible for the mode, which has less kinetic energy than does the barrier, to quantum mechanically "tunnel" through the barrier. The resulting inversion, as shown in the figure, is the passage from the left well, which has the wavefunction ψ_L, to the right well, which has the wavefunction ψ_R. The total wavefunction for each vibrational state of the mode, ψ_υ, must be either symmetrical or antisymmetrical to inversion – so by making an analogy to the discussions eqns 14.7 and 14.29 we conclude that there are actually two nearly degenerate states for each vibrational state υ. Neglecting overlap in the normalization constant, they are:

$$\psi_{\upsilon,\text{plus}} = \frac{1}{\sqrt{2}}(\psi_{L\upsilon}(s) + \psi_{R\upsilon}(s)) \qquad \psi_{\upsilon,\text{minus}} = \frac{1}{\sqrt{2}}(\psi_{L\upsilon}(s) + \psi_{R\upsilon}(s))$$

The energy difference between these two states is $0.8\,\text{cm}^{-1}$ for $\upsilon = 0$ and $36\,\text{cm}^{-1}$ for $\upsilon = 1$. These are very small differences which appear in the microwave. The relative energies of these states are shown in the figure.

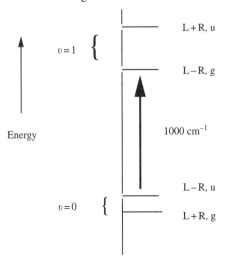

(a) The criteria that $V(s_e) = V(-s_e) = 0$ and $dV(s_e)/ds = dV(-s_e)/ds = 0$ provide two equations that may be solved for $C(B)$ and $D(B)$.

$$\frac{dV(s)}{ds} = \frac{-2A}{D} \operatorname{sech}^2 \left(\frac{s}{D} \right) \times \tanh \left(\frac{s}{D} \right) \times \left(2B \operatorname{sech}^2 \left(\frac{s}{D} \right) - C \right)$$

Only the last factor can equal zero at $s = s_e$, so $\operatorname{sech}^2 \left(\frac{s_e}{D} \right) = \frac{C}{2B}$

additionally, $V(s_e) = 0 = A \left(1 + B \operatorname{sech}^4 \left(\frac{s_e}{D} \right) - C \operatorname{sech}^2 \left(\frac{s_e}{D} \right) \right) = A \left(1 + \frac{C^2}{4B} - \frac{C^2}{2B} \right)$ or

$$\boxed{C = 2\sqrt{B}}$$

Substitution of the latter equation into the former gives:

$$\operatorname{sech}^2 \left(\frac{s_e}{D} \right) = \frac{C}{2B} = \frac{1}{\sqrt{B}} \quad \text{or} \quad D = \frac{s_e}{\operatorname{arcsech}(B^{-1/4})} = \boxed{\frac{s_e}{\operatorname{arccosh}(B^{1/4})} = C}$$

(b) Two equations are required for the determination of the potential energy parameters A and B. Parameters C and D are evaluated with the equations of part **(a)**. The equations are:

$$E_{1,\text{minus}} - E_{0,\text{minus}} = 1000 \, \text{cm}^{-1} hc \quad \text{and}$$

$$E_{1,\text{plus}} - E_{1,\text{minus}} = 36 \, \text{cm}^{-1} hc$$

Use mathematical software to setup integrals that will calculate each of the energy terms of the above two expressions. The integrals will be functions of A and B. A Given/Find solve block (Mathcad) or a numerical solver can then be used to adjust the values of A and B until the above two constraints are satisfied.

$$\text{For example, } E_{1,\text{plus}} = \frac{\int \psi_{1,\text{plus}} H \psi_{1,\text{plus}} \, d\tau}{\int \psi_{1,\text{plus}}^2 \, d\tau}$$

where $H = \dfrac{-h^2}{8\pi^2 m_{\text{eff}}} \dfrac{d^2}{ds^2} + V(s, A, B)$

and $V(s, A, B) = A \left(1 + B \operatorname{sech}^4 \left(\frac{s}{D(B)} \right) - C(B) \operatorname{sech}^2 \left(\frac{s}{D(B)} \right) \right)$

Use harmonic oscillator wavefuntions (eqn 12.35) centered at $s = s_e (= 38.1 \, \text{pm})$ and $s = -s_e$ to construct an approximate $\psi_{1,\text{plus}}$.

The method yields: $\boxed{A = 46119 \, \text{cm}^{-1}, B = 1.4275, C = 2.3896, \text{ and } D = 88.99 \, \text{pm}}$

The height of the inversion barrier is given by

$$\varepsilon = V(0, A, B) = \boxed{1750 \, \text{cm}^{-1} = \varepsilon}$$

Solutions to applications

P16.30   **(a)** In order to absorb infrared radiation, a molecule must either have a permanent dipole moment or its dipole must change during vibration (Sections 16.2 and 16.10). $N_2(g)$ and $O_2(g)$ have neither of these properties and are infrared inactive. Water has a permanent dipole (Section 16.15) and three infrared active normal modes at $1595 \, \text{cm}^{-1}$, $3652 \, \text{cm}^{-1}$, and $3756 \, \text{cm}^{-1}$. Carbon dioxide has no permanent dipole (Fig. 16.48) but it does have three normal modes which change the dipole moment during vibration. There is the antisymmetrical stretch at $2349\,\text{cm}^{-1}$ and a doubly degenerate bending mode at $667\,\text{cm}^{-1}$.

(b) The absorption bands are broad because vibration–rotation transitions with lifetime or pressure broadening (Section 16.3).

(c)

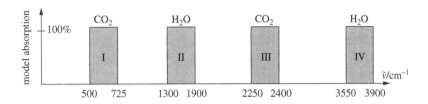

In Application Problem 11.27 it is shown that the wavenumber distribution of black-body radiation is given by

$$f(T, \tilde{\nu}) = \frac{8\pi hc\tilde{\nu}^3}{e^{hc\tilde{\nu}/kT} - 1}$$

The total energy flux emitted by a black-body at temperature T in a band from $\tilde{\nu}_{\text{low}}$ to $\tilde{\nu}_{\text{high}}$ is

$$J_{\text{band}}(T) = \int_{\text{band}} dJ = \frac{C}{4} \int_{\text{band}} dE = \frac{C}{4} \int_{\tilde{\nu}_{\text{low}}}^{\tilde{\nu}_{\text{high}}} f(T, \tilde{\nu})\, d\tilde{\nu}$$

The above equation gives the energy absorbed by each one of the four absorption bands in the model. The total energy absorbed by the gases per unit surface area of the Earth is

$$J_{\text{abs}}(T) = \sum_{i=1}^{4} J_{\text{band i}}(T)$$

(c) Black-body radiation absorption by the atmosphere decreases the radiative energy loss by the Earth's surface. The energy balance between the Earth's (radius R) absorption of solar radiation and the Earth's energy emission to space is:

total solar energy flux absorption by Earth = total black-body emission at T by Earth

$$\pi R^2(1 - \text{albedo})\,(\text{total solar energy flux}) = 4\pi R^2(\sigma T^4 - J_{\text{abs}}(T))$$

$$\boxed{(1 - \text{albedo})\,(\text{total solar energy flux}) = 4(\sigma T^4 - J_{\text{abs}}(T))}$$

where albedo = 0.29 and solar flux = $0.1353\,\text{W cm}^{-2}$. Mathematical software can be used to find the value of T that makes the left and right sides of the above equation equal. A numerical solver or the Given/Find solve block of Mathcad can be used to find that $\boxed{T = 281.88\text{K}}$. Taking the Earth's average temperature to be 288.16 K and theoretical temperature in the absence of the greenhouse effect to be 255.1 K (Problem 11.27), this model explains 81.1% of the greenhouse effect. $\dfrac{(281.9 - 255.1)100}{(288.16 - 255.1)} = \boxed{81.1\%}$

P16.32 According to Problem 13.28(a), the Doppler effect obeys

$$\nu_{\text{receding}} = \nu f \quad \text{where } f = \left(\frac{1 - s/c}{1 + s/c}\right)^{1/2}.$$

This can be rearranged to yield:

$$s = \frac{1 - f^2}{1 + f^2} c.$$

We are given wavelength data, so we use:

$$f = \frac{\nu_{\text{star}}}{\nu} = \frac{\lambda}{\lambda_{\text{star}}}.$$

The ratio is:

$$f = \frac{654.2\,\text{nm}}{706.5\,\text{nm}} = 0.9260,$$

so $s = \dfrac{1 - 0.9260^2}{1 + 0.9260^2} c = \boxed{0.0768c} = 2.30 \times 10^7\,\text{ms}^{-1}$

The broadening of the line is due to local events (collisions) in the distant star. It is temperature dependent and hence yields the surface temperature of the star. Eqn 16.24 relates the observed linewidth to temperature:

$$\delta\lambda_{\text{obs}} = \frac{2\lambda}{c}\left(\frac{2kT\ln 2}{m}\right)^{1/2} \quad \text{so} \quad T = \left(\frac{c\,\delta\lambda}{2\lambda}\right)^2 \frac{m}{2k\ln 2},$$

$$T = \left(\frac{(2.998 \times 10^8\,\text{m s}^{-1})(61.8 \times 10^{-12}\,\text{m})}{2(654.2 \times 10^{-9})}\right)^2 \left[\frac{(47.95\,\text{u})(1.661 \times 10^{-27}\,\text{kg u}^{-1})}{2(1.381 \times 10^{-23}\,\text{J K}^{-1})\ln 2}\right],$$

$$T = \boxed{8.34 \times 10^5\,\text{K}}.$$

P16.33

$$E_J = J(J+1)hcB, \quad g_J = 2J + 1$$

$$E_1 - E_0 = 2hcB = hc\left(\frac{1}{\lambda_{\text{shorter}}} - \frac{1}{\lambda_{\text{longer}}}\right)$$

$$B = \frac{1}{2}\left(\frac{1}{\lambda_{\text{shorter}}} - \frac{1}{\lambda_{\text{longer}}}\right) = \frac{1}{2}\left(\frac{1}{\lambda_{\text{shorter}}} - \frac{1}{\lambda_{\text{shorter}} + \Delta\lambda}\right)$$

$$= \frac{1}{2}\left(\frac{1}{\lambda_{\text{shorter}}}\right) \times \left(1 - \frac{1}{1 + \frac{\Delta\lambda}{\lambda_{\text{shorter}}}}\right)$$

$$= \frac{1}{2}\left(\frac{1}{387.5\,\text{nm}}\right) \times \left(1 - \frac{1}{1 + \frac{0.061}{387.5}}\right) \times \left(\frac{10^9\,\text{nm}}{10^2\,\text{cm}}\right)$$

$$B = \boxed{2.031\,\text{cm}^{-1}}$$

$$\frac{E_1 - E_0}{k} = \frac{2hcB}{k} = \frac{2(6.626 \times 10^{-34}\,\text{J s}) \times (3.00 \times 10^{10}\,\text{cm s}^{-1}) \times (2.031\,\text{cm}^{-1})}{1.381 \times 10^{-23}\,\text{J K}^{-1}}$$

$$= 5.84\overline{7}\,\text{K}$$

Intensity of $J' \leftarrow J$ absorption line $I_J \propto g J e^{-E_J/kT}$

$$\frac{I_{\lambda_{\text{longer}}}}{I_{I_{\lambda_{\text{shorter}}}}} \simeq \frac{g_1 e^{-E_1/kT}}{g_0 e^{-E_0/kT}} = \frac{g_1}{g_0} e^{-(E_1-E_0)/kT}$$

Solve for T

$$T = \left(\frac{E_1 - E_0}{k}\right) \times \left(\frac{1}{\ln\left(\frac{g_1 \, I_{\lambda_{\text{shorter}}}}{g_0 I_{\lambda_{\text{longer}}}}\right)}\right) = 5.84\bar{7}\,\text{K} \left(\frac{1}{\ln(3 \times 4)}\right) = \boxed{2.35\,\text{K}}$$

17 Spectroscopy 2: electronic transitions

Solutions to exercises

Discussion questions

E17.1(a) The process of the determination of the term symbol for dioxygen, $^3\Sigma_g^-$, is described in Section 17.1(b) and will not be repeated here. The interpretation of the symbol follows: the letter Σ means that the magnitude of the total orbital angular momentum about the internuclear axis is 0; the left superscript 3 means that the component of the total spin angular momentum about the internuclear axis is $1 (2 \times 1 + 1 = 3)$; the subscript g means that the parity of the term is even; and the superscript $-$ means that the molecular wavefunction for O_2 changes sign upon reflection in the plane containing the nuclei.

E17.2(a) A band head is the convergence of the frequencies of electronic transitions with increasing rotational quantum number, J. They result from the rotational structure superimposed on the vibrational structure of the electronic energy levels of the diatomic molecule. See Figs 17.8 and 17.11. To understand how a band head arises, one must examine the equations describing the transition frequencies (eqns 17.5). As seen from the analysis in Section 17.1(f), convergence can only arise when terms in both $(B' - B)$ and $(B' + B)$ occur in the equation. Since only a term in $(B' - B)$ occurs for the Q branch, no band head can arise for that branch.

E17.3(a) The overall process associated with fluorescence involves the following steps. The molecule is first promoted from the vibrational ground state of a lower electronic level to a higher vibrational-electronic energy level by absorption of energy from a radiation field. Because of the requirements of the Franck-Condon principle, the transition is to excited vibrational levels of the upper electronic state. See Fig. 17.16. Therefore, the absorption spectrum shows a vibrational structure characteristic of the upper state. The excited state molecule can now lose energy to the surroundings through radiationless transitions and decay to the lowest vibrational level of the upper state. A spontaneous radiative transition now occurs to the lower electronic level and this fluorescence spectrum has a vibrational structure characteristic of the lower state. The fluorescence spectrum is not the mirror image of the absorption spectrum because the vibrational frequencies of the upper and lower states are different due to the difference in their potential energy curves.

E17.4(a) See Section 17.5 for a detailed description of both the theory and experiment involved in laser action. Here we restrict our discussion to only the most fundamental concepts. The basic requirement for a laser is that it have at least three energy levels. Of these levels, the highest lying state must be capable of being efficiently populated above its thermal equilibrium value by a pulse of radiation. A second state, lower in energy, must be a metastable state with a long enough lifetime for it to accumulate a population greater than its thermal equilibrium value by spontaneous transitions from the higher overpopulated state.

The metastable state must than be capable of undergoing stimulated transitions to a third lower lying state. This last requirements implies not only that the metastable state have more than its thermal equilibrium population, but also that it have a higher population than the third lower lying state, namely that it achieve population inversion. See Figs 17.23 and 17.24 for a description of the three- and four-level laser. The amplification process occurs when low intensity radiation of frequency equal to the transition frequency between the metastable state and the lower lying state stimulates the transition to the lower lying state and many more photons (higher intensity of the radiation) of that frequency are created. Examples of practical lasers are listed and discussed in Section 17.6.

Numerical exercises

E17.5(a) Refer to Fig. 17.2 of the text.

(a) π^* is gerade, $\boxed{g}$

(b) g, u is inapplicable to a heteronuclear molecule, for it has no centre of inversion.

(c) A δ orbital (Fig. 17.1(a)) is gerade, $\boxed{g}$

(d) A δ^* orbital (Fig. 17.1(b)) is ungerade, $\boxed{u}$

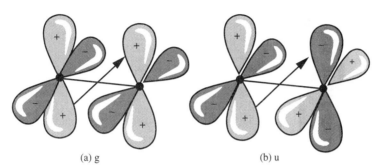

(a) g (b) u **Figure 17.1**

E17.6(a) The left superscript is the value of $2S + 1$, so $2S + 1 = 2$ implies that $\boxed{S = \frac{1}{2}}$. The symbol Σ indicates that the total orbital angular momentum around the molecular axis is $\boxed{\text{zero}}$. The latter implies that the unpaired electron must be in a σ orbital. From Fig. 14.30 of the text, we predict the configuration of the ion to be $1\sigma^2 2\sigma^{*2} 1\pi^4 3\sigma^1$, which is in accord with the $^2\Sigma_g$ term symbol since 3σ is an even function (Fig. 17.2) and all lower energy orbitals are filled, leaving one unpaired electron; thus $S = \frac{1}{2}$.

E17.7(a) The reduction in intensity obeys the Beer–Lambert law introduced in Chapter 16. It applies equally well to the spectroscopic methods of this chapter.

$$\log \frac{I}{I_0} = -\varepsilon[\text{J}]l \ [16.10 \text{ and } 16.11]$$

$$= (-855 \text{ L mol}^{-1} \text{ cm}^{-1}) \times (3.25 \times 10^{-3} \text{ mol L}^{-1}) \times (0.25 \text{ cm})$$

$$= -0.69\bar{5}$$

Hence, $\frac{I}{I_0} = 0.20$, and the reduction in intensity is $\boxed{80 \text{ per cent}}$

E17.8(a) $\log \dfrac{I}{I_0} = -\varepsilon[\text{J}]l \ [16.10, 16.11]$

Hence, $\varepsilon = -\dfrac{1}{[\text{J}]l} \log \dfrac{I}{I_0} = -\dfrac{\log 0.201}{(1.11 \times 10^{-4} \text{ mol L}^{-1}) \times (1.00 \text{ cm})} = \boxed{6.28 \times 10^3 \text{ L mol}^{-1} \text{ cm}^{-1}}$

E17.9(a) $[\text{J}] = -\dfrac{1}{\varepsilon l} \log \dfrac{I}{I_0} [16.10, 16.11] = \dfrac{-1}{(286 \text{ L mol}^{-1} \text{ cm}^{-1}) \times (0.65 \text{ cm})} \log (1 - 0.465)$

$= \boxed{1.5 \text{ mmol L}^{-1}}$

E17.10(a) $A = \int \varepsilon \, d\tilde{\nu}$ [16.12]

The integral can be approximated by the area under the triangle [area = $\frac{1}{2}$ × base × height]

$$A = \tfrac{1}{2} \times (43480 - 34480)\,\text{cm}^{-1} \times (1.21 \times 10^4\,\text{L mol}^{-1}\,\text{cm}^{-1})$$

$$= \boxed{5.44 \times 10^7\,\text{L mol}^{-1}\,\text{cm}^{-2}}.$$

E17.11(a) π-electrons in polyenes may be considered as particles in a one-dimensional box. Applying the Pauli exclusion principle, the N conjugated electrons will fill the levels, two electrons at a time, up to the level $n = \dfrac{N}{2}$. Since N is also the number of alkene carbon atoms. Nd is the length of the box, with d the carbon–carbon interatomic distance. Hence

$$E_n = \frac{n^2 h^2}{8m N^2 d^2}$$

where, for the lowest energy transition ($\Delta n = +1$)

$$\Delta E = h\nu = \frac{hc}{\lambda} = E_{(N/2)+1} - E_{(N/2)} = \frac{h^2(N+1)}{8md^2 N^2}$$

Therefore, the larger N, the larger λ. Hence the absorption at 243 nm is due to the diene and that at 192 nm to the butene.

Question. How accurate is the formula derived above in predicting the wavelengths of the absorption maxima in these two compounds?

E17.12(a) $\varepsilon = -\dfrac{1}{[\text{J}]l} \log \dfrac{I}{I_0}$ [16.10, 16.11]   with $l = 0.20\,\text{cm}$

We use this formula to draw up the following table

$[\text{Br}_2]/\text{mol L}^{-1}$	0.0010	0.0050	0.0100	0.0500	
I/I_0	0.814	0.356	0.127	3.0×10^{-5}	
$\varepsilon/(\text{L mol}^{-1}\,\text{cm}^{-1})$	447	449	448	452	mean: $44\overline{9}$

Hence, the molar absorption coefficient is $\varepsilon = \boxed{450\,\text{L mol}^{-1}\,\text{cm}^{-1}}$

E17.13(a) $\varepsilon = -\dfrac{1}{[\text{J}]l} \log \dfrac{I}{I_0}$ [16.10, 16.11] $= \dfrac{-1}{(0.010\,\text{mol L}^{-1}) \times (0.20\,\text{cm})} \log 0.48 = \boxed{159\,\text{L mol}^{-1}\,\text{cm}^{-1}}$

$$T = \frac{I}{I_0} = 10^{-[\text{J}]\varepsilon l}$$

$$= 10^{(-0.010\,\text{mol L}^{-1}) \times (159\,\text{L mol}^{-1}\,\text{cm}^{-1}) \times (0.40)} = 10^{-0.63\overline{6}} = 0.23, \text{ or } \boxed{23\,\text{per cent}}.$$

E17.14(a) $l = \dfrac{-1}{\varepsilon[\text{J}]} \log \dfrac{I}{I_0}$

For water, $[\text{H}_2\text{O}] \approx \dfrac{1.00\,\text{kg/L}}{18.02\,\text{g mol}^{-1}} = 55.5\,\text{mol L}^{-1}$

and $\varepsilon[\text{J}] = (55.5\,\text{M}) \times (6.2 \times 10^{-5}\,\text{M}^{-1}\,\text{cm}^{-1}) = 3.4 \times 10^{-3}\,\text{cm}^{-1} = 0.34\,\text{m}^{-1}$, so $\dfrac{1}{\varepsilon[\text{J}]} = 2.9\,\text{nm}$

Hence, $l/\text{m} = -2.9 \times \log \dfrac{I}{I_0}$

(a) $\dfrac{I}{I_0} = 0.5$, $l = -2.9\,\text{m} \times \log 0.5 = \boxed{0.9\,\text{m}}$ **(b)** $\dfrac{I}{I_0} = 0.1$, $l = -2.9\,\text{m} \times \log 0.1 = \boxed{3\,\text{m}}$

E17.15(a) We will make the same assumption as in Exercise 17.10(a) namely that the absorption curve can be approximated by a triangle. Refer to Fig. 17.2.

$$\mathcal{A} = \int \varepsilon \, d\tilde{\nu} \; [16.12]$$

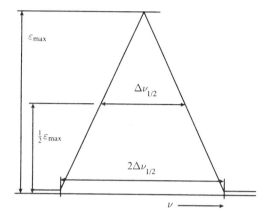

Figure 17.2

From the illustration,

$$\mathcal{A} = \tfrac{1}{2} \times \varepsilon_{\max} \times 2\Delta\tilde{\nu}_{1/2} \; [\text{area} = \tfrac{1}{2} \times \text{height} \times \text{base}] = \varepsilon_{\max}\Delta\tilde{\nu}_{1/2}$$

$$\mathcal{A} = 5000\,\text{cm}^{-1} \times \varepsilon_{\max}$$

(a) $\mathcal{A} = 5000\,\text{cm}^{-1} \times 1 \times 10^4\,\text{L mol}^{-1}\,\text{cm}^{-1} = \boxed{5 \times 10^7\,\text{L mol}^{-1}\,\text{cm}^{-2}}$

(b) $\mathcal{A} = (5000\,\text{cm}^{-1}) \times (5 \times 10^2\,\text{L mol}^{-1}\,\text{cm}^{-1}) = \boxed{25 \times 10^5\,\text{L mol}^{-1}\,\text{cm}^{-2}}$

E17.16(a) The internuclear distance in H_2^+ is greater than that in H_2. The change in bond length and the corresponding shift in the molecular potential energy curves reduces the Franck–Condon factor for transitions between the two ground vibrational states. It creates a better overlap between $v = 2$ of H_2^+ and $v = 0$ of H_2, and so increases the Franck–Condon factor of that transition.

Solutions to problems

Solutions to numerical problems

P17.1 The potential energy curves for the $X^3\Sigma_g^-$ and $B^3\Sigma_u^-$ electronic states of O_2 are represented schematically in Fig. 17.3 along with the notation used to represent the energy separation of this problem. Curves for the other electronic state of O_2 are not shown. Ignoring rotational structure and anharmonicity we may write

$$\tilde{\nu}_{00} \approx T_e + \tfrac{1}{2}(\tilde{\nu}' - \tilde{\nu}) = 6.175\,\text{eV} \times \left(\frac{8065.5\,\text{cm}^{-1}}{1\,\text{eV}}\right) + \tfrac{1}{2}(700 - 1580)\,\text{cm}^{-1}$$

$$= \boxed{49\,364\,\text{cm}^{-1}}$$

Comment. Note that the selection rule $\Delta v = \pm 1$ does not apply to vibrational transitions between different electronic states.

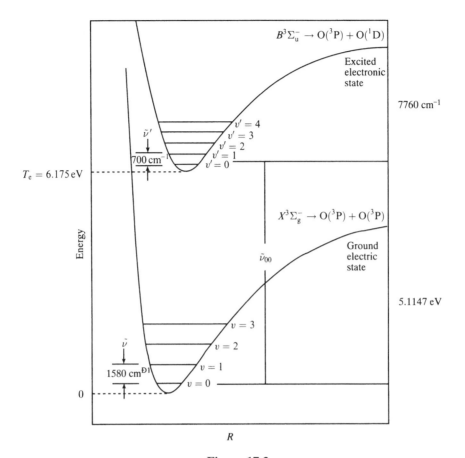

Figure 17.3

Question. What is the percentage change in $\tilde{\nu}_{00}$ if the anharmonicity constants $x_e\tilde{\nu}$ (Section 16.11) 12.0730 cm^{-1} and 8.002 cm^{-1} for the ground and excited states, respectively, are included in the analysis?

P17.2 The energy of the dissociation products of the B state, $O(^3P)$ and $O(^1D)$, above the $v = 0$ state of the ground state is $7760\,\text{cm}^{-1} + 49\,363\,\text{cm}^{-1} = 57\,123\,\text{cm}^{-1}$. One of these products, $O(^1D)$, has energy $15\,870\,\text{cm}^{-1}$ above the energy of the ground-state atom, $O(^3P)$. Hence, the energy of two ground-state atoms, $2O(^3P)$, above the $v = 0$ state of the ground electronic state is $57\,123\,\text{cm}^{-1} - 15\,870\,\text{cm}^{-1} = 41\,253\,\text{cm}^{-1} = \boxed{5.1147\,\text{eV}}$. These energy relations are indicated (not to scale) in Fig. 17.3.

P17.4 We write $\varepsilon = \varepsilon_{\max}e^{-x^2} = \varepsilon_{\max}e^{-\tilde{\nu}^2/2\Gamma}$, the variable being $\tilde{\nu}$ and Γ being a constant. $\tilde{\nu}$ is measured from the band centre, at which $\tilde{\nu} = 0$, $\varepsilon = \frac{1}{2}\varepsilon_{\max}$ when $\tilde{\nu}^2 = 2\Gamma \ln 2$.

Therefore, the width at half height is

$$\Delta\tilde{\nu}_{1/2} = 2 \times (2\Gamma \ln 2)^{1/2}, \quad \text{implying that} \quad \Gamma = \frac{\Delta\tilde{\nu}_{1/2}^2}{8 \ln 2}$$

Now we carry out the integration

$$\mathcal{A} = \int \varepsilon \, d\tilde{\nu} = \varepsilon_{\max} \int_{-\infty}^{\infty} e^{-\tilde{\nu}^2/2\Gamma} \, d\tilde{\nu} = \varepsilon_{\max}(2\Gamma\pi)^{1/2} \quad \left[\int_{-\infty}^{\infty} e^{-x^2} \, dx = \pi^{1/2}\right]$$

$$= \varepsilon_{\max} \left(\frac{2\pi \Delta\tilde{\nu}_{1/2}^2}{8 \ln 2}\right)^{1/2} = \left(\frac{\pi}{4 \ln 2}\right)^{1/2} \varepsilon_{\max} \Delta\tilde{\nu}_{1/2} = 1.0645\,\varepsilon_{\max} \Delta\tilde{\nu}_{1/2}$$

From Fig. 17.51 of the text we estimate $\varepsilon_{max} \approx 9.5 \, L \, mol^{-1} \, cm^{-1}$ and $\Delta \tilde{\nu}_{1/2} \approx 4760 \, cm^{-1}$. Then

$$A = 1.0645 \times (9.5 \, L \, mol^{-1} \, cm^{-1}) \times (4760 \, cm^{-1}) = \boxed{4.8 \times 10^4 \, L \, mol^{-1} \, cm^{-2}}$$

The area under the curve on the printed page is about $1288 \, mm^2$, each mm^2 corresponds to about $190.5 \, cm^{-1} \times 0.189 \, L \, mol^{-1} \, cm^{-1}$, and so $\int \varepsilon \, d\tilde{\nu} \approx 4.64 \times 10^4 \, L \, mol^{-1} \, cm^{-2}$. The agreement with the calculated value above is good.

P17.6   For a photon to induce a spectroscopic transition, the transition moment $\langle \mu \rangle$ must be nonzero. The Laporte selection rule forbids transitions that involve no change in parity. So transitions to the Π_u states are forbidden. (Note. These states may not even be reached by a vibronic transition, for these molecules have only one vibrational mode and it is centrosymmetric.)

We will judge transitions to the other states with the assistance of the $D_{\infty h}$ character table. The transition moment is the integral $\int \psi_f^* \mu \psi_i \, d\tau$, where the dipole moment operator has components proportional to the Cartesian coordinates. The integral vanishes unless the integrand, or at least some part of it, belongs to the totally symmetric representation of the molecule's point group. To find the character of the integrand, we multiply together the characters of its factors. Note that the μ_z has the same symmetry species as the ground state, namely A_{1u}, and the product of the ground state and μ_z has the A_{1g} symmetry species; since the symmetry species are mutually orthogonal, only a state with A_{1g} symmetry can be reached from the ground state with z-polarized light. The $^2\Sigma_g^+$ state is such a state, so $\boxed{^2\Sigma_g^+ \leftarrow \, ^2\Sigma_u^+ \text{ is allowed.}}$ That leaves x- or y-polarized transitions to the $^2\Pi_g$ states to consider.

	E	$\infty C_2'$	$2C_\phi$	i	$\infty \sigma_v$	$2S_\phi$
$\Sigma_u^+(A_{1u})$	1	-1	1	-1	1	-1
μ_x or $_y(E_{1u})$	2	0	$2\cos\phi$	-2	0	$2\cos\phi$
$\Pi_g(E_{1g})$	2	0	$2\cos\phi$	2	0	$-2\cos\phi$
Integrand	4	0	$4\cos^2\phi$	4	0	$4\cos^2\phi$

The little orthogonality theorem (see the solution to Problem 15.18) gives the coefficient of A_{1g} in the integrand as

$$c_{A_{1g}} = (1/h) \Sigma_c g(C) \chi(C) = [4 + 0 + 2(4\cos^2\phi) + 4 + 0 + 2(4\cos^2\phi)]/\infty = 0.$$

So the integrand does not contain A_{1g}, and the $\boxed{\text{transition to } ^2\Pi_g \text{ would be forbidden.}}$

P17.7   We use the technique described in Example 16.5, the Birge–Sponer extrapolation method, and plot the difference $\Delta \tilde{\nu}_v$ against $v + \frac{1}{2}$.

We then draw up the following table:

$\Delta \tilde{\nu}_v$	688.0	665.1	641.5	617.6	591.8	561.2	534.0
$v + \dfrac{1}{2}$	$\dfrac{1}{2}$	$\dfrac{3}{2}$	$\dfrac{5}{2}$	$\dfrac{7}{2}$	$\dfrac{9}{2}$	$\dfrac{11}{2}$	$\dfrac{13}{2}$

$\Delta \tilde{\nu}_v$	502.1	465.5	428.9	388.2	343.1	300.9	255.0
$v + \dfrac{1}{2}$	$\dfrac{15}{2}$	$\dfrac{17}{2}$	$\dfrac{19}{2}$	$\dfrac{21}{2}$	$\dfrac{23}{2}$	$\dfrac{25}{2}$	$\dfrac{27}{2}$

The data are plotted in Fig. 17.4. Each square corresponds to $25 \, cm^{-1}$. The area under the non-linear extrapolated line is 295 squares; therefore the dissociation energy is $7375 \, cm^{-1}$. The $^3\Sigma_u^- \leftarrow X$

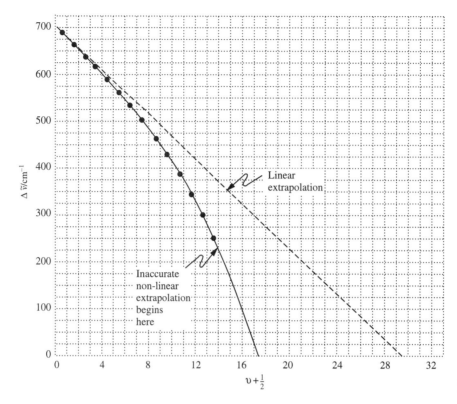

Figure 17.4

excitation energy (where X denotes the ground state) to $v = 0$ is $49\,357.6\,\text{cm}^{-1}$ which corresponds to $6.12\,\text{eV}$. The $^3\Sigma_u^-$ dissociation energy for

$$O_2(^3\Sigma_u^-) \rightarrow O + O^*$$

is $7375\,\text{cm}^{-1}$, or $0.91\,\text{eV}$. Therefore, the energy of

$$O_2(X) \rightarrow O + O^*$$

is $6.12\,\text{eV} + 0.91\,\text{eV} = 7.03\,\text{eV}$. Since $O^* \rightarrow$ is $-190\,\text{kJ mol}^{-1}$, corresponding to $-1.97\,\text{eV}$, the energy of

$$O_2(X) \rightarrow 2O$$

is $7.03\,\text{eV} - 1.97\,\text{eV} = \boxed{5.06\,\text{eV}}$.

Comment. This value of the dissociation energy is close to the experimental value of $5.08\,\text{eV}$ quoted by Herzberg [Further reading, Chapters 16 and 17], but differs somewhat from the value obtained in Problem 17.2. The difficulty arises from the Brige–Sponer extrapolation which works best when the experimental data fit a linear extrapolation curve as in Example 16.5. A glance at Figure 17.4 shows that the plot of the data is far from linear; hence, it is not surprising that the extrapolation here does not compare well to the extrapolation quoted in Problem 17.2. The extrapolation can be improved by using quadratic of higher terms in the formula for ΔG [Chapter 16].

P17.9 The ratio of the transition probabilities of spontaneous emission to stimulated emission at a frequency ν is given by

$$A = \left(\frac{8\pi h\nu^3}{c^3}\right) B[16.18] = \frac{k}{\lambda^3} B, \quad \text{where } k \text{ is a constant and we have used } \nu = \frac{c}{\lambda}.$$

Thus at 400 nm

$$A(400) = \frac{k}{(400)^3} B(400), \quad \text{and at 500 nm} \quad A(500) = \frac{k}{(500)^3} B(500)$$

Then, $\dfrac{A(500)}{A(400)} = \left(\dfrac{(400)^3}{(500)^3}\right) \times \left(\dfrac{B(500)}{B(400)}\right) = \left(\dfrac{64}{125}\right) \times 10^{-5} = 5 \times 10^{-6}$

Lifetimes and half-lives are inversely proportional to transition probabilities (rate constants) and hence

$$t_{1/2}(\text{T} \to \text{S}) = \frac{1}{5 \times 10^{-6}} t_{1/2}(\text{S}^* \to \text{S}) = (2 \times 10^5) \times (1.0 \times 10^{-9}\,\text{s}) = \boxed{2 \times 10^{-4}\,\text{s}}$$

P17.10 $V = \text{volume of crystal} = \pi r^2 \cdot l = \pi \left(\dfrac{0.50\,\text{cm}}{2}\right)^2 \times (5.0\,\text{cm}) = 0.98\,\text{cm}^3$

$$\text{Maximum mass of excited ions} \simeq \left(\frac{0.050\,\text{g Cr}^{3+}}{100\,\text{g crystal}}\right) \times \left(\frac{3.97\,\text{g crystal}}{\text{cm}^3}\right) \times (0.98\,\text{cm}^3)$$

$$\simeq 1.95 \times 10^{-3}\,\text{g Cr}^{3+}$$

$$\text{Maximum number of photon emissions} \simeq (1.95 \times 10^{-3}\,\text{g Cr}^{3+}) \times \left(\frac{1\,\text{mol}}{52.0\,\text{g Cr}}\right)$$

$$\times \left(\frac{6.022 \times 10^{23}}{\text{mol}}\right) \simeq 2.26 \times 10^{19}$$

$$\text{Maximum total energy of emissions} \simeq 2.26 \times 10^{19} \frac{hc}{\lambda}$$

$$\simeq \frac{2.26 \times 10^{19}\,(6.626 \times 10^{-34}\,\text{J s}) \times (3.00 \times 10^8\,\text{m s}^{-1})}{694.3 \times 10^{-9}\,\text{m}}$$

$$\simeq 6.46\,\text{J}$$

$$\text{Maximum power} = \frac{\text{Maximum emitted energy}}{\Delta t} = \frac{6.46\,\text{J}}{100 \times 10^{-9}\,\text{s}}$$

$$\text{Maximum power} \approx \boxed{65\,\text{MW}}$$

P17.12 The valence electron configuration of NO is $\boxed{(1\sigma)^2(2\sigma^*)^2(1\pi)^4(3\sigma)^2(2\pi^*)^1}$. The data refer to the kinetic energies of the ejected electrons, and so the ionization energies are 16.52 eV, 15.65 eV, and 9.21 eV. The 16.52 eV line refers to ionization of 3σ electron, and the 15.65 eV line (with its long vibrational progression) to ionization of a 1π electron. The 9.21 eV line refers to the ionization of the least strongly attached electron, that is, $2\pi^*$.

Solutions to theoretical problems

P17.15 **(a)** Ethene (ethylene) belongs to D_{2h}. In this group the x, y, and z components of the dipole moment transform as B_{3u}, B_{2u}, and B_{1u} respectively. [See a more extensive set of character tables than in the text.] The π orbital is B_{1u} (like z, the axis perpendicular to the plane) and π^* is B_{3g}. Since $B_{3g} \times B_{1u} = B_{2u}$ and $B_{2u} \times B_{2u} = A_{1g}$, the transition is $\boxed{\text{allowed}}$ (and is y-polarized).

(b) Regard the CO group with its attached groups as locally C_{2v}. The dipole moment has components that transform as $A_1(z)$, $B_1(x)$, and $B_2(y)$, with the z-axis along the C=O direction and x perpendicular to the R_2CO plane. The n orbital is p_y (in the R_2CO plane), and hence transforms as B_2. The π^* orbital is p_x (perpendicular to the R_2CO plane), and hence transforms as B_1. Since $\Gamma_f \times \Gamma_i = B_1 \times B_2 = A_2$, but no component of the dipole moment transforms as A_2, the transition is $\boxed{\text{forbidden}}$.

P17.18 $\mu = -eSR$ [given]

$$S = \left[1 + \frac{R}{a_0} + \frac{1}{3}\left(\frac{R}{a_0}\right)^2\right] e^{-R/a_0} \text{ [Problem 14.3]}$$

$$f = \frac{8\pi^2 m_e \nu}{3he^2}\mu^2 = \frac{8\pi^2 m_e \nu}{3h}R^2 S^2 = \frac{8\pi^2 m_e \nu a_0^2}{3h}\left(\frac{RS}{a_0}\right)^2 = \boxed{\left(\frac{RS}{a_0}\right)^2 f_0}$$

We then draw up the following table

R/a_0	0	1	2	3	4	5	6	7	8
f/f_0	0	0.737	1.376	1.093	0.573	0.233	0.080	0.024	0.007

These points are plotted in Fig. 17.5.

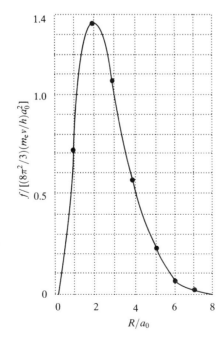

Figure 17.5

The maximum in f occurs at the maximum of RS

$$\frac{d}{dR}(RS) = S + R\frac{dS}{dR} = \left[1 + \frac{R}{a_0} - \frac{1}{3}\left(\frac{R}{a_0}\right)^3\right]e^{-R/a_0} = 0 \quad \text{at } R = R^*$$

That is, $1 + \dfrac{R^*}{a_0} - \dfrac{1}{3}\left(\dfrac{R^*}{a_0}\right)^3 = 0$.

This equation may be solved either numerically or analytically (see Abramowitz and Stegun, *Handbook of mathematical functions*, Section 3.8.2), and $R^* = 2.10380a_0$

As $R \to 0$, the transition becomes $s \to s$, which is forbidden. As $R \to \infty$, the electron is confined to a single atom because its wavefunction does not extend to the other.

P17.20   The fluorescence spectrum gives the vibrational splitting of the lower state. The wavelengths stated correspond to the wavenumbers 22 730, 24 390, 25 640, 27 030 cm^{-1}, indicating spacings of 1660, 1250, and 1390 cm^{-1}. The absorption spectrum spacing gives the separation of the vibrational levels of the upper state. The wavenumbers of the absorption peaks are 27 800, 29 000, 30 300, and 32 800 cm^{-1}. The vibrational spacings are therefore 1200, 1300, and 2500 cm^{-1}.

P17.23   Refer to *Justification* 17.4. If a substance responds non-linearly to an electric field E, then it induces a dipole moment:

$$\mu = \alpha E + \beta E^2 .$$

If the electric field is oscillating at two frequencies, we can write the electric field as

$$E = E_1 \cos \omega_1 t + E_2 \cos \omega_2 t,$$

and the non-linear response as

$$\beta E^2 = \beta(E_1 \cos \omega_1 t + E_2 \cos \omega_2 t)^2,$$
$$\beta E^2 = \beta(E_1^2 \cos^2 \omega_1 t + E_2^2 \cos^2 \omega_2 t + 2E_1 E_2 \cos \omega_1 t \cos \omega_2 t).$$

Application of trigonometric identities allows a product of cosines to be re-written as a sum:

$$\cos A \cos B = \tfrac{1}{2}\cos(A - B) + \tfrac{1}{2}\cos(A + B).$$

Using this result (a special case of which applies to the $\cos^2$ terms), yields:

$$\beta E^2 = \tfrac{1}{2}\beta[E_1^2(1 + \cos 2\omega_1 t) + E^2(1 + \cos 2\omega_2 t) + 2E_1 E_2(\cos(\omega_1 + \omega_2)t + \cos(\omega_1 - \omega_2)t].$$

This expression includes responses at twice the original frequencies as well as at the sum and difference frequencies.

P17.26   **(a)** The integrated absorption coefficient is (specializing to a triangular lineshape)

$$\mathcal{A} = \int \varepsilon(\tilde{\nu})\, d\tilde{\nu} = (1/2)\varepsilon_{\max}\Delta\tilde{\nu}$$

$$= (1/2) \times (150\,\text{L mol}^{-1}\,\text{cm}^{-1}) \times (34\,483 - 31\,250)\,\text{cm}^{-1},$$

$$\mathcal{A} = \boxed{2.42 \times 10^5\,\text{L mol}^{-1}\,\text{cm}^{-2}}.$$

(b) The concentration of gas under these conditions is

$$c = \frac{n}{V} = \frac{p}{RT} = \frac{2.4\,\text{Torr}}{(62.364\,\text{Torr L mol}^{-1}\,\text{K}^{-1}) \times (373\,\text{K})} = 1.03 \times 10^{-4}\,\text{mol L}^{-1}.$$

Over 99 per cent of these gas molecules are monomers, so we take this concentration to be that of CH_3I. (If 1 of every 100 of the original monomers turned to dimers, each produces 0.5 dimers; remaining monomers represent 99 of 99.5 molecules.) Beer's law states

$$A = \varepsilon c l = (150\,\text{L mol}^{-1}\,\text{cm}^{-1}) \times (1.03 \times 10^{-4}\,\text{mol L}^{-1}) \times (12.0\,\text{cm}) = \boxed{0.185}.$$

(c) The concentration of gas under these conditions is

$$c = \frac{n}{V} = \frac{p}{RT} = \frac{100\,\text{Torr}}{(62.364\,\text{Torr L mol}^{-1}\,\text{K}^{-1}) \times (373\,\text{K})} = 4.30 \times 10^{-3}\,\text{mol L}^{-1}$$

Since 18 per cent of these CH_3I units are in dimers (forming 9 per cent as many molecules as were originally present as monomers), the monomer concentration is only 82/91 of this value or $3.87 \times 10^{-3}\,\text{mol L}^{-1}$. Beer's law is

$$A = \varepsilon c l = (150\,\text{L mol}^{-1}\,\text{cm}^{-1}) \times (3.87 \times 10^{-3}\,\text{mol L}^{-1}) \times (12.0\,\text{cm}) = \boxed{6.97}.$$

If this absorbance were measured, the molar absorption coefficient inferred from it without consideration of the dimerization would be

$$\varepsilon = A/cl = 6.97/((4.30 \times 10^{-1}\,\text{mol L}^{-1}) \times (12.0\,\text{cm}))$$

$$= \boxed{135\,\text{L mol}^{-1}\,\text{cm}^{-1}}$$

an apparent drop of 10 per cent compared to the low-pressure value.

P17.28   In Fig. 17.6

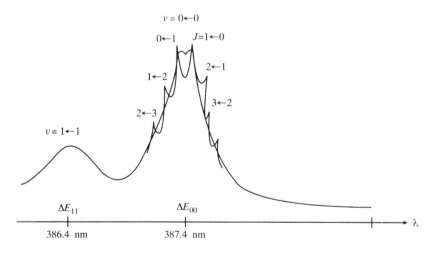

Figure 17.6

$$\Delta E_{11} = \frac{hc}{\lambda_{11}} = \frac{hc}{386.4\,\text{nm}} = 5.1409 \times 10^{-19}\,\text{J} = 3.2087\,\text{eV}$$

and

$$\Delta E_{00} = \frac{hc}{\lambda_{00}} = \frac{hc}{387.6\,\text{nm}} = 5.1250 \times 10^{-19}\,\text{J} = 3.1987\,\text{eV}$$

Energy of excited singlet, S_1: $E_1(v, J) = V_1 + (v + 1/2)\tilde{v}_1 hc + J(J+1)\tilde{B}_1 hc$

Energy of ground singlet, S_0: $E_0(v, J) = V_0 + (v + 1/2)\tilde{v}_0 hc + J(J+1)\tilde{B}_0 hc$

The midpoint of the 0–0 band corresponds to the forbidden Q branch ($\Delta J = 0$) with $J = 0$ and $v = 0 \leftarrow 0$.

$$\Delta E_{00} = E_1(0, 0) - E_0(0, 0) = (V_1 - V_0) + \tfrac{1}{2}(\tilde{v}_1 - \tilde{v}_0)hc \tag{1}$$

The midpoint of the 1–1 band corresponds to the forbidden Q branch ($\Delta J = 0$) with $J = 0$ and $v = 1 \leftarrow 1$.

$$\Delta E_{11} = E_1(1, 0) - E_0(1, 0) = (V_1 - V_0) + \tfrac{3}{2}(\tilde{v}_1 - \tilde{v}_0)hc \tag{2}$$

Multiplying eqn 1 by three and subtracting eqn 2 gives

$$3\Delta E_{00} - \Delta E_{11} = 2(V_1 - V_0)$$

$$\begin{aligned}
V_1 - V_0 &= \tfrac{1}{2}(3\Delta E_{00} - \Delta E_{11}) \\
&= \tfrac{1}{2}\{3(5.1250) - (5.1409)\}10^{-19}\,\text{J} \\
&= 5.1171 \times 10^{-19}\,\text{J} = \boxed{3.1938\,\text{eV}}
\end{aligned} \tag{3}$$

This is the potential energy difference between S_0 and S_1.

Equations (1) and (3) may be solved for $\tilde{v}_1 - \tilde{v}_0$.

$$\begin{aligned}
\tilde{v}_1 - \tilde{v}_0 &= 2\{\Delta E_{00} - (V_1 - V_0)\} \\
&= 2\{5.1250 - 5.1171\}10^{-19}\,\text{J}/hc \\
&= 1.5800 \times 10^{-21}\,\text{J} = 0.009\,861\,5\,\text{eV} = \boxed{79.538\,\text{cm}^{-1}}
\end{aligned}$$

The $\tilde{v}_1$ value can be determined by analyzing the band head data for which $J + 1 \leftarrow J$.

$$\begin{aligned}
\Delta E_{10}(J) &= E_1(0, J) - E_0(1, J+1) \\
&= V_1 - V_0 + \tfrac{1}{2}(\tilde{v}_1 - 3\tilde{v}_0)hc + J(J+1)\tilde{B}_1 hc - (J+1) \times (J+2)\tilde{B}_0 hc
\end{aligned}$$

$$\Delta E_{00}(J) = V_1 - V_0 + \tfrac{1}{2}(\tilde{v}_1 - \tilde{v}_0)hc + J(J+1)\tilde{B}_1 hc - (J+1) \times (J+2)\tilde{B}_0 hc$$

Therefore,

$$\Delta E_{00}(J) - \Delta E_{10}(J) = \tilde{v}_0 hc$$

$$\Delta E_{00}(J_{\text{head}}) = \frac{hc}{388.3\,\text{nm}} = 5.1158 \times 10^{-19}\,\text{J}$$

$$\Delta E_{10}(J_{\text{head}}) = \frac{hc}{421.6\,\text{nm}} = 4.7117 \times 10^{-19}\,\text{J}$$

$$\tilde{v}_0 = \frac{\Delta E_{00}(J) - \Delta E_{10}(J)}{hc}$$

$$= \frac{(5.1158 - 4.7117) \times 10^{-19} \text{ J}}{hc}$$

$$= \frac{4.0410 \times 10^{-20} \text{ J}}{hc} = 0.25222 \text{ eV} = \boxed{2034.3 \text{ cm}^{-1}}$$

$$\tilde{v}_1 = \tilde{v}_0 + 79.538 \text{ cm}^{-1}$$

$$= (2034.3 + 79.538) \text{ cm}^{-1} = \boxed{2113.8 \text{ cm}^{-1} = \frac{4.1990 \times 10^{-20} \text{ J}}{hc}}$$

$$\frac{I_{1-1}}{I_{0-0}} \approx \frac{e^{-E_1(1,0)/kT_{\text{eff}}}}{e^{-E_1(0,0)/kT_{\text{eff}}}} = e^{-(E_1(1,0) - E_1(0,0))/kT_{\text{eff}}}$$

$$\approx e^{-hc\tilde{v}_1/kT_{\text{eff}}}$$

$$\ln\left(\frac{I_{1-1}}{I_{0-0}}\right) = -\frac{hc\tilde{v}_1}{kT_{\text{eff}}}$$

$$T_{\text{eff}} = \frac{hc\tilde{v}_1}{k \ln\left(\frac{I_{0-0}}{I_{1-1}}\right)} = \frac{4.1990 \times 10^{-20} \text{ J}}{(1.38066 \times 10^{-23} \text{ J K}^{-1}) \ln(10)} = \boxed{1321 \text{ K}}$$

The relative population of the $v = 0$ and $v = 1$ vibrational states is the inverse of the relative intensities of the transitions from those states; hence $\frac{1}{0.1} = \boxed{10}$

It would seem that with such a high effective temperature more than eight of the rotational levels of the S_1 state should have a significant population. But the spectra of molecules in comets are never as clearly resolved as those obtained in the laboratory and that is most probably the reason why additional rotational structure does not appear in these spectra.

18 Spectroscopy 3: magnetic resonance

Solutions to exercises

Discussion questions

E18.1(a) Detailed discussions of the origins of the local, neighbouring group, and solvent contributions to the shielding constant can be found in Sections 17.5(d), (e), and (f) as well as the books listed under *Further reading*. Here we will merely summarize the major features.

The local contribution is essentially the contribution of the electrons in the atom that contains the nucleus being observed. It can be expressed as a sum of a diamagnetic and paramagnetic parts, that is $\sigma\,(\text{local}) = \sigma_d + \sigma_p$. The diamagnetic part arises because the applied field generates a circulation of charge in the ground state of the atom. In turn, the circulating charge generates a magnetic field. The direction of this field can be found through Lenz's law which states that the induced magnetic field must be opposite in direction to the field producing it. Thus it shields the nucleus. The diamagnetic contribution is roughly proportional to the electron density on the atom and it is the only contribution for closed shell free atoms and for distributions of charge that have spherical or cylindrical symmetry. The local paramagnetic contribution is somewhat harder to visualize since there is no simple and basic principle analogous to Lenz's law that can be used to explain the effect. The applied field adds a term to the hamiltonian of the atom which mixes in excited electronic states into the ground state and any theoretical calculation of the effect requires detailed knowledge of the excited state wave functions. It is to be noted that the paramagnetic contribution does not require that the atom or molecule be paramagnetic. It is paramagnetic only in the sense in that it results in an induced field in the same direction as the applied field.

The neighbouring group contributions arise in a manner similar to the local contributions. Both diamagnetic and paramagnetic currents are induced in the neighbouring atoms and these currents results in sheilding contributions to the nucleus of the atom being observed. However, there are some differences: The magnitude of the effect is much smaller because the induced currents in neighbouring atoms are much farther away. It also depends on the anisotropy of the magnetic susceptibility (see Chapter 23) of the neighbouring group as shown in eqn 18.30. Only anisotropic susceptibilities result in a contribution.

Solvents can influence the local field in many different ways. Detailed theoretical calculations of the effect are difficult due to the complex nature of the solute-solvent interaction. Polar solvent–polar solute interactions are an electric field effect that usually causes deshielding of the solute protons. Solvent magnetic antisotropy can cause shielding or deshielding, for example, for solutes in benzene solution. In addition, there are a variety of specific chemical interactions between solvent and solute that can affect the chemical shift. See the references listed under *Further reading* for more details.

E18.2(a) Both spin–lattice and spin–spin relaxation are caused by fluctuating magnetic and electric fields at the nucleus in question and these fields result from the random thermal motions present in the solution or other form of matter. These random motions can be a result of a number of processes and it is hard to summarize all that could be important. In theory every known nuclear interaction coupled with every type of motion can contribute to relaxation and detailed treatments can be exceedingly complex. However, they all depend on the magnetogyric ratio of the atom in question and the magnetogyric ratio of the proton is much larger than that of [13]C. Hence the interaction of the proton with fluctuating local magnetic fields caused by the presence of neighboring magnetic nuclei will be greater, and the relaxation will be quicker, corresponding to a shorter relaxation time for protons. Another consideration is the structure of compounds containing carbon and hydrogen. Typically the C atoms are in the interior of the molecule bonded to other C atoms, 99% of which are nonmagnetic, so the primary relaxation effects are due to bonded protons. Protons are on the outside of the molecule and are subject to many more interactions and hence faster relaxation.

E18.3(a) Spin–spin couplings in NMR are due to a polarization mechanism which is transmitted through bonds. The following description applies to the coupling between the protons in a H_X—C—H_Y group as is typically found in organic compounds. See Fig. 18.32 of the text. On H_X, the Fermi contact interaction causes the spins of its proton and electron to be aligned antiparallel. The spin of the electron from C in the H_X—C bond is then aligned antiparallel to the electron from H_X due to the Pauli exclusion principle. The spin of the C electron in the bond H_Y is then aligned parallel to the electron from H_X because of Hund's rule. Finally the alignment is transmitted through the second bond in the same manner as the first. This progression of alignments (antiparallel × antiparallel × parallel × antiparallel × antiparallel) yields an overall energetically favourable parallel alignment of the two proton nuclear spins. Therefore, in this case the coupling constant, $^2J_{HH}$ is negative in sign.

The hyperfine structure in the ESR spectrum of an atomic or molecular system is a result of two interactions: an anisotrophic dipolar coupling between the net spin of the unpaired electrons and the nuclear spins and also an isotropic coupling due to the Fermi contact interaction. In solution, only the Fermi contact interaction contributes to the splitting as the dipolar contribution averages to zero in a rapidly tumbling system. In the case of π-electron radicals, such as $C_6H_6^-$, no hyperfine interaction between the unpaired electron and the ring protons might have been expected. The protons lie in the nodal plane of the molecular orbital occupied by the unpaired electron, so any hyperfine structure cannot be explained by a simple Fermi contact interaction which requires an unpaired electron density at the proton. However, an indirect spin polarization mechanism, similar to that used to explain spin–spin couplings in NMR, can account for the existence of proton hyperfine interactions in the ESR spectra of these systems. Refer to Fig. 18.4 of the text. Because of Hund's rule, the unpaired electron and the first electron in the C—H bond (the one from the C atom), will tend to align parallel to each other. The second electron in the C—H bond (the one from H) will then align antiparallel to the first by the Pauli principle, and finally the Fermi contact interaction will align the proton and electron on H antiparallel. The net result (parallel × antiparallel × antiparallel) is that the spins of the unpaired electron and the proton are aligned parallel and effectively they have detected each other.

Numerical exercises

E18.4(a) The resonance frequency is equal to the Larmor frequency of the proton and is given by

$$\nu = \nu_L = \frac{\gamma \mathcal{B}_0}{2\pi} \; [18.17] \quad \text{with } \gamma = \frac{g_I \mu_N}{\hbar} \; [18.14]$$

hence $\nu = \dfrac{g_I \mu_N \mathcal{B}_0}{h} = \dfrac{(5.5857) \times (5.0508 \times 10^{-27} \, \text{J T}^{-1}) \times (14.1 \, \text{T})}{6.626 \times 10^{-34} \, \text{J s}} = \boxed{600 \, \text{MHz}}$

E18.5(a) $E_{m_I} = -\gamma \hbar \mathcal{B}_0 m_I \; [18.16] = -g_I \mu_N \mathcal{B}_0 m_I \; [18.14, \gamma \hbar = g_I \mu_N]$

$m_I = \dfrac{3}{2}, \dfrac{1}{2}, -\dfrac{1}{2}, -\dfrac{3}{2}$

$E_{m_I} = (-0.4289) \times (5.051 \times 10^{-27} \, \text{J T}^{-1}) \times (7.500 \, \text{T} \times m_I) = \boxed{-1.625 \times 10^{-26} \, \text{J} \times m_I}$

E18.6(a) The energy level separation is

$$\Delta E = h\nu \quad \text{where } \nu = \frac{\gamma \mathcal{B}_0}{2\pi} \; [18.17]$$

So

$$\nu = \frac{(6.73 \times 10^7 \, \text{T}^{-1} \, \text{s}^{-1}) \times 14.4 \, \text{T}}{2\pi} = 1.54 \times 10^8 \, \text{s}^{-1}$$

$$= 1.54 \times 10^8 \, \text{Hz} = \boxed{154 \, \text{MHz}}$$

E18.7(a) **(a)** As shown in Exercise 18.4(a) a 600-MHz NMR spectrometer operates in a magnetic field of 14.1 T. Thus

$$\Delta E = \gamma \hbar \mathcal{B}_0 = h\nu_{\mathrm{L}} = h\nu \quad \text{at resonance}$$

$$= (6.626 \times 10^{-34}\,\mathrm{J\,s}) \times (6.00 \times 10^8\,\mathrm{s}^{-1}) = \boxed{3.98 \times 10^{-25}\,\mathrm{J}}$$

(b) A 600-MHz NMR spectrometer means 600-MHz is the resonance frequency for protons for which the magnetic field is 14.1 T. In high-field NMRs it is the field not the frequency which is fixed, so for the deuteron

$$\nu = \frac{g_I \mu_{\mathrm{N}} \mathcal{B}_0}{h} \quad [\text{Exercise 18.4(a)}]$$

$$= \frac{(0.8575) \times (5.051 \times 10^{-27}\,\mathrm{J\,T}^{-1}) \times (14.1\,\mathrm{T})}{6.626 \times 10^{-34}\,\mathrm{J\,s}} = 9.22 \times 10^7\,\mathrm{Hz} = 92.2\,\mathrm{MHz}$$

$$\Delta E = h\nu = (6.626 \times 10^{-34}\,\mathrm{J\,s}) \times (9.21\bar{6} \times 10^7\,\mathrm{s}^{-1}) = \boxed{6.11 \times 10^{-26}\,\mathrm{J}}$$

Thus the separation in energy is larger for the proton **(a)**.

E18.8(a) This is similar to Exercise 18.7**(a)(b)**. There the energy difference was calculated for $|\Delta m_I| = 1$, here it is for $|\Delta m_I| = 2$. That is

$$\Delta E = 2g_I \mu_{\mathrm{N}} \mathcal{B}_0 = (2) \times (0.4036) \times (5.051 \times 10^{-27}\,\mathrm{J\,T}^{-1}) \times (15.00\,\mathrm{T})$$

$$= \boxed{6.116 \times 10^{-26}\,\mathrm{J}}$$

E18.9(a) In all cases the selection rule $\Delta m_I = \pm 1$ is applied; hence

$$\mathcal{B}_0 = \frac{h\nu}{g_I \mu_{\mathrm{N}}} = \frac{6.626 \times 10^{-34}\,\mathrm{J\,Hz}^{-1}}{5.0508 \times 10^{-27}\,\mathrm{J\,T}^{-1}} \times \frac{\nu}{g_I}$$

$$= (1.3119 \times 10^{-7}) \times \frac{(\nu/\mathrm{Hz})}{g_I}\mathrm{T} = (0.13119) \times \frac{(\nu/\mathrm{MHz})}{g_I}\mathrm{T}$$

We can draw up the following table

$\mathcal{B}_0/\mathrm{T}$	**(a)** $^1\mathrm{H}$	**(b)** $^2\mathrm{H}$	**(c)** $^{13}\mathrm{C}$
g_I	5.5857	0.85745	1.4046
(i) 250 MHz	5.87	38.3	23.4
(ii) 500 MHz	11.7	76.6	46.8

Comment. Magnetic fields above 30 T have not yet been obtained for use in NMR spectrometers. As discussed in the solution to Exercise 18.7**(a)(b)**, it is the field, not the frequency, that is fixed in high-field NMR spectrometers. Thus an NMR spectrometer that is called a 500-MHz spectrometer refers to the resonance frequency for protons and has a magnetic field fixed at 11.7 T.

Question. What are the resonance frequencies of these nuclei in 250-MHz and 500-MHz spectrometers? See Exercise 18.7**(a)(b)**.

E18.10(a) The ground state has

$$m_I = +\tfrac{1}{2} = \alpha \text{ spin}, \qquad m_I = -\tfrac{1}{2} = \beta \text{ spin} \quad [18.16]$$

Hence, with

$$\delta N = N_\alpha - N_\beta$$

$$\frac{\delta N}{N} = \frac{N_\alpha - N_\beta}{N_\alpha + N_\beta} = \frac{N_\alpha - N_\alpha e^{-\Delta E/kT}}{N_\alpha + N_\alpha e^{-\Delta E/kT}} \quad \text{[\textit{Justification} 18.1]}$$

$$= \frac{1 - e^{-\Delta E/kT}}{1 + e^{-\Delta E/kT}} \approx \frac{1 - (1 - \Delta E/kT)}{1 + 1} \approx \frac{\Delta E}{2kT} = \frac{g_I \mu_N \mathcal{B}_0}{2kT} \quad \text{[for } \Delta E \ll kT\text{]}$$

That is, $\dfrac{\delta N}{N} \approx \dfrac{g_I \mu_N \mathcal{B}_0}{2kT} = \dfrac{(5.5857) \times (5.0508 \times 10^{-27}\,\text{J T}^{-1}) \times (\mathcal{B}_0)}{(2) \times (1.38066 \times 10^{-23}\,\text{J K}^{-1}) \times (298\,\text{K})} \approx 3.43 \times 10^{-6}\,\mathcal{B}_0/\text{T}$

(a) $\mathcal{B}_0 = 0.3\,\text{T}, \qquad \delta N/N = \boxed{1 \times 10^{-6}}$

(b) $\mathcal{B}_0 = 1.5\,\text{T}, \qquad \delta N/N = \boxed{5.1 \times 10^{-6}}$

(c) $\mathcal{B}_0 = 10\,\text{T}, \qquad \delta N/N = \boxed{3.4 \times 10^{-5}}$

E18.11(a) $\quad \delta N \approx \dfrac{N g_I \mu_N \mathcal{B}_0}{2kT} \text{[Exercise 18.10(a)]} = \dfrac{Nh\nu}{2kT}$

Thus, $\delta N \propto \nu$

$$\frac{\delta N(800\,\text{MHz})}{\delta N(60\,\text{MHz})} = \frac{800\,\text{MHz}}{60\,\text{MHz}} = \boxed{13}$$

This ratio is not dependent on the nuclide as long as the approximation $\Delta E \ll kT$ holds (Exercise 18.10(a)).

E18.12 $\quad \mathcal{B}_{\text{loc}} = (1 - \sigma)\mathcal{B}_0 \text{ [18.23]}$

$$|\Delta \mathcal{B}_{\text{loc}}| = |(\Delta \sigma)|\mathcal{B}_0 \approx |[\delta(CH_3) - \delta(CHO)]|\mathcal{B}_0 \qquad \left[|\Delta\sigma| \approx \left|\frac{\nu - \nu_0}{\nu_0}\right|\right]$$

$$= |(2.20 - 9.80)| \times 10^{-6}\mathcal{B}_0 = 7.60 \times 10^{-6}\mathcal{B}_0$$

(a) $\mathcal{B}_0 = 1.5\,\text{T}, \qquad |\Delta \mathcal{B}_{\text{loc}}| = 7.60 \times 10^{-6} \times 1.5\,\text{T} = \boxed{11\,\mu\text{T}}$

(b) $\mathcal{B}_0 = 15\,\text{T}, \qquad |\Delta \mathcal{B}_{\text{loc}}| = \boxed{110\,\mu\text{T}}$

E18.13(a) $\quad \nu - \nu_0 = \nu_0 \delta \times 10^{-6} \text{ [18.25]}$

$$|\Delta \nu| \quad \equiv (\nu - \nu_0)(CHO) - (\nu - \nu_0)(CH_3)$$

$$= \nu(CHO) - \nu(CH_3)$$

$$= \nu_0[\delta(CHO) - \delta(CH_3)] \times 10^{-6}$$

$$= (9.80 - 2.20) \times 10^{-6}\nu_0 = 7.60 \times 10^{-6}\nu_0$$

(a) $\nu_0 = 250\,\text{MHz}, \qquad |\Delta \nu| = 7.60 \times 10^{-6} \times 250\,\text{MHz} = 1.90\,\text{kHz}$

(b) $\nu_0 = 500\,\text{MHz}, \qquad |\Delta \nu| = 3.80\,\text{kHz}$

(a) The spectrum is shown in Fig. 18.1 with the value of $|\Delta \nu|$ as calculated above.

(b) When the frequency is changed to 500 MHz, the $|\Delta \nu|$ changes to 3.80 kHz. The fine structure (the splitting within groups) remains the same as spin–spin splitting is unaffected by the strength of the applied field. However, the intensity of the lines increases by a factor of 2 because $\delta N/N \propto \nu$ (Exercise 18.11(a)).

The observed splitting pattern is that of an AX_3 (or A_3X) species, the spectrum of which is described in Section 18.6.

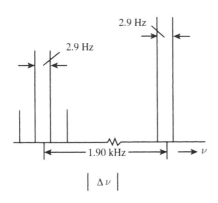

Figure 18.1

E18.14(a) $\tau \approx \dfrac{\sqrt{2}}{\pi \Delta \nu}$ [18.44, with $\delta\nu$ written as $\Delta\nu$]

$\Delta\nu = \nu_0(\delta' - \delta) \times 10^{-6}$ [Exercise 18.13(**a**)]

Then $\tau \approx \dfrac{\sqrt{2}}{\pi\nu_0(\delta' - \delta) \times 10^{-6}}$

$\approx \dfrac{\sqrt{2}}{(\pi) \times (250 \times 10^6 \,\text{Hz}) \times (5.2 - 4.0) \times 10^{-6}} \approx 1.5 \times 10^{-3} \,\text{s}$

Therefore, the signals merge when the lifetime of each isomer is less than about 1.5 ms, corresponding to a conversion rate of about $\boxed{6.7 \times 10^2 \,\text{s}^{-1}}$

E18.15(a) The four equivalent ^{19}F nuclei ($I = \frac{1}{2}$) give a single line. However, the ^{10}B nucleus ($I = 3$, 19.6 per cent abundant) splits this line into $2 \times 3 + 1 = 7$ lines and the ^{11}B nucleus ($I = \frac{3}{2}$, 80.4 per cent abundant) into $2 \times \frac{3}{2} + 1 = 4$ lines. The splitting arising from the ^{11}B nucleus will be larger than that arising from the ^{10}B nucleus (since its magnetic moment is larger, by a factor of 1.5, Table 18.1). Moreover, the total intensity of the four lines due to the ^{11}B nuclei will be greater (by a factor of $80.4/19.6 \approx 4$) than the total intensity of the seven lines due to the ^{10}B nuclei. The individual line intensities will be in the ratio $\frac{7}{4} \times 4 = 7$ ($\frac{4}{7}$ the number of lines and about four times as abundant). The spectrum is sketched in Fig. 18.2.

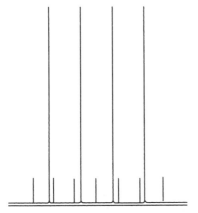

Figure 18.2

E18.16(a) The A, M, and X resonances lie in distinctively different groups. The A resonance is split into a 1 : 2 : 1 triplet by the M nuclei, and each line of that triplet is split into a 1 : 4 : 6 : 4 : 1 quintet by the X nuclei, (with $J_{AM} > J_{AX}$). The M resonance is split into a 1 : 3 : 3 : 1 quartet by the A nuclei and each line is split into a quintet by the X nuclei (with $J_{AM} > J_{MX}$). The X resonance is split into a quartet by the A nuclei and then each line is split into a triplet by the M nuclei (with $J_{AX} > J_{MX}$). The spectrum is sketched in Fig. 18.3.

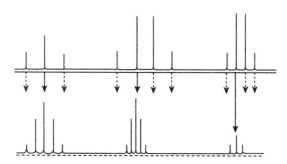

Figure 18.3

E18.17(a) (a) If there is rapid rotation about the axis, the H nuclei are both chemically and magnetically equivalent.

(b) Since $J_{cis} \neq J_{trans}$, the H nuclei are chemically but not magnetically equivalent.

E18.18(a) Analogous to precession of the magnetization vector in the laboratory frame due to the presence of $\mathcal{B}_0$ that is

$$\nu_L = \frac{\gamma \mathcal{B}_0}{2\pi} \ [18.17],$$

there is a precession in the rotating frame, due to the presence of $\mathcal{B}_1$, namely

$$\nu_L = \frac{\gamma \mathcal{B}_1}{2\pi} \quad \text{or} \quad \omega_1 = \gamma \mathcal{B}_1 \quad [\omega = 2\pi\nu]$$

Since ω is an angular frequency, the angle through which the magnetization vector rotates is

$$\theta = \gamma \mathcal{B}_1 t = \frac{g_I \mu_N}{\hbar} \mathcal{B}_1 t$$

and $\mathcal{B}_1 = \dfrac{\theta \hbar}{g_I \mu_N t} = \dfrac{\left(\frac{\pi}{2}\right) \times (1.055 \times 10^{-34}\,\text{J s})}{(5.586) \times (5.051 \times 10^{-27}\,\text{J T}^{-1}) \times (1.0 \times 10^{-5}\,\text{s})} = \boxed{5.9 \times 10^{-4}\,\text{T}}$

A 180° pulse requires $2 \times 10\,\mu s = \boxed{20\,\mu s}$

E18.19(a) (a) $\mathcal{B}_0 = \dfrac{h\nu}{g_I \mu_N} = \dfrac{(6.626 \times 10^{-34}\,\text{J Hz}^{-1}) \times (9 \times 10^9\,\text{Hz})}{(5.5857) \times (5.051 \times 10^{-27}\,\text{J T}^{-1})} = \boxed{2 \times 10^2\,\text{T}}$

(b) $\mathcal{B}_0 = \dfrac{h\nu}{g_e \mu_B} = \dfrac{(6.626 \times 10^{-34}\,\text{J Hz}^{-1}) \times (300 \times 10^6\,\text{Hz})}{(2.0023) \times (9.274 \times 10^{-24}\,\text{J T}^{-1})} = \boxed{10\,\text{mT}}$

Comment. Because of the sizes of these magnetic fields neither experiment seems feasible.

Question. What frequencies are required to observe electron resonance in the magnetic field of a 300 MHz NMR magnet and nuclear resonance in the field of a 9 GHz ($g = 2.00$) ESR magnet? Are these experiments feasible?

E18.20(a) $g = \dfrac{h\nu}{\mu_B \mathcal{B}_0}$ [18.48]

We shall often need the value

$$\frac{h}{\mu_B} = \frac{6.62608 \times 10^{-34}\,\text{J Hz}^{-1}}{9.27402 \times 10^{-24}\,\text{J T}^{-1}} = 7.14478 \times 10^{-11}\,\text{T Hz}^{-1}$$

Then, in this case

$$g = \frac{(7.14478 \times 10^{-11}\,\text{T Hz}^{-1}) \times (9.2231 \times 10^9\,\text{Hz})}{329.12 \times 10^{-3}\,\text{T}} = \boxed{2.0022}$$

E18.21(a) $a = \mathcal{B}(\text{line 3}) - \mathcal{B}(\text{line 2}) = \mathcal{B}(\text{line 2}) - \mathcal{B}(\text{line 1})$

$$\left.\begin{array}{l} \mathcal{B}_3 - \mathcal{B}_2 = (334.8 - 332.5)\,\text{mT} = 2.3\,\text{mT} \\ \mathcal{B}_2 - \mathcal{B}_1 = (332.5 - 330.2)\,\text{mT} = 2.3\,\text{mT} \end{array}\right\} \quad a = \boxed{2.3\,\text{mT}}$$

Use the centre line to calculate g

$$g = \frac{h\nu}{\mu_B \mathcal{B}_0} = (7.14478 \times 10^{-11}\,\text{T Hz}^{-1}) \times \frac{9.319 \times 10^9\,\text{Hz}}{332.5 \times 10^{-3}\,\text{T}} = \boxed{2.002\bar{5}}$$

E18.22(a)   The centre of the spectrum will occur at 332.5 mT. Proton 1 splits the line into two components with separation 2.0 mT and hence at 332.5 ± 1.0 mT. Proton 2 splits these two hyperfine lines into two, each with separation 2.6 mT, and hence the lines occur at $332.5 \pm 1.0 \pm 1.3$ mT. The spectrum therefore consists of four lines of $\boxed{\text{equal intensity}}$ at the fields $\boxed{330.2\,\text{mT}, \ 332.2\,\text{mT}, \ 332.8\,\text{mT}, \ 334.8\,\text{mT}}$

E18.23(a)   We construct Fig. 18.4(a) for CH_3 and Fig. 18.4(b) for CD_3. The predicted intensity distribution is determined by counting the number of overlapping lines of equal intensity from which the hyperfine line is constructed.

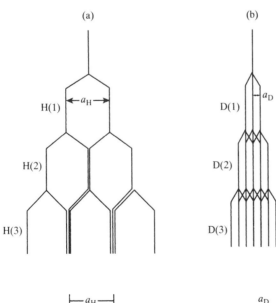

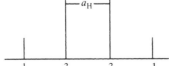

Figure 18.4

E18.24(a) $\quad \mathcal{B}_0 = \dfrac{h\nu}{g\mu_B} = \dfrac{7.14478 \times 10^{-11}}{2.0025}\, \text{T Hz}^{-1} \times \nu\ [\text{Exercise } 18.20(\text{a})] = 35.68\, \text{mT} \times (\nu/\text{GHz})$

    (a) $\quad \nu = 9.302\, \text{GHz}, \qquad \mathcal{B}_0 = \boxed{331.9\, \text{mT}}$

    (b) $\quad \nu = 33.67\, \text{GHz}, \qquad \mathcal{B}_0 = 1201\, \text{mT} = \boxed{1.201\, \text{T}}$

E18.25(a) Since the number of hyperfine lines arising from a nucleus of spin I is $2I + 1$, we solve $2I + 1 = 4$ and find that $\boxed{I = \tfrac{3}{2}}$.

    Comment. Four lines of equal intensity could also arise from two inequivalent nuclei with $I = \tfrac{1}{2}$.

E18.26(a) The X nucleus produces six lines of equal intensity. The pair of H nuclei in XH_2 split each of these lines into a $1 : 2 : 1$ triplet (Fig. 18.5(a)). The pair of D nuclei ($I = 1$) in XD_2 split each line into a $1 : 2 : 3 : 2 : 1$ quintet (Fig. 18.5(b)). The total number of hyperfine lines observed is then $6 \times 3 = 18$ in XH_2 and $6 \times 5 = 30$ in XD_2.

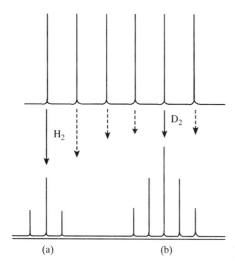

    Figure 18.5

Solutions to problems

Solutions to numerical problems

P18.1 $\quad g_I = -3.8260 \quad$ (Table 18.2)

$$\mathcal{B}_0 = \frac{h\nu}{g_I \mu_N} = \frac{(6.626 \times 10^{-34}\, \text{J Hz}^{-1}) \times \nu}{(-)(3.8260) \times (5.0508 \times 10^{-27}\, \text{J T}^{-1})} = 3.429 \times 10^{-8}(\nu/\text{Hz})\, \text{T}$$

Therefore, with $\nu = 300\, \text{MHz}$,

$$\mathcal{B}_0 = (3.429 \times 10^{-8}) \times (300 \times 10^6\, \text{T}) = \boxed{10.3\, \text{T}}$$

$$\frac{\delta N}{N} \approx \frac{g_I \mu_N \mathcal{B}_0}{2kT}\quad [\text{Exercise } 18.7(\text{a})]$$

$$= \frac{(-3.8260) \times (5.0508 \times 10^{-27}\, \text{J T}^{-1}) \times (10.3\, \text{T})}{(2) \times (1.381 \times 10^{-23}\, \text{J K}^{-1}) \times (298\, \text{K})} = \boxed{2.42 \times 10^{-5}}$$

Since $g_I < 0$ (as for an electron, the magnetic moment is antiparallel to its spin), the $\boxed{\beta}$ state $\left(m_I = -\frac{1}{2}\right)$ lies lower.

P18.3   The envelopes of maxima and minima of the curve are determined by T_2 through eqn 25, but the time interval between the maxima of this decaying curve corresponds to the reciprocal of the frequency difference $\Delta \nu$ between the pulse frequency ν_0 and the Larmor frequency ν_L, that is $\Delta \nu = |\nu_0 - \nu_L|$

$$\Delta \nu = \frac{1}{0.10\,\text{s}} = 10\,\text{s}^{-1} = 10\,\text{Hz}$$

Therefore the Larmor frequency is $\boxed{300 \times 10^6\,\text{Hz} \pm 10\,\text{Hz}}$

According to eqn 25 the intensity of the maxima in the FID curve decays exponentially as e^{-t/T_2}

Therefore T_2 corresponds to the time at which the intensity has been reduced to $1/e$ of the original value. In the text figure, this corresponds to a time slightly before the fourth maximum has occurred, or about $\boxed{0.29\,\text{s}}$

P18.4   The three rotational conformations of $F_2BrC\text{--}CBrCl_2$ are shown in Fig. 18.6. In conformation I, the two F atoms are equivalent. However, in conformations II and III they are non-equivalent. At low temperature, the molecular residence time in conformation I is longer (because this conformation has the lowest repulsive energy of the large bromine atoms) than that of conformations II and III, which have equal residence times. With its longer residence time, we expect that the NMR signal intensity of conformation I should be stronger and we can conclude that it is the low-temperature singlet. It is a singlet because equivalent atoms do not have detectable spin–spin couplings.

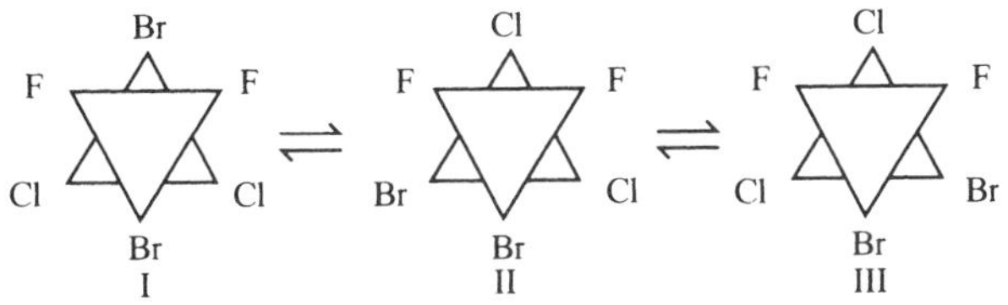

Figure 18.6

The fluorines of conformations II and III are non-equivalent, so their coupling is observed at low-temperature. Fluorine has a nuclear spin of $1/2$, so we expect a doublet for each fluorine. These are observed with strong geminal coupling of 160 Hz. As temperature increases, the rate of rotation between II and III increases and the two fluorines become equivalent in these conformations, though remaining distinct from I. The doublets collapse to singlets. With a further temperature increase to $-30°C$, and above, the rate of rotation about the $C\text{--}C$ bond becomes so rapid that the residence times of the three conformations become equal. The very short residence times produce an average NMR signal that is a singlet and the fluorines appear totally equivalent.

The spectra shown in text Fig. 18.55 for conformations II and III show both spin–spin coupling and differences in chemical shift. The spin–spin splitting is 160 Hz. The difference in chemical shift can be estimated from the separation between the doublet centres, Δ

$$\Delta = (J^2 + \delta\nu^2)^{1/2}$$

Δ is estimated from the figure to be 210 Hz. This yields for $\delta\nu$, the chemical shift,

$$\delta\nu = (\Delta^2 - J^2)^{1/2}$$

$$= (210^2 - 160^2)^{1/2}\,\text{Hz} \approx 140\,\text{Hz}$$

Collapse to a single line will occur when the rate of interconversion satisfies

$$k \approx \frac{1}{\tau} \approx \frac{\pi \Delta}{\sqrt{2}} \quad [18.44]$$

$$k = \frac{\pi \times 200 \, \text{s}^{-1}}{\sqrt{2}} \approx \boxed{4 \times 10^2 \, \text{s}^{-1}}$$

The relative intensities, I, of the lines at $-80°C$ can be used to estimate the energy difference $(E_{II} - E_I)$ between conformation I and conformations II and III. We assume that the relative intensities of the lines are proportional to the populations of conformers and that these populations follow the Boltzmann distribution (Chapters 2 and 19). Then

$$\frac{I_I}{I_{II}} = \frac{e^{-E_I/RT}}{e^{-E_{II}/RT}} = e^{(E_{II}-E_I)/RT}$$

$$E_{II} - E_I = RT \ln\left(\frac{I_I}{I_{II}}\right) = 8.314 \, \text{J K}^{-1} \, \text{mol}^{-1} \times (273 - 80) \, \text{K} \ln(10)$$

$$= 3.7 \times 10^3 \, \text{J mol}^{-1} = \boxed{3.7 \, \text{kJ mol}^{-1}}$$

This energy difference is not, however, the rotational energy barrier between the rotational isomers. The latter can be estimated from the rate of interconversion between the isomers as a function of temperature. That rate of interconversion is roughly $4 \times 10^2 \, \text{s}^{-1}$ at $-30°C$. At $-60°C$, as estimated from the line width at that temperature [16.26], it is roughly $1/3$ of that value, or $\sim 1.3 \times 10^2 \, \text{s}^{-1}$. Assuming that the rate of interconversion satisfies an Arrhenius type of behaviour, $k \propto e^{-E_a/RT}$, where E_a is the rotational energy barrier,

$$\frac{k(-30°C)}{k(-60°C)} = 3 = e^{\left\{-\frac{E_a}{R}\left(\frac{1}{243\,\text{K}} - \frac{1}{213\,\text{K}}\right)\right\}}$$

$$E_a = \frac{R \ln 3}{\left(\frac{1}{213\,\text{K}} - \frac{1}{243\,\text{K}}\right)} = 1.6 \times 10^4 \, \text{J mol}^{-1} = \boxed{16 \, \text{kJ mol}^{-1}}$$

This value is typical of the rotational barriers observed in compounds of this kind.

P18.6 (a) The Karplus equation [18.35] for $^3J_{HH}$ is a linear equation in $\cos\phi$ and $\cos 2\phi$. The experimentally determined equation for $^3J_{SnSn}$ is a linear equation in $^3J_{HH}$. In general, if $F(f)$ is linear in f, and if $f(x)$ is linear in x, then $F(x)$ is linear. So we expect $^3J_{SnSn}$ to be linear in $\cos\phi$ and $\cos 2\phi$. This is demonstrated in (b).

(b) $\quad ^3J_{SnSn}/\text{Hz} = 78.86(^3J_{HH}/\text{Hz}) + 27.84$

Inserting the Karplus equation for $^3J_{HH}$ we obtain

$^3J_{SnSn}/\text{Hz} = 78.86\{A + B\cos\phi + C\cos 2\phi\} + 27.84$. Using $A = 7$, $B = -1$, and $C = 5$, we obtain

$$^3J_{SnSn}/\text{Hz} = \boxed{580 - 79\cos\phi + 395\cos 2\phi}$$

The plot of $^3J_{SnSn}$ is shown in Fig. 18.7.

(c) A staggered configuration (Fig. 18.8) with the $SnMe_3$ groups *trans* to each other is the preferred configuration. The $SnMe_3$ repulsions are then at a minimum.

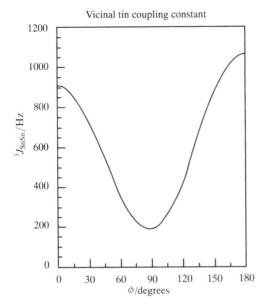

Vicinal tin coupling constant

Figure 18.7

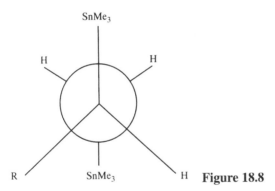

SnMe$_3$

H   H

R   SnMe$_3$ H   **Figure 18.8**

P18.7

$$g = \frac{h\nu}{\mu_B \mathcal{B}_0} [18.48] = \frac{(7.14478 \times 10^{-11}\,\text{T}) \times (\nu/\text{Hz})}{\mathcal{B}_0}$$

$$= \frac{(7.14478 \times 10^{-11}\,\text{T}) \times (9.302 \times 10^9)}{\mathcal{B}_0} = \frac{0.66461}{\mathcal{B}_0/T}$$

$$g_\parallel = \frac{0.66461}{0.33364} = \boxed{1.992} \qquad g_\perp = \frac{0.66461}{0.33194} = \boxed{2.002}$$

P18.9   Construct the spectrum by taking into account first the two equivalent ^{14}N splitting (producing a $\boxed{1:2:3:2:1 \text{ quintet}}$) and then the splitting of each of these lines into a $\boxed{1:4:6:4:1 \text{ quintet}}$ by the four equivalent protons. The resulting 25-line spectrum is shown in Fig. 18.9. Note that Pascal's triangle does not apply to the intensities of the quintet due to ^{14}N, but does apply to the quintet due to the protons.

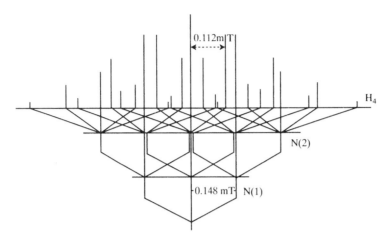

Figure 18.9

Solutions to theoretical problems

P18.12   Use eqn 18.29 and illustration 18.2. For hydrogen itself, we have:

$$\sigma_d = \frac{e^2 \mu_0}{12\pi m_e} \left\langle \frac{1}{r} \right\rangle = \frac{e^2 \mu_0}{12\pi_e a_0}.$$

The only difference in wavefunction (and therefore in the expectation value of $1/r$) between hydrogen and a more general hydrogenic ion is that the latter has a_0/Z where the former has a_0, so:

$$\boxed{\sigma_d = \frac{e^2 \mu_0 Z}{12\pi m_e a_0} = 1.78 \times 10^{-5} Z}.$$

P18.13   (a) The table displays experimental ^{13}C chemical shifts and computed* atomic charges on the carbon atom *para* to a number of substituents in substituted benzenes. Two sets of charges are shown, one derived by fitting the electrostatic potential and the other by Mulliken population analysis.

Substituent	OH	CH$_3$	H	CF$_3$	CN	NO$_2$
δ	130.1	128.4	128.5	128.9	129.1	129.4
Electrostatic charge/e	−0.1305	−0.1273	−0.0757	−0.0227	−0.0152	−0.0541
Mulliken charge/e	−0.1175	−0.1089	−0.1021	−0.0665	−0.0805	−0.0392

*Semi-empirical, PM3 level, PC Spartan Pro^{TM}

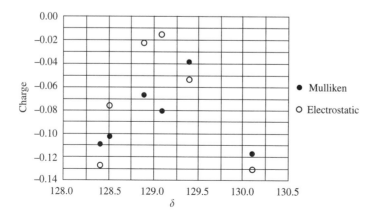

(b) Neither set of charges correlates well to the chemical shifts. If one removes phenol from the data set, a correlation would be apparent, particularly for the Mulliken Charges.

(c) The diamagnetic local contribution to shielding is roughly proportional to the electron density on the atom. The extent to which the *para*-carbon atom is affected by electron-donating or-withdrawing groups on the other side of the benzene ring is reflected in the net charge on the atom. If the diamagnetic local contribution dominated, then the more positive the atom, the greater the deshielding and the greater the chemical shift δ would be. That no such correlation is observed leads to several possible hypotheses: for example, the diamagnetic local contribution is not the dominant contribution in these molecules (or not in all of these molecules), or the computation is not sufficiently accurate to provide meaningful atomic charges.

P18.15 Equation 18.47 may be written

$$\mathcal{B} = k(1 - 3\cos^2\theta)$$

where k is a constant independent of angle.

Thus, $\langle\mathcal{B}\rangle \propto \displaystyle\int_0^\pi (1 - 3\cos^2\theta)\sin\theta\,d\theta \int_0^{2\pi} d\phi$

$$\propto \int_1^{-1} (1 - 3x^2)\,dx \times 2\pi \quad [x = \cos\theta, dx = -\sin\theta\,d\theta]$$

$$\propto (x - x^3)\Big|_1^{-1} = 0$$

P18.16 The shape of spectral line $\mathcal{I}(\omega)$ is related to the free induction decay signal $G(t)$ by

$$\mathcal{I}(\omega) = A\,\mathrm{Re}\int_0^\infty G(t)e^{i\omega t}\,dt$$

where A is a constant and Re means take the real part of what follows. Calculate the lineshape corresponding to an oscillating, decaying function

$$G(t) = \cos\omega_0 t\, e^{-t/\tau}$$

$$\mathcal{I}(\omega) = A\,\mathrm{Re}\int_0^\infty G(t)e^{i\omega t}\,dt$$

$$= A\,\mathrm{Re}\int_0^\infty \cos\omega_0 t\, e^{-t/\tau + i\omega t}\,dt$$

$$= \frac{1}{2}A\,\mathrm{Re}\int_0^\infty (e^{-i\omega_0 t} + e^{i\omega_0 t})e^{-t/\tau + i\omega t}\,dt$$

$$= \frac{1}{2}A\,\mathrm{Re}\int_0^\infty \{e^{i(\omega_0 + \omega + i/\tau)t} + e^{-i(\omega_0 - \omega - i/\tau)t}\}\,dt$$

$$= -\frac{1}{2}A\,\mathrm{Re}\left[\frac{1}{i(\omega_0 + \omega + i/\tau)} - \frac{1}{i(\omega_0 - \omega - i/\tau)}\right]$$

when ω and ω_0 are similar to magnetic resonance frequencies (or higher), only the second term in brackets in significant $\left(\text{because } \dfrac{1}{(\omega_0 + \omega)} \ll 1 \text{ but } \dfrac{1}{(\omega_0 - \omega)} \text{ may be large if } \omega \approx \omega_0\right)$. Therefore

$$\mathcal{I}(\omega) \approx \frac{1}{2} A \operatorname{Re} \frac{1}{i(\omega_0 - \omega)^2 + 1/\tau}$$

$$= \frac{1}{2} A \operatorname{Re} \frac{1}{(\omega_0 - \omega)^2 + 1/\tau^2}$$

$$\boxed{= \frac{1}{2} \frac{A\tau}{1 + (\omega_0 - \omega)^2 \tau^2}}$$

which is Lorentzian line centred on ω_0, of amplitude $\frac{1}{2} A\tau$ and width $\dfrac{2}{\tau}$ at half-height.

P18.18   (a) $\mu = g_I \mu_N |I|$ $[\mu_N = 5.05079 \times 10^{-27} \, \text{JT}^{-1}]$

Using the formulas

$$\text{Sensitivity ratio } (\nu) = \frac{R_\nu(\text{nuclide})}{R_\nu(^1\text{H})} = \frac{2}{3}(I+1)\left[\frac{\mu(\text{nuclide})}{\mu(^1\text{H})}\right]$$

$$\text{Sensitivity ratio}(\mathcal{B}) = \frac{R_\mathcal{B}(\text{nuclide})}{R_\mathcal{B}(^1\text{H})} = \frac{1}{6}\left(\frac{I+1}{I^2}\right)\left[\frac{\mu(\text{nuclide})}{\mu(^1\text{H})}\right]^3$$

we construct the following table

Nuclide	Spin I	μ/μ_N	Sensitivity ratio(ν)	Sensitivity ratio ($\mathcal{B}$)
^{2}H	1	0.85745	0.409	0.00965
^{13}C	$\frac{1}{2}$	0.7023	0.251	0.01590
^{14}N	1	0.40356	0.193	0.00101
^{19}F	$\frac{1}{2}$	2.62835	0.941	0.83350
^{31}P	$\frac{1}{2}$	1.1317	0.405	0.06654
^{1}H	$\frac{1}{2}$	2.79285		

(b) $\mu = \gamma \hbar I = g_I \mu_N |I|$

Hence $\gamma = \dfrac{\mu}{\hbar I}$

At constant frequency

$$R_\nu \propto (I+1)\mu\omega_0^2 \text{ or } R_\nu \propto (I+1)\mu \quad [\omega_0 \text{ is constant between the nuclei}]$$

Thus

$$\text{Sensitivity ratio}(\nu) = \frac{R_\nu(\text{nuclide})}{R_\nu(^1\text{H})}$$

$$= \tfrac{2}{3}(I+1)\left[\frac{\mu(\text{nuclide})}{\mu(^1\text{H})}\right] = \tfrac{2}{3}(I+1)\left[\frac{\mu(\text{nuclide})/\mu_N}{\mu(^1\text{H})/\mu_N}\right]$$

as above. Substituting $\omega_0 = \gamma \mathcal{B}_0$ and $\gamma = \dfrac{\mu}{\hbar I}$, $\omega_0 = \dfrac{\mu \mathcal{B}_0}{\hbar I}$ so

$$R_\mathcal{B} \propto \frac{(I+1)\mu^3 \mathcal{B}_0^2}{I^2}$$

$$\text{Sensitivity ratio}(\mathcal{B}) = \frac{R_\mathcal{B}(\text{nuclide})}{R_\mathcal{B}(^1\text{H})} = \tfrac{1}{6}\left(\frac{I+1}{I^2}\right)\left[\frac{\mu(\text{nuclide})}{\mu(^1\text{H})}\right]^3$$

$$= \tfrac{1}{6}\left(\frac{I+1}{I^2}\right)\left[\frac{\mu(\text{nuclide})/\mu_N}{\mu(^1\text{H})/\mu_N}\right]^3$$

as in part (a)

Solutions to applications

P18.19 $$\langle \mathcal{B}_{nucl} \rangle = \frac{-g_I \mu_N \mu_0 m_I}{4\pi R^3} \frac{\int_0^{\theta_{max}} (1 - 3\cos^2 \theta) \sin \theta \, d\theta}{\int_0^{\theta_{max}} \sin \theta \, d\theta}$$

The denominator is the normalization constant, and ensures that the total probability of being between 0 and θ_{max} is 1.

$$= \frac{-g_I \mu_N \mu_0 m_I}{4\pi R^3} \frac{\int_1^{x_{max}} (1 - 3x^2) dx}{\int_1^{x_{max}} dx} [x_{max} = \cos \theta_{max}]$$

$$= \frac{-g_I \mu_N \mu_0 m_I}{4\pi R^3} \times \frac{x_{max}(1 - x_{max}^2)}{x_{max} - 1} = \boxed{\frac{-g_I \mu_N \mu_0 m_I}{4\pi R^3} (\cos^2 \theta_{max} + \cos \theta_{max})}$$

If $\theta_{max} = \pi$ (complete rotation), $\cos \theta_{max} = -1$ and $\langle \mathcal{B}_{nucl} \rangle = 0$. If $\theta_{max} = 30°$, $\cos^2 \theta_{max} + \cos \theta_{max} = 1.616$, and

$$\langle \mathcal{B}_{nucl} \rangle = \frac{(5.5857) \times (5.0508 \times 10^{-27} \, \text{J T}^{-1}) \times (4\pi \times 10^{-7} \, \text{T}^2 \, \text{J}^{-1} \, \text{m}^3) \times (1.616)}{(4\pi) \times (1.58 \times 10^{-10} \, \text{m})^3 \times (2)}$$

$$= \boxed{0.58 \, \text{mT}}$$

P18.20 At, say, room temperature, the tumbling rate of benzene, the small molecule, in a mobile solvent, may be close to the Larmor frequency, and hence its spin-lattice relaxation time will be short. As the temperature increases, the tumbling rate may increase well beyond the Larmor frequency, resulting in an increased spin–lattice relaxation time.

For the large oligopeptide at room temperature, the tumbling rate may be well below the Larmor frequency, but with increasing temperature it will approach the Larmor frequency due to the increased thermal motion of the molecule combined with the decreased viscosity of the solvent. Therefore, the spin-lattice relaxation time may decrease.

P18.22 **(a)** The first figure displays spin densities computed by molecular modeling software (*ab initio*, density functional theory, Gaussian 98^{TM}).

(b) First, note that the software assigned slightly different values to the two protons *ortho* to the oxygen and to the two protons *meta* to the oxygen. This is undoubtedly a computational artifact, a result of the minimum-energy structure having one methyl proton in the plane of the ring, which makes the right and left side of the ring slightly non-equivalent. (See second figure.) In fact, fast internal rotation makes the two halves of the ring equivalent. We will take the spin density at the *ortho* carbons to be 0.285 and those of the *meta* carbons to be −0.132. Predict the form of the spectrum by using the McConnell equation (18.52) for the splittings. The two *ortho* protons give rise to a $1:2:1$ triplet with splittings 0.285×2.25 mT $= 0.64$ mT; these will in turn be split by the two *meta* protons into $1:2:1$ triplets with splitting 0.132×2.25 mT $= 0.297$ mT. And finally, these lines will be seen to be further split by the three methyl protons into $1:3:3:1$ quartets with splittings 1.045 mT. Note that the McConnel relation cannot be applied to calculate these latter splittings, but the software generates them directly from calculated spin densities on the methyl hydrogens. The computed splittings agree well with experiment at the *ortho* positions $(0.60$ mT) and at the methyl hydrogens $(1.19$ mT), but less well at the *meta* positions $(0.145$ mT).

19 Statistical thermodynamics: the concepts

Solutions to exercises

Discussion questions

E19.1(a) The Boltzmann distribution applies to particles which can be thought of as distinguishable from one another and for which there is no restriction on the number of them which can occupy a given energy state. For such particles, the number of ways a specified set of population numbers (a configuration) can be achieved is given by eqn 19.1. This number is referred to as the statistical weight, W, of the configuration. It turns out that for all practical purposes the properties of a system consisting of, say a mole of particles, can be determined solely from the configuration with the largest statistical weight. Therefore, the procedure for determining the Boltzmann distribution reduces to a mathematical process of finding the set of population numbers, that is, the configuration or distribution, which maximizes W. The maximum value of W obeys $d \ln W = 0$, so it might seem that we should be able to obtain W_{max} by simple differentiation of W with respect to the population numbers. But it is not quite that simple because we are looking for a maximum subject to two constraints on the distribution numbers: that the total number of particles, N, and the total energy, E, of the system must be constants. The mathematical technique for finding this constrained maximum is described in detail in *Justification* 19.3 and will not be repeated here. It involves the method of undetermined multipliers described in *Further information 1*. After the elimination of one of the undetermined multipliers, α, and the identification of the other, β, as $1/kT$, the Boltzmann distribution, eqn 19.6a, follows.

E19.2(a) See Figures 19.8 and 19.11, *Illustration* 19.3, Self-test 19.6, and the solution to Exercise 19.12(a)

E19.3(a) We evaluate β by comparing calculated and experimental values for thermodynamic properties. The calculated values are obtained from the theoretical formulas for these properties, all of which are expressed in terms of the parameter β. So there can be many ways of identifying β, as many as there are thermodynamic properties. One way is through the energy as shown in Section 19.3(b). Another is through the pressure as demonstrated in Example 20.1. Another yet is through the entropy, and this approach to the identification may be the most fundamental. See *Further reading* for elaboration of this method.

E19.4(a) An ensemble is a set of a large number of imaginary replications of the actual system. These replications are identical in some respects, but not in all respects. For example, in the canonical ensemble, all replications have the same number of particles, the same volume, and the same temperature, but not the same energy. Ensembles are useful in statistical thermodynamics because it is mathematically more tractable to perform an ensemble average to determine the (time averaged) thermodynamic properties than it is to perform an average over time to determine these properties. Recall that macroscopic thermodynamic properties are averages over the time dependent properties of the particles that compose the macroscopic system. In fact, it is taken as a fundamental principle of statistical thermodynamics that the (sufficiently long) time average of every physical observable is equal to its ensemble average. This principle is connected to a famous assumption of Boltzmann's called the ergodic hypothesis. A thorough discussion of these topics would take us far beyond what we need here. See the references under *Further reading*.

Numerical exercises

E19.5(a)
$$n_i = \frac{Ne^{-\beta\varepsilon_i}}{q} \qquad \left[19.6a, \text{ with } q = \sum_i e^{-\beta\varepsilon_i}\right]$$

Hence, $\dfrac{n_2}{n_1} = \dfrac{e^{-\beta \varepsilon_2}}{e^{-\beta \varepsilon_1}} = e^{-\beta(\varepsilon_2 - \varepsilon_1)} = e^{-\beta \Delta \varepsilon} = e^{-\Delta \varepsilon / kT}$ $\left[\beta = \dfrac{1}{kT} \right]$

as $T \to \infty$, $\dfrac{n_2}{n_1} = e^{-0} = \boxed{1}$

E19.6(a) $q = \dfrac{V}{\Lambda^3} = \left(\dfrac{2\pi m}{h^2 \beta} \right)^{3/2} V \, [19.22] = \left(\dfrac{2\pi m k T}{h^2} \right)^{3/2} V$

$$= \left(\dfrac{(2\pi) \times (120 \times 10^{-3} \, \mathrm{kg\,mol^{-1}}) \times (1.381 \times 10^{-23} \, \mathrm{J\,K^{-1}}) \times T}{(6.022 \times 10^{23} \, \mathrm{mol^{-1}}) \times (6.626 \times 10^{-34} \, \mathrm{J\,s})^2} \right)^{3/2} \times (2.00 \times 10^{-6} \, \mathrm{m^3})$$

(a) $T = 300 \, \mathrm{K}$, $q = (4.94 \times 10^{23}) \times (300)^{3/2} = \boxed{2.57 \times 10^{27}}$

(b) $T = 600 \, \mathrm{K}$, $q = (4.94 \times 10^{23}) \times (600)^{3/2} = \boxed{7.26 \times 10^{27}}$

E19.7(a) $q = \dfrac{V}{\Lambda^3} \, [19.22]$,   implying that $\dfrac{q}{q'} = \left(\dfrac{\Lambda'}{\Lambda} \right)^3$

However, as $\Lambda \propto \dfrac{1}{m^{1/2}}$, $\dfrac{q}{q'} = \left(\dfrac{m}{m'} \right)^{3/2}$

Therefore, $\dfrac{q(\mathrm{D_2})}{q(\mathrm{H_2})} = 2^{3/2} = \boxed{2.83}$ $(m(\mathrm{D_2}) = 2m(\mathrm{H_2}))$

E19.8(a) $q = \sum\limits_{\text{levels}} g_j e^{-\beta \varepsilon_j} \, [19.12] = 3 + (e^{-\beta \varepsilon_1}) + (3 e^{-\beta \varepsilon_2})$

$\beta \varepsilon = \dfrac{hc\tilde{\nu}}{kT} = \dfrac{1.4388(\tilde{\nu}/\mathrm{cm^{-1}})}{T/\mathrm{K}}$ [inside front cover]

Therefore, $q = 3 + (e^{-(1.4388) \times (3500)/1900}) + (3 e^{-(1.4388) \times (4700)/1900}) = 3 + 0.0706 + 0.085 = \boxed{3.156}$

E19.9(a) $E = -\dfrac{N}{q} \dfrac{dq}{d\beta} \, [19.25] = -\dfrac{N}{q} \dfrac{d}{d\beta} (3 + e^{-\beta \varepsilon_1} + 3 e^{-\beta \varepsilon_2}) = -\dfrac{N}{q} (-\varepsilon_1 e^{-\beta \varepsilon_1} - 3 \varepsilon_2 e^{-\beta \varepsilon_2})$

$\quad = \dfrac{Nhc}{q} (\tilde{\nu}_1 e^{-\beta hc \tilde{\nu}_1} + 3 \tilde{\nu}_2 e^{-\beta hc \tilde{\nu}_2})$

$\quad = \left(\dfrac{N_A hc}{3.156} \right) \times \left(\tilde{\nu}_1 e^{(-hc\tilde{\nu}_1)/kT} + 3 \tilde{\nu}_2 e^{-(hc\tilde{\nu}_2)/kT} \right)$

$\quad = \left(\dfrac{N_A hc}{3.156} \right) \times \left(0 + 3500 \, \mathrm{cm^{-1}} \times e^{-(1.4388 \times 3500)/1900} + 3 \times 4700 \, \mathrm{cm^{-1}} \times e^{-(1.4388 \times 4700)/1900} \right)$

$\quad = N_A hc \times (204.9 \, \mathrm{cm^{-1}}) = \boxed{2.45 \, \mathrm{kJ\,mol^{-1}}}$

E19.10(a) $\dfrac{n_i}{N} = \dfrac{e^{-\beta \varepsilon_i}}{q} \, [19.6a]$

Therefore, $\dfrac{n_{\mathrm{ex}}}{n_{\mathrm{g}}} = \dfrac{e^{-\beta \varepsilon_{\mathrm{ex}}}}{e^{-\beta \varepsilon_{\mathrm{g}}}} = e^{-\beta \varepsilon}$ $[\varepsilon = \varepsilon_{\mathrm{ex}} - \varepsilon_{\mathrm{g}}]$

Solving for β, $\beta = \dfrac{1}{\varepsilon} \ln \dfrac{n_{\mathrm{g}}}{n_{\mathrm{ex}}}$ or $T = \dfrac{\varepsilon/k}{\ln\left(\dfrac{n_{\mathrm{g}}}{n_{\mathrm{ex}}} \right)}$

and $T = \dfrac{\left(\frac{hc\tilde{\nu}}{k}\right)}{\ln\left(\frac{n_g}{n_{ex}}\right)} = \dfrac{(1.4388\,\text{cm K}) \times (540\,\text{cm}^{-1})}{\ln\left(\frac{0.90}{0.10}\right)} = \boxed{354\,\text{K}}$

E19.11(a) $q = \sum_i e^{-\beta\varepsilon_i} = \boxed{1 + e^{-2\mu_B\beta B}}$ [energies measured from lower state]

$$\langle\varepsilon\rangle = \frac{E}{N} = -\frac{1}{q}\frac{dq}{d\beta}\,[19.25] = \boxed{\frac{2\mu_B B\,e^{-2\mu_B\beta B}}{1 + e^{-2\mu_B\beta B}}}$$

We write $x = 2\mu_B\beta B$, then $\dfrac{\langle\varepsilon\rangle}{2\mu_B B} = \dfrac{e^{-x}}{1 + e^{-x}} = \dfrac{1}{e^x + 1}$

This function is plotted in Fig. 19.1. For the partition function we plot

$q = 1 + e^{-x}$

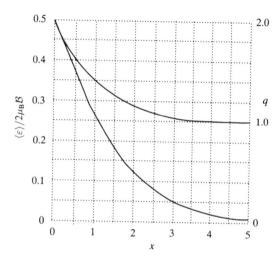

Figure 19.1

The relative populations are

$\dfrac{n_+}{n_-} = e^{-\beta\Delta\varepsilon}\,[\text{Exercise 19.5}] = e^{-x}$

$$x = 2\mu_B\beta B = \frac{(2) \times (9.274 \times 10^{-24}\,\text{J T}^{-1}) \times (1.0)\text{T}}{(1.381 \times 10^{-23}\,\text{J K}^{-1})T} = 1.343/(T/\text{K})$$

(a) $T = 4\,\text{K},\quad \dfrac{n_+}{n_-} = e^{-1.343/4} = \boxed{0.71}$ **(b)** $T = 298\,\text{K},\quad \dfrac{n_+}{n_-} = e^{-1.343/298} = \boxed{0.996}$

E19.12(a)   The energy separation is $\varepsilon = k \times (10\,\text{K})$

(a) $\dfrac{n_1}{n_0} = e^{-\beta(\varepsilon_1 - \varepsilon_0)}\,[\text{Exercise 19.5}] = e^{-\beta\varepsilon} = e^{-10/(T/\text{K})}$

(1) $T = 1.0\,\text{K};\quad \dfrac{n_1}{n_0} = e^{-10} = \boxed{5 \times 10^{-5}}$

(2) $T = 10\,\text{K};\quad \dfrac{n_1}{n_0} = e^{-1.0} = \boxed{0.4}$

(3) $T = 100\,\text{K};\quad \dfrac{n_1}{n_0} = e^{-0.100} = \boxed{0.905}$

(b) $q = \sum_j g_j e^{-\varepsilon_j/kT} = e^0 + e^{-1.0} = \boxed{1.4}$

(c) $E = -\dfrac{N_A}{q}\dfrac{dq}{d\beta}$ [19.25]

$q = 1 + e^{-10\,\text{K}\times k\beta}$

$E = -\dfrac{N_A}{q}\left\{-(10\,\text{K}) \times k e^{-(10\,\text{K}\times k\beta)}\right\} = \dfrac{(10\,\text{K}) \times R}{1 + e^{(10\,\text{K}\times k\beta)}} = \dfrac{(10\,\text{K}) \times R}{1 + e^{10/(T/\text{K})}}$

At $T = 10\,\text{K},\quad E = \dfrac{(10\,\text{K}) \times R}{1 + e} = \dfrac{(10\,\text{K}) \times (8.314\,\text{J K}^{-1}\,\text{mol}^{-1})}{3.718} = \boxed{22\,\text{J mol}^{-1}}$

(d) $C_V = \left(\dfrac{\partial U}{\partial T}\right)_V = \dfrac{dE}{dT}\qquad \dfrac{d}{dT} = -\dfrac{1}{kT^2}\dfrac{d}{d\beta}$

$= -\dfrac{1}{kT^2}\dfrac{d}{d\beta}\left(\dfrac{(10\,\text{K}) \times R}{1 + e^{(10\,\text{K}\times k\beta)}}\right) = R\left(\dfrac{e}{(1+e)^2}\right) = 0.19\bar{7}R = \boxed{1.6\,\text{J K}^{-1}\,\text{mol}^{-1}}$

(e) $S = \dfrac{U - U(0)}{T} + N_A k \ln q = \dfrac{E}{T} + R \ln q$

$= \left(\dfrac{22.3\bar{6}\,\text{J mol}^{-1}}{10\,\text{K}}\right) + (R \ln 1.3\bar{6}8) = \boxed{4.8\,\text{J K}^{-1}\,\text{mol}^{-1}}$

E19.13(a) $\dfrac{n_1}{n_0} = e^{-\beta\varepsilon} = e^{-hc\tilde{\nu}/kT}$

$\tilde{\nu} = 2991\,\text{cm}^{-1}$ [Table 16.2, assume $^1\text{H}^{35}\text{Cl}$]

$\dfrac{1}{e} = e^{-hc\tilde{\nu}/kT}$

$-1 = \dfrac{-hc\tilde{\nu}}{kT}$

$T = \dfrac{hc\tilde{\nu}}{k} = (1.4388\,\text{cm K}) \times (2991\,\text{cm}^{-1}) = \boxed{4303\,\text{K}}$

Comment. Vibrational energy level separations are large compared to kT at room temperature which is $207\,\text{cm}^{-1}$ at 298 K. Thus high temperatures are required to achieve substantial population in excited vibrational states.

Question. If thermal decomposition of HCl occurs when 1 per cent of HCl molecules find themselves in a vibrational state of energy corresponding to the bond dissociation energy $(431\,\text{kJ mol}^{-1})$, what temperature is required? Assume ε is constant at $2991\,\text{cm}^{-1}$ and do not take the result too seriously.

E19.14(a) $S_m^{\ominus} = R \ln\left(\dfrac{e^{5/2}kT}{p^{\ominus}\Lambda^3}\right)$ [19.45b with $p = p^{\ominus}$]

$\Lambda = \dfrac{h}{(2\pi mkT)^{1/2}} = \dfrac{6.626 \times 10^{-34}\,\text{J s}}{[(2\pi) \times (20.18) \times (1.6605 \times 10^{-27}\,\text{kg}) \times (1.381 \times 10^{-23}\,\text{J K}^{-1}T)]^{1/2}}$

$= \dfrac{3.886 \times 10^{-10}\,\text{m}}{(T/\text{K})^{1/2}}$

$S_m^{\ominus} = R \ln\left(\dfrac{(e^{5/2}) \times (1.381 \times 10^{-23}\,\text{J K}^{-1}T)}{(1 \times 10^5\,\text{Pa}) \times (3.886 \times 10^{-10}\,\text{m})^3}\right) \times \left(\dfrac{T}{\text{K}}\right)^{3/2} = R \ln(28.67) \times (T/\text{K})^{5/2}$

(a) $T = 200\,\text{K},\quad S_m^{\ominus} = (8.314\,\text{J K}^{-1}\,\text{mol}^{-1}) \times \ln(28.67) \times (200)^{5/2} = \boxed{138\,\text{J K}^{-1}\,\text{mol}^{-1}}$

(b) $\quad T = 298.15 \, \text{K}, \quad S_m^{\ominus} = (8.314 \, \text{J K}^{-1} \, \text{mol}^{-1}) \times \ln(28.67) \times (298.15)^{5/2}$

$$= \boxed{146 \, \text{J K}^{-1} \, \text{mol}^{-1}}$$

E19.15(a) $\quad q = \dfrac{1}{1 - e^{-\beta\varepsilon}}$ [Example 19.2] $= \dfrac{1}{1 - e^{-hc\beta\tilde{\nu}}}$

$$hc\beta\tilde{\nu} = \frac{(1.4388 \, \text{cm K}) \times (560 \, \text{cm}^{-1})}{500 \, \text{K}} = 1.611$$

Therefore, $q = \dfrac{1}{1 - e^{-1.611}} = 1.249$

The internal energy due to vibrational excitation is

$$U - U(0) = \frac{N\varepsilon e^{-\beta\varepsilon}}{1 - e^{-\beta\varepsilon}} [19.27a \text{ and Example } 19.2] = \frac{Nhc\tilde{\nu}e^{-hc\tilde{\nu}\beta}}{1 - e^{-hc\tilde{\nu}\beta}} = \frac{Nhc\tilde{\nu}}{e^{hc\tilde{\nu}\beta} - 1}$$

$$= (0.249) \times (Nhc) \times (560 \, \text{cm}^{-1})$$

and hence

$$\frac{S_m}{N_A k} = \frac{U - U(0)}{N_A kT} + \ln q \, [19.34] = (0.249) \times \left(\frac{hc}{kT}\right) \times (560 \, \text{cm}^{-1}) + \ln(1.249)$$

$$= \left(\frac{(0.249) \times (1.4388 \, \text{K cm}) \times (560 \, \text{cm}^{-1})}{500 \, \text{K}}\right) + \ln(1.249) = 0.401 + 0.222 = 0.623$$

Hence, $S_m = 0.623R = \boxed{5.18 \, \text{J K}^{-1} \, \text{mol}^{-1}}$

E19.16(a) $\boxed{\text{(a)}}$ Yes; He atoms indistinguishable and non-localized.

$\boxed{\text{(b)}}$ Yes; CO molecules indistinguishable and non-localized.

(c) No; CO molecules can be identified by their locations.

$\boxed{\text{(d)}}$ Yes; H_2O molecules indistinguishable and non-localized.

Solutions to problems

Solutions to numerical problems

P19.1 $\quad$ Number of configurations of combined system, $W = W_1 W_2$

$$W = (10^{20}) \times (2 \times 10^{20}) = \boxed{2 \times 10^{40}}$$

$$S = k \ln W \, [19.30]; \qquad S_1 = k \ln W_1; \qquad S_2 = k \ln W_2$$

$$S = k \ln(2 \times 10^{40}) = k\{\ln 2 + 40 \ln 10\} = 92.8 \, k$$

$$= 92.8 \times (1.381 \times 10^{-23} \, \text{J K}^{-1}) = \boxed{1.282 \times 10^{-21} \, \text{J K}^{-1}}$$

$$S_1 = k \ln(10^{20}) = k\{20 \ln 10\} = 46.1 \, k$$

$$= 46.1 \times (1.381 \times 10^{-23} \, \text{J K}^{-1}) = \boxed{0.637 \times 10^{-21} \, \text{J K}^{-1}}$$

$$S_2 = k \ln(2 \times 10^{20}) = k\{\ln 2 + 20 \ln 10\} = 46.7\,k$$

$$= 46.7 \times (1.381 \times 10^{-23}\,\mathrm{J\,K^{-1}}) = \boxed{0.645 \times 10^{-21}\,\mathrm{J\,K^{-1}}}$$

These results are significant in that they show that the statistical mechanical entropy is an additive property consistent with the thermodynamic result. That is, $S = S_1 + S_2 = (0.637 \times 10^{-21} + 0.645 \times 10^{-21})\,\mathrm{J\,K^{-1}} = 1.282 \times 10^{-21}\,\mathrm{J\,K^{-1}}$

P19.2   Although He is a liquid at these temperatures ($T_b = 4.22\,\mathrm{K}$), we will test it as if it were a perfect gas with no interaction potential.

$$p_i = \frac{N_i}{N} = g_i e^{-\beta \varepsilon_i}/q \quad [19.6a]$$

$$\varepsilon_i = \frac{h^2}{8mX^2}\{n_x^2 + n_y^2 + n_z^2\} \quad [19.19]; \qquad q = \frac{V}{\Lambda^3}; \quad \Lambda = h\left(\frac{\beta}{2\pi m}\right)^{1/2} \quad [19.22]$$

Ground state $n_x = n_y = n_z = 1;\ g = 1$

First excited state

$$\left.\begin{array}{ll} n_x = n_y = 1; & n_z = 2 \\ n_x = n_z = 1; & n_y = 2 \\ n_y = n_z = 1; & n_x = 2 \end{array}\right\} g = 3$$

$$q = \frac{V}{\Lambda^3} = \frac{V}{h^3}(2\pi m k T)^{3/2}$$

$$= \frac{(1\,\mathrm{cm}^3) \times \left(\frac{1\,\mathrm{m}^3}{10^6\,\mathrm{cm}^3}\right) \times \left[2\pi(1.381 \times 10^{-23}\,\mathrm{J\,K^{-1}})\right]^{+3/2} \times (mT)^{+3/2}}{(6.626 \times 10^{-34}\,\mathrm{J\,s})^3}$$

$$= 2.28 \times 10^{60}\,\mathrm{kg}^{-3/2}\,\mathrm{K}^{-3/2}(mT)^{3/2}$$

$$\beta\varepsilon_{1\text{st excited}} = \left(\frac{1}{kT}\right) \times \left(\frac{h^2}{8mX^2}\right) \quad (6)$$

$$= \frac{6(6.626 \times 10^{-34}\,\mathrm{J\,s})^2}{8(1.381 \times 10^{-23}\,\mathrm{J\,K^{-1}}) \times (0.01\,\mathrm{m})^2}\frac{1}{mT}$$

$$= \frac{2.38 \times 10^{-40}\,\mathrm{kg\,K}}{mT}$$

$$p_{1\text{st excited}} = \frac{3e^{-\left(\frac{2.38 \times 10^{-40}\,\mathrm{kg\,K}}{mT}\right)}}{(2.78 \times 10^{60}\,\mathrm{kg}^{-3/2}\,\mathrm{K}^{-3/2}) \times (mT)^{3/2}}$$

Isotope	m/kg	T/K	$p_{1\text{st excited}}$	Occupancy $= pN = 10^{22}p$
^{4}He	6.64×10^{-27}	0.0010	6.30×10^{-17}	6.30×10^5
		2.0	7.04×10^{-22}	7
		4.0	2.49×10^{-22}	2
^{3}He	5.01×10^{-27}	0.0010	9.63×10^{-17}	9.63×10^5
		2.0	1.08×10^{-21}	11
		4.0	3.81×10^{-22}	4

These results may at first seem to contradict the expected common sense result that the populations of excited states increase as the temperature increases, but the energy separations of these states is so small that even a slight increase in temperature promotes the particles to much higher quantum states.

P19.3 $S = k \ln W$ [19.30]

Therefore,

$$\left(\frac{\partial S}{\partial U}\right)_V = \frac{k}{W}\left(\frac{\partial W}{\partial U}\right)_V$$

or

$$\left(\frac{\partial W}{\partial U}\right)_V = \frac{W}{k}\left(\frac{\partial S}{\partial U}\right)_V$$

But from eqn 5.4

$$\left(\frac{\partial U}{\partial S}\right)_V = T$$

So,

$$\left(\frac{\partial S}{\partial U}\right)_V = \frac{1}{T}$$

then

$$\left(\frac{\partial W}{\partial U}\right)_V = \frac{W}{k}\left(\frac{1}{T}\right)$$

Therefore,

$$\frac{\Delta W}{W} \approx \frac{\Delta U}{kT}$$

$$= \frac{100 \times 10^3 \, \text{J}}{(1.381 \times 10^{-23} \, \text{J K}^{-1}) \times 298 \, \text{K}}$$

$$= \boxed{2.4 \times 10^{25}}$$

P19.5 $q = \frac{V}{\Lambda^3}, \qquad \Lambda = \frac{h}{(2\pi mkT)^{1/2}} \qquad \left[19.22, \beta = \frac{1}{kT}\right]$

and hence

$$T = \left(\frac{h^2}{2\pi mk}\right) \times \left(\frac{q}{V}\right)^{2/3}$$

$$= \left(\frac{(6.626 \times 10^{-34} \, \text{J s})^2}{(2\pi) \times (39.95) \times (1.6605 \times 10^{27} \, \text{kg}) \times (1.381 \times 10^{-23} \, \text{J K}^{-1})}\right)$$

$$\times \left(\frac{10}{1.0 \times 10^{-6} \, \text{m}^3}\right)^{2/3}$$

$$= \boxed{3.5 \times 10^{-15} \, \text{K}} \text{ [a very low temperature]}$$

The exact partition function in one dimension is

$$q = \sum_{n=1}^{\infty} e^{-(n^2-1)h^2\beta/8mL^2}$$

For an Ar atom in a cubic box of side 1.0 cm,

$$\frac{h^2\beta}{8mL^2}$$

$$= \frac{(6.626 \times 10^{-34}\,\text{J s})^2}{(8) \times (39.95) \times (1.6605 \times 10^{-27}\,\text{kg}) \times (1.381 \times 10^{-23}\,\text{J K}^{-1}) \times (3.5 \times 10^{-15}\,\text{K}) \times (1.0 \times 10^{-2}\,\text{m})^2}$$

$$= 0.17\bar{1}$$

Then $q = \sum_{n=1}^{\infty} e^{-0.17\bar{1}(n^2-1)} = 1.00 + 0.60 + 0.25 + 0.08 + 0.02 + \cdots = 1.95$

The partition function for motion in three dimensions is therefore $q = (1.95)^3 = \boxed{7.41}$

Comment. Temperatures as low as 3.5×10^{-15} K have never been achieved. However, a temperature of 2×10^{-8} K has been attained by adiabatic nuclear demagnetization (Box 4.2).

Question. Does the integral approximation apply at 2×10^{-8} K?

P19.7 (a) $q = \sum_j g_j e^{-\beta\varepsilon_j}$ [19.12] $= \sum_j g_j e^{-hc\beta\tilde{\nu}_j}$

We use $hc\beta = \dfrac{1}{207\,\text{cm}^{-1}}$ at 298 K and $\dfrac{1}{3475\,\text{cm}^{-1}}$ at 5000 K. Therefore,

(i) $q = 5 + e^{-4707/207} + 3e^{-4751/207} + 5e^{-10559/207}$

$\quad = (5) + (1.3 \times 10^{-10}) + (3.2 \times 10^{-10}) + (3.5 \times 10^{-22}) = \boxed{5.00}$

(ii) $q = 5 + e^{-4707/3475} + 3e^{-4751/3475} + 5e^{-10559/3475}$

$\quad = (5) + (0.26) + (0.76) + (0.24) = \boxed{6.26}$

(b) $p_j = \dfrac{g_j e^{-\beta\varepsilon_j}}{q} = \dfrac{g_j e^{-hc\beta\tilde{\nu}_j}}{q}$ [19.10, with degeneracy g_j included]

Therefore, $p_0 = \dfrac{5}{q} = \boxed{1.00}$ at 298 K and $\boxed{0.80}$ at 5000 K

$$p_2 = \frac{3e^{-4751/207}}{5.00} = \boxed{6.5 \times 10^{-11}} \text{ at 298 K}$$

$$p_2 = \frac{3e^{-4751/3475}}{6.26} = \boxed{0.12} \text{ at 5000 K}$$

(c) $S_m = \dfrac{U_m - U_m(0)}{T} + Nk \ln q$ [19.34]

We need $U_m - U_m(0)$, and evaluate it by explicit summation

$$U_m - U_m(0) = E = \frac{N_A}{q} \sum_j g_j \varepsilon_j e^{-\beta\varepsilon_j} \quad [\text{19.24 with degeneracy } g_j \text{ included}]$$

In terms of wavenumber units

(i) $\dfrac{U_m - U_m(0)}{N_A hc} = \dfrac{1}{5.00}\{0 + 4707\,\text{cm}^{-1} \times e^{-4707/207} + \cdots\} = 4.32 \times 10^{-7}\,\text{cm}^{-1}$

(ii) $\dfrac{U_m - U_m(0)}{N_A hc} = \dfrac{1}{6.26}\{0 + 4707\,\text{cm}^{-1} \times e^{-4707/3475} + \cdots\} = 1178\,\text{cm}^{-1}$

Hence, at 298 K

$$U_m - U_m(0) = 5.17 \times 10^{-6} \,\text{J mol}^{-1}$$

and at 5000 K

$$U_m - U_m(0) = 14.10 \,\text{kJ mol}^{-1}$$

It follows that

(i) $S_m = \left(\dfrac{5.17 \times 10^{-6} \,\text{J mol}^{-1}}{298 \,\text{K}} \right) + (8.314 \,\text{J K}^{-1} \,\text{mol}^{-1}) \times (\ln 5.00)$

$= \boxed{13.38 \,\text{J K}^{-1} \,\text{mol}^{-1}}$ [essentially $R \ln 5$]

(ii) $S_m = \left(\dfrac{14.09 \times 10^3 \,\text{J mol}^{-1}}{5000 \,\text{K}} \right) + (8.314 \,\text{J K}^{-1} \,\text{mol}^{-1}) \times (\ln 6.26) = \boxed{18.07 \,\text{J K}^{-1} \,\text{mol}^{-1}}$

P19.9 $q = \sum_j g_j e^{-\beta \varepsilon_j}$ [19.12] $= \sum_j g_j e^{-hc\beta \tilde{\nu}_j}$

$$p_i = \frac{g_i e^{-\beta \varepsilon_i}}{q} \text{ [19.10]} = \frac{g_i e^{-hc\beta \tilde{\nu}_i}}{q}$$

We measure energies from the lower states, and write

$$q = 2 + 2e^{-hc\beta \tilde{\nu}} = 2 + 2e^{-(1.4388 \times 121.1)/(T/\text{K})} = 2 + 2e^{-174.2/(T/\text{K})}$$

This function is plotted in Fig. 19.2.

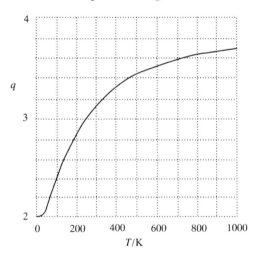

$$q$$

$$T/\text{K}$$

Figure 19.2

(a) At 300 K

$$p_0 = \frac{2}{q} = \frac{1}{1 + e^{-174.2/300}} = \boxed{0.64}$$

$$p_1 = 1 - p_0 = \boxed{0.36}$$

(b) The electronic contribution to U_m in wavenumber units is

$$\frac{U_m - U_m(0)}{N_A hc} = -\frac{1}{hcq} \frac{dq}{d\beta} \text{ [19.25]} = \frac{2\tilde{\nu} e^{-hc\beta\tilde{\nu}}}{q}$$

$$= \frac{(121.1 \,\text{cm}^{-1}) \times (e^{-174.2/300})}{1 + e^{-174.2/300}} = 43.45 \,\text{cm}^{-1}$$

which corresponds to $\boxed{0.52 \,\text{kJ mol}^{-1}}$

For the electronic contribution to the molar entropy, we need q and $U_m - U_m(0)$ at 500 K as well as at 300 K. These are

	300K	500K
$U_m - U_m(0)$	$0.518\,\text{kJ mol}^{-1}$	$0.599\,\text{kJ mol}^{-1}$
q	3.120	3.412

Then we form

$$S_m = \frac{U_m - U_m(0)}{T} + R \ln q \quad [19.34]$$

$$\text{At 300 K} \quad S_m = \left(\frac{518\,\text{J mol}^{-1}}{300\,\text{K}}\right) + (8.314\,\text{J K}^{-1}\,\text{mol}^{-1}) \times (\ln 3.120) = 11.2\,\text{J K}^{-1}\,\text{mol}^{-1}$$

$$\text{At 500 K} \quad S_m = \left(\frac{599\,\text{J mol}^{-1}}{500\,\text{K}}\right) + (8.314\,\text{J K}^{-1}\,\text{mol}^{-1}) \times (\ln 3.412) = 11.4\,\text{J K}^{-1}\,\text{mol}^{-1}$$

P19.12 (a) Total entropy, $S = S_1 + S_2 = (5.69 + 11.63)\,\text{J K}^{-1} = 17.32\,\text{J K}^{-1}$

$$W = e^{S/k} = e^{17.32\,\text{J K}^{-1}/1.381 \times 10^{-23}\,\text{J K}^{-1}} \quad [19.30]$$
$$= e^{1.254 \times 10^{24}} = 10^{5.44 \times 10^{23}}$$

(b) Total entropy, $S = 2\,\text{mol}(9.03\,\text{J K}^{-1}\,\text{mol}^{-1}) = 18.06\,\text{J K}^{-1}$

$$W = e^{S/k} = e^{18.06\,\text{J K}^{-1}/1.381 \times 10^{-23}\,\text{J K}^{-1}}$$
$$= e^{1.31 \times 10^{24}} = 10^{5.69 \times 10^{23}}$$

The final temperature is not the average because the molar heat capacity of graphite increases with temperature. At 298 K it is $8.54\,\text{J K}^{-1}\,\text{mol}^{-1}$, whereas at 498 K it is $14.64\,\text{J K}^{-1}\,\text{mol}^{-1}$.

(c) At constant internal energy and volume the condition for spontaneity is $\boxed{\Delta S_{U,V} > 0}$. Since $W_{(b)} > W_{(a)}$, the process part (b) is $\boxed{\text{spontaneous}}$.

Solutions to theoretical problems

P19.14 (a) $\quad W = \dfrac{N!}{n_1!n_2!\cdots}\,[19.1] = \dfrac{5!}{0!5!0!0!0!} = \boxed{1}$

(b) We draw up the following table

0	ε	2ε	3ε	4ε	5ε	$W = \dfrac{N!}{n_1!n_2!\cdots}$
4	0	0	0	0	1	5
3	1	0	0	1	0	20
3	0	1	1	0	0	20
2	2	0	1	0	0	30
2	1	2	0	0	0	30
1	3	1	0	0	0	20
0	5	0	0	0	0	1

The most probable configurations are $\boxed{\{2, 2, 0, 1, 0, 0\}}$ and $\boxed{\{2, 1, 2, 0, 0, 0\}}$ jointly.

P19.17 **(a)** $q = \sum_j g_j e^{-\beta \varepsilon_j} = 1 + 3e^{-\beta \varepsilon} = \boxed{1 + 3e^{-\varepsilon/kT}}$

at $T = \dfrac{\varepsilon}{k}$, $q = 1 + 3e^{-1} = 2.104$

(b) $U_m - U_m(0) = E = -\dfrac{N_A}{q}\dfrac{dq}{d\beta} = \dfrac{N_A}{q}(3\varepsilon e^{-\beta \varepsilon})$

$$= \dfrac{N_A}{q}(3RTe^{-1}) = \dfrac{3RT}{2.104\,e} = \boxed{0.5245RT}$$

A numerical value cannot be obtained for the energy without specific knowledge of the temperature, but that is not required for the heat capacity or the entropy.

$$C_V = \left(\dfrac{\partial U_m}{\partial T}\right)_V = \left(\dfrac{\partial E}{\partial T}\right)_V$$

Since $\dfrac{d}{dT} = \dfrac{d\beta}{dT} \times \dfrac{d}{d\beta} = -\dfrac{1}{kT^2}\dfrac{d}{d\beta} = -k\beta^2\dfrac{d}{d\beta}$

$$C_V = -k\beta^2\left(\dfrac{\partial E}{\partial \beta}\right)_V = -k\beta^2(3\varepsilon N_A)\dfrac{\partial}{\partial \beta}\left(\dfrac{e^{-\beta\varepsilon}}{q}\right)$$

$$= -k\beta^2(3\varepsilon N_A)\dfrac{\partial}{\partial \beta}\left(\dfrac{e^{-\beta\varepsilon}}{1 + 3e^{-\beta\varepsilon}}\right)$$

$$= -k\beta^2(3\varepsilon N_A)\left[\dfrac{(1 + 3e^{-\beta\varepsilon}) \times (-\varepsilon)e^{-\beta\varepsilon} - e^{-\beta\varepsilon}(-3\varepsilon e^{-\beta\varepsilon})}{(1 + 3e^{-\beta\varepsilon})^2}\right]$$

$$= -k\beta^2(3\varepsilon N_A)\left[\dfrac{-\varepsilon e^{-\beta\varepsilon} - 3\varepsilon e^{-2\beta\varepsilon} + 3\varepsilon e^{-2\beta\varepsilon}}{(1 + 3e^{-\beta\varepsilon})^2}\right]$$

$$= -k\beta^2(3\varepsilon N_A)\left[\dfrac{-\varepsilon e^{-\beta\varepsilon}}{(1 + 3e^{-\beta\varepsilon})^2}\right]$$

$$= \dfrac{3R\varepsilon^2 e^{-\beta\varepsilon}}{(kT)^2(1 + 3e^{-\beta\varepsilon})^2}$$

For $\varepsilon = kT$, $C_V = \dfrac{3Re^{-1}}{(1 + 3e^{-1})^2} = \dfrac{3R}{e(1 + \frac{3}{e})^2} = \boxed{2.074\,\text{J K}^{-1}\,\text{mol}^{-1}}$

Note that taking the derivative of $0.5245RT$ with regard to T does not give the correct answer. That is because the temperature dependence of q is not taken into account by that process.

$$\dfrac{\partial}{\partial T}(0.5245RT) = 0.5245R = 4.361\,\text{J K}^{-1}\,\text{mol}^{-1}$$

and this is not the correct value.

The calculation of S does not require taking another derivative, so we can use $E = 0.5245RT$

$$S_m = \dfrac{E}{T} + R\ln q = 0.5245\,R + R\ln(2.104) = \boxed{10.55\,\text{J K}^{-1}\,\text{mol}^{-1}}.$$

P19.18 (a) $U - U(0) = -N\dfrac{\mathrm{d}\ln q}{\mathrm{d}\beta}$ [19.27b], with $q = \dfrac{1}{1 - \mathrm{e}^{-\beta\varepsilon}}$ [19.15]

$$\frac{\mathrm{d}\ln q}{\mathrm{d}\beta} = \frac{1}{q}\frac{\mathrm{d}q}{\mathrm{d}\beta} = \frac{-\varepsilon\mathrm{e}^{-\beta\varepsilon}}{1 - \mathrm{e}^{-\beta\varepsilon}}$$

$$a\varepsilon = \frac{U - U(0)}{N} = \frac{\varepsilon\mathrm{e}^{(-\beta\varepsilon)}}{1 - \mathrm{e}^{-\beta\varepsilon}} = \frac{\varepsilon}{\mathrm{e}^{\beta\varepsilon} - 1}$$

Hence, $\mathrm{e}^{\beta\varepsilon} = \dfrac{1 + a}{a}$, implying that, $\beta = \dfrac{1}{\varepsilon}\ln\left(1 + \dfrac{1}{a}\right)$

For a mean energy of ε, $a = 1$, $\beta = \dfrac{1}{\varepsilon}\ln 2$, implying that

$$T = \frac{\varepsilon}{k\ln 2}\ln 2 = (50\,\mathrm{cm}^{-1}) \times \left(\frac{hc}{k\ln 2}\right) = \boxed{104\,\mathrm{K}}$$

(b) $q = \dfrac{1}{1 - \mathrm{e}^{-\beta\varepsilon}} = \dfrac{1}{1 - \left(\frac{a}{1+a}\right)} = \boxed{1 + a}$

(c) $\dfrac{S}{Nk} = \dfrac{U - U(0)}{NkT} + \ln q$ [19.34] $= a\beta\varepsilon + \ln q$

$$= a\ln\left(1 + \frac{1}{a}\right) + \ln(1 + a) = a\ln(1 + a) - a\ln a + \ln(1 + a)$$

$$= \boxed{(1 + a)\ln(1 + a) - a\ln a}$$

When the mean energy is ε, $a = 1$ and then $\boxed{\dfrac{S}{Nk} = 2\ln 2}$.

Solutions to applications

P19.20 (a) By Archimede's principle

$$F_{\mathrm{net}} = F_{\mathrm{gravity}} - F_{\mathrm{buoyancy}} = mg - m_{\mathrm{fluid}}g$$
$$= v\rho gh - v\rho_0 gh = v(\rho - \rho_0)g$$

By the definition of potential energy

$$F_{\mathrm{net}} = \frac{\mathrm{d}V}{\mathrm{d}h}\quad\text{or}\quad \mathrm{d}V = F_{\mathrm{net}}\,\mathrm{d}h$$

$$V = \int_0^h v(\rho - \rho_0)g\,\mathrm{d}h = \boxed{v(\rho - \rho_0)gh}$$

(b) $\dfrac{\mathcal{N}(h)}{\mathcal{N}(0)} = \mathrm{e}^{-\{v(\rho - \rho_0)gh/kT\}}$ [19.6a]

$$\frac{v(\rho - \rho_0)gh}{kT} = \ln\frac{\mathcal{N}(0)}{\mathcal{N}(h)}$$

(c) $k = \dfrac{v(\rho - \rho_0)gh}{T \ln\left(\frac{\mathcal{N}(0)}{\mathcal{N}(h)}\right)}$

$= \dfrac{(1.03 \times 10^{-19}\,\mathrm{m}^3) \times [(1.21 - 1.00) \times 10^3\,\mathrm{kg\,m}^{-3}]}{(277\,\mathrm{K})\ln 2}$

$\times (9.81\,\mathrm{m\,s}^{-2}) \times (1.23 \times 10^{-5}\mathrm{m})$

$= \boxed{1.36 \times 10^{-23}\,\mathrm{J\,K}^{-1}}$

$N_A = \dfrac{R}{k} = \dfrac{8.315\,\mathrm{J\,K}^{-1}\,\mathrm{mol}^{-1}}{1.36 \times 10^{-23}\,\mathrm{J\,K}^{-1}}$

$= \boxed{6.11 \times 10^{23}\,\mathrm{mol}^{-1}}$

These are remarkably close to the modern values.

P19.23   At equilibrium $\dfrac{N(r)/V}{N(r_0)/V} = e^{-\{V(r)-V(r_0)\}/kT}$ [19.6a]

Since $V(r) = -GMm/r$, $V(\infty) = 0$ and [Note: $V(r)$ is potential energy, V is volume]

$\dfrac{N(\infty)/V}{N(r_0)/V} = e^{V(r_0)/kT}$

which says that $N(\infty)/V \propto e^{V(r_0)/kT} = $ constant. This is obviously not the current distribution for planetary atmospheres where $\lim\limits_{r\to\infty} N(r)/V = 0$. Consequently, we may conclude that the earth's atmosphere, or any other planetary atmosphere, cannot be at equilibrium.

20 Statistical thermodynamics: the machinery

Solutions to exercises

Discussion questions

E20.1(a) An approximation involved in the derivation of all of these expressions is the assumption that the contributions from the different modes of motion are separable. The expression $q^R = kT/hcB$ is the high temperature approximation to the rotational partition function for nonsymmetrical linear rotors. The expression $q^V = kT/hc\tilde{v}$ is the high temperature form of the partition function for one vibrational mode of the molecule in the harmonic approximation. The expression $q^E = g^E$ for the electronic partition function applies at normal temperatures to atoms and molecules with no low lying electronic energy levels.

E20.2(a) Residual entropy is due to the presence of some disorder in the system even at $T = 0$. It is observed in systems where there is very little energy difference between alternative arrangements of the molecules at very low temperatures. Consequently, the molecules cannot lock into a preferred orderly arrangement and some disorder persists.

E20.3(a) Equations of state can be thought of as expressions for the pressure of a gas in terms of the state functions, n, V, and T. They are obtained from the expression for the pressure in terms of the canonical partition function given in eqn 20.4.

$$p = kT \left(\frac{\partial \ln Q}{\partial V} \right)_T \qquad [20.4]$$

Partition functions for perfect and imperfect gases are different. That for the perfect gas is given by $Q = q^N/N!$ with $q = V/\Lambda^3$. There is no one form for imperfect gases. One example is shown in Self-test 20.1, another which can be shown to lead to the van der Waals equation of state is

$$Q = \frac{1}{N!} \left(\frac{2\pi mkT}{h^2} \right)^{3N/2} (V - Nb)\, e^{aN^2/kTV}$$

For the case of the perfect gas there are no molecular features in the partition function, but for imperfect gases there are repulsive and attractive features in the partition function which are related to the structure of the molecules.

Numerical exercises

E20.4(a) $C_{V,m} = \frac{1}{2}(3 + v_R^* + 2v_V^*)R$ [20.40]

with a mode active if $T > \theta_M$.

(a) $v_R^* = 2$, $v_V^* \approx 0$; hence $C_{V,m} = \frac{1}{2}(3 + 2)R = \boxed{\frac{5}{2}R}$ [experimental: $3.4R$]

(b) $v_R^* = 3$, $v_V^* \approx 0$; hence $C_{V,m} = \frac{1}{2}(3 + 3)R = \boxed{3R}$ [experimental: $3.2R$]

(c) $v_R^* = 3$, $v_V^* \approx 0$; hence $C_{V,m} = \frac{1}{2}(3 + 3)R = \boxed{3R}$ [experimental: $8.8R$]

Comment. Data from the books by Herzberg (see *Further reading*, Chapters 16 and 17) give for the vibrational wavenumbers

I_2 $\tilde{\nu} = 214 \, \text{cm}^{-1} \approx 207 \, \text{cm}^{-1} \, [kT \text{ at } T = 298 \, \text{K}]$

CH_4 all greater than $1300 \, \text{cm}^{-1}$

C_6H_6 4 less than $207 \, \text{cm}^{-1} \, [kT \text{ at } T = 298 \, \text{K}]$

Thus, we expect the vibrational mode of I_2 to contribute significantly to $C_{V,m}$, and hence $C_{V,m} > \dfrac{5}{2}R$.

We expect little vibrational contribution to $C_{V,m}$ for methane; hence $C_{V,m} \approx 3R$. For benzene, there is a lot of vibrational contribution; hence $C_{V,m} \gg 3R$.

E20.5(a)   Assuming that all rotational modes are active we can draw up the following table for $C_{V,m}$, $C_{p,m}$, and γ with and without active vibrational modes.

	$C_{V,m}$	$C_{p,m}$	γ	Exptl	
$NH_3(\nu_V^* = 0)$	$3R$	$4R$	1.33	1.31	closer
$NH_3(\nu_V^* = 6)$	$9R$	$10R$	1.11		
$CH_4(\nu_V^* = 0)$	$3R$	$4R$	1.33	1.31	closer
$CH_4(\nu_V^* = 9)$	$12R$	$13R$	1.08		

The experimental values are obtained from Table 2.6 assuming $C_{p,m} = C_{V,m} + R$. It is clear from the comparison in the above table that the vibrational modes are not active. This is confirmed by the experimental vibrational wavenumbers (see Herzberg references in *Further reading*, Chapters 16 and 17) all of which are much greater than kT at 298 K.

E20.6(a) $q^R = \dfrac{0.6950}{\sigma} \times \dfrac{T/K}{(B/\text{cm}^{-1})}$ [Table 20.3]

$$= \dfrac{(0.6950) \times (T/\text{K})}{10.59} \, [\sigma = 1] = 0.06563(T/\text{K})$$

(a) $q^R = (0.06563) \times (298) = \boxed{19.6}$ **(b)** $q^R = (0.06563) \times (523) = \boxed{34.3}$.

E20.7(a)   Look for the rotational subgroup of the molecule (the group of the molecule composed only of the identity and the rotational elements, and assess its order).

(a) CO. Full group $C_{\infty v}$; subgroup C_1; hence $\sigma = \boxed{1}$

(b) O_2. Full group $D_{\infty h}$; subgroup C_2; hence $\sigma = \boxed{2}$

(c) H_2S. Full group C_{2v}; subgroup C_2; hence $\sigma = \boxed{2}$

(d) SiH_4. Full group T_d; subgroup T; hence $\sigma = \boxed{12}$

(e) $CHCl_3$. Full group C_{3v}; subgroup C_3; hence $\sigma = \boxed{3}$

See the references in the *Further reading* for Chapter 15 for a more complete set of character tables including those of the rotational subgroups.

E20.8(a) $q^R = \dfrac{1.0270}{\sigma} \dfrac{(T/K)^{3/2}}{(ABC/\text{cm}^{-3})^{1/2}}$ [Table 20.3]

$$= \dfrac{1.0270 \times 298^{3/2}}{(2) \times (27.878 \times 14.509 \times 9.287)^{1/2}} \, [\sigma = 2] = \boxed{43.1}$$

The high-temperature approximation is valid if $T > \theta_R$, where

$$\theta_R = \frac{hc(ABC)^{1/3}}{k},$$

$$= \frac{(6.626 \times 10^{-34}\,\mathrm{J\,s}) \times (2.998 \times 10^{10}\,\mathrm{cm\,s^{-1}}) \times [(27.878) \times (14.509) \times (9.287)\,\mathrm{cm^{-3}}]^{1/3}}{1.38 \times 10^{-23}\,\mathrm{J\,K^{-1}}}$$

$$= \boxed{22.36\,\mathrm{K}}$$

E20.9(a) $\quad q^R = 43.1$ [Exercise 20.8]

All the rotational modes of water are fully active at 25°C (Example 20.6 and Exercise 20.8); therefore

$$U_m^R - U_m^R(0) = E^R = \tfrac{3}{2}RT$$

$$S_m^R = \frac{E^R}{T} + R\,\ln q^R$$

$$= \tfrac{3}{2}R + R\,\ln 43.1 = \boxed{43.76\,\mathrm{J\,K^{-1}\,mol^{-1}}}$$

Comment. Division of q^R by N_A! is not required for the internal contributions; internal motions may be thought of as localized (distinguishable). It is the overall canonical partition function, which is a product of internal and external contributions, that is divided by N!

E20.10(a) (a) For a spherical rotor (Section 16.5)

$$E = hcBJ(J+1)\ [16.31]\quad [B = 5.2412\,\mathrm{cm^{-1}}\ \text{for } CH_4]$$

and the degeneracy is $g(J) = (2J+1)^2$. Hence

$$q \approx \frac{1}{\sigma} \sum_J (2J+1)^2 e^{-\beta hcBJ(J+1)}$$

which is analogous for spherical rotors to eqn 14 for q^R for linear rotors.

$$hcB\beta = \frac{(1.4388\,\mathrm{K}) \times (5.2412)}{T} = \frac{7.5410}{T/\mathrm{K}}, \qquad \sigma = 12$$

$$q = \frac{1}{12} \sum_J (2J+1)^2 e^{-7.5410\,J(J+1)/(T/\mathrm{K})}$$

$$= \frac{1}{12}(1.0000 + 8.5561 + 21.480 + 36.173 + \cdots)$$

$$= \frac{1}{12} \times 443.427 = \boxed{36.95}\ \text{at } 298\,\mathrm{K}$$

The sum converged after 20 terms.
Similarly, at 500 K

$$q = \frac{1}{12}(1.0000 + 8.7326 + 22.8370 + 40.8880 + \cdots) = \frac{1}{12} \times 960.96 = \boxed{80.08}$$

The sum converged after 24 terms.

(Note that the results are still approximate because the symmetry number is a valid corrector only at high temperatures. To get exact values of q we should do a detailed analysis of the rotational states allowed by the Pauli principle.)

(b) $\quad q \approx \dfrac{1.0270}{\sigma} \times \dfrac{(T/K)^{3/2}}{(B/cm^{-1})^{3/2}}$ [Table 20.3, $A = B = C$]

$\qquad = \dfrac{1.0270}{12} \times \dfrac{(T/K)^{3/2}}{(5.2412)^{3/2}} = 7.133 \times 10^{-3} \times (T/K)^{3/2}$

$\qquad$ At 298 K, $q = 7.133 \times 10^{-3} \times 298^{3/2} = \boxed{36.7}$

$\qquad$ At 500 K, $q = 7.133 \times 10^{-3} \times 500^{3/2} = \boxed{79.7}$

$\qquad$ The difference in this case is small.

E20.11(a) $\quad q^R = \dfrac{kT}{\sigma hcB}$ [20.17], $\quad B = \dfrac{\hbar}{4\pi cI}, \quad I = \mu R^2$

$\qquad$ Hence $q = \dfrac{8\pi^2 kTI}{\sigma h^2} = \dfrac{8\pi^2 kT\mu R^2}{\sigma h^2}$

$\qquad$ For O_2, $\mu = \frac{1}{2}m(O) = \frac{1}{2} \times 16.00\,u = 8.00\,u$ and $\sigma = 2$; therefore

$$q = \dfrac{(8\pi^2) \times (1.381 \times 10^{-23}\,J\,K^{-1}) \times (300\,K) \times (8.00) \times (1.6605 \times 10^{-27}\,kg) \times (1.21 \times 10^{-10}\,m)^2}{(2) \times (6.626 \times 10^{-34}\,J\,s)^2}$$

$\qquad = \boxed{72.5}$

E20.12(a) $\quad C_{V,m}/R = f^2, \quad f = \left(\dfrac{\theta_V}{T}\right) \times \left(\dfrac{e^{-\theta_V/2T}}{1 - e^{-\theta_V/T}}\right)$ [20.39]; $\quad \theta = \dfrac{hc\tilde{\nu}}{k}$

$\qquad$ We write $x = \dfrac{\theta_V}{T}$; then $C_{V,m}/R = \dfrac{x^2 e^{-x}}{(1 - e^{-x})^2}$

$\qquad$ This function is plotted in Fig. 20.1. For the acetylene (ethyne) calculation, use the expression above for each mode. We draw up the following table using $kT/hc = 207\,cm^{-1}$ at 298 K and $348\,cm^{-1}$ at 500 K, and $\theta_V/T = hc\tilde{\nu}/kT$.

	x		$C_{V,m}/R$	
$\tilde{\nu}/cm^{-1}$	298 K	500 K	298 K	500 K
612	2.96	1.76	0.505	0.777
612	2.96	1.76	0.505	0.777
729	3.52	2.09	0.389	0.704
729	3.52	2.09	0.389	0.704
1974	9.54	5.67	0.007	0.112
3287	15.88	9.45	3.2×10^{-5}	0.007
3374	16.30	9.70	2.2×10^{-5}	0.006

$\qquad$ The heat capacity of the molecule is the sum of these contributions, namely

$\qquad$ **(a)** $1.796R = \boxed{14.93\,J\,K^{-1}\,mol^{-1}}$ at 298 K and   **(b)** $3.086R = \boxed{25.65\,J\,K^{-1}\,mol^{-1}}$ at 500 K.

E20.13(a) $\quad$ In each case the contribution to G is given by

$$G - G(0) = -nRT \ln q \text{ [See Comment to Exercise 20.6]}$$

$\qquad$ Therefore, we first evaluate q^R and q^V.

$\qquad q^R = \dfrac{0.6950}{\sigma} \dfrac{T/K}{B/cm^{-1}}$ [Table 20.3, $\sigma = 2$]

$\qquad\quad = \dfrac{(0.6950) \times (298)}{(2) \times (0.3902)} = 265$

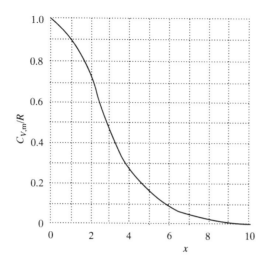

Figure 20.1

$$q^V = \left(\frac{1}{1 - e^{-a}}\right) \times \left(\frac{1}{1 - e^{-b}}\right)^2 \times \left(\frac{1}{1 - e^{-c}}\right) \text{ [Table 20.3]}$$

with

$$a = \frac{(1.4388) \times (1388.2)}{298} = 6.70\overline{2}$$

$$b = \frac{(1.4388) \times (667.4)}{298} = 3.22\overline{2}$$

$$c = \frac{(1.4388) \times (2349.2)}{298} = 11.3\overline{4}$$

Hence

$$q^V = \frac{1}{1 - e^{-6.702}} \times \left(\frac{1}{1 - e^{-3.222}}\right)^2 \times \frac{1}{1 - e^{-11.34}} = 1.08\overline{6}$$

Therefore, the rotational contribution to the molar Gibbs energy is

$$-RT \ln q^R = -8.314 \, \text{J K}^{-1} \, \text{mol}^{-1} \times 298 \, \text{K} \times \ln 265$$

$$= \boxed{-13.8 \, \text{kJ mol}^{-1}}$$

and the vibrational contribution is

$$-RT \ln q^V = -8.314 \, \text{J K}^{-1} \, \text{mol}^{-1} \times 298 \, \text{K} \times \ln 1.08\overline{6} = \boxed{-0.20 \, \text{kJ mol}^{-1}}$$

E20.14(a) $q = \sum_j g_j e^{-\beta \varepsilon_j}, \quad g_j = 2J + 1$

$$= 4 + 2e^{-\beta \varepsilon} \ [g(^2P_{3/2}) = 4, \ g(^2P_{1/2}) = 2]$$

$$U - U(0) = -\frac{N}{q} \frac{dq}{d\beta} = \frac{N \varepsilon e^{-\beta \varepsilon}}{2 + e^{-\beta \varepsilon}}$$

$$C_V = \left(\frac{\partial U}{\partial T}\right)_V = -k\beta^2 \left(\frac{\partial U}{\partial \beta}\right)_V = \frac{2R(\varepsilon \beta)^2 e^{-\beta \varepsilon}}{(2 + e^{-\beta \varepsilon})^2} \ [N = N_A]$$

(a) Therefore, since at 500 K $\beta\varepsilon = 2.53\overline{5}$

$$C_{V,m}/R = \frac{(2) \times (2.53\overline{5})^2 \times (e^{-2.53\overline{5}})}{(2 + e^{-2.53\overline{5}})^2} = \boxed{0.236}$$

(b) At 900 K, when $\beta\varepsilon = 1.408$,

$$C_{V,m}/R = \frac{(2) \times (1.408)^2 \times (e^{-1.408})}{(2 + e^{-1.408})^2} = \boxed{0.193}$$

Comment. $C_{V,m}$ is smaller at 900 K than at 500 K, for then the temperature is higher than the peak in the "two-level" heat capacity curve.

E20.15(a) We assume that the upper eight of the $\left(2 \times \frac{9}{2} + 1\right) = 9$ spin–orbit states of the ion lies at an energy much greater than kT at 1 K; hence, since the spin degeneracy of Co^{2+} is 4 (the ion is a spin quartet), $q = 4$. The contribution to the entropy is

$$R \ln q = (8.314\,\text{J K}^{-1}\,\text{mol}^{-1}) \times (\ln 4) = \boxed{11.5\,\text{J K}^{-1}\,\text{mol}^{-1}}.$$

E20.16(a) In each case $S_m = R \ln s$ [20.52]. Therefore,

(a) $S_m = R \ln 3 = 1.1R = \boxed{9\,\text{J K}^{-1}\,\text{mol}^{-1}}$ **(b)** $S_m = R \ln 5 = 1.6R = \boxed{13\,\text{J K}^{-1}\,\text{mol}^{-1}}$

(c) $S_m = R \ln 6 = 1.8R = \boxed{15\,\text{J K}^{-1}\,\text{mol}^{-1}}$.

E20.17 We use eqn 20.57 with $X = I$, $X_2 = I_2$, $\Delta E_0 = D_0$.

$$D_0 = D_e - \frac{1}{2}\tilde{\nu} = 1.5422\,\text{eV} \times \frac{8065.5\,\text{cm}^{-1}}{1\,\text{eV}} - 107.18\,\text{cm}^{-1}$$

$$= 1.2331 \times 10^4\,\text{cm}^{-1} = 1.475 \times 10^5\,\text{J mol}^{-1}$$

$$K = \left(\frac{q_{I,m}^{\ominus 2}}{q_{I_2,m}^{\ominus} N_A}\right) e^{-\Delta E_0/RT} \quad [20.57]$$

$$q_{I,m}^{\ominus} = q_m^T(I)q^E(I), \quad q^E(I) = 4$$

$$q_{I_2,m}^{\ominus} = q_m^T(I_2)q^R(I_2)q^V(I_2)q^E(I_2), \quad q^E(I_2) = 1$$

$$\frac{q_m^T}{N_A} = 2.561 \times 10^{-2}(T/K)^{5/2} \times (M/\text{g mol}^{-1})^{3/2} \quad [\text{Table 20.3}]$$

$$\frac{q_m^T(I_2)}{N_A} = 2.561 \times 10^{-2} \times 1000^{5/2} \times 253.8^{3/2} = 3.27 \times 10^9$$

$$\frac{q_m^T(I)}{N_A} = 2.561 \times 10^{-2} \times 1000^{5/2} \times 126.9^{3/2} = 1.16 \times 10^9$$

$$q^R(I_2) = \frac{0.6950}{\sigma} \times \frac{T/K}{B/\text{cm}^{-1}} = \frac{1}{2} \times 0.6950 \times \frac{1000}{0.0373} = 931\overline{6}$$

$$q^V(I_2) = \frac{1}{1 - e^{-a}}, \quad a = 1.4388 \frac{\tilde{\nu}/\text{cm}^{-1}}{T/\text{K}} \quad [\text{Table 20.3}]$$

$$= \frac{1}{1 - e^{-1.4388 \times 214.36/1000}} = 3.77$$

$$K = \frac{(1.16 \times 10^9 \times 4)^2 e^{-17.741}}{(3.27 \times 10^9) \times (9316) \times (3.77)} = \boxed{3.70 \times 10^{-3}}$$

Solutions to problems

Solutions to numerical problems

P20.1

$$q^E = \sum_j g_j e^{-\beta \varepsilon_j} = 2 + 2e^{-\beta \varepsilon}, \quad \varepsilon = \Delta\varepsilon = 121.1 \, \text{cm}^{-1}$$

$$U_m - U_m(0) = -\frac{N_A}{q^E} \left(\frac{\partial q^E}{\partial \beta} \right)_V \quad [20.26] = \frac{2 N_A \varepsilon e^{-\beta \varepsilon}}{q^E}$$

$$C_{V,m} = -k\beta^2 \left(\frac{\partial U_m}{\partial \beta} \right)_V \quad [20.35]$$

Let $x = \beta \varepsilon$, then $d\beta = \frac{1}{\varepsilon} dx$

$$C_{V,m} = -N_A k \left(\frac{x}{\varepsilon} \right)^2 \times (\varepsilon)^2 \frac{\partial}{\partial x} \left(\frac{e^{-x}}{1 + e^{-x}} \right) = R \left(\frac{x^2 e^{-x}}{(1 + e^{-x})^2} \right)$$

Therefore

$$C_{V,m}/R = \frac{x^2 e^{-x}}{(1 + e^{-x})^2}, \quad x = \beta \varepsilon$$

We then draw up the following table

T/K	50	298	500
$(kT/hc)/\text{cm}^{-1}$	34.8	207	348
x	3.48	0.585	0.348
$C_{V,m}/R$	0.351	0.079	0.029
$C_{V,m}/(\text{J K}^{-1}\,\text{mol}^{-1})$	2.91	0.654	0.244

Comment. Note that the double degeneracies do not affect the results because the two factors of 2 in q cancel when U is formed. In the range of temperature specified, the electronic contribution to the heat capacity decreases with increasing temperature.

P20.3   The energy expression for a particle on a ring is

$$E = \frac{\hbar^2 m_l^2}{2I} \quad [12.49]$$

Therefore

$$q = \sum_{m=-\infty}^{\infty} e^{-m^2 \hbar^2/2IkT} = \sum_{m=-\infty}^{\infty} e^{-\beta \hbar^2 m^2/2I}$$

The summation may be approximated by an integration

$$q \approx \frac{1}{\sigma} \int_{-\infty}^{\infty} e^{-m^2\hbar^2/2IkT} \, dm = \frac{1}{\sigma} \left(\frac{2IkT}{\hbar^2} \right)^{1/2} \int_{-\infty}^{\infty} e^{-x^2} \, dx$$

$$\approx \frac{1}{\sigma} \left(\frac{2\pi IkT}{\hbar^2} \right)^{1/2}$$

$$U - U(0) = -\frac{N}{q} \frac{\partial q}{\partial \beta} = \frac{N}{2\beta} = \frac{1}{2} NkT = \frac{1}{2} RT \quad (N = N_A)$$

$$C_{V,m} = \left(\frac{\partial U_m}{\partial T} \right)_V = \frac{1}{2} R = \boxed{4.2 \, \text{J K}^{-1} \, \text{mol}^{-1}}$$

$$S_m = \frac{U_m - U_m(0)}{T} + R \ln q$$

$$= \frac{1}{2} R + R \ln \frac{1}{\sigma} \left(\frac{2\pi IkT}{\hbar^2} \right)^{1/2}$$

$$= \frac{1}{2} R + R \ln \frac{1}{3} \left(\frac{(2\pi) \times (5.341 \times 10^{-47} \, \text{kg m}^2) \times (1.381 \times 10^{-23} \, \text{J K}^{-1}) \times (298)}{(1.055 \times 10^{-34} \, \text{J s})^2} \right)^{1/2}$$

$$= \frac{1}{2} R + 1.31R = 1.81R, \text{ or } \boxed{15 \, \text{J K}^{-1} \, \text{mol}^{-1}}$$

P20.5 The absorption lines are the values of $\{E(J+1) - E(J)\}/hc$ for $J = 0, 1, \ldots$. Therefore, we can reconstruct the energy levels from the data; they are

$$\frac{E_J}{hc} = \sum_{J'=0}^{J-1} \{E(J'+1) - E(J')\}/hc$$

Using $hc\beta = \frac{hc}{kT} = 207.223 \, \text{cm}^{-1}$ [inside front cover]

$$q = \sum_{J=0}^{\infty} (2J+1)e^{-\beta E(J)}$$

$$= 1 + 3e^{-21.19/207.223} + 5e^{-(21.19+42.37)/207.223} + 7e^{-(21.19+42.37+63.56)/207.223} + \cdots$$

$$= 1 + 2.708 + 3.679 + 3.790 + 3.237 + \cdots = \boxed{19.89}$$

P20.6
$$\frac{q_m^T}{N_A} = 2.561 \times 10^{-2} \times (T/\text{K})^{5/2} \times (M/\text{g mol}^{-1}) \text{ [Table 20.3]}$$

$$= (2.561 \times 10^{-2}) \times (298)^{5/2} \times (28.02)^{3/2} = 5.823 \times 10^6$$

$$q^R = \frac{1}{2} \times 0.6950 \times \frac{298}{1.9987} \text{ [Table 20.3]} = 51.81$$

$$q^V = \frac{1}{1 - e^{-2358/207.2}} \text{ [Table 20.3]} = 1.00$$

Therefore

$$\frac{q_m^{\ominus}}{N_A} = (5.82\bar{3} \times 10^6) \times (51.8\bar{1}) \times (1.00) = 3.02 \times 10^8$$

$$U_m - U_m(0) = \frac{3}{2}RT + RT = \frac{5}{2}RT \qquad [T \gg \theta_T, \theta_R]$$

Hence

$$S_m^{\ominus} = \frac{U_m - U_m(0)}{T} + R\left(\ln\frac{q_m^{\ominus}}{N_A} + 1\right)$$

$$= \frac{5}{2}R + R\{\ln 3.02 \times 10^8 + 1\} = 23.03R = \boxed{191.4\,\text{J K}^{-1}\,\text{mol}^{-1}}$$

The difference between the experimental and calculated values is negligible, indicating that the residual entropy is negligible.

P20.7   The molar entropy is given by

$$S_m = \frac{U_m - U_m(0)}{T} + R\left(\ln\frac{q_m}{N_A} - 1\right) \quad \text{where} \quad \frac{U_m - U_m(0)}{T} = -N_A\left(\frac{\partial \ln q}{\partial \beta}\right)_V$$

and $\dfrac{q_m}{N_A} = \dfrac{q_{m,\text{tr}}}{N_A}q_{\text{rot}}\,q_{\text{vib}}\,q_{\text{elec}}$

The energy term $U_m - U_m(0)$ works out to be

$$U_m - U_m(0) = N_A[\langle\varepsilon_{\text{tr}}\rangle + \langle\varepsilon_{\text{rot}}\rangle + \langle\varepsilon_{\text{vib}}\rangle + \langle\varepsilon_{\text{elec}}\rangle]\ \text{[Table 20.3]}$$

Translation:

$$\frac{q_{m,\text{tr}}^{\circ}}{N_A} = \frac{kT}{p^{\circ}\Lambda^3} = 2.561 \times 10^{-2}(T/\text{K})^{5/2} \times (M/\text{g mol}^{-1})^{3/2}\ \text{[Table 20.3]}$$

$$= 2.561 \times 10^{-2} \times (298)^{5/2} \times (38.00)^{3/2}$$

$$= 9.20 \times 10^6 \quad \text{and} \quad \langle\varepsilon_{\text{tr}}\rangle = 3/2kT$$

Rotation of a linear molecule

$$q_{\text{rot}} = \frac{kT}{\sigma hcB} = \frac{0.6950}{\sigma} \times \frac{T/\text{K}}{B/\text{cm}^{-1}}\ \text{[Table 20.3]}$$

The rotational constant is

$$B = \frac{\hbar}{4\pi cI} = \frac{\hbar}{4\pi c\mu R^2}$$

$$= \frac{(1.0546 \times 10^{-34}\,\text{J s}) \times (6.022 \times 10^{23}\,\text{mol}^{-1})}{4\pi(2.998 \times 10^{10}\,\text{cm s}^{-1}) \times (1/2 \times 19.00 \times 10^{-3}\,\text{kg mol}^{-1}) \times (190.0 \times 10^{-12}\,\text{m})^2}$$

$$= 0.4915\,\text{cm}^{-1} \quad \text{so} \quad q_{\text{rot}} = \frac{0.6950}{2} \times \frac{298}{0.4915} = 210.7$$

Also $\langle\varepsilon_{\text{rot}}\rangle = kT$

Vibration

$$q_{\text{vib}} = \frac{1}{1 - e^{-hc\tilde{v}/kT}} = \frac{1}{1 - \exp\left(\frac{-1.4388(\tilde{v}/\text{cm}^{-1})}{T/K}\right)} = \frac{1}{1 - \exp\left(\frac{-1.4388(450.0)}{298}\right)}$$

$$= 1.129$$

$$\langle \varepsilon_{\text{vib}} \rangle = \frac{hc\tilde{v}}{e^{hc\tilde{v}/kT} - 1} = \frac{(6.626 \times 10^{-34}\,\text{J s}) \times (2.998 \times 10^{10}\,\text{cm s}^{-1}) \times (450.0\,\text{cm}^{-1})}{\exp\left(\frac{1.4388(450.0)}{298}\right) - 1}$$

$$= 1.149 \times 10^{-21}\,\text{J}$$

The Boltzmann factor for the lowest-lying electronic excited state is

$$\exp\left(\frac{-(1.609\,\text{eV}) \times (1.602 \times 10^{-19}\,\text{J eV}^{-1})}{(1.381 \times 10^{-23}\,\text{J K}^{-1}) \times (298\,\text{K})}\right) = 6 \times 10^{-28}$$

so we may take q_{elec} to equal the degeneracy of the ground state, namely 2 and $\langle \varepsilon_{\text{elec}} \rangle$ to be zero. Putting it all together yields

$$\frac{U_m - U_m(0)}{T} = \frac{N_A}{T}\left(\frac{3}{2}kT + kT + 1.149 \times 10^{-21}\,\text{J}\right) = \frac{5}{2}R + \frac{N_A(1.149 \times 10^{-21}\,\text{J})}{T}$$

$$= (2.5) \times (8.3145\,\text{J mol}^{-1}\,\text{K}^{-1}) + \frac{(6.022 \times 10^{23}\,\text{mol}^{-1}) \times (1.149 \times 10^{-21}\,\text{J})}{298\,\text{K}}$$

$$= 23.11\,\text{J mol}^{-1}\,\text{K}^{-1}$$

$$R\left(\ln\frac{q_m}{N_A} - 1\right) = (8.3145\,\text{J mol}^{-1}\,\text{K}^{-1}) \times \{\ln[(9.20 \times 10^6) \times (210.7) \times (1.129) \times (2)] - 1\}$$

$$= 176.3\,\text{J mol}^{-1}\,\text{K}^{-1} \quad \text{and} \quad S_m^\circ = \boxed{199.4\,\text{J mol}^{-1}\,\text{K}^{-1}}$$

P20.8 The vibrational temperature is defined by

$$k\theta_V = hc\tilde{v},$$

so a vibration with θ_V less than 1000 K has a wavenumber less than

$$\tilde{v} = \frac{k\theta_V}{hc} = \frac{(1.381 \times 10^{-23}\,\text{J K}^{-1}) \times (1000\,\text{K})}{(6.626 \times 10^{-34}\,\text{J s}) \times (2.998 \times 10^{10}\,\text{cm s}^{-1})} = 695.2\,\text{cm}^{-1}$$

There are seven such wavenumbers listed among those for C_{60} : two T_{1u}, a T_{2u}, a G_u, and three H_u. The number of *modes* involved, v_V^*, must take into account the degeneracy of these vibrational energies

$$v_V^* = 2(3) + 1(3) + 1(4) + 3(5) = \boxed{28}$$

The molar heat capacity of a molecule is roughly

$$C_{V,m} = \tfrac{1}{2}(3 + v_R^* + 2v_V^*)R \;\; [42] = \tfrac{1}{2}(3 + 3 + 2 \times 28)R = 31R = 31(8.3145\,\text{J mol}^{-1}\,\text{K}^{-1})$$

$$= \boxed{258\,\text{J mol}^{-1}\,\text{K}^{-1}}$$

P20.11 $H_2O + DCl \rightleftharpoons HDO + HCl$

$$K = \frac{q^{\ominus}(HDO)q^{\ominus}(HCl)}{q^{\ominus}(H_2O)q^{\ominus}(DCl)} e^{-\beta \Delta_r E_0} \quad [20.54, \text{ with } \Delta_r E_0 \text{ in joules}]$$

The ratio of translational partition functions (Table 20.3) is

$$\frac{q_m^T(HDO)q_m^T(HCl)}{q_m^T(H_2O)q_m^T(DCl)} = \left(\frac{M(HDO)M(HCl)}{M(H_2O)M(DCl)} \right)^{3/2}$$

$$= \left(\frac{19.02 \times 36.46}{18.02 \times 37.46} \right)^{3/2} = 1.041$$

The ratio of rotational partition functions is

$$\frac{q^R(HDO)q^R(HCl)}{q^R(H_2O)q^R(DCl)} = 2 \times \frac{(27.88 \times 14.51 \times 9.29)^{1/2} \times 5.449}{(23.38 \times 9.102 \times 6.417)^{1/2} \times 10.59} = 1.707$$

($\sigma = 2$ for H_2O; $\sigma = 1$ for the other molecules).

The ratio of vibrational partition functions is

$$\frac{q^V(HDO)q^V(HCl)}{q^V(H_2O)q^V(DCl)} = \frac{q(2726.7)q(1402.2)q(3707.5)q(2991)}{q(3656.7)q(1594.8)q(3755.8)q(2145)} = Q$$

where

$$q(x) = \frac{1}{1 - e^{-1.4388 \times x/(T/K)}}$$

$$\frac{\Delta_r E_0}{hc} = \frac{1}{2}\{(2726.7 + 1402.2 + 3707.5 + 2991) - (3656.7 + 1594.8 + 3755.8 + 2145)\} \text{ cm}^{-1}$$

$$= -162 \text{ cm}^{-1}$$

Therefore, $K = 1.041 \times 1.707 \times Q \times e^{1.4388 \times 162/(T/K)} = 1.777 \, Q e^{233/(T/K)}$

We then draw up the following table (using a computer)

T/K	100	200	300	400	500	600	700	800	900	1000
K	18.3	5.70	3.87	3.19	2.85	2.65	2.51	2.41	2.34	2.29

and specifically $K = \boxed{3.89}$ at **(a)** 298 K and $\boxed{2.41}$ at **(b)** 800 K.

Solutions to theoretical problems

P20.12   A Sackur–Tetrode type of equation describes the translational entropy of the gas. Here

$$q^T = q_x^T q_y^T \quad \text{with } q_x^T = \left(\frac{2\pi m X^2}{\beta h^2} \right)^{1/2} \quad [19.18]$$

Therefore,

$$q^T = \left(\frac{2\pi m}{\beta h^2} \right) XY = \frac{2\pi m \sigma}{\beta h^2}, \quad \sigma = XY$$

$$U_m - U_m(0) = -\frac{N_A}{q} \left(\frac{\partial q}{\partial \beta} \right) = RT \quad [\text{or by equipartition}]$$

$$S_m = \frac{U_m - U_m(0)}{T} + R(\ln q_m - \ln N_A + 1) \quad \left[q_m = \frac{q}{n}\right]$$

$$= R + R \ln\left(\frac{eq_m}{N_A}\right) = R \ln\left(\frac{e^2 q_m}{N_A}\right)$$

$$= \boxed{R \ln\left(\frac{2\pi e^2 m \sigma_m}{h^2 N_A \beta}\right)} \quad \left[\sigma_m = \frac{\sigma}{n}\right]$$

Since in three dimensions

$$S_m = R \ln\left\{e^{5/2}\left(\frac{2\pi m}{h^2 \beta}\right)^{3/2} \frac{V_m}{N_A}\right\} \quad \text{[Sackur–Tetrode equation]}$$

The entropy of condensation is the difference

$$\Delta S_m = R \ln \frac{e^2 (2\pi m / h^2 \beta) \times (\sigma_m / N_A)}{e^{5/2} (2\pi m / h^2 \beta)^{3/2} \times (V_m / N_A)}$$

$$= \boxed{R \ln\left\{\left(\frac{\sigma_m}{V_m}\right) \times \left(\frac{h^2 \beta}{2\pi me}\right)^{1/2}\right\}}$$

P20.16   (a) $$U - U(0) = -\frac{N}{q}\frac{\partial q}{\partial \beta} = -\frac{N}{q}\sum_j \varepsilon_j e^{-\beta \varepsilon_j} = \frac{NkT}{q}\dot{q}$$

$$= \boxed{nRT\left(\frac{\dot{q}}{q}\right)}$$

$$C_V = \left(\frac{\partial U}{\partial T}\right)_V = \frac{d\beta}{dT}\left(\frac{\partial U}{\partial \beta}\right)_V = \frac{1}{kT^2}\frac{\partial}{\partial \beta}\left(\frac{N}{q}\sum_j \varepsilon_j e^{-\beta \varepsilon_j}\right)$$

$$= \left(\frac{N}{kT^2}\right) \times \left[\frac{1}{q}\sum_j \varepsilon_j^2 e^{-\beta \varepsilon_j} + \frac{1}{q^2}\left(\frac{\partial q}{\partial \beta}\right)\sum_j \varepsilon_j e^{-\beta \varepsilon_j}\right]$$

$$= \left(\frac{N}{kT^2}\right) \times \left[\frac{1}{q}\sum_j \varepsilon_j^2 e^{-\beta \varepsilon_j} - \frac{1}{q^2}\left(\sum_j \varepsilon_j e^{-\beta \varepsilon_j}\right)^2\right]$$

$$= \left(\frac{N}{kT^2}\right) \times \left[\frac{k^2 T^2 \ddot{q}}{q} - \frac{k^2 T^2}{q^2}\dot{q}^2\right]$$

$$= \boxed{nR\left\{\frac{\ddot{q}}{q} - \left(\frac{\dot{q}}{q}\right)^2\right\}}$$

$$S = \frac{U - U(0)}{T} + nR\ln\left(\frac{q}{N} + 1\right) = \boxed{nR\left(\frac{\dot{q}}{q} + \ln\frac{eq}{N}\right)}$$

(b) At 5000 K, $\dfrac{kT}{hc} = 3475\,\text{cm}^{-1}$. We form the sums

$$q = \sum_j e^{-\beta\varepsilon_j} = 1 + e^{-21870/3475} + 3e^{-21870/3475} + \cdots = 1.0167$$

$$\dot{q} = \sum_j \frac{\varepsilon_j}{kT} e^{-\beta\varepsilon_j} = \frac{hc}{kT} \sum_j \tilde{\nu}_j e^{-\beta\varepsilon_j}$$

$$= \left(\frac{1}{3475}\right) \times \{0 + 21850\,e^{-21850/3475} + 3 \times 21870\,e^{-21870/3475} + \cdots\} = 0.1057$$

$$\ddot{q} = \sum_j \left(\frac{\varepsilon_j}{kT}\right)^2 e^{-\beta\varepsilon_j} = \left(\frac{hc}{kT}\right)^2 \sum_j \tilde{\nu}_j^2 e^{-\beta\varepsilon_j}$$

$$= \left(\frac{1}{3475}\right)^2 \times \{0 + 21850^2\,e^{-21850/3475} + 3 \times 21870^2\,e^{-21870/3475} + \cdots\} = 0.6719$$

The electronic contribution to the molar constant-volume heat capacity is

$$C_{V,m} = R\left\{\frac{\ddot{q}}{q} - \left(\frac{\dot{q}}{q}\right)^2\right\}$$

$$= 8.314\,\text{J K}^{-1}\,\text{mol}^{-1} \times \left\{\frac{0.6719}{1.0167} - \left(\frac{0.1057}{1.0167}\right)^2\right\} = \boxed{5.41\,\text{J K}^{-1}\,\text{mol}^{-1}}$$

P20.18 $c_s = \left(\dfrac{\gamma RT}{M}\right)^{1/2}, \quad \gamma = \dfrac{C_{p,m}}{C_{V,m}}, \quad C_{p,m} = C_{V,m} + R$

(a) $C_{V,m} = \frac{1}{2}R(3 + \nu_R^* + 2\nu_V^*) = \frac{1}{2}R(3 + 2) = \frac{5}{2}R$

$C_{p,m} = \frac{5}{2}R + R = \frac{7}{2}R$

$\gamma = \dfrac{7}{5} = 1.40; \quad$ hence $\boxed{c_s = \left(\dfrac{1.40RT}{M}\right)^{1/2}}$

(b) $C_{V,m} = \dfrac{1}{2}R(3+2) = \dfrac{5}{2}R, \quad \gamma = 1.40, \quad \boxed{c_s = \left(\dfrac{1.40RT}{M}\right)^{1/2}}$

(c) $C_{V,m} = \frac{1}{2}R(3+3) = 3R$

$C_{p,m} = 3R + R = 4R, \quad \gamma = \dfrac{4}{3}, \quad \boxed{c_s = \left(\dfrac{4RT}{3M}\right)^{1/2}}$

For air, $M \approx 29\,\text{g mol}^{-1}$, $T \approx 298\,\text{K}$, $\gamma = 1.40$

$$c_s = \left(\frac{(1.40) \times (2.48\,\text{kJ mol}^{-1})}{29 \times 10^{-3}\,\text{mol}^{-1}}\right)^{1/2} = \boxed{350\,\text{m s}^{-1}}$$

Solutions to applications

P20.19 $S = k \ln W$ [19.30]

$\qquad = k \ln 4^N = Nk \ln 4$

$\qquad = (5 \times 10^8) \times (1.38 \times 10^{-23}\,\mathrm{J\,K^{-1}}) \times \ln 4 = \boxed{9.57 \times 10^{-15}\,\mathrm{J\,K^{-1}}}$

Comment. Even for a molecule as large as DNA the residual molecular entropy is small compared to normal entropies for macroscopic systems.

P20.21   The standard molar Gibbs energy is given by

$$G_m^\circ - G_m^\circ(0) = RT \ln \frac{q_m^\circ}{N_A} \; [20.10] \quad \text{where} \quad \frac{q_m^\circ}{N_A} = \frac{q_{m,\mathrm{tr}}^\circ}{N_A} q_{\mathrm{rot}}\, q_{\mathrm{vib}}\, q_{\mathrm{elec}}$$

Translation

$$\frac{q_{m,\mathrm{tr}}^\circ}{N_A} = 2.561 \times 10^{-2} \times (T/\mathrm{K})^{5/2} \times (M/\mathrm{g\,mol^{-1}})^{3/2}$$

$$\qquad = 2.561 \times 10^{-2} \times (200.0)^{5/2} \times (102.9)^{3/2} = 1.512 \times 10^7$$

Rotation of a nonlinear molecule

$$q_{\mathrm{rot}} = \frac{1}{\sigma}\left(\frac{kT}{hc}\right)^{3/2}\left(\frac{\pi}{ABC}\right)^{1/2} = \frac{1.0270}{\sigma} \times \frac{(T/\mathrm{K})^{3/2}}{(ABC/\mathrm{cm^{-3}})^{1/2}}$$

$$\qquad = \frac{1.0270}{2} \times \frac{[(200.0) \times (2.998 \times 10^{10}\,\mathrm{cm\,s^{-1}})]^{3/2}}{[(13109.4) \times (2409.8) \times (2139.7) \times (10^6\,\mathrm{s^{-1}})^3/\mathrm{cm^{-3}}]^{1/2}} = 2.900 \times 10^4$$

Vibration

$$q_{\mathrm{vib}}^{(1)} = \frac{1}{1 - \exp\left(\frac{-1.4388(\tilde{\nu}/\mathrm{cm^{-1}})}{T/\mathrm{K}}\right)} = \frac{1}{1 - \exp\left(\frac{-1.4388(753)}{200.0}\right)} = 1.004$$

$$q_{\mathrm{vib}}^{(2)} = \frac{1}{1 - \exp\left(\frac{-1.4388(542)}{200.0}\right)} = 1.021$$

$$q_{\mathrm{vib}}^{(3)} = \frac{1}{1 - \exp\left(\frac{-1.4388(310)}{200.0}\right)} = 1.120$$

$$q_{\mathrm{vib}}^{(4)} = \frac{1}{1 - \exp\left(\frac{-1.4388(127)}{200.0}\right)} = 1.670$$

$$q_{\mathrm{vib}}^{(5)} = \frac{1}{1 - \exp\left(\frac{-1.4388(646)}{200.0}\right)} = 1.010$$

$$q_{\mathrm{vib}}^{(6)} = \frac{1}{1 - \exp\left(\frac{-1.4388(419)}{200.0}\right)} = 1.052$$

$$q_{\mathrm{vib}} = \prod_{i=1}^{6} q_{\mathrm{vib}}^{(i)} = 2.037$$

Putting it all together yields

$$G_m^\circ - G_m^\circ(0) = (8.3145 \, \text{J} \, \text{mol}^{-1} \, \text{K}^{-1}) \times (200.0 \, \text{K}) \times \ln(1.512 \times 10^7)$$
$$\times (2.900 \times 10^4) \times (2.037) \times (1)$$

$$G_m^\circ - G_m^\circ(0) = 4.576 \times 10^4 \, \text{J} \, \text{mol}^{-1} = \boxed{45.76 \, \text{kJ} \, \text{mol}^{-1}}$$

21 Molecular interactions

Solutions to exercises

Discussion questions

E21.1(a) Molecules with a permanent separation of electric charge have a permanent dipole moment. In molecules containing atoms of differing electronegativity, the bonding electrons may be displaced in such a way as to produce a net separation of charge in the molecule. Separation of charge may also arise from a difference in atomic radii of the bonded atoms as illustrated in Fig. 21.1. The separation of charges in the bonds in usually, though not always, in the direction of the more electronegative atom but depends on the precise bonding situation in the molecule as described in Section 21.1(a). A heteronuclear diatomic molecule necessarily has a dipole moment if there is a difference in electronegativity between the atoms, but the situation in polyatomic molecules is more complex. A polyatomic molecule has a permanent dipole moment only if it fulfills certain symmetry requirements as discussed in Section 15.3(a).

An external electric field can distort the electron density in both polar and nonpolar molecules and this results in an induced dipole moment that is proportional to the field. The constant of proportionality is called the polarizability.

E21.2(a) Dipole moments are not measured directly, but are calculated from a measurement of the relative permittivity, ε_r (dielectric constant) of the medium. Equation 21.16 implies that the dipole moment can be determined from a measurement of ε_r as a function of temperature. This approach is illustrated in Example 21.1. In another method, the relative permittivity of a solution of the polar molecule is measured as a function of concentration. The calculation is again based on the Debye equation, but in a modified form. The values obtained by this method are accurate only to about 10%. See the references listed under *Further reading* for the details of this approach, in particular, the article by C. P. Smyth. A third method is based on the relation between relative permittivity and refractive index, eqn 21.19, and thus reduces to a measurement of the refractive index. Accurate values of the dipole moments of gaseous molecules can be obtained from the Stark effect in their microwave spectra.

E21.3(a) See Fig. 21.15. If the A—H bond in the A—H$\cdots$B arrangement is regarded as formed from the overlap of an orbital on A, ψ_A, and a hydrogen 1s orbital ψ_H, and if the lone pair on B occupies an orbital on B, ψ_B, then when the two molecules are close together, we can build three molecular orbitals from the three basis orbitals:

$$\psi = C_A \psi_A + C_H \psi_H + C_B \psi_B$$

One of the molecular orbitals is bonding, one almost nonbonding, and the third antibonding. These three orbitals need to accommodate four electrons, two from the A—H bond and two from the lone pair on B. Two enter the bonding orbital and two the nonbonding orbital, so the net effect is a lowering of the energy, that is, a bond has formed.

E21.4(a) The increase in entropy of a solution when hydrophobic molecules or groups in molecules cluster together and reduce their structural demands on the solvent (water) is the origin of the hydrophobic interaction that tends to stabilize clustering of hydrophobic groups in solution. A manifestation of the hydrophobic interaction is the clustering together of hydrophobic groups in biological macromolecules. For example, the side chains of amino acids that are used to form the polypeptide chains of proteins are hydrophobic, and the hydrophobic interaction is a major contributor to the tertiary structure of polypeptides. At first thought, this clustering would seem to be a nonspontaneous process as the clustering of the solute results in a decrease in entropy of the solute. However, the clustering of the solute results in greater freedom of movement of the solvent molecules and an accompanying increase in disorder and entropy of the solvent. The total entropy of the system has increased and the process is spontaneous.

Numerical exercises

E21.5(a) A molecule with a centre of symmetry may not be polar. Therefore $ClF_3 (D_{3h})$ may not be polar as the group D_{3h} contains C_2 and σ_h (equivalent to i). Molecules belonging to the groups C_n and C_{nv} may be polar (Section 15.3); therefore $\boxed{O_3}$ (C_{2v}) is polar as well as $\boxed{H_2O_2}$ (C_2) except in one configuration when the two O—H bonds are at $180°$ to each other. But nearly free rotation about the O—O bond makes the average dipole zero.

E21.6(a) Polarizability, dipole moment, and molar polarization are related by

$$P_m = \left(\frac{N_A}{3\varepsilon_0}\right) \times \left(\alpha + \frac{\mu^2}{3kT}\right) \quad [21.16]$$

In order to solve for α, it is first necessary to obtain μ from the temperature variation of P_m.

$$\alpha + \frac{\mu^2}{3kT} = \frac{3\varepsilon_0 P_m}{N_A}$$

Therefore, $\left(\dfrac{\mu^2}{3k}\right) \times \left(\dfrac{1}{T} - \dfrac{1}{T'}\right) = \left(\dfrac{3\varepsilon_0}{N_A}\right) \times (P - P')$ $(P$ at T, P' at $T')$

and hence

$$\mu^2 = \frac{\left(\frac{9\varepsilon_0 k}{N_A}\right) \times (P - P')}{\frac{1}{T} - \frac{1}{T'}}$$

$$= \frac{(9) \times (8.854 \times 10^{-12}\,\mathrm{J^{-1}\,C^2\,m^{-1}}) \times (1.381 \times 10^{-23}\,\mathrm{J\,K^{-1}}) \times (70.62 - 62.47) \times 10^{-6}\,\mathrm{m^3\,mol^{-1}}}{(6.022 \times 10^{23}\,\mathrm{mol^{-1}}) \times \left(\frac{1}{351.0\,\mathrm{K}} - \frac{1}{423.2\,\mathrm{K}}\right)}$$

$$= 3.06\overline{4} \times 10^{-59}\,\mathrm{C^2 m^2}$$

and hence $\mu = \boxed{5.5 \times 10^{-30}\,\mathrm{C\,m}}$, or $1.7\,\mathrm{D}$

Then $\alpha = \dfrac{3\varepsilon_0 P_m}{N_A} - \dfrac{\mu^2}{3kT} = \dfrac{(3) \times (8.854 \times 10^{-12}\,\mathrm{J^{-1}\,C^2\,m^{-1}}) \times (70.62 \times 10^{-6}\,\mathrm{m^3\,mol^{-1}})}{6.022 \times 10^{23}\,\mathrm{mol^{-1}}}$

$$- \frac{3.06\overline{4} \times 10^{-59}\,\mathrm{C^2\,m^2}}{(3) \times (1.381 \times 10^{-23}\,\mathrm{J\,K^{-1}}) \times (351.0\,\mathrm{K})}$$

$$= \boxed{1.01 \times 10^{-39}\,\mathrm{J^{-1}\,C^2 m^2}}$$

Corresponding to $\alpha' = \dfrac{\alpha}{4\pi\varepsilon_0}$ [21.10] $= \boxed{9.1 \times 10^{-24}\,\mathrm{cm^3}}$

E21.7(a) $\dfrac{\varepsilon_r - 1}{\varepsilon_r + 2} = \dfrac{\rho P_m}{M}$ [21.15] $= \dfrac{(1.89\,\mathrm{g\,cm^{-3}}) \times (27.18\,\mathrm{cm^3\,mol^{-1}})}{92.45\,\mathrm{g\,mol^{-1}}} = 0.556$

Hence, $\varepsilon_r = \dfrac{(1) + (2) \times (0.556)}{1 - 0.556} = \boxed{4.8}$

E21.8(a) $n_r = (\varepsilon_r)^{1/2}$ [21.19]

$$\frac{\varepsilon_r - 1}{\varepsilon_r + 2} = \frac{N\alpha}{3\varepsilon_0} \quad [21.17]; \qquad N = \frac{\rho N_A}{M}$$

Therefore,

$$\alpha = \left(\frac{3\varepsilon_0 M}{\rho N_A}\right) \times \left(\frac{n_r^2 - 1}{n_r^2 + 2}\right) = \left(\frac{(3) \times (8.854 \times 10^{-12}\,\text{J}^{-1}\,\text{C}^2\,\text{m}^{-1}) \times (267.8\,\text{g mol}^{-1})}{(3.32 \times 10^6\,\text{g m}^{-3}) \times (6.022 \times 10^{23}\,\text{mol}^{-1})}\right)$$

$$\times \left(\frac{1.732^2 - 1}{1.732^2 + 2}\right) = \boxed{1.42 \times 10^{-39}\,\text{J}^{-1}\,\text{C}^2\,\text{m}^2}$$

and $\alpha' = \boxed{1.28 \times 10^{-23}\,\text{cm}^3}$

E21.9(a) $\mu = qR$ ($q = be$, $b = $ bond order)

For example, $\mu_{\text{ionic}}(\text{C–O}) = (1.602 \times 10^{-19}\,\text{C}) \times (1.43 \times 10^{-10}\,\text{m}) = 22.9 \times 10^{-30}\,\text{C m} = 6.86\,\text{D}$

Then, percentage ionic character $= \dfrac{\mu_{\text{obs}}}{\mu_{\text{ionic}}} \times 100$ per cent

$\Delta\chi$ values are based on Pauling electronegativities.

We draw up the following table

Bond	$\mu_{\text{obs}}/\text{D}$	$\mu_{\text{ionic}}/\text{D}$	per cent	$\Delta\chi$
C—O	1.2	6.86	17	0.8
C=O	2.7	11.72	23	0.8

There is no correlation based on this set of bonds between the same two atoms, but in general there is a qualitative correlation for bonds between different atoms.

Comment. There are other contributions to the observed dipole moment besides the term qR. These are a result of the delocalization of the charge distribution in the bond orbitals.

E21.10(a) Refer to Fig. 21.2 of the text, and add moments vectorially.

Use $\mu = 2\mu_1 \cos \dfrac{1}{2}\theta$ [21.3b].

(a) *p*-xylene: the resultant is zero, so $\mu = \boxed{0}$

(b) *o*-xylene: $\mu = (2) \times (0.4\,\text{D}) \times \cos 30° = \boxed{0.7\,\text{D}}$

(c) *m*-xylene: $\mu = (2) \times (0.4\,\text{D}) \times \cos 60° = \boxed{0.4\,\text{D}}$

The *p*-xylene molecule belongs to the group D_{2h}, and so it is necessarily nonpolar.

E21.11(a)

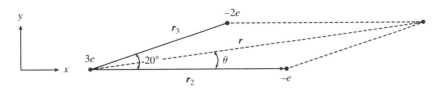

Figure 21.1

The dipole moment is the vector sum (see Fig. 21.1)

$$\boldsymbol{\mu} = \sum_i q_i \boldsymbol{r}_i = 3e(0) - e\boldsymbol{r}_2 - 2e\boldsymbol{r}_3$$

$$\boldsymbol{r}_2 = \mathbf{i}x_2, \qquad \boldsymbol{r}_3 = \mathbf{i}x_3 + \mathbf{j}y_3$$

$$x_2 = +0.32 \, \text{nm}$$

$$x_3 = r_3 \cos 20° = (+0.23 \, \text{nm}) \times (0.940) = 0.21\overline{6} \, \text{nm}$$

$$y_3 = r_3 \sin 20° = (+0.23 \, \text{nm}) \times (0.342) = 0.078\overline{7} \, \text{nm}$$

The components of the vector sum are the sums of the components. That is (with all distances in nm)

$$\mu_x = -ex_2 - 2ex_3 = -(e) \times \{(0.32) + (2) \times (0.21\overline{6})\} = -(e) \times (0.752 \, \text{nm})$$

$$\mu_y = -2ey_3 = -(e) \times (2) \times (0.078\overline{7}) = -(e) \times (0.1574 \, \text{nm})$$

$$\mu = (\mu_x^2 + \mu_y^2)^{1/2} = (e) \times (0.76\overline{8} \, \text{nm}) = (1.602 \times 10^{-19} \, \text{C}) \times (0.768 \times 10^{-9} \, \text{m})$$

$$= 1.2\overline{3} \times 10^{-28} \, \text{C m} = \boxed{37 \, \text{D}}$$

The angle that μ makes with x-axis is given by

$$\cos\theta = \frac{|\mu_x|}{\mu} = \frac{0.752}{0.768}; \qquad \boxed{\theta = 11.7°}$$

E21.12(a) $\mu^* = \alpha\mathcal{E} [21.8] = 4\pi\varepsilon_0\alpha'\mathcal{E} \, [21.10]$

$$= (4\pi) \times (8.854 \times 10^{-12} \, \text{J}^{-1} \, \text{C}^2 \, \text{m}^{-1}) \times (1.48 \times 10^{-30} \, \text{m}^3) \times (1.0 \times 10^5 \, \text{Vm}^{-1})$$

$$= 1.6 \times 10^{-35} \, \text{C m} \quad [1 \, \text{J} = 1 \, \text{C V}]$$

which corresponds to $\boxed{4.9 \, \mu\text{D}}$.

E21.13(a) The solution to Exercise 21.8(a) showed that

$$\alpha = \left(\frac{3\varepsilon_0 M}{\rho N_A}\right) \times \left(\frac{n_r^2 - 1}{n_r^2 + 2}\right) \quad \text{or} \quad \alpha' = \left(\frac{3M}{4\pi\rho N_A}\right) \times \left(\frac{n_r^2 - 1}{n_r^2 + 2}\right)$$

which may be solved for n_r to yield

$$n_r = \left(\frac{\beta' + 2\alpha'}{\beta' - \alpha'}\right)^{1/2} \quad \text{with} \quad \beta' = \frac{3M}{4\pi\rho N_A}$$

$$\beta' = \frac{(3) \times (18.02 \, \text{g mol}^{-1})}{(4\pi) \times (0.99707 \times 10^6 \, \text{g m}^{-3}) \times (6.022 \times 10^{23} \, \text{mol}^{-1})} = 7.165 \times 10^{-30} \, \text{m}^3$$

$$n_r = \left(\frac{(7.165) + (2) \times (1.5)}{(7.165) - (1.5)}\right)^{1/2} = \boxed{1.34}$$

There is little or no discrepancy to be explained!

E21.14(a) $\dfrac{\varepsilon_r - 1}{\varepsilon_r + 2} = \left(\dfrac{\rho N_A}{3\varepsilon_0 M}\right) \times \left(\alpha + \dfrac{\mu^2}{3kT}\right)$ [21.16, with 21.15]

Hence, $\varepsilon_r = \dfrac{1 + 2x}{1 - x}$ with $x = \left(\dfrac{\rho N_A}{3\varepsilon_0 M}\right) \times \left(\alpha + \dfrac{\mu^2}{3kT}\right)$

$$x = \left(\frac{(1.173 \times 10^6 \, \text{g m}^{-3}) \times (6.022 \times 10^{23} \, \text{mol}^{-1})}{(3) \times (8.854 \times 10^{-12} \, \text{J}^{-1} \, \text{C}^2 \, \text{m}^{-1}) \times (112.6 \, \text{g mol}^{-1})} \right)$$

$$\times \left[(4\pi) \times (8.854 \times 10^{-12} \, \text{J}^{-1} \, \text{C}^2 \, \text{m}^{-1}) \times (1.23 \times 10^{-29} \, \text{m}^3) \right.$$

$$\left. + \left(\frac{[(1.57) \times (3.336 \times 10^{-30} \, \text{C m})]^2}{(3) \times (1.381 \times 10^{-23} \, \text{J K}^{-1}) \times (298.15 \, \text{K})} \right) \right]$$

$$= 0.848$$

Therefore, $\varepsilon_r = \dfrac{(1) + (2) \times (0.848)}{1 - 0.848} = \boxed{18}$

E21.15(a) We start with the equation derived in *Justification* 21.4

$$\Delta\theta = (n_R - n_L) \times \left(\frac{2\pi l}{\lambda} \right)$$

From the construction in Fig. 21.2 we see that the angle of rotation of the plane of polarization is $\delta = \dfrac{\Delta\theta}{2}$.

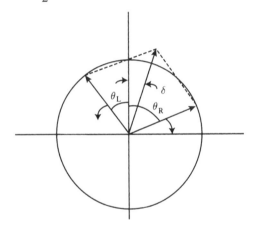

Figure 21.2

$$\delta = (n_R - n_L) \times \left(\frac{\pi l}{\lambda} \right)$$

$$(n_R - n_L) = \frac{\delta\lambda}{\pi l} = \frac{\delta\lambda}{180° \times l} = \frac{(250°) \times (5.00 \times 10^{-7} \, \text{m})}{(180°) \times (0.10 \, \text{m})} = \boxed{6.9 \times 10^{-6}}$$

Solutions to problems

Solutions to numerical problems

P21.1 The positive (H) end of the dipole will lie closer to the (negative) anion. The electric field generated by a dipole is

$$\mathcal{E} = \left(\frac{\mu}{4\pi\varepsilon_0} \right) \times \left(\frac{2}{r^3} \right) \quad [21.23]$$

$$= \frac{(2) \times (1.85) \times (3.34 \times 10^{-30} \, \text{C m})}{(4\pi) \times (8.854 \times 10^{-12} \, \text{J}^{-1} \, \text{C}^2 \, \text{m}^{-1}) \times r^3} = \frac{1.11 \times 10^{-19} \, \text{V m}^{-1}}{(r/\text{m})^3} = \frac{1.11 \times 10^8 \, \text{V m}^{-1}}{(r/\text{nm})^3}$$

(a) $\mathcal{E} = \boxed{1.1 \times 10^8 \text{ V m}^{-1}}$ when $r = 1.0$ nm

(b) $\mathcal{E} = \dfrac{1.11 \times 10^8 \text{ V m}^{-1}}{0.3^3} = \boxed{4 \times 10^9 \text{ V m}^{-1}}$ for $r = 0.3$ nm

(c) $\mathcal{E} = \dfrac{1.11 \times 10^8 \text{ V m}^{-1}}{30^3} = \boxed{4 \text{ kV m}^{-1}}$ for $r = 30$ nm.

P21.3   The equations relating dipole moment and polarizability volume to the experimental quantities ε_r and ρ are

$$P_m = \left(\frac{M}{\rho}\right) \times \left(\frac{\varepsilon_r - 1}{\varepsilon_r + 2}\right) \text{ [21.15] and } P_m = \frac{4\pi}{3} N_A \alpha' + \frac{N_A \mu^2}{9\varepsilon_0 kT} \text{ [21.16, with } \alpha = 4\pi\varepsilon_0\alpha']$$

Therefore, we draw up the following table (with $M = 119.4 \text{ g mol}^{-1}$)

$\theta/^\circ$C	−80	−70	−60	−40	−20	0	20
T/K	193	203	213	233	253	273	293
$\dfrac{1000}{T/K}$	5.18	4.93	4.69	4.29	3.95	3.66	3.41
ε_r	3.1	3.1	7.0	6.5	6.0	5.5	5.0
$\dfrac{\varepsilon_r - 1}{\varepsilon_r + 2}$	0.41	0.41	0.67	0.65	0.63	0.60	0.57
$\rho/\text{g cm}^{-3}$	1.65	1.64	1.64	1.61	1.57	1.53	1.50
$P_m/(\text{cm}^3 \text{ mol}^{-1})$	29.8	29.9	48.5	48.0	47.5	46.8	45.4

P_m is plotted against $\dfrac{1}{T}$ in Fig. 21.3.

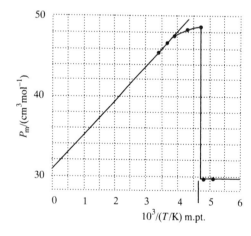

Figure 21.3

The (dangerously unreliable) intercept is ≈ 30 and the slope is $\approx 4.5 \times 10^3$. It follows that

$$\alpha' = \frac{(3) \times (30 \text{ cm}^3 \text{ mol}^{-1})}{(4\pi) \times (6.022 \times 10^{23} \text{ mol}^{-1})} = \boxed{1.2 \times 10^{-23} \text{ cm}^3}$$

To determine μ we need

$$\mu = \left(\frac{9\varepsilon_0 k}{N_A}\right)^{1/2} \times (\text{slope} \times \text{cm}^3\,\text{mol}^{-1}\,\text{K})^{1/2}$$

$$= \left(\frac{(9) \times (8.854 \times 10^{-12}\,\text{J}^{-1}\,\text{C}^2\,\text{m}^{-1}) \times (1.381 \times 10^{-23}\,\text{J}\,\text{K}^{-1})}{6.022 \times 10^{-23}\,\text{mol}^{-1}}\right)^{1/2}$$

$$\times (\text{slope} \times \text{cm}^3\,\text{mol}^{-1}\text{K})^{1/2}$$

$$= (4.275 \times 10^{-29}\,\text{C}) \times \left(\frac{\text{mol}}{\text{K}\,\text{m}}\right)^{1/2} \times (\text{slope} \times \text{cm}^3\,\text{mol}^{-1}\,\text{K})^{1/2}$$

$$= (4.275 \times 10^{-29}\,\text{C}) \times (\text{slope} \times \text{cm}^3\,\text{m}^{-1})^{1/2} = (4.275 \times 10^{-29}\,\text{C}) \times (\text{slope} \times 10^{-6}\,\text{m}^2)^{1/2}$$

$$= (4.275 \times 10^{-32}\,\text{C}\,\text{m}) \times (\text{slope})^{1/2} = (1.282 \times 10^{-2}\,\text{D}) \times (\text{slope})^{1/2}$$

$$= (1.282 \times 10^{-2}\,\text{D}) \times (4.5 \times 10^3)^{1/2} = \boxed{0.86\,\text{D}}$$

The sharp decrease in P_m occurs at the freezing point of chloroform ($-63°C$), indicating that the dipole reorientation term no longer contributes. Note that P_m for the solid corresponds to the extrapolated, dipole-free, value of P_m, so the extrapolation is less hazardous than it looks.

P21.5 $$P_m = \frac{4\pi}{3} N_A \alpha' + \frac{N_A \mu^2}{9\varepsilon_0 k T} \quad [21.16,\ \text{with } \alpha = 4\pi\varepsilon_0\alpha']$$

Therefore, draw up the following table

T/K	292.2	309.0	333.0	387.0	413.0	446.0
$\dfrac{1000}{T/\text{K}}$	3.42	3.24	3.00	2.58	2.42	2.24
$P_m/(\text{cm}^3\,\text{mol}^{-1})$	57.57	55.01	51.22	44.99	42.51	39.59

The points are plotted in Fig. 21.4.

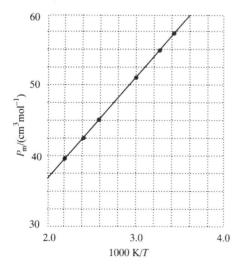

Figure 21.4

The extrapolated (least squares) intercept lies at $5.65 \, \text{cm}^3 \, \text{mol}^{-1}$ (not shown in the figure), and the least squares slope is $1.52 \times 10^4 \, \text{cm}^3 \, \text{K}^{-1} \, \text{mol}^{-1}$. It follows that

$$\alpha' = \frac{3P_m \, (\text{at intercept})}{4\pi N_A} = \frac{3 \times 5.65 \, \text{cm}^3 \, \text{mol}^{-1}}{4\pi \times 6.022 \times 10^{23} \, \text{mol}^{-1}}$$

$$= \boxed{2.24 \times 10^{-24} \, \text{cm}^3}$$

$\mu = 1.282 \times 10^{-2} \, \text{D} \times (1.52 \times 10^4)^{1/2}$ [from Problem 21.3] $= \boxed{1.58 \, \text{D}}$.

The high-frequency contribution to the molar polarization, P_m', at 273 K may be calculated from the refractive index

$$P_m' = \left(\frac{M}{\rho} \right) \times \left(\frac{\varepsilon_r - 1}{\varepsilon_r + 2} \right) \, [21.15] = \left(\frac{M}{\rho} \right) \times \left(\frac{n_r^2 - 1}{n_r^2 + 2} \right)$$

Assuming that ammonia under these conditions (1.00 atm pressure assumed) can be considered a perfect gas, we have

$$\rho = \frac{pM}{RT}$$

and $\dfrac{M}{\rho} = \dfrac{RT}{p} = \dfrac{82.06 \, \text{cm}^3 \, \text{atm} \, \text{K}^{-1} \, \text{mol}^{-1} \times 273 \, \text{K}}{1.00 \, \text{atm}} = 2.24 \times 10^4 \, \text{cm}^3 \, \text{mol}^{-1}$

Then $P_m' = 2.24 \times 10^4 \, \text{cm}^3 \, \text{mol}^{-1} \times \left\{ \dfrac{(1.000379)^2 - 1}{(1.000379)^2 + 2} \right\} = \boxed{5.66 \, \text{cm}^3 \, \text{mol}^{-1}}$.

If we assume that the high-frequency contribution to P_m remains the same at 292.2 K then we have

$$\frac{N_A \mu^2}{q \varepsilon_0 kT} = P_m - P_m' = (57.57 - 5.66) \, \text{cm}^3 \, \text{mol}^{-1}$$

$$= 51.91 \, \text{cm}^3 \, \text{mol}^{-1} = 5.191 \times 10^{-5} \, \text{m}^3 \, \text{mol}^{-1}$$

Solving for μ we have

$$\mu = \left(\frac{9\varepsilon_0 k}{N_A} \right)^{1/2} T^{1/2} (P_m - P_m')^{1/2}$$

The factor $\left(\dfrac{9\varepsilon_0 k}{N_A} \right)^{1/2}$ has been calculated in Problem 21.3 and is $4.275 \times 10^{-29} \, \text{C} \times \left(\dfrac{\text{mol}}{\text{K m}} \right)^{1/2}$

Therefore $\mu = 4.275 \times 10^{-29} \, \text{C} \times \left(\dfrac{\text{mol}}{\text{K m}} \right)^{1/2} \times (292.2 \, \text{K})^{1/2} \times (5.191 \times 10^{-5})^{1/2} \left(\dfrac{\text{m}^3}{\text{mol}} \right)^{1/2}$

$$= 5.26 \times 10^{-30} \, \text{C m} = \boxed{1.58 \, \text{D}}$$

The agreement is exact!

P21.8 The rotational constant is related to the moment of inertia, which in turn is related to the internuclear separation

$$B = \frac{h}{4\pi cI} = \frac{h}{4\pi cm_{\text{eff}} R^2} \quad \text{so} \quad R = \left(\frac{h}{4\pi c B m_{\text{eff}}} \right)^{1/2}$$

The effective mass is given by

$$m_{\text{eff}}^{-1} = m_1^{-1} + m_2^{-1} = (39.963 \, \text{u})^{-1} + (19.992 \, \text{u})^{-1} = 7.5043 \times 10^{-2} \, \text{u}^{-1}$$

or $\quad m_{\text{eff}} = \dfrac{1.66054 \times 10^{-27} \, \text{kg} \, \text{u}^{-1}}{7.5043 \times 10^{-2} \, \text{u}^{-1}} = 2.2128 \times 10^{-26} \, \text{kg}$

then $R = \left(\dfrac{1.0546 \times 10^{-34} \, \text{J s}}{4\pi \, (2914.9 \times 10^6 \, \text{s}^{-1}) \times (2.2128 \times 10^{-26} \, \text{kg})} \right)^{1/2} = \boxed{3.6071 \times 10^{-10} \, \text{m}}$

The distortion constant is related to the fundamental vibrational wavenumber by

$$D_J = \dfrac{4B^3}{\tilde{v}^2} \quad \text{so} \quad \tilde{v} = \left(\dfrac{4B^3}{D_J} \right)^{1/2} = \left(\dfrac{4(cB)^3}{c^2(cD_J)} \right)^{1/2}$$

$$\tilde{v} = \left(\dfrac{4(2914.9 \times 10^6 \, \text{s}^{-1})^3}{(2.998 \times 10^{10} \, \text{cm s}^{-1})^2 \times (231.01 \times 10^3 \, \text{s}^{-1})} \right)^{1/2} = \boxed{21.84 \, \text{cm}^{-1}}$$

The force constant is related to the vibrational frequency by

$$w = \left(\dfrac{k}{m_{\text{eff}}} \right)^{1/2} = 2\pi \tilde{v} = 2\pi c \tilde{v} \quad \text{so} \quad k = (2\pi c \tilde{v})^2 m_{\text{eff}}$$

$$k = [2\pi (2.998 \times 10^{10} \, \text{cm s}^{-1}) \times (21.84 \, \text{cm}^{-1})]^2 \times (2.2128 \times 10^{-26} \, \text{kg}) = \boxed{0.3746 \, \text{kg s}^{-2}}$$

P21.10   (a) The depth of the well in energy units is

$$\varepsilon = hcD_e = \boxed{1.51 \times 10^{-23} \, \text{J}}$$

The distance at which the potential is zero is given by

$$R_e = 2^{1/6} r_0 \quad \text{so} \quad r_0 = R_e 2^{-1/6} = 2^{-1/6}(297 \, \text{pm}) = \boxed{265 \, \text{pm}}.$$

(b) In Fig. 21.5 both potentials were plotted with respect to the bottom of the well, so the Lennard-Jones potential is the usual L-J potential plus ε.

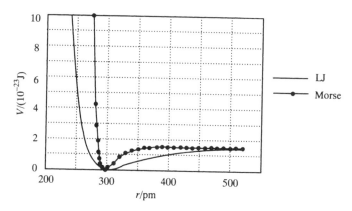

Figure 21.5

Note that the Lennard-Jones potential has a much softer repulsive branch than the Morse.

Solution to theoretical problems

P21.13 (a) Consider the arrangement shown in Fig. 21.6(a). There are a total of $3 \times 3 = 9$ Coulombic interactions at the distances shown. The total potential energy of interaction of the two quadrupoles is

$$V = \frac{q_1 q_2}{4\pi\varepsilon_0} \times \left[\left(\frac{1}{r} - \frac{2}{r-l} + \frac{1}{r-2l} \right) - 2\left(\frac{1}{r+l} - \frac{2}{r} + \frac{1}{r-l} \right) \right.$$
$$\left. + \left(\frac{1}{r+2l} - \frac{2}{r+l} + \frac{1}{r} \right) \right]$$
$$= \frac{q_1 q_2}{4\pi\varepsilon_0 r} \times \left[\left(1 - \frac{2}{1-\lambda} + \frac{1}{1-2\lambda} \right) - 2\left(\frac{1}{1+\lambda} - 2 + \frac{1}{1-\lambda} \right) \right.$$
$$\left. + \left(\frac{1}{1+2\lambda} - \frac{2}{1+\lambda} + 1 \right) \right] \quad \left(\lambda = \frac{l}{r} \ll 1 \right)$$

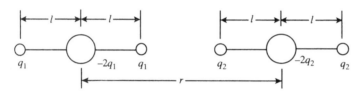

Figure 21.6(a)

Expand each term using

$$\frac{1}{1+x} = 1 - x + x^2 - x^3 + x^4 - \cdots$$

and keep up to λ^4 (the preceding terms cancel). The result is

$$V = \frac{q_1 q_2}{4\pi\varepsilon_0 r} \times 24\lambda^4 = \frac{6 q_1 q_2 l^4}{\pi\varepsilon_0 r^5}$$

Define the quadrupole moments of the two distributions as

$$Q_1 = q_1 l^2, \qquad Q_2 = q_2 l^2$$

and hence obtain $\boxed{V = \dfrac{6 Q_1 Q_2}{\pi\varepsilon_0} \times \dfrac{1}{r^5}}$

(b) Consider Fig. 21.6(b). There are three different distances, r, r', and r''. Three interactions are at r, four at r', and two at r''.

$$r' = (r^2 + l^2)^{1/2} = r(1 + \lambda^2)^{1/2} \approx r\left(1 + \frac{\lambda^2}{2} - \frac{\lambda^4}{8} + \cdots \right)$$
$$r'' = (r^2 + 4l^2)^{1/2} = r(1 + 4\lambda^2)^{1/2} \approx r(1 + 2\lambda^2 - 2\lambda^4 + \cdots)$$

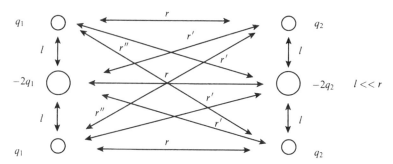

Figure 21.6(b)

$$V = \frac{q_1 q_2}{4\pi\varepsilon_0} \times \left[\left(\frac{1}{r} - \frac{2}{r'} + \frac{1}{r''}\right) - 2\left(\frac{2}{r'} - \frac{4}{r} + \frac{2}{r'}\right) + \left(\frac{1}{r''} - \frac{2}{r'} + \frac{1}{r}\right)\right]$$

$$= \left(\frac{2q_1 q_2}{4\pi\varepsilon_0}\right) \times \left(\frac{3}{r} - \frac{4}{r'} + \frac{1}{r''}\right) = \left(\frac{2q_1 q_2}{4\pi\varepsilon_0 r}\right) \times \left(3 - 4\frac{r}{r'} + \frac{r}{r''}\right)$$

Substituting for r' and r'' in terms of r and λ from above we obtain (dropping terms beyond λ^4)

$$V = V_0 \left(3 - \frac{4}{\left(1 + \frac{\lambda^2}{2} - \frac{\lambda^4}{8}\right)} + \frac{1}{(1 + 2\lambda^2 - 2\lambda^4)}\right) \quad \left[V_0 = \frac{2q_1 q_2}{4\pi\varepsilon_0 r}\right]$$

$$= V_0 \left[3 - 4\left(1 - \frac{\lambda^2}{2} + \frac{\lambda^4}{8} + \frac{\lambda^4}{4}\right) + (1 - 2\lambda^2 + 2\lambda^4 + 4\lambda^4)\right]$$

The terms in λ^0 and λ^2 cancel leaving

$$V = V_0 \left(6 - \frac{3}{2}\right)\lambda^4 = \frac{9}{2} V_0 \lambda^4 = \frac{9 q_1 q_2 \lambda^4}{4\pi\varepsilon_0 r} = \frac{9 q_1 q_2 l^4}{4\pi\varepsilon_0 r^5} = \boxed{\frac{9 Q_1 Q_2}{4\pi\varepsilon_0 r^5}}.$$

P21.14   Exercise 21.8 showed

$$\alpha = \left(\frac{3\varepsilon_0 M}{\rho N_A}\right) \times \left(\frac{n_r^2 - 1}{n_r^2 + 2}\right) \quad \text{or} \quad \alpha' = \left(\frac{3M}{4\pi\rho N_A}\right) \times \left(\frac{n_r^2 - 1}{n_r^2 + 2}\right)$$

Therefore, $\dfrac{n_r^2 - 1}{n_r^2 + 2} = \dfrac{4\pi\alpha' N_A \rho}{3M}$

Solving for n_r, $n_r = \left(\dfrac{1 + \frac{8\pi\alpha'\rho N_A}{3M}}{1 - \frac{4\pi\alpha'\rho N_A}{3M}}\right)^{1/2} = \left(\dfrac{1 + \frac{8\pi\alpha' p}{3kT}}{1 - \frac{4\pi\alpha' p}{3kT}}\right)^{1/2} \quad \left[\text{for a gas, } \rho = \frac{M}{V_m} = \frac{Mp}{RT}\right]$

$$\approx \left[\left(1 + \frac{8\pi\alpha' p}{3kT}\right) \times \left(1 + \frac{4\pi\alpha' p}{3kT}\right)\right]^{1/2} \quad \left[\frac{1}{1-x} \approx 1 + x\right]$$

$$\approx \left(1 + \frac{12\pi\alpha' p}{3kT} + \cdots\right)^{1/2} \approx 1 + \frac{2\pi\alpha' p}{kT} \quad \left[(1+x)^{1/2} \approx 1 + \frac{1}{2}x\right]$$

Hence, $\boxed{n_r = 1 + \text{const.} \times p}$, with constant $= \boxed{\dfrac{2\pi\alpha'}{kT}}$. From the first line above,

$$\alpha' = \left(\frac{3M}{4\pi N_A \rho}\right) \times \left(\frac{n_r^2 - 1}{n_r^2 + 2}\right) = \boxed{\left(\frac{3kT}{4\pi p}\right) \times \left(\frac{n_r^2 - 1}{n_r^2 + 2}\right)}$$

P21.16 The dimers should have a zero dipole moment. The strong molecular interactions in the pure liquid probably break up the dimers and produce hydrogen-bonded groups of molecules with a chain-like structure. In very dilute benzene solutions, the molecules should behave much like those in the gas and should tend to form planar dimers. Hence the relative permittivity $\boxed{\text{should decrease}}$ as the dilution increases.

P21.17 Consider a single molecule surrounded by $N - 1(\approx N)$ others in a container of volume V. The number of molecules in a spherical shell of thickness dr at a distance r is $4\pi r^2 \times \dfrac{N}{V}\, dr$. Therefore, the interaction energy is

$$u = \int_a^R 4\pi r^2 \times \left(\frac{N}{V}\right) \times \left(\frac{-C_6}{r^6}\right) dr = \frac{-4\pi N C_6}{V} \int_a^R \frac{dr}{r^4}$$

where R is the radius of the container and d the molecular diameter (the distance of closest approach). Therefore,

$$u = \left(\frac{4\pi}{3}\right) \times \left(\frac{N}{V}\right)(C_6) \times \left(\frac{1}{R^3} - \frac{1}{d^3}\right) \approx \frac{-4\pi N C_6}{3V d^3}$$

because $d \ll R$. The mutual pairwise interaction energy of all N molecules is $U = \frac{1}{2}Nu$ (the $\frac{1}{2}$ appears because each pair must be counted only once, i.e. A with B but not A with B and B with A). Therefore,

$$U = \boxed{\frac{-2\pi N^2 C_6}{3V d^3}}$$

For a van der Waals gas, $\dfrac{n^2 a}{V^2} = \left(\dfrac{\partial U}{\partial V}\right)_T = \dfrac{2\pi N^2 C_6}{3V^2 d^3}$

and therefore $a = \boxed{\dfrac{2\pi N_A^2 C_6}{3d^3}}$ $[N = nN_A]$

P21.20 Refer to Fig. 21.7(a).

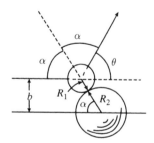

Figure 21.7(a)

The scattering angle is $\theta = \pi - 2\alpha$ if specular reflection occurs in the collision (angle of impact equal to angle of departure from the surface). For $b \le R_1 + R_2$, $\sin\alpha = \dfrac{b}{R_1 + R_2}$.

$$\theta = \begin{cases} \pi - 2\arcsin\left(\dfrac{b}{R_1 + R_2}\right) & b \le R_1 + R_2 \\ 0 & b > R_1 + R_2 \end{cases}$$

The function is plotted in Fig. 21.7(b).

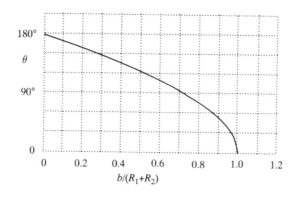

Figure 21.7(b)

P21.22 The interaction is a dipole-induced-dipole interaction. The energy is given by eqn 21.26:

$$V = -\frac{\mu 1^2 \alpha 2'}{4\pi \varepsilon_0 r^6} = -\frac{[(1.26\,D)(3.336 \times 10^{-30}\,C\,m\,D^{-1})]^2(1.04 \times 10^{-29}\,m^3)}{4\pi(8.854 \times 10^{-12}\,J^{-1}\,C^2\,m^{-1})(1.0 \times 10^{-9}\,m)^6}$$

$$V = \boxed{-1.6 \times 10^{-24}\,J = -1.0\,J\,mol^{-1}}.$$

P21.23 **(a)** The energy of induced-dipole–induced-dipole interactions can be approximated by the London formula (eqn 21.27):

$$V = -\frac{C}{r^6} = -\frac{3\alpha_1'\alpha_2'}{2r^6}\frac{I_1 I_2}{I_1 + I_2} = -\frac{3\alpha'^2 I}{4r^6}$$

where the second equality uses the fact that the interaction is between two of the same molecule. For two phenyl groups, we have:

$$V = -\frac{3(1.04 \times 10^{-29}m^3)^2(5.0\,eV)(1.602 \times 10^{-19}\,J\,eV^{-1})}{4(1.0 \times 10^{-9}m)^6} = 6.6 \times 10^{-23}J$$

or $\boxed{-39\,J\,mol^{-1}}$.

(b) The potential energy is everywhere negative. We can obtain the distance dependence of the force by taking

$$F = -\frac{dV}{dr} = -\frac{6C}{r^7}.$$

This force is everywhere attractive (*i.e.*, it works against increasing the distance between interacting groups). The force $\boxed{\text{approches zero as the distance becomes very large}}$; there is no finite distance at which the dispersion force is zero. (Ofcourse, if one takes into account repulsive forces, then the net force is zero at a distance at which the attractive and repulsive forces balance.)

22 Macromolecules and aggregates

Solutions to exercises

Discussion questions

E22.1(a) Contour length: the length of the macromolecule measured along its backbone from atom to atom. This is the stretched-out length of the macromolecule, but with bond angles maintained within the monomer units. It is directly proportional to the number of monomer units, N, and to the length of each unit (eqn 22.4).

Root mean square separation: one measure of the average separation of the ends of a random coil. It is the square root of the mean value of R^2, where R is the separation of the two ends of the coil. This mean value is calculated by weighting each possible value of R^2 with the probability, f (eqn 22.2), of that value of R occurring. It is proportional to $N^{1/2}$ and the length of each unit (eqn 22.5).

Radius of gyration: the radius of a thin hollow spherical shell of the same mass and moment of inertia as the macromolecule. In general, it is not easy to visualize this distance geometrically. However, for the simple case of a molecule consisting of a chain of identical atoms this quantity can be visualized as the root mean square distance of the atoms from the centre of mass. It also depends on $N^{1/2}$, but is smaller than the root mean square separation by a factor of $(1/6)^{1/2}$ (eqn 22.7).

E22.2(a) For a molecular mechanics calculation, potential energy functions are chosen for all the interactions between the atoms in the molecule; the calculation itself is a mathematical procedure that locates the energy minima (local and global) of the molecule as a function of bond distances and bond angles. Because only the potential energy is included in the calculation, contributions to the total energy from the kinetic energy are excluded in the result. The global minimum of a molecular mechanics calculation is a snapshot of the molecular structure at $T = 0$. No equations of motion are solved in a molecular mechanics calculation. The structure of a macromolecule (or any molecule, for that matter) can, in principle, be determined by solving the time independent Schrodinger equation for the molecule with methods similar to those described in Chapter 14. But due to the very large size of macromolecules, these methods may be impractical and inaccurate.

In a molecular dynamics calculation, equations of motion are integrated to determine the trajectories of all atoms in the molecule. The equations of motion can, in principle, be either classical (Newton's laws of motion) or quantum mechanical. But, in practice, due to the very large number of atoms in a macromolecule, Newton's equations of motion are used. Quantum mechanical methods are too time consuming, complicated, and at this stage too inaccurate to be popular in the field of polymer chemistry.

E22.3(a) Number average: the value obtained by weighting each molar mass by the number of molecules with that mass

$$\overline{M}_n = \frac{1}{N} \sum_i N_i M_i$$

In this expression, N_i is the number of molecules of Molar mass M_i and N is the total number of molecules. Measurements of the osmotic pressures of macromolecular solutions yield the number average molar mass.

Weight average: the value obtained by weighting each molar mass by the mass of each one present

$$\overline{M}_W = \frac{1}{m} \sum_i m_i M_i$$

In this expression, m_i is the total mass of molecules with molar mass M_i and m is the total mass of the sample. Light scattering experiments give the weight average molar mass.

Viscosity average: the value obtained from measurements of the intrinsic viscosities of solutions of the macromolecule

$$\overline{M}_V = \left(\frac{1}{m}\sum_i m_i M_i\right)^a$$

The symbols have the same meaning as above, but a is the Mark–Kuhn–Houwink–Sakurada equation. In terms of this equation, the weight average molar mass corresponds to $a = 1$, the number average molar mass to $a = -1$. Experimentally, a is found to be in the range 0.5–1.0. Therefore, M_V is closer to M_W than M_n.

Z-average molar mass: this is defined through the formula

$$\overline{M}_Z = \frac{\sum_i N_i M_i^3}{\sum_i N_i M_i^2}$$

The Z-average molar mass is obtained from sedimentation equilibria experiments.

E22.4(a) (a) The increase in temperature with the hydrophobic chain length is a result of the increased strength of the van der Waals interaction between long unsaturated portions of the chains that can interlock well with each other. The introduction of double bonds in the chains can affect the interlocking of the parallel chains by putting kinks in the chains, thereby decreasing the strength of the van der Waals interactions between chains. Double bonds can be either cis or trans. Only cis-double bonds produce a kink, but most fatty acids are the cis-isomer. So we expect that the transition temperatures will decrease in rough proportion to the number of C=C bonds.

(b) The addition of cholesterol is expected to increase the temperature of the transition from the liquid crystalline state to the liquid state by altering the conformations of the hydrocarbon chains. Cholesterol stabilizes extended chain conformations of adjacent hydrocarbon sections by van der Waals interactions relative to the coiled conformations that predominate when cholesterol is absent. The extended chains can pack better than coiled arrangements. However the lower transition temperature, that from the solid crystalline state to the liquid crystalline form, is probably decreased upon addition of cholesterol; its presence prevents the hydrophobic chains from freezing into a solid array by disrupting their packing. This will also spread the melting point over a range of temperatures.

Numerical exercises

E22.5(a) $R_{rms} = N^{1/2}l[22.5] = (700)^{1/2} \times (0.90\,\text{nm}) = \boxed{24\,\text{nm}}$

E22.6(a) The repeating unit (monomer) of polyethylene is ($-CH_2-CH_2-$) which has a molar mass of $28\,\text{g mol}^{-1}$. The number of repeating units, N, is therefore

$$N = \frac{280\,000\,\text{g mol}^{-1}}{28\,\text{g mol}^{-1}}$$

$$= 1.00 \times 10^4; \qquad l = 2R(C-C)\left[\text{Add } \tfrac{1}{2} \text{ bond on each side of monomer}\right]$$

$$R_c = Nl[22.4] = 2 \times (1.00 \times 10^4) \times (154\,\text{pm}) = 3.08 \times 10^6\,\text{pm} = \boxed{3.08 \times 10^{-6}\,\text{m}}$$

$$R_{rms} = N^{1/2} \times l[22.5] = 2 \times (1.00 \times 10^4)^{1/2} \times (154\,\text{pm})$$

$$= 3.08 \times 10^4\,\text{pm} = \boxed{3.08 \times 10^{-8}\,\text{m}}$$

E22.7(a) Equal amounts imply equal numbers of molecules; hence

$$\overline{M}_n = \frac{N_1 M_1 + N_2 M_2}{N} \; [22.17] = \frac{n_1 M_1 + n_2 M_2}{n} = \frac{1}{2}(M_1 + M_2) \quad \left[n_1 = n_2 = \frac{1}{2}n\right]$$

$$= \frac{62 + 78}{2} \, \text{kg mol}^{-1} = \boxed{70 \, \text{kg mol}^{-1}}$$

$$\overline{M}_W = \frac{m_1 M_1 + m_2 M_2}{m} \; [22.18] = \frac{n_1 M_1^2 + n_2 M_2^2}{n_1 M_1 + n_2 M_2} = \frac{M_1^2 + M_2^2}{M_1 + M_2} \quad [n_1 = n_2]$$

$$= \frac{62^2 + 78^2}{62 + 78} \, \text{kg mol}^{-1} = \boxed{71 \, \text{kg mol}^{-1}}$$

E22.8(a) $R_g = \dfrac{N^{1/2} l}{\sqrt{6}} \; [22.7] \quad N = 6 \left(\dfrac{R_g}{l}\right)^2 = (6) \times \left(\dfrac{7.3 \, \text{nm}}{0.154 \, \text{nm}}\right)^2 = \boxed{1.4 \times 10^4}$

E22.9(a) (a) Osmometry gives the number-average molar mass, so

$$\overline{M}_n = \frac{N_1 M_1 + N_2 M_2}{N_1 + N_2} = \frac{\left(\frac{m_1}{M_1}\right) M_1 + \left(\frac{m_2}{M_2}\right) M_2}{\left(\frac{m_1}{M_1}\right) + \left(\frac{m_2}{M_2}\right)} = \frac{m_1 + m_2}{\left(\frac{m_1}{M_1}\right) + \left(\frac{m_2}{M_2}\right)}$$

$$= \frac{100 \, \text{g}}{\left(\frac{30 \, \text{g}}{30 \, \text{kg mol}^{-1}}\right) + \left(\frac{70 \, \text{g}}{15 \, \text{kg mol}^{-1}}\right)} \; [\text{assume 100 g of solution}] = \boxed{18 \, \text{kg mol}^{-1}}$$

(b) Light-scattering gives the mass-average molar mass, so

$$\overline{M}_W = \frac{m_1 M_1 + m_2 M_2}{m_1 + m_2} = (0.30) \times (30) + (0.70) \times (15) \, \text{kg mol}^{-1} = \boxed{20 \, \text{kg mol}^{-1}}$$

E22.10(a) The formula for the rotational correlation time is:

$$\tau = \frac{4\pi a^3 \eta}{3kT}$$

With $\eta(H_2O) = 0.8909 \times 10^{-3} \, \text{kg m}^{-1} \text{s}^{-1}$ and $a(\text{SA}) = 3.0 \, \text{nm}$,

$$\tau = \frac{4\pi \times (3.0 \times 10^{-9} \, \text{m})^3 \times 0.8909 \times 10^{-3} \, \text{kg m}^{-1} \text{s}^{-1}}{3 \times 1.381 \times 10^{-23} \, \text{J K}^{-1} \times 298 \, \text{K}} = \boxed{2.4 \times 10^{-8} \, \text{s}}$$

With $\eta(CCl_4) = 0.895 \times 10^{-3} \, \text{kg m}^{-1} \text{s}^{-1}$ and $a(CCl_4) = 250 \, \text{pm}$,

$$\tau = \frac{4\pi \times (2.50 \times 10^{-10} \, \text{m})^3 \times 0.895 \times 10^{-3} \, \text{kg m}^{-1} \text{s}^{-1}}{3 \times 1.381 \times 10^{-23} \, \text{J K}^{-1} \times 298 \, \text{K}} = \boxed{1.4 \times 10^{-11} \, \text{s}}$$

E22.11(a) The effective mass of the particles is

$$m_{\text{eff}} = bm = (1 - \rho v_s)m \, [22.36] = m - \rho v_s m = v\rho_p - v\rho = v(\rho_p - \rho)$$

where v is the particle volume, ρ_p is the particle density. Equating the forces

$$m_{\text{eff}} r\omega^2 = fs = 6\pi \eta a s \quad [22.37, \; a = \text{particle radius}]$$

or $v(\rho_p - \rho) r\omega^2 = 6\pi \eta a s$ or $\frac{4}{3}\pi a^3 (\rho_p - \rho) r\omega^2 = 6\pi \eta a s$

Solving for s, $s = \dfrac{2a^2(\rho_p - \rho)r\omega^2}{9\eta}$

Thus, the relative rates of sedimentation are $\dfrac{s_2}{s_1} = \dfrac{a_2^2}{a_1^2} = 10^2 = \boxed{100}$

E22.12(a) The data yield the number-average molar mass using

$$\overline{M}_n = \frac{SRT}{bD}[22.41] = \frac{SRT}{(1-\rho v_s)D} \quad [22.36, \text{ for } b]$$

$$= \frac{(4.48 \times 10^{-13}\,\text{s}) \times (8.314\,\text{J K}^{-1}\,\text{mol}^{-1}) \times (293\,\text{K})}{[(1) - (0.9982 \times 10^3\,\text{kg m}^3) \times (0.749 \times 10^{-3}\,\text{m}^3\,\text{kg}^{-1})] \times (6.9 \times 10^{-11}\,\text{m}^2\,\text{s}^{-1})}$$

$$= \boxed{63\,\text{kg mol}^{-1}}$$

E22.13(a) See the solution to Exercise 22.11(a). In place of force $= m_{\text{eff}}r\omega^2$ we have force $= m_{\text{eff}}g$
The rest of the analysis is similar, leading to

$$s = \frac{2a^2(\rho_p - \rho)g}{9\eta} = \frac{(2) \times (2.0 \times 10^{-5}\,\text{m})^2 \times (1750 - 1000)\,\text{kg m}^{-3} \times (9.81\,\text{m s}^{-2})}{(9) \times (8.9 \times 10^{-4}\,\text{kg m}^{-1}\,\text{s}^{-1})}$$

$$= \boxed{7.3 \times 10^{-4}\,\text{m s}^{-1}}$$

E22.14(a) $\overline{M}_n = \dfrac{SRT}{bD}[22.41] = \dfrac{(3.2 \times 10^{-13}\,\text{s}) \times (8.314\,\text{J K}^{-1}\,\text{mol}^{-1}) \times (293\,\text{K})}{[(1) - (0.656) \times (1.06)] \times (8.3 \times 10^{-11}\,\text{m}^2\,\text{s}^{-1})} = \boxed{31\,\text{kg mol}^{-1}}$

E22.15(a) The number of solute molecules with potential energy E is proportional to $e^{-E/kT}$; hence

$$c \propto N \propto e^{-E/kT} \quad E = \tfrac{1}{2}m_{\text{eff}}r^2\omega^2$$

Therefore, $c \propto e^{Mb\omega^2 r^2/2RT}$ $[m_{\text{eff}} = bm, M = mN_A]$

$$\ln c = \text{const.} + \frac{Mb\omega^2 r^2}{2RT} \quad [b = 1 - \rho v_s]$$

and slope of $\ln c$ against r^2 is equal to $\dfrac{Mb\omega^2}{2RT}$. Therefore

$$M = \frac{2RT \times \text{slope}}{b\omega^2} = \frac{(2) \times (8.314\,\text{J K}^{-1}\,\text{mol}^{-1}) \times (300\,\text{K}) \times (729 \times 10^4\,\text{m}^{-2})}{(1 - 0.997 \times 0.61) \times \left(\frac{(2\pi) \times (50000)}{60\,\text{s}}\right)^2}$$

$$= \boxed{3.4 \times 10^3\,\text{kg mol}^{-1}}$$

E22.16(a) The centrifugal force acting is $F = mr\omega^2$, and by Newton's second law of motion, $F = ma$; hence

$$a = r\omega^2 = 4\pi^2 r\nu^2 = 4\pi^2 \times (6.0 \times 10^{-2}\,\text{m}) \times \left(\frac{80 \times 10^3}{60\,\text{s}}\right)^2 = 4.2\bar{1} \times 10^6\,\text{m s}^{-2}$$

Then, since $g = 9.81\,\text{m s}^{-2}$, $a = \boxed{4.3 \times 10^5 g}$

Solutions to problems

Solutions to numerical problems

P22.2 $S = \dfrac{s}{r\omega^2}$ [22.38]

Since $s = \dfrac{dr}{dt}$, $\dfrac{s}{r} = \dfrac{1}{r}\dfrac{dr}{dt} = \dfrac{d\ln r}{dt}$

and if we plot $\ln r$ against t, the slope gives S through

$$S = \frac{1}{\omega^2}\frac{d\ln r}{dt}$$

The data are as follows

t/min	15.5	29.1	36.4	58.2
r/cm	5.05	5.09	5.12	5.19
$\ln(r/\text{cm})$	1.619	1.627	1.633	1.647

The points are plotted in Fig. 22.1.

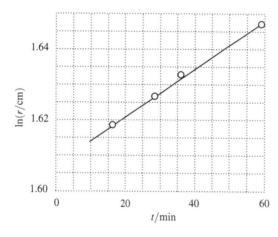

Figure 22.1

The least-squares slope is 6.62×10^{-4}, so

$$S = \frac{6.62 \times 10^{-4}\,\text{min}^{-1}}{\omega^2} = \frac{(6.62 \times 10^{-4}) \times \left(\frac{1}{60}\right)\text{s}^{-1}}{\left(2\pi \times \frac{4.5 \times 10^4}{60\,\text{s}}\right)^2} = 4.9\overline{7} \times 10^{-13}\,\text{s} \quad \text{or} \quad \boxed{5.0\,\text{Sv}}$$

P22.5 $[\eta] = \lim_{c \to 0} \left(\dfrac{\eta/\eta_0 - 1}{c}\right)$ [22.45]

We see that the intercept of a plot of the right-hand side against c, extrapolated to $c = 0$, gives $[\eta]$. We begin by constructing the following table using $\eta_0 = 0.985\,\text{g m}^{-1}\,\text{s}^{-1}$

$c/(\text{g L}^{-1})$	1.32	2.89	5.73	9.17
$\left(\dfrac{\eta/\eta_0 - 1}{c}\right)\Big/(\text{L g}^{-1})$	0.0731	0.0755	0.0771	0.0825

The points are plotted in Fig. 22.2. The least-squares intercept is at 0.0716, so $[\eta] = \boxed{0.0716\,\mathrm{L\,g^{-1}}}$

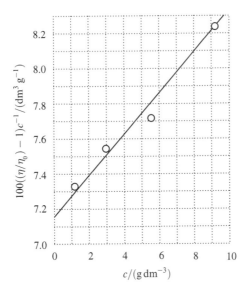

Figure 22.2

P22.6   We need to determine the intrinsic viscosity from a plot of $\dfrac{\left(\frac{\eta}{\eta_0}\right) - 1}{c/(\mathrm{g\,L^{-1}})}$ against c, extrapolated to $c = 0$ as in Example 22.7. Then from the relation

$$[\eta] = K\overline{M}_V^{\,a} \quad [22.47]$$

with K and a from Table 23.3, the viscosity average molar mass $\overline{M}_V$ may be calculated. η/η_0 values are determined from the times of flow using the relation

$$\frac{\eta}{\eta_0} = \frac{t}{t_0} \times \frac{\rho}{\rho_0} \approx \frac{t}{t_0}$$

noting that in the limit as $c \to 0$ it becomes exact. As explained in Example 23.5, $[\eta]$ can also be determined from the limit of $\dfrac{1}{c}\ln\left(\dfrac{\eta}{\eta_0}\right)$ as $c \to 0$.

We draw up the following table

$c/(\mathrm{g\,L^{-1}})$	0.000	2.22	5.00	8.00	10.00
t/s	208.2	248.1	303.4	371.8	421.3
$\dfrac{\eta}{\eta_0}$	—	1.192	1.457	1.786	2.024
$\dfrac{100\left[\left(\frac{\eta}{\eta_0}\right) - 1\right]}{c/(\mathrm{g\,L^{-1}})}$	—	8.63	9.15	9.82	10.24
$\ln\left(\dfrac{\eta}{\eta_0}\right)$	—	0.1753	0.3766	0.5799	0.7048
$\dfrac{100\ln\left(\frac{\eta}{\eta_0}\right)}{c/(\mathrm{g\,L^{-1}})}$	—	7.89	7.52	7.24	7.05

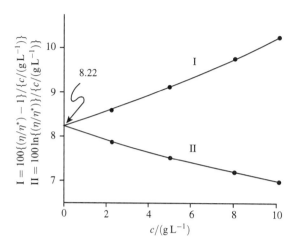

Figure 22.3

The points are plotted in Fig. 22.3.

The intercept as determined from the simultaneous extrapolation of both plots is $0.0822\,\text{L}\,\text{g}^{-1}$.

$$\overline{M}_V = \left(\frac{[\eta]}{K}\right)^{1/a} = \left(\frac{0.0822\,\text{L}\,\text{g}^{-1}}{9.5 \times 10^{-6}\,\text{L}\,\text{g}^{-1}}\right)^{1/0.74} = \boxed{2.1 \times 10^5\,\text{g}\,\text{mol}^{-1}}$$

Comment. This value differs markedly in molar mass from the sample of polystyrene in toluene described in Example 22.7.

P22.7   We follow the procedure of Example 22.7. Also compare to Problems 22.5 and 22.6.

$$[\eta] = \lim_{c \to 0}\left(\frac{\eta/\eta_0 - 1}{c}\right) \quad \text{and} \quad [\eta] = K\overline{M}_V^a \quad \text{[with K and a from Table 22.4]}$$

We draw up the following table using $\eta_0 = 0.647 \times 10^{-3}\,\text{kg}\,\text{m}^{-1}\,\text{s}^{-1}$

$c/(\text{g}/100\,\text{cm}^3)$	0	0.2	0.4	0.6	0.8	1.0
$\eta/(10^{-3}\,\text{kg}\,\text{m}^{-1}\,\text{s}^{-1})$	0.647	0.690	0.733	0.777	0.821	0.865
$\left(\dfrac{\eta/\eta_0 - 1}{c}\right)\Big/(100\,\text{cm}^3\,\text{g}^{-1})$		0.333	0.332	0.335	0.336	0.337

The values are plotted in Fig. 22.4, and extrapolated to 0.330.

Hence $[\eta] = (0.330) \times (100\,\text{cm}^3\,\text{g}^{-1}) = 33.0\,\text{cm}^3\,\text{g}^{-1}$

and $M_V = \left(\dfrac{33.0\,\text{cm}^3\,\text{g}^{-1}}{8.3 \times 10^{-2}\,\text{cm}^3\,\text{g}^{-1}}\right)^{1/0.50} = 158 \times 10^3$

That is, $M = \boxed{158\,\text{kg}\,\text{mol}^{-1}}$

P22.8   The relationship [22.47] between $[\eta]$ and $\overline{M}_V$ can be transformed into a linear one

$$\ln[\eta] = \ln K + a \ln M_V$$

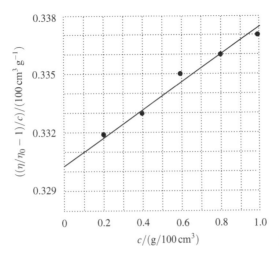

Figure 22.4

so a plot of $\ln [\eta]$ versus $\ln M_V$ will have a slope of a and a y-intercept of $\ln K$. The transformed data and plot are shown below (Fig. 22.5).

$\overline{M}_V/(\text{kg mol}^{-1})$	10.0	19.8	106	249	359	860	1800	5470	9720	56 800
$[\eta]/(\text{cm}^3\,\text{g}^{-1})$	8.90	11.9	28.1	44.0	51.2	77.6	113.9	195	275	667
$\ln \overline{M}_V/(\text{kg mol}^{-1})$	2.30	2.99	4.66	5.52	5.88	6.76	7.50	8.61	9.18	10.9
$\ln [\eta]/(\text{cm}^3\,\text{g}^{-1})$	2.19	2.48	3.34	3.78	3.94	4.35	4.74	5.27	5.62	6.50

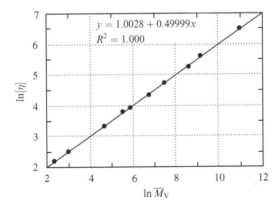

Figure 22.5

Thus $a = \boxed{0.500}$ and $K = e^{1.0028}\,\text{cm}^3\,\text{g}^{-1}\,\text{kg}^{-1/2}\,\text{mol}^{1/2} = \boxed{2.73\,\text{cm}^3\,\text{g}^{-1}\,\text{kg}^{-1/2}\,\text{mol}^{1/2}}$

Solving for $\overline{M}_V$ yields

$$\overline{M}_V = \left(\frac{[\eta]}{K}\right)^{1/a} = \left(\frac{100\,\text{cm}^3\,\text{g}^{-1}}{2.73\,\text{cm}^3\,\text{g}^{-1}\,\text{kg}^{-1/2}\,\text{mol}^{1/2}}\right)^2 = \boxed{1.34 \times 10^3\,\text{kg mol}^{-1}}$$

P22.9   The empirical Mark–Kuhn–Houwink–Sakurada equation [22.47] is

$$[\eta] = K\overline{M}_V^{a}$$

As the constant a may be non-integral the molar mass here is to be interpreted as unitless, that is, as $\overline{M}_V/(\text{g mol}^{-1})$. The units of K are then the same as $[\eta]$.

We fit the data to the above equation and obtain K and a from the fitting procedure. The plot is shown in Fig. 22.6.

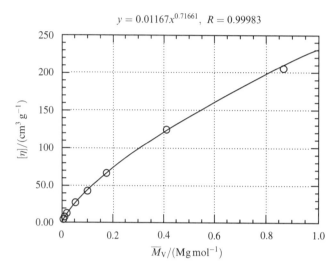

$$y = 0.01167x^{0.71661}, \quad R = 0.99983$$

Figure 22.6

$$\boxed{K = 0.0117\,\text{cm}^3\,\text{g}^{-1}}$$

$$\boxed{a = 0.717}$$

This value for a is not much different from that for polystyrene in benzene listed in Table 22.4. This is somewhat surprising as one would expect both the K and a values to be solvent-dependent. THF is not chemically similar to benzene. On the other hand, benzene and toluene are very much alike, yet the values of K and a as determined in Example 22.7 are markedly different from those in Table 22.4 for polystyrene in benzene.

P22.10   See section 7.5(e) and Example 7.5.

$$\frac{\Pi}{c} = \frac{RT}{\overline{M}_n}\left(1 + B\frac{c}{\overline{M}_n} + \cdots\right) \quad \text{[Example 7.5, with } \Pi = \rho g h]$$

Therefore, to determine $\overline{M}_n$ and B we need to plot Π/c against c. We draw up the following table

$c/(\text{g L}^{-1})$	1.21	2.72	5.08	6.60
$(\Pi/c)/(\text{Pa/g L}^{-1})$	111	118	129	136

The points are plotted in Fig. 22.7.

A least-squares analysis gives an intercept of $105.\overline{4}$ and a slope of 4.64. It follows that

$$\frac{RT}{\overline{M}_n} = 105.\overline{4}\,\text{Pa g}^{-1}\,\text{L} = 105.\overline{4}\,\text{Pa kg}^{-1}\,\text{m}^3$$

and hence that $\overline{M}_n = \dfrac{(8.314\,\text{J K}^{-1}\,\text{mol}^{-1}) \times (293\,\text{K})}{105.\overline{4}\,\text{Pa kg}^{-1}\,\text{m}^3} = \boxed{23.1\,\text{kg mol}^{-1}}$

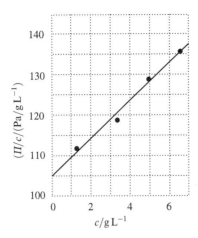

Figure 22.7

The slope of the graph is equal to $\dfrac{RTB}{M_n^2}$, so

$$\frac{RTB}{M_n^2} = 4.64\,\mathrm{Pa\,g^{-2}\,L^2} = 4.64\,\mathrm{Pa\,kg^{-2}\,m^6}$$

Therefore, $B = \dfrac{(23.1\,\mathrm{kg\,mol^{-1}})^2 \times (4.64\,\mathrm{Pa\,kg^{-2}\,m^6})}{(8.314\,\mathrm{J\,K^{-1}\,mol^{-1}}) \times (293\,\mathrm{K})} = \boxed{1.02\,\mathrm{m^3\,mol^{-1}}}$

Solutions to theoretical problems

P22.13 **(a)** $R_{\mathrm{rms}}^2 = \displaystyle\int_0^\infty R^2 f\, \mathrm{d}R$ [22.5]

$$f = 4\pi \left(\frac{a}{\pi^{1/2}}\right)^3 R^2 e^{-a^2 R^2} \ [22.2], \quad a = \left(\frac{3}{2Nl^2}\right)^{1/2}$$

Therefore,

$$R_{\mathrm{rms}}^2 = 4\pi \left(\frac{a}{\pi^{1/2}}\right)^3 \int_0^\infty R^4 e^{-a^2 R^2}\, \mathrm{d}R = 4\pi \left(\frac{a}{\pi^{1/2}}\right)^3 \times \left(\frac{3}{8}\right) \times \left(\frac{\pi}{a^{10}}\right)^{1/2}$$

$$= \frac{3}{2a^2} = Nl^2$$

Hence, $R_{\mathrm{rms}} = \boxed{lN^{1/2}}$

(b) $R_{\mathrm{mean}} = \displaystyle\int_0^\infty R f\, \mathrm{d}r = 4\pi \left(\frac{a}{\pi^{1/2}}\right)^3 \int_0^\infty R^3 e^{-a^2 R^2}\, \mathrm{d}R$

$$= 4\pi \left(\frac{a}{\pi^{1/2}}\right)^3 \times \left(\frac{1}{2a^4}\right) = \frac{2}{a\pi^{1/2}} = \boxed{\left(\frac{8N}{3\pi}\right)^{1/2} l}$$

(c) Set $\dfrac{\mathrm{d}f}{\mathrm{d}R} = 0$ and solve for R

$$\frac{\mathrm{d}f}{\mathrm{d}R} = 4\pi \left(\frac{a}{\pi^{1/2}}\right)^3 \{2R - 2a^2 R^3\} e^{-a^2 R^2} = 0 \quad \text{when } a^2 R^2 = 1$$

Therefore, the most probable separation is

$$R^* = \frac{1}{a} = \boxed{l\left(\frac{2}{3}N\right)^{1/2}}$$

When $N = 4000$ and $l = 154\,\text{pm}$,

(a) $R_{\text{rms}} = \boxed{9.74\,\text{nm}}$ **(b)** $R_{\text{mean}} = \boxed{8.97\,\text{nm}}$ **(c)** $R^* = \boxed{7.95\,\text{nm}}$

P22.16 $B = \frac{1}{2}N_A v_p = 4N_A v_{\text{mol}}\ [\text{Problem 22.19}] = \frac{16\pi}{2}N_A a_{\text{eff}}{}^3 = \frac{16\pi}{3}N_A \gamma^3 R_{\text{g}}{}^3\ [a_{\text{eff}} = \gamma R_{\text{g}}]$

(a) $R_{\text{g}} = \dfrac{N^{1/2}l}{\sqrt{6}}$ [22.7]

$$B = \frac{16\pi}{3 \times 6^{3/2}}\gamma^3 l^3 N^{3/2} N_A = \boxed{4.22 \times 10^{23}\,\text{mol}^{-1} \times (l\sqrt{N})^3}$$

$$= (4.22 \times 10^{23}\,\text{mol}^{-1}) \times [(154 \times 10^{-12}\,\text{m}) \times \sqrt{4000}]^3 = \boxed{0.39\,\text{m}^3\,\text{mol}^{-1}}$$

(b) $R_{\text{g}} = 2^{1/2} \times R_{\text{g}}(\text{free})$ [22.10]

$$B = 2^{3/2} \times B(\text{free}) = \boxed{1.19 \times 10^{24}\,\text{mol}^{-1} \times (l\sqrt{N})^3} = (2^{3/2}) \times (0.39\,\text{m}^3\,\text{mol}^{-1})$$

$$= \boxed{1.1\,\text{m}^3\,\text{mol}^{-1}}$$

Solutions to applications

P22.18   The table below lists initial and optimized torsional angles and optimized conformational potential energies for model dipeptides (molecular mechanics, MMFF force field, PC Spartan Pro^{TM}):

	initial		optimized		
	ϕ	ψ	ϕ	ψ	energy (kJ mol^{-1})
(a) R = H	75	−65	165	−50	−37.9
	180	180	180	−178	−54.2
	65	35	179	−178	−48.6
(b) R = CH$_3$	75	−65	149	−57	−29.0
	180	180	151	−168	−40.7
	65	35	75	−32	−41.8

The computations were set up by minimizing the energy subject to two constraints, namely the fixed initial values of ϕ and ψ. Then the constraints were removed, and the entire structure was allowed to relax to a minimum energy. In the R = H case, two of the three initial conformers converged to the same final values of ϕ and ψ, but not quite to the same conformation; the two conformations differed in the orientation of the methyl and hydrogen on the nitrogen of the N-terminal amino acid residue. The different conformations appear to represent local energy minima. It ought not to be surprising that there are several such minima in even a short peptide chain that contains so many nearly free internal rotations. The potential energy surface even of the model compound contains multiple fairly shallow minima. The conformations of the R = CH$_3$ model all keep the central methyl group away from the ends of the chain.

P22.19 The center of the spheres cannot approach more closely than $2a$; hence the excluded volume is

$$v_P = \frac{4}{3}\pi(2a)^3 = 8\left(\frac{4}{3}\pi a^3\right) = \boxed{8v_{mol}}$$

where v_{mol} is the molecular volume.

The osmotic virial coefficient, B (see eqn 7.41), arises largely from the effect of excluded volume. If we imagine a solution of a macromolecule being built by the successive addition of macromolecules to the solvent, each one being excluded by the ones that preceded it, then the value of B turns out to be

$$B = \tfrac{1}{2}N_A v_P$$

where v_P is the excluded volume due to a single molecule.

$$B(\text{BSV}) = \frac{1}{2}N_A \times \frac{32}{3}\pi a^3 = \frac{16}{3}\pi a^3 N_A$$

$$= \left(\frac{16\pi}{3}\right) \times (6.022 \times 10^{23}\,\text{mol}^{-1}) \times (14.0 \times 10^{-9}\,\text{m})^3 = \boxed{28\,\text{m}^3\,\text{mol}^{-1}}$$

$$B(\text{Hb}) = \left(\frac{16\pi}{3}\right) \times (6.022 \times 10^{23}\,\text{mol}^{-1}) \times (3.2 \times 10^{-9}\,\text{m})^3 = \boxed{0.33\,\text{m}^3\,\text{mol}^{-1}}$$

Since $\Pi = RT[P] + BRT[P]^2$ [7.41], if we write $\Pi^\circ = RT[P]$,

$$\frac{\Pi - \Pi^\circ}{\Pi^\circ} = \frac{BRT[P]^2}{RT[P]} = B[P]$$

For BSV,

$$[P] = \left(\frac{1.0\,\text{g}}{M}\right) \times (10\,\text{L}^{-1}) = \frac{10\,\text{g}\,\text{L}^{-1}}{1.07 \times 10^7\,\text{g}\,\text{mol}^{-1}} = 9.35 \times 10^{-7}\,\text{mol}\,\text{L}^{-1} = 9.35 \times 10^{-4}\,\text{mol}\,\text{m}^{-3}$$

and $\dfrac{\Pi - \Pi^\circ}{\Pi^\circ} = (28\,\text{m}^3\,\text{mol}^{-1}) \times (9.35 \times 10^{-4}\,\text{mol}\,\text{m}^{-3}) = 2.6 \times 10^{-2}$ corresponding to $\boxed{2.6\text{ percent}}$

For Hb, $[P] = \dfrac{10\,\text{g}\,\text{L}^{-1}}{66.5 \times 10^3\,\text{g}\,\text{mol}^{-1}} = 0.15\,\text{mol}\,\text{m}^{-3}$

and $\dfrac{\Pi - \Pi^\circ}{\Pi^\circ} = (0.15\,\text{mol}\,\text{m}^{-3}) \times (0.33\,\text{m}^3\,\text{mol}^{-1}) = 5.0 \times 10^{-2}$

which corresponds to $\boxed{5\text{ percent}}$.

P22.20 Assume the solute particles are solid spheres and see how well R_g calculated on the basis of that assumption agrees with experimental values.

$$R_g = (0.05690) \times \{(v_s/\text{cm}^3\,\text{g}^{-1}) \times (M/\text{g}\,\text{mol}^{-1})\}^{1/3}\,\text{nm} \quad \text{[Problem 22.15]}$$

and draw up the following table

	$M/(\text{g}\,\text{mol}^{-1})$	$v_s/(\text{cm}^3\,\text{g}^{-1})$	$(R_g/\text{nm})_{\text{calc}}$	$(R_g/\text{nm})_{\text{expt}}$
Serum albumin	66×10^3	0.752	2.09	2.98
Busy stunt virus	10.6×10^6	0.741	11.3	12.0
DNA	4×10^6	0.556	7.43	117.0

Therefore, serum albumin and bushy stunt virus resemble solid spheres, but DNA does not.

P22.23
$$\overline{M}_n = \frac{SRT}{bD}[22.41] = \frac{SRT}{(1 - \rho v_s)D}[22.36, \text{ for } b]$$

$$= \frac{(4.5 \times 10^{-13}\text{s}) \times (8.314\,\text{J K}^{-1}\text{mol}^{-1}) \times (293\,\text{K})}{(1 - 0.75 \times 0.998) \times (6.3 \times 10^{-11}\,\text{m}^2\,\text{s}^{-1})} = \boxed{69\,\text{kg mol}^{-1}}$$

Now combine $f = 6\pi a\eta$ [22.34] with $f = \dfrac{kT}{D}$ [22.33]

$$a = \frac{kT}{6\pi\eta D} = \frac{(1.381 \times 10^{-23}\,\text{J K}^{-1}) \times (293\,\text{K})}{(6\pi) \times (1.00 \times 10^{-3}\,\text{kg m}^{-1}\text{s}^{-1}) \times (6.3 \times 10^{-11}\,\text{m}^2\,\text{s}^{-1})} = \boxed{3.4\,\text{nm}}$$

23 The solid state

Solutions to exercises

Discussion questions

E23.1(a) Lattice planes are labelled by their Miller indices h, k, and l, where h, k, and l refer respectively to the reciprocals of the smallest intersection distances (in units of the lengths of the unit cell, a, b and c) of the plane along the x, y, and z axes.

E23.2(a) If the overall amplitude of a wave diffracted by planes (hkl) is zero, that plane is said to be absent in the diffraction pattern.

When the phase difference between adjacent planes in the set of planes (hkl) is π, destructive interference between the waves diffracted from the planes can occur and this will diminish the intensity of the diffracted wave. This is illustrated in Fig. 23.21. The overall intensity of a diffracted wave from a plane (hkl) is determined from a calculation of the structure factor, F_{hkl}, which is a function of the positions (hence, of the Miller indices) and of the scattering factors of the atoms in the crystal (see eqn 23.7). If F_{hkl} is zero for the plane (hkl), that plane is absent. See Example 23.3.

E23.3(a) The majority of metals crystallize in structures which can be interpreted as the closest packing arrangements of hard spheres. These are the cubic close-packed (ccp) and hexagonal close-packed (hcp) structures. In these models, 74% of the volume of the unit cell is occupied by the atoms (packing fraction = 0.74). Most of the remaining metallic elements crystallize in the body-centered cubic (bcc) arrangement which is not too much different from the close-packed structures in terms of the efficiency of the use of space (packing fraction 0.68 in the hard sphere model). Polonium is an exception; it crystallizes in the simple cubic structure which has a packing fraction of 0.52. See the solution to Problem 23.16 for a derivation of all the packing fractions in cubic systems. If atoms were truly hard spheres, we would expect that all metals would crystallize in either the ccp or hcp close-packed structures. The fact that a significant number crystallize in other structures is proof that a simple hard sphere model is an inaccurate representation of the interactions between the atoms. Covalent bonding between the atoms may influence the structure.

E23.4(a) Because enantiomers give almost identical diffraction patterns it is difficult to distinguish between them. But absolute configurations can be obtained from an analysis of small differences in diffraction intensities by a method developed by J. M. Bijvoet. The method makes use of extra phase shifts that occur when the frequency of the X-rays approaches an absorption frequency of atoms in the compound. The phase shifts are called anomalous scattering and result in different intensities in the diffraction patterns of different enantiomers. See Fig. 23.41 and Section 23.7(b) of the text for an illustration of the origin of this anomalous phase shift. The incorporation of heavy atoms into the compound makes the observation of the extra phase shift easier to observe, but with very sensitive modern diffractometers this is no longer strictly necessary.

E23.5(a) The most obvious difference is that there is no magnetic analog of electric charge; hence, there are no magnetic 'ions.' Both electric and magnetic moments exist and these can be either permanent or induced. Induced magnetic moments in the entire sample can be either parallel or antiparallel to the applied field producing them (paramagnetic or diamagnetic moments), whereas in the electric case they are always parallel. Magnetization, M, is the analog of polarization, P. Although both magnetization and induced dipole moment are proportional to the fields producing them, they are not analogous quantities, neither are volume magnetic susceptibility, χ, and electric polarizability, α. The magnetic quantities refer to the sample as a whole, the electric quantities to the molecules. Molar magnetic susceptibility is analogous to molar polarization as can be seen by comparing equations 23.30 and 21.16 and magnetizability is analogous to electric polarizability.

Numerical exercises

E23.6(a) There are four equivalent lattice points in the fcc unit cell. One way of choosing them is shown by the positions of the Cl$^-$ ions in Fig. 23.1 (which is similar to Fig. 23.23 in the text). The three lattice points equivalent to $\left(\frac{1}{2}, 0, 0\right)$ are $\boxed{\left(1, \frac{1}{2}, 0\right)}$, $\boxed{\left(1, 0, \frac{1}{2}\right)}$, and $\boxed{\left(\frac{1}{2}, \frac{1}{2}, \frac{1}{2}\right)}$. Figure 23.1 shows location of the atoms in the fcc unit cell of NaCl. The tinted circles are Na$^+$; the open circles are Cl$^-$.

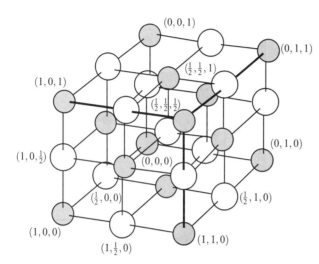

Figure 23.1

Comment. The positions of the other Cl$^-$ ions in Fig. 23.1 do not correspond to lattice points of the unit cell shown, as they are generated by full unit cell translations, and hence belong to neighbouring unit cells.

Question. What Na$^+$ positions define the unit cell of NaCl in Fig. 23.1? What lattice points are equivalent to $(0, 0, 0)$?

E23.7(a) The planes are sketched in Fig. 23.2. Expressed in multiples of the unit cell distances the planes are labelled $(2, 3, 2)$ and $(2, 2, \infty)$. Their Miller indices are the reciprocals of these multiples with all fractions cleared, thus

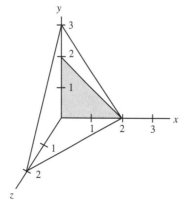

Figure 23.2

$$(2, 3, 2) \rightarrow \left(\tfrac{1}{2}, \tfrac{1}{3}, \tfrac{1}{2}\right) \rightarrow (3, 2, 3) \quad \text{[multiply by 6]}$$

$$(2, 2, \infty) \rightarrow \left(\tfrac{1}{2}, \tfrac{1}{2}, 0\right) \rightarrow (1, 1, 0) \quad \text{[multiply by 2]}$$

Dropping the commas, the planes are written $\boxed{(3\,2\,3)}$ and $\boxed{(1\,1\,0)}$

E23.8(a) $\quad d_{khl} = \dfrac{a}{(h^2 + k^2 + l^2)^{1/2}}$ [23.2]

Therefore, $d_{111} = \dfrac{a}{3^{1/2}} = \dfrac{432\,\text{pm}}{3^{1/2}} = \boxed{249\,\text{pm}} \qquad d_{211} = \dfrac{a}{6^{1/2}} = \dfrac{432\,\text{pm}}{6^{1/2}} = \boxed{176\,\text{pm}}$

$d_{100} = a = \boxed{432\,\text{pm}}$

E23.9(a) $\quad \lambda = 2d \sin\theta\,[23.5] = (2) \times (99.3\,\text{pm}) \times (\sin 20.85°) = \boxed{70.7\,\text{pm}}$

Comment. Knowledge of the type of crystal is not needed to complete this exercise.

E23.10(a) Refer to Fig. 23.22 of the text. Systematic absences correspond to $h + k + l = $ odd. Hence the first three lines are from planes $(1\,1\,0)$, $(2\,0\,0)$, and $(2\,1\,1)$.

$$\sin\theta_{hkl} = \dfrac{\lambda}{2d_{hkl}}\,[23.5], \quad d_{hkl} = \dfrac{a}{(h^2 + k^2 + l^2)^{1/2}}\,[23.2], \text{ then}$$

$$\sin\theta_{hkl} = (h^2 + k^2 + l^2)^{1/2} \times \left(\dfrac{\lambda}{2a}\right)$$

In a bcc unit cell, the body diagonal of the cube is $4R$ where R is the atomic radius. The relationship of the side of the unit cell to R is therefore (using the Pythagorean theorem twice)

$$(4R)^2 = a^2 + 2a^2 = 3a^2 \quad \text{or} \quad a = \dfrac{4R}{3^{1/2}}$$

This can be seen from Fig. 23.3.

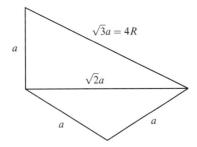

Figure 23.3

$$a = \dfrac{4 \times 126\,\text{pm}}{3^{1/2}} = 291\,\text{pm}$$

$$\dfrac{\lambda}{2a} = \dfrac{58\,\text{pm}}{(2) \times (291\,\text{pm})} = 0.099\overline{7}$$

$\sin\theta_{110} = \sqrt{2} \times (0.099\overline{7}) = 0.14\overline{1} \qquad 2\theta_{110} = \boxed{16°}$

$\sin\theta_{200} = (2) \times (0.099\overline{7}) = 0.19\overline{9} \qquad 2\theta_{200} = \boxed{23°}$

$\sin\theta_{211} = \sqrt{6} \times (0.099\overline{7}) = 0.24\overline{4} \qquad 2\theta_{211} = \boxed{28°}$

E23.11(a) $\theta = \arcsin \dfrac{\lambda}{2d}$ [23.5, $\arcsin \equiv \sin^{-1}$]

$$\Delta\theta = \arcsin \frac{\lambda_1}{2d} - \arcsin \frac{\lambda_2}{2d} = \arcsin \left(\frac{154.051 \text{ pm}}{(2) \times (77.8 \text{ pm})} \right) - \arcsin \left(\frac{154.433 \text{ pm}}{(2) \times (77.8 \text{ pm})} \right)$$

$$= -1.07° = -0.0187 \text{ rad}$$

The angle θ in radians is related to the distances D of the reflection line from the centre of the pattern by $\theta = \dfrac{D}{2R}$; hence

$$D = 2R\theta = (2) \times (5.74 \text{ cm}) \times (0.0187) = \boxed{0.215 \text{ cm}}$$

E23.12(a) A tetragonal unit cell, as shown in Fig. 23.8 of the text, has $a = b \neq c$. Therefore

$$V = (651 \text{ pm}) \times (651 \text{ pm}) \times (934 \text{ pm}) = \boxed{3.96 \times 10^{-28} \text{ m}^3}$$

E23.13(a) $\rho = \dfrac{\text{mass of unit cell}}{\text{volume of unit cell}} = \dfrac{m}{V}$

$m = nM = \dfrac{N}{N_A} M$ [N is the number of formula units per unit cell]

Then, $\rho = \dfrac{NM}{VN_A}$

and $N = \dfrac{\rho V N_A}{M}$

$$= \frac{(3.9 \times 10^6 \text{ g m}^{-3}) \times (634) \times (784) \times (516 \times 10^{-36} \text{ m}^3) \times (6.022 \times 10^{23} \text{ mol}^{-1})}{154.77 \text{ g mol}^{-1}} = 3.9$$

Therefore, $\boxed{N = 4}$ and the true calculated density (in the absence of defects) is

$$\rho = \frac{(4) \times (154.77 \text{ g mol}^{-1})}{(634) \times (784) \times (516 \times 10^{-30} \text{ cm}^3) \times (6.022 \times 10^{23}) \text{ mol}^{-1}} = \boxed{4.01 \text{ g cm}^{-3}}$$

E23.14(a) $d_{hkl} = \left[\left(\dfrac{h}{a} \right)^2 + \left(\dfrac{k}{b} \right)^2 + \left(\dfrac{l}{c} \right)^2 \right]^{-1/2}$ [23.3]

$$d_{411} = \left[\left(\frac{4}{812} \right)^2 + \left(\frac{1}{947} \right)^2 + \left(\frac{1}{637} \right)^2 \right]^{-1/2} \text{ pm} = \boxed{190 \text{ pm}}$$

E23.15(a) Since the reflection at $32.6°$ is (220), we know that

$$d_{220} = \frac{\lambda}{2 \sin \theta} [23.5] = \frac{154 \text{ pm}}{2 \sin 32.6} = 143 \text{ pm}$$

and hence, since $d_{220} = \dfrac{a}{(2^2 + 2^2)^{1/2}} [23.1] = \dfrac{a}{8^{1/2}}$

it follows that $a = (8^{1/2}) \times (143 \text{ pm}) = 404 \text{ pm}$

The indices of the other reflections are obtained from

$$(h^2 + k^2 + l^2) = \left(\frac{a}{d_{hkl}} \right)^2 [1] = \left(\frac{(a) \times 2 \sin \theta}{\lambda} \right)^2 \quad \text{[using eqn 23.5]}$$

We draw up the following table

θ	$a^2\left(\dfrac{2\sin\theta}{\lambda}\right)^2$	$h^2 + k^2 + l^2$	(hkl)	a/pm
19.4	3.04	3	(111)	402
22.5	4.03	4	(200)	402
32.6	7.99	8	(220)	404
39.4	11.09	11	(311)	402

The values of a in the final column are obtained from

$$a = \left(\frac{\lambda}{2\sin\theta}\right) \times (h^2 + k^2 + l^2)^{1/2}$$

and average to 402 pm.

E23.16(a) $\theta_{hkl} = \arcsin\dfrac{\lambda}{2d_{hkl}}$ [from eqn 23.5] $= \arcsin\left\{\dfrac{\lambda}{2}\left[\left(\dfrac{h}{2}\right)^2 + \left(\dfrac{k}{b}\right)^2 + \left(\dfrac{l}{c}\right)^2\right]^{1/2}\right\}$ [from eqn 23.3]

$$= \arcsin\left\{77\left[\left(\frac{h}{542}\right)^2 + \left(\frac{k}{917}\right)^2 + \left(\frac{l}{645}\right)^2\right]^{1/2}\right\}$$

Therefore,

$$\theta_{100} = \arcsin\left(\frac{77}{542}\right) = \boxed{8.17°} \qquad \theta_{010} = \arcsin\left(\frac{77}{917}\right) = \boxed{4.82°}$$

$$\theta_{111} = \arcsin\left\{77 \times \left[\left(\frac{1}{542}\right)^2 + \left(\frac{1}{917}\right)^2 + \left(\frac{1}{645}\right)^2\right]^{1/2}\right\} = \arcsin\frac{77}{378} = \boxed{11.75°}$$

E23.17(a) From the discussion of systematic absences (Section 23.3 and Fig. 23.24 of the text) we can conclude that the unit cell is $\boxed{\text{face-centred cubic}}$

E23.18(a) $F_{hkl} = \displaystyle\sum_j f_j e^{2\pi i(hx_j + ky_j + lz_j)}$ [23.7]

with $f_j = \frac{1}{8}f$ (each atom is shared by eight cells). Therefore,

$$F_{hkl} = \tfrac{1}{8}f\{1 + e^{2\pi ih} + e^{2\pi ik} + e^{2\pi il} + e^{2\pi i(h+k)} + e^{2\pi i(h+l)} + e^{2\pi i(k+l)} + e^{2\pi i(h+k+l)}\}$$

However all the exponential terms are unity since, h, k, and l are all integers and

$$e^{i\theta} = \cos\theta + i\sin\theta\,[\theta = 2\pi h, 2\pi k, \ldots] = \cos\theta = 1$$

Therefore, $F_{hkl} = \boxed{f}$

E23.19(a) The hatched area is $h \times 2R = 3^{1/2}R \times 2R = 2\sqrt{3}R^2$ where $h = 2R\cos 30°$. The net number of cylinders in a hatched area is 1, and the area of the cylinder's base is πR^2. The volume of the prism (of which the hatched area is the base) is $2\sqrt{3}R^2L$, and the volume occupied by the cylinders is $\pi R^2 L$. Hence, the packing fraction is

$$f = \frac{\pi R^2 L}{2\sqrt{3}R^2 L} = \frac{\pi}{2\sqrt{3}} = \boxed{0.9069}$$

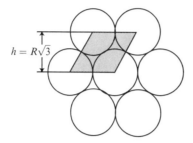

Figure 23.4

E23.20(a) For sixfold coordination see Fig. 23.5.

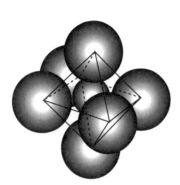 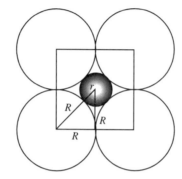

Figure 23.5

We assume that the larger spheres of radius R touch each other and that they also touch the smaller interior sphere. Hence, by the Pythagorean theorem

$$(R + r)^2 = 2(R)^2 \quad \text{or} \quad \left(1 + \frac{r}{R}\right)^2 = 2$$

Thus, $\dfrac{r}{R} = \boxed{0.414}$

E23.21(a) The radius ratios determined in Exercises 23.20(a) and 23.20(b) correspond to the smallest value of the radius of the interior cation, since any smaller value would tend to bring the anions closer and increase their interionic repulsion and at the same time decrease the attractions of cation and anion.

(a) $\dfrac{r_+}{r_-} = 0.414$ [result of Exercise 23.20(a)]

$r_+(\text{smallest}) = (0.414) \times (140 \, \text{pm})[\text{Table 23.3}] = \boxed{58.0 \, \text{pm}}$

(b) $\dfrac{r_+}{r_-} = 0.732$ [result of Exercise 23.20(b)]

$r_+(\text{smallest}) = (0.732) \times (140 \, \text{pm}) = \boxed{102 \, \text{pm}}$

Comment. As is evident from the data in Table 23.3 larger values than these do not preclude the occurrence of coordination number 6.

E23.22(a) Figure 23.39 in the text shows the diamond structure. Figure 23.6(a) below is an easier to visualize form of the structure which shows the unit cell of diamond.

The number of carbon atom in the unit cell is $\left(8 \times \frac{1}{8}\right) + \left(6 \times \frac{1}{2}\right) + (4 \times 1) = 8$ ($\frac{1}{8}$ for a corner atom, $\frac{1}{2}$ for a face-centred atom, and 1 for an atom entirely in the cell). The positions of the atoms are

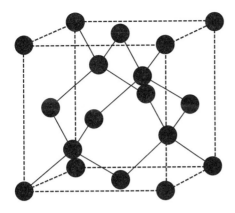

Figure 23.6(a)

$(0, 0, 0)$, $\left(\frac{1}{2}, \frac{1}{2}, 0\right)$, $\left(\frac{1}{2}, 0, \frac{1}{2},\right)$, $\left(0, \frac{1}{2}, \frac{1}{2}\right)$, $\left(\frac{1}{4}, \frac{1}{4}, \frac{1}{4}\right)$, $\left(\frac{1}{4}, \frac{3}{4}, \frac{3}{4}\right)$, $\left(\frac{3}{4}, \frac{1}{4}, \frac{3}{4}\right)$, and $\left(\frac{3}{4}, \frac{3}{4}, \frac{1}{4}\right)$ as indicated in Fig. 23.6(b).

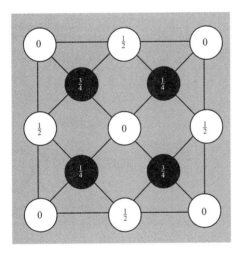

Figure 23.6(b)

The fractions in Fig. 23.6(b) denote height above the base in units of the cube edge, a. Two atoms that touch lie along the body diagonal at $(0, 0, 0)$ and $\left(\frac{1}{4}, \frac{1}{4}, \frac{1}{4}\right)$. Hence the distance $2r$ is one-fourth of the body diagonal which is $\sqrt{3}a$ in a cube. That is $2r = \dfrac{\sqrt{3}a}{4}$

The packing fraction is $\dfrac{\text{volume of atoms}}{\text{volume of unit cell}} = \dfrac{8V_a}{a^3} = \dfrac{(8) \times \frac{4}{3}\pi r^3}{\left(\frac{8r}{\sqrt{3}}\right)^3} = \boxed{0.340}$

E23.23(a) The volume change is a result of two partially counteracting factors: (1) different packing fraction (f), and (2) different radii.

$$\frac{V(\text{bcc})}{V(\text{hcp})} = \frac{f(\text{hcp})}{f(\text{bcc})} \times \frac{v(\text{bcc})}{v(\text{hcp})}$$

$$f(\text{hcp}) = 0.7404, \quad f(\text{bcc}) = 0.6802 \quad [\textit{Justification } 23.3 \text{ and Problem } 23.16]$$

$$\frac{V(\text{bcc})}{V(\text{hcp})} = \frac{0.7405}{0.6802} \times \frac{(142.5)^3}{(145.8)^3} = 1.016$$

Hence there is an $\boxed{\text{expansion}}$ of 1.6 per cent.

E23.24(a) Draw points corresponding to the vectors joining each pair of atoms. Heavier atoms give more intense contribution than light atoms. Remember that there are two vectors joining any pair of atoms ($\overrightarrow{AB}$ and $\overleftarrow{AB}$); don't forget the AA zero vectors for the centre point of the diagram. See Fig. 23.7.

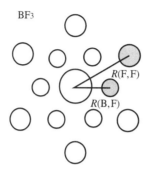

Figure 23.7

E23.25(a)
$$\lambda = \frac{h}{p} = \frac{h}{mv}$$

Hence, $v = \dfrac{h}{m\lambda} = \dfrac{6.626 \times 10^{-34}\,\text{J s}}{(1.675 \times 10^{-27}\,\text{kg}) \times (50 \times 10^{-12}\,\text{m})} = \boxed{7.9\,\text{km s}^{-1}}$

E23.26(a) Combine $E = \frac{1}{2}kT$ and $E = \frac{1}{2}mv^2$ with $m^2v^2 = \dfrac{h^2}{\lambda^2}$ [de Broglie relation]; then

$$E = \frac{h^2}{2m\lambda^2} \quad \text{and} \quad E = e\Delta\phi$$

Therefore, $\Delta\phi = \dfrac{h^2}{2me\lambda^2} = \dfrac{(6.626 \times 10^{-34}\,\text{J s})^2}{(2) \times (9.109 \times 10^{-31}\,\text{kg}) \times (1.602 \times 10^{-19}\,\text{C}) \times (18 \times 10^{-12}\,\text{m})^2}$

$$= \boxed{4.6\,\text{kV}} \quad (1\,\text{J} = 1\,\text{C V})$$

E23.27(a)
$$E_p = -A \times \frac{|z_1 z_2| N_A e^2}{4\pi\epsilon_0 d} \quad [23.13]$$

$$E_p^* = N_A C' e^{-d/d^*} \quad [23.14]$$

The sum of these two expressions gives the total potential energy and the minimum in this quantity is obtained by differentiation with respect to d.

$$\frac{\text{d}(E_p + E_p^*)}{\text{d}d} = 0 = \frac{A|z_1 z_2| N_A e^2}{4\pi\epsilon_0 d^2} - N_A \, C' \left(\frac{1}{d^*}\right) e^{-d/d^*}$$

N_A can be eliminated and we can write

$$C' e^{-d/d^*} = \frac{A|z_1 z_2| e^2}{4\pi\epsilon_0 d} \left(\frac{d^*}{d}\right)$$

Then after substitution into eqn 23.14 we obtain

$$E_{p,\text{min}} = -A \frac{|z_1 z_2| N_A e^2}{4\pi\epsilon_0 d} + A N_A \frac{|z_1 z_2| e^2}{4\pi\epsilon_0 d} \left(\frac{d^*}{d}\right)$$

$$= -A N_A \left(\frac{|z_1 z_2| e^2}{4\pi\epsilon_0 d}\right) \left(1 - \frac{d^*}{d}\right)$$

which is eqn 23.15

E23.28(a) Washing a cotton shirt disturbs the secondary structure of the cellulose. The water tends to break the hydrogen bonds between the cellulose chains by forming its own hydrogen bonds to the chains. Upon drying, the hydrogen bonds between chains reform, but in a random manner, causing wrinkles. The wrinkles are removed by moistening the shirt, which again breaks the hydrogen bonds, making the fibre more flexible and plastically deformable. A hot iron shapes the cloth and causes the water to evaporate so new hydrogen bonds are formed between the chains while they are held in place by the pressure of the iron.

E23.29(a) $E = \dfrac{\text{normal stress}}{\text{normal strain}}$ [23.16a] $= 1.26\,\text{Pa} = 1.2 \times 10^9\,\text{kg m}^{-1}\,\text{s}^{-2}$

We solve this relation for the normal strain after calculating the normal stress from the data provided.

$$\text{normal stress} = \text{force per unit area} = F/A$$

$$\text{normal strain} = \text{relative elongation} = \Delta L/L$$

$$\Delta L/L = \frac{F/A}{E} = \frac{mg/A}{E} = \frac{mg}{AE} = \frac{mg}{\pi \left(\frac{d}{2}\right)^2 E}$$

$$= \frac{1.0\,\text{kg} \times 9.8\,\text{m s}^{-2}}{\pi \left(\frac{1.0 \times 10^{-3}\,\text{m}}{2}\right)^2 \times 1.2 \times 10^9\,\text{kg m}^{-1}\,\text{s}^{-2}}$$

$$= \boxed{0.010} \text{ or about 1\% elongation}$$

E23.30(a) Poisson's ratio: $\nu_p = \dfrac{\text{transverse strain}}{\text{normal strain}} = 0.45$

We note that the transverse strain is usually a contraction and that it is usually evenly distributed in both transverse directions. That is, if $(\Delta L/L)_z$ is the normal strain, then the transverse strains, $(\Delta L/L)_x$ and $(\Delta L/L)_y$, are equal. In this case

$$(\Delta L/L)_z = +0.010, \qquad \left(\frac{\Delta L}{L}\right)_x = -0.0045 = \left(\frac{\Delta L}{L}\right)_y$$

$$\text{New volume} = (1 - 0.0045)^2 \times (1 + 0.010) \times 1.0\,\text{cm}^3$$
$$= 1.00093\,\text{cm}^3$$

The change in volume is $\boxed{9.3 \times 10^{-4}\,\text{cm}^3}$

E23.31(a) $m = g_e\{S(S+1)\}^{1/2}\mu_B$ [23.34, with S in place of s]

Therefore, since $m = 3.81\mu_B$

$$S(S+1) = \left(\tfrac{1}{4}\right) \times (3.81)^2 = 3.63, \text{ implying that } S = 1.47$$

Since $S \approx \tfrac{3}{2}$, there must be $\boxed{\text{three}}$ unpaired spins.

E23.32(a) $\chi_m = \chi V_m$ [23.28] $= \dfrac{\chi M}{\rho} = \dfrac{(-7.2 \times 10^{-7}) \times (78.11\,\text{g mol}^{-1})}{0.879\,\text{g cm}^{-3}} = \boxed{-6.4 \times 10^{-5}\,\text{cm}^3\,\text{mol}^{-1}}$

E23.33(a) We need to compare the experimentally determined expression for χ_m to the theoretical expression

$$\chi_m = \frac{N_A g_e^2 \mu_0 \mu_B^2 S(S+1)}{3kT} \quad [23.35]$$

where in making the comparison we are assuming spin-only magnetism. Inserting the constants we obtain (*Illustration* 23.1)

$$\chi_m = (6.3001 \times 10^{-6}\,\text{m}^3\,\text{K}\,\text{mol}^{-1}) \times \left(\frac{S(S+1)}{T}\right) = \frac{1.22 \times 10^{-5}\,\text{m}^3\,\text{K}\,\text{mol}^{-1}}{T}$$

Therefore, $S(S+1) = \dfrac{1.22 \times 10^{-5}}{6.3001 \times 10^{-6}} = 1.94 \approx 2$ or $S = 1$

and the number of unpaired electrons is $\boxed{2}$.

The problem of the Lewis structure is resolved in molecular orbital theory which shows that it is possible to have simultaneously a double bond and two unpaired electrons. See Section 14.5(f).

Comment. The discrepancy between 1.94 and 2 in $S(S+1)$ can probably be accounted for by allowing for some orbital contribution to the magnetic moment of O_2. The assumption of spin-only magnetism is not exact.

E23.34(a) $\chi_m(\text{theor}) = \dfrac{N_A g_e^2 \mu_0 \mu_B^2 S(S+1)}{3kT} \quad [23.35]$

The molar susceptibility is given by

$$\chi_m = \frac{N_A g_e^2 \mu_0 \mu_B^2 S(S+1)}{3kT} \qquad \text{so} \qquad S(S+1) = \frac{3kT \chi_m}{N_A g_e^2 \mu_0 \mu_B^2}$$

$$S(S+1) = \frac{3(1.381 \times 10^{-23}\,\text{J}\,\text{K}^{-1}) \times (294.53\,\text{K}) \times (0.1463 \times 10^{-6}\,\text{m}^3\,\text{mol}^{-1})}{(6.022 \times 10^{23}\,\text{mol}^{-1}) \times (2.0023)^2 \times (4\pi \times 10^{-7}\,\text{T}^2\,\text{J}^{-1}\,\text{m}^3) \times (9.27 \times 10^{-24}\,\text{J}\,\text{T}^{-1})^2}$$

$$= 6.84$$

so

$$S^2 + S - 6.841 = 0 \quad \text{and} \quad S = \frac{-1 + \sqrt{1 + 4(6.841)}}{2} = 2.163$$

corresponding to $\boxed{4.326}$ effective unpaired spins. The theoretical number is $\boxed{5}$ corresponding to the $3d^5$ electronic configuration of Mn^{2+}.

Comment. The discrepancy between the two values is accounted for by an antiferromagnetic interaction between the spins which alters χ_m from the form of eqn 23.35.

E23.35(a) $\chi_m = (6.3001) \times \left(\dfrac{S(S+1)}{T/K}\,\text{cm}^3\,\text{mol}^{-1}\right)$ [*Illustration* 23.1]

Since Cu(II) is a d^9 species, it has one unpaired spin, and so $S = s = \frac{1}{2}$. Therefore,

$$\chi_m = \frac{(6.3001) \times \left(\frac{1}{2}\right) \times \left(\frac{3}{2}\right)}{298}\,\text{cm}^3\,\text{mol}^{-1} = \boxed{+0.016\,\text{cm}^3\,\text{mol}^{-1}}$$

E23.36(a) The magnitude of the orientational energy is given by

$$g_e \mu_B M_S \mathcal{B} \quad \text{with } M_S = S = 1$$

Setting this equal to kT and solving for $\mathcal{B}$

$$\mathcal{B} = \frac{kT}{g_e \mu_B} = \frac{(1.38 \times 10^{-23}\,\text{J K}^{-1}) \times (298\,\text{K})}{(2.00) \times (9.27 \times 10^{-24}\,\text{J T}^{-1})} = \boxed{222\,\text{T}}$$

Comment. This is an enormous magnetic field and it is a measure of the strength of the internal magnetic fields required for spin alignment in ferromagnetic and antiferromagnetic materials in which such alignments occur.

Solutions to problems

Solutions to numerical problems

P23.2 A large separation between the sixth and seventh lines relative to the separation between the fifth and sixth lines is characteristic of a simple (primitive) cubic lattice . This is readily seen without indexing the lines. The conclusion that the unit cell is simple cubic is then confirmed by the presence of reflections from (100) planes.

$$d_{100} = a[23.1] = \frac{\lambda}{2\sin\theta}\ [23.5]$$

$$a = \frac{154\,\text{pm}}{(2) \times (0.225)} = \boxed{342\,\text{pm}}$$

P23.3 See Fig. 23.23 of the text or Fig. 23.1 of this manual. The length of an edge in the fcc lattice of these compounds is

$$a = 2(r_+ + r_-)$$

Then
(1) $a(\text{NaCl}) = 2(r_{\text{Na}^+} + r_{\text{Cl}^-}) = 562.8\,\text{pm}$ (2) $a(\text{KCl}) = 2(r_{\text{K}^+} + r_{\text{Cl}^-}) = 627.7\,\text{pm}$
(3) $a(\text{NaBr}) = 2(r_{\text{Na}^+} + r_{\text{Br}^-}) = 596.2\,\text{pm}$ (4) $a(\text{KBr}) = 2(r_{\text{K}^+} + r_{\text{Br}^-}) = 658.6\,\text{pm}$

If the ionic radii of all the ions are constant then

$$(1) + (4) = (2) + (3)$$

$$(1) + (4) = (562.8 + 658.6)\,\text{pm} = 1221.4\,\text{pm}$$

$$(2) + (3) = (627.7 + 596.2)\,\text{pm} = 1223.9\,\text{pm}$$

The difference is slight; hence the data support the constancy of the radii of the ions.

P23.5 For the three given reflections

$$\sin 19.076° = 0.32682 \qquad \sin 22.171° = 0.37737 \qquad \sin 32.256° = 0.53370$$

For cubic lattices $\sin\theta_{hkl} = \dfrac{\lambda(h^2 + k^2 + l^2)^{1/2}}{2a}$ [23.5 with 23.2]

First consider the possibility of simple cubic; the first three reflections are (100), (110), and (111). (See Fig. 23.22 of the text.)

$$\frac{\sin\theta(100)}{\sin\theta(110)} = \frac{1}{\sqrt{2}} \neq \frac{0.32682}{0.37737} \quad \text{[not simple cubic]}$$

Consider next the possibility of body-centred cubic; the first three reflections are (110), (200), and (211).

$$\frac{\sin\theta(110)}{\sin\theta(200)} = \frac{\sqrt{2}}{\sqrt{4}} = \frac{1}{\sqrt{2}} \neq \frac{0.32682}{0.37737} \quad \text{[not bcc]}$$

Consider finally face-centred cubic; the first three reflections are (111), (200), and (220)

$$\frac{\sin\theta(111)}{\sin\theta(200)} = \frac{\sqrt{3}}{\sqrt{4}} = 0.86603$$

which compares very favourably to $\dfrac{0.32682}{0.37737} = 0.86605$. Therefore, the lattice is $\boxed{\text{face-centred cubic}}$

This conclusion may easily be confirmed in the same manner using the second and third reflection.

$$a = \frac{\lambda}{2\sin\theta}(h^2 + k^2 + l^2)^{1/2} = \left(\frac{154.18\,\text{pm}}{(2)\times(0.32682)}\right) \times \sqrt{3} = \boxed{408.55\,\text{pm}}$$

$$\rho = \frac{NM}{N_A V}[\text{Exercise 23.13(a)}] = \frac{(4)\times(107.87\,\text{g mol}^{-1})}{(6.0221\times10^{23}\,\text{mol}^{-1})\times(4.0855\times10^{-8}\,\text{cm})^3}$$

$$= \boxed{10.507\,\text{g cm}^{-3}}$$

This compares favourably to the value listed in the *Data section*.

P23.7 $\lambda = 2a\sin\theta_{100}$ as $d_{100} = a$

Therefore, $a = \dfrac{\lambda}{2\sin\theta_{100}}$ and

$$\frac{a(\text{KCl})}{a(\text{NaCl})} = \frac{\sin\theta_{100}(\text{NaCl})}{\sin\theta_{100}(\text{KCl})} = \frac{\sin 6°0'}{\sin 5°23'} = 1.114$$

Therefore, $a(\text{KCl}) = (1.114)\times(564\,\text{pm}) = \boxed{628\,\text{pm}}$

The relative densities calculated from these unit cell dimensions are

$$\frac{\rho(\text{KCl})}{\rho(\text{NaCl})} = \left(\frac{M(\text{KCl})}{M(\text{NaCl})}\right) \times \left(\frac{a(\text{NaCl})}{a(\text{KCl})}\right)^3 = \left(\frac{74.55}{58.44}\right) \times \left(\frac{564\,\text{pm}}{628\,\text{pm}}\right)^3 = 0.924$$

Experimentally

$$\frac{\rho(\text{KCl})}{\rho(\text{NaCl})} = \frac{1.99\,\text{g cm}^{-3}}{2.17\,\text{g cm}^{-3}} = 0.917$$

and the measurements $\boxed{\text{are broadly consistent}}$

P23.9   When there is only one pair of identical atoms, the Wierl equation reduces to

$$I(\theta) = f^2 \frac{\sin sR}{sR}$$

Extrema occur at

$$sR = \frac{\sin sR}{\cos sR} = \tan sR$$

$sR = 0$ and $sR = 4.49$ correspond to maxima and minima respectively. To find the scattering angles, we need

$$sR = \frac{4\pi R}{\lambda} \sin \frac{1}{2}\theta \quad \text{so} \quad \theta = 2\sin^{-1}\left(\frac{sR\lambda}{4\pi R}\right)$$

The first maximum occurs at $\boxed{\theta = 0}$ for both neutrons and electrons. The minimum are

Neutron : $\theta = 2 \sin^{-1}\left(\dfrac{(4.49) \times (78 \text{ pm})}{4\pi (229.0 \text{ pm})}\right) = \boxed{14.0°}$

Electron : $\theta = 2 \sin^{-1}\left(\dfrac{(4.49) \times (4.0 \text{ pm})}{4\pi (229.0 \text{ pm})}\right) = \boxed{0.72°}$

P23.11   The volume per unit cell is

$$V = abc = (3.6881 \text{ nm}) \times (0.9402 \text{ nm}) \times (1.7652 \text{ nm}) = 6.121 \text{ nm}^3 = 6.121 \times 10^{-21} \text{ cm}^3$$

The mass per unit cell is 8 times the mass of the formula unit, $RuN_2C_{28}H_{44}S_4$, for which the molar mass is

$$M = \{101.07 + 2(14.007) + 28(12.011) + 44(1.008) + 4(32.066)\} \text{ g mol}^{-1} = 638.01 \text{ g mol}^{-1}$$

The density is

$$\rho = \frac{m}{V} = \frac{8M}{N_A V} = \frac{8(638.01 \text{ g mol}^{-1})}{(6.022 \times 10^{23} \text{ mol}^{-1}) \times (6.121 \times 10^{-21} \text{ cm}^3)} = \boxed{1.385 \text{ g cm}^{-3}}$$

The osmium analogue has a molar mass of 727.1 g mol^{-1}. If the volume of the crystal changes negligibly with the substitution, then the densities of the complexes are in proportion to their molar masses

$$\rho_{Os} = \frac{727.1}{638.01}(1.385 \text{ g cm}^{-3}) = \boxed{1.578 \text{ g cm}^{-3}}$$

P23.13   The molar magnetic susceptibility is given by

$$\chi_m = \frac{N_A g_e^2 \mu_0 \mu_B^2 S(S+1)}{3kT} = 6.3001 \times \frac{S(S+1)}{T/K} \text{ cm}^3 \text{ mol}^{-1} \text{ [}Illustration \text{ 23.1]}$$

For $S = 2$, $\chi_m = \dfrac{(6.3001) \times (2) \times (2+1)}{298} \text{ cm}^3\text{mol}^{-1} = \boxed{0.127 \text{ cm}^3\text{mol}^{-1}}$

For $S = 3$, $\chi_m = \dfrac{(6.3001) \times (3) \times (3+1)}{298} \text{ cm}^3\text{mol}^{-1} = \boxed{0.254 \text{ cm}^3\text{mol}^{-1}}$

For $S = 4$, $\chi_m = \dfrac{(6.3001) \times (4) \times (4+1)}{298} \text{ cm}^3\text{mol}^{-1} = \boxed{0.423 \text{ cm}^3\text{mol}^{-1}}$

Instead of a single value of S, we use an average weighted by the Boltzmann factor

$$\exp\left(\frac{-50 \times 10^3 \,\text{J mol}^{-1}}{(8.3145\,\text{J mol}^{-1}\text{K}^{-1}) \times (298\,\text{K})}\right) = 1.7 \times 10^{-9}$$

Thus the $S = 2$ and $S = 4$ forms are present in negligible quantities compared to the $S = 3$ form. The compound's susceptibility, then, is that of the $S = 3$ form, namely $\boxed{0.254\,\text{cm}^3\,\text{mol}^{-1}}$

Solutions to theoretical problems

P23.14   Consider, for simplicity, the two-dimensional lattice and planes shown in Fig. 23.8.

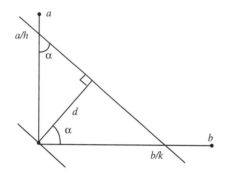

Figure 23.8

P23.19 $G = U - TS - tl$ [given]

Hence $dG = dU - T\,dS - S\,dT - l\,dt - t\,dl = T\,dS + t\,dl - T\,dS - S\,dT - l\,dt - t\,dl = \boxed{-S\,dT - l\,dt}$

$A = U - TS = G + tl$

Hence $dA = dG + t\,dl + l\,dt = -S\,dT - l\,dt + t\,dl + l\,dt = \boxed{-S\,dT + t\,dl}$

Since dG and dA are both exact differentials

$$\left(\frac{\partial S}{\partial l}\right)_T = -\left(\frac{\partial t}{\partial T}\right)_l \quad \text{and} \quad \left(\frac{\partial S}{\partial t}\right)_T = \left(\frac{\partial l}{\partial T}\right)_t$$

Since $dU = T\,dS + t\,dl$ [given],

$$\left(\frac{\partial U}{\partial l}\right)_T = T\left(\frac{\partial S}{\partial l}\right)_T + t = \boxed{-T\left(\frac{\partial t}{\partial T}\right)_l + t} \quad \text{[Maxwell relation, above]}$$

P23.21   According to eqn 23.19:

$$G = \frac{E}{2(1 + \nu_P)} \quad \text{and} \quad K = \frac{E}{3(1 - 2\nu_P)}.$$

Substituting the Lamé-constant expressions for E and v_P into the right-hand side of these relationships yields:

$$G = \frac{\frac{\mu(3\lambda+2\mu)}{\lambda+\mu}}{2\left[1 + \frac{\lambda}{2(\lambda+\mu)}\right]} \quad \text{and} \quad K = \frac{\frac{\mu(3\lambda+2\mu)}{\lambda+\mu}}{3\left[1 - \frac{\lambda}{\lambda+\mu}\right]}.$$

Expanding leads to:

$$G = \frac{\frac{\mu(3\lambda+2\mu)}{\lambda+\mu}}{2\left[\frac{2\lambda+2\mu+\lambda}{2(\lambda+\mu)}\right]} = \frac{\mu(3\lambda+2\mu)}{3\lambda+2\mu} = \boxed{\mu}$$

and

$$K = \frac{\frac{\mu(3\lambda+2\mu)}{\lambda+\mu}}{3\left[\frac{\lambda+\mu-\lambda}{\lambda+\mu}\right]} = \frac{\mu(3\lambda+2\mu)}{3\mu} = \boxed{\frac{3\lambda+2\mu}{3}},$$

as the problem asks us to prove.

P23.24

$$N_2O_4(g) \overset{K}{\rightleftharpoons} 2NO_2(g)$$

$$(1-\alpha)n \qquad 2\alpha n \qquad \text{amounts}$$

$$\frac{1-\alpha}{1+\alpha} \qquad \frac{2\alpha}{1+\alpha} \qquad \text{mole fractions}$$

$$\left(\frac{1-\alpha}{1+\alpha}\right)p \qquad \left(\frac{2\alpha}{1+\alpha}\right)p \qquad \text{partial pressures } [p \equiv p/p^{\ominus} \text{ here}]$$

$$K = \frac{\left(\frac{2\alpha}{1+\alpha}\right)^2 p}{\left(\frac{1-\alpha}{1+\alpha}\right)} = \frac{4\alpha^2}{1-\alpha^2}p$$

Now solve for α.

$$\alpha^2 = \frac{K}{4p+K}, \qquad \alpha = \left(\frac{K}{4p+K}\right)^{1/2}$$

The degree of dimerization is $d = 1 - \alpha = 1 - \left(\frac{K}{4p+K}\right)^{1/2} = \boxed{1 - \left(\frac{1}{4(p/K)+1}\right)^{1/2}}$

The susceptibility varies in proportion to $\alpha = 1 - d$. As pressure increases, α decreases, and the susceptibility $\boxed{\text{decreases}}$.

To determine the effect of temperature we need $\Delta_r H \approx \Delta_r H^{\ominus}$ for the reaction above.

$$\Delta_r H^{\ominus} = 2 \times (33.18\,\text{kJ mol}^{-1}) - 9.16\,\text{kJ mol}^{-1} = +57.2\,\text{kJ mol}^{-1}$$

A positive $\Delta_r H^{\ominus}$ indicates that $NO_2(g)$ is favoured as the temperature increases; hence the susceptibility $\boxed{\text{increases}}$ with temperature.

P23.26 The density of a face-centred cubic crystal is $4m/V$ where m is the of the unit hung on each lattice point and V is the volume of the unit cell. (The 4 comes from the fact that each of the cell's 8 vertices is shared by 8 cells, and each of the cell's 6 faces is shared by 2 cells.)

So $\rho = \dfrac{4m}{a^3} = \dfrac{4M}{N_A a^3}$ and $M = \dfrac{1}{4}\rho N_A a^3$

$M = \frac{1}{4}(1.287\,\text{g cm}^{-3}) \times (6.022 \times 10^{23}\,\text{mol}^{-1}) \times (12.3 \times 10^{-7}\,\text{cm})^3$

$= \boxed{3.61 \times 10^5 \text{ g mol}^{-1}}$

Solutions to applications

P23.27

$$\rho = \frac{m(\text{unit cell})}{V(\text{unit cell})} = \frac{(2) \times \frac{M(CH_2CH_2)}{N_A}}{abc}$$

$$= \frac{(2) \times (28.05\,\text{g mol}^{-1})}{(6.022 \times 10^{23}\,\text{mol}^{-1}) \times [(740 \times 493 \times 253) \times 10^{-36}]\,\text{m}^3}$$

$$= 1.01 \times 10^6\,\text{g m}^{-3} = \boxed{1.01\,\text{g cm}^{-3}}$$

P23.28 The glass transition temperature T_g is the temperature at which internal bond rotations stop. In effect, the easier such rotations are, the lower T_g. Internal rotations are more difficult for polymers that have bulky side chains than for polymers without such chains because the side chains of neighbouring molecules can impede each others' motion. Of the four polymers in this problem, polystyrene has the largest side chain (phenyl) and the largest T_g. The chlorine atoms in poly(vinyl chloride) interfere with each other's motion more than the smaller hydrogen atoms that hang from the carbon backbone of polyethylene. Poly(oxymethylene), like polyethylene, has only hydrogen atoms protruding from its backbone; however, poly(oxymethylene) has fewer hydrogen protrusions and a still lower T_g than polyethylene.

P23.30 As demonstrated in *Justification* 23.3 of the text, close-packed spheres fill 0.7404 of the total volume of the crystal. Therefore 1 cm^3 of close-packed carbon atoms would contain

$$\frac{0.74040\,\text{cm}^3}{\left(\frac{4}{3}\pi r^3\right)} = 3.838 \times 10^{23}\,\text{atoms}$$

$$\left(r = \left(\frac{154.45}{2}\right)\text{pm} = 77.225\,\text{pm} = 77.225 \times 10^{-10}\,\text{cm}\right)$$

Hence the close-packed density would be

$$\rho = \frac{\text{mass in 1 cm}^3}{1\,\text{cm}^3} = \frac{(3.838 \times 10^{23}\,\text{atom}) \times (12.01\,\text{u/atom}) \times (1.6605 \times 10^{-24}\,\text{g u}^{-1})}{1\,\text{cm}^3}$$

$$= \boxed{7.654\text{g cm}^{-3}}$$

The diamond structure (solution to Exercise 23.22) is a very open structure which is dictated by the tetrahedral bonding of the carbon atoms. As a result many atoms that would be touching each other in a normal fcc structure do not in diamond; for example, the C atom in the centre of a face does not touch the C atoms at the corners of the face.

Part 3: Change

24 Molecules in motion

Solutions to exercises

Discussion questions

E24.1(a) **(a)** See Section 27.1 which discusses the collision theory of gas phase reactions. Their rate depends on the number of collisions having a relative kinetic energy above a certain critical value ε_a. The relative kinetic energy, in turn, depends on the relative velocities of the colliding molecules. The rate of reaction also depends on the number of collisions per unit volume per unit time, or collision density, Z_{AB}, the formula for which is derived in *Justification* 27.1. Z_{AB} also depends on the average relative velocity of the colliding molecules.

(b) A complete analysis of the composition of planetary atmospheres is a complicated process. See the solution to Problem 24.33 for detailed calculations on the depletion of the Earth's atmosphere and on the atmosphere of planets in general. Also see the solution to Problem 19.22 which deals with the inherent instability of planetary atmospheres. The simple answer to this question, though, is that light molecules are more likely to have velocities in excess of the escape velocity than are heavy molecules. Therefore, heavy molecules will remain in the atmosphere much longer than light molecules, though all will eventually escape, unless there is a source of replenishment.

E24.2(a) Gases are very dilute systems and on average the molecules are very far apart from each other except when they collide. So what little resistance there is to flow in a gaseous fluid is almost entirely due to the collisions between molecules. The frequency of collisions increases with increasing temperature (see eqns 24.11 and 27.6); hence the viscosity of gases increases with temperature. In liquids, on the other hand, the molecules are very close to each other, which results in there being strong forces of attraction between them which resist their movement relative to each other. However, as the temperature increases, more and more molecules are likely to have sufficient kinetic energy to overcome the forces of attraction, resulting in decreased viscosity.

E24.3(a) **(a)** This is Fick's first law of diffusion in one dimension written in terms of concentrations rather than activities; hence, it applies strictly only to ideal solutions.

(b) In addition to the restriction to ideal solutions as in **(a)**, the derivation of this expression uses the additional approximation that the frictional retarding force on a moving particle is proportional to the first power of the speed of the particle (as opposed to a more general functional relation).

(c) The restrictions of parts **(a)** and **(b)** still apply, as well as a third, which is the assumption that the particle is spherical.

Numerical exercises

E24.4(a) **(a)** The formula for the mean speed is derived in Example 24.1 and is

$$\bar{c} = \left(\frac{8RT}{\pi M}\right)^{1/2}$$

Thus $\dfrac{\bar{c}(H_2)}{\bar{c}(Hg)} = \left[\dfrac{M(Hg)}{M(H_2)}\right]^{1/2} = \left(\dfrac{200.6\,u}{2.016\,u}\right)^{1/2} = \boxed{9.975}$

(b) The average kinetic energy involves the root mean square speed since the kinetic energy, ε, is given by $\bar{\varepsilon} = \dfrac{1}{2}m\langle v^2 \rangle = \dfrac{1}{2}mc^2$ where $c = \sqrt{\langle v^2 \rangle}$ and $c = \left(\dfrac{3RT}{M}\right)^{1/2}$ [24.3]

Thus, $\dfrac{\bar{\varepsilon}(H_2)}{\bar{\varepsilon}(Hg)} = \dfrac{\frac{1}{2}m(H_2)\left[\frac{3RT}{M(H_2)}\right]}{\frac{1}{2}m(Hg)\left[\frac{3RT}{M(Hg)}\right]} = \boxed{1}$ since the masses, m, are proportional to the molar masses, M, $M = N_A m$.

Comment. Neither ratio is dependent on temperature and the ratio of energies is independent of both temperature and mass.

E24.5(a) (a) On the assumption that the gas is perfect, the temperature is easily calculated from eqn 1.2 after solving for T. $T = \dfrac{pV}{nR}$

$$n = \dfrac{1.0 \times 10^{23}\ \text{molecules}}{6.02 \times 10^{23}\ \text{molecules mol}^{-1}} = 0.16\bar{6}\ \text{mol}$$

$$T = \dfrac{(1.00 \times 10^5\ \text{Pa}) \times (1.0\,\text{L}) \times \left(\frac{1\,\text{m}^3}{10^3\,\text{L}}\right)}{(0.16\bar{6}\,\text{mol}) \times (8.314\,\text{J K}^{-1}\,\text{mol}^{-1})} = \boxed{72\,\text{K}}$$

(b) $c = \left(\dfrac{3RT}{M}\right)^{1/2}$ [24.3] $= \left(\dfrac{3 \times (8.314\,\text{J K}^{-1}\,\text{mol}^{-1}) \times (72.\overline{46}\,\text{K})}{2.016 \times 10^{-3}\,\text{kg mol}^{-1}}\right)^{1/2} = \boxed{9.5 \times 10^2\ \text{m s}^{-1}}$

(c) The temperature would not be different if they were O_2 molecules and exerted the same pressure in the same volume, but their root mean square speed would be different.

Comment. This exercise could have been solved by first obtaining the root mean square speed from $pV = \frac{1}{3}nMc^2$ [24.1] and then using eqn 24.3 to solve for the temperature. The results should be identical.

E24.6(a) This solution to this exercise is similar to that of Exercise 24.5 (b)(b). Here p is calculated from the mean free path, rather than the mean free path from p, as in Exercise 24.5 (b)(b).

$$\lambda = \dfrac{kT}{2^{1/2}\sigma p}\ \text{[24.14] implies that}\ p = \dfrac{kT}{2^{1/2}\sigma\lambda}$$

with $\lambda \approx 10\,\text{cm} = \sqrt[3]{1000\,\text{cm}^3}$

$$p = \dfrac{(1.381 \times 10^{-23}\,\text{J K}^{-1}) \times (298.15\,\text{K})}{(2^{1/2}) \times (0.36 \times 10^{-18}\,\text{m}^2) \times (0.10\,\text{m})} = \boxed{0.081\,\text{Pa}}$$

This pressure corresponds to 8.0×10^{-7} atm and to 6.1×10^{-4} Torr, a pressure much larger than that of Exercise 24.5 (b)(b).

E24.7(a) This exercise is similar to Exercise 24.5(b)(b) and the solution involves the same procedure.

$$\lambda = \dfrac{kT}{2^{1/2}\sigma p}\text{[24.14]} = \dfrac{(1.381 \times 10^{-23}\,\text{J K}^{-1}) \times (217\,\text{K})}{(2^{1/2}) \times (0.43 \times 10^{-18}\,\text{m}^2) \times (0.050) \times (1.013 \times 10^5\,\text{Pa})}$$

$$= \boxed{9.7 \times 10^{-7}\,\text{m}}$$

E24.8(a) The collision frequency, z, is given by $z = \dfrac{2^{1/2}\sigma\bar{c}p}{kT}$[24.12, 24.9], which becomes after substitution for

$$\bar{c} = \left(\dfrac{8RT}{\pi M}\right)^{1/2}\ \text{[Example 24.1]} = \left(\dfrac{8kT}{\pi m}\right)^{1/2}$$

$$z = \left(2^{1/2}\right) \times \sigma \times \left(\frac{8kT}{\pi m}\right)^{1/2} \times \left(\frac{p}{kT}\right) = \left(\frac{16}{\pi mkT}\right)^{1/2} \times \sigma p$$

$$= \left(\frac{16}{\pi \times (39.95) \times (1.6605 \times 10^{-27}\,\text{kg}) \times (1.381 \times 10^{-23}\,\text{J K}^{-1}) \times (298\,\text{K})}\right)^{1/2}$$

$$\times (0.36 \times 10^{-18}\,\text{m}^2) \times (p)$$

$$= (4.92 \times 10^4\,\text{s}^{-1}) \times (p/\text{Pa}) = (4.92 \times 10^4\,\text{s}^{-1}) \times (1.0133 \times 10^5) \times (p/\text{atm})$$

$$= (4.98 \times 10^9\,\text{s}^{-1}) \times (p/\text{atm})$$

Therefore

(a) $z = \boxed{5 \times 10^{10}\,\text{s}^{-1}}$ when $p = 10\,\text{atm}$,

(b) $z = \boxed{5 \times 10^{9}\,\text{s}^{-1}}$ when $p = 1\,\text{atm}$, and

(c) $z = \boxed{5 \times 10^{3}\,\text{s}^{-1}}$ when $p = 10^{-6}\,\text{atm}$, z is directly proportional to p at constant T.

E24.9(a) $\quad \lambda = \frac{kT}{2^{1/2}\sigma p} [24.14] = \frac{(1.381 \times 10^{-23}\,\text{J K}^{-1}) \times (298.15\,\text{K})}{(2^{1/2}) \times (0.43 \times 10^{-18}\,\text{m}^2) \times (p)}$

$$= \frac{6.8 \times 10^{-3}\,\text{m}}{(p/\text{Pa})} = \frac{6.7 \times 10^{-8}\,\text{m}}{p/\text{atm}}$$

(a) When $p = 10\,\text{atm}$, $\lambda = 6.7 \times 10^{-9}\,\text{m}$, or $\boxed{6.7\,\text{nm}}$

(b) When $p = 1\,\text{atm}$, $\lambda = \boxed{67\,\text{nm}}$

(c) When $p = 10^{-6}\,\text{atm}$, $\lambda = \boxed{6.7\,\text{cm}}$

The mean free path is inversely proportional to p and to z (Exercise 1.17(a)).

E24.10(a) The Maxwell distribution of speeds is $f(v) = 4\pi \left(\frac{M}{2\pi RT}\right)^{3/2} v^2 e^{-Mv^2/2RT}$ [24.4]

The factor, $\frac{M}{2RT}$, can be evaluated as

$$\frac{M}{2RT} = \frac{28.02 \times 10^{-3}\,\text{kg mol}^{-1}}{2 \times (8.314\,\text{J K}^{-1}\,\text{mol}^{-1}) \times (500\,\text{K})} = 3.37 \times 10^{-6}\,\text{m}^{-2}\,\text{s}^2$$

Though $f(v)$ varies over the range 290 to 300 m s^{-1}, the variation is small over this small range and its value at the centre of the range can be used.

$$f(295\,\text{m s}^{-1}) = (4\pi) \times \left(\frac{3.37 \times 10^{-6}\,\text{m}^{-2}\text{s}^2}{\pi}\right)^{3/2} \times (295\,\text{m s}^{-1})^2 \times e^{(-3.37 \times 10^{-6}) \times (295)^2}$$

$$= 9.06 \times 10^{-4}\,\text{m}^{-1}\text{s}$$

Therefore, the fraction of molecules in the specified range is

$$f \times \Delta v = (9.06 \times 10^{-4}\,\text{m}^{-1}\text{s}) \times (10\,\text{m s}^{-1}) = \boxed{9.06 \times 10^{-3}}$$

corresponding to 0.91 per cent.

Comment. This is a rather small percentage and suggests that the approximation of constancy of $f(v)$ over the range is adequate. To test the approximation $f(290 \, \text{m s}^{-1})$ and $f(300 \, \text{m s}^{-1})$ could be evaluated.

E24.11(a) We first calculate, Z_W, the number of collisions per unit area per unit time; the number of collisions is then Z_W multiplied by the area of the surface and the time.

$$Z_W = \frac{p}{(2\pi mkT)^{1/2}} \, [24.15]$$

$$= \frac{90 \, \text{Pa}}{[(2\pi) \times (39.95) \times (1.6605 \times 10^{-27} \, \text{kg}) \times (1.381 \times 10^{-23} \, \text{J K}^{-1}) \times (500 \, \text{K})]^{1/2}}$$

$$= 1.7 \times 10^{24} \, \text{m}^{-2} \, \text{s}^{-1}$$

Therefore, the number of collisions is

$$N = (1.7 \times 10^{24} \, \text{m}^{-2} \, \text{s}^{-1}) \times (2.5 \times 3.0 \times 10^{-6} \, \text{m}^2) \times (15 \, \text{s}) = \boxed{1.9 \times 10^{20}}$$

Comment. Equation 24.15 in the form $p = Z_W(2\pi mkT)^{1/2}$ is considered the molecular explanation of pressure and is used in this form in Example 24.3.

Question. How many collisions are there per second on the walls of a room with dimensions $3 \, \text{m} \times 5 \, \text{m} \times 5 \, \text{m}$ with 'air' molecules at $25°\text{C}$ and $1.00 \, \text{atm}$?

E24.12(a) $$\Delta m = Z_W A_0 m \Delta t \text{ [Example 24.3]} = \frac{p A_0 m \Delta t}{(2\pi mkT)^{1/2}} = p A_0 \Delta t \left(\frac{m}{2\pi kT}\right)^{1/2}$$

$$= p A_0 \Delta t \left(\frac{M}{2\pi RT}\right)^{1/2}$$

From the data, with $A_0 = \pi r^2$,

$$\Delta m = (0.835 \, \text{Pa}) \times (\pi) \times (1.25 \times 10^{-3} \, \text{m})^2 \times (7.20 \times 10^3 \, \text{s})$$

$$\times \left(\frac{260 \times 10^{-3} \, \text{kg mol}^{-1}}{(2\pi) \times (8.314 \, \text{J K}^{-1} \, \text{mol}^{-1}) \times (400 \, \text{K})}\right)^{1/2}$$

$$= 1.04 \times 10^{-4} \, \text{kg}, \quad \text{or} \quad \boxed{104 \, \text{mg}}$$

Question. For the same solid shaped in the form of a sphere of radius $0.050 \, \text{m}$ and suspended in a vacuum, what will be the mass loss in $2.00 \, \text{h}$? *Hint*. Make any reasonable approximations.

E24.13(a) $$J_z = -\kappa \frac{dT}{dz} \, [24.21] = \left(\frac{-0.163 \, \text{mJ cm}^{-2} \, \text{s}^{-1}}{\text{K cm}^{-1}}\right) \times (-2.5 \, \text{K m}^{-1}) \text{ [Table 24.2]}$$

$$= (0.41 \, \text{mJ cm}^{-2} \, \text{s}^{-1}) \times (\text{cm/m}) = 0.41 \times 10^{-2} \, \text{mJ cm}^{-2} \, \text{s}^{-1} = \boxed{4.1 \times 10^{-2} \, \text{J m}^{-2} \, \text{s}^{-1}}$$

E24.14(a) The thermal conductivity, κ, is a function of the mean free path, λ, which in turn is a function of the collision cross-section, σ. Hence, reversing the order, σ can be obtained from κ.

$$\kappa = \tfrac{1}{3}\lambda \bar{c} C_{V,\text{m}}[A] \, [24.28]$$

$$\bar{c} = \left(\frac{8RT}{\pi M}\right)^{1/2} [24.35] \quad \text{and} \quad \lambda = \frac{kT}{2^{1/2}\sigma p}[24.14] = \frac{V}{2^{1/2}\sigma n N_A} = \frac{1}{2^{1/2}\sigma N_A[A]}$$

Hence,

$$[A]\lambda \bar{c} = \left(\frac{8RT}{\pi M}\right)^{1/2} \times \left(\frac{1}{2^{1/2}\sigma N_A}\right) = \left(\frac{4RT}{\pi M}\right)^{1/2} \times \left(\frac{1}{\sigma N_A}\right)$$

and so

$$\kappa = \left(\frac{1}{3\sigma N_A}\right) \times \left(\frac{4RT}{\pi M}\right)^{1/2} C_{V,m} = \left(\frac{1}{3\sigma N_A}\right) \times \left(\frac{4RT}{\pi M}\right)^{1/2}$$

$$\times \frac{3}{2}R\left[C_{V,m} = \tfrac{3}{2}R\right] = \left(\frac{k}{2\sigma}\right) \times \left(\frac{4RT}{\pi M}\right)^{1/2}$$

$$\sigma = \left(\frac{k}{2\kappa}\right) \times \left(\frac{4RT}{\pi M}\right)^{1/2} = \left(\frac{1.318 \times 10^{-23}\,\mathrm{J\,K^{-1}}}{(2) \times (0.0465\,\mathrm{J\,s^{-1}\,K^{-1}\,m^{-1}})}\right)$$

$$\times \left(\frac{(4) \times (8.314\,\mathrm{J\,K^{-1}\,mol^{-1}}) \times (273\,\mathrm{K})}{(\pi) \times (20.2 \times 10^{-3}\,\mathrm{kg\,mol^{-1}})}\right)^{1/2}$$

$$= \boxed{5.6 \times 10^{-20}\,\mathrm{m^2}} \quad \text{or} \quad 0.056\,\mathrm{nm^2}$$

The experimental value is $0.24\,\mathrm{nm^2}$.

Question. What approximations inherent in the equations used in the solution to this exercise are the likely cause of the factor of 4 difference between the experimental and calculated values of the collision cross-section for neon?

E24.15(a) $J_z(\text{energy}) = -\kappa \dfrac{\mathrm{d}T}{\mathrm{d}z}$ [24.21]

This is the rate of energy transfer per unit area. For an area A

Rate of energy transfer: $\dfrac{\mathrm{d}E}{\mathrm{d}t} = AJ_z = \kappa A \dfrac{\mathrm{d}T}{\mathrm{d}z}$

Therefore, with $\kappa \approx 0.241\,\mathrm{mJ\,cm^{-2}\,s^{-1}}/(\mathrm{K\,cm^{-1}})$ [Table 24.2]

$$\frac{\mathrm{d}E}{\mathrm{d}t} \approx \left(\frac{0.241\,\mathrm{mJ\,cm^{-2}\,s^{-1}}}{\mathrm{K\,cm^{-1}}}\right) \times (1.0 \times 10^4\,\mathrm{cm^2}) \times \left(\frac{35\,\mathrm{K}}{5.0\,\mathrm{cm}}\right)$$

$$\approx 17 \times 10^3\,\mathrm{mJ\,s^{-1}} = 17\,\mathrm{J\,s^{-1}}, \quad \text{or} \quad \boxed{17\,\mathrm{W}}$$

Therefore, a $\boxed{17\,\mathrm{W}}$ heater is required.

E24.16(a) The pressure change follows the equation

$$p = p_0 e^{-t/\tau}, \quad \tau = \left(\frac{2\pi m}{kT}\right)^{1/2} \times \left(\frac{V}{A_0}\right) \quad \text{[Example 24.2]}$$

Therefore, the time required for the pressure to fall from p_0 to p is

$$t = \tau \ln \frac{p_0}{p}$$

Consequently for two different gases at the same initial and final pressures

$$\frac{t'}{t} = \frac{\tau'}{\tau} = \left(\frac{M'}{M}\right)^{1/2}$$

and hence $M' = \left(\dfrac{t'}{t}\right)^2 \times M = \left(\dfrac{52}{42}\right)^2 \times (28.02 \text{ g mol}^{-1}) = \boxed{43 \text{ g mol}^{-1}}$

Comment. The actual value of CO_2 is 44.01 g mol^{-1}

E24.17(a) $t = \tau \ln \dfrac{p_0}{p}, \quad \tau = \left(\dfrac{2\pi m}{kT}\right)^{1/2} \times \left(\dfrac{V}{A_0}\right)$ [Example 24.2 and Exercise 24.16(a)]

Since $\tau = \left(\dfrac{2\pi M}{RT}\right)^{1/2} \times \left(\dfrac{V}{A_0}\right) = \left(\dfrac{(2\pi) \times (32.0 \times 10^{-3} \text{ kg mol}^{-1})}{(8.314 \text{ J K}^{-1} \text{ mol}^{-1}) \times (298 \text{ K})}\right)^{1/2}$

$$\times \left(\frac{3.0 \text{ m}^3}{\pi \times (1.0 \times 10^{-4} \text{ m})^2}\right) = 8.6 \times 10^5 \text{ s}$$

we find that $t = (8.6 \times 10^5) \times \ln\left(\dfrac{0.80}{0.70}\right) = \boxed{1.1 \times 10^5 \text{ s}}$ (30 h)

E24.18(a) $\eta = \frac{1}{3} m \lambda \bar{c} N_A [A]$ [24.33, $M = m N_A$]

$$\lambda \bar{c} [A] = \left(\frac{4RT}{\pi M}\right)^{1/2} \times \left(\frac{1}{\sigma N_A}\right) \text{ [Exercise 24.14(a)]}$$

Therefore, $\eta = \left(\dfrac{m}{3\sigma}\right) \times \left(\dfrac{4RT}{\pi M}\right)^{1/2}$

and $\sigma = \left(\dfrac{m}{3\eta}\right) \times \left(\dfrac{4RT}{\pi M}\right)^{1/2}$

$$= \left(\frac{(20.2) \times (1.6605 \times 10^{-27} \text{ kg})}{(3) \times (2.98 \times 10^{-5} \text{ kg m}^{-1} \text{ s}^{-1})}\right) \times \left(\frac{(4) \times (8.314 \text{ J K}^{-1} \text{ mol}^{-1}) \times (273 \text{ K})}{\pi \times (20.2 \times 10^{-3} \text{ kg mol}^{-1})}\right)^{1/2}$$

$$= \boxed{1.42 \times 10^{-19} \text{ m}^2} \quad \text{or} \quad 0.142 \text{ nm}^2$$

E24.19(a) $\dfrac{dV}{dt} = \dfrac{(p_1^2 - p_2^2)\pi r^4}{16 l \eta p_0}$ [24.35] which rearranges to

$$p_1^2 = p_2^2 + \left(\frac{16 l \eta p_0}{\pi r^4}\right) \times \left(\frac{dV}{dt}\right)$$

$$= p_2^2 + \left(\frac{(16) \times (8.50 \text{ m}) \times (1.76 \times 10^{-5} \text{ kg m}^{-1} \text{ s}^{-1}) \times (1.00 \times 10^5 \text{ Pa})}{\pi \times (5.0 \times 10^{-3} \text{ m})^4}\right) \times \left(\frac{9.5 \times 10^2 \text{ m}^3}{3600 \text{ s}}\right)$$

$$= p_2^2 + (3.22 \times 10^{10} \text{ Pa}^2) = (1.00 \times 10^5)^2 \text{ Pa}^2 + (3.22 \times 10^{10} \text{ Pa}^2) = 4.22 \times 10^{10} \text{ Pa}^2$$

Hence, $p_1 = \boxed{205 \text{ kPa}}$ (2.05 bar)

E24.20(a) $\eta = \frac{1}{3}m\lambda\bar{c}N_A[A]$ [24.33, with $M = mN_A$] $= \left(\frac{m}{3\sigma}\right) \times \left(\frac{4RT}{\pi M}\right)^{1/2}$

$$= \left(\frac{(29) \times (1.6605 \times 10^{-27}\,\text{kg})}{(3) \times (0.40 \times 10^{-18}\,\text{m}^2)}\right) \times \left(\frac{(4) \times (8.314\,\text{J K}^{-1}\,\text{mol}^{-1}) \times T}{\pi \times (29 \times 10^{-3}\,\text{kg mol}^{-1})}\right)^{1/2}$$

$$= (7.7 \times 10^{-7}\,\text{kg m}^{-1}\,\text{s}^{-1}) \times (T/\text{K})^{1/2}$$

(a) At $T = 273\,\text{K}$, $\eta = 1.3 \times 10^{-5}\,\text{kg m}^{-1}\,\text{s}^{-1}$, or $\boxed{130\,\mu\text{P}}$

(b) At $T = 298\,\text{K}$, $\eta = \boxed{130\,\mu\text{P}}$ (c) At $T = 1000\,\text{K}$, $\eta = \boxed{240\,\mu\text{P}}$

E24.21(a) $\kappa = \frac{1}{3}\lambda\bar{c}C_{V,m}[A]$ [24.28] $= \left(\frac{k}{2\sigma}\right) \times \left(\frac{4RT}{\pi M}\right)^{1/2}$ [Exercise 24.14(a)]

$$= \left(\frac{1.381 \times 10^{-23}\,\text{J K}^{-1}}{(2) \times (\sigma/\text{nm}^2) \times 10^{-18}\,\text{m}^{-2}}\right) \times \left(\frac{(4) \times (8.314\,\text{J K}^{-1}\,\text{mol}^{-1}) \times (300\,\text{K})}{\pi \times (M/\text{g mol}^{-1}) \times 10^{-3}\,\text{kg mol}^{-1}}\right)^{1/2}$$

$$= \frac{1.23 \times 10^{-2}\,\text{J K}^{-1}\,\text{m}^{-1}\,\text{s}^{-1}}{(\sigma/\text{nm}^2) \times (M/\text{g mol}^{-1})^{1/2}}$$

(a) For Ar, $\kappa = \dfrac{1.23 \times 10^{-2}\,\text{J K}^{-1}\,\text{m}^{-1}\,\text{s}^{-1}}{(0.36) \times (39.95)^{1/2}} = \boxed{5.4\,\text{mJ K}^{-1}\,\text{m}^{-1}\,\text{s}^{-1}}$

(b) For He, $\kappa = \dfrac{1.23 \times 10^{-2}\,\text{J K}^{-1}\,\text{m}^{-1}\,\text{s}^{-1}}{(0.21) \times (4.00)^{1/2}} = \boxed{29\,\text{mJ K}^{-1}\,\text{m}^{-1}\,\text{s}^{-1}}$

The rate of flow of energy as heat is [Exercise 24.15(a)]

$$kA\frac{dT}{dz} = \kappa \times (100 \times 10^{-4}\,\text{m}^2) \times (150\,\text{K m}^{-1}) = (1.50\,\text{K m}) \times \kappa$$

$$= 8.1\,\text{mJ s}^{-1} = \boxed{8.1\,\text{mW}}\text{ for Ar, } 44\,\text{mJ s}^{-1} = \boxed{44\,\text{mW}}\text{ for He}$$

E24.22(a) $\dfrac{dV}{dt} \propto \dfrac{1}{\eta}$ [24.35]

which implies

$$\frac{\eta(CO_2)}{\eta(Ar)} = \frac{\tau(CO_2)}{\tau(Ar)} = \frac{55\,\text{s}}{83\,\text{s}} = 0.66\bar{3}$$

Therefore, $\eta(CO_2) = 0.66\bar{3} \times \eta(Ar) = \boxed{13\bar{8}\,\mu\text{P}}$

For the molecular diameter of CO_2 we use

$$\sigma = \left(\frac{m}{3\eta}\right) \times \left(\frac{3RT}{\pi M}\right)^{1/2} \text{ [Exercise 24.18(a)]}$$

$$= \left(\frac{(44.01) \times (1.6605 \times 10^{-27}\,\text{kg})}{(3) \times (1.38 \times 10^{-5}\,\text{kg m}^{-1}\,\text{s}^{-1})}\right) \times \left(\frac{(4) \times (8.314\,\text{J K}^{-1}\,\text{mol}^{-1}) \times (298\,\text{K})}{\pi \times (44.01 \times 10^{-3}\,\text{kg mol}^{-1})}\right)^{1/2}$$

$$= 4.7 \times 10^{-19}\,\text{m}^2 \approx \pi d^2$$

therefore $d \approx \left(\dfrac{1}{\pi} \times (4.7 \times 10^{-19}\,\text{m}^2)\right)^{1/2} = \boxed{390\,\text{pm}}$

E24.23(a) $\kappa = \frac{1}{3}\lambda \bar{c} C_{V,m}[\mathrm{A}]$ [24.28], $\bar{c} = \left(\dfrac{8RT}{\pi M}\right)^{1/2}$

$\lambda = \dfrac{kT}{2^{1/2}\sigma p}$ [24.14] $= \dfrac{1}{2^{1/2}\sigma N_A[\mathrm{A}]}$ $\left[\dfrac{p}{kT} = N_A[\mathrm{A}]\right]$

Therefore, $\kappa = \dfrac{\bar{c} C_{V,m}}{(3)\times(2^{1/2})\sigma N_A}$

For argon $M \approx 39.95\ \mathrm{g\ mol^{-1}}$

$$\bar{c} = \left(\frac{(8)\times(8.314\ \mathrm{J\,K^{-1}\,mol^{-1}})\times(298\ \mathrm{K})}{\pi\times(39.95\times10^{-3}\ \mathrm{kg\,mol^{-1}})}\right)^{1/2} = 397\ \mathrm{m\,s^{-1}}$$

$$\kappa = \frac{(397\ \mathrm{m\,s^{-1}})\times(12.5\ \mathrm{J\,K^{-1}\,mol^{-1}})}{(3)\times(2^{1/2})\times(0.36\times10^{-18}\ \mathrm{m^2})\times(6.022\times10^{23}\ \mathrm{mol^{-1}})} = \boxed{5.4\times10^{-3}\ \mathrm{J\,K^{-1}\,m^{-1}\,s^{-1}}}$$

Comment. This calculated value does not agree well with the value of κ listed in Table 24.22.

Question. Can the differences between the calculated and experimental values of κ be accounted for by the difference in temperature (298 K here, 273 K in Table 24.22)? If not, what might be responsible for the difference?

E24.24(a) $D = \frac{1}{3}\lambda\bar{c}$ [24.23] $= \left(\dfrac{2}{3p\sigma}\right)\times\left(\dfrac{k^3 T^3}{\pi m}\right)^{1/2}$

$$= \left(\frac{2}{(3p)\times(0.36\times10^{-18}\ \mathrm{m^2})}\right)\times\left(\frac{(1.381\times10^{-23}\ \mathrm{J\,K^{-1}})^3\times(298\ \mathrm{K})^3}{\pi\times(39.95)\times(1.6605\times10^{-27}\ \mathrm{kg})}\right)^{1/2}$$

$$= \frac{1.07\ \mathrm{m^2\,s^{-1}}}{(p/\mathrm{Pa})}$$

Therefore, **(a)** at 1 Pa, $D = \boxed{1.1\ \mathrm{m^2\,s^{-1}}}$, **(b)** at 100 kPa, $D = \boxed{1.1\times10^{-5}\ \mathrm{m^2\,s^{-1}}}$, and **(c)** at 10 MPa, $D = \boxed{1.1\times10^{-7}\ \mathrm{m^2\,s^{-1}}}$

The flux due to diffusion is

$$J = -D\left(\frac{dN}{dz}\right)\ [24.20]$$

Dividing both sides by the Avogadro constant converts the flux to number of moles per unit area per second. Thus

$$J = -D\frac{dc}{dx}\ [24.75] = -D\frac{d}{dx}\left(\frac{n}{V}\right)$$

$$= -\left(\frac{D}{RT}\right)\frac{dp}{dx}\quad\text{[perfect gas law]}$$

The negative sign indicates flow from high pressure to low. For a pressure gradient of $0.10\ \mathrm{atm\,cm^{-1}}$

$$J = \left[\frac{D/(\mathrm{m^2\,s^{-1}})}{(8.3145\ \mathrm{J\,K^{-1}\,mol^{-1}}\times298\ \mathrm{K})}\right]\times(0.10\ \mathrm{atm\,cm^{-1}}\times100\ \mathrm{cm\,m^{-1}}\times1.01\times10^5\ \mathrm{Pa\,atm^{-1}})$$

$$= (4.1\times10^2\ \mathrm{mol\,m^{-2}\,s^{-1}})\times(D/(\mathrm{m^2\,s^{-1}}))$$

(a) $J = (4.1 \times 10^2 \, \text{mol} \, \text{m}^{-2} \, \text{s}^{-1}) \times 1.07 = \boxed{4.4 \times 10^2 \, \text{mol} \, \text{m}^{-2} \, \text{s}^{-1}}$

(b) $J = (4.1 \times 10^2 \, \text{mol} \, \text{m}^{-2} \, \text{s}^{-1}) \times 1.07 \times 10^{-5} = \boxed{4.4 \times 10^{-3} \, \text{mol} \, \text{m}^{-2} \, \text{s}^{-1}}$

(c) $J = (4.1 \times 10^2 \, \text{mol} \, \text{m}^{-2} \, \text{s}^{-1}) \times 1.07 \times 10^{-7} = \boxed{4.4 \times 10^{-5} \, \text{mol} \, \text{m}^{-2} \, \text{s}^{-1}}$

E24.25(a) Molar ionic conductivity is related to mobility by

$$\lambda = zuF \, [24.54]$$
$$= 1 \times 7.91 \times 10^{-8} \, \text{m}^2 \, \text{s}^{-1} \, \text{V}^{-1} \times 96\,485 \, \text{C} \, \text{mol}^{-1}$$
$$= \boxed{7.63 \times 10^{-3} \, \text{S} \, \text{m}^2 \, \text{mol}^{-1}}$$

E24.26(a) $\quad s = u\mathcal{E} \, [24.52] \quad \mathcal{E} = \dfrac{\Delta\phi}{l}$

Therefore,

$$s = u\left(\frac{\Delta\phi}{l}\right) = (7.92 \times 10^{-8} \, \text{m}^2 \, \text{s}^{-1} \, \text{V}^{-1}) \times \left(\frac{35.0 \, \text{V}}{8.00 \times 10^{-3} \, \text{m}}\right)$$
$$= 3.47 \times 10^{-4} \, \text{m} \, \text{s}^{-1}, \quad \text{or} \quad \boxed{347 \, \mu\text{m} \, \text{s}^{-1}}$$

E24.27(a) $\quad t_+^\circ = \dfrac{u_+}{u_+ + u_-} \, [24.61] = \dfrac{4.01 \times 10^{-4} \, \text{cm}^2 \, \text{s}^{-1} \, \text{V}^{-1}}{(4.01 + 8.09) \times 10^{-4} \, \text{cm}^2 \, \text{s}^{-1} \, \text{V}^{-1}} \, \text{(Table 24.6)} = \boxed{0.331}$

E24.28(a) The basis for the solution is Kohlrausch's law of independent migration of ions [32]. Switching counterions does not affect the mobility of the remaining other ion at infinite dilution.

$$\Lambda_m^\circ = \nu_+ \lambda_+ + \nu_- \lambda_- \, [24.40]$$
$$\Lambda_m^\circ(\text{KCl}) = \lambda(\text{K}^+) + \lambda(\text{Cl}^-) = 14.99 \, \text{mS} \, \text{m}^2 \, \text{mol}^{-1}$$
$$\Lambda_m^\circ(\text{KNO}_3) = \lambda(\text{K}^+) + \lambda(\text{NO}_3^-) = 14.50 \, \text{mS} \, \text{m}^2 \, \text{mol}^{-1}$$
$$\Lambda_m^\circ(\text{AgNO}_3) = \lambda(\text{Ag}^+) + \lambda(\text{NO}_3^-) = 13.34 \, \text{mS} \, \text{m}^2 \, \text{mol}^{-1}$$

Hence, $\Lambda_m^\circ(\text{AgCl}) = \Lambda_m^\circ(\text{AgNO}_3) + \Lambda_m^\circ(\text{KCl}) - \Lambda_m^\circ(\text{KNO}_3)$

$$= (13.34 + 14.99 - 14.50) \, \text{mS} \, \text{m}^2 \, \text{mol}^{-1} = \boxed{13.83 \, \text{mS} \, \text{m}^2 \, \text{mol}^{-1}}$$

Question. How well does this result agree with the value calculated directly from the data of Table 24.5?

E24.29(a) $\quad u = \dfrac{\lambda}{z\mathcal{F}} \, [24.54]$

$$u(\text{Li}^+) = \frac{3.87 \, \text{mS} \, \text{m}^2 \, \text{mol}^{-1}}{9.6485 \times 10^4 \, \text{C} \, \text{mol}^{-1}} = 4.01 \times 10^{-5} \, \text{mS} \, \text{C}^{-1} \, \text{m}^2$$

$$= \boxed{4.01 \times 10^{-8} \, \text{m}^2 \, \text{V}^{-1} \, \text{s}^{-1}} \, (1 \, \text{C} \, \Omega = 1 \, \text{A} \, \text{s} \, \Omega = 1 \, \text{V} \, \text{s})$$

$$u(\text{Na}^+) = \frac{5.01 \, \text{mS} \, \text{m}^2 \, \text{mol}^{-1}}{9.6485 \times 10^4 \, \text{C} \, \text{mol}^{-1}} = \boxed{5.19 \times 10^{-8} \, \text{m}^2 \, \text{V}^{-1} \, \text{s}^{-1}}$$

$$u(\text{K}^+) = \frac{7.35 \, \text{mS} \, \text{m}^2 \, \text{mol}^{-1}}{9.6485 \times 10^4 \, \text{C} \, \text{mol}^{-1}} = \boxed{7.62 \times 10^{-8} \, \text{m}^2 \, \text{V}^{-1} \, \text{s}^{-1}}$$

E24.30(a) $D = \dfrac{uRT}{z\mathcal{F}}$ [24.79]

$$= \dfrac{(7.40 \times 10^{-8}\,\text{m}^2\,\text{s}^{-1}\,\text{V}^{-1}) \times (8.314\,\text{J}\,\text{K}^{-1}) \times (298\,\text{K})}{9.6485 \times 10^4\,\text{C}\,\text{mol}^{-1}} = \boxed{1.90 \times 10^{-9}\,\text{m}^2\,\text{s}^{-1}}$$

E24.31(a)   Equation [24.91] gives the mean square distance travelled in any one dimension. We need the distance travelled from a point in any direction. The distinction here is the distinction between one-dimensional and three-dimensional diffusion. The mean square three-dimensional distance can be obtained from the one-dimensional mean square distance since motions in the three directions are independent. Since

$$r^2 = x^2 + y^2 + z^2 \quad \text{[Pythagorean theorem]}$$

$$\langle r^2 \rangle = \langle x^2 \rangle + \langle y^2 \rangle + \langle z^2 \rangle = 3\langle x^2 \rangle \quad \text{[independent motion]}$$

$$= 3 \times 2Dt \quad [24.91 \text{ for } \langle x^2 \rangle] = 6Dt$$

Therefore, $t = \dfrac{\langle r^2 \rangle}{6D} = \dfrac{(5.0 \times 10^{-3}\,\text{m})^2}{(6) \times (3.17 \times 10^{-9}\,\text{m}^2\,\text{s}^{-1})} = \boxed{1.3 \times 10^3\,\text{s}}$

E24.32(a) $a = \dfrac{kT}{6\pi\eta D}$ [24.83 and Example 24.5]

$$= \dfrac{(1.381 \times 10^{-23}\,\text{J}\,\text{K}^{-1}) \times (298\,\text{K})}{(6\pi) \times (1.00 \times 10^{-3}\,\text{kg}\,\text{m}^{-1}\,\text{s}^{-1}) \times (5.2 \times 10^{-10}\,\text{m}^2\,\text{s}^{-1})} = 4.2 \times 10^{-10}\,\text{m}, \quad \text{or} \quad \boxed{420\,\text{pm}}$$

E24.33(a)   The Einstein–Smoluchowski equation [24.93] relates the diffusion constant to the unit jump distance and time

$$D = \dfrac{\lambda^2}{2\tau} \quad [24.93] \quad \text{so} \quad \tau = \dfrac{\lambda^2}{2D}$$

If the jump distance is about one molecular diameter, or two effective molecular radii, then the jump distance can be obtained by use of the Stokes–Einstein equation [24.83]

$$D = \dfrac{kT}{6\pi\eta a} = \dfrac{kT}{3\pi\eta\lambda} \quad \text{so} \quad \lambda = \dfrac{kT}{3\pi\eta D}$$

and $\tau = \dfrac{(kT)^2}{18(\pi\eta)^2 D^3} = \dfrac{[(1.381 \times 10^{-23}\,\text{J}\,\text{K}^{-1}) \times (298\,\text{K})]^2}{18[\pi(0.601 \times 10^{-3}\,\text{kg}\,\text{m}^{-1}\,\text{s}^{-1})]^2 \times (2.13 \times 10^{-9}\,\text{m}^2\,\text{s}^{-1})^3}$

$$= 2.73 \times 10^{-11}\,\text{s} = \boxed{27\,\text{ps}}$$

Comment. In the strictest sense we are again (cf. Exercise 24.31(a)) dealing with three-dimensional diffusion here. However, since we are assuming that only one jump occurs, it is probably an adequate approximation to use an equation derived for one-dimensional diffusion. For three-dimensional diffusion the equation analogous to eqn 24.93 is

$$\tau = \dfrac{\lambda^2}{6D}$$

Question. Can you derive this equation? *Hint.* Use an analysis similar to that described in the solution to Exercise 24.31(a).

E24.34(a) For three-dimensional diffusion we use the equation analogous to eqn 24.91 derived in Exercise 24.31(a), that is

$$\langle r^2 \rangle = 6Dt$$

For iodine in benzene [Data from Table 24.8]

$$\langle r^2 \rangle^{1/2} = [(6) \times (2.13 \times 10^{-9}\,\text{m}^2\,\text{s}^{-1}) \times (1.0\,\text{s})]^{1/2} = \boxed{113\,\mu\text{m}}$$

For sucrose in water

$$\sqrt{\langle r^2 \rangle} = \sqrt{6Dt} = [6 \times (0.5216 \times 10^{-9}\,\text{m}^2\,\text{s}^{-1}) \times (1.0\text{s})]^{1/2} = \boxed{5.594 \times 10^{-5}\,\text{m}}$$

Solutions to problems

Solutions to numerical problems

P24.1 The time in seconds for a disk to rotate $360°$ is the inverse of the frequency. The time for it to advance $2°$ is $\dfrac{\left(\frac{2°}{360°}\right)}{\nu}$. This is the time required for slots in neighbouring disks to coincide. For an atom to pass through all neighbouring slots it must have the speed $v_x = \dfrac{1.0\,\text{cm}}{\left(\frac{2}{360}\right)} = 180\,\nu\,\text{cm} = 180(\nu/\text{Hz})\,\text{cm}\,\text{s}^{-1}$

Hence, the distributions of the x-component of velocity are

ν/Hz	20	40	80	100	120
$v_x/(\text{cm}\,\text{s}^{-1})$	3600	7200	14400	18000	21600
$\mathcal{I}(40\,\text{K})$	0.846	0.513	0.069	0.015	0.002
$\mathcal{I}(100\,\text{K})$	0.592	0.485	0.217	0.119	0.057

Theoretically, the velocity distribution in the x-direction is

$$f(v_x) = \left(\frac{m}{2\pi kT}\right)^{1/2} e^{-mv_x^2/2kT} \quad [24.6, \text{with } M/R = m/k]$$

Therefore, as $\mathcal{I} \propto f$, $\mathcal{I} \propto \left(\dfrac{1}{T}\right)^{1/2} e^{-mv_x^2/2kT}$

Since $\dfrac{mv_x^2}{2kT} = \dfrac{83.8 \times (1.6605 \times 10^{-27}\,\text{kg}) \times \{1.80(\nu/\text{Hz})\text{m}\,\text{s}^{-1}\}^2}{(2) \times (1.381 \times 10^{-23}\,\text{J}\,\text{K}^{-1}) \times (T)} = \dfrac{1.63 \times 10^{-2}(\nu/\text{Hz})^2}{T/\text{K}}$

We can write $\mathcal{I} \propto \left(\dfrac{1}{T/\text{K}}\right)^{1/2} e^{-1.63 \times 10^{-2}(\nu/\text{Hz})^2/(T/\text{K})}$ and draw up the following table, obtaining the constant of proportionality by fitting $\mathcal{I}$ to the value at $T = 40\,\text{K}$, $\nu = 80\,\text{Hz}$

ν/Hz	20	40	80	100	120
$\mathcal{I}(40\,\text{K})$	0.80	0.49	(0.069)	0.016	0.003
$\mathcal{I}(100\,\text{K})$	0.56	0.46	0.209	0.116	0.057

in fair agreement with the experimental data.

P24.2   For discrete rather than continuous variables the equation analogous to the equation for obtaining $\bar{c}$
(Example 24.1) is $\langle v_x \rangle = \sum_i v_{i,x} \left(\dfrac{N_i}{N} \right) = \dfrac{1}{N} \sum_i N_i v_{i,x}$ with $\left(\dfrac{N_i}{N} \right)$ the analogue of $f(v)$.

$$N = 40 + 62 + 53 + 12 + 2 + 38 + 59 + 60 + 2 = 328$$

(a) $\langle v_x \rangle = \dfrac{1}{328} \{40 \times 80 + 62 \times 85 + \cdots + 2 \times 100 + 38 \times (-80)$

$+ 59 \times (-85) + \cdots + 2 \times (-100)\} \, \text{km h}^{-1}$

$= \boxed{2.8 \, \text{km h}^{-1}} \, \text{East}$

(b) $\langle |v_x| \rangle = \dfrac{1}{328} \{40 \times 80 + 62 \times 85 + \cdots + 2 \times 100 + 38 \times 80$

$+ 59 \times 85 + \cdots + 2 \times 100\} \, \text{km h}^{-1}$

$= \boxed{86 \, \text{km h}^{-1}}$

(c) $\langle v_x^2 \rangle = \dfrac{1}{328} \{40 \times 80^2 + 62 \times 85^2 + \cdots + 2 \times 100^2\}(\text{km h}^{-1})^2 = 7430 \, (\text{km h}^{-1})^2$

$\sqrt{\langle v_x^2 \rangle} = \boxed{86 \, \text{km h}^{-1}}$ $\left[\text{that } \sqrt{\langle v_x^2 \rangle} = \langle |v_x| \rangle \text{ in this case is coincidental.} \right]$

P24.5   The number of molecules that escape in unit time is the number per unit time that would have collided
with a wall section of area A equal to the area of the small hole. That is,

$$\frac{\mathrm{d}N}{\mathrm{d}t} = -Z_W A = \frac{-Ap}{(2\pi mkT)^{1/2}} \, [24.15]$$

where p is the (constant) vapour pressure of the solid. The change in the number of molecules inside
the cell in an interval Δt is therefore $\Delta N = -Z_W A \Delta t$, and so the mass loss is

$$\Delta w = \Delta N m = -Ap \left(\frac{m}{2\pi kT} \right)^{1/2} \Delta t = -Ap \left(\frac{M}{2\pi RT} \right)^{1/2} \Delta t$$

Therefore, the vapour pressure of the substance in the cell is

$$p = \left(\frac{-\Delta w}{A \Delta t} \right) \times \left(\frac{2\pi RT}{M} \right)^{1/2}$$

For the vapour pressure of germanium

$$p = \left(\frac{4.3 \times 10^{-8} \, \text{kg}}{\pi \times (5.0 \times 10^{-4} \, \text{m})^2 \times (7200 \, \text{s})} \right) \times \left(\frac{(2\pi) \times (8.314 \, \text{J K}^{-1} \, \text{mol}^{-1}) \times (1273 \, \text{K})}{72.6 \times 10^{-3} \, \text{kg mol}^{-1}} \right)^{1/2}$$

$$= 7.3 \times 10^{-3} \, \text{Pa}, \quad \text{or} \quad \boxed{7.3 \, \text{mPa}}$$

P24.6   Radioactive decay follows first-order kinetics (Chapter 25); hence the two contributions to the rate
of change of the number of helium atoms are

$$\frac{\mathrm{d}N}{\mathrm{d}t} = k_r[\text{Bk}] \, (\text{Radioactive decay}) \qquad \frac{\mathrm{d}N}{\mathrm{d}t} = -Z_W[\text{A}] \quad [\text{Problem 24.5}]$$

Therefore, the total rate of change is

$$\frac{\mathrm{d}N}{\mathrm{d}t} = k_r[\text{Bk}] - Z_W A \quad \text{with } Z_W = \frac{p}{(2\pi mkT)^{1/2}}$$

$$[\text{Bk}] = [\text{Bk}]_0 e^{-k_r t} \quad \text{and} \quad p = \frac{nRT}{V} = \frac{nN_A kT}{V} = \frac{NkT}{V}$$

Therefore, the pressure of helium inside the container obeys

$$\frac{dp}{dt} = \frac{kT}{V}\frac{dN}{dt} = \frac{kk_r T}{V}[Bk]_0 e^{-k_r t} - \frac{\left(\frac{pAkT}{V}\right)}{(2\pi mkT)^{1/2}}$$

If we write $a = \dfrac{kk_r T[Bk]_0}{V}$, $b = \left(\dfrac{A}{V}\right) \times \left(\dfrac{kT}{2\pi m}\right)^{1/2}$, the rate equation becomes

$$\frac{dp}{dt} = ae^{-k_r t} - bp, \quad p = 0 \text{ at } t = 0$$

which is a first-order linear differential equation with the solution

$$p = \left(\frac{a}{k_r - b}\right) \times (e^{-bt} - e^{-k_r t})$$

Since $[Bk] = \frac{1}{2}[Bk]_0$ when $t = 4.4\,h$, it follows from the radioactive decay law ($[Bk] = [Bk]_0 e^{-k_r t}$) that (Chapter 25)

$$k_r = \frac{\ln 2}{(4.4) \times (3600\,s)} = 4.4 \times 10^{-5}\,s^{-1}$$

We also know that $[Bk]_0 = \left(\dfrac{1.0 \times 10^{-3}\,g}{244\,g\,mol^{-1}}\right) \times (6.022 \times 10^{23}\,mol^{-1}) = 2.5 \times 10^{18}$

Then, $a = \dfrac{kk_r T[Bk]_0}{V} = \dfrac{(1.381 \times 10^{-23}\,J\,K^{-1}) \times (4.4 \times 10^{-5}\,s^{-1}) \times (298\,K) \times (2.5 \times 10^{18})}{1.0 \times 10^{-6}\,m^3}$

$$= 0.45\,Pa\,s^{-1}$$

and $b = \left(\dfrac{\pi \times (2.0 \times 10^{-6}\,m)^2}{1.0 \times 10^{-6}\,m^3}\right) \times \left(\dfrac{(1.381 \times 10^{-23}\,J\,K^{-1}) \times (298\,K)}{(2\pi) \times (4.0) \times (1.6605 \times 10^{-27}\,kg)}\right)^{1/2} = 3.9 \times 10^{-3}\,s^{-1}$

Hence, $p = \left(\dfrac{0.45\,Pa\,s^{-1}}{[(4.4 \times 10^{-5}) - (3.9 \times 10^{-3})]\,s^{-1}}\right) \times (e^{-3.9 \times 10^{-3}(t/s)} - e^{-4.4 \times 10^{-5}(t/s)})$

$$= (120\,Pa) \times (e^{-4.4 \times 10^{-5}(t/s)} - e^{-3.9 \times 10^{-3}(t/s)})$$

(a) $t = 1\,h$, $p = (120\,Pa) \times (e^{-0.16} - e^{-14}) = \boxed{100\,Pa}$

(b) $t = 10\,h$, $p = (120\,Pa) \times (e^{-1.6} - e^{140}) = \boxed{24\,Pa}$

P24.8 $\kappa \propto \dfrac{1}{R}$ [24.37, and the discussion above 24.37]

Because both solutions are aqueous their conductivities include a contribution of $76\,mS\,m^{-1}$ from the water. Therefore,

$$\frac{\kappa(acid\ soln)}{\kappa(KCl\ soln)} = \frac{\kappa(acid) + \kappa(water)}{\kappa(KCl) + \kappa(water)} = \frac{R(KCl\ soln)}{R(acid\ soln)} = \frac{33.21\,\Omega}{300.0\,\Omega}$$

Hence, $\kappa(acid) = \{\kappa(KCl) + \kappa(water)\} \times \left(\dfrac{33.21}{300.0}\right) - \kappa(water) = 53\,mS\,m^{-1}$

$$\Lambda_m = \frac{\kappa}{c} = \frac{53\,mS\,m^{-1}}{1.00 \times 10^5\,mol\,m^{-3}} = \boxed{5.3 \times 10^{-4}\,mS\,m^2\,mol^{-1}}$$

P24.9 $\Lambda_m = \Lambda_m^\circ - \mathcal{K}c^{1/2}$ [24.39], $\Lambda_m = \dfrac{C}{cR}$ [24.38]

where $C = 20.63 \text{ m}^{-1}$ ($C = \kappa^* R^*$, where κ^* and R^* are the conductivity and resistance, of a standard solution respectively)

Therefore, we draw up the following table

c/M	0.0005	0.001	0.005	0.010	0.020	0.050
$(c/\text{M})^{1/2}$	0.224	0.032	0.071	0.100	0.141	0.224
R/Ω	3314	1669	342.1	174.1	89.08	37.14
$\Lambda_m/(\text{mS m}^2 \text{ mol}^{-1})$	12.45	12.36	12.06	11.85	11.58	11.11

The values of Λ_m are plotted against $c^{1/2}$ in Fig. 24.1.

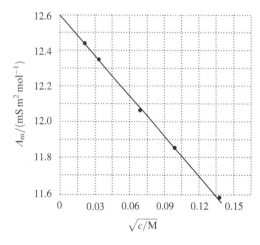

Figure 24.1

The limiting value is $\Lambda_m^\circ = \boxed{12.6 \text{ mS m}^2 \text{ mol}^{-1}}$. The slope is -7.30; hence

$\mathcal{K} = \boxed{7.30 \text{ mS m}^2 \text{ mol}^{-1} \text{ M}^{-1/2}}$.

(a) $\Lambda_m = (5.01 + 7.68) \text{ mS m}^2 \text{ mol}^{-1} - (+7.30 \text{ mS m}^2 \text{ mol}^{-1}) \times (0.010)^{1/2}$

     $= \boxed{11.96 \text{ mS m}^2 \text{ mol}^{-1}}$

(b) $\kappa = c\Lambda_m = (10 \text{ mol m}^{-3}) \times (11.96 \text{ mS m}^2 \text{ mol}^{-1}) = 119.6 \text{ mS m}^2 \text{ m}^{-3} = \boxed{119.6 \text{ mS m}^{-1}}$

(c) $R = \dfrac{C}{\kappa} = \dfrac{20.63 \text{ m}^{-1}}{119.6 \text{ mS m}^{-1}} = \boxed{172.5 \, \Omega}$

P24.11 $s = u\mathcal{E}$ [24.52]   with $\mathcal{E} = \dfrac{10 \text{ V}}{1.00 \text{ cm}} = 10 \text{ V cm}^{-1}$

     $s(\text{Li}^+) = (4.01 \times 10^{-4} \text{ cm}^2 \text{ s}^{-1} \text{ V}^{-1}) \times (10 \text{ V cm}^{-1}) = \boxed{4.0 \times 10^{-3} \text{ cm s}^{-1}}$

     $s(\text{Na}^+) = (5.19 \times 10^{-4} \text{ cm}^2 \text{ s}^{-1} \text{ V}^{-1}) \times (10 \text{ V cm}^{-1}) = \boxed{5.2 \times 10^{-3} \text{ cm s}^{-1}}$

     $s(\text{K}^+) = (7.62 \times 10^{-4} \text{ cm}^2 \text{ s}^{-1} \text{ V}^{-1}) \times (10 \text{ V cm}^{-1}) = \boxed{7.6 \times 10^{-3} \text{ cm s}^{-1}}$

$$t = \frac{d}{s} \quad \text{with } d = 1.0 \, \text{cm}:$$

$$t(\text{Li}^+) = \frac{1.0 \, \text{cm}}{4.0 \times 10^{-3} \, \text{cm s}^{-1}} = \boxed{250 \, \text{s}}, \qquad t(\text{Na}^+) = \boxed{190 \, \text{s}}, \qquad t(\text{K}^+) = \boxed{130 \, \text{s}}$$

(a) For the distance moved during a half-cycle, write

$$d = \int_0^{1/2v} s \, dt = \int_0^{1/2v} u\mathcal{E} \, dt = u\varepsilon_0 \int_0^{1/2v} \sin(2\pi v t) \, dt \quad [\mathcal{E} = \mathcal{E}_0 \sin(2\pi v t)]$$

$$= \frac{u\mathcal{E}_0}{\pi v} = \frac{u \times (10 \, \text{V cm}^{-1})}{\pi \times (1.0 \times 10^3 \, \text{s}^{-1})} \quad [\text{assume } \mathcal{E}_0 = 10 \, \text{V}] = 3.18 \times 10^{-3} u \, \text{V s cm}^{-1}$$

That is, $d/\text{cm} = (3.18 \times 10^{-3}) \times (u/\text{cm}^2 \, \text{V}^{-1} \, \text{s}^{-1})$. Hence,

$$d(\text{Li}^+) = (3.18 \times 10^{-3}) \times (4.0 \times 10^{-4} \, \text{cm}) = \boxed{1.3 \times 10^{-6} \, \text{cm}}$$

$$d(\text{Na}^+) = \boxed{1.7 \times 10^{-6} \, \text{cm}}, \qquad d(\text{K}^+) = \boxed{2.4 \times 10^{-6} \, \text{cm}}$$

(b) These correspond to about $\boxed{43}$, $\boxed{55}$, and $\boxed{81}$ solvent molecule diameters respectively.

P24.13

$$t = \frac{zcVF}{I\Delta t} = \frac{zcAFl}{I\Delta t} \quad [24.64]$$

$$= \left(\frac{(21 \, \text{mol m}^{-3}) \times (\pi) \times (2.073 \times 10^{-3} \, \text{m})^2 \times (9.6485 \times 10^4 \, \text{C mol}^{-1})}{18.2 \times 10^{-3} \, \text{A}} \right) \times \left(\frac{l}{\Delta t} \right)$$

$$= (1.50 \times 10^3 \, \text{m}^{-1} \, \text{s}) \times \left(\frac{l}{\Delta t} \right) = (1.50) \times \left(\frac{l/\text{mm}}{\Delta t/\text{s}} \right)$$

Then we draw up the following table

$\Delta t / \text{s}$	200	400	600	800	1000
l/mm	64	128	192	254	318
t_+	0.48	0.48	0.48	0.48	0.48
$t_- = 1 - t_+$	0.52	0.52	0.52	0.52	0.52

Hence, we conclude that $t_+ = \boxed{0.48}$ and $t_- = \boxed{0.52}$. For the mobility of K^+ we use

$$t_+ = \frac{\lambda_+}{\Lambda_{\text{m}}^\circ} \, [24.62] = \frac{u_+ F}{\Lambda_{\text{m}}^\circ} \, [24.54]$$

to obtain

$$u_+ = \frac{t_+ \Lambda_{\text{m}}^\circ}{F} = \frac{(0.48) \times (149.9 \, \text{S cm}^2 \, \text{mol}^{-1})}{9.6485 \times 10^4 \, \text{C mol}^{-1}} = \boxed{7.5 \times 10^{-4} \, \text{cm}^2 \, \text{s}^{-1} \, \text{V}^{-1}}$$

$$\lambda_+ = t_+ \Lambda_{\text{m}}^\circ [24.62] = (0.48) \times (149.9 \, \text{S cm}^2 \, \text{mol}^{-1}) = \boxed{72 \, \text{S cm}^2 \, \text{mol}^{-1}}$$

P24.15

$$\mathcal{F} = -\frac{RT}{c} \times \frac{dc}{dx} \quad [24.74]$$

$$\frac{dc}{dx} = \frac{(0.05 - 0.10) \, \text{M}}{0.10 \, \text{m}} = -0.50 \, \text{M m}^{-1} \quad [\text{linear gradation}]$$

$$RT = 2.48 \times 10^3 \, \text{J mol}^{-1} = 2.48 \times 10^3 \, \text{N m mol}^{-1}$$

(a) $\mathcal{F} = \left(\dfrac{-2.48 \,\text{kN m mol}^{-1}}{0.10 \,\text{M}} \right) \times (-0.50 \,\text{M m}^{-1}) = \boxed{12 \,\text{kN mol}^{-1}}, \; \boxed{2.1 \times 10^{-20} \,\text{N molecule}^{-1}}$

(b) $\mathcal{F} = \left(\dfrac{-2.48 \,\text{kN m mol}^{-1}}{0.075 \,\text{M}} \right) \times (-0.50 \,\text{M m}^{-1}) = \boxed{17 \,\text{kN mol}^{-1}}, \; \boxed{2.8 \times 10^{-20} \,\text{N molecule}^{-1}}$

(c) $\mathcal{F} = \left(\dfrac{-2.48 \,\text{kN m mol}^{-1}}{0.05 \,\text{M}} \right) \times (-0.50 \,\text{M m}^{-1}) = \boxed{25 \,\text{kN mol}^{-1}}, \; \boxed{4.1 \times 10^{-20} \,\text{N molecule}^{-1}}$

P24.16 $D = \dfrac{uRT}{zF}$ [24.79]   and $a = \dfrac{ze}{6\pi\eta u}$ [24.53]

$$D = \frac{(8.314 \,\text{J K}^{-1}\,\text{mol}^{-1}) \times (298.15 \,\text{K}) \times u}{9.6485 \times 10^4 \,\text{C mol}^{-1}} = 2.569 \times 10^{-2} \,\text{V} \times u$$

so $D/(\text{cm}^2\,\text{s}^{-1}) = (2.569 \times 10^{-2}) \times u/(\text{cm}^2\,\text{s}^{-1}\,\text{V}^{-1})$

$$a = \frac{1.602 \times 10^{-19} \,\text{C}}{(6\pi) \times (0.891 \times 10^{-3} \,\text{kg m}^{-1}\,\text{s}^{-1}) \times u}$$

$$= \frac{9.54 \times 10^{-18} \,\text{C kg}^{-1}\,\text{m s}}{u} = \frac{9.54 \times 10^{-18} \,\text{V}^{-1}\,\text{m}^3\,\text{s}^{-1}}{u} \quad (1\,\text{J} = 1\,\text{C V}, \; 1\,\text{J} = 1\,\text{kg m}^2\,\text{s}^{-2})$$

and so $a/\text{m} = \dfrac{9.54 \times 10^{-14}}{u/\text{cm}^2\,\text{s}^{-1}\,\text{V}^{-1}}$

and therefore $a/\text{pm} = \dfrac{9.54 \times 10^{-2}}{u/\text{cm}^2\,\text{s}^{-1}\,\text{V}^{-1}}$

We can now draw up the following table using data from Table 24.6

	Li^+	Na^+	K^+	Rb^+
$10^4 u/(\text{cm}^2\,\text{s}^{-1}\,\text{V}^{-1})$	4.01	5.19	7.62	7.92
$10^5 D/\text{cm}^2$	1.03	1.33	1.96	2.04
a/pm	238	184	125	120

The ionic radii themselves (i.e. their crystallographic radii) are

	Li^+	Na^+	K^+	Rb^+
r_+/pm	59	102	138	149

and it would seem that K^+ and Rb^+ have effective hydrodynamic radii that are smaller than their ionic radii. The effective hydrodynamic and ionic volumes of Li^+ and Na^+ are $\dfrac{4\pi}{3}\pi a^3$ and $\dfrac{4\pi}{3}\pi r_+^3$ respectively, and so the volumes occupied by hydrating water molecules are

(a) Li^+: $\Delta V = \left(\dfrac{4\pi}{3} \right) \times (212^3 - 59^3) \times 10^{-36}\,\text{m}^3 = 5.5\bar{6} \times 10^{-29}\,\text{m}^3$

(b) Na^+: $\Delta V = \left(\dfrac{4\pi}{3} \right) \times (164^3 - 102^3) \times 10^{-36}\,\text{m}^3 = 2.1\bar{6} \times 10^{-29}\,\text{m}^3$

The volume occupied by a single H_2O molecule is approximately
$\left(\dfrac{4\pi}{3} \right) \times (150\,\text{pm})^3 = 1.4 \times 10^{-29}\,\text{m}^3$.

Therefore, Li^+ has about $\boxed{\text{four}}$ firmly attached H_2O molecules whereas Na^+ has only $\boxed{\text{one to two}}$ (according to this analysis).

P24.18   This is essentially one-dimensional diffusion and therefore eqn 24.88 applies.

$$c = \frac{n_0 e^{-x^2/4Dt}}{A(\pi D t)^{1/2}} \quad [24.88]$$

and we know that $n_0 = \left(\dfrac{10\,g}{342\,g\,mol^{-1}}\right) = 0.0292\,mol$

$$A = \pi R^2 = 19.6\,cm^2, \qquad D = 5.21 \times 10^{-6}\,cm^2\,s^{-1} \text{ [Table 24.8]}$$

$$A(\pi D t)^{1/2} = (19.6\,cm^2) \times [(\pi) \times (5.21 \times 10^{-6}\,cm^2\,s^{-1}) \times (t)]^{1/2}$$

$$= 7.93 \times 10^{-2}\,cm^3 \times (t/s)^{1/2}$$

$$\frac{x^2}{4Dt} = \frac{25\,cm^2}{(4) \times (5.21 \times 10^{-6}\,cm^2\,s^{-1}) \times t} = \frac{1.20 \times 10^6}{(t/s)}$$

Therefore, $c = \left(\dfrac{0.0292\,mol \times 10^{22}}{(7.93 \times 10^{-2}\,cm^3) \times (t/s)^{1/2}}\right) \times e^{-1.20 \times 10^6/(t/s)}$

$$= (369\,M) \times \left(\frac{e^{-1.20 \times 10^6/(t/s)}}{(t/s)^{1/2}}\right)$$

(a) $t = 10\,s, \qquad c = (369\,M) \times \left(\dfrac{e^{-1.2 \times 10^5}}{10^{1/2}}\right) \approx \boxed{0}$

(b) $t = 1\,yr = 3.16 \times 10^7\,s, \qquad c = (369\,M) \times \left(\dfrac{e^{-0.038}}{(3.16 \times 10^7)^{1/2}}\right) = \boxed{0.063\,M}$

Comment. This problem illustrates the extreme slowness of diffusion through typical macroscopic distances; however, it is rapid enough through distances comparable to the dimensions of a cell. Compare to Problem 24.36.

P24.20   Kohlrausch's law states that the molar conductance of a strong electrolyte varies with the square root of concentration

$$\Lambda_m = \Lambda_m^\circ - \mathcal{K} c^{1/2}$$

Therefore, a plot of Λ_m versus $c^{1/2}$ should be a straight line with y-intercept Λ_m°. The data and plot (Fig. 24.2) are shown below

NaI			KI		
$c/(mmol\,L^{-1})$	$c^{1/2}$	$\Lambda_m/(S\,cm^2\,mol^{-1})$	$c/(mmol\,L^{-1})$	$c^{1/2}$	$\Lambda_m/(S\,cm^2\,mol^{-1})$
32.02	5.659	50.26	17.68	4.205	42.45
20.28	4.503	51.99	10.88	3.298	45.91
12.06	3.473	54.01	7.19	2.68	47.53
8.64	2.94	55.75	2.67	1.63	51.81
2.85	1.69	57.99	1.28	1.13	54.09
1.24	1.11	58.44	0.83	0.91	55.78
0.83	0.91	58.67	0.19	0.44	57.42

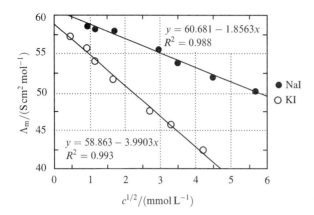

Figure 24.2

Thus $\Lambda_m^\circ(\text{NaI}) = \boxed{60.7\,\text{S cm}^2\,\text{mol}^{-1}}$ and $\Lambda_m^\circ(\text{KI}) = \boxed{58.9\,\text{S cm}^2\,\text{mol}^{-1}}$

Since these two electrolytes have a common anion, the difference in conductances is due to the cations

$$\lambda^\circ(\text{Na}^+) - \lambda^\circ(\text{K}^+) = \Lambda_m^\circ(\text{NaI}) - \Lambda_m^\circ(\text{KI}) = \boxed{1.8\,\text{S cm}^2\,\text{mol}^{-1}}$$

The analogous quantities in water are

$$\Lambda_m^\circ(\text{NaI}) = \lambda(\text{Na}^+) + \lambda(\text{I}^-) = (50.10 + 76.8)\,\text{S cm}^2\,\text{mol}^{-1} = \boxed{126.9\,\text{S cm}^2\,\text{mol}^{-1}}$$

$$\Lambda_m^\circ(\text{KI}) = \lambda(\text{K}^+) + \lambda(\text{I}^-) = (73.50 + 76.8)\,\text{S cm}^2\,\text{mol}^{-1} = \boxed{150.3\,\text{S cm}^2\,\text{mol}^{-1}}$$

$$\lambda^\circ(\text{Na}^+) - \lambda^\circ(\text{K}^+) = (50.10 - 73.50)\,\text{S cm}^2\,\text{mol}^{-1} = \boxed{-23.4\,\text{S cm}^2\,\text{mol}^{-1}}$$

The ions are considerably more mobile in water than in this solvent. Also, the differences between Na^+ and K^+ are minimized and even inverted compared to water.

Solutions to theoretical problems

P24.23   The most probable speed of a gas molecule corresponds to the condition that the Maxwell distribution be a maximum (it has no minimum); hence we find it by setting the first derivative of the function to zero and solve for be value of v for which this condition holds.

$$f(v) = 4\pi \left(\frac{m}{2\pi kT}\right)^{3/2} v^2 e^{-mv^2/2kT} = \text{const} \times v^2 e^{-mv^2/2kT} \quad \left[\frac{M}{R} = \frac{m}{k}\right]$$

$$\frac{df(v)}{ds} = 0 \quad \text{when} \left(2 - \frac{mv^2}{kT}\right) = 0$$

So, $\boxed{v(\text{most probable}) = c^* = \left(\dfrac{2kT}{m}\right)^{1/2} = \left(\dfrac{2RT}{M}\right)^{1/2}}$

The average kinetic energy corresponds to the average of $\frac{1}{2}mv^2$. The average is obtained by

determining $\langle v^2 \rangle = \displaystyle\int_0^\infty v^2 f(v)\,dv = 4\pi \left(\frac{m}{2\pi}\right)^{3/2} \times \left(\frac{1}{kT}\right)^{3/2} \int_0^\infty v^4 e^{-mv^2/2kT}\,dv$

The integral evaluates to $\dfrac{3}{8}\pi^{1/2}\left(\dfrac{m}{2kT}\right)^{-5/2}$. Then

$$\langle v^2 \rangle = 4\pi \left(\frac{m}{2\pi}\right)^{3/2} \times \left(\frac{1}{kT}\right)^{3/2} \times \left(\frac{3}{8}\pi^{1/2}\right) \times \left(\frac{2kT}{m}\right)^{5/2} = \frac{3kT}{m}$$

thus $\langle \varepsilon \rangle = \frac{1}{2}m\langle v^2 \rangle = \frac{3}{2}kT$

P24.24 We proceed as in Section 24.1(a) except that, instead of taking a product of three one-dimensional distributions in order to get the three-dimensional distribution, we make a product of two one-dimensional distributions.

$$f(v_x, v_y)dv_x\, dv_y = f(v_x^2)f(v_y^2)\, dv_x\, dv_y = \left(\frac{m}{2\pi kT}\right) e^{-mv^2/2kT}\, dv_x\, dv_y$$

where $v^2 = v_x^2 + v_y^2$. The probability $f(v)\, dv$ that the molecules have a two-dimensional speed, v, in the range $v, v + dv$ is sum of the probabilities that it is in any of the area elements $dv_x\, dv_y$ in the circular shell of raidus v. The sum of the area elements is the area of the circular shell of radius v and thickness dv which is $\pi(v + dv)^2 - \pi v^2 = 2\pi v\, dv$. Therefore

$$\boxed{f(v) = 2\pi \left(\frac{m}{2\pi kT}\right) v e^{-mv^2/2kT}} \qquad \boxed{\left[\frac{M}{R} = \frac{m}{k}\right]}$$

The mean speed is determined as $\bar{c} = \displaystyle\int_0^\infty v f(v)\, dv = \int_0^\infty \frac{m}{kT} v^2 e^{-mv^2/2kT}\, dv$

Using standard integrals this evaluates to $\boxed{\bar{c} = \left(\dfrac{\pi kT}{2m}\right)^{1/2} = \left(\dfrac{\pi RT}{2M}\right)^{1/2}}$

Comment. The two-dimensional gas serves as a model of the motion of molecules of surfaces. See Chapter 28.

P24.26 Rewriting eqn 24.4 with $\left(\dfrac{M}{R}\right) = \left(\dfrac{m}{k}\right)$

$$f(v) = 4\pi \left(\frac{m}{2\pi kT}\right)^{3/2} v^2 e^{-mv^2/2kT}$$

The proportion of molecules with speeds less than c is

$$P = \int_0^c f(v)\, dv = 4\pi \left(\frac{m}{2\pi kT}\right)^{3/2} \int_0^c v^2 e^{-mv^2/2kT}\, dv$$

Defining $a \equiv \dfrac{m}{2kT}$

$$P = 4\pi \left(\frac{a}{\pi}\right)^{3/2} \int_0^c v^2 e^{-av^2}\, dv = -4\pi \left(\frac{a}{\pi}\right)^{3/2} \frac{d}{da} \int_0^c e^{-av^2}\, dv$$

Defining $x^2 \equiv av^2$, $dv = a^{-1/2}\, dx$

$$P = -4\pi \left(\frac{a}{\pi}\right)^{3/2} \frac{d}{da} \left\{ \frac{1}{a^{1/2}} \int_0^{ca^{1/2}} e^{-x^2}\, dx \right\}$$

$$= -4\pi \left(\frac{a}{\pi}\right)^{3/2} \left\{ -\frac{1}{2} \left(\frac{1}{a}\right)^{3/2} \int_0^{ca^{1/2}} e^{-x^2}\, dx + \left(\frac{1}{a}\right)^{1/2} \frac{d}{da} \int_0^{ca^{1/2}} e^{-x^2}\, dx \right\}$$

Then we use $\displaystyle\int_0^{ca^{1/2}} e^{-x^2}\, dx = \left(\frac{\pi^{1/2}}{2}\right) \operatorname{erf}(ca^{1/2})$

$$\frac{d}{da} \int_0^{ca^{1/2}} e^{-x^2}\, dx = \left(\frac{dca^{1/2}}{da}\right) \times (e^{-c^2 a}) = \frac{1}{2} \left(\frac{c}{a^{1/2}}\right) e^{-c^2 a}$$

where we have used $\dfrac{d}{dz} \displaystyle\int_0^z f(y)dy = f(z)$

Substituting and cancelling we obtain $P = \text{erf}(ca^{1/2}) - \dfrac{2ca^{1/2}}{\pi^{1/2}} e^{-c^2 a}$

Now, $c = \left(\dfrac{3kT}{m}\right)^{1/2}$, so $ca^{1/2} = \left(\dfrac{3kT}{m}\right)^{1/2} \times \left(\dfrac{m}{2kT}\right)^{1/2} = \left(\dfrac{3}{2}\right)^{1/2}$, and

$$P = \text{erf}\left(\sqrt{\dfrac{3}{2}}\right) - \left(\dfrac{6}{\pi}\right)^{1/2} e^{-3/2} = 0.92 - 0.31 = \boxed{0.61}$$

Therefore (b) $\boxed{61 \text{ per cent}}$ of the molecules have a speed less than the root mean square speed and
(a) $\boxed{39 \text{ per cent}}$ have a speed greater than the root mean square speed. (c) For the proportions in
terms of the mean speed $\bar{c}$, replace c by $\bar{c} = \left(\dfrac{8kT}{\pi m}\right)^{1/2} = \left(\dfrac{8}{3\pi}\right)^{1/2} c$, so $\bar{c}a^{1/2} = \dfrac{2}{\pi^{1/2}}$.

Then $P = \text{erf}(\bar{c}a^{1/2}) - \left(\dfrac{2\bar{c}a^{1/2}}{\pi^{1/2}}\right) \times (e^{-\bar{c}^2 a}) = \text{erf}\left(\dfrac{2}{\pi^{1/2}}\right) - \dfrac{4}{\pi} e^{-4/\pi} = 0.889 - 0.356 = \boxed{0.533}$

That is, $\boxed{53 \text{per cent}}$ of the molecules have a speed less than the mean, and $\boxed{47 \text{per cent}}$ have a speed
greater than the mean.

P24.30 $p(x) = \dfrac{N!}{\left\{\frac{1}{2}(N+s)\right\}! \left\{\frac{1}{2}(N-s)\right\}! 2^N}$ [*Justification* 24.12], $s = \dfrac{x}{\lambda}$

$$p(6d) = \dfrac{N!}{\left\{\frac{1}{2}(N+6)\right\}! \left\{\frac{1}{2}(N-6)\right\}! 2^N}$$

(a) $N = 4$, $p(6\lambda) = \boxed{0}$ $(m! = \infty \text{ for } m < 0)$

(b) $N = 6$, $p(6\lambda) = \dfrac{6!}{6!0!2^6} = \dfrac{1}{2^6} = \dfrac{1}{64} = \boxed{0.016}$

(c) $N = 12$, $p(6\lambda) = \dfrac{12!}{9!3!2^{12}} = \dfrac{12 \times 11 \times 10}{3 \times 2 \times 2^{12}} = \boxed{0.054}$

(NB $0! = 1$)

P24.32 $AB \rightleftharpoons A^+ + B^-$; $\gamma_{AB} \simeq 1$ because AB interacts weakly with ions.

$$K = \dfrac{a_{A^+} a_{B^-}}{a_{AB}} = \left(\dfrac{\gamma_A - \gamma_{B^+}}{\gamma_{AB}}\right) \times \left(\dfrac{c_A - c_{B^+}}{c_{AB}}\right)$$

$$K = \gamma_\pm^2 \left[\dfrac{(\alpha c)(\alpha c)}{(1-\alpha)c}\right] = \gamma_\pm^2 \left(\dfrac{\alpha^2 c}{1-\alpha}\right) \quad \text{or} \quad \dfrac{\gamma_\pm^2 c}{K} = \dfrac{1-\alpha}{\alpha^2}$$

$$\Lambda_m = \dfrac{\kappa}{c} = \dfrac{(\lambda_+ + \lambda_-)c_{ion}}{c} = \dfrac{(\lambda_+ + \lambda_-)\alpha c}{c} = (\lambda_+ + \lambda_-)\alpha$$

Let $\Lambda_m^\circ = \lambda_+ + \lambda_-$ be the molar conductivity when the solution is infinitely dilute and $\alpha = 1$ (eqn
24.40). Then, $\alpha = \dfrac{\Lambda_m}{\lambda_+ + \lambda_-} = \dfrac{\Lambda_m}{\Lambda_m^\circ}$. Substitution into equilibrium expression gives:

$$K = \gamma_\pm^2 c \left(\dfrac{\Lambda_m}{\Lambda_m^\circ}\right)^2 \left(\dfrac{1}{1 - \frac{\Lambda_m}{\Lambda_m^\circ}}\right)$$

$$1 - \frac{\Lambda_m}{\Lambda_m^\circ} = \left(\frac{\Lambda_m}{\Lambda_m^\circ}\right)^2 \frac{\gamma_\pm^2 c}{K} = \left(\frac{\Lambda_m}{\Lambda_m^\circ}\right)^2 \left(\frac{1-\alpha}{\alpha^2}\right)$$

Division by Λ_m gives:

$$\frac{1}{\Lambda_m} - \frac{1}{\Lambda_m^\circ} = \left(\frac{1-\alpha}{\alpha^2}\right) \frac{\Lambda_m}{(\Lambda_m^\circ)^2}$$

$$\frac{1}{\Lambda_m} = \frac{1}{\Lambda_m^\circ} + \left(\frac{1-\alpha}{\alpha^2}\right) \frac{\Lambda_m}{(\Lambda_m^\circ)^2}$$

Solutions to applications

P24.34   The diffusion coefficient for a perfect gas is

$$D = \tfrac{1}{3}\lambda\bar{c} \quad \text{where } \lambda = (2^{1/2}\sigma\mathcal{N})^{-1} \text{ where } \mathcal{N} \text{ is number density.}$$

The mean speed is

$$\bar{c} = \left(\frac{8kT}{\pi m}\right)^{1/2} = \left(\frac{8(1.381 \times 10^{-23}\,\text{J K}^{-1}) \times (10^4\,\text{K})}{\pi(1\,\text{u}) \times (1.66 \times 10^{-27}\,\text{kg u}^{-1})}\right)^{1/2} = 1.46 \times 10^4\,\text{m s}^{-1}$$

So $D = \dfrac{\bar{c}}{3\sigma\mathcal{N}2^{1/2}} = \dfrac{1.46 \times 10^4\,\text{m s}^{-1}}{3(0.21 \times 10^{-18}\,\text{m}^2) \times (1 \times (10^{-2}\,\text{m})^{-3})2^{1/2}} = \boxed{1.\bar{6} \times 10^{16}\,\text{m}^2\,\text{s}^{-1}}$

The thermal conductivity is

$$\kappa = \frac{\bar{c}C_{V,m}}{3\sigma N_A 2^{1/2}} = \frac{(1.46 \times 10^4\,\text{m s}^{-1}) \times (20.784 - 8.3145)\,\text{J K}^{-1}\,\text{mol}^{-1}}{3(0.21 \times 10^{-18}\,\text{m}^2) \times (6.022 \times 10^{23}\,\text{mol}^{-1})2^{1/2}}$$

$$\kappa = \boxed{0.3\underline{4}\,\text{J K}^{-1}\,\text{m}^{-1}\,\text{s}^{-1}}$$

Comment. The validity of these calculations is in doubt because the kinetic theory of gases assumes the Maxwell–Boltzmann distribution, essentially an equilibrium distribution. In such a dilute medium, the timescales on which particles exchange energy by collision make an assumption of equilibrium unwarranted. It is especially dubious considering that atoms are more likely to interact with photons from stellar radiation than with other atoms.

P24.35   For order of magnitude calculations we restrict our assumed values to powers of 10 of the base units. Thus

$$\rho = 1\,\text{g cm}^{-3} = 1 \times 10^3\,\text{kg m}^{-3}$$

$$\eta(\text{air}) = 1 \times 10^{-5}\,\text{kg m}^{-1}\,\text{s}^{-1} \text{ [See comment and question below.]}$$

We need the diffusion constant

$$D = \frac{kT}{6\pi\eta a}$$

a is calculated from the volume of the virus which is assumed to be spherical

$$V = \frac{m}{\rho} \approx \frac{(1 \times 10^5\,\text{u}) \times (1 \times 10^{-27}\,\text{kg u}^{-1})}{1 \times 10^3\,\text{kg m}^3} \approx 1 \times 10^{-25}\,\text{m}^3$$

$$V = \tfrac{4}{3}\pi a^3$$

$$a \approx \left(\frac{V}{4}\right)^{1/3} \approx \left(\frac{1 \times 10^{-25}\,\text{m}^3}{4}\right)^{1/3} \approx 1 \times 10^{-8}\,\text{m}$$

$$D \approx \left(\frac{(1 \times 10^{-23}\,\text{J K}^{-1}) \times (300\,\text{K})}{(6\pi) \times (1 \times 10^{-5}\,\text{kg m}^{-1}\,\text{s}^{-1}) \times (1 \times 10^{-8}\,\text{m})}\right) \approx 1 \times 10^{-9}\,\text{m}^2\,\text{s}^{-1}$$

For three-dimensional diffusion

$$t = \frac{\langle r^2 \rangle}{6D} \approx \frac{1\,\text{m}^2}{1 \times 10^{-8}\,\text{m}^2\,\text{s}^{-1}} \approx \boxed{10^8\,\text{s}}$$

Therefore it does not seem likely that a cold could be caught by the process of diffusion.

Comment. In a Fermi calculation only those values of physical quantities that can be determined by scientific common sense should be used. Perhaps the value for $\eta(\text{air})$ used above does not fit that description.

Question. Can you obtain the value of $\eta(\text{air})$ by a Fermi calculation based on the relation in Table 24.3?

P24.36   The mean square displacement is (from Exercise 24.31(b))

$$\langle r^2 \rangle = 6Dt \quad \text{so} \quad t = \frac{\langle r^2 \rangle}{6D} = \frac{(1.0 \times 10^{-6}\,\text{m})^2}{6(1.0 \times 10^{-11}\,\text{m}^2\,\text{s}^{-1})} = \boxed{1.7 \times 10^{-2}\,\text{s}}$$

P24.37 $c(x, t) = c_0 + (c_s - c_0)\{1 - \text{erf}(\xi)\}$ where $\xi(x, t) = \dfrac{x}{(4Dt)^{1/2}}$

In order for $c(x, t)$ to be the correct solution of this diffusion problem it must satisfy the boundary condition, the initial condition, and the diffusion equation (eqn 24.84). According to *Justification* 12.3,

$$\text{erf}(\xi) = 1 - \frac{2}{\pi^{1/2}} \int_{\xi}^{\infty} e^{y^2}\,dy$$

at the boundary $x = 0$, $\xi = 0$, and $\text{erf}(0) = 1 - \dfrac{2}{\pi^{1/2}} \displaystyle\int_{0}^{\infty} e^{-y^2}\,dy = 1 - \left(\dfrac{2}{\pi^{1/2}}\right) \times \left(\dfrac{\pi^{1/2}}{2}\right) = 0.$

Thus, $c(0, t) = c_0 + (c_s - c_0)\{1 - 0\} = c_s$. The boundary condition is satisfied. At the initial time $(t = 0)$, $\xi(x, 0) = \infty$ and $\text{erf}(\infty) = 1$. Thus, $c(x, 0) = c_0 + (c_s - c_0)\{1 - 1\} = c_0$. The initial condition is satisfied. We must find the analytical forms for $\partial c/\partial t$ and $\partial^2 c/\partial x^2$. If they are proportional with a constant of proportionality equal to D, $c(x, t)$ satisfies the diffusion equation.

$$\frac{\partial c(x, t)}{\partial x} = D\left[\frac{1}{2}\frac{(c_s - c_0)x}{\sqrt{\pi}(Dt)^{3/2}}e^{-\frac{x^2}{4Dt}}\right]$$

$$\frac{\partial^2 c(x, t)}{\partial x^2} = \left[\frac{1}{2}\frac{(c_s - c_0)x}{\sqrt{\pi}(Dt)^{3/2}}e^{-\frac{x^2}{4Dt}}\right]$$

The constant of proportionality between the partials equals D and we conclude that the suggested solution satisfies the diffusion equation.

Diffusion through alveoli sites (about 1 cell thick) of oxygen and carbon dioxide between lungs and blood capillaries (also about 1 cell thick) occurs through about 0.075 mm (the diameter of a red blood cell). So we will examine diffusion profiles for $0 \leq x \leq 0.1$ mm. The largest distance suggests that the longest time that must be examined is estimated with eqn 24.90.

$$t_{max} \simeq \frac{\pi x_{max}^2}{4D} = \frac{\pi (1 \times 10^{-4}\,\text{m})^2}{4(2.10 \times 10^{-9}\,\text{m}^2\,\text{s}^{-1})} = 3.74\,\text{s}$$

The following graph shows oxygen concentration distributions for times between 0.01 s and 4.0 s.

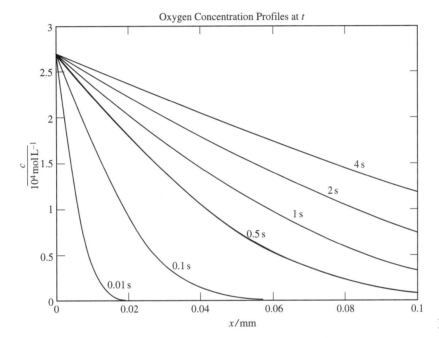

Figure 24.3

Example 7.4 uses Henry's law to show that the equilibrium concentration of oxygen in water equals $2.7 \times 10^{-4}\,\text{mol L}^{-1}$. We use this as an estimate for c_s and take c_0 to equal zero.

25 The rates of chemical reactions

Solutions to exercises

Discussion questions

E25.1(a) No solution.

E25.2(a) The overall reaction order is the sum of the powers of the concentrations of all of the substances appearing in the *experimental* rate law for the reaction (eqn 25.5); hence, it is the sum of the individual orders (exponents) associated with a given reactant (or product). Reaction order is an experimentally determined, *not theoretical*, quantity, although theory may attempt to predict it. *Molecularity* is the number of reactant molecules participating in an elementary reaction. This concept has meaning only for an elementary reaction, but reaction order applies to any reaction. In general, reaction order bears no necessary relation to the stoichiometry of the reaction, with the exception of elementary reactions, where the order of the reaction corresponds to the number of molecules participating in the reaction; that is, to its molecularity. Thus for an elementary reaction, overall order and molecularity are the same and are determined by the stoichiometry.

E25.3(a) The steady-state approximation is the assumption that the rate of change of the concentrations of intermediates in consecutive chemical reactions is negligibly small. It is a good approximation when at least one of the reaction steps involving the intermediate is very fast, that is, has a large rate constant relative to other steps. See Section 25.7(c). A pre-equilibrium approximation is similar in that it is a good approximation when the rate of formation of the intermediate from the reactants and the rate of its reversible decay back to the reactions are both very fast in comparison to the rate of formation of the product from the intermediate. This results in the intermediate being in approximate equilibrium with the reactants over relatively long time periods (though short compared to the overall time scale of the reaction). Hence the concentration of the intermediate remains approximately constant over the time period that the equilibrium can be considered to be maintained. This allows one to relate the rate constants and concentrations to each other through a constant (the pre-equilibrium constant). See Section 25.7(d).

E25.4(a) The primary isotope effect is the change in rate constant of a reaction in which the breaking of a bond involving the isotope occurs. The reaction coordinate in a C—H bond breaking process corresponds to the stretching of that bond. The vibrational energy of the stretching depends upon the effective mass of the C and H atoms. See eqn 16.57. Upon deuteration, the zero point energy of the bond is lowered due to the greater mass of the deuterium atom. However the height of the energy barrier is not much changed because the relevant vibration in the activated complex has a very low force constant (bonding in the complex is very weak), so there is little zero point energy associated with the complex and little change in its zero point energy upon deuteration. The net effect is an increase in the activation energy of the reaction. We then expect that the rate constant for the reaction will be lowered in the deuterated molecule and that is what is observed. See the derivation leading to eqns 25.47–25.49 for a quantitative description of the effect.

A secondary kinetic isotope effect is the reduction in the rate of a reaction involving the bonded isotope even though the bond is not broken in the reaction. The cause is again related to the change in zero point energy that occurs upon replacement of an atom with its isotope, but in this case it arises from the differences in zero point energies between reactants and an activated complex with significantly different structure. See *Illustration* 25.3 for an example of the estimation of the magnitude of the effect in a heterolytic dissociation reaction.

If the rate of a reaction is altered by isotopic substitution it implies that the substituted site plays an important role in the mechanism of the reaction. For example, an observed effect on the rate can identify bond breaking events in the rate determining step of the mechanism. On the other hand, if no isotope effect is observed, the site of the isotopic substitution may play no critical role in the mechanism of the reaction.

Numerical exercises

E25.5(a) $v = \dfrac{1}{v_J} \dfrac{d[J]}{dt}$ [25.1c] so $\dfrac{d[J]}{dt} = v_J v$

The reaction has the form

$$0 = 3C + D - A - 2B$$

Rate of formation of C $= 3v = \boxed{3.0\,\text{mol}\,\text{L}^{-1}\,\text{s}^{-1}}$

Rate of formation of D $= v = \boxed{1.0\,\text{mol}\,\text{L}^{-1}\text{s}^{-1}}$

Rate of consumption of A $= v = \boxed{1.0\,\text{mol}\,\text{L}^{-1}\text{s}^{-1}}$

Rate of consumption of B $= 2v = \boxed{2.0\,\text{mol}\,\text{L}^{-1}\text{s}^{-1}}$

E25.6(a) $v = \dfrac{1}{v_J} \dfrac{d[J]}{dt}$ [25.1c]

For the reaction $2A + B \rightarrow 2C + 3D$, $v_C = +2$; hence

$$v = \tfrac{1}{2} \times (1.0\,\text{M}\,\text{s}^{-1}) = \boxed{0.50\,\text{mol}\,\text{L}^{-1}\text{s}^{-1}}$$

Rate of formation of D $= 3v = \boxed{1.5\,\text{mol}\,\text{L}^{-1}\,\text{s}^{-1}}$

Rate of consumption of A $= 2v = \boxed{1.0\,\text{mol}\,\text{L}^{-1}\,\text{s}^{-1}}$

Rate of consumption of B $= v = \boxed{0.50\,\text{mol}\,\text{L}^{-1}\,\text{s}^{-1}}$

E25.7(a) The rate is expressed in $\text{mol}^{-1}\text{L}^{-1}\text{s}^{-1}$; therefore

$$\text{mol}\,\text{L}^{-1}\text{s}^{-1} = [k] \times (\text{mol}\,\text{L}^{-1}) \times (\text{mol}\,\text{L}^{-1}) \quad [[k] = \text{units of } k]$$

requires the units to be $\boxed{\text{L}\,\text{mol}^{-1}\,\text{s}^{-1}}$

(a) Rate of formation of A $= v = \boxed{k[A][B]}$ **(b)** Rate of consumption of C $= 3v = \boxed{3k[A][B]}$

E25.8(a) $\dfrac{d[C]}{dt} = k[A][B][C]$

$$v = \dfrac{1}{v_J} \dfrac{d[J]}{dt} \quad \text{with } v_J = v_C = 2$$

Therefore $v = \dfrac{1}{2}\dfrac{d[C]}{dt} = \boxed{\dfrac{1}{2}k[A][B][C]}$

The units of k, $[k]$, must satisfy

$$\text{mol}\,\text{L}^{-1}\text{s}^{-1} = [k] \times (\text{mol}\,\text{L}^{-1}) \times (\text{mol}\,\text{L}^{-1}) \times (\text{mol}\,\text{L}^{-1})$$

Therefore, $[k] = \boxed{\text{L}^2\,\text{mol}^{-2}\,\text{s}^{-1}}$

E25.9(a) For A $\rightarrow$ Products

$$v_A = -\dfrac{d[A]}{dt} = k[A]^a$$

Since concentration and partial pressure are proportional to each other we may write

$$v_A = -\frac{dp_A}{dt} = kp_A^a$$

and $\dfrac{v_{A,1}}{v_{A,2}} = \dfrac{p_{A,1}^a}{p_{A,2}^a} = \left(\dfrac{p_{A,1}}{p_{A,2}}\right)^a$

Taking logarithms

$$\log\left(\frac{v_{A,1}}{v_{A,2}}\right) = a\log\left(\frac{p_{A,1}}{p_{A,2}}\right)$$

$$a = \frac{\log\left(\frac{v_{A,1}}{v_{A,2}}\right)}{\log\left(\frac{p_{A,1}}{p_{A,2}}\right)} = \frac{\log\left(\frac{1.07}{0.76}\right)}{\log\left(\frac{0.95}{0.80}\right)} = 1.9\overline{9}$$

Hence, the reaction is $\boxed{\text{second-order}}$

Comment. Knowledge of the initial pressure is not required for the solution to this exercise.

E25.10(a) The general expression for the half-life of a reaction of the type $A \rightarrow P$ is

$$t_{1/2} = \frac{2^{n-1} - 1}{(n-1)k[A]_0^{n-1}} \text{ [Table 25.3]} = f(n,k)[A]_0^{1-n}$$

where $f(n,k) = \dfrac{2^{n-1} - 1}{(n-1)k}$. Then

$$\log t_{1/2} = \log f + (1-n)\log p_0 \quad [p_0 \propto [A]_0]$$

Hence, $\log\left(\dfrac{t_{1/2}(p_{0,1})}{t_{1/2}(p_{0,2})}\right) = (1-n)\log\left(\dfrac{p_{0,1}}{p_{0,2}}\right) = (n-1)\log\left(\dfrac{p_{0,2}}{p_{0,1}}\right)$

or $(n-1) = \dfrac{\log\left(\frac{410}{880}\right)}{\log\left(\frac{169}{363}\right)} = 0.999 \approx 1$

Therefore, $\boxed{n = 2}$ in agreement with the result of Exercise 25.9(a).

E25.11(a) $2N_2O_5 \rightarrow 4NO_2 + O_2 \quad v = k[N_2O_5]$

Therefore, rate of consumption of $N_2O_5 = 2v = 2k[N_2O_5]$ [1]

$$\frac{d[N_2O_5]}{dt} = -2k[N_2O_5]$$

$$[N_2O_5] = [N_2O_5]_0 e^{-2kt}$$

which implies that $t = \dfrac{1}{2k}\ln\dfrac{[N_2O_5]_0}{[N_2O_5]}$

and therefore that $t_{1/2} = \dfrac{1}{2k}\ln 2 = \dfrac{\ln 2}{(2)\times(3.38\times 10^{-5}\,\text{s}^{-1})} = \boxed{1.03\times 10^4\,\text{s}}$

Since the partial pressure of N_2O_5 is proportional to its concentration,

$$p(N_2O_5) = p_0(N_2O_5)e^{-2kt}$$

(a) $\quad p(N_2O_5) = (500\,\text{Torr}) \times \left(e^{-(6.76\times10^{-5})\times10}\right) = \boxed{499.\overline{7}\,\text{Torr}}$

(b) $\quad p(N_2O_5) = (500\,\text{Torr}) \times \left(e^{-(6.76\times10^{-5})\times600}\right) = \boxed{480\,\text{Torr}}$

E25.12(a) We use $kt = \dfrac{1}{[B]_0 - [A]_0} \ln\left\{\left(\dfrac{[B]}{[B]_0}\right) \bigg/ \left(\dfrac{[A]}{[A]_0}\right)\right\}$ [25.16]

(a) The stoichiometry of the reaction requires that when $\Delta[A] = (0.050 - 0.020)\,\text{mol L}^{-1} = 0.030$ mol L^{-1}, $\Delta[B] = 0.030\,\text{mol L}^{-1}$ as well. Thus $[B] = 0.080\,\text{mol L}^{-1} - 0.030\,\text{mol L}^{-1} = 0.050\,\text{mol L}^{-1}$ when $[A] = 0.20\,\text{mol L}^{-1}$. Thus,

$$kt = \left(\frac{1}{(0.080 - 0.050)\,\text{mol L}^{-1}}\right) \ln\left\{\left(\frac{0.050}{0.080}\right) \bigg/ \left(\frac{0.020}{0.050}\right)\right\}$$

$$k \times 1.0\,\text{h} = 14.88\,\text{L mol}^{-1}$$

$$k = 14.\overline{9}\,\text{L mol}^{-1}\,\text{h}^{-1} = \boxed{4.1 \times 10^{-3}\,\text{L mol}^{-1}\,\text{s}^{-1}}$$

(b) The half-life with respect to A is the time required for $[A]$ to fall to $0.025\,\text{mol L}^{-1}$. We solve eqn 15 for t

$$t_{1/2}(A) = \left(\frac{1}{(14.\overline{9}\,\text{L mol}^{-1}\,\text{h}^{-1}) \times (0.030\,\text{mol L}^{-1})}\right) \times \ln\left\{\left(\frac{0.055}{0.080}\right)\bigg/0.50\right\}$$

$$= 0.71\overline{2}\,\text{h} = \boxed{2.6 \times 10^3\,\text{s}}$$

Similarly, $t_{1/2}(B) = \left(\dfrac{1}{0.44\overline{7}\,\text{h}^{-1}}\right) \ln\left\{0.50\bigg/\left(\dfrac{0.010}{0.050}\right)\right\} = 2.0\overline{5}\,\text{h} = \boxed{7.4 \times 10^3\,\text{s}}$

Comment. This exercise illustrates that there is no unique half-life for reactions other than those of the type $A \rightarrow P$.

E25.13(a) **(a)** For a second-order reaction, denoting the units of k by $[k]$

$$\text{mol L}^{-1}\,\text{s}^{-1} = [k] \times (\text{mol L}^{-1})^2, \quad \text{therefore} \quad [k] = \boxed{\text{L mol}^{-1}\,\text{s}^{-1}}$$

For a third-order reaction

$$\text{mol L}^{-1}\,\text{s}^{-1} = [k] \times (\text{mol L}^{-1})^3, \quad \text{therefore} \quad [k] = \boxed{\text{L}^2\,\text{mol}^{-2}\,\text{s}^{-1}}$$

(b) For a second-order reaction

$$\text{kPa s}^{-1} = [k] \times \text{kPa}^2, \quad \text{therefore} \quad [k] = \boxed{\text{kPa}^{-1}\,\text{s}^{-1}}$$

For a third-order reaction

$$\text{kPa s}^{-1} = [k] \times \text{kPa}^3, \quad \text{therefore} \quad [k] = \boxed{\text{kPa}^{-2}\,\text{s}^{-1}}$$

E25.14(a) For a reaction of the type $A + B \rightarrow$ products we use

$$kt = \left(\frac{1}{[B]_0 - [A]_0}\right) \ln\left\{\left(\frac{[B]}{[B]_0}\right) \Big/ \left(\frac{[A]}{[A]_0}\right)\right\} \quad [25.16]$$

Introducing $[B] = [B]_0 - x$ and $[A] = [A]_0 - x$ and rearranging we obtain

$$kt = \left(\frac{1}{[B]_0 - [A]_0}\right) \ln\left(\frac{[A]_0([B]_0 - x)}{([A]_0 - x)[B]_0}\right)$$

Solving for x

$$x = \frac{[A]_0[B]_0\left\{e^{k([B]_0-[A]_0)t} - 1\right\}}{[B]_0 e^{([B]_0-[A]_0)kt} - [A]_0} = \frac{(0.050) \times (0.100\,\text{mol L}^{-1}) \times \left\{e^{(0.100-0.050)\times 0.11 \times t/s} - 1\right\}}{(0.100) \times \left\{e^{(0.100-0.050)\times 0.11 \times t/s}\right\} - 0.050}$$

$$= \frac{(0.100\,\text{mol L}^{-1}) \times (e^{5.5\times 10^{-3}\,t/s} - 1)}{2e^{5.5\times 10^{-3}\,t/s} - 1}$$

(a) $$x = \frac{(0.100\,\text{mol L}^{-1}) \times (e^{0.055} - 1)}{2e^{0.055} - 1} = 5.1 \times 10^{-3}\,\text{mol L}^{-1}$$

which implies that $[\text{NaOH}] = (0.050 - 0.0051)\,\text{mol L}^{-1} = \boxed{0.045\,\text{mol L}^{-1}}$ and

$[\text{CH}_3\text{COOC}_2\text{H}_5] = (0.100 - 0.0051)\,\text{mol L}^{-1} = \boxed{0.095\,\text{mol L}^{-1}}$

(b) $$x = \frac{(0.100\,\text{mol L}^{-1}) \times (e^{3.3} - 1)}{2e^{3.3} - 1} = 0.049\,\text{mol L}^{-1}$$

Hence, $[\text{NaOH}] = (0.050 - 0.049)\,\text{mol L}^{-1} = \boxed{0.001\,\text{mol L}^{-1}}$

$[\text{CH}_3\text{COOC}_2\text{H}_5] = (0.100 - 0.049)\,\text{mol L}^{-1} = \boxed{0.051\,\text{mol L}^{-1}}$

E25.15(a) The rate of consumption of A is

$$\frac{d[A]}{dt} = -2k[A]^2 \quad [\nu_A = -2]$$

which integrates to $\dfrac{1}{[A]} - \dfrac{1}{[A]_0} = 2kt$ [25.13b with k replaced by $2k$]

Therefore, $t = \dfrac{1}{2k}\left(\dfrac{1}{[A]} - \dfrac{1}{[A]_0}\right) = \left(\dfrac{1}{(2) \times (3.50 \times 10^{-4}\,\text{L mol}^{-1}\,\text{s}^{-1})}\right)$

$$\times \left(\frac{1}{0.011\,\text{mol L}^{-1}} - \frac{1}{0.260\,\text{mol L}^{-1}}\right) = \boxed{1.24 \times 10^5\,\text{s}}$$

E25.16(a) $\ln k = \ln A - \dfrac{E_a}{RT}$ [25.25] $\ln k' = \ln A - \dfrac{E_a}{RT'}$

Hence, $E_a = \dfrac{R \ln\left(\frac{k'}{k}\right)}{\left(\frac{1}{T} - \frac{1}{T'}\right)} = \dfrac{(8.314\,\text{J K}^{-1}\,\text{mol}^{-1}) \times \ln\left(\frac{1.38\times 10^{-2}}{2.80\times 10^{-3}}\right)}{\frac{1}{303\,\text{K}} - \frac{1}{323\,\text{K}}} = \boxed{64.9\,\text{kJ mol}^{-1}}$

For A, we use

$$A = k \times e^{E_a/RT} [25.27] = (2.80 \times 10^{-3}\,\text{mol L}^{-1}\,\text{s}^{-1}) \times e^{64.9\times 10^3/(8.314\times 303)}$$

$$= \boxed{4.32 \times 10^8\,\text{mol L}^{-1}\,\text{s}^{-1}}$$

E25.17(a) If cleavage of a C—D or C—H bond is involved in the rate-determining step then use

$$\frac{k(C-D)}{k(C-H)} = e^{-\lambda}, \quad \lambda = \left(\frac{\hbar k_f^{1/2}}{2kT}\right) \times \left(\frac{1}{\mu_{CH}^{1/2}} - \frac{1}{\mu_{CD}^{1/2}}\right)$$

$$\left[25.49 \text{ with } \tilde{\nu} = \frac{1}{2\pi C}\left(\frac{k_f}{\mu_{CH}}\right)^{1/2} \text{ (C-H)}\right]$$

and see if this accounts for the difference.

$$\mu_{CD} \approx \frac{2 \times 12}{2 + 12}\,u = 1.71\,u$$

$$\mu_{CH} \approx \frac{1 \times 12}{1 + 12}\,u = 0.92\,u$$

$$\lambda \approx \left(\frac{(1.054 \times 10^{-34}\,\text{J s}) \times (450\,\text{N m}^{-1})^{-1/2}}{(2) \times (1.381 \times 10^{-23}\,\text{J K}^{-1}) \times (298\,\text{K})}\right) \times \left(\frac{1}{0.92^{1/2}} - \frac{1}{1.71^{1/2}}\right)$$

$$\times \left(\frac{1}{1.6605 \times 10^{-27}\,\text{kg})^{1/2}}\right)$$

$$\approx 1.85$$

Hence, $\dfrac{k_2(D)}{k_2(H)} = e^{-1.85} = \boxed{0.156}$

That is, $k_2(H) \approx 6.4 \times k_2(D)$, in reasonable accord with the data.

E25.18(a) The rate of change of [A] is

$$\frac{d[A]}{dt} = -k[A]^n$$

Hence, $\displaystyle\int_{[A]_0}^{[A]} \frac{d[A]}{[A]^n} = -k \int_0^t dt = -kt$

Therefore, $kt = \left(\dfrac{1}{n-1}\right) \times \left(\dfrac{1}{[A]^{n-1}} - \dfrac{1}{[A]_0^{n-1}}\right)$

and $kt_{1/2} = \left(\dfrac{1}{n-1}\right) \times \left(\dfrac{2^{n-1}}{[A]_0^{n-1}} - \dfrac{1}{[A]_0^{n-1}}\right) = \left(\dfrac{2^{n-1}-1}{n-1}\right) \times \left(\dfrac{1}{[A]_0^{n-1}}\right)$ [as in Table 25.3]

Hence, $\boxed{t_{1/2} \propto \dfrac{1}{[A]_0^{n-1}}}$

E25.19(a) $\dfrac{1}{k} = \dfrac{1}{k_a p_A} + \dfrac{k_a'}{k_a k_b}$ [analogous to eqn 25.63]

Therefore, for two different pressures we have

$$\frac{1}{k} - \frac{1}{k'} = \frac{1}{k_a}\left(\frac{1}{p} - \frac{1}{p'}\right)$$

and hence $k_a = \dfrac{\left(\dfrac{1}{p} - \dfrac{1}{p'}\right)}{\left(\dfrac{1}{k} - \dfrac{1}{k'}\right)} = \dfrac{\left(\dfrac{1}{12\,\text{Pa}} - \dfrac{1}{1.30 \times 10^3\,\text{Pa}}\right)}{\left(\dfrac{1}{2.10 \times 10^{-5}\,\text{s}^{-1}} - \dfrac{1}{2.50 \times 10^{-4}\,\text{s}^{-1}}\right)} = 1.9 \times 10^{-6}\,\text{Pa}^{-1}\,\text{s}^{-1},$

or $\boxed{1.9\,\text{MPa}^{-1}\,\text{s}^{-1}}$

E25.20(a) $NH_4^+(aq) + H_2O(l) \rightleftharpoons NH_3(aq) + H_3O^+(aq)$ $pK_a = 9.25$

$NH_3(aq) + H_2O(l) \underset{k'}{\overset{k}{\rightleftharpoons}} NH_4^+(aq) + OH^-(aq)$

$pK_b = pK_w - pK_a = 14.00 - 9.25 = 4.75$

Therefore, $K_b = \dfrac{k}{k'} = 10^{-4.75} = 1.78 \times 10^{-5}\,mol\,L^{-1}$

and $k = k'K_b = (1.78 \times 10^{-5}\,mol\,L^{-1}) \times (4.0 \times 10^{10}\,L\,mol^{-1}s^{-1}) = \boxed{7.1 \times 10^5\,s^{-1}}$

$\dfrac{1}{\tau} = k + k'([NH_4^+] + [OH^-])$ [Example 25.4]

$= k + 2k'K_b^{1/2}[NH_3]^{1/2}$ $[[NH_4^+] = [OH^-] = (K_b[NH_3])^{1/2}]$

$= (7.1 \times 10^5\,s^{-1}) + (2) \times (4.0 \times 10^{10}\,L\,mol^{-1}\,s^{-1}) \times (1.78 \times 10^{-5}\,mol\,L^{-1})^{1/2}$

$\times\,(0.15\,mol\,L^{-1})^{1/2}$

$= 1.31 \times 10^8\,s^{-1};$ hence $\boxed{\tau = 7.61\,ns}$

Comment. The rate constant k corresponds to the pseudo-first-order protonation of NH_3 in excess water and hence has the units s^{-1}. Therefore, K_b in this problem must be assigned the units $mol\,L^{-1}$ to obtain proper cancellation of units in the equation for $\dfrac{1}{\tau}$.

Solutions to problems

Solutions to numerical problems

P25.1   A simple but practical approach is to make an initial guess at the order by observing whether the half-life of the reaction appears to depend on concentration. If it does not, the reaction is first-order; if it does, it may be second-order. Examination of the data shows that the first half-life is roughly 45 minutes, but that the second is about double the first. (Compare the $0 \to 50.0$ minute data to the $50.0 \to 150$ minute data.) Therefore, assume second-order and confirm by a plot of $\dfrac{1}{[A]}$ against time.

We draw up the following table ($A = NH_4CNO$)

t/min	0	20.0	50.0	65.0	150
$m(urea)/g$	0	7.0	12.1	13.8	17.7
$m(A)/g$	22.9	15.9	10.8	9.1	5.2
$[A]/(mol\,L^{-1})$	0.382	0.265	0.180	0.152	0.0866
$\dfrac{1}{[A]}/(L\,mol^{-1})$	2.62	3.77	5.56	6.59	11.5

The data are plotted in Fig. 25.1 and fit closely to a straight line. Hence, the reaction is $\boxed{\text{second-order}}$

The rate constant is the slope. A least-squares fit gives $\boxed{k = 0.059\overline{4}\,L\,mol^{-1}\,min^{-1}}$ At 300 min [A] = $0.049\,mol\,L^{-1}$. These calculations were performed on an inexpensive hand-held calculator which is pre-programmed to do linear regression (and other kinds too). The value of [A] at 300 min is provided

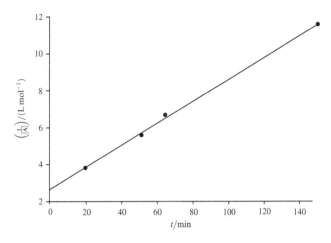

Figure 25.1

automatically by the calculator. It could be obtained by

$$\frac{1}{[A]} = kt + \frac{1}{[A]_0} \quad [25.13b]$$

or $[A] = \dfrac{[A]_0}{kt[A]_0 + 1} = \dfrac{0.382 \text{ mol L}^{-1}}{(0.059\overline{4}) \times (300) \times (0.382) + 1} = 0.049 \text{ mol L}^{-1}$

The mass of NH_4CNO left after 300 minutes is

$$\text{mass} = (0.048\overline{9} \text{ mol L}^{-1}) \times (1.00 \text{ L}) \times (60.06 \text{ g mol}^{-1}) = \boxed{2.94 \text{ g}}$$

P25.3 The procedure adopted in the solutions to Problems 25.1 and 25.2 is employed here. Examination of the data indicates that the half-life is independent of concentration and that the reaction is therefore first-order. That is confirmed by a plot of $\ln\left(\dfrac{[A]}{[A]_0}\right)$ against time.

We draw up the following table ($A = $ nitrile)

$t/(10^3 \text{ s})$	0	2.00	4.00	6.00	8.00	10.00	12.00
$[A]/(\text{mol L}^{-1})$	1.10	0.86	0.67	0.52	0.41	0.32	0.25
$\dfrac{[A]}{[A]_0}$	1	0.78	0.61	0.47	0.37	0.29	0.23
$\ln\left(\dfrac{[A]}{[A]_0}\right)$	0	-0.246	-0.496	-0.749	-0.987	-1.235	-1.482

A least-squares fit to a linear equation gives $k = \boxed{1.2\overline{3} \times 10^{-4} \text{ s}^{-1}}$ with a correlation coefficient of 1.000.

P25.5 As described in Example 25.5, if the rate constant obeys the Arrhenius equation [25.25], a plot of $\ln k$ against $\dfrac{1}{T}$ should yield a straight line with slope $\dfrac{-E_a}{R}$. However, since data are available only at three temperatures, we use the two-point method, that is

$$\ln \frac{k_2}{k_1} = -\frac{E_a}{R}\left(\frac{1}{T_2} - \frac{1}{T_1}\right)$$

which yields $E_a = \dfrac{-R \ln\left(\frac{k_2}{k_1}\right)}{\left(\frac{1}{T_2} - \frac{1}{T_1}\right)}$

For the pair $\theta = 0°C$ and $40°C$,

$$E_a = \frac{-R \ln\left(\frac{576}{2.46}\right)}{\left(\frac{1}{313\,K} - \frac{1}{273\,K}\right)} = 9.69 \times 10^4 \, J\,mol^{-1}$$

For the pair $\theta = 20°C$ and $40°C$,

$$E_a = \frac{-R \ln\left(\frac{576}{45.1}\right)}{\left(\frac{1}{313\,K} - \frac{1}{293\,K}\right)} = 9.71 \times 10^4 \, J\,mol^{-1}$$

The agreement of these values of E_a indicates that the rate constant data fits the Arrhenius equation and that the activation energy is $\boxed{9.70 \times 10^4 \, J\,mol^{-1}}$

P25.6   We have, since both reactions are first-order

$$-\frac{d[A]}{dt} = k_1[A] + k_2[A]$$

or $\dfrac{dx}{dt} = k_1([A]_0 - x) + k_2([A]_0 - x) \; [x = [A]_0 - [A]] = (k_1 + k_2) \times ([A]_0 - x)$

which integrates to $(k_1 + k_2)t = \ln \dfrac{[A]_0}{[A]_0 - x}$

Solving for x, $x = [A]_0 \left(1 - e^{-(k_1+k_2)t}\right)$

For reaction (2) above $x_2 = [A]_0(1 - e^{-k_2 t})$

At the start of the reaction, $t = 0$ and x and x_2 are both zero as well. When $t \to \infty$ we may expand both exponentials

$$x = [A]_0(k_1 + k_2)t \qquad x_2 = [A]_0 k_2 t$$

The fraction of the ketene formed is

$$\frac{x_2}{x} = \frac{k_2}{k_1 + k_2} = \frac{4.65\,s^{-1}}{(3.74\,s^{-1}) + (4.65\,s^{-1})} = 0.554$$

The maximum percentage yield is then $\boxed{55.4 \text{ per cent}}$

Comment. If a substance reacts by parallel processes of the same order, then the ratio of the amounts of products will be constant and independent of the extent of the reaction, no matter what the order.

Question. Can you demonstrate the truth of the statement made in the above comment?

P25.9   If the reaction is first-order the concentrations obey

$$\ln\left(\frac{[A]}{[A]_0}\right) = -kt \quad [25.10b]$$

and, since pressures and concentrations of gases are proportional, the pressures should obey

$$\ln \frac{p_0}{p} = kt$$

and $\frac{1}{t} \ln \frac{p_0}{p}$ should be a constant. We test this by drawing up the following table

p_0/Torr	200	200	400	400	600	600
t/s	100	200	100	200	100	200
p/Torr	186	173	373	347	559	520
$10^4 \left(\dfrac{1}{t/s} \right) \ln \dfrac{p_0}{p}$	7.3	7.3	7.0	7.1	7.1	7.2

The values in the last row of the table are virtually constant, and so (in the pressure range spanned by the data) the reaction has $\boxed{\text{first-order kinetics}}$ with $k = \boxed{7.2 \times 10^{-4}\,\text{s}^{-1}}$

P25.10 $\quad A + B \rightarrow P, \quad \dfrac{d[P]}{dt} = k[A]^m[B]^n$

and for a short interval δt,

$$\delta[P] \approx k[A]^m[B]^n \delta t$$

Therefore, since $\delta[P] = [P]_t - [P]_0 = [P]_t$,

$$\frac{[P]}{[A]} = k[A]^{m-1}[B]^n \delta t$$

$\dfrac{[\text{Chloropropane}]}{[\text{Propene}]}$ independent of [Propene] implies that $m = 1$.

$$\frac{[\text{Chloropropane}]}{[\text{HCl}]} = \begin{cases} p(\text{HCl}) & 10 & 7.5 & 5.0 \\ & 0.05 & 0.03 & 0.01 \end{cases}$$

These results suggest that the ratio is roughly proportional to p^2, and therefore that $m = 3$ when A is identified with HCl. The rate law is therefore

$$\frac{d[\text{Chloropropane}]}{dt} = k[\text{Propane}][\text{HCl}]^3$$

and the reaction is $\boxed{\text{first-order}}$ in propene and $\boxed{\text{third-order}}$ in HCl.

P25.12 $\quad 2\text{HCl} \rightleftharpoons (\text{HCl})_2, \quad K_1 \quad [(\text{HCl})_2] = K_1[\text{HCl}]^2$

$\text{HCl} + \text{CH}_3\text{CH}{=}\text{CH}_2 \rightleftharpoons \text{Complex} \quad K_2$

$[\text{Complex}] = K_2[\text{HCl}][\text{CH}_3\text{CH}{=}\text{CH}_2]$

$(\text{HCl})_2 + \text{Complex} \rightarrow \text{CH}_3\text{CHClCH}_3 + 2\text{HCl} \quad k$

$\text{rate} = k[(\text{HCl})_2][\text{Complex}] = kK_2[(\text{HCl})_2][\text{HCl}][\text{CH}_3\text{CH}{=}\text{CH}_2]$

$$= \boxed{kK_2K_1[\text{HCl}]^3[\text{CH}_3\text{CH}{=}\text{CH}_2]}$$

Use infrared spectroscopy to search for $(\text{HCl})_2$.

P25.14 $E_a = \dfrac{R \ln\left(\dfrac{k'}{k}\right)}{\left(\dfrac{1}{T} - \dfrac{1}{T'}\right)}$ [Exercise 25.16(a) from eqn 25.25]

We then draw up the following table

T/K		300.3	300.3	341.2
T'/K		341.2	392.2	392.2
$10^{-7}k/(\text{L mol}^{-1}\,\text{s}^{-1})$		1.44	1.44	3.03
$10^{-7}k'/(\text{L mol}^{-1}\,\text{s}^{-1})$		3.03	6.9	6.9
$E_a/(\text{kJ mol}^{-1})$		15.5	16.7	18.0

The mean is $\boxed{16.7\,\text{kJ mol}^{-1}}$ For A, use

$A = k e^{E_a/RT}$

and draw up the following table

T/K	300.3	341.2	392.2
$10^{-7}k/(\text{L mol}^{-1}\,\text{s}^{-1})$	1.44	3.03	6.9
E_a/RT	6.69	5.89	5.12
$10^{-10}A/(\text{L mol}^{-1}\,\text{s}^{-1})$	1.16	1.10	1.16

The mean is $\boxed{1.14 \times 10^{10}\,\text{L mol}^{-1}\,\text{s}^{-1}}$

P25.16   The relation between the equilibrium constant and the rate constants is obtained from

$\Delta_r G^{\ominus} = -RT \ln K$ with $\Delta_r G^{\ominus} = \Delta_r H^{\ominus} - T\Delta_r S^{\ominus}$ and $K = \dfrac{k}{k'}$.

$$K = \frac{k}{k'} = \exp\left(\frac{-\Delta_r H^{\ominus}}{RT}\right)\exp\left(\frac{\Delta_r S^{\ominus}}{R}\right) = \left(\frac{A}{A'}\right)\exp\left(\frac{E_a' - E_a}{RT}\right)$$

Setting the temperature-dependent parts equal yields

$\Delta_r H^{\ominus} = E_a - E_a' = [-4.2 - (53.3)]\,\text{kJ mol}^{-1} = -57.5\,\text{kJ mol}^{-1}$

Setting the temperature-independent parts equal yields

$$\exp\left(\frac{\Delta_r S^{\ominus}}{R}\right) = \left(\frac{A}{A'}\right) \quad \text{so} \quad \Delta_r S^{\ominus} = R\ln\left(\frac{A}{A'}\right) = (8.3145\,\text{J K}^{-1}\,\text{mol}^{-1})\ln\left(\frac{1.0 \times 10^9}{1.4 \times 10^{11}}\right)$$

$\Delta_r S^{\ominus} = -41.1\,\text{J K}^{-1}\,\text{mol}^{-1}$

The thermodynamic quantities of the reaction are related to standard molar quantities

$\Delta_r H^{\ominus} = \Delta_f H^{\ominus}(C_2H_6) + \Delta_f H^{\ominus}(Br) - \Delta_f H^{\ominus}(C_2H_5) - \Delta_f H^{\ominus}(HBr)$

so $\Delta_f H^{\ominus}(C_2H_5) = \Delta_f H^{\ominus}(C_2H_6) + \Delta_f H^{\ominus}(Br) - \Delta_f H^{\ominus}(HBr) - \Delta_r H^{\ominus}$

$\Delta_f H^{\ominus}(C_2H_5) = [(-84.68) + 111.88 - (-36.40) - (-57.5)]\,\text{kJ mol}^{-1} = \boxed{121.2\,\text{kJ mol}^{-1}}$

$S_m^{\ominus}(C_2H_5) = [229.60 + 175.02 - 198.70 - (-41.1)]\,\text{J mol}^{-1}\,\text{K}^{-1} = \boxed{247.0\,\text{J K}^{-1}\,\text{mol}^{-1}}$

$$\Delta_f G^{\ominus}(C_2H_5) = [-32.82 + 82.396 - (-53.45)] \text{ kJ mol}^{-1} - \Delta_r G^{\ominus}$$

$$= 103.03 \text{ kJ mol}^{-1} - \Delta_r G^{\ominus}$$

$$\Delta_r G^{\ominus} = \Delta_r H^{\ominus} - T\Delta_r S^{\ominus} = -57.5 \text{ kJ mol}^{-1} - (298 \text{ K}) \times (-41.1 \times 10^{-3} \text{ kJ K}^{-1} \text{ mol}^{-1})$$

$$= -45.3 \text{ kJ mol}^{-1}$$

$$\Delta_f G^{\ominus}(C_2H_5) = [103.03 - (-45.3)] \text{ kJ mol}^{-1} = \boxed{148.3 \text{ kJ mol}^{-1}}$$

P25.17   **(a)** Let A represent $SiCl_3Br$ and B represent Br_2.

$$\frac{d[A]}{dt} = k[SiCl_3H]^a[B]^b$$

But $[A] = \alpha[A]_0$ and the equilibrium constant for the dissociation of dibromine is so small that

$$[B] = [B]_0 - \alpha[A]_0 \qquad [SiCl_3H] = (1-\alpha)[A]_0.$$

substitution of these relationships into the rate law gives:

$$\frac{d\alpha}{dt} = k[A]_0^{a-1}(1-\alpha)^a([B]_0 - \alpha[A]_0)^b$$

$$\int_0^\alpha \frac{d\alpha}{(1-\alpha)^a([B]_0 - \alpha[A]_0)^b} = \int_0^t k[A]_0^{a-1} dt$$

We make the reasonable assumption that $a = 1$ because it seems very unlikely that the reaction mechanism would involve a binary reaction between two trichlorosilane molecules.

$$\underbrace{\int_0^\alpha \frac{d\alpha}{(1-\alpha)([B]_0 - \alpha[A]_0)^b}}_{I(\alpha,b)} = kt$$

Data at 180°C can be used to evaluate the value of b. Use $i = 0, 1, 2, 3$ to index the data points of the set. Then, use mathematical software to systematically adjust the value of both b and k until the sum of the squares of errors (SSE) is minimized. The Given/Minerr solve block of Mathcad performs the task very nicely.

$$SSE(b, k) = \sum_{i=0}^{3} \{I(\alpha_i, b) - k t_i\}^2$$

It is found that $b = 0.497$.

Single term expressions for rate laws often have reactant reaction orders that are integers or half integers. Recognizing that the above value of b is very close to $\frac{1}{2}$, we conclude that $b = 0.500$ and that the observed deviation from the half integer is due to experimental error.

$$\boxed{a = 1 \quad \text{and} \quad b = 1/2}$$

(b) Each data set is used to determine k at the temperature of the data. The technique of part (a) is repeated but with fixed values for a and b. Only k is systematically adjusted to minimize $SSE(k)$.

The table summarizes results.

T/K	$k/\mathrm{cm}^{3/2}\,\mathrm{mol}^{-1/2}\,\mathrm{s}^{-1}$
433.15	0.077
453.15	0.504
473.15	2.26
493.15	12.0

The activation parameters are determined with a linear regression analysis of results plotted as a $\ln(k)$ versus $1/T$ graph (eqn 25.25 and Example 25.5). The regression fit gives:

$$\text{slope} = -1.776 \times 10^4\,\mathrm{K}$$

$$\text{intercept} = 38.45$$

$$E_a = -R \times \text{slope} = \boxed{148\,\mathrm{kJ\,mol}^{-1}}$$

$$A = \mathrm{e}^{\text{intercept}}\,\mathrm{cm}^{3/2}\,\mathrm{mol}^{-1/2}\,\mathrm{s}^{-1} = \boxed{5.00 \times 10^{16}\,\mathrm{cm}^{3/2}\,\mathrm{mol}^{-1/2}\mathrm{s}^{-1}}$$

(c) We assume that the dissociation of dibromine is at equilibrium.

$$\mathrm{Br_2(g)} \overset{K}{\rightleftharpoons} 2 \cdot \mathrm{Br(g)} \quad [\mathrm{Br}] = \sqrt{K[\mathrm{Br_2}]}$$

$$\mathrm{SiCl_3H(g)} + \cdot\mathrm{Br(g)} \xrightarrow{k_1} \cdot\mathrm{SiCl_3(g)} + \mathrm{HBr(g)}$$

$$\cdot\mathrm{SiCl_3(g)} + \mathrm{Br_2(g)} \xrightarrow{k_2} \mathrm{SiCl_3Br(g)} + \cdot\mathrm{Br(g)}$$

Applying the steady-state approximation to the $\cdot\mathrm{SiCl_3}$ concentration in the proposed mechanism gives:

$$\frac{d[\mathrm{SiCl_3}]}{dt} = k_1[\mathrm{SiCl_3H}][\mathrm{Br}] - k_2[\mathrm{SiCl_3}][\mathrm{Br_2}] = 0$$

$$k_1 K^{1/2}[\mathrm{SiCl_3H}][\mathrm{Br_2}]^{1/2} - k_2[\mathrm{SiCl_3}][\mathrm{Br_2}] = 0$$

$$[\mathrm{SiCl_3}] = \frac{k_1 K^{1/2}[\mathrm{SiCl_3H}]}{k_2[\mathrm{Br_2}]^{1/2}}$$

The above equation is substituted into the rate expression for $\mathrm{SiCl_3Br}$.

$$\frac{d[\mathrm{SiCl_3Br}]}{dt} = k_2[\mathrm{SiCl_3}][\mathrm{Br_2}] = k_1 K^{1/2}[\mathrm{SiCl_3H}][\mathrm{Br_2}]^{1/2}$$

This is the form of the rate law found in part **(a)** (i.e., $a = 1$ and $b = 1/2$). Since the proposed mechanism gives a rate law that matches the experimental rate law, the mechanism might possibly be correct.

V. F. Kochubei and M. V. Karmagin, *Kinetics and Catalysis*, *37*(1), 13 (1996).

Solutions to theoretical problems

P25.19

$$\mathrm{A} \rightleftharpoons \mathrm{B}$$

$$\frac{d[\mathrm{A}]}{dt} = -k[\mathrm{A}] + k'[\mathrm{B}] \quad \frac{d[\mathrm{B}]}{dt} = -k'[\mathrm{B}] + k[\mathrm{A}]$$

$[\mathrm{A}] + [\mathrm{B}] = [\mathrm{A}]_0 + [\mathrm{B}]_0$ at all times.

Therefore, $[B] = [A]_0 + [B]_0 - [A]$

$$\frac{d[A]}{dt} = -k[A] + k'\{[A]_0 + [B]_0 - [A]\} = -(k + k')[A] + k'([A]_0 + [B]_0)$$

The solution is $[A] = \dfrac{k'([A]_0 + [B]_0) + (k[A]_0 - k'[B]_0)e^{-(k+k')t}}{k + k'}$

The final composition is found by setting $t = \infty$

$$[A]_\infty = \left(\frac{k'}{k + k'}\right) \times ([A]_0 + [B]_0)$$

$$[B]_\infty = [A]_0 + [B]_0 - [A]_\infty = \left(\frac{k}{k + k'}\right) \times ([A]_0 + [B]_0)$$

Note that $\boxed{\dfrac{[B]_\infty}{[A]_\infty} = \dfrac{k}{k'}}$

P25.21 $\qquad \dfrac{d[A]}{dt} = -2k[A]^2[B], \quad 2A + B \rightarrow P$

(a) Let $[P] = x$ at t, then $[A] = A_0 - 2x$ and $[B] = B_0 - x$. Therefore,

$$\frac{d[A]}{dt} = -2\frac{dx}{dt} = -2k(A_0 - 2x)^2 \times (B_0 - x)$$

$$\frac{dx}{dt} = k(A_0 - 2x)^2 \times \left(\frac{1}{2}A_0 - x\right) = \frac{1}{2}k(A_0 - 2x)^3$$

$$\frac{1}{2}kt = \int_0^x \frac{dx}{(A_0 - 2x)^3} = \frac{1}{4} \times \left[\left(\frac{1}{A_0 - 2x}\right)^2 - \left(\frac{1}{A_0}\right)^2\right]$$

Therefore, $\boxed{kt = \dfrac{2x(A_0 - x)}{A_0^2(A_0 - 2x)^2}}$

(b) $\qquad \dfrac{dx}{dt} = k(A_0 - 2x)^2 \times (B_0 - x) = k(A_0 - 2x)^2 \times (A_0 - x) \quad \boxed{B_0 = 2 \times \dfrac{1}{2}A_0 = A_0}$

$$kt = \int_0^x \frac{dx}{(A_0 - 2x)^2 \times (A_0 - x)}$$

We proceed by the method of partial fractions (which is employed in the general case too), and look for the concentrations α, β, and γ in

$$\frac{1}{(A_0 - 2x)^2 \times (A_0 - x)} = \frac{\alpha}{(A_0 - 2x)^2} + \frac{\beta}{A_0 - 2x} + \frac{\gamma}{A_0 - x}$$

which requires that

$$\alpha(A_0 - x) + \beta(A_0 - 2x) \times (A_0 - x) + \gamma(A_0 - 2x)^2 = 1$$

$$(A_0\alpha + A_0^2\beta + A_0^2\gamma) - (\alpha + 3\beta A_0 + 4\gamma A_0)x + (2\beta + 4\gamma)x^2 = 1$$

This must be true for all x; therefore

$$A_0\alpha + A_0^2\beta + A_0^2\gamma = 1$$

$$\alpha + 3A_0\beta + 4A_0\gamma = 0$$

$$2\beta + 4\gamma = 0$$

These solve to give $\alpha = \dfrac{2}{A_0}$, $\beta = \dfrac{-2}{A_0^2}$, and $\gamma = \dfrac{1}{A_0^2}$

Therefore,

$$kt = \int_0^x \left(\frac{(2/A_0)}{(A_0 - 2x)^2} - \frac{(2/A_0^2)}{A_0 - 2x} + \frac{(1/A_0^2)}{A_0 - x} \right) dx$$

$$= \left(\frac{(1/A_0)}{A_0 - 2x} + \frac{1}{A_0^2} \ln(A_0 - 2x) - \frac{1}{A_0^2} \ln(A_0 - x) \right) \Bigg|_0^x$$

$$= \boxed{\left(\frac{2x}{A_0^2(A_0 - 2x)} \right) + \left(\frac{1}{A_0^2} \right) \ln\left(\frac{A_0 - 2x}{A_0 - x} \right)}$$

P25.22   The rate equations are

$$\frac{d[A]}{dt} = -k_a[A] + k_a'[B]$$

$$\frac{d[B]}{dt} = k_a[A] - k_a'[B] - k_b[B] + k_b'[C]$$

$$\frac{d[C]}{dt} = k_b[B] - k_b'[C]$$

These equations are a set of coupled differential equations and, though this is not immediately apparent, do admit of an analytical general solution. However, we are looking for specific circumstances under which the mechanism reduces to the second form given. Since the reaction involves an intermediate, let us explore the result of applying the steady-state approximation to it. Then

$$\frac{d[B]}{dt} = k_a[A] - k_a'[B] - k_b[B] + k_b'[C] = 0$$

and $[B] = \dfrac{k_a[A] + k_b'[C]}{k_a' + k_b}$

Therefore, $\dfrac{d[A]}{dt} = -\dfrac{k_a k_b}{k_a' + k_b}[A] + \dfrac{k_a' k_b'}{k_a' + k_b}[C]$

This rate expression may be compared to that given in the text [Section 25.4] for the mechanism

$$A \underset{k}{\overset{k}{\rightleftharpoons}} B \quad [\text{Here } A \underset{k_{\text{eff}}'}{\overset{k_{\text{eff}}}{\rightleftharpoons}} C]$$

Hence, $k_{\text{eff}} = \dfrac{k_a k_b}{k_a' + k_b}$ $k_{\text{eff}}' = \dfrac{k_a' k_b'}{k_a' + k_b}$

The solutions are $[A] = \left(\dfrac{k_{\text{eff}}' + k_{\text{eff}} e^{-(k_{\text{eff}} + k_{\text{eff}}')t}}{k_{\text{eff}} + k_{\text{eff}}'} \right) \times [A]_0$

and $[C] = [A_0] - [A]$

Thus, the conditions under which the first mechanism given reduces to the second are the conditions under which the steady-state approximation holds, namely, when B is a reactive intermediate .

Solutions to applications

P25.27
$$[^{14}C] = [^{14}C]_0 e^{-kt} \ [25.10b], \quad k = \frac{\ln 2}{t_{1/2}}$$

Solving for t, $t = \frac{1}{k} \ln \frac{[^{14}C]_0}{[^{14}C]} = \frac{t_{1/2}}{\ln 2} \ln \frac{[^{14}C]_0}{[^{14}C]} = \left(\frac{5730 \text{ y}}{\ln 2}\right) \times \ln\left(\frac{1.00}{0.72}\right) = \boxed{2720 \text{ y}}$

P25.29
The data for this experiment do not extend much beyond one half-life. Therefore the half-life method of predicting the order of the reaction as described in the solutions to Problems 25.1 and 25.2 cannot be used here. However, a similar method based on three-quarters lives will work. Analogous to the derivation leading to eqn 25.11, we may write

$$kt_{3/4} = -\ln \frac{\frac{3}{4}[A]_0}{[A]_0} = -\ln \frac{3}{4} = \ln \frac{4}{3} = 0.288$$

or $t_{3/4} = \dfrac{0.288}{k}$

and we see that the three-quarters life is also independent of concentration for a first-order reaction. Examination of the data shows that the first three-quarters life is about 80 min (0.237 mol L^{-1}) and by interpolation the second is also about 80 min (0.178 mol L^{-1}). Therefore the reaction is first-order and the rate constant is approximately

$$k = \frac{0.288}{t_{3/4}} \approx \frac{0.288}{80 \text{ min}} = 3.6 \times 10^{-3} \text{ min}^{-1}$$

A least-squares fit of the data to the first-order integrated rate law [25.10b] gives the slightly more accurate result, $\boxed{k = 3.65 \times 10^{-3} \text{ min}^{-1}}$ The average lifetime is calculated from

$$\frac{[A]}{[A]_0} = e^{-kt} \ [25.10b]$$

which has the form of a distribution function. The ratio $\dfrac{[A]}{[A]_0}$ is the fraction of sucrose molecules which have lived to time t. The average lifetime is then

$$\langle t \rangle = \frac{\int_0^\infty t e^{-kt} \, dt}{\int_0^\infty e^{-kt} \, dt} = \frac{1}{k} = \boxed{274 \text{ min}}$$

The denominator ensures normalization of the distribution function.

Comment. The average lifetime is also called the relaxation time. Compare to eqn 25.24. Note that the average lifetime is not the half-life. The latter is 190 minutes. Also note that $2 \times t_{3/4} \neq t_{1/2}$

P25.31
$$E_a = -R \frac{d \ln(k)}{d(1/T)} \ [25.26]$$

$$= -R \ln(10) \frac{d \log(k)}{d(1/T)} = -R \ln(10) \frac{d}{d(1/T)} \left\{ 11.75 - \frac{5488}{T/K} \right\}$$

$$= -R \ln(10)\{-5488 \text{ K}\} = \{8.3145 \text{ J K}^{-1} \text{ mol}^{-1}\} \ln(10)\{5488 \text{ K}\}$$

$$\boxed{E_a = 105 \text{ kJ mol}^{-1}}$$

$$\Delta_r G^{\oplus} = -RT \ln K \quad [9.19]$$

$$= -RT \ln(10) \log K$$

At 298.15 K

$$\Delta_r G^{\oplus} = -(8.3145 \, \text{J K}^{-1} \, \text{mol}^{-1}) \times (298.15 \, \text{K}) \ln(10) \left\{ -1.36 + \frac{1794}{298.15} \right\}$$

$$\boxed{\Delta_r G^{\oplus} = -26.6 \, \text{kJ mol}^{-1}}$$

$$\Delta_r H^{\oplus} = -R \frac{d \ln(K)}{d(1/T)} \quad [9.26]$$

$$= -R \ln(10) \frac{d \log(K)}{d(1/T)}$$

$$= -R \ln(10) \frac{d}{d(1/T)} \left\{ -1.36 + \frac{1794}{T/\text{K}} \right\}$$

$$= -R \ln(10)\{1794 \, \text{K}\} = -\{8.3145 \, \text{J K}^{-1} \, \text{mol}^{-1}\} \ln(10)\{1794 \, \text{K}\}$$

$$\boxed{\Delta_r H^{\oplus} = -34.3 \, \text{kJ mol}^{-1}}$$

The reaction is

The equations for the rate constant k and the equilibrium constant K were obtained under conditions corresponding to the biological standard state (pH $= 7$, $p = 1$ bar; Section 9.5). Thus the values of $\Delta_r G$ calculated from the equation for K are $\Delta_r G^{\oplus}$ values which differ significantly from $\Delta_r G^{\ominus}$ (pH $= 1$, $p = 1$ bar). Prebiotic conditions are more likely to be near pH $= 7$ than pH $= 1$, so we expect that the reaction will still be favourable ($K \gg 1$) thermodynamically.

Because $\Delta_r G = \Delta_r G^{\oplus} + RT \ln Q$ [9.11] and since we might expect $Q < 1$ in a prebiotic environment, $\Delta_r G < \Delta_r G^{\oplus}$. But, as shown in the calculation above, $\Delta_r G^{\oplus}$ is rather large and negative ($-26.6 \, \text{kJ mol}^{-1}$), so we expect it will still be large and negative under the prebiotic conditions; hence the reaction will be spontaneous for these conditions. We expect that $\Delta_r H \approx \Delta_r H^{\oplus}$ because enthalpy changes largely reflect bond breakage and bond formation energies.

A plot of the equation for the rate constant k is shown in Fig. 25.2(a) and that for the equilibrium constant in Fig. 25.2(b). From a kinetic point of view the reaction becomes more favourable at higher temperatures; from a thermodynamic point of view it becomes less favourable, but $K \gg 1$ at all temperatures.

P25.32   The Arrhenius expression for the rate constant is

$$k = A e^{-E_a/RT} \quad [25.27] \quad \text{so} \quad \ln k = \ln A - E_a/RT \quad [25.25]$$

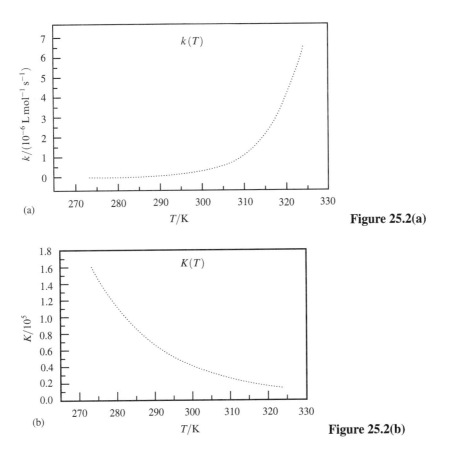

(a)

Figure 25.2(a)

(b)

Figure 25.2(b)

A plot of $\ln k$ versus $1/T$ will have slope $-E_a/R$ and y-intercept $\ln A$. The transformed data and plot (Fig. 25.3) follow

T/K	295	295	223	218	213	206	200	195
$10^{-6}k/(\mathrm{L\,mol^{-1}\,s^{-1}})$	3.70	3.55	0.494	0.452	0.379	0.295	0.241	0.217
$\ln k/(\mathrm{L\,mol^{-1}\,s^{-1}})$	15.12	15.08	13.11	13.02	12.85	12.59	12.39	12.29
K/T		0.00339	0.00339	0.00448	0.00459	0.00469	0.00485	0.00500	0.00513

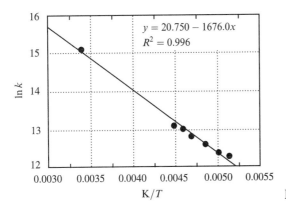

$$y = 20.750 - 1676.0x$$
$$R^2 = 0.996$$

Figure 25.3

So $E_a = -(8.3145\,\mathrm{J\,K^{-1}\,mol^{-1}}) \times (-1676\,\mathrm{K}) = 1.39 \times 10^4\,\mathrm{J\,mol^{-1}} = \boxed{13.9\,\mathrm{kJ\,mol^{-1}}}$

and $A = e^{20.750}\,\mathrm{L\,mol^{-1}\,s^{-1}} = \boxed{1.03 \times 10^9\,\mathrm{L\,mol^{-1}\,s^{-1}}}$

P25.34   The rate constants are:

$$k = A \exp\left(\frac{-E_a}{RT}\right) \text{[25.27]}$$

$$k_1 = (1.13 \times 10^9\,\mathrm{L\,mol^{-1}\,s^{-1}}) \exp\left(\frac{-14.1 \times 10^3\,\mathrm{J\,mol^{-1}}}{(8.3145\,\mathrm{J\,mol^{-1}\,K^{-1}}) \times (298\,\mathrm{K})}\right)$$

$$= \boxed{3.82 \times 10^6\,\mathrm{L\,mol^{-1}\,s^{-1}}}$$

$$k_2 = (6.0 \times 10^8\,\mathrm{L\,mol^{-1}\,s^{-1}}) \exp\left(\frac{-17.5 \times 10^3\,\mathrm{J\,mol^{-1}}}{(8.3145\,\mathrm{J\,mol^{-1}\,K^{-1}}) \times (298\,\mathrm{K})}\right)$$

$$= \boxed{5.1 \times 10^5\,\mathrm{L\,mol^{-1}\,s^{-1}}}$$

$$k_3 = (1.01 \times 10^9\,\mathrm{L\,mol^{-1}\,s^{-1}}) \exp\left(\frac{-13.6 \times 10^3\,\mathrm{J\,mol^{-1}}}{(8.3145\,\mathrm{J\,mol^{-1}\,K^{-1}}) \times (298\,\mathrm{K})}\right)$$

$$= \boxed{4.17 \times 10^6\,\mathrm{L\,mol^{-1}\,s^{-1}}}$$

Compared to reaction 1, reaction 2 shows a significant kinetic isotope effect whereas reaction 3 shows practically none. This difference should not be surprising: in reaction 2 a C—D bond is broken, whereas in reaction 3 the D atom is simply along for the ride already attached to the O atom. Compare the measured isotope effect of 0.13 to that expected in reaction 2.

$$\frac{k_2}{k_1} = \exp\left(\frac{-\hbar k_f^{1/2}}{2kT}\left(\frac{1}{\mu_{CH}^{1/2}} - \frac{1}{\mu_{CD}^{1/2}}\right)\right) \left[25.49 \text{ with } \tilde{\nu} = \frac{1}{2\pi c}\left(\frac{k_f}{\mu_{CH}}\right)^{1/2} (\text{C—H})\right]$$

We take $\mu_{CH} \approx m_H$ and $\mu_{CD} \approx m_D \approx 2m_H$, so

$$\frac{k_2}{k_1} = \exp\left(\left(\frac{-(1.0546 \times 10^{-34}\,\mathrm{J\,s}) \times (500\,\mathrm{kg\,s^{-2}})^{1/2}}{2(1.381 \times 10^{-23}\,\mathrm{J\,K^{-1}}) \times (298\,\mathrm{K})}\right) \times \left(1 - \frac{1}{2^{1/2}}\right)\right.$$

$$\left. \times \left(\frac{6.022 \times 10^{23}\,\mathrm{mol^{-1}}}{1 \times 10^{-3}\,\mathrm{kg\,mol^{-1}}}\right)^{1/2}\right)$$

$$= \boxed{0.13}$$

in agreement with the experimental value.

26 The kinetics of complex reactions

Solutions to exercises

Discussion questions

E26.1(a)   **(a)**

(1) $AH \rightarrow A\cdot + H\cdot$	Initiation [radicals formed]
(2) $A\cdot \rightarrow B\cdot + C$	Propagation [new radicals formed]
(3) $AH + B\cdot \rightarrow A\cdot + D$	Propagation [new radicals formed]
(4) $A\cdot + B\cdot \rightarrow P$	Termination [non-radical product formed]

         (b)

(1) $A_2 \rightarrow A\cdot + A\cdot$	Initiation [radicals formed]
(2) $A\cdot \rightarrow B\cdot + C$	Propagation [radicals formed]
(3) $A\cdot + P \rightarrow B\cdot$	Retardation [product destroyed, but chain not terminated]
(4) $A\cdot + B\cdot \rightarrow P$	Termination [non-radical product formed]

E26.2(a)   The Michaelis–Menten mechanism of enzyme activity models the enzyme with one active site, that weakly and reversibly, binds a substrate in homogeneous solution. It is a three-step mechanism. The first and second steps are the reversible formation of the enzyme–substrate complex (ES). The third step is the decay of the complex into the product. The steady-state approximation is applied to the concentration of the intermediate (ES) and its use simplifies the derivation of the final rate expression. However, the justification for the use of the approximation with this mechanism is suspect, in that both rate constants for the reversible steps may not be as large, in comparison to the rate constant for the decay to products, as they need to be for the approximation to be valid. The simplest form of the mechanism applies only when $k_b \gg k_a'$. Nevertheless, the form of the rate equation obtained does seem to match the principal experimental features of enzyme catalyzed reactions; it explains why there is a maximum in the reaction rate and provides a mechanistic understanding of the turnover number. The model may be expanded to include multisubstrate reaction rate and provides a mechanistic understanding of the turnover number. The model may be expanded to include multisubstrate reactions and inhibition.

E26.3(a)   In an autocatalytic mechanism, products (or intermediates) are involved in the steps leading to their formation. The presence of autocatalytic steps in a mechanism is a necessary condition for oscillation to occur, but not all autocatalytic mechanisms can lead to oscillations. When they do, it is the concentrations of the intermediates that oscillate. For a chemical system to be capable of sustained oscillation, it must be open, far from equilibrium, and be able to exist in two steady states (bistability). In an open system, there is no requirement that entropy must increase, and the apparent conflict of sustainable oscillations with the second law of thermodynamics is removed. In order to achieve oscillation in a chemical system, the mechanism of the reaction must involve feedback through autocatalytic steps describable by non-linear ordinary differential equations. An easy to understand model of how oscillations can be sustained is the Lotka–Volterra mechanism discussed in Section 26.8(a).

E26.4(a)   The primary quantum yield is associated with the primary photochemical event in the overall photochemical process which may involve secondary events as well. An example that illustrates both kind of events is the photolysis of HI described in Section 26.12(a). The primary quantum yield is defined as the ratio of the number of primary events to the number of photons absorbed (eqn 26.30) and its value can never exceed one. However, in reactions described by complex mechanisms, the overall quantum yield, which is the number of reactant molecules consumed in both primary and secondary processes per photon absorbed, can easily exceed one. Experimental procedures for the determination of the overall quantum yield involve measurements of the intensity of the radiation used, defined here as the number of photons generated and directed at the reacting sample, and of the amount of product formed. This ratio is the overall quantum yield. See Example 26.6. In addition to chemical reactions, the concept of the quantum yield enters into the description of other kinds of photochemical processes, such as fluorescence and phosphorescence, and in each case there are techniques specific to the process for the determination of the quantum yield.

Numerical exercises

In the following exercises and problems, it is recommended that rate constants are labelled with the number of the step in the proposed reaction mechanism and that any reverse steps are labelled similarly but with a prime.

E26.5(a) We assume that the steady-state approximation applies to [O] (but see the question below). Then

$$\frac{d[O]}{dt} = 0 = k_1[O_3] - k_1'[O][O_2] - k_2[O][O_3]$$

Solving for [O],

$$[O] = \frac{k_1[O_3]}{k_1'[O_2] + k_2[O_3]}$$

$$\text{Rate} = -\frac{1}{2}\frac{d[O_3]}{dt}$$

$$\frac{d[O_3]}{dt} = -k_1[O_3] + k_1'[O][O_2] - k_2[O][O_3]$$

Substituting for [O] from above

$$\frac{d[O_3]}{dt} = -k_1[O_3] + \frac{k_1[O_3](k_1'[O_2] - k_2[O_3])}{k_1'[O_2] + k_2[O_3]}$$

$$= \frac{-k_1[O_3](k_1'[O_2] + k_2[O_3]) + k_1[O_3](k_1'[O_2] - k_2[O_3])}{k_1'[O_2] + k_2[O_3]} = \frac{-2k_1k_2[O_3]^2}{k_1'[O_2] + k_2[O_3]}$$

$$\boxed{\text{Rate} = \frac{k_1k_2[O_3]^2}{k_1'[O_2] + k_2[O_3]}}$$

Question. Can you determine the rate law expression if the first step of the proposed mechanism is a rapid pre-equilibrium? Under what conditions does the rate expression above reduce to the latter?

E26.6(a) The steady-state expressions are now

$$k_2[NO_2][NO_3] - k_3[NO][NO_3] = 0$$

$$k_1[N_2O_5] - k_1'[NO_2][NO_3] - k_2[NO_2][NO_3] - k_3[NO][NO_3] = 0$$

$$\frac{d[N_2O_5]}{dt} = -k_1[N_2O_5] + k_1'[NO_2][NO_3]$$

From the steady-state equations

$$k_3[NO][NO_3] = k_2[NO_2][NO_3]$$

$$[NO_2][NO_3] = \frac{k_1}{k_1' + 2k_2}[N_2O_5]$$

Substituting, $\dfrac{d[N_2O_5]}{dt} = -k_1[N_2O_5] + \dfrac{k_1k_1'}{k_1' + 2k_2}[N_2O_5] = \dfrac{-2k_1k_2}{k_1' + 2k_2}[N_2O_5]$

$$\text{Rate} = \frac{k_1k_2}{k_1' + 2k_2}[N_2O_5] = k[N_2O_5]$$

E26.7(a)   At 800 K, the branching-chain explosion occurs between $\boxed{0.16\,\text{kPa and } 4.0\,\text{kPa}}$

E26.8(a) $$\frac{d[A^-]}{dt} = k_1[AH][B] - k_2[A^-][BH^+] - k_3[A^-][AH] = 0$$

Therefore, $[A^-] = \boxed{\dfrac{k_1[AH][B]}{k_2[BH^+] + k_3[AH]}}$

and the rate of formation of product is

$$\frac{d[P]}{dt} = k_3[AH][A^-] = \boxed{\frac{k_1 k_3[AH]^2[B]}{k_2[BH^+] + k_3[AH]}}$$

E26.9(a) $$\frac{d[AH]}{dt} = -k_a[AH] - k_c[AH][B]$$

(i) $\dfrac{d[A]}{dt} = k_a[AH] - k_b[A] + k_c[AH][B] - k_d[A][B] \approx 0$

(ii) $\dfrac{d[B]}{dt} = k_b[A] - k_c[AH][B] - k_d[A][B] \approx 0$

$(i + ii)\ [A][B] = \left(\dfrac{k_a}{2k_d}\right)[AH]$ ⎫
$(i - ii)\ [A] = \left(\dfrac{k_a + 2k_c[B]}{2k_b}\right)[AH]$ ⎬

Then, solving for $[A]$

$$[A] = k[AH], \quad k = \left(\frac{k_a}{4k_b}\right) \times \left[1 + \left(1 + \frac{8k_b k_c}{k_a k_d}\right)^{1/2}\right]$$

from which it follows that

$$[B] = \frac{k_a[AH]}{2k_d[A]} = \frac{k_a}{2k k_d}$$

and hence that $\dfrac{d[AH]}{dt} = -k_a[AH] - \left(\dfrac{k_a k_c}{2k k_d}\right)[AH] = \boxed{-k_{\text{eff}}[AH]}$

with $\boxed{k_{\text{eff}} = k_a + \dfrac{k_a k_c}{2k k_d}}$

E26.10(a)   Maximum velocity $= k_b[E]_0$ [26.20b]

Therefore, since $v = \dfrac{k_b[S][E]_0}{K_M + [S]}$ [26.20a]

we know that

$$v_{\text{max}} = k_b[E]_0 = \left(\frac{K_M + [S]}{[S]}\right)v = \left(\frac{0.035 + 0.110}{0.110}\right) \times (1.15 \times 10^{-3}\,\text{mol}\,\text{L}^{-1}\,\text{s}^{-1})$$

$$= \boxed{1.52 \times 10^{-3}\,\text{mol}\,\text{L}^{-1}\,\text{s}^{-1}}$$

E26.11(a)   Number of photons absorbed $= \phi^{-1} \times$ number of molecules that react [Sections 26.11 and 26.12].
Therefore,

$$\text{Number absorbed} = \frac{(1.14 \times 10^{-3}\,\text{mol}) \times (6.022 \times 10^{23}\,\text{einstein}^{-1})}{2.1 \times 10^2\,\text{mol einstein}^{-1}} = \boxed{3.3 \times 10^{18}}$$

E26.12(a) For a source of power P and wavelength λ, the amount of photons (n_λ) generated in a time t is

$$n_\lambda = \frac{Pt}{h\nu N_A} = \frac{P\lambda t}{hc N_A} = \frac{(100\,\text{W}) \times (45) \times (60\,\text{s}) \times (490 \times 10^{-9}\,\text{m})}{(6.626 \times 10^{-34}\,\text{J s}) \times (2.998 \times 10^8\,\text{m s}^{-1}) \times (6.022 \times 10^{23}\,\text{mol}^{-1})}$$

$$= 1.11\,\text{mol}$$

The amount of photons absorbed is 60 per cent of this incident flux, or 0.664 mol. Therefore,

$$\phi = \frac{0.344\,\text{mol}}{0.664\,\text{mol}} = \boxed{0.518}$$

Alternatively, expressing the amount of photons in einsteins [1 mol photons $=$ 1 einstein], $\phi = 0.518\,\text{mol einstein}^{-1}$.

Solutions to problems

Solutions to numerical problems

P26.1

$$H + NO_2 \rightarrow OH + NO \quad k_1 = 2.9 \times 10^{10}\,\text{L mol}^{-1}\,\text{s}^{-1}$$

$$OH + OH \rightarrow H_2O + O \quad k_2 = 1.55 \times 10^9\,\text{L mol}^{-1}\,\text{s}^{-1}$$

$$O + OH \rightarrow O_2 + H \quad k_3 = 1.1 \times 10^{10}\,\text{L mol}^{-1}\,\text{s}^{-1}$$

$$[H]_0 = 4.5 \times 10^{-10}\,\text{mol cm}^{-3} \qquad [NO_2]_0 = 5.6 \times 10^{-10}\,\text{mol cm}^{-3}$$

$$\frac{d[O]}{dt} = k_2[OH]^2 + k_3[O][OH] \qquad \frac{d[O_2]}{dt} = k_3[O][OH]$$

$$\frac{d[OH]}{dt} = k_1[H][NO_2] - 2k_2[OH]^2 - k_3[O][OH] \qquad \frac{d[NO_2]}{dt} = -k_1[H][NO_2]$$

$$\frac{d[H]}{dt} = k_3[O][OH] - k_1[H][NO_2]$$

These equations serve to show how even a simple sequence of reactions leads to a complicated set of nonlinear differential equations. Since we are interested in the time behaviour of the composition we may not invoke the steady-state assumption. The only thing left is to use a computer and to integrate the equations numerically. The outcome of this is the set of curves shown in Fig. 26.1 (they have been

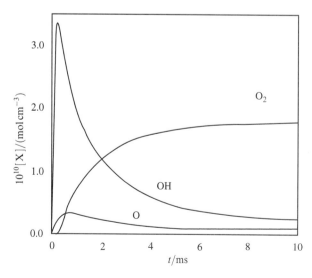

Figure 26.1

sketched from the original reference). The similarity to an A $\rightarrow$ B $\rightarrow$ C scheme should be noticed (and expected), and the general features can be analysed quite simply in terms of the underlying reactions.

P26.3   The roles are

(1) $N_2O \rightarrow N_2 + O$ initiation
(2) $O + SiH_4 \rightarrow SiH_3 + OH$ propagation [or transfer]
(3) $OH + SiH_4 \rightarrow SiH_3 + H_2O$ propagation [or transfer]
(4) $SiH_3 + N_2O \rightarrow SiH_3O + N_2$ propagation
(5) $SiH_3O + SiH_4 \rightarrow SiH_3OH + SiH_3$ propagation
(6) $SiH_3 + SiH_3O \rightarrow (H_3Si)_2O$ termination

The rate of silane consumption is

$$\frac{d[SiH_4]}{dt} = -k_2[SiH_4][O] - k_3[SiH_4][OH] - k_5[SiH_3O][SiH_4]$$

Steady-state approximation (SSA) for O

$$\frac{d[O]}{dt} = k_1[N_2O] - k_2[SiH_4][O] \approx 0 \quad \text{so} \quad [O] = \frac{k_1[N_2O]}{k_2[SiH_4]}$$

SSA for OH

$$\frac{d[OH]}{dt} = k_2[SiH_4][O] - k_3[OH][SiH_4] \approx 0 = k_1[N_2O] - k_3[OH][SiH_4]$$

so $[OH] = \dfrac{k_1[N_2O]}{k_3[SiH_4]}$

SSA for SiH_3O and SiH_3

$$\frac{d[SiH_3O]}{dt} = k_4[SiH_3][N_2O] - k_5[SiH_3O][SiH_4] - k_6[SiH_3O][SiH_3] = 0$$

$$\frac{d[SiH_3]}{dt} = k_2[SiH_4][O] + k_3[SiH_4][OH] - k_4[SiH_3][N_2O]$$
$$+ k_5[SiH_3O][SiH_4] - k_6[SiH_3O][SiH_3]$$
$$= 2k_1[N_2O] - k_4[SiH_3][N_2O] + k_5[SiH_3O][SiH_4] - k_6[SiH_3O][SiH_3] \approx 0$$

Adding these expressions together yields

$$0 = 2k_1[N_2O] - 2k_6[SiH_3O][SiH_3] \quad \text{so} \quad [SiH_3] = \frac{k_1[N_2O]}{k_6[SiH_3O]}$$

and subtracting them gives

$$0 = 2k_1[N_2O] - k_4[SiH_3][N_2O] + k_5[SiH_3O][SiH_4]$$

Solve for $[SiH_3O]$

$$0 = 2k_1[N_2O] - \frac{k_1 k_4 [N_2O]^2}{k_6[SiH_3O]} + k_5[SiH_3O][SiH_4]$$
$$= 2k_1 k_6 [SiH_3O][N_2O] - k_1 k_4 [N_2O]^2 + k_5 k_6 [SiH_3O]^2[SiH_4]$$

$$[SiH_3O] = \frac{-2k_1k_6[N_2O] \pm (4k_1^2k_6^2[N_2O]^2 + 4k_1k_4k_5k_6[N_2O]^2[SiH_4])^{1/2}}{2k_5k_6[SiH_4]}$$

$$= \frac{k_1[N_2O]}{k_5[SiH_4]}\left[-1 + \left(1 + \frac{k_4k_5[SiH_4]}{k_1k_6}\right)^{1/2}\right]$$

If k_1 is small, then

$$[SiH_3O] \approx \frac{k_1[N_2O]}{k_5[SiH_4]}\left(\frac{k_4k_5[SiH_4]}{k_1k_6}\right)^{1/2} = [N_2O]\left(\frac{k_1k_4}{k_5k_6[SiH_4]}\right)^{1/2}$$

Putting it all together yields

$$\frac{d[SiH_4]}{dt} = -2k_1[N_2O] - k_5[SiH_4][N_2O]\left(\frac{k_1k_4}{k_5k_6[SiH_4]}\right)^{1/2} \approx \boxed{-\left(\frac{k_1k_4k_5}{k_6}\right)^{1/2}[N_2O][SiH_4]^{1/2}}$$

P26.4   (a) Note that $[A]_0 - x = \dfrac{[A]_0}{1 + \left(\frac{x}{[A]_0 - x}\right)}$

as can be seen by dividing both sides by $[A]_0 - x$. The ratio $x/([A]_0 - x)$ is given in the problem.

Test of first-order rate law. The plot of $\ln([A]_0 - x)$ against t is linear with slope $-k$ and intercept $\ln[A]_0$. The regression fit [not shown] of $\ln([A]_0 - x)$ against t gives the following results for $X = H$ in the XC_6H_4HgCl formula.

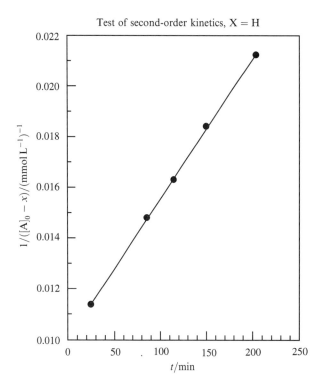

Figure 26.2(a)

$R = 0.986\,324$

Slope $= -3.44 \times 10^{-3}\,\text{min}^{-1}$ standard deviation $= 2.34 \times 10^{-4}\,\text{min}^{-1}$

Intercept $= \ln([\text{A}]_0/(\text{mmol}\,\text{L}^{-1})) = 4.532$ standard deviation $= 0.031$

Test of second-order rate law. The plot of $1/([\text{A}]_0 - x)$ against t [Fig. 26.2(a)] is linear with slope k and intercept $1/[\text{A}]_0$. The regression fit of $1/([\text{A}]_0 - x)$ against t gives the following results for $\text{X} = \text{H}$ in the $\text{XC}_6\text{H}_4\text{HgCl}$ formula.

$R = 0.999\,691$

Slope $= 5.46 \times 10^{-5}\,(\text{mmol}\,\text{L}^{-1})^{-1}\,\text{min}^{-1}$

standard deviation $= 7.84 \times 10^{-7}\,(\text{mmol}\,\text{L}^{-1})^{-1}\,\text{min}^{-1}$

Intercept $= 1.01 \times 10^{-2}(\text{mmol}\,\text{L}^{-1})^{-1}$

standard deviation $= 1.02 \times 10^{-4}\,(\text{mmol}\,\text{L}^{-1})^{-1}$

Both the correlation coefficients and the standard deviations of the slopes indicate that the regression plot of $1/([\text{A}]_0 - x)$ against t is the better fit. We conclude that the rate law is $\boxed{\text{second-order}}$ when $\text{X} = \text{H}$.

For $\text{X} = \text{Cl}$ the regression fit [not shown] of $\ln([\text{A}]_0 - x)$ against t gives

$R = 0.995\,421$

Slope $= -1.807 \times 10^{-3}\,\text{min}^{-1}$ standard deviation $= 1.23 \times 10^{-4}\,\text{min}^{-1}$

Intercept $= \ln([\text{A}]_0/(\text{mmol}\,\text{L}^{-1})) = 4.597$ standard deviation $= 0.030$

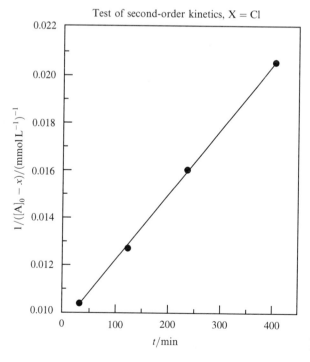

Test of second-order kinetics, $\text{X} = \text{Cl}$

Figure 26.2(b)

while the regression fit of $1/([A]_0 - x)$ against t [Fig. 26.2(b)] gives

$R = 0.999\,697$

$\text{Slope} = 2.716 \times 10^{-5}\,(\text{mmol L}^{-1})^{-1}\,\text{min}^{-1}$

$\text{standard deviation} = 4.72 \times 10^{-7}\,(\text{mmol L}^{-1})\,\text{min}^{-1}$

$\text{Intercept} = 9.51 \times 10^{-3}\,(\text{mmol L}^{-1})^{-1}$

$\text{standard deviation} = 1.15 \times 10^{-4}\,(\text{mmol L}^{-1})^{-1}$

Once again the $\boxed{\text{second-order}}$ fit is the better.

(b) Since the third step is very fast while the second is slow, we may consider the second step of the proposed mechanism to be rate-determining.

$$\frac{d[\text{ArHgCl}]}{dt} = -k_2[(\text{ArRh(CO)}_2)_2][\text{ArHgCl}] + k_{-2}[\text{Ar}_2\text{Rh(CO)}_2\text{HgCl}]$$

The second term to the right is very small because the fast step 3 prevents the build-up of $[\text{Ar}_2\text{Rh(CO)}_2\text{HgCl}]$. Thus,

$$\frac{d[\text{ArHgCl}]}{dt} = -k_2[(\text{ArRh(CO)}_2)_2][\text{ArHgCl}]$$

If we assume that step 1 is fast enough so as to be at quasi-equilibrium, we may write that

$$K_1 \approx \frac{[(\text{ArRh(CO)}_2)_2][\text{HgCl}_2]}{[(\text{ClRh(CO)}_2)_2][\text{ArHgCl}]}$$

or $(\text{ArRh(CO)}_2)_2 \approx \dfrac{K_1[(\text{ClRh(CO)}_2)_2][\text{ArHgCl}]}{[\text{HgCl}_2]}$

Substitution into the rate expression gives

$$\frac{d[\text{ArHgCl}]}{dt} = -\frac{k_2 K_1[(\text{ClRh(CO)}_2)_2][\text{ArHgCl}]^2}{[\text{HgCl}_2]}$$

If the concentrations of both the catalyst and HgCl_2 remain constant, the rate will be second-order in $[\text{ArHgCl}]$ as observed. Step 3 assumes that the catalyst concentration remains constant. The limited solubility of HgCl_2 keeps this concentration constant.

P26.6 $$\frac{d[\text{HI}]}{dt} = 2k_b[\text{I}\cdot]^2[\text{H}_2] \tag{1}$$

$$\frac{d[\text{I}\cdot]}{dt} = 2k_a[\text{I}_2] - 2k_a'[\text{I}\cdot]^2 - 2k_b[\text{I}\cdot]^2[\text{H}_2]$$

In the steady-state approximation for $[\text{I}\cdot]$

$$\frac{d[\text{I}\cdot]}{dt} = 0 = 2k_a[\text{I}_2] - 2k_a'[\text{I}\cdot]_{SS}^2 - 2k_b[\text{I}\cdot]_{SS}^2[\text{H}_2]$$

$$[\text{I}\cdot]_{SS}^2 = \frac{k_a}{k_a' + k_b[\text{H}_2]}[\text{I}_2] \tag{2}$$

Substitution of (2) into (1) gives

$$\frac{d[HI]}{dt} = \frac{2k_b k_a [I_2][H_2]}{k_a' + k_b[H_2]}$$

This simple rate law is observed when step (b) is rate-determining so that step (a) is a rapid equilibrium and [I·] is in an approximate steady state. This is equivalent to $k_b[H_2] \ll k_a'$ and hence,

$$\frac{d[HI]}{dt} = 2k_b K [I_2][H_2]$$

P26.7

$$UO_2^{2+} + h\nu \rightarrow (UO_2^{2+})^*$$

$$(UO_2^{2+})^* + (COOH)_2 \rightarrow UO_2^{2+} + H_2O + CO_2 + CO$$

$$2MnO_4^- + 5(COOH)_2 + 6H^+ \rightarrow 10CO_2 + 8H_2O + 2Mn^{2+}$$

$17.0 \, cm^3$ of 0.212 M KMnO$_4$ is equivalent to

$$\tfrac{5}{2} \times (17.0 \, cm^3) \times (0.212 \, mol \, L^{-1}) = 9.01 \times 10^{-3} \, mol \, (COOH)_2$$

The initial sample contained 5.232 g (COOH)$_2$, corresponding to

$$\frac{5.232 \, g}{90.04 \, g \, mol^{-1}} = 5.81 \times 10^{-2} \, mol \, (COOH)_2$$

Therefore, $(5.81 \times 10^{-2} \, mol) - (9.01 \times 10^{-3} \, mol) = 4.91 \times 10^{-2} \, mol$ of the acid has been consumed. A quantum efficiency 0.53 implies that the amount of photons absorbed must have been

$$\frac{4.91 \times 10^{-2} \, mol}{0.53} = 9.3 \times 10^{-2} \, mol$$

Since the exposure was for 300 s, the rate of incidence of photons was

$$\frac{9.3 \times 10^{-2} \, mol}{300 \, s} = 3.1 \times 10^{-4} \, mol \, s^{-1}$$

Since 1 mol photons = 1 einstein, the incident rate is $\boxed{3.1 \times 10^{-4} \, einstein \, s^{-1}}$ or $\boxed{1.9 \times 10^{20} \, s^{-1}}$

P26.9 Since $I_f = k_f[S^*]_t = k_f[S^*]_0 \, e^{t/\tau_0}$, we surmise that a graph of $\ln(I_f/I_0)$ against t should be linear with a slope equal to $-1/\tau_0$ in the absence of a quencher. The plot is in fact linear with a regression slope equal to $-1.004 \times 10^5 \, s^{-1}$

$$\tau_0 = \frac{1}{1.004 \times 10^5 \, s^{-1}} = 9.96 \, \mu s$$

In the presence of a quencher, a graph of $\ln(I_f/I_0)$ against t is still linear but with a slope equal to $-1/\tau$. This plot is found to be linear with a regression slope equal to $-1.788 \times 10^5 s^{-1}$.

$$\tau = \frac{1}{1.788 \times 10^5 \, s^{-1}} = 5.59 \, \mu s$$

$$\frac{1}{\tau} = \frac{1}{\tau_0} + k_Q[Q] \quad [26.35]$$

$$k_q = \frac{\tau^{-1} - \tau_0^{-1}}{[N_2]} = \frac{RT(\tau^{-1} - \tau_0^{-1})}{p_{N_2}}$$

$$= \frac{(0.08206 \, \text{L atm K}^{-1} \, \text{mol}^{-1})(300 \, \text{K})(1.788 - 1.004)10^5 \, \text{s}^{-1}}{9.74 \times 10^{-4} \, \text{atm}}$$

$$k_q = \boxed{1.98 \times 10^9 \, \text{L mol}^{-1} \, \text{s}^{-1}}$$

P26.10 $$E_T = \frac{R_0^6}{R_0^6 + R^6} \quad \text{or} \quad \frac{1}{E_T} = 1 + (R/R_0)^6 \quad [26.36]$$

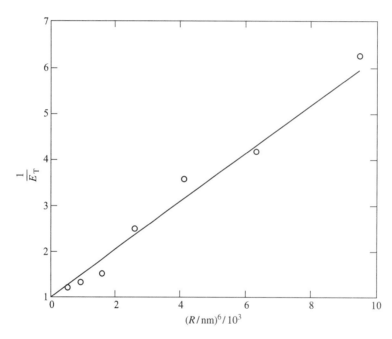

Figure 26.3

Since a plot of E_T^{-1} values against R^6 appears to be linear with an intercept equal to 1, we conclude that eqn 26.36 adequately describes the data. Solving eqn 26.36 for R_0 gives $R_0 = R(E_T^{-1} - 1)^{1/6}$. R_0 may be evaluated by taking the mean of experimental data in this expression. The two data points at lowest R must be excluded from the mean as they are highly uncertain. $\boxed{R_0 = 3.5\bar{2} \, \text{nm}}$ with a standard deviation of $0.17\bar{3} \, \text{nm}$.

Solutions to theoretical problems

P26.13 $$\langle \overline{M} \rangle_N = \frac{M}{1 - p} \quad [\text{Eqn 26.8a with } \langle \overline{M} \rangle_N = \langle n \rangle M]$$

The probability P_n that a polymer consists of n monomers is equal to the probability that it has $n - 1$ reacted end groups and one unreacted end group. The former probability is p^{n-1}; the latter $1 - p$. Therefore, the total probability of finding an n-mer is

$$P_n = p^{n-1}(1 - p)$$

$$\langle M^2 \rangle_N = M^2 \langle n^2 \rangle = M^2 \sum_n n^2 P_n = M^2 (1-p) \sum_n n^2 p^{n-1} = M^2 (1-p) \frac{d}{dp} p \frac{d}{dp} \sum_n p^n$$

$$= M^2 (1-p) \frac{d}{dp} p \frac{d}{dp} (1-p)^{-1} = \frac{M^2 (1+p)}{(1-p)^2}$$

We see that $\langle n^2 \rangle = \dfrac{1+p}{(1-p)^2}$

and that $\langle M^2 \rangle_N - \langle \overline{M} \rangle_{N^2} = M^2 \left(\dfrac{1+p}{(1-p)^2} - \dfrac{1}{(1-p)^2} \right) = \dfrac{pM^2}{(1-p)^2}$

Hence, $\boxed{\delta M = \dfrac{p^{1/2} M}{1-p}}$

The time dependence is obtained from

$$p = \frac{kt[A]_0}{1 + kt[A]_0} \quad [26.7]$$

and $\dfrac{1}{1-p} = 1 + kt[A]_0$ [26.8b]

Hence $\dfrac{p^{1/2}}{1-p} = p^{1/2}(1 + kt[A]_0) = \{kt[A]_0(1 + kt[A]_0)\}^{1/2}$

and $\delta M = \boxed{M\{kt[A]_0(1 + kt[A]_0)\}^{1/2}}$

P26.15 In termination by disproportionation, the radicals do not combine. The average number of monomers in a polymer molecule equals the number in the radical, the kinetic chain length, ν.

$$\langle n \rangle = \nu = \boxed{k[\cdot M][I]^{1/2}} \quad [26.14] \tag{26.14}$$

P26.17 $\dfrac{d[P]}{dt} = k[A]^2[P]$

$[A] = A_0 - x, \quad [P] = P_0 + x, \quad \dfrac{d[P]}{dt} = \dfrac{dx}{dt} = k(A_0 - x)^2 (P_0 + x)$

$$\int_0^x \frac{dx}{(A_0 - x)^2 (P_0 + x)} = kt$$

Solve the integral by partial fractions

$$\frac{1}{(A_0 - x)^2 (P_0 + x)} = \frac{\alpha}{(A_0 - x)^2} + \frac{\beta}{A_0 - x} + \frac{\gamma}{P_0 + x}$$

$$= \frac{\alpha(P_0 + x) + \beta(A_0 - x)(P_0 + x) + \gamma(A_0 - x)^2}{(A_0 - x)^2 (P_0 + x)}$$

$$\left. \begin{array}{r} P_0 \alpha + A_0 P_0 \beta + A_0^2 \gamma = 1 \\ \alpha + (A_0 - P_0)\beta - 2A_0 \gamma = 0 \\ -\beta + \gamma = 0 \end{array} \right\}$$

This set of simultaneous equations solves to

$$\alpha = \frac{1}{A_0 + P_0}, \qquad \beta = \gamma = \frac{\alpha}{A_0 + P_0}$$

Therefore,

$$
\begin{aligned}
kt &= \left(\frac{1}{A_0 + P_0}\right) \int_0^x \left[\left(\frac{1}{A_0 - x}\right)^2 + \left(\frac{1}{A_0 + P_0}\right)\left(\frac{1}{A_0 - x} + \frac{1}{P_0 + x}\right)\right] dx \\
&= \left(\frac{1}{A_0 + P_0}\right)\left\{\left(\frac{1}{A_0 - x}\right) - \left(\frac{1}{A_0}\right) + \left(\frac{1}{A_0 + P_0}\right)\left[\ln\left(\frac{A_0}{A_0 - x}\right) + \ln\left(\frac{P_0 + x}{P_0}\right)\right]\right\} \\
&= \left(\frac{1}{A_0 + P_0}\right)\left[\left(\frac{x}{A_0(A_0 - x)}\right) + \left(\frac{1}{A_0 + P_0}\right)\ln\left(\frac{A_0(P_0 + x)}{(A_0 - x)P_0}\right)\right]
\end{aligned}
$$

Therefore, with $y = \dfrac{x}{A_0}$ and $p = \dfrac{P_0}{A_0}$,

$$A_0(A_0 + P_0)kt = \boxed{\left(\frac{y}{1 - y}\right) + \left(\frac{1}{1 + p}\right)\ln\left(\frac{p + y}{p(1 - y)}\right)}$$

The maximum rate occurs at

$$\frac{dv_P}{dt} = 0, \qquad v_P = k[A]^2[P]$$

and hence at the solution of

$$2k\left(\frac{d[A]}{dt}\right)[A][P] + k[A]^2\frac{d[P]}{dt} = 0$$

$$-2k[A][P]v_P + k[A]^2 v_P = 0 \quad [\text{as } v_A = -v_P]$$

$$k[A]([A] - 2[P])v_P = 0$$

That is, the rate is a maximum when $[A] = 2[P]$, which occurs at

$$A_0 - x = 2P_0 + 2x, \quad \text{or} \quad x = \tfrac{1}{3}(A_0 - 2P_0); \quad y = \tfrac{1}{3}(1 - 2p)$$

Substituting this condition into the integrated rate law gives

$$A_0(A_0 + P_0)kt_{max} = \left(\frac{1}{1 + p}\right)\left(\frac{1}{2}(1 - 2p) + \ln\frac{1}{2p}\right)$$

or $\boxed{(A_0 + P_0)^2 kt_{max} = \tfrac{1}{2} - p - \ln 2p}$

P26.19 Write the differential equations for [X] and [Y]

(i) $\dfrac{d[X]}{dt} = k_a[A][X] - k_b[X][Y]$

(ii) $\dfrac{d[Y]}{dt} = k_b[X][Y] - k_c[Y]$

and express them as finite-difference equations

(i) $X(t_{i+1}) = X(t_i) + k_a[A]X(t_i)\Delta t - k_b X(t_i)Y(t_i)\Delta t$

(ii) $Y(t_{i+1}) = Y(t_i) - k_c Y(t_i)\Delta t + k_b X(t_i)Y(t_i)\Delta t$

and iterate for different values of [A], X(0), and Y(0). For the steady state,

(i) $\dfrac{d[X]}{dt} = k_a[A][X] - k_b[X][Y] = 0$

(ii) $\dfrac{d[Y]}{dt} = k_b[X][Y] - k_c[Y] = 0$

which solve to

(i) $k_b[X] = k_c$ (ii) $k_a[A] = k_b[Y]$

Hence, $\boxed{[X] = \dfrac{k_c}{k_b}}$ $\boxed{[Y] = \dfrac{k_a[A]}{k_b}}$

P26.22 $A \rightarrow 2R$ $\mathcal{I}$

$A + R \rightarrow R + B$ k_2

$R + R \rightarrow R_2$ k_3

$\dfrac{d[A]}{dt} = \boxed{-\mathcal{I} - k_2[A][R]}$, $\dfrac{d[R]}{dt} = 2\mathcal{I} - 2k_3[R]^2 = 0$

The latter implies that $[R] = \left(\dfrac{\mathcal{I}}{k_3}\right)^{1/2}$, and so

$\dfrac{d[A]}{dt} = \boxed{-\mathcal{I} - k_2\left(\dfrac{\mathcal{I}}{k_3}\right)^{1/2}}[A]$

$\dfrac{d[B]}{dt} = k_2[A][R] = k_2\left(\dfrac{\mathcal{I}}{k_3}\right)^{1/2}[A]$

Therefore, only the combination $\dfrac{k_2}{k_3^{1/2}}$ may be determined if the reaction attains a steady state.

Comment. If the reaction can be monitored at short enough times so that termination is negligible compared to initiation, then $[R] \approx 2\mathcal{I}t$ and $\dfrac{d[B]}{dt} \approx 2k_2\mathcal{I}t\,[A]$. So monitoring B sheds light on just k_2.

P26.24 $\dfrac{d[Cr(CO)_5]}{dt} = I - k_2[Cr(CO)_5][CO] - k_3[Cr(CO)_5][M] + k_4[Cr(CO)_5M] = 0$ [steady state]

Hence, $[Cr(CO)_5] = \dfrac{I + k_4[Cr(CO)_5M]}{k_2[CO] + k_3[M]}$

$\dfrac{d[Cr(CO)_5M]}{dt} = k_3[Cr(CO)_5][M] - k_4[Cr(CO)_5M]$

Substituting for $[Cr(CO)_5]$ from above,

$\dfrac{d[Cr(CO)_5M]}{dt} = \dfrac{k_3I[M] - k_2k_4[Cr(CO)_5M][CO]}{k_2[CO] + k_3[M]} = -f[Cr(CO)_5M]$

if $f = \boxed{\dfrac{k_2k_4[CO]}{k_2[CO] + k_3[M]}}$

and we have taken $k_3 I[M] \ll k_2 k_4[\text{Cr(CO)}_5 M][CO]$. Therefore,

$$\frac{1}{f} = \frac{1}{k_4} + \frac{k_3[M]}{k_2 k_4[CO]}$$

and a graph of $\dfrac{1}{f}$ against $[M]$ should be a straight line.

Solutions to applications

P26.25   The rate equation is

$$\frac{dN}{dt} = bN - dN$$

which has the solution

$$\boxed{N(t) = N_0 e^{(b-d)t} = N_0 e^{kt}}$$

A least-squares fit to the above data gives

$$N_0 = 0.484 \times 10^9 \approx 0.5 \times 10^9$$

$$k = 9.19 \times 10^{-3}\,\text{y}^{-1}$$

$$R^2 = (\text{coefficient of determination}) = 0.983$$

$$\text{Standard error of estimate} = 0.130 \times 10^9$$

Thus, this model of population growth for the planet as a whole fits the data fairly well.

Comment. Despite the fact that the Malthusian model seems to fit the (admittedly crude) population data it has been much criticized. An alternative rate equation that takes into account the carrying capacity K of the planet is due to Verhulst (1836). This rate equation is

$$\frac{dN}{dt} = kN\left(1 - \frac{N}{K}\right)$$

Question. Does the Verhulst model fit our limited data any better?

P26.27 $\dfrac{d[P]}{dt} = \dfrac{k_b[E]_0[S]}{K_M + [S]}$ [26.19 with $[S] = [S]_0$]

Write $v = \dfrac{d[P]}{dt}$, then $\dfrac{1}{v} = \left(\dfrac{1}{k_b[E]_0}\right) + \left(\dfrac{K_M}{k_b[E]_0}\right) \times \left(\dfrac{1}{[S]}\right)$

We therefore draw up the following table

$10^3[S]/\text{mol L}^{-1}$	50	17	10	5	2
$\dfrac{1}{[S]/\text{mol L}^{-1}}$	20.0	58.8	100	200	500
$v/(\text{mm}^3\,\text{min}^{-1})$	16.6	12.4	10.1	6.6	3.3
$\dfrac{1}{v/\text{mm}^3\,\text{min}^{-1}}$	0.0602	0.0806	0.0990	0.152	0.303

The points are plotted in Fig 26.4

The intercept lies at 0.050, which implies that $\dfrac{1}{k_b[E]_0} = 0.050\,\text{mm}^{-3}\,\text{min}$. The slope is 5.06×10^{-4}, which implies that

$$\frac{K_M}{k_b[E]_0} = 5.06 \times 10^{-4}\,\text{mm}^{-3}\,\text{min mol L}^{-1}$$

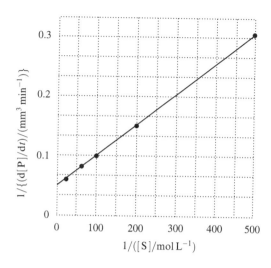

Figure 26.4

and therefore that $K_M = \dfrac{5.06 \times 10^{-4}\ \text{mm}^{-3}\ \text{min mol L}^{-1}}{0.050\ \text{mm}^{-3}\ \text{min}} = \boxed{0.010\ \text{mol L}^{-1}}$

P26.29 **(a)** The dissociation equilbrium may be rearranged to give the following relationships.

$$[E^-] = K_{E,a}[EH]/[H^+] \quad [EH_2^+] = [EH][H^+]/K_{E,b}$$

$$[ES^-] = K_{ES,a}[ESH]/[H^+] \quad [ESH_2] = [ESH][H^+]/K_{ES,b}$$

Mass balance provides an equation for [EH].

$$[E]_0 = [E^-] + [EH] + [EH_2^+] + [ES^-] + [ESH] + [ESH_2]$$

$$= \frac{K_{E,a}[EH]}{[H^+]} + [EH] + \frac{[EH][H^+]}{K_{E,b}} + \frac{K_{ES,a}[ESH]}{[H^+]} + [ESH] + \frac{[ESH][H^+]}{K_{ES,b}}$$

$$[EH] = \frac{[E]_0 - \left\{1 + \frac{[H^+]}{K_{ES,b}} + \frac{K_{ES,a}}{[H^+]}\right\}[ESH]}{1 + \frac{[H]^+}{K_{E,b}} + \frac{K_{E,a}}{[H^+]}}$$

$$= \frac{[E]_0 - c_1[ESH]}{c_2}$$

The steady-state approximation provides an equation for [ESH],

$$\frac{d[ESH]}{dt} = k_a[EH][S] - k_a'[ESH] - k_b[ESH] = 0$$

$$[ESH] = \frac{k_a}{k_a' + k_b}[EH][S] = K_M^{-1}[EH][S]$$

$$= K_M^{-1}[S]\left\{\frac{[E]_0 - c_1[ESH]}{c_2}\right\}$$

$$[ESH] = \frac{K_M^{-1}[S][E]_0/c_2}{1 + \frac{K_M^{-1}[S]c_1}{c_2}} = \frac{[E]_0/c_1}{1 + \frac{K_M(c_2/c_1)}{[S]}}$$

The rate law becomes:

$$v = d[P]/dt = k_b[ESH]$$

$$v = \frac{v'_{max}}{1 + k'_{M/[S]}}$$

where $\quad v'_{max} = \dfrac{k_b[E]_0}{\left\{1 + \frac{[H^+]}{K_{ES,b}} + \frac{K_{ES,a}}{[H^+]}\right\}}$

$$K'_M = K_M \left\{ \frac{1 + \frac{[H^+]}{K_{E,b}} + \frac{K_{E,a}}{[H^+]}}{1 + \frac{[H^+]}{K_{ES,b}} + \frac{K_{ES,a}}{[H^+]}} \right\}$$

(b) $v_{max} = 1.0 \times 10^{-6}\,mol\,L^{-1}\,s^{-1}$

$K_{ES,b} = 1.0 \times 10^{-6}\,mol\,L^{-1}$

$K_{ES,a} = 1.0 \times 10^{-8}\,mol\,L^{-1}$

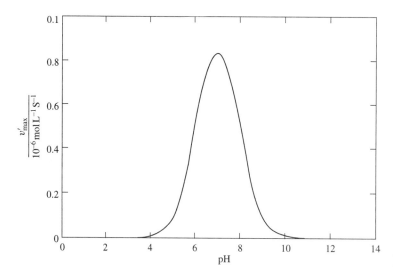

Figure 26.5(a)

The graph indicates a maximum value of v'_{max} at pH = 7.0 for this set of equilibrium and kinetic constants. A formula for the pH of the maximum can be derived by finding the point at which $\dfrac{dv'_{max}}{d[H^+]} = 0$. This gives:

$$[H^+]_{max} = (K_{ES,a}\,K_{ES,b})^{1/2}$$

Inserting constants, $[H^+]_{max} = \sqrt{(1.0 \times 10^{-8}\,mol\,L^{-1})(1.0 \times 10^{-6}\,mol\,L^{-1})}$

$$= 1.0 \times 10^{-7}\,mol\,L^{-1}$$

which corresponds to $\boxed{pH = 7.0}$

(c) $v_{max} = 1.0 \times 10^{-6}\,mol\,L^{-1}\,s^{-1}$
$K_{ES,b} = 1.0 \times 10^{-4}\,mol\,L^{-1}$
$K_{ES,a} = 1.0 \times 10^{-10}\,mol\,L^{-1}$

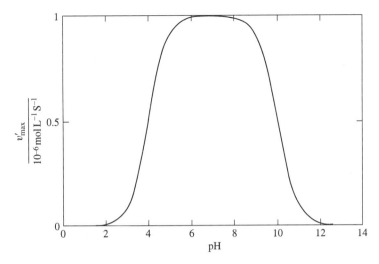

Figure 26.5(b)

The constants of part (**c**) give a much broader curve than do the constants of part (**b**). This reflects the behaviour of the term $1 + [H^+]/K_{ES,b} + K_{ES,a}/[H^+]$ in the denominator of the v'_{max} expression. When $K_{ES,b}$ is relatively large, large $[H^+]$ values (low pH) cause growth in the values of v'_{max}. However, when $K_{ES,a}$ is relatively small, very small $[H^+]$ values (high pH) cause a decline in the v'_{max} values.

P26.30   We construct Lineweaver–Burk plots for $[I] = 0$ and for $[I] = 15\,\mu mol\,L^{-1}$. If there is common vertical axis intercept, the inhibition is competitive. If there is a common horizontal axis intercept is noncompetitive. Draw up the following table.

(a) $\dfrac{1}{v}\Big/(\mu mol\,L^{-1}s^{-1})^{-1}$	2.04	1.05	0.77	0.67	0.62
(b) $\dfrac{1}{v}\Big/(\mu mol\,L^{-1}s^{-1})^{-1}$	3.7	1.92	1.41	1.23	1.16
$\dfrac{1}{[S]}\Big/(\mu mol\,L^{-1})^{-1}$		100	33	14	8.33	5.56

These data are plotted in Fig. 26.6 below. There is a common intercept on the horizontal axis. Therefore, the inhibition is non-competitive.

P26.31   The description of the progress of infectious diseases can be represented by the mechanism

$$S \rightarrow I \rightarrow R$$

Only the first step is autocatalytic as indicated in the first rate expression. If the three rate equations are added

$$\frac{dS}{dt} + \frac{dI}{dt} + \frac{dR}{dt} = 0$$

and, hence, there is no change with time of the total population, that is

$$S(t) + I(t) + R(t) = N$$

Whether the infection spreads or dies out is determined by

$$\frac{dI}{dt} = rSI - aI$$

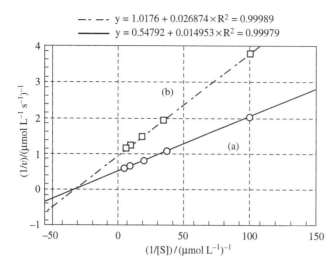

Figure 26.6

At $t = 0$, $I = I(0) = I_0$. Since the process is autocatalytic $I(0) \neq 0$.

$$\left(\frac{dI}{dt}\right)_{t=0} = I_0(rS_0 - a)$$

If $a > rS_0$, $\left(\dfrac{dI}{dt}\right)_{t=0} < 0$, and the infection dies out. If $a < rS$, $\left(\dfrac{dI}{dt}\right)_{t=0} > 0$ and the infection spreads (an epidemic). Thus

$$\boxed{\dfrac{a}{r} < S_0}\ \text{[infection spreads]}$$

$$\boxed{\dfrac{a}{r} > S_0}\ \text{[infection dies out]}$$

P26.33 **(a)**

$2NO \rightarrow N_2O + O$	k_a	initiation
$O + NO \rightarrow O_2 + N$	k_b	propagation
$N + NO \rightarrow N_2 + O$	k_c	propagation
$2O + M \rightarrow O_2 + M$	k_d	termination
$O_2 + M \rightarrow 2O + M$	k_{-d}	initiation

(b) $$\frac{d[NO]}{dt} = -2k_a[NO]^2 - k_b[O][NO] - k_c[N][NO]$$

To determine the steady-state concentration of N, $[N]_{SS}$, write the rate expression for $d[N]/dt$ and set it equal to zero.

$$\frac{d[N]}{dt} = k_b[O][NO] - k_c[N]_{SS}[NO] = 0$$

$$[N]_{SS} = \frac{k_b}{k_c}[O]$$

Substitution of $[N]_{SS}$ into the expression of $d[NO]/dt$ indicates that, under steady-state conditions for [N], $v_b = v_c$ and

$$\boxed{\frac{d[NO]}{dt} = -2k_a[NO]^2 - 2k_b[O][NO]}$$

[N] steady-state conditions.

If the propagation step is much more rapid than initiation, the last term predominates

$$\frac{d[NO]}{dt} = -2k_b[O][NO]$$

[N] is in steady-state and initiation is very slow.

If oxygen atoms and molecules are in equilibrium

$$2O + M \underset{k_{-d}}{\overset{k_d}{\rightleftharpoons}} O_2 + M$$

$$K_{O/O_2} = \frac{k_d}{k_{-d}} = \frac{[O_2]}{[O]^2}$$

$$[O] = \left(\frac{k_{-d}[O_2]}{k_d}\right)^{1/2}$$

Substitution into the previous rate expression yields

$$\frac{d[NO]}{dt} = -2k_b\left(\frac{k_{-d}}{k_d}\right)^{1/2}[O_2]^{1/2}[NO]$$

[N] is in steady-state; initiation is very slow; atomic and molecular oxygen are in equilibrium.

(c)   Since $k \propto e^{-E_a/RT}$ where E_a is the activation energy, we may write the individual rate constants in the form $k_i \propto e^{-E_i/RT}$ where the subscript 'a' has been dropped and 'i' represents the ith elementary step with activation energy E_i. Substitution of such expressions into the last equation of part **(b)** yields

$$\frac{d[NO]}{dt} \propto e^{-E_b/RT}\left(\frac{e^{-E_{-d}/RT}}{e^{-E_d/RT}}\right)^{1/2}[O_2]^{1/2}[NO]$$

$$\propto e^{-\left(\frac{-\left(E_b + \frac{1}{2}E_{-d} - \frac{1}{2}E_d\right)}{RT}\right)}[O_2]^{1/2}[NO]$$

We conclude that the effective activation energy, $E_{a,\text{eff}}$, is given by

$$E_{a,\text{eff}} = E_b + \tfrac{1}{2}E_{-d} - \tfrac{1}{2}E_d$$

(d)   Using the estimate that activation energies are approximately equal to the bond energies that must be broken

$$E_{a,\text{eff}} \approx B(NO) + \tfrac{1}{2}B(O_2) - \tfrac{1}{2}B(O)$$

$$\approx 630.57\,\text{kJ mol}^{-1} + \tfrac{1}{2}(498.36\,\text{kJ mol}^{-1}) - \tfrac{1}{2}(0)$$

$$\approx 879.75\,\text{kJ mol}^{-1}$$

where this is the unimolecular bond-breakage estimate of activation energies.

The previous estimate of $E_{a,\text{eff}}$ may be much too high because the activation energy of step (**b**) is probably being greatly overestimated. A more realistic estimate of E_b would be that difference between the energies of the NO bond that must be broken and the O_2 bond that is formed. Then,

$$E_{a,\text{eff}} \approx \{B(\text{NO}) - B(O_2)\} + \tfrac{1}{2}B(O_2) - \tfrac{1}{2}B(O)$$

$$\approx B(\text{NO}) - \tfrac{1}{2}B(O_2)$$

$$\approx 630.57\,\text{kJ}\,\text{mol}^{-1} - \tfrac{1}{2}(498.36\,\text{kJ}\,\text{mol}^{-1})$$

$$\approx \boxed{381.39\,\text{kJ}\,\text{mol}^{-1}}$$

The energy of the activated complex of step (**b**) is the difference between bond-breakage and bond-formation energies.

It is interesting to compare these estimates with the value based upon E_i values that have been determined by experiment

$$E_{a,\text{eff}} = (161\,\text{kJ}\,\text{mol}^{-1}) + \tfrac{1}{2}(493\,\text{kJ}\,\text{mol}^{-1}) - \tfrac{1}{2}(14\,\text{kJ}\,\text{mol}^{-1})$$

$$= 401\,\text{kJ}\,\text{mol}^{-1}$$

This value is based upon experimental activation energies for the elementary steps.

(**e**)   We now eliminate the assumption of O/O_2 equilibrium and assume that both [N] and [O] are at steady-state values. From part (b), $[N_{SS}] = k_b[O]_{SS}/k_c$

$$\frac{d[\text{O}]}{dt} = k_a[\text{NO}]^2 - k_b[\text{O}]_{SS}[\text{NO}] + k_c[\text{N}]_{SS}[\text{NO}] - 2k_d[\text{O}]_{SS}^2[\text{M}] + 2k_{-d}[O_2][\text{M}] = 0$$

$$k_a[\text{NO}]^2 - k_b[\text{O}]_{SS}[\text{NO}] + k_b[\text{O}]_{SS}[\text{NO}] - 2k_d[\text{O}]_{SS}^2[\text{M}] + 2k_{-d}[O_2][\text{M}] = 0$$

$$k_a[\text{NO}]^2 - 2k_d[\text{O}]_{SS}^2[\text{M}] + 2k_{-d}[O_2][\text{M}] = 0$$

At very low values of $[O_2]$ the last term is negligible so that

$$[\text{O}]_{SS} \approx \left(\frac{k_a}{2k_d[\text{M}]}\right)^{1/2}[\text{NO}]$$

Substitution of the expressions for $[N]_{SS}$ and $[O]_{SS}$ into the expression for $d[\text{NO}]/dt$ (top of part (**b**)) gives

$$\frac{d[\text{NO}]}{dt} = -2k_a[\text{NO}]^2 - k_b[\text{O}]_{SS}[\text{NO}] - k_c\left(\frac{k_b[\text{O}]_{SS}}{k_c}\right)[\text{NO}]$$

$$= -2k_a[\text{NO}]^2 - 2k_b[\text{O}]_{SS}[\text{NO}]$$

$$= -2k_a[\text{NO}]^2 - 2k_b\left(\frac{k_a}{2k_d[\text{M}]}\right)^{1/2}[\text{NO}]^2$$

If propagation is much more rapid than initiation so that $k_b\left(\dfrac{k_a}{2k_d[\text{M}]}\right)^{1/2} \gg k_a$, this expression becomes

$$\boxed{\frac{d[\text{NO}]}{dt} = -2k_b\left(\frac{k_a}{2k_d[\text{M}]}\right)^{1/2}[\text{NO}]^2}$$

(f) $NO + O_2 \rightarrow O + NO_2 \quad k_e$ initiation

$$\frac{d[NO]}{dt} = -2k_a[NO]^2 - k_b[O][NO] - k_c[N][NO] - k_e[NO][O_2]$$

if the conversion has proceeded to the extent that $[O_2]$ has become significant and $k_e[NO][O_2] \gg 2k_a[NO]^2$

$$\frac{d[NO]}{dt} = -k_b[O][NO] - k_c[N][NO] - k_e[NO][O_2]$$

Applying the steady-state approximation to both [N] and [O] gives $[N]_{SS} = k_b[O]_{SS}/k_c$ and

$$\frac{d[O]}{dt} = k_a[NO]^2 - k_b[O]_{SS}[NO] + k_c[N]_{SS}[NO] - 2k_d[O]^2_{SS}[M] + 2k_{-d}[O^2][M]$$

$$+ k_e[O_2][NO] = 0 \quad \text{and} \quad -k_b[O]_{SS}[NO] + k_c[N]_{SS}[NO] = 0$$

Thus $2k_d[O]^2_{SS}[M] - 2k_{-d}[O_2][M] - k_e[O_2][NO] = 0$

At high concentrations of O_2, species 'M' is likely to be O_2 and $k_d[O]^2_{SS}[M] \gg k_b[O]_{SS}[NO]$. The value of k_{-d} is so small that it can be neglected.

$$2k_d[O]^2_{SS}[M] \approx k_e[O_2][NO]$$

$$[O]_{SS} = \left(\frac{k_e}{2k_d[M]}\right)^{1/2} [O_2]^{1/2}[NO]^{1/2}$$

Substitution of $[N]_{SS}$ and $[O]_{SS}$ into the expression for $d[NO]/dt$ gives

$$\frac{d[NO]}{dt} = -k_b[O]_{SS}[NO] - k_c[N]_{SS}[NO] - k_e[NO][O_2]$$

$$= -2k_b[O]_{SS}[NO] - k_e[NO][O_2]$$

$$= -2k_b \left(\frac{k_e}{2k_d[M]}\right)^{1/2} [O_2]^{1/2}[NO]^{3/2} - k_e[NO][O_2]$$

If propagation is much more rapid than initiation, the expression becomes

$$\boxed{\frac{d[NO]}{dt} = -2k_b \left(\frac{k_e}{2k_d[M]}\right)^{1/2} [O_2]^{1/2}[NO]^{3/2}}$$

$$E_{a,eff} = E_b + \tfrac{1}{2}E_e - \tfrac{1}{2}E_d$$

Using the experimental values of E_i, $E_{a,eff}$ is estimated to be given by

$$E_{a,eff} = 161 \, kJ \, mol^{-1} + \tfrac{1}{2}(198 \, kJ \, mol^{-1}) - \tfrac{1}{2}(14 \, kJ \, mol^{-1})$$

$$= \boxed{253 \, kJ \, mol^{-1}}$$

This value is consistent with the low range of the experimental values of $E_{a,eff}$, whereas the value found in part **(b)** is consistent with the high experimental values.

27 Molecular reaction dynamics

Solutions to exercises

Discussion questions

E27.1(a) The harpoon mechanism accounts for the large steric factor of reactions of the kind $K + Br_2 \rightarrow KBr + Br$ in beams. It is supposed that an electron hops across from K to Br_2 when they are within a certain distance, and then the two resulting ions are drawn together by their mutual Coulombic attraction.

E27.2(a) The Eyring equation (eqn 27.53) results from activated complex theory which is an attempt to account for the rate constants of bimolecular reactions of the form $A + B \rightleftharpoons C^{\ddagger} \rightarrow P$ in terms of the formation of an activated complex. In the formulation of the theory, it is assumed that the activated complex and the reactants are in equilibrium, and the concentration of activated complex is calculated in terms of an equilibrium constant, which in turn is calculated from the partition functions of the reactants and a postulated form of the activated complex. It is further supposed that one normal mode of the activated complex, the one corresponding to displacement along the reaction coordinate, has a very low force constant and displacement along this normal mode leads to products provided that the complex enters a certain configuration of its atoms, which is known as the transition state. The derivation of the equilibrium constant from the partition functions leads to eqn 27.51 and in turn to eqn 27.53, the Eyring equation. See Section 27.4 for a more complete discussion of a complicated subject.

E27.3(a) *Infrared chemiluminescence.* Chemical reactions may yield products in excited states. The emission of radiation as the molecules decay to lower energy states is called chemiluminescence. If the emission is from vibrationally excited states, then it is infrared chemiluminescence. The vibrationally excited product molecule in the example of Figure 27.13 in the text is CO. By studying the intensities of the infrared emission spectrum, the populations of the vibrational states in the product CO may be determined and this information allows us to determine the relative rates of formation of CO in these excited states.

  Multi-photon ionization (MPI). Multi-photon absorption is the absorption of two or more photons by the molecule in its transition to a higher electronic state. The frequencies of the photons satisfy the condition

$$\Delta E = h\nu_1 + h\nu_2 + \cdots$$

which is similar to the frequency condition for one-photon absorption. However, multi-photon selection rules are different from one-photon selection rules. Therefore, multi-photon processes allows examination of energy states that otherwise could not be reached. In multi-photon ionization, the second or third photon takes the molecule into the energy continuum above its highest lying energy state. This technique is especially useful for the study of weakly fluorescing molecules.

  Resonant multi-photon ionization (REMPI). This is a variant of MPI described above, in which one or more photons promote a molecule to an electronically excited state and then additional photons generate ions from the excited state. The power of this method in the study of chemical reactions is its selectivity. In a chemically reacting system, individual reactants and products can be chosen by tuning the frequency of the laser generating the radiation to the electronic absorption band of specific molecules.

  Reaction product imaging. In this technique, product ions are accelerated by an electric field toward a phosphorescent screen and the light emitted from the screen is imaged by a charge-coupled device. The significance of this experiment to the study of chemical reactions is that it allows for a detailed analysis of the angular distribution of products.

Femtosecond spectroscopy. See Box 27.1 for a more detailed discussion. Until recently, because of their exceedingly short lifetimes, there have been no direct observations of the activated complexes postulated to exist in the transition state of chemical reactions. But, after the development of femtosecond pulsed lasers, species resembling activated complexes can now be studied spectroscopically. Transitions to and from the activated complex have been observed and such experiments have greatly extended our knowledge of the dynamics of chemical reactions.

Numerical exercises

E27.4(a)
$$z = \frac{2^{1/2}\sigma\bar{c}p}{kT} \quad [24.12 \text{ with } 24.9]$$

and $\bar{c} = \left(\dfrac{8RT}{\pi M}\right)^{1/2}$ [Example 24.1] $= \left(\dfrac{8kT}{\pi m}\right)^{1/2}$

Therefore, $z = \dfrac{4\sigma p}{(\pi mkT)^{1/2}}$ with $\sigma \approx \pi d^2 \approx 4\pi R^2$

The collision frequency z gives the number of collisions made by a single molecule. We can obtain the *total* collision frequency, the rate of collisions between all the molecules in the gas, by multiplying z by $\frac{1}{2}N$ (the factor $\frac{1}{2}$ ensures that the $A \ldots A'$ and $A' \ldots A$ collisions are counted as one). Therefore the collision density, Z, the total number of collisions per unit time per unit volume, is

$$Z_{AA} = \frac{\frac{1}{2}zN}{V} = \frac{\sigma\bar{c}}{2^{1/2}}\left(\frac{N}{V}\right)^2$$

Introducing the expression for $\bar{c}$ above

$$Z_{AA} = \sigma\left(\frac{4kT}{\pi m}\right)^{1/2} \times \left(\frac{N}{V}\right)^2 = \sigma\left(\frac{4kT}{\pi m}\right)^{1/2} \times \left(\frac{p}{kT}\right)^2 \quad [N/V = p/kT]$$

We express these equations in the form

$$z = \frac{(16\pi R^2) \times (1.00 \times 10^5 \text{ Pa})}{\{(\pi) \times (M/\text{g mol}^{-1}) \times (1.6605 \times 10^{-27} \text{ kg}) \times (1.381 \times 10^{-23} \text{ J K}^{-1}) \times (298.15 \text{ K})\}^{1/2}}$$

$$= \frac{(1.08 \times 10^{30} \text{ m}^{-2} \text{ s}^{-1}) \times R^2}{(M/\text{g mol}^{-1})^{1/2}} = \frac{1.08 \times 10^6 \times (R/\text{pm})^2 \text{ s}^{-1}}{(M/\text{g mol}^{-1})^{1/2}}$$

$$Z_{AA} = 4\pi R^2 \left(\frac{(4) \times (1.381 \times 10^{-23} \text{ J K}^{-1}) \times (298.15 \text{ K})}{(\pi) \times (M/\text{g mol}^{-1}) \times (1.6605 \times 10^{-27} \text{ kg})}\right)^{1/2}$$

$$\times \left(\frac{1.00 \times 10^5 \text{ Pa}}{(1.381 \times 10^{-23} \text{ J K}^{-1}) \times (298.15 \text{ K})}\right)^2$$

$$= \frac{(1.32 \times 10^{55} \text{ m}^{-5} \text{ s}^{-1}) \times R^2}{(M/\text{g mol}^{-1})^{1/2}} = \frac{1.35 \times 10^{31}(R/\text{pm})^2}{(M/\text{g mol}^{-1})^{1/2}} \text{ m}^{-3} \text{ s}^{-1}$$

For NH_3, $R = 190 \text{ pm}$, $M = 17 \text{ g mol}^{-1}$

$$z = \frac{(1.08 \times 10^6) \times (190^2 \text{ s}^{-1})}{17^{1/2}} = \boxed{9.5 \times 10^9 \text{ s}^{-1}}$$

$$Z_{AA} = \frac{(1.32 \times 10^{31}) \times (190^2\,\text{m}^{-3}\,\text{s}^{-1})}{17^{1/2}} = \boxed{1.2 \times 10^{35}\,\text{m}^{-3}\,\text{s}^{-1}}$$

For the percentage increase at constant volume, use

$$\frac{1}{z}\frac{dz}{dT} = \frac{1}{\bar{c}}\frac{d\bar{c}}{dT} = \frac{1}{2T}, \qquad \left(\frac{1}{Z}\right)\left(\frac{dZ}{dT}\right) = \frac{1}{2T}$$

Therefore, $\dfrac{\delta z}{z} \approx \dfrac{\delta T}{2T}$ and $\dfrac{\delta T}{Z} \approx \dfrac{\delta T}{2T}$

and since $\dfrac{\delta T}{T} = \dfrac{10\,\text{K}}{298\,\text{K}} = 0.034$, both z and Z increase by about $\boxed{1.7\ \text{per cent}}$

E27.5(a)　In each case use $f = e^{-E_a/RT}$

(a) $\dfrac{E_a}{RT} = \dfrac{10 \times 10^3\,\text{J mol}^{-1}}{(8.314\,\text{J K}^{-1}\,\text{mol}^{-1}) \times (300\,\text{K})} = 4.01,$　$f = e^{-4.01} = \boxed{0.018}$

$\dfrac{E_a}{RT} = \dfrac{10 \times 10^3\,\text{J mol}^{-1}}{(8.314\,\text{J K}^{-1}\,\text{mol}^{-1}) \times (1000\,\text{K})} = 1.20,$　$f = e^{-1.20} = \boxed{0.30}$

(b) $\dfrac{E_a}{RT} = \dfrac{100 \times 10^3\,\text{J mol}^{-1}}{(8.314\,\text{J K}^{-1}\,\text{mol}^{-1}) \times (300\,\text{K})} = 40.1,$　$f = e^{-40.1} = \boxed{3.9 \times 10^{-18}}$

$\dfrac{E_a}{RT} = \dfrac{100 \times 10^3\,\text{J mol}^{-1}}{(8.314\,\text{J K}^{-1}\,\text{mol}^{-1}) \times (1000\,\text{K})} = 12.0,$　$f = e^{-12.0} = \boxed{6.0 \times 10^{-6}}$

E27.6(a)　The percentage increase is

$$(100) \times \left(\frac{\delta f}{f}\right) \approx (100) \times \left(\frac{df}{dT}\right) \times \left(\frac{\delta T}{f}\right) \approx \frac{100 E_a}{RT^2}\delta T$$

(a)　$E_a = 10\,\text{kJ mol}^{-1}, \qquad \delta T = 10\,\text{K}$

$$(100)\left(\frac{\delta f}{f}\right) = \frac{(100) \times (10 \times 10^3\,\text{J mol}^{-1}) \times (10\,\text{K})}{(8.314\,\text{J K}^{-1}\,\text{mol}^{-1}) \times (T^2)}$$

$$= \frac{1.20 \times 10^6}{(T/\text{K})^2} = \begin{cases} \boxed{13\ \text{per cent}}\ \text{at}\ 300\,\text{K} \\ \boxed{1.2\ \text{per cent}}\ \text{at}\ 1000\,\text{K} \end{cases}$$

(b)　$E_a = 100\,\text{kJ mol}^{-1}, \qquad \delta T = 10\,\text{K}$

$$(100)\left(\frac{\delta f}{f}\right) = \frac{1.20 \times 10^7}{(T/\text{K})^2} = \begin{cases} \boxed{130\ \text{per cent}}\ \text{at}\ 300\,\text{K} \\ \boxed{12\ \text{per cent}}\ 1000\,\text{K} \end{cases}$$

E27.7(a)　$k_2 = \sigma \left(\dfrac{8kT}{\pi\mu}\right)^{1/2} N_A e^{-E_a/RT}$ [27.16 with 27.9]

The activation energy E_a to be used in this formula is related to the experimental activation by

$$E_a = E_a^{\text{exp}} - \tfrac{1}{2}RT \text{ [Footnote 2, p. 948]}$$

$$= (1.71 \times 10^5\,\text{J mol}^{-1}) - \left(\tfrac{1}{2}\right) \times (8.314\,\text{J K}^{-1}\,\text{mol}^{-1}) \times (650\,\text{K})$$

$$= 1.68\bar{3} \times 10^5\,\text{J mol}^{-1}$$

$$e^{-E_a/RT} = e^{-1.68\overline{3}\times 10^5 \,\mathrm{J\,mol^{-1}}/(8.314\,\mathrm{J\,K^{-1}\,mol^{-1}}\times 650\,\mathrm{K})} = 2.9\overline{9} \times 10^{-14}$$

$$\left(\frac{8kT}{\pi\mu}\right)^{1/2} = \left(\frac{(8)\times(1.381\times 10^{-23}\,\mathrm{J\,K^{-1}})\times(650\,\mathrm{K})}{(\pi)\times(3.32\times 10^{-27}\,\mathrm{kg})}\right)^{1/2} = 2.62\overline{3} \times 10^3\,\mathrm{m\,s^{-1}}$$

$$k_2 = (0.36\times 10^{-18}\,\mathrm{m^2})\times(2.62\overline{3}\times 10^3\,\mathrm{m\,s^{-1}})\times(6.022\times 10^{23}\,\mathrm{mol^{-1}})\times(2.9\overline{9}\times 10^{-14})$$

$$= 1.7\times 10^{-5}\,\mathrm{m^3\,mol^{-1}\,s^{-1}} = \boxed{1.7\times 10^{-2}\,\mathrm{L\,mol^{-1}\,s^{-1}}}$$

Comment. Estimates of collision cross-sections are notoriously variable. For the $H_2 + I_2$ reaction they have ranged from $0.28\,\mathrm{nm^2}$ to $0.50\,\mathrm{nm^2}$. However, that factor alone will not account for the differences between theoretical and experimental values of k_2. See Example 27.1.

E27.8(a) $k_d = 4\pi R^* D N_A$ [27.27]

$D = D_A + D_B = 2\times 5\times 10^{-9}\,\mathrm{m^2\,s^{-1}} = 1\times 10^{-8}\,\mathrm{m^2\,s^{-1}}$

$k_d = (4\pi)\times(0.4\times 10^{-9}\,\mathrm{m})\times(1\times 10^{-8}\,\mathrm{m^2\,s^{-1}})\times(6.02\times 10^{23}\,\mathrm{mol^{-1}}) = 3\times 10^7\,\mathrm{m^3\,mol^{-1}\,s^{-1}}$

$= \boxed{3\times 10^{10}\,\mathrm{L\,mol^{-1}\,s^{-1}}}$

E27.9(a) $k_d = \dfrac{8RT}{3\eta}$ [27.35] $= \dfrac{(8)\times(8.314\times\mathrm{J\,K^{-1}\,mol^{-1}})\times(298\,\mathrm{K})}{3\eta}$

$= \dfrac{6.61\times 10^3\,\mathrm{J\,mol^{-1}}}{\eta} = \dfrac{6.61\times 10^3\,\mathrm{kg\,m^2\,s^{-2}\,mol^{-1}}}{(\eta/\mathrm{kg\,m^{-1}s^{-1}})\times\mathrm{kg\,m^{-1}\,s^{-1}}} = \dfrac{6.61\times 10^3\,\mathrm{m^3\,mol^{-1}\,s^{-1}}}{(\eta/\mathrm{kg\,m^{-1}\,s^{-1}})}$

$= \dfrac{6.61\times 10^6\,\mathrm{M^{-1}\,s^{-1}}}{(\eta/\mathrm{kg\,m^{-1}\,s^{-1}})} = \dfrac{6.61\times 10^9\,\mathrm{M^{-1}\,s^{-1}}}{(\eta/\mathrm{cP})}$

(a) Water, $\eta = 1.00\,\mathrm{cP}$,

$$k_d = \frac{6.61\times 10^9}{1.00}\,\mathrm{L\,mol^{-1}\,s^{-1}} = 6.61\times 10^9\,\mathrm{L\,mol^{-1}\,s^{-1}}$$

$$= \boxed{6.61\times 10^6\,\mathrm{m^3\,mol^{-1}\,s^{-1}}}$$

(b) Pentane, $\eta = 0.22\,\mathrm{cP}$,

$$k_d = \frac{6.61\times 10^9}{0.22}\,\mathrm{L\,mol^{-1}\,s^{-1}} = 3.0\times 10^{10}\,\mathrm{L\,mol^{-1}\,s^{-1}}$$

$$= \boxed{3.0\times 10^7\,\mathrm{m^3\,mol^{-1}\,s^{-1}}}$$

E27.10(a) $k_d = \dfrac{8RT}{3\eta} = \dfrac{(8)\times(8.314\,\mathrm{J\,K^{-1}\,mol^{-1}})\times(298\,\mathrm{K})}{(3)\times(0.89\times 10^{-3}\,\mathrm{kg\,m^{-1}\,s^{-1}})}$

$= 7.4\times 10^6\,\mathrm{m^3\,mol^{-1}\,s^{-1}} = \boxed{7.4\times 10^9\,\mathrm{L\,mol^{-1}\,s^{-1}}}$

Since this reaction is elementary bimolecular it is second-order; hence

$$t_{1/2} = \frac{1}{k_d[A]_0}\ \text{[Table 25.3]}$$

$$= \frac{1}{(7.4\times 10^9\,\mathrm{L\,mol^{-1}\,s^{-1}})\times(1.0\times 10^{-3}\,\mathrm{mol\,L^{-1}})} = \boxed{1.4\times 10^{-7}\,\mathrm{s}}$$

E27.11(a) $P = \dfrac{\sigma^*}{\sigma}$ [Section 27.1(c)]

For the mean collision cross-section, write $\sigma_A = \pi d_A{}^2$, $\sigma_B = \pi d_B{}^2$, and $\sigma = \pi d^2$, with $d = \frac{1}{2}(d_A + d_B)$

$$\sigma = \tfrac{1}{4}\pi(d_A + d_B)^2 = \tfrac{1}{4}\pi(d_A{}^2 + d_B{}^2 + 2d_A d_B)$$

$$= \tfrac{1}{4}(\sigma_A + \sigma_B + 2\sigma_A{}^{1/2}\sigma_B{}^{1/2})$$

$$= \tfrac{1}{4}\{0.95 + 0.65 + 2 \times (0.95 \times 0.65)^{1/2}\}\,\text{nm}^2 = 0.793\,\text{nm}^2$$

Therefore, $P \approx \dfrac{9.2 \times 10^{-22}\,\text{m}^2}{0.793 \times 10^{-18}\,\text{m}^2} = \boxed{1.2 \times 10^{-3}}$

E27.12(a) Since the reaction is assumed to be elementary bimolecular, it is necessarily second-order; hence

$$\frac{d[P]}{dt} = k_2[A][B]$$

$$k_2 = 4\pi R^* D N_A\,[27.27] = 4\pi R^*(D_A + D_B)N_A$$

$$= \frac{2kTN}{3\eta}(R_A + R_B) \times \left(\frac{1}{R_A} + \frac{1}{R_B}\right)$$

$$= \frac{2RT}{3\eta}(R_A + R_B) \times \left(\frac{1}{R_A} + \frac{1}{R_B}\right)$$

$$= \frac{(2) \times (8.314\,\text{J K}^{-1}\,\text{mol}^{-1}) \times (313\,\text{K})}{(3) \times (2.37 \times 10^{-3}\,\text{kg m}^{-1}\,\text{s}^{-1})} \times (294 + 825) \times \left(\frac{1}{294} + \frac{1}{825}\right)$$

$$= 3.8 \times 10^6\,\text{mol}^{-1}\,\text{m}^3\,\text{s}^{-1} = 3.8 \times 10^9\,\text{L mol}^{-1}\,\text{s}^{-1}$$

Therefore, the initial rate is

$$\frac{d[P]}{dt} = (3.8 \times 10^9\,\text{L mol}^{-1}\,\text{s}^{-1}) \times (0.150\,\text{mol L}^{-1}) \times (0.330\,\text{mol L}^{-1})$$

$$= \boxed{1.9 \times 10^8\,\text{mol L}^{-1}\,\text{s}^{-1}}$$

Comment. If eqn 27.35 is used in place of eqn 27.27, $k_2 = 2.9 \times 10^9\,\text{L mol}^{-1}\,\text{s}^{-1}$ which yields $\dfrac{d[P]}{dt} = 1.4 \times 10^8\,\text{mol L}^{-1}\,\text{s}^{-1}$. In this case the approximation that led to eqn 27.35 results in a difference of ~ 30 per cent.

E27.13(a) For reactions in solution the relation between energy and enthalpy of activation is $\Delta^{\ddagger}H = E_a - RT$ [Footnote 7, p 961]

$$k_2 = B\,e^{\Delta^{\ddagger}S/R}e^{-\Delta^{\ddagger}H/RT}, \quad B = \left(\frac{kT}{h}\right) \times \left(\frac{RT}{p^{\ominus}}\right) \text{ [27.60]}$$

$$= B\,e^{\Delta^{\ddagger}S/R}e^{-E_a/RT}\,e = A\,e^{-E_a/RT}$$

Therefore, $A = e\,B\,e^{\Delta^{\ddagger}S/R}$, implying that $\Delta^{\ddagger}S = R\left(\ln\dfrac{A}{B} - 1\right)$

Therefore, since $E_a = 8681\,\text{K} \times R$

$$\Delta^{\ddagger}H = E_a - RT = (8681\,\text{K} - 303\,\text{K})R$$

$$= (8378\,\text{K}) \times (8.314\,\text{J K}^{-1}\,\text{mol}^{-1}) = \boxed{69.7\,\text{kJ mol}^{-1}}$$

$$B = \frac{(1.381 \times 10^{-23}\,\text{J K}^{-1}) \times (303\,\text{K})}{6.626 \times 10^{-34}\,\text{J s}} \times \frac{(8.314\,\text{J K}^{-1}\,\text{mol}^{-1}) \times (303\,\text{K})}{10^5\,\text{Pa}}$$

$$= 1.59 \times 10^{11}\,\text{m}^3\,\text{mol}^{-1}\,\text{s}^{-1} = 1.59 \times 10^{14}\,\text{L mol}^{-1}\,\text{s}^{-1}$$

and hence $\Delta^{\ddagger}S = R\left[\ln\left(\dfrac{2.05 \times 10^{13}\,\text{L mol}^{-1}\,\text{s}^{-1}}{1.59 \times 10^{14}\,\text{L mol}^{-1}\,\text{s}^{-1}}\right) - 1\right]$

$$= 8.314\,\text{J K}^{-1}\,\text{mol}^{-1} \times (-3.05) = \boxed{-25\,\text{J K}^{-1}\,\text{mol}^{-1}}$$

E27.14(a) $\Delta^{\ddagger}H = E_a - RT$ [Exercise 27.13(a)]

$\Delta^{\ddagger}H = (9134\,\text{K} - 303\,\text{K}) \times (8.314\,\text{J K}^{-1}\,\text{mol}^{-1}) = +73.4\,\text{kJ mol}^{-1}$

$\Delta^{\ddagger}S = R\left(\ln\dfrac{A}{B} - 1\right)$ [Exercise 27.13(a)]

with $B = \left(\dfrac{kT}{h}\right) \times \left(\dfrac{RT}{p^{\ominus}}\right)$ [27.60] $= 1.59 \times 10^{14}\,\text{L mol}^{-1}\,\text{s}^{-1}$ at 30°C

Therefore, $\Delta^{\ddagger}S = 8.314\,\text{J K}^{-1}\,\text{mol}^{-1} \times \left[\ln\left(\dfrac{7.78 \times 10^{14}}{1.59 \times 10^{14}}\right) - 1\right] = +4.9\,\text{J K}^{-1}\,\text{mol}^{-1}$

Hence, $\Delta^{\ddagger}G = \Delta^{\ddagger}H - T\Delta^{\ddagger}S = \{(73.4) - (303) \times (4.9 \times 10^{-3})\}\,\text{kJ mol}^{-1} = \boxed{+71.9\,\text{kJ mol}^{-1}}$

E27.15(a) $\Delta^{\ddagger}H = E_a - 2RT$

$$= \{(56.8) - (2) \times (8.314 \times 10^{-3}) \times (338)\}\,\text{kJ mol}^{-1} = 51.2\,\text{kJ mol}^{-1}$$

$k_2 = A\,e^{-E_a/RT}$ implies that

$A = k_2\,e^{E_a/RT} = 7.84 \times 10^{-3}\,\text{kPa}^{-1}\,\text{s}^{-1} \times e^{58.6 \times 10^3/(8.314 \times 338)}$

$$= 4.70\overline{5} \times 10^6\,\text{kPa}^{-1}\,\text{s}^{-1} = 4.70\overline{5} \times 10^3\,\text{Pa}^{-1}\,\text{s}^{-1}$$

In terms of molar concentrations

$$v = k_2\,p_A\,p_B = k_2(RT)^2[A][B]$$

and instead of $\dfrac{dp_A}{dt} = -k_2\,p_A\,p_B$

we have $\dfrac{d[A]}{dt} = -k_2\,RT[A][B]$

and hence use

$A = (4.70\overline{5} \times 10^3\,\text{Pa}^{-1}\,\text{s}^{-1}) \times (8.314\,\text{J K}^{-1}\,\text{mol}^{-1}) \times (338\,\text{K}) = 1.32\overline{2} \times 10^7\,\text{m}^3\,\text{mol}^{-1}\,\text{s}^{-1}$

Then $B = \dfrac{kT}{h} \times \dfrac{RT}{p^{\ominus}}$ [27.60] $= \dfrac{(1.381 \times 10^{-23}) \times (338\,\text{K})}{6.626 \times 10^{-34}\,\text{J s}} \times \dfrac{(8.314\,\text{J K}^{-1}\,\text{mol}^{-1}) \times (338\,\text{K})}{10^5\,\text{Pa}}$

$$= 1.98 \times 10^{11}\,\text{m}^3\,\text{s}^{-1}\,\text{mol}^{-1}$$

and $\Delta^{\ddagger}S = R\left[\ln\left(\dfrac{A}{B}\right) - 2\right]$ [27.62] $= (8.314\,\text{J K}^{-1}\,\text{mol}^{-1}) \times \left\{\ln\left(\dfrac{1.32\overline{2} \times 10^7}{1.98 \times 10^{11}}\right) - 2\right\}$

$$= \boxed{-96.6\,\text{J K}^{-1}\,\text{mol}^{-1}}$$

and hence $\Delta^{\ddagger}G = \Delta^{\ddagger}H - T\,\Delta^{\ddagger}S = [(51.2) - (338) \times (-96.6 \times 10^{-3})]\,\text{kJ mol}^{-1}$

$$= +83.9\,\text{kJ mol}^{-1}$$

E27.16(a) $k_2 = N_A \sigma^* \left(\dfrac{8kT}{\pi\mu}\right)^{1/2} e^{-\Delta E_0/RT}$ [27.56]

The pre-exponential factor is

$$A = N_A \sigma^* \left(\frac{8kT}{\pi\mu}\right)^{1/2}$$

Therefore, $\dfrac{A}{B} = \left(\dfrac{N_A \sigma^* hp^{\ominus}}{kT \times RT}\right) \times \left(\dfrac{8kT}{\pi\mu}\right)^{1/2} = \dfrac{8^{1/2}\sigma^* hp^{\ominus}}{(\pi\mu k^3 T^3)^{1/2}}$

For identical particles, $\mu = \frac{1}{2}m$, so

$$\frac{A}{B} = \frac{4\sigma^* hp^{\ominus}}{(\pi m k^3 T^3)^{1/2}}$$

$$= \frac{(4) \times (0.4 \times 10^{-18}\,\text{m}^2) \times (6.626 \times 10^{-34}\,\text{J s}) \times (10^5\,\text{Pa})}{\{(\pi) \times (50) \times (1.6605 \times 10^{-27}\,\text{kg}) \times (1.381 \times 10^{-23}\,\text{J K}^{-1} \times 300\,\text{K})^3\}^{1/2}}$$

$$= 7.78 \times 10^{-4}$$

and hence $\Delta^{\ddagger}S = R\left[\ln\left(\dfrac{A}{B}\right) - 2\right]$ [from eqn 27.62] $= 8.314\,\text{J K}^{-1}\,\text{mol}^{-1}\{\ln 7.78 \times 10^{-4} - 2\}$

$$= \boxed{-76\,\text{J K}^{-1}\,\text{mol}^{-1}}$$

E27.17(a) $B = \left(\dfrac{kT}{h}\right) \times \left(\dfrac{RT}{p^{\ominus}}\right)$ [27.60]

$$= \left(\frac{(1.381 \times 10^{-23}\,\text{J K}^{-1}) \times (298.15\,\text{K})}{6.626 \times 10^{-34}\,\text{J s}}\right) \times \left(\frac{(8.314\,\text{J K}^{-1}\,\text{mol}^{-1}) \times (298.15\,\text{K})}{10^5\,\text{Pa}}\right)$$

$$= 1.540 \times 10^{11}\,\text{m}^3\,\text{mol}^{-1}\,\text{s}^{-1} = 1.540 \times 10^{14}\,\text{L mol}^{-1}\,\text{s}^{-1}$$

Therefore

(a) $\Delta^{\ddagger}S = R\left[\ln\left(\dfrac{4.6 \times 10^{12}}{1.540 \times 10^{14}}\right) - 2\right] = \boxed{-45.8\,\text{J K}^{-1}\,\text{mol}^{-1}}$

(b) $\Delta^{\ddagger}H = E_a - 2RT = \{(10.0) - (2) \times (2.48)\}\,\text{kJ mol}^{-1} = \boxed{+5.0\,\text{kJ mol}^{-1}}$

(c) $\Delta^{\ddagger}G = \Delta^{\ddagger}H - T\Delta^{\ddagger}S = \{(5.0) - (298.15) \times (-45.8 \times 10^{-3})\}\,\text{kJ mol}^{-1} = \boxed{+18.7\,\text{kJ mol}^{-1}}$

E27.18(a) $\log k_2 = \log k_2^{\circ} + 2Az_A z_B I^{1/2}$ [27.69]

Hence

$$\log k_2^{\circ} = \log k_2 - 2Az_A z_B I^{1/2}$$

$$= (\log 12.2) - (2) \times (0.509) \times (1) \times (-1) \times (0.0525^{1/2}) = 1.32$$

$$k_2^{\circ} = \boxed{20.9\,\text{L}^2\,\text{mol}^{-2}\,\text{min}^{-1}}$$

Solutions to problems

Solutions to numerical problems

P27.2 Draw up the following table as the basis of an Arrhenius plot

T/K	600	700	800	1000
$10^3\,\text{K}/T$	1.67	1.43	1.25	1.00
$k/(\text{cm}^3\,\text{mol}^{-1}\,\text{s}^{-1})$	4.6×10^2	9.7×10^3	1.3×10^5	3.1×10^6
$\ln(k/\text{cm}^3\,\text{mol}^{-1}\,\text{s}^{-1})$	6.13	9.18	11.8	14.9

The points are plotted in Fig. 27.1.

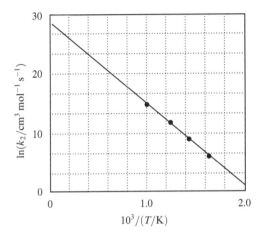

Figure 27.1

The least-squares intercept is at 28.3, which implies that

$$A/(\text{cm}^3\,\text{mol}^{-1}\,\text{s}^{-1}) = e^{28.3} = 2.0 \times 10^{12}$$

From $A = N_A \sigma^* \left(\dfrac{8kT}{\pi \mu}\right)^{1/2}$ [Exercise 27.16(a)]

$$\sigma^* = \frac{A_{\text{exptl}}}{N_A(8kT/\pi\mu)^{1/2}} \quad \text{with } \mu = \tfrac{1}{2}m(\text{NO}_2)$$

$$= \left(\frac{A_{\text{exptl}}}{4N_A}\right)\left(\frac{\pi m}{kT}\right)^{1/2} = \left(\frac{2.0 \times 10^6\,\text{m}^3\,\text{mol}^{-1}\,\text{s}^{-1}}{(4) \times (6.022 \times 10^{23}\,\text{mol}^{-1})}\right)$$

$$\times \left(\frac{(\pi) \times (46\,\text{u}) \times (1.6605 \times 10^{-27}\,\text{kg}\,\text{u}^{-1})}{(1.381 \times 10^{-23}\,\text{J}\,\text{K}^{-1}) \times (750\,\text{K})}\right)^{1/2}$$

$$= 4.0 \times 10^{-21}\,\text{m}^2 \quad \text{or} \quad \boxed{4.0 \times 10^{-3}\,\text{nm}^2}$$

$$P = \frac{\sigma^*}{\sigma} = \frac{4.0 \times 10^{-3}\,\text{nm}^2}{0.60\,\text{nm}^2} = \boxed{0.007}$$

P27.4   Draw up the following table for an Arrhenius plot

$\theta/°C$	-24.82	-20.73	-17.02	-13.00	-8.95
T/K	248.33	252.42	256.13	260.15	264.20
$10^3/(T/K)$	4.027	3.962	3.904	3.844	3.785
$\ln(k/s^{-1})$	-9.01	-8.37	-7.73	-7.07	-6.55

The points are plotted in Fig. 27.2.

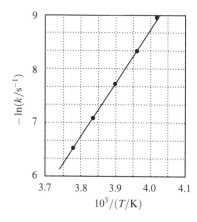

Figure 27.2

A least-squares fit of the data yields the intercept $+32.6$ at $\dfrac{1}{T} = 0$ and slope -10.33×10^3 K.
The former implies that $\ln\left(\dfrac{A}{s^{-1}}\right) = 32.6$, and hence that $A = 1.4 \times 10^{14}\,s^{-1}$. The slope yields
$\dfrac{E_a}{R} = 10.33 \times 10^3$ K, and hence $E_a = \boxed{85.9\,\text{kJ mol}^{-1}}$

In solution $\Delta^{\ddagger}H = E_a - RT$, so at $-20°C$

$$\Delta^{\ddagger}H = (85.9\,\text{kJ mol}^{-1}) - (8.314\,\text{J K}^{-1}\,\text{mol}^{-1}) \times (253\,\text{K})$$

$$= \boxed{83.8\,\text{kJ mol}^{-1}}$$

We assume that the reaction is first-order for which, by analogy to Section 27.4

$$K^{\ddagger} = K = \frac{kT}{h\nu}\overline{K}^{\ddagger}$$

and $k_1 = k^{\ddagger}\,K^{\ddagger} = \nu \times \dfrac{kT}{h\nu} \times \overline{K}^{\ddagger}$

with $\Delta^{\ddagger}G = -RT\,\ln\overline{K}^{\ddagger}$

Therefore, $k_1 = A\,e^{-E_a/RT} = \dfrac{kT}{h}\,e^{-\Delta^{\ddagger}G/RT} = \dfrac{kT}{h}\,e^{\Delta^{\ddagger}S/R}\,e^{-\Delta^{\ddagger}H/RT}$

and hence we can identify $\Delta^{\ddagger}S$ by writing

$$k_1 = \frac{kT}{h}\,e^{\Delta^{\ddagger}S/R}\,e^{-E_a/RT}\,e = A\,e^{-E_a/RT}$$

and hence obtain

$$\Delta^{\ddagger}S = R\left[\ln\left(\frac{hA}{kT}\right) - 1\right]$$

$$= 8.314 \, \text{J K}^{-1} \, \text{mol}^{-1} \times \left[\ln\left(\frac{(6.626 \times 10^{-34} \, \text{J s}) \times (1.4 \times 10^{14} \, \text{s}^{-1})}{(1.381 \times 10^{-23} \, \text{J K}^{-1}) \times (253 \, \text{K})}\right) - 1\right]$$

$$= \boxed{+19.1 \, \text{J K}^{-1} \text{mol}^{-1}}$$

Therefore, $\Delta^{\ddagger}G = \Delta^{\ddagger}H - T\Delta^{\ddagger}S = 83.8 \, \text{kJ mol}^{-1} - 253 \, \text{K} \times 19.1 \, \text{J K}^{-1} \, \text{mol}^{-1}$

$$= \boxed{+79.0 \, \text{kJ mol}^{-1}}$$

P27.5 $\log k_2 = \log k_2^{\circ} + 2Az_A z_B I^{1/2}$ with $A = 0.509 \, (\text{mol L}^{-1})^{-1/2}$ [27.69]

This expression suggests that we should plot $\log k$ against $I^{1/2}$ and determine z_B from the slope, since we know that $|z_A| = 1$. We draw up the following table

$I/(\text{mol L}^{-1})$	0.0025	0.0037	0.0045	0.0065	0.0085
$(I/(\text{mol L}^{-1}))^{1/2}$	0.050	0.061	0.067	0.081	0.092
$\log(k_2/(\text{L mol}^{-1} \text{s}^{-1}))$	0.021	0.049	0.064	0.072	0.100

These points are plotted in Fig. 27.3.

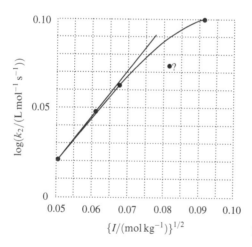

Figure 27.3

The slope of the limiting line in Fig. 27.3 is ≈ 2.5. Since this slope is equal to $2Az_A z_B \times (\text{mol L}^{-1})^{1/2} = 1.018 \, z_A z_B$, we have $z_A z_B \approx 2.5$. But $|z_A| = 1$, and so $|z_B| = 2$. Furthermore, z_A and z_B have the same sign because $z_A z_B > 0$. (The data refer to I^- and $S_2O_8^{2-}$.)

P27.8   Both approaches involve plots of $\log k$ versus $\log \gamma$, where γ is the activity coefficient. The limiting law has $\log \gamma$ proportional to $I^{1/2}$ (where I is ionic strength), so a plot of $\log k$ versus $I^{1/2}$ should give a straight line whose y-intercept is $\log k^{\circ}$ and whose slope is $2Az_A z_B$, where z_A and z_B are charges involved in the activated complex. The extended Debye–Hückel law has $\log \gamma$ proportional to $\left(\dfrac{I^{1/2}}{1 + BI^{1/2}}\right)$, so it requires plotting $\log k$ versus $\left(\dfrac{I^{1/2}}{1 + BI^{1/2}}\right)$, and it also has a slope of $2Az_A z_B$ and a y-intercept of $\log k^{\circ}$. The ionic strength in a 2 : 1 electrolyte solution is three times the molar concentration. The transformed data and plot (Fig. 27.4) follow

$[Na_2SO_4]/(mol\,kg^{-1})$	0.2	0.15	0.1	0.05	0.025	0.0125	0.005
$k/(L^{1/2}\,mol^{-1/2}\,s^{-1})$	0.462	0.430	0.390	0.321	0.283	0.252	0.224
$I^{1/2}$	0.775	0.671	0.548	0.387	0.274	0.194	0.122
$I^{1/2}/(1+BI^{1/2})$	0.436	0.401	0.354	0.279	0.215	0.162	0.109
$\log k$	-0.335	-0.367	-0.409	-0.493	-0.548	-0.599	-0.650

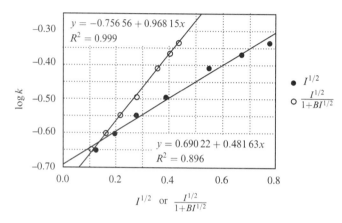

Figure 27.4

The line based on the limiting law appears curved. The zero-ionic-strength rate constant based on it is

$$k^\circ = 10^{-0.690}\,L^{1/2}\,mol^{-1/2}\,s^{-1} = 0.204\,L^{1/2}\,mol^{-1/2}\,s^{-1}$$

The slope is positive, so the complex must overcome repulsive interactions. The product of charges, however, works out to be 0.5, not easily interpretable in terms of charge numbers. The line based on the extended law appears straighter and has a better correlation coefficient. The zero-ionic-strength rate constant based on it is

$$k^\circ = 10^{-0.757}\,L^{1/2}\,mol^{-1/2}\,s^{-1} = 0.175\,L^{1/2}\,mol^{-1/2}\,s^{-1}$$

The product of charges works out to be 0.9, nearly 1, interpretable in terms of

a complex of two univalent ions of the same sign

P27.9 (a) $\dfrac{d[F_2O]}{dt} = -k_1[F_2O]^2 - k_2[F][F_2O]$ (1)

$\dfrac{d[F]}{dt} = k_1[F_2O]^2 - k_2[F][F_2O] + 2k_3[OF]^2 - 2k_4[F]^2[F_2O]$ (2)

$\dfrac{d[OF]}{dt} = k_1[F_2O]^2 + k_2[F][F_2O] - 2k_3[OF]^2$ (3)

applying the steady-state approximation to both [F] and [OF] and adding the resulting equations gives

$$k_1[F_2O]^2 - k_2[F]_{SS}[F_2O] + 2k_3[OF]_{SS}^2 - 2k_4[F]_{SS}^2[F_2O] = 0$$
$$\underline{k_1[F_2O]^2 + k_2[F]_{SS}[F_2O] - 2k_3[OF]_{SS}^2 \hspace{4.5cm} = 0}$$
$$2k_1[F_2O]^2 \hspace{5.5cm} - 2k_4[F]_{SS}^2[F_2O] = 0$$

solving for $[F]_{SS}$ gives

$$[F]_{SS} = \left(\frac{k_1}{k_4} [F_2O] \right)^{1/2} \tag{4}$$

substituting (4) into (1)

$$\frac{d[F_2O]}{dt} = -k_1 [F_2O]^2 - k_2 \left(\frac{k_1}{k_4} \right)^{1/2} [F_2O]^{3/2}$$

or

$$-\frac{d[F_2O]}{dt} = k_1 [F_2O]^2 + k_2 \left(\frac{k_1}{k_4} \right)^{1/2} [F_2O]^{3/2} \tag{5}$$

Comparison with the experimental rate law reveals that they are consistent when we make the following identifications.

$$k = k_1 = 7.8 \times 10^{13} e^{-E_1/RT} \, \text{L mol}^{-1} \, \text{s}^{-1}$$

$$E_1 = (19350 \, \text{K}) R = 160.9 \, \text{kJ mol}^{-1}$$

$$k' = k_2 \left(\frac{k_1}{k_4} \right)^{1/2} = 2.3 \times 10^{10} e^{-E'/RT} \, \text{L mol}^{-1} \, \text{s}^{-1}$$

$$E' = (16910 \, \text{K}) R = 140.6 \, \text{kJ mol}^{-1}$$

(b) $\quad \frac{1}{2} O_2 + F_2 \rightarrow F_2O \quad \Delta_f H(F_2O) = 24.41 \, \text{kJ mol}^{-1}$

$$\qquad\qquad 2F \rightarrow F_2 \qquad \Delta H = -D(F\text{—}F) = -160.6 \, \text{kJ mol}^{-1}$$

$$\qquad\qquad O \rightarrow \tfrac{1}{2} O_2 \quad \Delta H = -\tfrac{1}{2} D(O\text{—}O) = -249.1 \, \text{kJ mol}^{-1}$$

$$\overline{\qquad\qquad 2F + O \rightarrow F_2O \qquad\qquad\qquad\qquad\qquad\qquad}$$

$$\Delta H(FO\text{—}F) + \Delta H(O\text{—}F) = -\left[\Delta_f H(F_2O) - D(F\text{—}F) - \tfrac{1}{2} D(O\text{—}O) \right]$$

$$= -(24.41 - 160.6 - 249.1) \, \text{kJ mol}^{-1}$$

$$= 385.3 \, \text{kJ mol}^{-1}$$

We estimate that $\boxed{\Delta H(FO\text{—}F) \approx E_1 = 160.9 \, \text{kJ mol}^{-1}}$
Then

$$\Delta H(O\text{—}F) = 385.3 \, \text{kJ mol}^{-1} - \Delta H(FO\text{—}F)$$

$$\approx (385.3 - 160.9) \, \text{kJ mol}^{-1}$$

$$\boxed{\Delta H(O\text{—}F) \approx 224.4 \, \text{kJ mol}^{-1}}$$

In order to determine the activation energy of reaction (2) we assume that each rate constant can be expressed in Arrhenius form, then

$$\ln k' = \ln k_2 + \tfrac{1}{2} \ln k_1 - \tfrac{1}{2} \ln k_4$$

or

$$\ln A' - \frac{E'}{RT} = \ln A_2 - \frac{E_2}{RT} + \frac{1}{2} \ln A_1 - \frac{1}{2} \frac{E_1}{RT} - \frac{1}{2} \ln A_4 + \frac{1}{2} \frac{E_4}{RT}$$

Differentiating with respect to T we obtain

$$E' = E_2 + \tfrac{1}{2}E_1 - \tfrac{1}{2}E_4 = 140.6\,\text{kJ mol}^{-1}$$

or

$$E_2 - \tfrac{1}{2}E_4 = E' - \tfrac{1}{2}E_1 = (140.6 - 80.4)\,\text{kJ mol}^{-1}$$

$$= 60.2\,\text{kJ mol}^{-1}$$

E_4 is expected to be small since reaction (4) is termolecular, so we set $E_4 \approx 0$, then

$$E_2 \approx \boxed{60\,\text{kJ mol}^{-1}}$$

P27.11   Linear regression analysis of ln(rate constant) against $1/T$ yields the following results

$$\ln(k/22.4\,\text{L mol}^{-1}\,\text{min}^{-1}) = C + B/T$$

where $C = 34.36$,   standard deviation $= 0.36$
        $B = -23\,227\,\text{K}$,   standard deviation $= 252\,\text{K}$
        $R = \boxed{0.999\,76}$ [good fit]

$$\ln(k'/22.4\,\text{L mol}^{-1}\,\text{min}^{-1}) = C_2 + B_2/T$$

where $C' = 28.30$,   standard deviation $= 0.84$
        $B' = -21065\,\text{K}$,   standard deviation $= 582\,\text{K}$
        $R = \boxed{0.998\,48}$ [good fit]

The regression parameters can be used in the calculation of the pre-experimental factor (A) and the activation energy (E_a) using $\ln k = \ln A - E_a/RT$.

$$\ln A = C + \ln(22.4) = 37.47$$

$$A = 1.87 \times 10^{16}\,\text{L mol}^{-1}\,\text{min}^{-1} = \boxed{3.12 \times 10^{14}\,\text{L mol}^{-1}\,\text{s}^{-1}}$$

$$E_a = -RB = -(8.3145\,\text{J K}^{-1}\,\text{mol}^{-1}) \times (-23\,227\,\text{K}) \times \left(\frac{10^{-3}\,\text{kJ}}{\text{J}}\right)$$

$$= \boxed{193\,\text{kJ mol}^{-1}}$$

$$\ln A' = C' + \ln(22.4) = 31.41$$

$$A' = 4.37 \times 10^{13}\,\text{L mol}^{-1}\,\text{min}^{-1} = \boxed{7.29 \times 10^{11}\,\text{L mol}^{-1}\,\text{s}^{-1}}$$

$$E_a' = -RB' = -(8.3145\,\text{J K}^{-1}\,\text{mol}^{-1}) \times (-21065\,\text{K}) \times \left(\frac{10^{-3}\,\text{kJ}}{\text{J}}\right)$$

$$= \boxed{175\,\text{kJ mol}^{-1}}$$

To summarize

	$A/(\text{L mol}^{-1}\,\text{s}^{-1})$	$E_a/(\text{kJ mol}^{-1})$
k	$3.12 \times 10^{14}(=A)$	193
k'	$7.29 \times 10^{11}(=A')$	175

Both sets of data, k and k', fit the Arrhenius equation very well and hence are consistent with the collision theory of bimolecular gas-phase reactions which provides an equation [27.18] compatible with the Arrhenius equation. The numerical values for k' and A may be compared to the results of Exercise 27.7(a) and are in rough agreement at 647 K, as is the value of E_a.

Solutions to theoretical problems

P27.13

$$[J]^* = k \int_0^t [J] e^{-kt} dt + [J] e^{-kt} \quad [27.40]$$

$$\frac{\partial [J]^*}{\partial t} = k[J] e^{-kt} + \frac{\partial [J]}{\partial t} e^{-kt} - k[J] e^{-kt} = \left(\frac{\partial [J]}{\partial t}\right) e^{-kt}$$

$$\frac{\partial^2 [J]^*}{\partial x^2} = k \int_0^t \left(\frac{\partial^2 [J]}{\partial x^2}\right) e^{-kt} dt + \left(\frac{\partial^2 [J]}{\partial x^2}\right) e^{-kt}$$

Then, since

$$D \frac{\partial^2 [J]}{\partial x^2} = \frac{\partial [J]}{\partial t} \quad [27.39, \ k = 0]$$

we find that

$$D \frac{\partial^2 [J]^*}{\partial x^2} = k \int_0^t \left(\frac{\partial [J]}{\partial t}\right) e^{-kt} dt + \left(\frac{\partial [J]}{\partial t}\right) e^{-kt}$$

$$= k \int_0^t \left(\frac{\partial [J]^*}{\partial t}\right) dt + \frac{\partial [J]^*}{\partial t} = k[J]^* + \frac{\partial [J]^*}{\partial t}$$

which rearranges to eqn 27.39. When $t = 0$, $[J]^* = [J]$, and so the same initial conditions are satisfied. (The same boundary conditions are also satisfied.)

P27.15

$$\frac{q_m^{\ominus T}}{N_A} = 2.561 \times 10^{-2} (T/K)^{5/2} (M/\text{g mol}^{-1})^{3/2} \quad [\text{Table 20.3}]$$

For $T \approx 300$ K, $M \approx 50$ g mol^{-1}, $\dfrac{q_m^{\ominus T}}{N_A} \approx \boxed{1.4 \times 10^7}$

$$q^R (\text{nonlinear}) = \frac{1.0270}{\sigma} \times \frac{(T/K)^{3/2}}{(ABC/\text{cm}^{-3})^{1/2}} \quad [\text{Table 20.3}]$$

For $T \approx 300$ K, $A \approx B \approx C = 2$ cm^{-1}, $\sigma \approx 2$ [Section 16.5], $q^R(\text{NL}) \approx \boxed{900}$

$$q^R (\text{linear}) = \frac{0.6950}{\sigma} \times \frac{(T/K)}{(B/\text{cm}^{-1})} \quad [\text{Table 20.3}]$$

For $T \approx 300$ K, $B \approx 1$ cm^{-1}, $\sigma \approx 1$ [Section 16.5], $q^R(\text{L}) \approx \boxed{200}$

$$q^V \approx \boxed{1} \quad \text{and} \quad q^E \approx \boxed{1} \quad [\text{Table 20.3}]$$

$$k_2 = \frac{\kappa kT}{h} \overline{K}^{\ddagger} \quad [27.53]$$

$$= \left(\frac{\kappa kT}{h}\right) \times \left(\frac{RT}{p}\right) \times \left(\frac{N_A \overline{q}_C^{\ominus}}{q_A^{\ominus} q_B^{\ominus}}\right) e^{-\Delta E_0/RT} \quad [27.51] \approx A e^{-E_a/RT}$$

We then use

$$\frac{q_A^{\ominus}}{N_A} = \frac{q_A^{\ominus T}}{N_A} \approx 1.4 \times 10^7 \ [\text{above}]$$

$$\frac{q_B^\ominus}{N_A} = \frac{q_B^{\ominus T}}{N_A} \approx 1.4 \times 10^7 \text{ [above]}$$

$$\frac{\overline{q_C^\ominus}}{N_A} = \frac{q_C^{\ominus T} q^{R}(L)}{N_A} \approx (2^{3/2}) \times (1.4 \times 10^7) \times (200 \text{ [above]}) = 7.9 \times 10^9$$

[The factor of $2^{3/2}$ comes from $m_C = m_A + m_B \approx 2m_A$ and $q^T \propto m^{3/2}$.]

$$\frac{RT}{p^\ominus} \approx \frac{(8.314 \, \text{J K}^{-1} \, \text{mol}^{-1}) \times (300 \, \text{K})}{10^5 \, \text{Pa}} = 2.5 \times 10^{-2} \, \text{m}^3 \, \text{mol}^{-1}$$

$$\frac{\kappa k T}{h} \approx \frac{kT}{h} = \frac{(1.381 \times 10^{-23} \, \text{J K}^{-1}) \times (300 \, \text{K})}{6.626 \times 10^{-34} \, \text{J s}} = 6.25 \times 10^{12} \, \text{s}^{-1}$$

Therefore, the pre-exponential factor

$$A \approx \frac{(6.25 \times 10^{12} \, \text{s}^{-1}) \times (2.5 \times 10^{-2} \, \text{m}^3 \, \text{mol}^{-1}) \times (7.9 \times 10^9)}{(1.4 \times 10^7)^2}$$

$$\approx 6.3 \times 10^6 \, \text{m}^3 \, \text{mol}^{-1} \, \text{s}^{-1} \quad \text{or} \quad \boxed{6.3 \times 10^9 \, \text{L mol}^{-1} \, \text{s}^{-1}}$$

If all three species are nonlinear,

$$\frac{q_A^\ominus}{N_A} \approx (1.4 \times 10^7) \times (900) = 1.3 \times 10^{10} \approx \frac{q_B^\ominus}{N_A}$$

$$\frac{\overline{q_C^\ominus}}{N_A} \approx (2^{3/2}) \times (1.4 \times 10^7) \times (900) = 3.6 \times 10^{10}$$

$$A \approx \frac{(6.25 \times 10^{12} \, \text{s}^{-1}) \times (2.5 \times 10^{-2} \, \text{m}^3 \, \text{mol}^{-1}) \times (3.6 \times 10^{10})}{(1.3 \times 10^{10})^2}$$

$$\approx 33 \, \text{m}^3 \, \text{mol}^{-1} \, \text{s}^{-1} \quad \text{or} \quad \boxed{3.3 \times 10^4 \, \text{L mol}^{-1} \, \text{s}^{-1}}$$

Therefore, $P = \dfrac{A(\text{NL})}{A(\text{L})} = \dfrac{3.3 \times 10^4}{6.3 \times 10^9} = \boxed{5.2 \times 10^{-6}}$

These numerical values may be compared to those given in Table 27.1 and in Example 27.1. They lie within the range found experimentally.

P27.17 We consider the y-direction to be the direction of diffusion. Hence, for the activated atom the vibrational mode in this direction is lost. Therefore,

$$q^\ddagger = q_z^{\ddagger V} q_x^{\ddagger V} \quad \text{for the activated atom, and}$$
$$q = q_x^{V} q_y^{V} q_z^{V} \quad \text{for an atom at the bottom of a well}$$

For classical vibration, $q^V \approx \dfrac{kT}{h\nu}$ [Section 27.4]

The diffusion process described is unimolecular, hence first-order, and therefore analogous to the second-order case of Section 27.4 [also see Problem 27.4] we may write

$$-\frac{d[x]}{dt} = k^\ddagger [x]^\ddagger = \nu [x]^\ddagger = \nu K^\ddagger [x] = k_1 [x] \qquad \left[K^\ddagger = \frac{[x]^\ddagger}{[x]} \right]$$

Thus

$$k_1 = \nu\, K^{\ddagger} = \nu \left(\frac{kT}{h\nu}\right) \times \left(\frac{q^{\ddagger}}{q}\right) e^{-\beta \Delta E_0} \quad \left[\beta = \frac{1}{RT} \text{here}\right]$$

where $q^{\ddagger}$ and q are the (vibrational) partition functions at the top and foot of the well respectively. Therefore

$$k_1 = \frac{kT}{h} \left(\frac{(kT/h\nu^{\ddagger})^2}{(kT/h\nu)^3}\right) e^{-\beta \Delta E_0} = \boxed{\frac{\nu^3}{\nu^{\ddagger 2}} e^{-\beta \Delta E_0}}$$

(a) $\nu^{\ddagger} = \nu; \quad k_1 = \nu e^{-\beta \Delta E_0}$. Assume $\Delta E_0 \approx E_a$; hence

$$k_1 \approx 10^{11}\,\text{Hz}\,e^{-60 \times 10^3/(8.314 \times 500)} = 5.4 \times 10^4\,\text{s}^{-1}$$

But $D = \dfrac{\lambda^2}{2\tau} \approx \dfrac{1}{2}\lambda^2 k_1 \left[24.93;\ \tau = \dfrac{1}{k_1}, \text{Problem 25.29}\right]$

$$= \tfrac{1}{2} \times (316\,\text{pm})^2 \times 5.4 \times 10^4\,\text{s}^{-1} = \boxed{2.7 \times 10^{-15}\,\text{m}^2\,\text{s}^{-1}}$$

(b) $\nu^{\ddagger} = \tfrac{1}{2}\nu; \quad k_1 = 4\nu\, e^{-\beta \Delta E_0} = 2.2 \times 10^5\,\text{s}^{-1}$

$$D = (4) \times (2.7 \times 10^{-15}\,\text{m}^2\,\text{s}^{-1}) = \boxed{1.1 \times 10^{-14}\,\text{m}^2\,\text{s}^{-1}}$$

P27.19   The change in intensity of the beam, dI, is proportional to the number of scatterers per unit volume, N_s, the intensity of the beam, I, and the path length dl. The constant of proportionality is defined as the collision cross-section σ. Therefore,

$$dI = -\sigma N_s I\, dl \quad \text{or} \quad d \ln I = -\sigma N_s\, dl$$

If the incident intensity (at $l = 0$) is I_0 and the emergent intensity is I, we can write

$$\ln \frac{I}{I_0} = -\sigma N_s l \quad \text{or} \quad \boxed{I = I_0\, e^{-\sigma N_s l}}$$

P27.21 $A + B \rightarrow C^{\ddagger} \rightarrow P$

$$k_2 = \left(\kappa \frac{kT}{h}\right) \times \left(\frac{N_A RT}{p^{\ominus}}\right) \frac{q_{C^{\ddagger}}^{\ominus}}{q_A^{\ominus} q_B^{\ominus}} e^{-\Delta E_0/RT} \quad [27.52]$$

We assume that the only factor that changes between the atomic and molecular case is the ratio of the partition functions.

(1) For collisions between atoms

$$q_A^{\ominus} = q_A^T \approx 10^{26}$$

$$q_B^{\ominus} = q_B^T \approx 10^{26}$$

$$q_C^{\ominus} = (q_C^R)^2 q_C^V q_C^T \approx (10^{1.5})^2 \times (1) \times (10^{26}) \approx 10^{29}$$

$$k_2(\text{atoms}) \propto \frac{10^{29}}{10^{26} \times 10^{26}} = 10^{-23}$$

(2) For collisions between nonlinear molecules

$$q_A^{\ominus} = (q_A^R)^3 (q_A^V)^{3N-6} (q_A^T) \approx (10^{1.5})^3 \times (1) \times (10^{26}) \approx 3 \times 10^{30}$$

$$q_B^{\ominus} = (q_B^R)^3 (q_B^V)^{3N'-6} (q_B^T) \approx 3 \times 10^{30}$$

$$q_C^{\ominus} = (q_C^R)^3 (q_C^V)^{3(N+N')-6} (q_C^T) \approx 3 \times 10^{30}$$

$$k_2(\text{molecules}) \propto \frac{3 \times 10^{30}}{1 \times 10^{61}} = 3 \times 10^{-31}$$

Therefore k_2 (atoms)$/k_2$(molecules)$\approx \dfrac{10^{-23}}{3 \times 10^{-31}} \approx \boxed{3 \times 10^7}$

Solutions to applications

P27.23   (a) The rate constant of a diffusion-limited reaction is

$$k = \frac{8RT}{3\eta} = \frac{8 \times (8.3145\,\text{J K}^{-1}\,\text{mol}^{-1}) \times (298\,\text{K}) \times (10^3\,\text{L m}^{-3})}{3 \times (1.06 \times 10^{-3}\,\text{kg m}^{-1}\,\text{s}^{-1})}$$

$$= \boxed{6.23 \times 10^9\,\text{L mol}^{-1}\,\text{s}^{-1}}$$

(b) The rate constant is related to the diffusion constants and reaction distance by

$$k = 4\pi R^* D N_A \quad \text{so} \quad R^* = \frac{k}{4\pi D N_A}$$

$$= \frac{(2.77 \times 10^9\,\text{L mol}^{-1}\,\text{s}^{-1}) \times (10^{-3}\,\text{m}^3\,\text{L}^{-1})}{4\pi \times (1 \times 10^{-9}\,\text{m}^2\,\text{s}^{-1}) \times (6.022 \times 10^{23}\,\text{mol}^{-1})}$$

$$R^* = \boxed{3.7 \times 10^{-10}\,\text{m or } 0.37\,\text{nm}}$$

28 Processes at solid surfaces

Solutions to exercises

Discussion questions

E28.1(a) A terrace is a flat layer of atoms on a surface. There can be more than one terrace on a surface, each at a different height. Steps are the joints between the terraces; the height of the step can be constant or variable.

E28.2(a) AES can provide a depth profile or fingerprint of the sample, since the Auger spectrum is characteristic of the material present. Information about the atoms present and their bonding can be obtained. The technique is limited to a depth of a about 100 nm.

EELS and HREELS can detect very tiny amounts of adsorbate. The incident beam can induce vibrational excitations in the adsorbate which is characteristic of the species and its environment.

RAIRS resolves the problem of the opacity of surfaces to infrared or visible radiation but the spectral bands observed are typically very weak.

SERS resolves the problem of the weak spectra observed in RAIRS. It generally gives a greatly enhanced resonance Raman intensity. The disadvantages are that it provides only a weak enhancement for flat single crystal surfaces and the technique works well only for certain metals.

SEXAFS can provide nearest neighbour distributions, giving the number and interatomic distances of surface atoms.

SHG provides information about adsorption and surface coverage and rapid surface changes.

UPS can provide detailed information about the chemisorption process, surface composition, and the oxidation state of the atoms. It can distinguish between chemical adsorbtion and physical adsorption.

XPS is similar to UPS in the information revealed.

See the references listed under *Further reading* for more information about these modern techniques for probing the properties of surfaces.

E28.3(a) *Langmuir isotherm*. This isotherm applies under the following conditions:

1. Adsorption cannot proceed beyond monolayer coverage.
2. All sites are equivalent and the surface is uniform.
3. The ability of a molecule to adsorb at a given site is independent of the occupation of neighbouring sites.

BET isotherm. Condition number 1 above is removed. This isotherm applies to multi-layer coverage.

Temkin isotherm. Condition number 2 is removed and it is assumed that the energetically most favourable sites are occupied first. The Temkin isotherm corresponds to supposing that the adsorption enthalpy changes linearly with pressure.

Freundlich isotherm. Condition 2 is again removed, but this isotherm corresponds to a logarithmic change in the adsorption enthalpy with pressure.

E28.4(a) Heterogeneous catalysis on a solid surface requires the reacting molecules or fragments to encounter each other by adsorption on the surface. Therefore, the rate of the catalysed reaction is determined by the sticking probabilities of the species on the surface as described by Fig. 28.30.

Numerical exercises

E28.5(a)
$$Z_W = (2.63 \times 10^{24}\,\text{m}^{-2}\,\text{s}^{-1}) \times \left(\frac{p/\text{Pa}}{\{(T/\text{K}) \times (M/\text{g mol}^{-1})\}^{1/2}} \right) \quad [28.1b]$$

$$= \left(\frac{(1.52 \times 10^{19}\,\text{cm}^{-2}\,\text{s}^{-1}) \times (p/\text{Pa})}{(M/\text{g mol}^{-1})^{1/2}} \right) \quad [T = 298\,\text{K}]$$

Another practical form of this equation at 298 K is

$$Z_W = \frac{(2.03 \times 10^{21} \, \text{cm}^{-2} \, \text{s}^{-1}) \times (p/\text{Torr})}{(M/\text{g mol}^{-1})^{1/2}} \quad [100 \, \text{Pa} = 0.750 \, \text{Torr}]$$

or

$$Z_W = \frac{(2.03 \times 10^{25} \, \text{m}^{-2} \, \text{s}^{-1}) \times (p/\text{Torr})}{(M/\text{g mol}^{-1})^{1/2}}$$

Hence, we can draw up the following table

	H_2	C_3H_8
$M/(\text{g mol}^{-1})$	2.02	44.09
$Z_W(\text{m}^{-2} \, \text{s}^{-1})$		
(i) 100 Pa	1.07×10^{25}	2.35×10^{24}
(ii) 10^{-7} Torr	1.4×10^{18}	3.1×10^{17}

E28.6(a)

$$p/\text{Pa} = \frac{\{Z_W/(\text{m}^{-2} \, \text{s}^{-1})\} \times \{(T/\text{K}) \times (M/\text{g mol}^{-1})\}^{1/2}}{2.63 \times 10^{24}} \quad [28.1b]$$

$$= \frac{\{Z_W/(\text{m}^{-2} \, \text{s}^{-1})\} \times (425 \times 39.95)^{1/2}}{2.63 \times 10^{24}}$$

$$= 4.95 \times 10^{-23} \times Z_W/(\text{m}^{-2} \, \text{s}^{-1})$$

The collision rate required is

$$Z_W = \frac{4.5 \times 10^{20} \, \text{s}^{-1}}{\pi \times (0.075 \, \text{cm})^2} = 2.5\bar{5} \times 10^{22} \, \text{cm}^{-2} \, \text{s}^{-1} = 2.5\bar{5} \times 10^{26} \, \text{m}^{-2} \, \text{s}^{-1}$$

Hence $p = (4.95 \times 10^{-23} \, \text{Pa}) \times (2.5\bar{5} \times 10^{26}) = \boxed{1.3 \times 10^4 \, \text{Pa}}$

E28.7(a)

$$Z_W = (2.63 \times 10^{24} \, \text{m}^{-2} \, \text{s}^{-1}) \times \left(\frac{p/\text{Pa}}{\{(T/\text{K}) \times (M/\text{g mol}^{-1})\}^{1/2}} \right) \quad [28.1b]$$

$$= (2.63 \times 10^{24} \, \text{m}^{-2} \, \text{s}^{-1}) \times \left(\frac{35}{(80 \times 4.00)^{1/2}} \right) = 5.1 \times 10^{24} \, \text{m}^{-2} \, \text{s}^{-1}$$

The area occupied by a Cu atom is $\left(\frac{1}{2}\right) \times (3.61 \times 10^{-10} \, \text{m})^2 = 6.52 \times 10^{-20} \, \text{m}^2$ (in an fcc unit cell, there is the equivalent of two Cu atoms per face). Therefore

$$\text{rate per Cu atom} = (5.2 \times 10^{24} \, \text{m}^{-2} \, \text{s}^{-1}) \times (6.52 \times 10^{-20} \, \text{m}^2) = \boxed{3.4 \times 10^5 \, \text{s}^{-1}}$$

E28.8(a) $V_{\text{mon}} = 2.86 \, \text{cm}^3$

$$n = \frac{pV}{RT} = \frac{(1.00 \, \text{atm}) \times (2.86 \times 10^{-3} \, \text{L})}{(0.0821 \, \text{L atm K}^{-1} \, \text{mol}^{-1}) \times (273 \, \text{K})} = 1.28 \times 10^{-4} \, \text{mol}$$

$$N = nN_A = 7.69 \times 10^{19}$$

$$A = (7.69 \times 10^{19}) \times (0.165 \times 10^{-18} \, \text{m}^2) = \boxed{12.7 \, \text{m}^2}$$

Comment. There is more than one method of estimating the effective cross-sectional area of an adsorbed molecule. One very simple method which is appropriate here is to obtain it from the density of the liquid.

Question. Given that the density of liquid nitrogen is $0.808 \, \text{g cm}^{-3}$, what is the effective cross-sectional area of a nitrogen molecule? How does this estimate compare with the value used above?

E28.9(a) $\theta = \dfrac{V}{V_\infty}$ [28.2] $= \dfrac{V}{V_{\text{mon}}} = \dfrac{Kp}{1 + Kp}$ [28.5]

which rearranges to [Example 28.1]

$$\frac{p}{V} = \frac{p}{V_{\text{mon}}} + \frac{1}{K V_{\text{mon}}}$$

Hence, $\dfrac{p_2}{V_2} - \dfrac{p_1}{V_1} = \dfrac{p_2}{V_{\text{mon}}} - \dfrac{p_1}{V_{\text{mon}}}$

Solving for V_{mon}

$$V_{\text{mon}} = \frac{p_2 - p_1}{\left(\frac{p_2}{V_2} - \frac{p_1}{V_1}\right)} = \frac{(760 - 142.4)\,\text{Torr}}{\left(\frac{760}{1.430} - \frac{142.4}{0.284}\right)\text{Torr cm}^{-3}} = \boxed{20.5\,\text{cm}^3}$$

E28.10(a) The enthalpy of adsorption is typical of $\boxed{\text{chemisorption}}$ (Table 28.2). The residence lifetime is

$$t_{1/2} = \tau_0 e^{E_d/RT}\,[28.18] \approx (1 \times 10^{-14}\,\text{s}) \times (e^{120\times 10^3/(8.314\times 400)})[E_d \approx -\Delta_{\text{ad}}H] \approx \boxed{50\,\text{s}}$$

E28.11(a) The average residence time for particle adsorbed on the surface is

$$t_{1/2} = \tau_0 e^{E_d/RT} \, [28.18]$$

$$E_d = \frac{R \ln\left(\frac{t'_{1/2}}{t_{1/2}}\right)}{\left(\frac{1}{T'} - \frac{1}{T}\right)} = \frac{(8.314\,\text{J K}^{-1}\,\text{mol}^{-1}) \times \ln\left(\frac{0.36}{3.49}\right)}{\left(\frac{1}{2548\,\text{K}} - \frac{1}{2362\,\text{K}}\right)} = \boxed{610\,\text{kJ mol}^{-1}}$$

$$\tau_0 = t_{1/2} e^{-E_d/RT} = (3.49\,\text{s}) \times e^{-610\times 10^3/(8.314\times 2362)} = \boxed{0.113 \times 10^{-12}\,\text{s}}$$

$$A = \ln 2/\tau_0 = 0.693/(0.113 \times 10^{-12}\,\text{s}) = \boxed{6.15 \times 10^{12}\,\text{s}^{-1}}$$

E28.12(a) $\theta = \dfrac{Kp}{1 + Kp}$ [28.5], which implies that $p = \left(\dfrac{\theta}{1-\theta}\right)\dfrac{1}{K}$

(a) $p = \left(\dfrac{0.15}{0.85}\right) \times \left(\dfrac{1}{0.85\,\text{kPa}^{-1}}\right) = \boxed{0.21\,\text{kPa}}$ **(b)** $p = \left(\dfrac{0.95}{0.05}\right) \times \left(\dfrac{1}{0.85\,\text{kPa}^{-1}}\right) = \boxed{22\,\text{kPa}}$

E28.13(a) $\dfrac{m_1}{m_2} = \dfrac{\theta_1}{\theta_2} = \dfrac{p_1}{p_2} \times \dfrac{1 + Kp_2}{1 + Kp_1}$

which solves to

$$K = \frac{\left(\frac{m_1 p_2}{m_2 p_1}\right) - 1}{p_2 - \left(\frac{m_1 p_2}{m_2}\right)} = \frac{\left(\frac{m_1}{m_2}\right) \times \left(\frac{p_2}{p_1}\right) - 1}{1 - \left(\frac{m_1}{m_2}\right)} \times \frac{1}{p_2}$$

$$= \frac{\left(\frac{0.44}{0.19}\right) \times \left(\frac{3.0}{26.0}\right) - 1}{1 - \left(\frac{0.44}{0.19}\right)} \times \frac{1}{3.0\,\text{kPa}} = 0.19\,\text{kPa}^{-1}$$

Therefore,

$$\theta_1 = \frac{(0.19\,\text{kPa}) \times (26.0\,\text{kPa})}{(1) + (0.19\,\text{kPa}^{-1}) \times (26.0\,\text{kPa})} = \boxed{0.83} \qquad \theta_2 = \frac{(0.19) \times (3.0)}{(1) + (0.19) \times (3.0)} = \boxed{0.36}$$

E28.14(a) $\quad t_{1/2} \approx \tau_0 e^{E_d/RT}\,[28.18] = (10^{-13}\,\text{s}) \times (e^{E_d/(2.48\,\text{kJ mol}^{-1})}) \quad$ [at 298 K]

(a) $\quad E_d = 15\,\text{kJ mol}^{-1}, \qquad t_{1/2} = (10^{-13}\,\text{s}) \times (e^{6.05}) = \boxed{4 \times 10^{-11}\,\text{s}}$

(b) $\quad E_d = 150\,\text{kJ mol}^{-1}, \qquad t_{1/2} = (10^{-13}\,\text{s}) \times (e^{6.05}) = \boxed{2 \times 10^{13}\,\text{s}}$

The latter corresponds to about 600 000 y. At 1000 K, $t_{1/2} = (10^{-13}\,\text{s}) \times (e^{E_d/8.314\,\text{kJ mol}^{-1}})$

(a) $t_{1/2} = \boxed{6 \times 10^{-13}\,\text{s}}$ (b) $t_{1/2} = \boxed{7 \times 10^{-6}\,\text{s}}$

E28.15(a) $\quad \theta = \dfrac{Kp}{1 + Kp}\,[28.5], \quad$ which implies that $\quad K = \left(\dfrac{\theta}{1-\theta}\right) \times \left(\dfrac{1}{p}\right)$

But $\ln \dfrac{K'}{K} = \dfrac{\Delta_r H}{R}\left(\dfrac{1}{T} - \dfrac{1}{T'}\right)$ [9.28]

Since θ at the new temperature is the same, $K \propto \dfrac{1}{p}$ and

$$\ln \frac{p}{p'} = \frac{\Delta_{ad} H}{R}\left(\frac{1}{T} - \frac{1}{T'}\right) = \left(\frac{-10.2\,\text{kJ mol}^{-1}}{8.314\,\text{J K}^{-1}\,\text{mol}^{-1}}\right) \times \left(\frac{1}{298\,\text{K}} - \frac{1}{313\,\text{K}}\right) = -0.197$$

which implies that $p' = (12\,\text{kPa}) \times (e^{0.197}) = \boxed{15\,\text{kPa}}$

E28.16(a) $\quad v = k\theta = \dfrac{kKp}{1 + Kp}$ [Example 28.4]

(a) On gold, $\theta \approx 1$, and $v = k\theta \approx$ constant, a $\boxed{\text{zeroth-order}}$ reaction.

(b) On platinum, $\theta \approx Kp$ (as $Kp \ll 1$), so $v = kKp$, and the reaction is $\boxed{\text{first-order}}$

E28.17(a) $\quad \theta = \dfrac{Kp}{1 + Kp} \quad$ and $\quad \theta' = \dfrac{K'p'}{1 + K'p}$

but $\theta = \theta'$, so

$$\frac{Kp}{1 + Kp} = \frac{K'p'}{1 + K'p'}$$

which requires $Kp = K'p'$. We also know that

$$\Delta_{ad} H^{\ominus} = RT^2 \left(\frac{\partial \ln K}{\partial T}\right)_{\theta} \quad [28.9]$$

and can therefore write

$$\Delta_{ad} H^{\ominus} \approx RT^2 \left(\frac{\ln K' - \ln K}{T' - T}\right) = \frac{RT^2 \ln\left(\frac{K'}{K}\right)}{T' - T} \approx \frac{RT^2 \ln\left(\frac{p}{p'}\right)}{T' - T}$$

$$\approx \frac{(8.314\,\text{J K}^{-1}\,\text{mol}^{-1}) \times (220\,\text{K})^2 \times \ln\left(\frac{4.9}{32}\right)}{60\,\text{K}} = \boxed{-13\,\text{kJ mol}^{-1}}$$

E28.18(a) The desorption time for a given volume is proportional to the half-life of the adsorbed species, and as

$$t_{1/2} = \tau_0 e^{E_d/RT} \quad [28.18]$$

we can write

$$E_d = \frac{R \ln\left(\frac{t_{1/2}}{t'_{1/2}}\right)}{\left(\frac{1}{T} - \frac{1}{T'}\right)} = \frac{R \ln\left(\frac{t}{t'}\right)}{\frac{1}{T} - \frac{1}{T'}}$$

where t and t' are the two desorption times. We evaluate E_d from the data for the two temperatures

$$E_d = \frac{8.314\,\text{J K}^{-1}\,\text{mol}^{-1}}{\left(\frac{1}{1856\,\text{K}} - \frac{1}{1978\,\text{K}}\right)} \times \ln\frac{27}{2.0} = \boxed{65\bar{0}\,\text{kJ mol}^{-1}}$$

We write

$$t = t_0\, e^{65\bar{0} \times 10^3/(8.314 \times 1856)} = t_0 \times (1.9\bar{6} \times 10^{18})$$

Therefore, since $t = 27\,\text{min}$, $t_0 = 1.3\bar{8} \times 10^{-17}\,\text{min}$. Consequently,

(a) At 298 K,

$$t = (1.3\bar{8} \times 10^{-17}\,\text{min}) \times e^{65\bar{0} \times 10^3/(8.314 \times 298)} = \boxed{1.1 \times 10^{97}\,\text{min}}$$

which is just about forever.

(b) At 3000 K,

$$t = (1.3\bar{8} \times 10^{-17}\,\text{min}) \times e^{65\bar{0} \times 10^3/(8.314 \times 3000)} = \boxed{2.9 \times 10^{-6}\,\text{min}}$$

Solutions to problems

Solutions to numerical problems

P28.1
$$Z_W = \frac{p}{(2\pi mkT)^{1/2}} \quad [28.1a]$$

$$= \frac{p/\text{Pa}}{[(2\pi) \times (32.0) \times (1.6605 \times 10^{-27}\,\text{kg}) \times (1.381 \times 10^{-23}\,\text{J K}^{-1}) \times (300\,\text{K})]^{1/2}}$$

$$= (2.69 \times 10^{22}\,\text{m}^{-2}\,\text{s}^{-1}) \times p/\text{Pa} = (2.69 \times 10^{18}\,\text{cm}^{-2}\,\text{s}^{-1}) \times p/\text{Pa}$$

Hence,

(a) at 100 kPa, $Z_W = \boxed{2.69 \times 10^{23}\,\text{cm}^{-2}\,\text{s}^{-1}}$ **(b)** at 1.000 Pa, $Z_W = \boxed{2.69 \times 10^{18}\,\text{cm}^{-2}\,\text{s}^{-1}}$

The nearest neighbour in titanium is 291 pm, so the number of atoms per cm^2 is approximately 1.4×10^{15} (the precise value depends on the details of the packing, which is hcp, and the identity of the surface). The number of collisions per exposed atom is therefore $\dfrac{Z_W}{1.4 \times 10^{15}\,\text{cm}^{-2}}$

(a) When $p = 100\,\text{kPa}$, $Z_{\text{atom}} = \boxed{2.0 \times 10^8\,\text{s}^{-1}}$ **(b)** When $p = 1.000\,\text{Pa}$, $Z_{\text{atom}} = \boxed{2.0 \times 10^3\,\text{s}^{-1}}$

P28.3   We follow Example 28.1 and draw up the following table

p/Torr	0.19	0.97	1.90	4.05	7.50	11.95
$\dfrac{p}{V}\Big/(\text{Torr cm}^{-3})$	4.52	5.95	8.60	12.6	18.3	25.4

$\dfrac{p}{V}$ is plotted against p in Fig. 28.1.

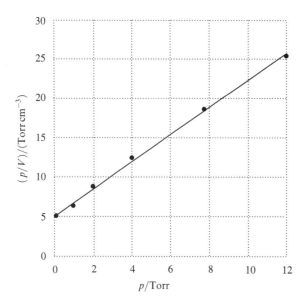

Figure 28.1

The low-pressure points fall on a straight line with intercept 4.7 and slope 1.8. It follows that $\dfrac{1}{V_\infty} =$ 1.8 Torr cm^{-3}/Torr $= 1.8\,\text{cm}^{-3}$, or $V_\infty = 0.57\,\text{cm}^3$ and $\dfrac{1}{KV_\infty} = 4.7\,\text{Torr cm}^{-3}$. Therefore,

$$K = \frac{1}{(4.7\,\text{Torr cm}^{-3}) \times (0.57\,\text{cm}^3)} = \boxed{0.37\,\text{Torr}^{-1}}$$

Comment. It is unlikely that low-pressure data can be used to obtain an accurate value of the volume corresponding to complete coverage. See Problem 28.5 for adsorption data at higher pressures.

P28.5   We assume that the data fit the Langmuir isotherm; to confirm this we plot $\dfrac{p}{V}$ against p and expect a straight line [Example 28.1]. We draw up the following table

p/atm	0.050	0.100	0.150	0.200	0.250
$\dfrac{p}{V}\Big/(10^{-2}\,\text{atm mL}^{-1})$	4.1	7.52	11.5	14.7	17.9

The data are plotted in Fig. 28.2.
They fit closely to a straight line with slope $0.720\,\text{mL}^{-1}$. Hence,

$$V_\infty = \boxed{1.3\bar{9}\,\text{mL}} = 1.3\bar{9} \times 10^{-3}\,\text{L} \approx V_{\text{mon}}$$

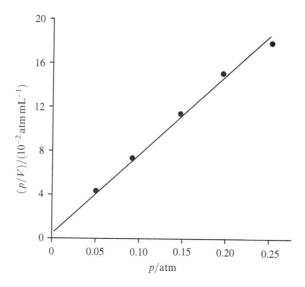

$$p/\text{atm}$$

Figure 28.2

The number of H_2 molecules corresponding to this volume is

$$N_{H_2} = \frac{pVN_A}{RT} = \frac{(1.00\,\text{atm}) \times (1.3\overline{9} \times 10^{-3}\,\text{L}) \times (6.02 \times 10^{23}\,\text{mol}^{-1})}{(0.0821\,\text{L atm K}^{-1}\,\text{mol}^{-1}) \times (273\,\text{K})} = 3.73 \times 10^{19}$$

The area occupied is the number of molecules times the area per molecule. The area per molecule can be estimated from the density of the liquid

$$A = \pi \left(\frac{3V}{4\pi}\right)^{2/3} \quad \left[V = \text{volume of molecule} = \frac{M}{\rho N_A}\right]$$

$$= \pi \left(\frac{3M}{4\pi p N_A}\right)^{2/3} = \pi \left(\frac{3 \times (2.02\,\text{g mol}^{-1})}{4\pi \times (0.0708\,\text{g cm}^{-3}) \times (6.02 \times 10^{23}\,\text{mol}^{-1})}\right)^{2/3}$$

$$= 1.58 \times 10^{-15}\,\text{cm}^2$$

Area occupied $= (3.73 \times 10^{19}) \times (1.58 \times 10^{-15}\,\text{cm}^2) = (5.9 \times 10^4\,\text{cm}^2) = \boxed{5.9\,\text{m}^2}$

Comment. The value for V_∞ calculated here may be compared to the value obtained in Problem 28.3. The agreement is not good and illustrates the point that these kinds of calculations provide only rough values of surface areas.

P28.6 $\qquad \theta = c_1 p^{1/c_2}$

We adapt this isotherm to a liquid by noting that $w_a \propto \theta$ and replacing p by [A], the concentration of the acid. Then $w_a = c_1[A]^{1/c_2}$ (with c_1, c_2 modified constants), and hence

$$\log w_a = \log c_1 + \frac{1}{c_2} \times \log[A]$$

We draw up the following table

$[A]/(\text{mol L}^{-1})$	0.05	0.10	0.50	1.0	1.5
$\log([A]/(\text{mol L}^{-1}))$	−1.30	−1.00	−0.30	−0.00	0.18
$\log(w_a/\text{g})$	−1.40	−1.22	−0.92	−0.80	−0.72

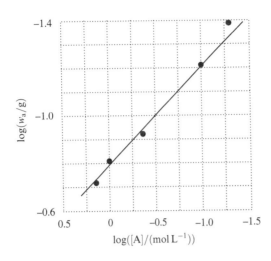

Figure 28.3

These points are plotted in Fig. 28.3.

They fall on a reasonably straight line with slope 0.42 and intercept -0.80. Therefore, $c_2 = \dfrac{1}{0.42} = \boxed{2.4}$ and $c_1 = \boxed{0.16}$. (The units of c_1 are bizarre: $c_1 = 0.16\,\text{g mol}^{-0.42}\,\text{dm}^{1.26}$.)

P28.8　**(a)** The volume of a spherical bubble is $V = \dfrac{4\pi r^3}{3}$. Using this expression in the perfect gas equation of state allows the evaluation of $\mathrm{d}N/\mathrm{d}t$

$$N = \frac{pV}{kT} = \frac{4\pi p r^3}{3kT}$$

$$\frac{\mathrm{d}N}{\mathrm{d}t} = \frac{4\pi p r^2}{kT}\frac{\mathrm{d}r}{\mathrm{d}t} = g(4\pi r^2)$$

$$\mathrm{d}r = \frac{gkT}{p}\,\mathrm{d}t$$

$$\int_{r_0}^{r}\mathrm{d}r = \frac{gkT}{p}\int_{0}^{t}\mathrm{d}t$$

assuming that the pressure of the bubble is constant

$$\boxed{r = r_0 + \left(\frac{gkT}{p}\right)t}$$

Thus, we find that $r = r_0 + vt$ where $\boxed{v = \dfrac{gkT}{p}}$.

(b) Linear regression analysis of the $r(t)$ data gives

$$r_0 = 0.018\,\text{cm}, \quad \text{standard deviation} = 0.0013\,\text{cm};$$

$$\boxed{v = 0.0041\,\text{cm s}^{-1}}, \quad \text{standard deviation} = 0.0005\,\text{cm s}^{-1}$$

with $\boxed{R = 0.959}$. About 96 per cent of the variation is explained by the linear regression.

$$g = \frac{vp}{kT} = \frac{(0.0041\,\text{cm s}^{-1}) \times (1 \times 10^5\,\text{Pa}) \times \left(\frac{10^{-2}\,\text{m}}{\text{cm}}\right)}{(1.381 \times 10^{-23}\,\text{J K}^{-1}) \times (298.15\,\text{K})}$$

$$\boxed{g = 1.0 \times 10^{21}\,\text{m}^{-3}\,\text{s}^{-1}}$$

The plot of $z(t)$ (Fig. 28.4) is seen to be nonlinear. To develop an empirical description of the data we will fit the data with two regression equations and compare goodness-of-fit for them. The two forms are

$$z(t)(\text{fit 1}) = at^b \quad \text{and} \quad z(t)(\text{fit 2}) = \boxed{a(e^{bt} - 1)}$$

where a and b are the regression parameters.

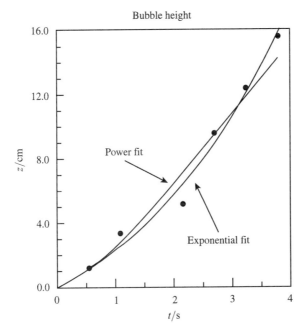

Figure 28.4

Performing the regression analysis, we find the following results,

Fit 1 (the power fit)

$a = 2.66\,\text{cm s}^{-b}$, standard deviation $= 0.098\,\text{cm s}^{-b}$
$b = 1.26$, standard deviation $= 0.11$
$R = 0.986$

Fit 2 (the exponential fit)

$\boxed{a = 5.71\,\text{cm}}$, standard deviation $= 2.24\,\text{cm}$

$\boxed{b = 0.35\,\text{s}^{-1}}$, standard deviation $= 0.08\,\text{s}^{-1}$

$\boxed{R = 0.994}$

The correlation coefficients suggest that the exponential fit is the better description of the data. However, standard deviations are relatively large and, although the exponential fit is to be preferred, we cannot totally reject the power fit possibility.

P28.10   Application of the van't Hoff equation [28.9] to adsorption equilibria yields

$$\left(\frac{\partial \ln K}{\partial T}\right)_\theta = \frac{\Delta_{ad} H^{\ominus}}{RT^2} \quad \text{or} \quad \left(\frac{\partial \ln K}{\partial (1/T)}\right)_\theta = \frac{-\Delta_{ad} H^{\ominus}}{R}$$

Hence, a plot (Fig. 28.5) of $\ln K$ against $1/T$ should be a straight line with slope $-\Delta_{ad} H^{\ominus}/R$. The transformed data and plot follow

T/K	283	298	308	318
$10^{-11} K$	2.642	2.078	1.286	1.085
$1000\,K/T$	3.53	3.36	3.25	3.14
$\ln K$	26.30	26.06	25.58	25.41

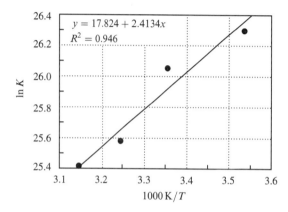

$y = 17.824 + 2.4134x$
$R^2 = 0.946$

Figure 28.5

The plot is not the straightest of lines. Still, we can extract

$$\Delta_{ad} H^{\ominus} = -(8.3145\,\text{J mol}^{-1}\,\text{K}^{-1}) \times (2.41 \times 10^3\,\text{K})$$

$$= -20.0 \times 10^3\,\text{J mol}^{-1} = \boxed{-20.1\,\text{kJ mol}^{-1}}$$

The Gibbs energy for absorption is

$$\Delta_{ad} G^{\ominus} = \Delta_{ad} H^{\ominus} - T\Delta_{ad} S^{\ominus} = -20.1\,\text{kJ mol}^{-1} - (298\,\text{K}) \times (0.146\,\text{kJ mol}^{-1}\,\text{K}^{-1})$$

$$= \boxed{-63.6\,\text{kJ mol}^{-1}}$$

P28.13   A linear regression fit, $Y(X) = a + bX$, is fitted to a set of N data pairs $\{X, Y\}_i$ by systematically adjusting the parameters a and b so as to minimize the sum of the squares of errors (SSE) between the experimental value Y_i and the regression value $Y(X_i)$. Parameter a is the intercept of the regression fit; parameter b is the slope.

$$\text{SSE} = \sum_i \{Y_i - Y(X_i)\}^2$$

The computational and statistical equations are:

$$X_{mean} = \frac{1}{N}\sum_i X_i, \quad Y_{mean} = \frac{1}{N}\sum_i Y_i, \quad (X^2)_{mean} = \frac{1}{N}\sum_i X_i^2$$

$$SS_X = \sum_i (X_i - X_{mean})^2, \quad SS_Y = \sum_i (Y_i - Y_{mean})^2$$

$$\text{slope, } b = \frac{\sum_i (X_i - X_{\text{mean}})(Y_i - Y_{\text{mean}})}{SS_X}$$

$$\text{intercept, } a = Y_{\text{mean}} - b X_{\text{mean}}$$

$$\text{standard deviation of the fit } Y \text{ on } X, S_{YX} = \sqrt{\frac{SSE}{N-2}}$$

$$\text{standard deviation of the slope, } S_{\text{slope}} = S_{YX} / \sqrt{SS_X}$$

$$\text{standard deviation of the intercept, } S_{\text{intercept}} = S_{YX} \sqrt{(X^2)_{\text{mean}} / SS_X}$$

$$\text{correlation coefficient, } r = b\, SS_x / \sqrt{SS_X\, SS_Y}$$

The correlation coefficient is a measure of the goodness of fit of the regression. If $|r|$ equals 1, the regression perfectly fits the data. If $|r|$ equals 0, X and Y are uncorrelated. The closer $|r|$ is to 1, the better the fit.

The Langmuir isotherm for adsorption from solution is constructed from the equation of Example 28.1 by substitution of the solution concentration (c) for pressure and substitution of the mass fraction adsorbed (x, the mass of solute adsorbed per unit mass of adsorbent) for gas volume.

$$\frac{c}{x} = \frac{c}{x_{\text{max}}} + \frac{1}{K x_{\text{max}}}$$

This has the linear form with $Y_i = \dfrac{c_i}{x_i}$, $x_i = c_i$, intercept $= \dfrac{1}{K x_{\text{max}}}$, slope $= \dfrac{1}{x_{\text{max}}}$. Application of the regression equations yields the following results:

$$x_{\text{max}} = 0.591; \qquad K = 0.155\, \text{L mg}^{-1}$$

$$r = 0.9240$$

$$\text{slope} = 1.693, \qquad S_{\text{slope}} = 0.313$$

$$\text{intercept} = 10.93\, \text{mg L}^{-1}, \qquad S_{\text{intercept}} = 2.76\, \text{mg L}^{-1}$$

The statistics indicate that the Langmuir isotherm fit has considerable uncertainty. The correlation coefficient indicates that about 92% of the $Y(X)$ variation is explained by the regression while the intercept and slope have 25% and 18% uncertainties, respectively.

The Freundlich isotherm for adsorption from solution is constructed from eqn 28.14.

$x = c_1\, x_{\text{max}} c^{1/c_2}$. Taking the equation logarithm gives

$$\ln(x) = \ln(c_1 x_{\text{max}}) + c_2^{-1} \ln(c)$$

This has the linear form with $Y_i = \ln(x_i)$, $x_i = \ln(c/\text{mg L}^{-1})$, intercept $= \ln(c_1 x_{\text{max}})$, slope $= c_2^{-1}$. Application of the regression equations yields the following results.

$$r = 0.9862$$

$$\text{slope} = 0.601, \qquad S_{\text{slope}} = 0.045$$

$$\text{intercept} = -2.497, \qquad S_{\text{intercept}} = 0.078$$

The Freundlich isotherm fits the data with less uncertainty than that of the Langmuir isotherm. The correlation coefficient indicates that about 99% of the $Y(X)$ variation is explained by the regression while the intercept and slope have 3% and 7% uncertainties, respectively. It is not possible to extract a value for x_{max} from the Freundlich equation fit because we know only that the $\ln(c_1 x_{max}) = -2.497$ and there is no method by which x_{max} alone can be known.

Division of the Langmuir equation by c yields

$$\frac{1}{x} = \frac{1}{x_{max}} + \frac{1}{K x_{max} c}$$

This has the linear form with $Y_i = \frac{1}{x_i}$, $X_i = \frac{1}{c_i}$, intercept $= \frac{1}{x_{max}}$, slope $= \frac{1}{K x_{max}}$. Application of the regression equations yields the following results.

$$\boxed{x_{max} = 0.380}; \qquad K = 0.351 \, \text{L mg}^{-1}$$

$r = 0.9984$

slope $= 7.506 \, \text{mg L}^{-1}$, $\qquad S_{slope} = 0.192 \, \text{mg L}^{-1}$

intercept $= 2.633$, $\qquad S_{intercept} = 0.386$

The correlation coefficient indicates that 99.8% of the $Y(x)$ variation is explained by the regression while the intercept and slope have 15% and 3% uncertainties respectively. This form of the Langmuir equation would seem to be the best fit of the data. However, it is always a good idea to carefully inspect the data plot and the goodness-of-fit of the regression.

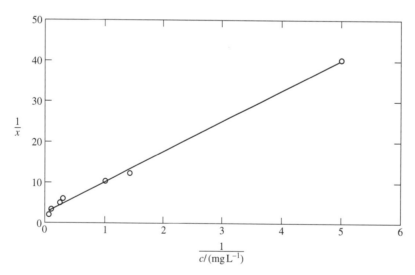

Figure 28.6

The regression line appears to fit the data very nicely. However, it is disquieting that there is a single point far from the origin and far outside the range of all other data. This point could be creating an illusion that the fit is very good. At the very least, any experimental uncertainty in that single point will disproportionally alter the calculated values of the intercept and x_{max}. It would be better to have several additional measurements in the range of $2 \, \text{L mg}^{-1} < \frac{1}{c} < 5 \, \text{L mg}^{-1}$.

In order to calculate the specific surface area of the activated carbon, we estimate that the surface area of a methylene blue molecule equals about $(1750\,\text{pm}) \times (900\,\text{pm}) = 1.58 \times 10^{-18}\,\text{m}^2$.

$$x_{max} = \frac{\text{adsorbed mass}}{\text{carbon mass}} = \frac{(\text{number of adsorbed molecules})M}{(\text{carbon mass})N_A}$$

$$= \left[\frac{\text{surface area of carbon}}{\text{surface area of molecule}}\right]\frac{M}{(\text{carbon mass})N_A}$$

$$= \left[\frac{\text{specific surface area of carbon}}{\text{surface area of molecule}}\right]\frac{M}{N_A}$$

$$\text{specific surface area of carbon} = \frac{(\text{surface area of molecule})N_A x_{max}}{M}$$

$$= \frac{\left(1.58 \times 10^{-18}\,\text{m}^2\right) \times \left(6.02 \times 10^{23}\,\text{mol}^{-1}\right) \times (0.380)}{391.85\,\text{g mol}^{-1}}$$

$$= \boxed{922\,\text{m}^2\,\text{mol}^{-1}}$$

P28.14   Refer to Fig. 28.7.

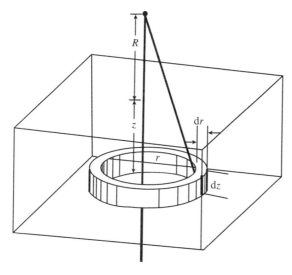

Figure 28.7

Let the number density of atoms in the solid be $\mathcal{N}$. Then the number in the annulus between r and $r + dr$ and thickness dz at a depth z below the surface is $2\pi\mathcal{N}r\,dr\,dz$. The interaction energy of these atoms and the single adsorbate atom at a height R above the surface is

$$dU = \frac{-2\pi\mathcal{N}r\,dr\,dz\,C_6}{\{(R+z)^2+r^2\}^3}$$

if the individual atoms interact as $\dfrac{-C_6}{d^6}$, with $d^2 = (R+z)^2 + r^2$. The total interaction energy of the atom with the semi-infinite slab of uniform density is therefore

$$U = -2\pi\mathcal{N}C_6\int_0^\infty dr\int_0^\infty dz\,\frac{r}{\{(R+z)^2+r^2\}^3}$$

We then use

$$\int_0^\infty \frac{r\,dr}{(a^2+r^2)^3} = \frac{1}{2}\int_0^\infty \frac{d(r^2)}{(a^2+r^2)^3} = \frac{1}{2}\int_0^\infty \frac{dx}{(a^2+x)^3} = \frac{1}{4a^4}$$

and obtain

$$U = -\frac{1}{2}\pi \mathcal{N}C_6 \int_0^\infty \frac{dz}{(R+z)^4} = \boxed{\frac{-\pi \mathcal{N}C_6}{6R^3}}$$

This result confirms that $U \propto \frac{1}{R^3}$. (A shorter procedure is to use a dimensional argument, but we need the explicit expression in the following.) When

$$V = 4\varepsilon\left[\left(\frac{\sigma}{R}\right)^{12} - \left(\frac{\sigma}{R}\right)^6\right] = \frac{C_{12}}{R^{12}} - \frac{C_6}{R^6}$$

we also need the contribution from C_{12}

$$U' = 2\pi \mathcal{N}C_{12}\int_0^\infty dr \int_0^\infty dz\frac{r}{\{(R+z)^2+r^2\}^6} = 2\pi \mathcal{N}C_{12} \times \frac{1}{10}\int_0^\infty \frac{dz}{(R+z)^{10}} = \frac{2\pi \mathcal{N}C_{12}}{90R^9}$$

and therefore the total interaction energy is

$$U = \frac{2\pi \mathcal{N}C_{12}}{90R^9} - \frac{\pi \mathcal{N}C_6}{6R^3}$$

We can express this result in terms of ε and σ by noting that $C_{12} = 4\varepsilon\sigma^{12}$ and $C_6 = 4\varepsilon\sigma^6$, for then

$$U = 8\pi \varepsilon \sigma^3 \mathcal{N}\left[\frac{1}{90}\left(\frac{\sigma}{R}\right)^9 - \frac{1}{12}\left(\frac{\sigma}{R}\right)^3\right]$$

For the position of equilibrium, we look for the value of R for which $\dfrac{dU}{dR} = 0$

$$\frac{dU}{dR} = 8\pi \varepsilon \sigma^3 \mathcal{N}\left[-\frac{1}{10}\left(\frac{\sigma^9}{R^{10}}\right) + \frac{1}{4}\left(\frac{\sigma^3}{R^4}\right)\right] = 0$$

Therefore, $\dfrac{\sigma^9}{10R^{10}} = \dfrac{\sigma^3}{4R^4}$ which implies that $R = \left(\dfrac{2}{5}\right)^{1/6}\sigma = \boxed{0.858\sigma}$. For $\sigma = 342\,\text{pm}$, $R \approx$ $\boxed{294\,\text{pm}}$

Solutions to theoretical problems

P28.15 A general change in the Gibbs function of a one-component system with a surface is

$$dG = -S\,dT + V\,dp + \gamma\,d\sigma + \mu\,dn$$

Let $G = G(g) + G(\sigma)$ and $n = n(g) + n(\sigma)$; then

$$dG(g) = -S(g)\,dT + V(g)\,dp + \mu(g)\,dn(g)$$

$$dG(\sigma) = -S(\sigma)\,dT + \gamma\,d\sigma + \mu(\sigma)\,dn(\sigma)$$

At equilibrium, $\mu(\sigma) = \mu(g) = \mu$. At constant temperature, $dG(\sigma) = \gamma\,d\sigma + \mu\,dn(\sigma)$. Since dG is an exact differential, this expression integrates to

$$G(\sigma) = \gamma\sigma + \mu n(\sigma)$$

Therefore, $dG(\sigma) = \sigma\,d\gamma + \gamma\,d\sigma + \mu\,dn(\sigma) + n(\sigma)\,d\mu$

But since $dG(\sigma) = \gamma \, d(\sigma) + \mu \, dn(\sigma)$

we conclude that $\sigma \, d\gamma + n(\sigma) \, d\mu = 0$

Since $d\mu = RT \, d \ln p$, this relation is equivalent to

$$n(\sigma) = -\frac{\sigma \, d\gamma}{d\mu} = -\left(\frac{\sigma}{RT}\right) \times \left(\frac{d\gamma}{d \ln p}\right)$$

Now express $n(\sigma)$ as an adsorbed volume using

$$n(\sigma) = \frac{p^{\ominus} V_a}{RT^{\ominus}}$$

and express $d\gamma$ as a kind of chemical potential through

$$d\mu' = \frac{RT^{\ominus}}{p^{\ominus}} \, d\gamma$$

evaluated at a standard temperature and pressure ($T^{\ominus}$ and $p^{\ominus}$), then

$$\boxed{-\left(\frac{\sigma}{RT}\right) \times \left(\frac{d\mu'}{d \ln p}\right) = V_a}$$

P28.16 $d\mu' = -c_2 \left(\dfrac{RT}{\sigma}\right) dV_a$

which implies that

$$\frac{d\mu'}{d \ln p} = \left(\frac{-c_2 RT}{\sigma}\right) \times \left(\frac{dV_a}{d \ln p}\right)$$

However, we established in Problem 28.15 that

$$\frac{d\mu'}{d \ln p} = \frac{-RT V_a}{\sigma}$$

Therefore,

$$-c_2 \left(\frac{RT}{\sigma}\right) \times \left(\frac{dV_a}{d \ln p}\right) = \frac{-RT V_a}{\sigma}, \quad \text{or} \quad c_2 \, d \ln V_a = d \ln p$$

Hence, $d \ln V_a^{c_2} = d \ln p$, and therefore $\boxed{V_a = c_1 p^{1/c_2}}$

P28.19   Equilibrium constants vary with temperature according to the van't Hoff equation [9.28] which can be written in the form

$$\frac{K_1}{K_2} = e^{-\left[\frac{\Delta_{ad} H^{\ominus}}{R}\left(\frac{1}{T_1} - \frac{1}{T_2}\right)\right]}$$

or

$$\frac{K_1}{K_2} = \exp\left[\frac{160 \times 10^3 \, \text{J mol}^{-1}}{8.3145 \, \text{J K}^{-1} \, \text{mol}^{-1}} \left(\frac{1}{673 \, \text{K}} - \frac{1}{773 \, \text{K}}\right)\right] = \boxed{40.4}$$

As measured by the equilibrium constant of adsorption, NO is about 40 times more strongly adsorbed at 500°C than at 400°C.

P28.20 **(a)** $\dfrac{1}{q_{VOC,RH=0}} = \dfrac{1+bp_{VOC}}{abc_{VOC}} = \dfrac{1}{abc_{VOC}} + \dfrac{1}{a}$

Parameters of regression fit

$\theta/°C$	$1/a$	$1/ab$	R	a	b/ppm^{-1}
33.6	9.07	709.8	0.9836	0.110	0.0128
41.5	10.14	890.4	0.9746	0.0986	0.0114
57.4	11.14	1599	0.9943	0.0898	0.00697
76.4	13.58	2063	0.9981	0.0736	0.00658
99	16.82	4012	0.9916	0.0595	0.00419

The linear regression fit is generally good at all temperatures with

R values in the range 0.975 to 0.991

(b) $\ln a = \ln k_a - \dfrac{\Delta_{ad}H}{R}\dfrac{1}{T}$

and $\ln b = \ln k_b - \dfrac{\Delta_b H}{R}\dfrac{1}{T}$

Linear regression analysis of $\ln a$ versus $1/T$ gives the intercept $\ln k_a$ and slope $-\Delta_{ad}H/R$ while a similar statement can be made for a $\ln b$ versus $1/T$ plot. The temperature must be in Kelvin.

For $\ln a$ versus $1/T$

$\ln k_a = -5.605$,   standard deviation $= 0.197$
$-\Delta_{ad}H/R = 1043.2\,\text{K}$,   standard deviation $= 65.4\,\text{K}$
$R = 0.9942$ [good fit]

$k_a = e^{-5.605} = \boxed{3.68 \times 10^{-3}}$

$\Delta_{ad}H = -(8.31451\,\text{J K}^{-1}\,\text{mol}^{-1}) \times (1043.2\,\text{K})$

$= \boxed{-8.67\,\text{kJ mol}^{-1}}$

For $\ln b$ versus $1/T$

$\ln(k_b/(\text{ppm}^{-1})) = -10.550$,   standard deviation $= 0.713$
$-\Delta_b H/R = 1895.4\,\text{K}$,   standard deviation $= 236.8$
$R = 0.9774$ [good fit]

$k_b = e^{-10.550}\,\text{ppm}^{-1} = \boxed{2.62 \times 10^{-5}\,\text{ppm}^{-1}}$

$\Delta_b H = -(8.31451\,\text{J K}^{-1}\,\text{mol}^{-1}) \times (1895.4\,\text{K})$

$\boxed{\Delta_b H = -15.7\,\text{kJ mol}^{-1}}$

(c) k_a may be interpreted to be the maximum adsorption capacity at an adsorption enthalpy of zero, while k_b is the maximum affinity in the case for which the adsorbant–surface bonding enthalpy is zero.

P28.21 **(a)** $q_{water} = k(\text{RH})^{1/n}$

With a power law regression analysis we find

$\boxed{k = 0.2289}$,   standard deviation $= 0.0068$
$1/n = 1.6182$,   standard deviation $= 0.0093$; $\boxed{n = 0.6180}$
$R = 0.999\,508$

A linear regression analysis may be performed by transforming the equation to the following form by taking the logarithm of the Freundlich type equation

$$\ln q_{water} = \ln k + \frac{1}{n} \ln(RH)$$

$\ln k = -1.4746$,   standard deviation $= 0.0068$; $\boxed{k = 0.2289}$

$\frac{1}{n} = 1.6183$,   standard deviation $= 0.0093$; $\boxed{n = 0.6180}$

$R = 0.999\,508$

The two methods give exactly the same result because the software package for performing the power law regression performs the transformation to linear form for you. Both methods are actually performing a linear regression.

The correlation coefficient indicates that 99.95 per cent of the data variation is explained with the Freundlich type isotherm. The Freundlich fit hypothesis looks very good.

(b) The Langmuir isotherm model describes adsorption sites that are independent and equivalent. This assumption seems to be valid for the VOC case in which molecules interact very weakly . However, water molecules interact much more strongly through forces such as hydrogen bonding and multilayers may readily form at the lower temperatures. The intermolecular forces of water apparently cause adsorption sites to become non-equivalent and dependent. In this particular case the Freundlich type isotherm becomes the better description.

(c) $r_{VOC} = 1 - q_{water}$ where $r_{VOC} \equiv q_{VOC}/q_{VOC,RH=0}$

$$r_{VOC} = 1 - k(RH)^{1/n}$$
$$1 - r_{VOC} = k(RH)^{1/n}$$

To determine the goodness-of-fit, k, and n we perform a power law regression fit of $1 - r_{VOC}$ against RH. Results are

$\boxed{k = 0.5227}$,   standard deviation $= 0.0719$

$\frac{1}{n} = 1.3749$,   standard deviation $= 0.0601$; $\boxed{n = 0.7273}$

$R = 0.996\,20$

Since 99.62 per cent of the variation is explained by the regression, we conclude that the hypothesis that $r_{VOC} = 1 - q_{water}$ may be very useful. The values of R and n differ significantly from those of part (a). It may be that water is adsorbing to some portions of the surface and VOC to others.

29 Dynamics of electron transfer

Solutions to exercises

Discussion questions

E29.1(a)   Donor (D) and acceptor (A) must collide before they can react. Consequently, the rate of their reaction in solution is initially determined by the rate of diffusion of the reacting species. After D and A have arrived at the critical reaction distance r^* (comparable to r, the edge-to-edge distance), the rate constant for electron transfer is a function of two factors. See eqn 29.2. The first is the tunnelling rate of the electron through an energy barrier that is a function of the ionization energies of the complexes DA and D^+A^-. The second is the Gibbs energy of activation.

Effective transfer can occur only when the electronic energies in the two complexes match. The electronic energies are a function of the internuclear separations in DA and D^+A^- as illustrated in Figs 29.1 and 29.2; therefore, the distance between D and A play a critical role in determining the rate of electron transfer. The tunnelling rate is determined by the matrix element of the coupling term in the Hamiltonian which exhibits an exponential dependence on the negative of r, as given by eqn 29.3.

Justification 29.1 shows how the Gibbs energy of activation is related to the reorganization energy associated with molecular rearrangements which include the relative reorientation of the D and A molecules and the relative reorientation of the solvent molecules surrounding DA.

E29.2(a)   There are three models of the structure of the electrical double layer. The Helmholtz model, the Gouy–Chapman model, and the Stern model. We will describe the Stern model which is a combination of the first two and illustrates most of the structural features associated with the double layer. The electrode surface is a rigid plane of, say, excess positive charge. Next to it is a plane of negatively charged ions with their solvating molecules, called the outer Helmholtz layer. Adjoining this region is a diffuse layer with perhaps only a slight excess of negative charge. This region fades away into the bulk neutral solution. At another level of sophistication, an inner Helmholtz plane is added, see Section 29.3(a) for a brief description of this layer.

E29.3(a)   The electrical double layer is present near the electrode surface whether or not current is flowing in the cell. The Nernst diffusion layer is invoked to explain polarization effects near a working electrode and is a region of linear variation in concentration between the bulk solution and outer Helmholtz plane. It is typically 0.1–0.5 mm in thickness without stirring or convection, but can be reduced to 0.001 mm with such agitation. The electrical double layer is unaffected by hydrodynamic flow and is typically about 1 nm in thickness.

E29.4(a)   The principles of operation of a fuel cell are very much the same as those of a conventional galvanic cell. Both employ a spontaneous electrochemical reaction to produce an electric current which can be used as a power source for external devices. The main difference between fuel cells and ordinary cells is that the reacting substance is a material that we normally classify as a fuel and it is continuously supplied to the cell from an external source. We wish to obtain large currents from fuel cells and in order to accomplish that goal a number of obstacles limiting the rate of reaction have to be overcome. One way to increase the rate of the reaction in the cell is to use a catalytic surface with a large effective surface area to increase the current density. Operating the cells at high temperatures can increase reaction rates and in some cases molten electrolytes and electrodes and employed.

Numerical exercises

E29.5(a)   Eqns 29.4 and 29.12 are two expressions that contain the given data, the desired quantity (reorganization energy, λ) and one other unknown quantity ($\Delta^\ddagger G$); thus, they constitute a system of two equations in two unknowns. Substituting eqn 29.4

$$\Delta^{\ddagger}G = \frac{(\Delta_r G^{\ominus} + \lambda)^2}{4\lambda},$$

into eqn 29.12 yields:

$$k_{et} = \frac{2\langle H_{DA}\rangle^2}{h}\left(\frac{\pi^3}{4\lambda RT}\right)^{1/2} \exp\left(\frac{-\Delta^{\ddagger}G}{RT}\right) = \frac{\langle H_{DA}\rangle^2}{h}\left(\frac{\pi^3}{\lambda RT}\right)^{1/2} \exp\left(\frac{-(\Delta_r G^{\circ} + \lambda)^2}{4\lambda RT}\right).$$

The only unknown in this equation is λ. Putting in the numbers, recognizing that

$$\langle H_{DA}\rangle = hcH = (6.626 \times 10^{-34}\,\text{J s})(2.998 \times 10^{10}\,\text{cm s}^{-1})(0.03\,\text{cm}^{-1}) = 6 \times 10^{-25}\,\text{J}$$

yields:

$$30.5\,\text{s}^{-1} = \frac{(6 \times 10^{-25}\,\text{J})^2}{6.626 \times 10^{-34}\,\text{J s}}\left(\frac{\pi^3}{\lambda(1.381 \times 10^{-23}\,\text{J K}^{-1})(298\,\text{K})}\right)^{1/2}$$

$$\times \exp\left(\frac{-[(-0.182\,\text{eV})(1.602 \times 10^{-19}\,\text{J eV}^{-1}) + \lambda]^2}{4\lambda(1.381 \times 10^{-23}\,\text{J K}^{-1})(298\,\text{K})}\right),$$

where the Boltzmann constant takes the place of the gas constant to put all energies on a molecular rather than a molar scale. One can solve this numerically using the root-finding command of a symbolic mathematics package, or graphically by plotting the right-hand side vs. the (constant) left-hand side and finding the value of λ at which the two lines cross. The reorganization energy turns out to be:

$$\boxed{\lambda = 1.\overline{9} \times 10^{-19}\,\text{J}}\ \text{or about}\ \boxed{1.2\,\text{eV}}.$$

E29.6(a) For the same donor and acceptor at different distances, eqn 29.13 applies:

$$\ln k_{et}/\text{s}^{-1} = -\beta r + \text{constant}.$$

The slope of a plot of k_{et} versus r is $-\beta$. The slope of a line defined by two points is:

$$\text{slope} = \frac{\Delta y}{\Delta x} = \frac{\ln k_{et,2}/\text{s} - \ln k_{et,1}/\text{s}}{r_2 - r_1} = -\beta = \frac{\ln 4.51 \times 10^4 - \ln 2.02 \times 10^5}{(1.23 - 1.11)\,\text{nm}},$$

$$\beta = \boxed{12\,\text{nm}^{-1}}$$

E29.7(a) $\quad \mathcal{E} = \dfrac{\Delta\phi}{l}$ [24.48]

$$= \frac{\sigma}{\varepsilon} = \frac{\sigma}{\varepsilon_r \varepsilon_0} = \frac{0.10\,\text{C m}^{-2}}{(48) \times (8.854 \times 10^{-12}\,\text{J}^{-1}\,\text{C}^2\,\text{m}^{-1})} = \boxed{2.4 \times 10^8\,\text{V m}^{-1}}$$

Comment. Surface electric fields are very large. Relative permittivities of solutions vary with concentration and temperature. The value for pure water at 20°C is 80.4.

E29.8(a) $\ln j = \ln j_0 + (1 - \alpha) f \eta$ $\left[29.39, f = \dfrac{F}{RT} \right]$ $\ln \dfrac{j'}{j} = (1 - \alpha) f (\eta' - \eta)$ which implies that for a current density j' we require an overpotential

$$\eta' = \eta + \frac{\ln \frac{j'}{j}}{(1 - \alpha) f} = (125\,\text{mV}) + \frac{\ln\left(\frac{75}{55}\right)}{(1 - 0.39) \times (25.69\,\text{mV})^{-1}} = \boxed{138\,\text{mV}}$$

E29.9(a) Take antilogarithms of eqn 29.39; then

$$j_0 = j\,e^{-(1-\alpha)\eta f} = (55.0\,\text{mA cm}^{-2}) \times e^{-0.61 \times 125\,\text{mV}/25.69\,\text{mV}} = \boxed{2.8\,\text{mA cm}^{-2}}$$

E29.10(a) $O_2(g)$ is produced at the anode in this electrolysis and $H_2(g)$ at the cathode. The net reaction is

$$2H_2O(l) \rightarrow 2H_2(g) + O_2(g)$$

For a large positive overpotential we use

$$\ln j = \ln j_0 + (1 - \alpha) f \eta \ [29.39]$$

$$\ln \frac{j'}{j} = (1 - \alpha) f (\eta' - \eta) = (0.5) \times \left(\frac{1}{0.02569\,\text{V}} \right) \times (0.6\,\text{V} - 0.4\,\text{V}) = 3.8\overline{9}$$

$$j' = j\,e^{3.8\overline{9}} = (1.0\,\text{mA cm}^2) \times (4\overline{9}) = 4\overline{9}\,\text{mA cm}^2$$

Hence, the anodic current density increases roughly by a factor of 50 with a corresponding increase in O_2 evolution.

E29.11(a) $j_0 = 6.3 \times 10^{-6}\,\text{A cm}^{-2}$, $\quad \alpha = 0.58$ [Table 29.1]

(a) $j = j_0 \{e^{(1-\alpha) f \eta} - e^{-\alpha f \eta}\}$ [29.35]

$$\frac{j}{j_0} = e^{\{(1-0.58) \times (1/0.02569) \times 0.20\}} - e^{\{-0.58 \times (1/0.02569) \times 0.2\}} = (26.\overline{3} - 0.011) \approx 26$$

$$j = (26) \times (6.3 \times 10^{-6}\,\text{A cm}^{-2}) = \boxed{1.7 \times 10^{-4}\,\text{A cm}^{-2}}$$

(b) The Tafel equation corresponds to the neglect of the second exponential above, which is very small for an overpotential of 0.2 V. Hence

$$j = \boxed{1.7 \times 10^{-4}\,\text{A cm}^{-2}}$$

The validity of the Tafel equation increases with higher overpotentials, but decreases at lower overpotentials. A plot of j against η becomes linear (non-exponential) as $\eta \rightarrow 0$.

E29.12(a) $j_{\text{lim}} = \dfrac{cRT\lambda}{zf\delta}$ [Example 29.3]

$$= \frac{(2.5 \times 10^{-3}\,\text{mol L}^{-1}) \times (25.69 \times 10^{-3}\,\text{V}) \times (61.9\,\text{S cm}^2\,\text{mol}^{-1})}{0.40 \times 10^{-3}\,\text{m}}$$

$$= 9.9\,\text{mol L}^{-1}\,\text{V S cm}^2\,\text{mol}^{-1}\,\text{m}^{-1}$$

$$= (9.9\,\text{mol m}^{-3}) \times (10^3) \times (\text{V}\,\Omega^{-1}) \times (10^{-4}\,\text{m}^2\,\text{mol}^{-1}\,\text{m}^{-1})$$

$$= \boxed{0.99\,\text{A m}^{-2}} \quad [1\,\text{V}\,\Omega^{-1} = 1\,\text{A}]$$

E29.13(a) For the cadmium electrode $E^{\ominus} = -0.40\,\text{V}$ [Table 10.7] and the Nernst equation for this electrode [Section 10.4(f)] is

$$E = E^{\ominus} - \frac{RT}{\nu F}\ln\left(\frac{1}{[\text{Cd}^{2+}]}\right) \qquad \nu = 2$$

Since the hydrogen overpotential is $0.60\,\text{V}$ evolution of H_2 will begin when the potential of the Cd electrode reaches $-0.60\,\text{V}$. Thus

$$-0.60\,\text{V} = -0.40\,\text{V} + \frac{0.02569\,\text{V}}{2}\ln[\text{Cd}^{2+}]$$

$$\ln[\text{Cd}^{2+}] = \frac{-0.20\,\text{V}}{0.0128\,\text{V}} = -15.\overline{6}$$

$$[\text{Cd}^{2+}] = \boxed{2 \times 10^{-7}\,\text{mol L}^{-1}}$$

Comment. Essentially all Cd^{2+} has been removed by deposition before evolution of H_2 begins.

E29.14(a) $\dfrac{j}{j_0} = e^{(1-\alpha)f\eta} - e^{-\alpha f\eta}\,[29.35] = e^{(1/2)f\eta} - e^{-(1/2)f\eta} \qquad [\alpha = 0.5]$

$$= 2\sinh\left(\frac{1}{2}f\eta\right) \qquad \left[\sinh x = \frac{e^x - e^{-x}}{2}\right]$$

and we use $\frac{1}{2}f\eta = \frac{1}{2} \times \dfrac{\eta}{25.69\,\text{mV}} = 0.01946(\eta/\text{mV})$

Or

$$j = 2j_0\sinh\left(\frac{1}{2}f\eta\right) = (1.58\,\text{mA cm}^{-2}) \times \sinh\left(\frac{0.01946\,\eta}{\text{mV}}\right)$$

(a) $\eta = 10\,\text{mV}$

$\qquad j = (1.58\,\text{mA cm}^{-2}) \times (\sinh 0.1946) = \boxed{0.31\,\text{mA cm}^{-2}}$

(b) $\eta = 100\,\text{mV}$

$\qquad j = (1.58\,\text{mA cm}^{-2}) \times (\sinh 1.946) = \boxed{5.41\,\text{mA cm}^{-2}}$

(c) $\eta = -0.5\,\text{V}$

$\qquad j = (1.58\,\text{mA cm}^{-2}) \times (\sinh -0.973) \approx \boxed{-2.19\,\text{A cm}^{-2}}$

E29.15(a) $E = E^{\ominus} + \dfrac{RT}{F}\ln\dfrac{a(\text{Fe}^{3+})}{a(\text{Fe}^{2+})}$ [Nernst equation]

$$E/\text{mV} = 770 + 25.7\ln\frac{a(\text{Fe}^{3+})}{a(\text{Fe}^{2+})}$$

$$\eta/\text{mV} = 1000 - E/\text{mV} = 229 - 25.7\ln\frac{a(\text{Fe}^{3+})}{a(\text{Fe}^{2+})}$$

and hence

$$j = 2j_0\sinh\left(\frac{0.01946\,\eta}{\text{mV}}\right) \quad \text{[Exercise 29.14(a)]}$$

$$= (5.0\,\text{mA cm}^{-2}) \times \sinh\left(4.46 - 0.50\ln\frac{a(\text{Fe}^{3+})}{a(\text{Fe}^{2+})}\right)$$

We can therefore draw up the following table

$\dfrac{a(Fe^{3+})}{a(Fe^{2+})}$	0.1	0.3	0.6	1.0	3.0	6.0	10.0
$j/(mA\,cm^{-2})$	684	395	278	215	124	88	68.0

The current density falls to zero when

$$4.46 = 0.50\ln\frac{a(Fe^{3+})}{a(Fe^{2+})}$$

which occurs when $a(Fe^{3+}) = 7480 \times a(Fe^{2+})$.

E29.16(a) $\quad I = 2j_0 S \sinh\left(\dfrac{0.01946\,\eta}{mV}\right)$ [Exercise 29.14(a)]

$$\eta = (51.39\,mV) \times \sinh^{-1}\left(\frac{I}{2j_0 S}\right)$$

$$= (51.39\,mV) \times \sinh^{-1}\left(\frac{20\,mA}{(2) \times (2.5\,mA\,cm^{-2}) \times (1.0\,cm^2)}\right)$$

$$= (51.39\,mV) \times (\sinh^{-1} 4.0) = \boxed{108\,mV}$$

E29.17(a) The current density of electrons is $\dfrac{j_0}{e}$ because each one carries a charge of magnitude e. Therefore,

    (a)   Pt|H_2|H^+; $j_0 = 0.79\,mA\,cm^{-2}$ [Table 29.1]

$$\frac{j_0}{e} = \frac{0.79\,mA\,cm^{-2}}{1.602 \times 10^{-19}\,C} = \boxed{4.9 \times 10^{15}\,cm^{-2}\,s^{-1}}$$

    (b)   Pt|Fe^{3+}, Fe^{2+}; $j_0 = 2.5\,mA\,cm^{-2}$

$$\frac{j_0}{e} = \frac{2.5\,mA\,cm^{-2}}{1.602 \times 10^{-19}\,C} = \boxed{1.6 \times 10^{16}\,cm^{-2}\,s^{-1}}$$

    (c)   Pb|H_2|H^+; $j_0 = 5.0 \times 10^{-12}\,A\,cm^{-2}$

$$\frac{j_0}{e} = \frac{5.0 \times 10^{-12}\,A\,cm^{-2}}{1.602 \times 10^{-19}\,C} = \boxed{3.1 \times 10^7\,cm^{-2}\,s^{-1}}$$

There are approximately $\dfrac{1.0\,cm^2}{(280\,pm)^2} = 1.3 \times 10^{15}$ atoms in each square centimetre of surface. The numbers of electrons per atom are therefore $\boxed{3.8\,s^{-1}}$, $\boxed{12\,s^{-1}}$, and $\boxed{2.4 \times 10^{-8}\,s^{-1}}$ respectively. The last corresponds to less than one event per year.

E29.18(a) $\quad \eta = \dfrac{RTj}{Fj_0}$ [29.37]

which implies that

$$I = Sj = \left(\frac{Sj_0 F}{RT}\right)\eta$$

An ohmic conductor of resistance r obeys $\eta = Ir$, and so we can identify the resistance as

$$r = \frac{RT}{Sj_0 F} = \frac{25.69 \times 10^{-3}\,V}{1.0\,cm^2 \times j_0} = \frac{25.69 \times 10^{-3}\,\Omega}{(j_0/A\,cm^{-2})} \quad [1\,V = 1\,A\,\Omega]$$

(a) $Pt|H_2|H^+$; $j_0 = 7.9 \times 10^{-4}\,A\,cm^{-2}$

$$r = \frac{25.60 \times 10^{-3}\,\Omega}{7.9 \times 10^{-4}} = \boxed{33\,\Omega}$$

(b) $Hg|H_2|H^+$; $j_0 = 0.79 \times 10^{-12}\,A\,cm^{-2}$

$$r = \frac{25.69 \times 10^{-3}\,\Omega}{0.79 \times 10^{-12}} = 3.3 \times 10^{10}, \quad or \quad \boxed{33\,G\Omega}$$

E29.19(a) For deposition of cations, a significant net current towards the electrodes is necessary. For copper and zinc, we have $E^\ominus \approx 0.34\,V$ and $-0.76\,V$, respectively. Therefore, deposition of copper occurs when the potential falls below $0.34\,V$ and continues until the copper ions are exhausted to the point that the limiting current density is reached. Then a further reduction in potential to below $-0.76\,V$ brings about the deposition of zinc.

Comment. The depositions will be very slow until E drops substantially below $E^\ominus$.

E29.20(a) See Exercise 29.13(a) for a related situation.

Hydrogen evolution occurs significantly (in the sense of having a current density of $1\,mA\,cm^{-2}$, which is 6.2×10^{15} electrons $cm^{-2}\,s^{-1}$, or $1.0 \times 10^{-8}\,mol\,cm^{-2}\,s^{-1}$, corresponding to about $1\,cm^3$ of gas per hour) when the overpotential is $\approx 0.60\,V$. Since $E = E^\ominus + \left(\dfrac{RT}{F}\right) \ln a(H^+) = -59\,mV \times$ pH, this rate of evolution occurs when the potential at the electrode is about $-0.66\,V$ (when pH ≈ 1). Ag^+ ($E^\ominus = 0.80\,V$) has a more positive deposition potential and so deposits first.

E29.21(a) We assume $\alpha \approx 0.5$; $E^\ominus(Zn^{2+}, Zn) = -0.76\,V$.

Zinc will deposit from a solution of unit activity when the potential is below $-0.76\,V$. The hydrogen ion current toward the zinc electrode is then

$$j(H^+) = j_0 e^{-\alpha f \eta} \text{ [29.34]}$$

$$j(H^+) = (5 \times 10^{-11}\,A\,cm^{-2}) \times (e^{760/51.4}) \quad \left[\eta = -760\,mV, \quad f = \frac{1}{25.7\,mV}\right]$$

$$= 1.3 \times 10^{-4}\,A\,cm^{-2}, \quad or \quad 0.13\,mA\,cm^{-2}$$

Using the criterion that $j > 1\,mA\,cm^{-2}$ [Exercise 29.20(a)] for significant evolution of hydrogen, this value of j corresponds to a negligible rate of evolution of hydrogen, and so $\boxed{\text{zinc may be deposited}}$ from the solution.

E29.22(a) Since $E^\ominus(Mg, Mg^{2+}) = -2.37\,V$, magnesium deposition will occur when the potential is reduced to below this value. The hydrogen ion current density is then (Exercise 29.21(a))

$$j(H^+) = (5 \times 10^{-11}\,A\,cm^{-2}) \times (e^{2370/51.4}) = 5.3 \times 10^9\,A\,cm^{-2}$$

which is a lot of hydrogen ($10^6\,L\,cm^{-2}\,s^{-1}$), and so magnesium $\boxed{\text{will not be plated out}}$

E29.23(a) The cell half-reactions are

$$Cd(OH)_2 + 2e^- \rightarrow Cd + 2OH^- \quad E^\ominus = -0.81\,V$$

$$NiO(OH) + H_2O + e^- \rightarrow Ni(OH)_2 + OH^- \quad E^\ominus = +0.49\,V$$

Therefore, the standard cell potential is $\boxed{+1.30\,V}$. If the cell is working reversibly yet producing $100\,mA$, the power it produces is

$$P = IE = (100 \times 10^{-3}\,A) \times (1.3\,V) = \boxed{0.13\,W}$$

E29.24(a) $E^{\ominus} = \frac{-\Delta_r G^{\ominus}}{\nu F}$

(a) $H_2 + \frac{1}{2}O_2 \rightarrow H_2O;\quad \Delta_r G^{\ominus} = -237\,\text{kJ}\,\text{mol}^{-1}$
Since $\nu = 2$,

$$E^{\ominus} = \frac{-(-237\,\text{kJ}\,\text{mol}^{-1})}{(2) \times (96.48\,\text{kC}\,\text{mol}^{-1})} = \boxed{+1.23\,\text{V}}$$

(b) $CH_4 + 2O_2 \rightarrow CO_2 + 2H_2O$
$\Delta_r G^{\ominus} = 2\Delta_f G^{\ominus}(H_2O) + \Delta_f G^{\ominus}(CO_2) - \Delta_f G^{\ominus}(CH_4)$
$\quad\quad = [(2) \times (-237.1) + (-394.4) - (-50.7)]\,\text{kJ}\,\text{mol}^{-1} = -817.9\,\text{kJ}\,\text{mol}^{-1}$

As written, the reaction corresponds to the transfer of eight electrons. It follows that, for the species in their standard states,

$$E^{\ominus} = \frac{-(-817.9\,\text{kJ}\,\text{mol}^{-1})}{(8) \times (96.48\,\text{kC}\,\text{mol}^{-1})} = \boxed{+1.06\,\text{V}}$$

E29.25(a)   The electrode potentials of half-reactions (a), (b), and (c) are (Section 29.8)

(a) $E(H_2, H^+) = -0.059\,\text{V}\,\text{pH} = (-7) \times (0.059\,\text{V}) = -0.41\,\text{V}$
(b) $E(O_2, H^+) = (1.23\,\text{V}) - (0.059\,\text{V})\text{pH} = +0.82\,\text{V}$
(c) $E(O_2, OH^-) = (0.40\,\text{V}) + (0.059\,\text{V})\text{pOH} = 0.81\,\text{V}$

$$E(M, M^+) = E^{\ominus}(M, M^+) + \left(\frac{0.059\,\text{V}}{z_+}\right)\log 10^{-6} = E^{\ominus}(M, M^+) - \frac{0.35\,\text{V}}{z_+}$$

Corrosion will occur if $E(a)$, $E(b)$, or $E(c) > E(M, M^+)$

(i) $E^{\ominus}(Fe, Fe^{2+}) = -0.44\,\text{V},\quad z_+ = 2$
$E(Fe, Fe^{2+}) = (-0.44 - 0.18)\,\text{V} = -0.62\,\text{V} < E(a, b, \text{and c})$

(ii) $E(Cu, Cu^+) = (0.52 - 0.35)\,\text{V} = 0.17\,\text{V} \begin{cases} > E(a) \\ < E(b \text{ and c}) \end{cases}$

$E(Cu, Cu^{2+}) = (0.34 - 0.18)\,\text{V} = 0.16\,\text{V} \begin{cases} > E(a) \\ < E(b \text{ and c}) \end{cases}$

(iii) $E(Pb, Pb^{2+}) = (-0.13 - 0.18)\,\text{V} = -0.31\,\text{V} \begin{cases} > E(a) \\ < E(b \text{ and c}) \end{cases}$

(iv) $E(Al, Al^{3+}) = (-1.66 - 0.12)\,\text{V} = -1.78\,\text{V} < E(a, b, \text{and c})$

(v) $E(Ag, Ag^+) = (0.80 - 0.35)\,\text{V} = 0.45\,\text{V} \begin{cases} > E(a) \\ < E(b \text{ and c}) \end{cases}$

(vi) $E(Cr, Cr^{3+}) = (-0.74 - 0.12)\,\text{V} = -0.86\,\text{V} < E(a, b, \text{and c})$

(vii) $E(Co, Co^{2+}) = (-0.28 - 0.15)\,\text{V} = -0.43\,\text{V} < E(a, b, \text{and c})$

Therefore, the metals with a thermodynamic tendency to corrode in moist conditions at pH $= 7$ are $\boxed{Fe, Al, Co, Cr}$ if oxygen is absent, but, if oxygen is present, all seven elements have a tendency to corrode.

E29.26(a) $\frac{(1.0\,\text{A}\,\text{m}^{-2}) \times (3.16 \times 10^7\,\text{s}\,\text{yr}^{-1})}{9.65 \times 10^4\,\text{C}\,\text{mol}^{-1}} = 32\overline{7}\,\text{mol}\,\text{e}^-\,\text{m}^{-2}\,\text{yr}^{-1} = 16\overline{4}\,\text{mol}\,\text{Fe}\,\text{m}^{-2}\,\text{yr}^{-1}$

$\frac{(16\overline{4}\,\text{mol}\,\text{m}^{-2}\,\text{yr}^{-1}) \times (55.85\,\text{g}\,\text{mol}^{-1})}{7.87 \times 10^6\,\text{g}\,\text{m}^{-3}} = 1.2 \times 10^{-3}\,\text{m}\,\text{yr}^{-1} = \boxed{1.2\,\text{mm}\,\text{yr}^{-1}}$

Solutions to problems

Solutions to numerical problems

P29.1 Estimate the bimolecular rate constant k_{12} for the reaction

$$\text{Ru(bpy)}_3{}^{3+} + \text{Fe(H}_2\text{O)}_6{}^{2+} \rightarrow \text{Ru(bpy)}_3{}^{2+} + \text{Fe(H}_2\text{O)}_6{}^{3+}$$

by using the approximate Marcus cross-relation:

$$k_{12} \approx (k_{11}k_{22}K)^{1/2}.$$

The standard cell potential for the reaction is:

$$E^{\ominus} = E_{\text{red}}^{\ominus}(\text{Ru(bpy)}_3{}^{3+}) - E_{\text{red}}^{\ominus}(\text{Fe(H}_2\text{O)}_6{}^{3+}) = (1.26 - 0.77)\,\text{V} = 0.49\,\text{V}.$$

The equilibrium constant is:

$$K = \exp\left(\frac{\nu F E^{\ominus}}{RT}\right) = \exp\left(\frac{(1)(96485\,\text{C mol}^{-1}\,\text{s}^{-1})(0.49\,\text{V})}{(8.3145\,\text{J mol}^{-1}\,\text{K}^{-1})(298\,\text{K})}\right) = 1.9 \times 10^8.$$

The rate constant is approximately:

$$k_{12} \approx [(4.0 \times 10^8\,\text{L mol}^{-1}\text{s}^{-1})(4.2\,\text{L mol}^{-1}\text{s}^{-1})(1.9 \times 10^8)]^{1/2},$$

$$k_{12} \approx \boxed{5.6 \times 10^8\,\text{L mol}^{-1}\text{s}^{-1}}.$$

P29.2 $$\ln j = \ln j_0 + (1 - \alpha)f\eta \quad [29.39]$$

Draw up the following table

η/mV	50	100	150	200	250
$\ln(j/\text{mA cm}^{-2})$	0.98	2.19	3.40	4.61	5.81

The points are plotted in Fig. 29.1.

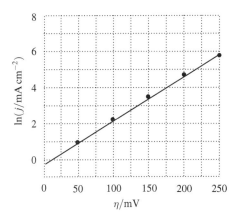

Figure 29.1

The intercept is at -0.25, and so $j_0/(\text{mA cm}^{-2}) = e^{-0.25} = \boxed{0.78}$. The slope is 0.0243, and so $\dfrac{(1-\alpha)F}{RT} = 0.0243\,\text{mV}^{-1}$. It follows that $1 - \alpha = 0.62$, and so $\boxed{\alpha = 0.38}$. If η were large but negative.

$$|j| \approx j_0 e^{-\alpha f \eta} [29.40] = (0.78\,\text{mA cm}^{-2}) \times (e^{-0.38\,\eta/25.7\,\text{mV}})$$
$$= (0.78\,\text{mA cm}^{-2}) \times (e^{-0.015(\eta/\text{mV})})$$

and we draw up the following table

η/mV	-50	-100	-150	-200	-250
$j/(\text{mA cm}^{-2})$	1.65	3.50	7.40	15.7	33.2

P29.4 $j_\text{lim} = \dfrac{zFDc}{\delta}$ [29.51],   and so $\delta = \dfrac{FDc}{j_\text{lim}}$ $[z = 1]$

Therefore,

$$\delta = \frac{(9.65 \times 10^4\,\text{C mol}^{-1}) \times (1.14 \times 10^{-9}\,\text{m}^2\,\text{s}^{-1}) \times (0.66\,\text{mol m}^{-3})}{28.9 \times 10^{-2}\,\text{A m}^{-2}}$$
$$= 2.5 \times 10^{-4}\,\text{m},   \text{or}   \boxed{0.25\,\text{mm}}$$

P29.5 $E' = E - IR_\text{s} - \dfrac{2RT}{zF} \ln g(I)$ [29.60]

$$g = \frac{\left(\dfrac{I}{A\bar{j}}\right)^{2z}}{\left[\left(1 - \dfrac{I}{Aj_{\text{lim,L}}}\right) \times \left(1 - \dfrac{I}{Aj_{\text{lim,R}}}\right)\right]^{1/2}}$$

with $j_\text{lim} = \dfrac{cRT\lambda}{zF\delta}$ [Example 29.3] $= a\lambda$

$$R_\text{s} = \frac{l}{\kappa A} = \frac{1}{cA\Lambda_\text{m}}   \text{with } \Lambda_\text{m} = \lambda_+ + \lambda_-$$

Therefore, $E' = E - \dfrac{Il}{cA\Lambda_\text{m}} - \dfrac{2RT}{zF} \ln g(I)$

with $g(I) = \dfrac{\left(\dfrac{I^2}{A^2 j_\text{LO} j_\text{RO}}\right)^z}{\left[1 - \left(\dfrac{I}{Aa_\text{L}\lambda_\text{L+}}\right)\right]^{1/2} \left[1 - \left(\dfrac{I}{Aa_\text{R}\lambda_\text{R+}}\right)\right]^{1/2}}$

with $a_\text{L} = \dfrac{RTc_\text{L}}{z_\text{L}F\delta_\text{L}}$ and $a_\text{R} = \dfrac{RTc_\text{R}}{z_\text{R}F\delta_\text{R}}$

For the cell $\text{Zn}|\text{ZnSO}_4(\text{aq})||\text{CuSO}_4(\text{aq})|\text{Cu}$, $l = 5\,\text{cm}$, $A = 5\,\text{cm}^2$, $c(\text{M}_\text{L}^+) = c(\text{M}_\text{R}^+) = 1\,\text{mol L}^{-1}$, $z_\text{L} = z_\text{R} = 2$, $\lambda_\text{L+} = 107\,\text{S cm}^2\,\text{mol}^{-1}$, $\lambda_\text{R+} = 106\,\text{S cm}^2\,\text{mol}^{-1} \approx \lambda_\text{L+}$, $\lambda_- = \lambda_{\text{SO}_4^{2-}} = 160\,\text{S cm}^2\,\text{mol}^{-1}$. $\Lambda_\text{m} \approx (107+160)\,\text{S cm}^2\,\text{mol}^{-1} = 267\,\text{S cm}^2\,\text{mol}^{-1}$ for both electrolyte solutions. We take $\delta \approx 0.25\,\text{mm}$ [Example 29.3] and $j_\text{LO} \approx j_\text{RO} \approx 1\,\text{mA cm}^{-2}$. We can also take

$$E^\ominus (a \approx 1) = E^\ominus(\text{Cu, Cu}^{2+}) - E^\ominus(\text{Zn, Zn}^{2+}) = [0.34 - (-0.76)]\,\text{V} = 1.10\,\text{V}$$
$$R_\text{s} = \frac{5\,\text{cm}}{(1\,\text{M}) \times (267\,\text{S cm}^2\,\text{mol}^{-1}) \times (5\,\text{cm}^2)} = 3.\overline{8}\,\Omega$$

$$j_{\lim} = j_{\lim}^+ = \frac{1}{2} \times \left(\frac{(0.0257\,\text{V}) \times (107\,\text{S cm}^2\,\text{mol}^{-1}) \times (1\,\text{M})}{0.25 \times 10^{-3}\,\text{m}} \right) \approx 5.5 \times 10^{-2}\,\text{S V cm}^{-2}$$

$$= 5.5 \times 10^{-2}\,\text{A cm}^{-2}$$

It follows that

$$E'/\text{V} = (1.10) - 3.7\overline{5}(I/\text{A}) - (0.0257)\ln\left(\frac{(I/5 \times 10^{-3}\,\text{A})^4}{1 - 3.6(I/\text{A})} \right)$$

$$= (1.10) - 3.7\overline{5}(I/\text{A}) - (0.0257)\ln\left(\frac{1.6 \times 10^9 (I/\text{A})^4}{1 - 3.6(I/\text{A})} \right)$$

This function is plotted in Fig. 29.2.

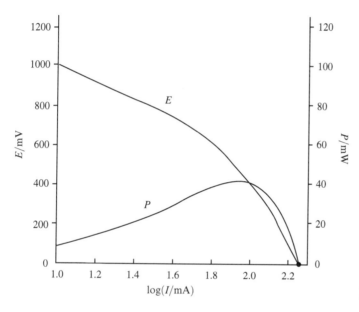

Figure 29.2

The power is

$$P = IE'$$

and so $P/\text{W} = 1.10(I/\text{A}) - 3.7\overline{5}(I/\text{A})^2 - 0.0257(I/\text{A})\ln\left(\frac{1.6 \times 10^9 (I/\text{A})^4}{1 - 3.6(I/\text{A})} \right)$

This function is also plotted in Fig. 29.2. Maximum power is delivered at about $\boxed{87\,\text{mA}}$ and 0.46 V, and is about 40 mW.

P29.6
$$E' = E - \left(\frac{4RT}{F} \right)\ln\left(\frac{I}{A\overline{j}} \right) - IR_s \quad [29.58]$$

$$P = IE' = IE - aI\ln\left(\frac{I}{I_0} \right) - I^2 R_s \quad \text{where } a = \frac{4RT}{F} \text{ and } I_0 = A\overline{j}. \text{ For maximum power,}$$

$$\frac{dP}{dI} = E - a\ln\left(\frac{I}{I_0} \right) - a - 2IR_s = 0$$

which requires

$$\ln\left(\frac{I}{I_0}\right) = \left(\frac{E}{a} - 1\right) - \frac{2IR_s}{a}$$

This expression may be written

$$\ln\left(\frac{I}{I_0}\right) = c_1 - c_2 I; \qquad c_1 = \frac{E}{a} - 1, \quad c_2 = \frac{2R_s}{a} = \frac{FR_s}{2RT}$$

For the present calculation, use the data in Problem 29.5. Then

$$I_0 = A\bar{j} = (5\,\text{cm}^2) \times (1\,\text{mA}\,\text{cm}^{-2}) = 5\,\text{mA}$$

$$c_1 = \frac{(1.10\,\text{V})}{(4) \times (0.0257\,\text{V})} - 1 = 10.7$$

$$c_2 = \frac{(3.7\bar{5}\,\Omega)}{(2) \times (0.0257\,\text{V})} = 73\,\Omega\,\text{V}^{-1} = 73\,\text{A}^{-1}$$

That is, $\ln(0.20I/\text{mA}) = 10.7 - 0.073(I/\text{mA})$

We then draw up the following table

I/mA	103	104	105	106	107
$\ln(0.20I/\text{mA})$	3.025	3.034	3.044	3.054	3.063
$10.7 - 0.073(I/\text{mA})$	3.181	3.108	3.035	2.962	2.889

The two sets of points are plotted in Fig. 29.3.

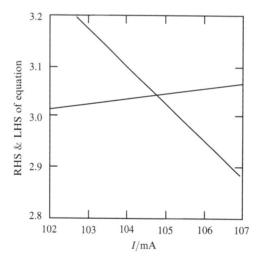

Figure 29.3

The lines intersect at $I = 105\,\text{mA}$, which therefore corresponds to the current at which maximum power is delivered. The power at this current is

$$P = (105\,\text{mA}) \times (1.10\,\text{V}) - (0.103\,\text{V}) \times (105\,\text{mA}) \times \ln\left(\frac{105}{5}\right) - (105\,\text{mA})^2 \times (3.7\bar{5}\,\Omega)$$

$$= 41\,\text{mW}$$

P29.7 $Fe^{2+} + 2e^- \rightarrow Fe$ $\nu = 2;$ $E^\ominus = -0.447\,V$

(a) $E_0 = E^\ominus - \dfrac{RT}{\nu F} \ln Q$ [10.34]

   $= E^\ominus - \dfrac{RT}{\nu F} \ln \dfrac{1}{[Fe^{2+}]}$ assuming $\gamma_{Fe^{2+}} = 1$

   $= -0.447\,V - \dfrac{25.693 \times 10^{-3}\,V}{2} \ln \left(\dfrac{mol\,L^{-1}}{1.70 \times 10^{-6}\,mol\,L^{-1}} \right)$

   $\boxed{E_0 = -0.618\,V}$

   $\eta = E' - E_0$ [29.32]

   η values are reported in the following table.

(b) $j = \dfrac{\nu F}{A} \dfrac{dn_{Fe}}{dt} = \dfrac{2(96485\,C\,mol^{-1})}{9.1\,cm^2} \dfrac{dn_{Fe}}{dt}$

   j values are reported in the following table.

   $j = j_0 \left(e^{(1-\alpha)f\eta} - e^{-\alpha f\eta} \right) = j_0 e^{-\alpha f\eta} \{e^{f\eta} - 1\}$

   $= -j_c \{e^{f\eta} - 1\}$ [29.34, 29.35]

   $j_c = \dfrac{-j}{e^{f\eta} - 1}$

   j_c values are reported in the following table

$\dfrac{dn_{Fe}}{dt} \Big/ (10^{-12}\,mol\,s^{-1})$	$-E'/mV$	$-\eta/mV$	$j/(\mu A\,cm^{-2})$	$j_c/(\mu A\,cm^{-2})$
1.47	702	84	0.0312	0.0324
2.18	727	109	0.0462	0.0469
3.11	752	134	0.0659	0.0663
7.26	812	194	0.154	0.154

(c) $j_c = j_0\,e^{-\alpha f\eta}$ [29.34]

   $\ln j_c = \ln j_0 - \alpha f\eta$

   Performing a linear regression analysis of the $\ln j_c$ versus η data, we find

   $\ln j_0 = -4.608,$ standard deviation $= 0.015$
   $\alpha f = 0.01413\,mV,$ standard deviation $= 0.00011$

   $\boxed{R = 0.999\,94}$

The correlation coefficient and the standard deviation indicate that the plot provides an excellent description of the data.

   $j_0 = e^{-4.608}$ or $\boxed{j_0 = 0.009\,97\,\mu A\,cm^{-2}}$

   $\alpha = \dfrac{0.01413}{f} = (0.01413\,mV^{-1}) \times (25.693\,mV)$

   $\boxed{\alpha = 0.363}$

P29.10   The accompanying Tafel plot (Fig. 29.4) of $\ln j$ against E shows no region of linearity so the Tafel equation cannot be used to determine j_0 and α.

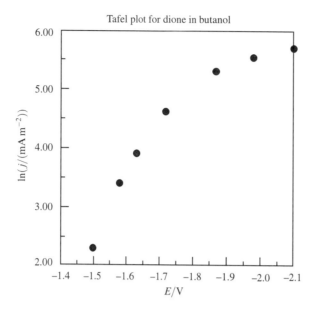

Figure 29.4

$$M_{sol} \overset{K_1}{\rightleftharpoons} M_{ads}$$

$$M_{ads} + H^+ + e^- \overset{K_2}{\rightleftharpoons} MH_{ads}$$

$$2MH_{ads} \xrightarrow[\text{rate-determining}]{k_3} HMMH$$

Assuming that the dimerization is rate-determining, two electrons are transferred per molecule of HMMH and $z = 2$. It is also reasonable to suppose that the first two reactions are at quasi-equilibrium. According to reaction 3, the current density is proportional to the square of the functional surface coverage by MH_{ads}, θ_{MH}.

$$j = zFk_3 \, \theta_{MH}^2$$

$$\ln j = \ln(zFk_3) + 2\ln\theta_{MH} \tag{1}$$

The characteristics of this equation differ from those of the Tafel equation at high negative overpotentials

$$\ln j = \ln j_0 - \alpha f \eta \quad [29.41]$$

at low concentrations of M the value of θ_{MH} changes with the overpotential in a non-exponential manner. This makes $\ln j$ non-linear throughout the potential range.

P29.11 (a) The polarization curve is described by the Butler–Volmer equation (eqns 29.34 and 29.35). The ratios j_a/j_0, $-j_c/j_0$ and j/j_0 are plotted in the following graph with $\alpha = 0.5$ at 298.15 K.

The polarization curve shows that if the overvoltage is greater than about 0.08 V, the net current is anodic. If the overvoltage is greater than about -0.08 V, the net current is cathodic. The anodic current density equals the cathodic current density at zero overvoltage.

The Butler–Volmer equation demonstrates contributions to the net current density. When the transfer coefficient is either zero or very small, the cathodic current makes a small contribution. When the transfer coefficient is equal to about 1, the anodic current makes a very small contribution.

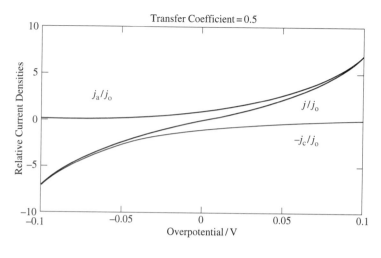

Figure 29.5(a)

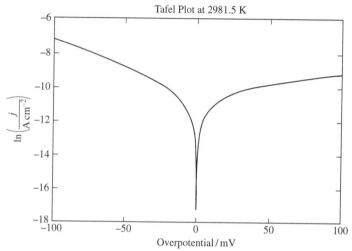

Figure 29.5(b)

(b) For $Ce^{4+} + e^- \rightleftharpoons Ce^{3+}$, $j_0 = 4.0 \times 10^{-5}\,\text{A cm}^{-2}$ and $\alpha = 0.75$ (Table 29.1).

In the limit of a very negative overpotential the slope of a Tafel plot equals $(1-\alpha)f$ while in the limit of very positive overpotential the slope equals $-\alpha f$ (eqns 29.39 and 29.41). Extrapolation of each linear extreme to an overpotential equal to zero gives the value of $\ln(j_0)$.

P29.13   **(a)** $Ti^{3+} + e^- \rightarrow Ti^{2+}$ $E^{\ominus} = -0.368\,\text{V}, z = 1$

We use $E^{\ominus}$ as the estimate of the standard formal potential at the ionic strength of the experiment, $E^{\ominus}$. Furthermore, it is reasonable to make the estimates

$$D_{Ox} = D_{Red} = 1 \times 10^{-5}\,\text{cm}^2\,\text{s}^{-1} \text{ and } \gamma = (D_{Ox}/D_{Red})^{1/2} = 1.$$

The simulated voltammogram at 298.15 K follows.

The voltammogram shows that, as the potential is scanned from E_{inital} set at -0.2 V at the rate of $10\,\text{m V s}^{-1}$, the cathodic current initially grows. It peaks at about -0.40 V, which is somewhat lower than the standard reduction potential of -0.368 V. At lower potentials the current declines as reduction depletes oxidant near the electrode. The decline would not occur at relatively low current density if diffusion from the bulk solution could more rapidly replenish reacted oxidant. This can be tested by redoing the plot with a larger diffusion coefficient.

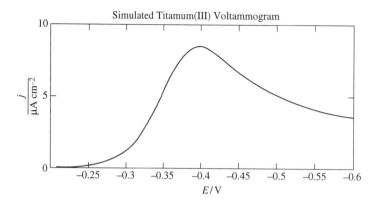

Figure 29.6

The current density is proportional to the square root of the scan rate. Doubling the scan rate causes an increase in the peak current density by a factor of $\sqrt{2}$.

(b) There is a $j(E)$ extremum when $\mathrm{d}j/\mathrm{d}E = 0$ and, consequently, $\mathrm{d}\chi(x)/\mathrm{d}x = 0$. The above plot indicates that $E_{\mathrm{pc}} \simeq -0.40\,\mathrm{V}$. A more accurate numerical determination of E_{pc} begins with numerical evaluation of $\mathrm{d}\chi/\mathrm{d}x$ and proceeds by systematically adjusting the value of x until the constraint $\mathrm{d}\chi/\mathrm{d}x = 0$ is satisfied. $x = x_{\mathrm{pc}}$ at that point. Mathematical software functions, such as the Given/Find solve block of mathcad, easily perform these tasks. We find that

$$x_{\mathrm{pc}} = 7.64632 \quad \text{and} \quad \boxed{E_{\mathrm{pc}} = E_{\mathrm{initial}} - \frac{x_{\mathrm{pc}}}{z\,f} = -0.39646\,\mathrm{V}}$$

$$Z(E_{\mathrm{pc}} - E^{\ominus}) = -28.46\,\mathrm{mV} \quad \text{or} \quad \boxed{E_{\mathrm{pc}} = E^{\ominus} - 28.5\,\mathrm{mV}/z}$$

The peak current density is $8.50\,\mu\mathrm{A\,cm}^{-2}\,(j_{\mathrm{p}})$. To find the potential at half-peak ($E_{\mathrm{p/2}}$), use a Given/Find solve block, or a solver of a mathematical software package, or the root function of a calculator to obtain the value of E for which

$$j(E) = j_{\mathrm{p}}/2$$

We find that $\boxed{E_{\mathrm{p/2}} = -0.3399\,\mathrm{V}}$ and $\boxed{E_{\mathrm{p/2}} - E_{\mathrm{pc}}} = -0.3399\,\mathrm{V} - (-0.3965\,\mathrm{V})$

$$\boxed{= 56.6\,\mathrm{mV}}$$

Criteria for reversibility include:

(i) Electroprocesses must be fast enough to maintain an equilibrium between oxidized and reduced species at the electrode.

(ii) $E_{\mathrm{p/2}} - E_{\mathrm{pc}} \simeq 56.5\,\mathrm{mV}$

(iii) $z(E_{\mathrm{pc}} - E^{\ominus}) \simeq -28.5\,\mathrm{mV}$

(iv) $j \propto c_{\mathrm{ox}}$ at fixed scan rate ν

(v) $j \propto \nu^{1/2}$

(vi) E_{pc} is independent of ν.

R.S. Nicholson and I. Shain, *Anal. Chem.*, **36**, 706 (1964).

D.H. Evans, K.M. O'Connell, R.A. Peterson, and M.J. Kelly, *J. Chem. Ed.*, **60**, 290 (1983).

Solutions to theoretical problems

P29.15

$$j = j_0 \left\{ e^{(1-\alpha)f\eta} - e^{-\alpha f\eta} \right\} \quad [29.35]$$

$$= j_0 \left\{ 1 + (1-\alpha)\eta f + \tfrac{1}{2}(1-\alpha)^2\eta^2 f^2 + \cdots - 1 + \alpha f\eta - \tfrac{1}{2}\alpha^2\eta^2 f^2 + \cdots \right\}$$

$$= j_0 \left\{ \eta f + \tfrac{1}{2}(\eta f)^2(1-2\alpha) + \cdots \right\}$$

$$\langle j \rangle = j_0 \left\{ \langle \eta \rangle f + \tfrac{1}{2}(1-2\alpha)f^2 \langle \eta^2 \rangle + \cdots \right\}$$

$$\langle \eta \rangle = 0 \quad \text{because} \quad \frac{\omega}{2\pi} \int_0^{2\pi/\omega} \cos \omega t \, dt = 0 \quad \left[\frac{2\pi}{\omega} \text{is the period} \right]$$

$$\langle \eta^2 \rangle = \frac{1}{2}\eta_0^2 \quad \text{because} \quad \frac{\omega}{2\pi} \int_0^{2\pi/\omega} \cos^2 \omega t \, dt = \frac{1}{2}$$

Therefore, $\boxed{ \langle j \rangle = \tfrac{1}{4}(1-2\alpha)f^2 j_0 \eta_0^2 }$

and $\langle j \rangle = 0$ when $\alpha = \tfrac{1}{2}$. For the mean current,

$$\langle I \rangle = \tfrac{1}{4}(1-2\alpha)f^2 j_0 S \eta_0^2$$

$$= \frac{1}{4} \times (1-0.76) \times \left(\frac{(7.90 \times 10^{-4}\,\mathrm{A\,cm^{-2}}) \times (1.0\,\mathrm{cm^2})}{(0.0257\,\mathrm{V})^2} \right) \times (10\,\mathrm{mV})^2$$

$$= \boxed{7.2\,\mathrm{\mu A}}$$

Solutions to applications

P29.18

For a series of reactions with a fixed edge-to-edge distance and reorganization energy, the log of the rate constant depends quadratically on the reaction free-energy; eqn 29.14 applies:

$$\ln k_{\mathrm{et}} = -\frac{(\Delta_r G^{\ominus})^2}{4\lambda kT} - \frac{\Delta_r G^{\ominus}}{2kT} + \text{constant},$$

where we have replaced RT by kT since the energies are given in molecular rather than molar units. Draw up the following table:

$\Delta_r G^{\ominus}/\mathrm{eV}$	$k_{\mathrm{et}}/(10^6\,\mathrm{s^{-1}})$	$\ln k_{\mathrm{et}}/\mathrm{s^{-1}}$
-0.665	0.657	13.4
-0.705	1.52	14.2
-0.745	1.12	13.9
-0.975	8.99	16.0
-1.015	5.76	15.6
-1.055	10.1	16.1

and plot $\ln k_{et}$ vs. $\Delta_r G^{\ominus}$:

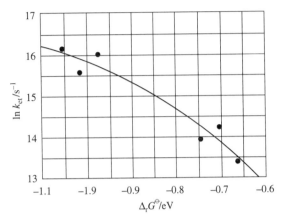

Figure 29.7

The least squares quadratic fit equation is:

$$\ln k_{et}/s^{-1} = 3.23 - 21.1(\Delta_r G^{\ominus}/eV) - 8.48 - (\Delta_r G^{\ominus}/eV)^2 \quad r^2 = 0.938$$

The coefficient of the quadratic term is:

$$-\frac{1}{4\lambda kT} = -\frac{8.48}{eV^2},$$

so $\quad \lambda = \dfrac{(eV)^2}{4(8.48)\,kT} = \dfrac{(1.602 \times 10^{-19}\,J\,eV^{-1})\,(eV)^2}{2(8.48)(1.381 \times 10^{-23}\,J\,K^{-1})(298\,K)},$

$\lambda = \boxed{2.30\,eV}$.

As a check on the reliability of the fit, note that according to eqn 29.14, the coefficient of the linear term is:

$$\frac{1}{2kT} = -\frac{21.1}{eV},$$

so $\quad T = \dfrac{eV}{2k(21.1)} = \dfrac{(1.602 \times 10^{-19}\,J\,eV^{-1})eV}{2(1.381 \times 10^{-23}\,J\,K^{-1})(21.1)} = 275\,K,$

which differs by about 8% from the stated temperature of 298 K.

P29.23   Corrosion occurs by way of the reaction

$$Fe + 2H^+ \rightarrow Fe^{2+} + H_2$$

The half-reactions at the anode and cathode are

Anode: $Fe \rightarrow Fe^{2+} + 2e^-$

Cathode: $2H^+ + 2e^- \rightarrow H_2$

$\Delta\phi_{corr} = (-0.720\,V) + (0.2802\,V) = -0.440\,V$

$\Delta\phi_{corr} = \eta(H) + \Delta\phi_e(H)$ [*Justification* 29.3]

$$\Delta\phi_e(H) = (-0.0592\,V) \times pH = (-0.0592\,V) \times 3 = -0.177\overline{6}\,V$$

$$\eta(H) = -\frac{1}{\alpha f}\ln\frac{j_{corr}}{j_0(H)} \quad [29.44]$$

Then, $\Delta\phi_{corr} = -0.440\,V = -\dfrac{1}{\alpha f}\ln\dfrac{j_{corr}}{j_0(H)} - 0.177\overline{6}\,V$

and $\ln\dfrac{j_{corr}}{j_0(H)} = (0.262\,V) \times \alpha f = (0.262\,V) \times (18\,V^{-1}) = 4.7\overline{16}$

$$j_{corr} = j_0(H) \times e^{4.71\overline{6}} = (1.0 \times 10^{-7}\,A\,cm^{-2}) \times (112) = 1.1\overline{2} \times 10^{-5}\,A\,cm^{-2}$$

Faraday's laws give the amount of iron corroded

$$n = \frac{I_{corr}t}{zF} = \frac{(1.1\overline{2} \times 10^{-5}\,A\,cm^{-2}) \times (8.64 \times 10^4\,s\,d^{-1})}{(2) \times (9.65 \times 10^4\,C\,mol^{-1})} = 5.0 \times 10^{-6}\,mol\,cm^{-2}\,d^{-1}$$

$$m = n \times (55.85\,g\,mol^{-1}) = (5.0 \times 10^{-6}\,mol\,cm^{-2}\,d^{-1}) \times (55.85 \times 10^3\,mg\,mol^{-1})$$

$$= \boxed{0.28\,mg\,cm^{-2}\,d^{-1}}$$

Solutions to box discussion questions

1.1.1 The pressure at the base of a column of height H is $p = \rho g H$ (Example 1.2). But the pressure at any altitude h within the atmospheric column of height H depends only on the air above it; therefore

$$p = \rho g(H - h) \quad \text{and} \quad \mathrm{d}p = -\rho g \, \mathrm{d}h$$

Since $\rho = \dfrac{pM}{RT}$ [Problem 1.5], $\mathrm{d}p = -\dfrac{pMg \, \mathrm{d}h}{RT}$, implying that $\dfrac{\mathrm{d}p}{p} = -\dfrac{Mg \, \mathrm{d}h}{RT}$

This relation integrates to $p = p_0 e^{-Mgh/RT}$

For air, $M \approx 29 \, \mathrm{g\,mol^{-1}}$ and at 298 K

$$\frac{Mg}{RT} \approx \frac{(29 \times 10^{-3} \, \mathrm{kg\,mol^{-1}}) \times (9.81 \, \mathrm{ms^{-2}})}{2.48 \times 10^3 \, \mathrm{J\,mol^{-1}}} = 1.1\overline{5} \times 10^{-4} \, \mathrm{m^{-1}} \quad [1\,\mathrm{J} = 1\,\mathrm{kg\,m^2\,s^{-2}}]$$

(a) $h = 15 \, \mathrm{cm}$

$$p = p_0 \times e^{(-0.15\,\mathrm{m}) \times (1.1\overline{5} \times 10^{-4}\mathrm{m^{-1}})} = 0.99\overline{998} \, p_0; \quad \frac{p - p_0}{p_0} = \boxed{0.00}$$

(b) $h = 1350 \, \mathrm{ft}$, which is equivalent to $412 \, \mathrm{m}[1 \, \mathrm{inch} = 2.54 \, \mathrm{cm}]$

$$p = p_0 \times e^{(-412\,\mathrm{m}) \times (1.1\overline{5} \times 10^{-4}\mathrm{m^{-1}})} = 0.95 \, p_0; \quad \frac{p - p_0}{p_0} = \boxed{-0.05}$$

2.1.1 The physical or chemical change associated with either exothermic or endothermic processes alters the heat capacity of the matter. This alters the thermogram baseline of the DSC.

2.2.1 Good approximate answers can be obtained from the data for the heat capacity and molar heat of vaporization of water at 25°C. [Table 2.6 and 2.3]

$$C_{p,\mathrm{m}}(\mathrm{H_2O, l}) = 75.3 \, \mathrm{J\,K^{-1}\,mol^{-1}} \quad \Delta_{\mathrm{vap}}H^{\ominus}(\mathrm{H_2O}) = 44.0 \, \mathrm{kJ\,mol^{-1}}$$

$$n(\mathrm{H_2O}) = \frac{65 \, \mathrm{kg}}{0.018 \, \mathrm{kg\,mol^{-1}}} = 3.6 \times 10^3 \, \mathrm{mol}$$

From $\Delta H = nC_{p,\mathrm{m}}\Delta T$, we obtain

$$\Delta T = \frac{\Delta H}{nC_{p,\mathrm{m}}}$$

$$= \frac{1.0 \times 10^4 \, \mathrm{kJ}}{(3.6 \times 10^3 \, \mathrm{mol}) \times (0.0753 \, \mathrm{kJ\,K^{-1}\,mol^{-1}})} = \boxed{+37 \, \mathrm{K}}$$

From $\Delta H = n\Delta_{\mathrm{vap}}H^{\ominus} = \dfrac{m}{M}\Delta_{\mathrm{vap}}H^{\ominus}$

$$m = \frac{M \times \Delta H}{\Delta_{\mathrm{vap}}H^{\ominus}} = \frac{(0.018 \, \mathrm{kg\,mol^{-1}}) \times (1.0 \times 10^4 \, \mathrm{kJ})}{44.0 \, \mathrm{kJ\,mol^{-1}}} = \boxed{4.09 \, \mathrm{kg}}$$

Comment. This estimate would correspond to about 30 glasses of water per day, which is much higher than the average consumption. The discrepancy may be a result of our assumption that evaporation of water is the main mechanism of heat loss. Heat is also lost by heat conduction between the hot body and cool environment and by inhalation of cool air accompanying exhalation of hot air.

4.1.1
$$c(T_c) = \frac{T_c}{T_h - T_c}; \quad c_0 = c(273\,K) = \frac{273\,K}{293\,K - 273\,K} = 13.65$$

$$dq = nC_{p,m}\,d\,T_c$$

$$|dw(\text{cooling})| = \frac{|dq|}{c(T_c)} = \frac{|nC_{p,m}dT|}{c(T_c)}$$

$$|w(\text{cooling})| = \left| \int_{T_i}^{T_f} \frac{nC_{p,m}\,dT}{c(T_c)} \right| = nC_{p,m} \left| \int_{T_i}^{T_f} \left(\frac{T_h}{T_c} - 1 \right) dT_c \right| \quad [T_i = T_h,\ T = T_c]$$

$$\approx \boxed{nC_{p,m} \left| T_f - T_i - T_i \ln\left(\frac{T_f}{T_i} \right) \right|}$$

$$|w(\text{total})| = |w(\text{cooling})| + |w(\text{freezing})|$$

$$|w(\text{cooling})| = \left(\frac{250\,g}{18.02\,g\,mol^{-1}} \right) \times (75.3\,J\,K^{-1}\,mol^{-1}) \times \left((-20\,K) - 293\,K \ln \frac{273}{293} \right) = 0.75\,kJ$$

$$|w(\text{freezing})| = \frac{n\Delta H_{fus}}{C_0} = \left(\frac{250\,g}{18.02\,g\,mol^{-1}} \right) \times \left(\frac{6.01\,kJ\,mol^{-1}}{13.65} \right) = 6.11\,kJ$$

Therefore, the total work is

$$|w(\text{total})| = 0.75\,kJ + 6.11\,kJ = \boxed{6.86\,kJ}$$

If the initial temperature were 25°C, no additional work would be needed because cooling from 25°C to 20°C is spontaneous.

$$t = \frac{6.86\,kJ}{100\,J\,s^{-1}} = \boxed{68.6\,s}$$

But this value for the time is meaningless as it does not take into account the rate of heat transfer.

4.2.1
Let T_i be the initial temperature and T_f be the final temperature. The entropy change for a demagnetization step of a magnetize-demagnetize cycle equals zero.

$$\Delta S_{demag} = 0 = a\,T_f^3 - b\,T_i^3 \quad \text{or} \quad T_f = \left(\frac{b}{a} \right)^{1/3} T_i$$

By induction we find that the final temperature after n cycles is given by

$$T_f = \left(\frac{b}{a} \right)^{n/3} T_i \quad \text{or} \quad n = \frac{3 \log\left(\frac{T_f}{T_i} \right)}{\log\left(\frac{b}{a} \right)}$$

$$= \left[\frac{3}{\log\left(\frac{5.1}{6.2} \right)} \right] \log\left(\frac{T_f}{T_i} \right)$$

$$= -35.37 \log\left(\frac{T_f}{T_i} \right)$$

(a) $n = -35.37 \log\left(\dfrac{10^{-3}\,K}{1\,K}\right) = 106$

(b) $n = -35.37 \log\left(\dfrac{10^{-9}\,K}{1\,K}\right) = 318$

6.1.1

(a) The Dieterici equation of state is purported to have good accuracy near the critical point. It does fail badly at high densities where V_m begins to approach the value of the Dieterici coefficient b. We will use it to derive a practical equation for the computations.

$$p_r = \frac{e^2\, T_r\, e^{-2/T_r V_r}}{2\,V_r - 1} \qquad \text{[Table 1.6]}$$

Substitution of the Dieterici eos derivative $(\partial p_r/\partial T_r)_{V_r} = (2 + T_r\, V_r) p_r / T_r^2\, V_r$ into the reduced form of eqn 5.8 gives

$$\left(\frac{\partial U_r}{\partial V_r}\right)_{T_r} = T_r \left(\frac{\partial p_r}{\partial T_r}\right)_{V_r} - p_r = \frac{2\,p_r}{T_r\,V_r} \qquad (U_r = U/p_c\,V_c)$$

Integration along the isotherm T_r from an infinite volume to V_r yields the practical computational equation.

$$\boxed{\Delta U_r(T_r,\,V_r) = -\int\limits_{\substack{V_r \\ T_r\,\text{constant}}}^{\infty} \frac{2\,p_r(T_r,\,V_r)}{T_r\,V_r}\, dV_r}$$

The integration is performed with mathematical software.

(b)

(c) $\quad \delta\,(T_r,\,V_r) = \sqrt{-p_c\,\Delta U_r\,/\,V_r} \quad$ where $\quad p_c = 72.9\,\text{atm}$

Carbon dioxide should have solvent properties similar to liquid carbon tetrachloride ($8 \leq \delta \leq 9$) when the reduced pressure is in the approximate range $\boxed{0.85 \text{ to } 0.90 \text{ when } T_r = 1}$

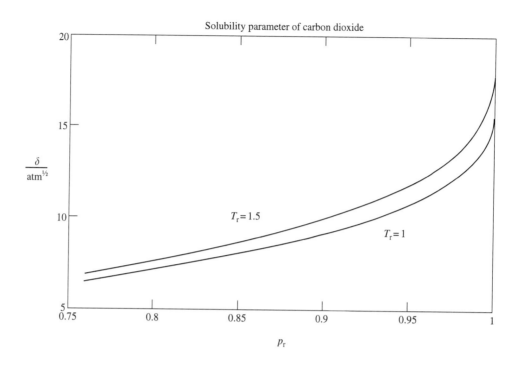

Solubility parameter of carbon dioxide

7.1.1   Imagine that the system of solutions is prepared by placing the aqueous polyelectrolyte $Na_\nu P$ on the "*in*side" of the membrane and a NaCl concentration equal to b on the "*out*side" of the membrane. Aqueous volumes on each side are equal. The polyelectrolyte is 100% dissociated.

$$Na_\nu P \xrightarrow{100\%} \nu Na^+(aq) + P^{\nu-}(aq)$$

The NaCl, moving from the high concentration outside the membrane to the low concentration inside, comes to equilibrium when $\mu_{Nacl,out} = \mu_{Nacl,in}$ assuming ideal solutions, eqn 7.55 gives:

$$\mu_{NaCl}^{\ominus} + RT \ln a_{NaCl,out} = \mu_{NaCl}^{\ominus} + RT \ln a_{NaCl,in}$$

$$a_{NaCl,out} = a_{NaCl,in}$$

$$[Na^+]_{in}[Cl^-]_{in} = [Na^+]_{out}[Cl^-]_{out} \quad \text{[ideal soln]}$$

Let $x = [Cl^-]_{in}$ at equilibrium and let $2[Cl^-] = b = [Cl^-]_{in} + [Cl^-]_{out}$. By the conservation of mass we may write that $[Na^+]_{in} = x + \nu[P^{\nu-}]$ and $[Na^+]_{out} = [Cl^-]_{out} = 2[Cl^-] - x$.

Substitution of the mass balance equations into the equilibrium criteria and solving for x gives

$$[Cl^-]_{in} = x = \frac{4\,[Cl^-]^2}{4\,[Cl^-] + \nu[P^{\nu-}]}$$

$$[Na^+]_{in} = \frac{(2[Cl^-] + \nu[P^{\nu-}])^2}{4\,[Cl^-] + \nu[P^{\nu-}]}$$

$$[Na^+]_{out} = [Cl^-]_{out} = \frac{2[Cl^-](2[Cl^-] + \nu[P^{\nu-}])}{4[Cl^-] + \nu[P^{\nu-}]}$$

$$[Na^+]_{in} - [Na^+]_{out} = \frac{\nu[P^{\nu-}](2[Cl^-] + \nu[P^{\nu-}])}{4[Cl^-] + \nu[P^{\nu-}]}$$

$$= \frac{\nu[P^{\nu-}](2[Cl^-] + \nu[P^{\nu-}])^2}{(2[Cl^-] + \nu[P^{\nu-}])(4[Cl^-] + \nu[P^{\nu-}])}$$

$$\boxed{[Na^+]_{in} - [Na^+]_{out} = \frac{\nu[P^{\nu-}][Na^+]_{in}}{2[Cl^-] + \nu[P^{\nu-}]}}$$

$$[Cl^-]_{in} - [Cl^-]_{out} = \frac{4[Cl^-]^2 - (4[Cl^-]^2 + 2\nu[P^{\nu-}][Cl^-])}{4[Cl^-] + \nu[p^{\nu-}]}$$

$$= \frac{-2\nu[P^{\nu-}][Cl^-](4[Cl^-]^2)}{(4[Cl^-]^2)(4[Cl^-] + \nu[P^{\nu-}])}$$

$$\boxed{[Cl^-]_{in} - [Cl^-]_{out} = -\frac{\nu[P^{\nu-}][Cl^-]_{in}}{2[Cl^-]}}$$

8.1.1   It is hard to imagine that the stacking of disk-like molecules (2) would result in "wires" capable of conducting electricity. Electrical conductivity mechanisms of metals are well known as is the conductivity of delocalized π-electrons found in the lamillar atomic planes of a graphite crystal. However, there is little electric conductivity normal to the covalent atomic planes of crystalline graphite. The nearest neighbor atomic distances are about 0.2–0.3 nm in metals and about 0.1–0.2 nm in molecules that have delocalized π-electrons. The lamillar planes of graphite are separated by 0.34 nm and held together by the weak van der Waals force. Stacks of disk-like organic molecules (2) seem to mimic the stacking of graphite planes but with greater separation (0.5 nm). No electric conductivity is expected along this type of discotic liquid crystal wire.

However, it may be possible to design a disk-like molecule that has

(a) filled, or partially filled, delocalized bonding molecular orbitals.

(b) relatively low energy, unfilled, delocalized antibonding molecular orbitals and

(c) a relatively small stacking distance.

A potential difference, when applied to the ends of a molecular stack, might cause a π-electron within one molecule to jump into an antibonding π-molecular orbital of an adjacent molecule. In such a case, the "wire" would conduct electricity. Relevant topics include molecular orbitals (Chapter 14) and electron tunnelling (Chapter 12).

8.2.1   To examine the process of zone levelling with the phase diagram below, consider a solid on the isopleth through a_1 and heat the sample without coming to overall equilibrium. If the temperature rises to a_2, a liquid of composition b_2 forms and the remaining solid is at a_2'. Heating that solid down an isopleth passing through a_2' forms a liquid of composition b_3 and leaves the solid at a_3'. This sequence of heater passes shows that in a pass the impurities at the end of a sample are reduced while being transfered to the liquid phase which moves with the heater down the length of the sample. With enough passes the dopant, which is initially at the end of the sample, is distributed evenly throughout.

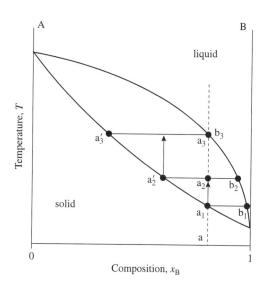

9.1.1 **(a)** At low pH the oxygen atoms of ATP are protonated; at high pH, acid-base neutralization reduces the degree of protonation. The electrostatic repulsions between the highly charged oxygen atoms of ATP^{4-} is expected to give it a high hydrolysis free energy – a trend that is compatible with the observed dependence upon pH. The repulsion makes $\Delta_r H$ more negative.

$$
\begin{array}{ccccc}
\text{OH} & & \text{OH} & & \text{OH} \\
| & & | & & | \\
\text{HO} - \text{P} - \text{O} - & \text{P} - \text{O} - & \text{P} - \text{O} - \text{ribose} - \text{adenine} \\
\| & & \| & & \| \\
\text{O} & & \text{O} & & \text{O}
\end{array}
$$

ATP, low pH form

$$
\begin{array}{ccccc}
\text{O}^- & & \text{O}^- & & \text{O}^- \\
| & & | & & | \\
\text{O}^- - \text{P} - \text{O} - & \text{P} - \text{O} - & \text{P} - \text{O} - \text{ribose} - \text{adenine} \\
\| & & \| & & \| \\
\text{O} & & \text{O} & & \text{O}
\end{array}
$$

ATP^{4-}, high pH form

(b) The phosphate species, $P_i(aq)$, produced in the hydrolysis of ATP has more resonance structures than ATP. Resonance lowers the energy of the dissociated phosphate making $\Delta_r H$ negative and contributing to the exergonicity of the hydrolysis. Double bonds are shorter than single bonds.

$$
\begin{array}{ccccc}
\text{O}^- & & \text{O} & & \text{O}^- \\
| & & \| & & | \\
\text{HO} - \text{P} - \text{O}^- & \longleftrightarrow & \text{HO} - \text{P} - \text{O}^- & \longleftrightarrow & \text{HO} - \text{P} = \text{O} \\
\| & & | & & | \\
\text{O} & & \text{O}^- & & \text{O}^-
\end{array}
$$

(c) triphosphate $\xrightarrow{\text{hydrolysis}}$ diphosphate $+$ phosphate

The triphosphate is a more highly ordered state than the state in which diphosphate and phosphate move independently, $\Delta_r S$ is positive. Analogously, the hydrolysis of ATP produces an increased number of independent chemical species so entropy must increase. The positive $\Delta_r S$ contributes to the hydrolysis exergonicity according to the relationship $\Delta_r G = \Delta_r H - T\Delta_r S$.

9.2.1 $\Delta_f G^{\ominus}(ZnO) = -318 \text{ kJ mol}^{-1}$ at $25\,^{\circ}$C (Table 2.1). Using the $\Delta_f G^{\ominus}$ value at $25\,^{\circ}$C and a slope that is common for all the oxides, we may add the approximate line for ZnO in the Ellingham diagram as shown in the following figure. Zinc oxide melts at 1975°C. The ZnO curve passes under the reaction (iii) curve at about $\boxed{1300^{\circ}\text{C}}$ so that is the estimate of the lowest temperature at which zinc oxide can be reduced to the metal by carbon.

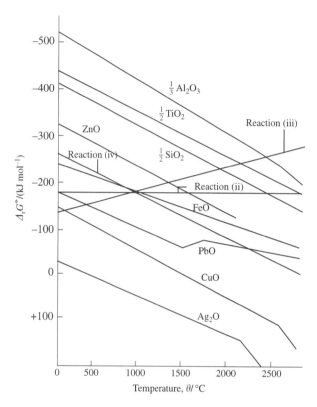

10.1.1 Yes, a bacterium can evolve to utilize the ethanol/nitrate pair to exergonically release the free energy needed for ATP synthesis. The ethanol reductant may yield any of the following products.

$$\begin{array}{ccccccc} CH_3CH_2OH & \longrightarrow & CH_3CHO & \longrightarrow & CH_3COOH & \longrightarrow & CO_2 + H_2O \\ \text{ethanol} & & \text{ethanal} & & \text{ethanoic acid} & & \end{array}$$

The nitrate oxidant may receive electrons to yield any of the following products.

$$\begin{array}{ccccccc} NO_3^- & \longrightarrow & NO_2^- & \longrightarrow & N_2 & \longrightarrow & NH_3 \\ \text{nitrate} & & \text{nitrite} & & \text{dinitrogen} & & \text{ammonia} \end{array}$$

Oxidation of two ethanol molecules to carbon dioxide and water can transfer 8 electrons to nitrate during the formation of ammonia. The half-reactions and net reaction are:

$$2\left[CH_3CH_2OH(l) \longrightarrow 2CO_2(g) + H_2O(l) + 4H^+(aq) + 4e^-\right]$$
$$NO_3^-(aq) + 9H^+(aq) + 8e^- \longrightarrow NH_3(aq) + 3H_2O$$

$$\rule{10cm}{0.4pt}$$

$$2CH_3CH_2OH(l) + H^+(aq) + NO_3^-(aq) \longrightarrow 4CO_2(g) + 5H_2O + NH_3(aq)$$

$\Delta_r G^{\ominus} = -2331.29\,\text{kJ}$ for the reaction as written (a Table 2.9 calculation). Of course, enzymes must evolve that couple this exergonic redox reaction to the production of ATP, which would then be available for carbohydrate, protein, lipid, and nucleic acid synthesis.

10.2.1

$$E = E_{ap} + \beta \frac{RT}{F} \ln(a_{Na^+} + ka_{K^+})$$

To determine the values of β and k for the Na^+-selective electrode, prepare a series of very low concentration K^+ solutions (free of Na^+) such that $a_{K^+} = [K^+]$ for each solution. Measure E for each solution, assuming that $E_{ap} = 0$ for the Na^+-free solutions,

$$E = \beta \frac{RT}{F} \ln(k[K^+])$$

$$= \beta \left(\frac{RT}{F}\right) \ln(k) + \beta \left(\frac{RT}{F}\right) \ln[K^+].$$

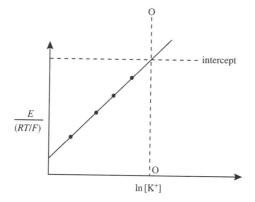

This equation indicates that a plot of $E/(RT/F)$ against $\ln[K^+]$ has a slope equal to β and intercept equal to $\beta \ln(k)$. A linear regression analysis of the plot yields values for the slope and intercept, β and k are calculated with the equations:

$$\beta = \text{slope} \quad \text{and} \quad k = e^{\text{intercept}/\beta}$$

Prepare a low concentration Na^+ solution such that $a_{Na^+} = [Na^+]$ and $a_{K^+} = [K^+] = 0$. Measure E and calculate the asymmetry potential with the equation:

$$E_{ap} = E - \beta \frac{RT}{F} \ln[Na^+]$$

11.1.1   The maxima and minima are determined by $\sin sR$. For the maxima, $\sin sR = 1$, and sR satisfies

$$sR = (4n+1)\frac{\pi}{2} \quad n = 0, 1, 2, 3, \ldots$$

Combining this relation with

$$s = \frac{4\pi}{\lambda} \sin \frac{1}{2}\theta$$

yields $\sin \dfrac{1}{2}\theta = \dfrac{(4n+1)\lambda}{8R}$

For $n = 0$ (the first maximum), for neutrons

$$\sin \frac{1}{2}\theta = \frac{\lambda}{8R} = \frac{80\,\text{pm}}{(8) \times (198.75\,\text{pm})} = 0.050\bar{3} \quad \theta = \boxed{5.8^\circ}$$

For electrons

$$\sin \frac{1}{2}\theta = \frac{\lambda}{8R} = \frac{4 \, \text{pm}}{(8) \times (198.75 \, \text{pm})} = 2.\overline{52} \times 10^{-3} \quad \theta = \boxed{0.3°}$$

For the minima, $\sin sR = -1$, and sR satisfies

$$sR = (4n + 3)\frac{\pi}{2} \quad n = 0, 1, 2, 3, \ldots$$

This yields

$$\sin \frac{1}{2}\theta = \frac{(4n + 3)\lambda}{8R}$$

For $n = 0$ (the first minimum), for neutrons

$$\sin \frac{1}{2}\theta = \frac{3\lambda}{8R} = \frac{(3) \times (80 \, \text{pm})}{(8) \times (198.75 \, \text{pm})} = 0.15\overline{1} \quad \boxed{\theta = 17°}$$

For electrons

$$\sin \frac{1}{2}\theta = \frac{3\lambda}{8R} = \frac{(3) \times (4 \, \text{pm})}{(8) \times (198.75 \, \text{pm})} = 7.\overline{5} \times 10^{-3} \quad \boxed{\theta = 0.9°}$$

Comment. The maxima and minima are widely separated in neutron diffraction but not in electron diffraction. Camera design is therefore different for neutron and electron diffraction.

12.1.1 $\quad \kappa L_2 = (2m_e (V - E)/h^2)^{1/2} L_2 \quad (V - E = 2.0 \, \text{eV})$

$$= (7.25 \times 10^9 \, \text{m}^{-1})(0.500 \times 10^{-9}) = 3.62$$

Since κL_2 is significantly greater than 1, we use eqn 12.28b to estimate the transmission probability, $T(L)$, for tunneling. The electrical current through the STM probe is proportional of $T(L)$.

$$\frac{\text{current } T(L_1)}{\text{current } T(L_2)} = \frac{16\varepsilon(1 - \varepsilon)e^{-2\kappa L_1}}{16\varepsilon(1 - \varepsilon)e^{-2\kappa L_2}} = e^{-2\kappa(L_1 - L_2)}$$

$$= e^{-2(7.25 \times 10^9 \, \text{m}^{-1})(0.60 - 0.50)10^{-9} \, \text{m}}$$

$$= \boxed{0.235}$$

13.1.1 A stellar surface temperature of 3000 K–4000 K, (a "red star") doesn't have the energetic particles and photons that are required for either the collisional or radiation excitation of a neutral hydrogen atom. Atomic hydrogen affects neither the absorption nor the emission lines of red stars in the absence of excitation. "Blue stars" have surface temperature of 15000 K–20000 K. Both the kinetic energy and the blackbody emissions display energies great enough to completely ionize hydrogen. Lacking an electron, the remaining proton cannot affect absorption and emission lines either.

In contrast, a star with a surface temperature of 8000 K–10000 K has a temperature low enough to avoid complete hydrogen ionization but high enough for blackbody radiation to cause electronic transitions of atomic hydrogen. Hydrogen spectral lines are intense for these stars.

Simple kinetic energy and radiation calculations confirm these assertions. For example, a plot of blackbody radiation against the radio photon energy and the ionization energy, I, is shown below. It

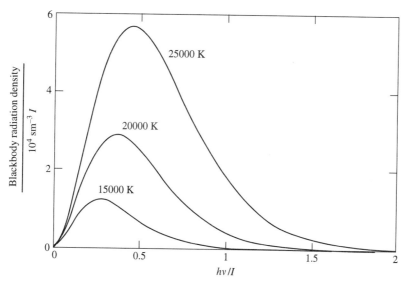

is clearly seen that at 25000 K a large fraction of the radiation is able to ionize the hydrogen ($h\nu/I$).
It is likely that at such high surface temperatures all hydrogen is ionized and, consequently, unable
to affect spectra.

Alternatively, consider the equilibrium between hydrogen atoms and their component charged
particles:

$$H = H^+ + e^-.$$

The equilibrium constant is:

$$K = \frac{p_+ p_-}{p_H p^\circ} = \exp\left(\frac{-\Delta G^\ominus}{RT}\right) = \exp\left(\frac{-\Delta H^\ominus}{RT}\right) = \exp\left(\frac{-\Delta S^\ominus}{RT}\right).$$

Clearly ΔS° is positive for ionization, which makes two particles out of one, and ΔH°, which is
close to the ionization energy, is also positive. At a sufficiently high temperature, ions will outnumber
neutral molecules. Using concepts developed in Chapters 19 and 20, one can compute the equilibrium
constant; it turns out to be 60. Hence, there are relatively few undissociated H atoms in the equilibrium
mixture which is consistent with the weak spectrum of neutral hydrogen observed.

The details of the calculation of the equilibrium constant based on the methods of Chapter 20 follows.
Consider the equilibrium between hydrogen atoms and their component charged particles:
$$H = H^+ + e^-.$$

The equilibrium constant is:
$$K = \frac{p_+ p_-}{p_H p^\ominus} = \exp\left(\frac{-\Delta G^\ominus}{RT}\right).$$

Jump ahead to Section 20.7(b) to use the statistical thermodynamic analysis of a dissociation
equilibrium:

$$K = \frac{q_+^\ominus q_-^\ominus}{q_H^\ominus N_A} e^{-\Delta_r E_0/RT}.$$

where $\quad q^\ominus = \dfrac{RT}{g p^\ominus \Lambda^3} \quad$ and $\quad \Lambda = \left(\dfrac{h^2}{2\pi k T m}\right)^{1/2}.$

and where g is the degeneracy of the species. Note that $g_+ = 2$, $g_- = 2$, and $g_H = 4$. Consequently, these factors cancel in the expression for K.

So $\quad K = \dfrac{RT}{p^{\ominus} N_A} \left(\dfrac{2\pi kT}{h^2} \right)^{3/2} \left(\dfrac{m_- m_+}{m_H} \right)^{3/2} e^{-\Delta_r E_0 / RT}$.

Note that the Boltzmann, Avogadro, and perfect gas constants are related ($R = N_A k$), and collect powers of kT; note also that the product of masses is the reduced mass, which is approximately equal to the mass of the electron; note finally that the molar energy $\Delta_r E_0$ divided by R is the same as the atomic ionization energy (2.179×10^{-18} J from Chapter 13.2(b)) divided by k:

$$K = \frac{(kT)^{5/2}(2\pi m_-)^{3/2}}{p^{\ominus} h^3} e^{-E/kT},$$

$$K = \frac{[(1.381 \times 10^{-23} \, \text{J K}^{-1})(25000 \, \text{K})]^{5/2} [2\pi (9.11 \times 10^{-31} \, \text{kg})]^{3/2}}{(10^5 \, \text{Pa})(6.626 \times 10^{-34} \, \text{J s})^3}$$

$$\times \exp \left(\frac{-2.179 \times 10^{-18} \, \text{J}}{(1.381 \times 10^{-23} \, \text{J K}^{-1})(25000 \, \text{K})} \right),$$

$$K = 60.$$

Thus, the equilibrium favors the ionized species, even though the ionization energy is greater than kT.

16.1.1 The question of whether to use CN or CH within the interstellar cloud of constellation Ophiuchus for the determination of the temperature of the cosmic background radiation depends upon which one has a rotational spectrum that best spans blackbody radiation of 2.726 K. Given $B_0(\text{CH}) = 14.190 \, \text{cm}^{-1}$, the rotational constant that is needed for the comparative analysis may be calculated from the 226.9 GHz spectral line of the Orion Nebula. Assuming that the line is for the $^{12}\text{C}^{14}\text{N}$ isotopic species and $J + 1 \leftarrow J = 1$, which gives a reasonable estimate of the CN bond length (117.4 pm), the CN rotational constant is calculated as follows.

$$B_0 = \nu/c = \frac{\nu}{2c(J+1)} = \frac{\nu}{4c} \quad [16.44] \tag{1}$$

$$= 1.892 \, \text{cm}^{-1} \tag{2}$$

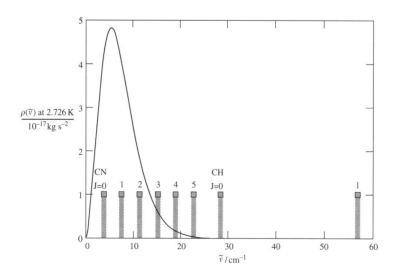

Blackbody radiation at 2.726 K may be plotted against radiation wavenumber with suitable transformation of eqn 11.5.

$$\rho(\tilde{\nu}) = \frac{8\pi hc\tilde{\nu}^3}{e^{hc\tilde{\nu}/kT} - 1}$$

Spectral absorption lines of $^{12}C^{14}N$ and $^{12}C^{1}H$ are calculated with eqn 16.44.

$$\tilde{\nu}(J+1 \leftarrow J) = 2B(J+1) \qquad J = 0, 1, 2, 3 \ldots .$$

The cosmic background radiation and molecular absorption lines are shown in the graph. It is evident that only CN spans the background radiation.

16.2.1   The basic principles of interference and diffraction are discussed in Chapter 9 of F. S. Crawford, Jr., *Waves* (Berkely Physics Course, V. 3), McGraw-Hill Book Co., 1968.

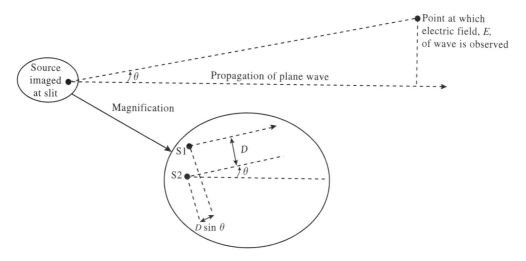

The phenomena of diffraction is observed as the increase in the cross sectional width of a light beam traveling through "slits" or spaces in matter. The angular width of the beam is said to be diffraction limited. By utilizing the Huygens' Principle the analysis of the passage of light through a slit of width D is simplified as the superposition of light sources within the slit. The smallest angle at which the sources destructively interfere gives the diffraction limit when the slit is viewed at a very great distance compared to the size of D. For example, if the slit is modeled with dual sources, as shown above, the beam of one source must travel the additional distance $D \sin \theta$ when observed at angle θ. The beams destructively interfere provided that the additional distance equals $\lambda/2$. For small angles $\sin \theta = \theta$ and the diffusion limit is approximately given by

$$\sin \theta = \frac{\lambda}{2D} \quad \text{or} \quad \theta = \frac{\lambda}{2D}.$$

The angular spread of a beam equals 2θ, or λ/D, in the diffraction limit. This describes the area for which the beam intensity is relatively large. Alternatively, the diffraction limit may be rearranged to the form $D = \lambda/2 \sin \theta$. In this form we may interpret θ as the half-angle of the widest cone of light collected by a microscope lens and D to be the minimum distance between two distinct objects viewed through the lens. This is an approximation of the Airy radius.

The Abbe diffraction limit of microscopy is applicable when the aperture of the objective lens is at a distance greater than the wavelength from the object being viewed. This is the so called "far-field scanning optical microscopy". The diffraction limit may be overcome by drawing out a very small aperture from optical fiber and carefully placing the aperture at a distance less than the wavelength from the object being viewed. This is called "near-field scanning optical microscopy" (NSOM or SNOM). SNOM techniques require careful control of the distance between the viewed object and the aperture using methodology analogous to scanning tunnelling microscopy (STM) and atomic force microscopy (AFM).

17.1.1 The following table summarizes AM1 calculations (an extended Hückel method) of the LUMO–HOMO separation in the 11-*cis* and 11-*trans* molecule (5) model of retinal. The $-46.0°$ torsional angle between the first two alternate double bonds indicates that they are not coplanar. In contrast, the C11C12C13C14 torsion angle shows that the C11=C12 double bond is close to coplanar with neighboring double bonds. The aromatic character of the alternating π-bond system is evidenced by contrasting the computed bond lengths at a single bond away from the π-system (C1–C2), a double bond (C11–C12), and a single bond between doubles (C12–C13) within the Lewis structure. We see a typical single bond length, a slightly elongated double bond length, and a bond length that is intermediate between a single and a double, respectively. The latter lengths are characteristic of aromaticity.

Conformation		11-*trans* (5)		11-*cis* (5)
$\Delta H_f^{\ominus}/\text{kJ mol}^{-1}$		725.07		738.1
$E_{\text{LUMO}}/\text{eV}$		-5.142		-5.138
$E_{\text{HOMO}}/\text{eV}$		-10.770		-10.888
$\Delta E/\text{eV}$	(a)	5.628	(b)	5.750
λ/nm	(a)	220.3	(b)	215.6
C5C6C7C8 torsion angle/°		-44.5		-46.0
C11C12C13C14 torsion angle/°		179.7		-165.5
C1–C2/pm		153.2		153.2
C11–C12/pm		137.3		136.7
C12–C13/pm		1.420		1.421

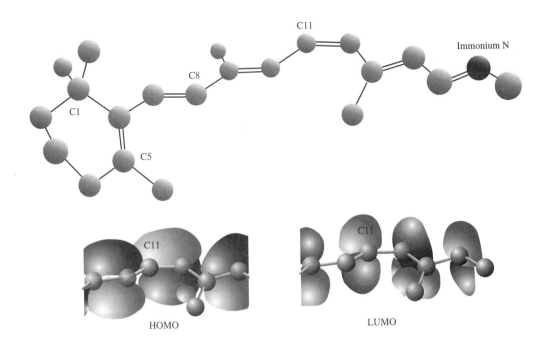

(c) The lowest $\pi^* \leftarrow \pi$ transition occurs in the ultraviolet with the 11-*cis* transition at higher energy (higher frequency, lower wavelength). It is apparent that important interactions between retinal and a surrounding opsin molecule are responsible for reducing the transition energy to the observed strong absorption in the 400 to 600 nm visible range.

17.2.1 (a) The molar concentration corresponding to 1 molecule per cubic μm is:

$$\frac{n}{V} = \frac{1}{6.022 \times 10^{23}\,\text{mol}^{-1}} \times \frac{(10^6\,\mu\text{m m}^{-1})^3}{(1.0\,\mu\text{m}^3)(10\,\text{dm m}^{-1})^3} = \boxed{1.7 \times 10^{-9}\,\text{mol dm}^{-3}},$$

i.e., nanomolar concentrations.

(b) An impurity of a compound of molar mass $100\,\text{g mol}^{-1}$ present at $1.0 \times 10^{-7}\,\text{kg}$ per $1.00\,\text{kg}$ water can be expected to be present at a level of N molecules per cubic μm where N is:

$$N = \frac{1.0 \times 10^{-7}\,\text{kg impurity}}{1.00\,\text{kg water}} \times \frac{6.022 \times 10^{23}\,\text{mol}^{-1}}{100 \times 10^{-3}\,\text{kg impurity mol}^{-1}}$$

$$\times (1.0 \times 10^3\,\text{kg water m}^{-3}) \times (10^{-6}\,\text{m})^3,$$

$$N = \boxed{6.0 \times 10^2}.$$

Pure as it seems, the solvent is much too contaminated for single-molecule spectroscopy.

18.1.1 We use $\nu = \dfrac{\gamma_N \mathcal{B}_{\text{loc}}}{2\pi} = \dfrac{\gamma_N}{2\pi}(1-\sigma)\mathcal{B}$ [18.24]

where $\mathcal{B}$ is the applied field.

Because shielding constants are quite small (a few parts per million) compared to 1, we may write for the purposes of this calculation

$$\nu = \frac{\gamma_N \mathcal{B}}{2\pi}$$

$$\nu_L - \nu_R = 100\,\text{Hz} = \frac{\gamma_N}{2\pi}(\mathcal{B}_L - \mathcal{B}_R)$$

$$\mathcal{B}_L - \mathcal{B}_R = \frac{2\pi \times 100\,\text{s}^{-1}}{\gamma_N}$$

$$= \frac{2\pi \times 100\,\text{s}^{-1}}{26.752 \times 10^7\,\text{T}^{-1}\,\text{s}^{-1}} - 2.35 \times 10^{-6}\,\text{T}$$

$$= 2.35\,\mu\text{T}$$

The field gradient required is then

$$\frac{2.35\,\mu\text{T}}{0.08\,\text{m}} = \boxed{29\,\mu\text{T m}^{-1}}$$

Note that knowledge of the spectrometer frequency, applied field, and the numerical value of the chemical shift (because constant) is not required.

18.2.1 The ESR spectra of a spin probe, such as the di-*tert*-butyl nitroxide radical, broadens with restricted motion of the probe. This suggests that the width of spectral lines may correlate with the depth to which a probe may enter into a biopolymer crevice. Deep crevices are expected to severely restrict probe motion and broaden the spectral lines. Additionally, the splitting and center of ESR spectra of an oriented sample can provide information about the shape of the biopolymer-probe environment because the probe ESR signal is anisotropic and depends upon the orientation of the probe with the external magnetic field. Oriented biopolymers occur in lipid membranes and in muscle fibers.

19.1.1 Heat flows spontaneously from a hot to a cold reservoir. For example, heat can always flow spontaneously from a high positive temperature to a low positive temperature. When a temperature scale includes the possibility of negative temperatures, but not the absolute zero of temperature, as it is unobtainable according to the third law of thermodynamics which we will continue to assume to hold, the ordering of the hot to cold reservoirs according to their temperatures proceeds in the sequence

$$|-0 \text{ K}, -1 \text{ K}, -2 \text{ K}, \ldots, -\infty \text{ K}|+\infty \text{ K}, \cdots, +2 \text{ K}, +1 \text{ K}, +0 \text{ K}|$$

$$| \text{hottest} \quad \longrightarrow \quad \text{colder} | \text{colder yet} \quad \longrightarrow \quad \text{coldest} |$$

$$\text{spontaneous} \qquad\qquad\qquad \text{spontaneous}$$
$$\text{heat flow} \qquad\qquad\qquad\quad \text{heat flow}$$

The sequence above has two discontinuities: that between $+ 0$ K and $- 0$ K, and that between $-\infty$ K and $+\infty$ K, the latter being a singularity as well. The former discontinuity cannot be traversed during refrigeration because of the Third Law; the question remains as to whether or not the latter can be traversed. Refrigeration requires the reversal of heat flow with respect to the spontaneous by doing work in order to transfer heat from a cold to a hot reservoir. The above sequence suggests that this is possible when either both reservoirs have a positive temperature or both have a negative temperature. The temperature of the refrigeration medium can change continuously in either case. However, when the hot reservoir has a negative temperature and the cold reservoir has a positive temperature, the refrigerant must be able to increase its temperature from T_c to $+ \infty$ K, pass to $-\infty$ K, and proceed to T_h. This is clearly impossible for a "normal" refrigerant. In order to achieve this kind of refrigeration, the refrigerant medium must have a finite number of energy states and work must be cleverly performed on the medium to invert the populations of its states, if it is to be able to transfer heat from a positive temperature to a negative temperature. However, it is possible to invert a population (as in the pumping of a laser) and it is possible to let an inverted population relax to a normal population. Both of these make the temperature of the system change sign and the latter corresponds to refrigeration. Imagine a process in a two-state system which takes the relative population of the upper state to 0.51 and imagine a process that relaxes its population to 0.49. Every physical property changes continuously with such a change except temperature. The temperature is discontinuous at $+/ - \infty$, but the system is indifferent to this discontinuity.

20.1.1 **(a)** $\quad q = 1 + \sum_{i=1}^{n} N_i K_i$

But, $\quad K_1 = \sigma e^{-\Delta G^{\ominus}/RT} = \sigma s$

$\qquad\quad K_2 = K_1 s = \sigma s^2$

$\qquad\quad K_3 = K_1 s^2 = \sigma s^3$

$\qquad\qquad \vdots$

$\qquad\quad K_i = K_1 s^{i-1} = \sigma s^i$

Therefore,

$$q = 1 + \sum_{i=1}^{n} N_i \sigma s^i$$

To show that $N_i = n - i + 1$ consider the following figure of n positions having an "X" label.

$$
\begin{array}{ccccccccc}
1 & 2 & 3 & \cdots & y_L & & & & \\
X & X & X & \cdots & X & \cdots & X & \cdots & X \quad X \\
& & & & & y_R & \cdots & n-1 & n
\end{array}
$$

Starting from the left, there are a total of y_L groups of i positions where y_L is limited because an additional group of i would extend beyond the nth position. There are more groups of i. We may start from the right and count off groups of i until reaching position y_R where y_R is limited because an additional group of i would be identical to the group starting at y_L. In fact, $y_R = n - \{y_L + (i - 1)\}$. Consequently,

$$N_i = y_L + y_R = y_L + n - \{y_L + (i - 1)\} = n - i + 1$$

(b) $\quad q = 1 + \sum_{i=1}^{n} N_i \, \sigma s^i$

$$\frac{dq}{ds} = \sum_{i=1}^{n} i \, N_i \sigma s^{i-1} = s^{-1} \sum_{i=1}^{n} i N_i \sigma s^i$$

and $\quad \displaystyle\sum_{i=1}^{n} i \, N_i \sigma s^i = \frac{dq}{ds}$

We may substitute the above expression into the equation for the degree of conversion, θ, that is given in the box.

$$\theta = \left(\frac{1}{nq}\right) \sum_{i=1}^{n} i \, N_i \sigma s^i = \frac{s}{nq} \frac{dq}{ds}$$

Since $dq/q = d(\ln q)$ and $ds/s = d(\ln s)$, the expression becomes

$$\theta = \left(\frac{1}{n}\right) \frac{d(\ln q)}{d(\ln s)}$$

21.1.1   **(a)** The table displays computed electrostatic charges (semi-empirical, PM3 level, PC Spartan Pro^{TM}) of the DNA bases, modified by addition of a methyl group to the position at which the base binds to the DNA backbone. (That is, R = methyl for the computations displayed, but R = DNA backbone in DNA.) See the first set of structures for numbering.

(b) and **(c)** On purely electrostatic grounds, one would expect the most positively charged hydrogen atoms of one molecule to bind to the most negatively charged atoms of another. The figure below depicts hydrogen atoms as black lines, and has thicker gray lines for the most positively charged hydrogens (those with a charge of at least 0.200); they also happen to be the hydrogens bound to electronegative atoms. The figure also has light gray type for the atoms with the greatest negative charges (more negative than -0.400), with a gray ball on the most negative carbon atoms. In principle, then, any of the thick gray lines of one molecule can line up next to any of the atoms in light gray type of its bonding partner. In practice, the carbon atoms are not good binding sites for steric reasons.

R-Adenine		R-Thymine		R-Guanine		R-Cytosine	
atom	charge	atom	charge	atom	charge	atom	charge
C1	0.905	C1	0.885	C1	0.720	C1	0.961
amino N	−0.656	O of C1	−0.580	O of C1	−0.524	amino N	−0.709
amino H†	0.288	C2	−0.554	N2	−0.473	amino H†	0.291
N2	−0.914	C2 methyl C	0.180	H of H2	0.233	N2	−0.901
C3	0.785	C2 methyl H†	−0.003	C3	0.794	C3	0.993
H of C3	−0.020	C3	0.173	amino N	−0.693	O of C3	−0.609
N4	−0.835	H of C3	0.111	amino H†	0.288	N4	−0.286
C5	0.639	N4	−0.390	N4	−0.757	methyl C*	0.119
N6	−0.183	N4 methyl C*	0.211	C5	0.325	methyl H*†	0.017
methyl C*	0.113	N4 methyl H*†	0.002	N6	0.079	C5	0.205
methyl H*†	0.022	C5	0.836	methyl C*	−0.008	H of C5	0.103
C7	0.320	O of C5	−0.596	methyl H*†	0.043	C6	−0.684
H of C7	0.056	N6	−0.540	C7	0.130	H of C6	0.174
N8	−0.584	H of N6	0.264	H of C7	0.086		
C9	−0.268			N8	−0.470		
				C9	−0.146		

*part of R group, so not really available for hydrogen bonding in DNA
†table displays average charge of atoms that are chemically equivalent

(d) The naturally occurring pairs are shown below. These configurations are quite accessible sterically, and they have the further advantage of multiple hydrogen bonds.

(e) See above

22.1.1 Single-walled carbon nanotubes (SWNT) may be either conductors or semiconductors depending upon the tube diameter and the chiral angle of the fused benzene rings with respect to the tube axis. Van der Waals forces cause SWNT to stick together in clumps, which are normally mixtures

large variety of chiral angles

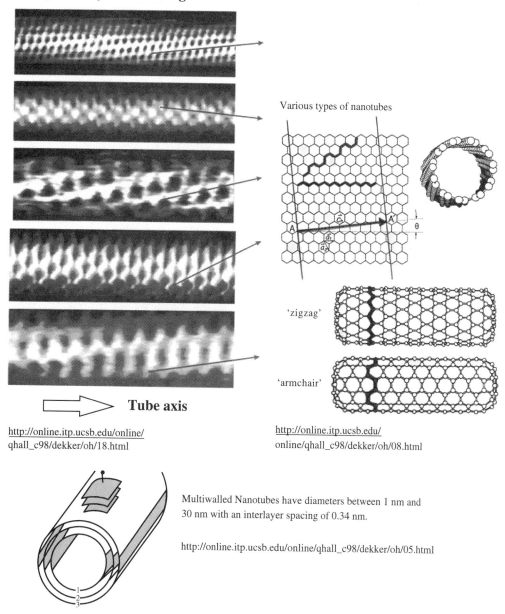

Tube axis

Various types of nanotubes

'zigzag'

'armchair'

http://online.itp.ucsb.edu/online/
qhall_c98/dekker/oh/18.html

http://online.itp.ucsb.edu/
online/qhall_c98/dekker/oh/08.html

Multiwalled Nanotubes have diameters between 1 nm and
30 nm with an interlayer spacing of 0.34 nm.

http://online.itp.ucsb.edu/online/qhall_c98/dekker/oh/05.html

of conductors and semiconductors. SWNT stick to many surfaces and they bend, or drape, around nano-sized features that are upon a surface.

Only the semiconductor SWNT are suitable for the preparation of field-effect transistors (FET) so IBM researchers (*Science*, April 27, 2001) have developed a destructive technique for eliminating conducting tubes from conductor/semiconductor clumps with a current burst. The technique can also be used to remove the outer layers of multi-walled tubes that consist of multiple concentric tubes about a common axis. Bandgaps increase as the diameter of multi-walled tubes is decreased which means that the destructive technique can be used to tailor a semiconductor tube to specific requirements.

Here is a list of ideas for producing transistors with SWNT.

Cees Dekker and students (S.J. Tans et al., *Nature*, **393**, 49 (1998)) have draped a semiconducting carbon nanotube over metal electrodes that are 400 nm apart atop a silicon surface coated with silicon dioxide. A bias voltage between the electrodes provides the source and drain of an FET. The silicon serves as a gate electrode. By adjusting the magnitude of an electric field applied to the gate, current flow across the nanotube may be turned on and off.

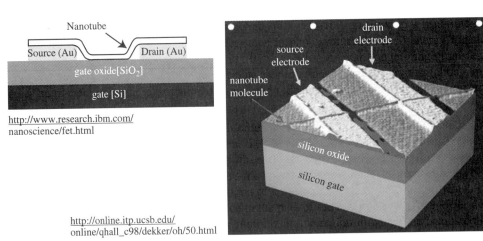

http://www.research.ibm.com/
nanoscience/fet.html

http://online.itp.ucsb.edu/
online/qhall_c98/dekker/oh/50.html

A section of a single nanotube may be exposed to potassium vapor to produce a p-n junction.

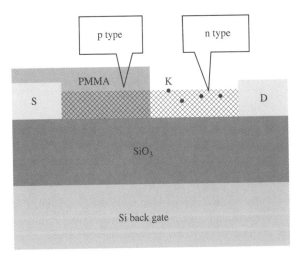

http://www.usc.edu/dept/ee/People/Faculty/Zhou/zhougroup/research.html

A single-electron transistor (SET) has been prepared by Cees Dekker and coworkers (*Science*, **293**, 76, (2001)) with a conducting nanotube. The SET is prepared by putting two bends in a tube with the tip of an AFM. Bending causes two buckles that, at a distance of 20 nm, serves as a conductance barrier. When an appropriate voltage is applied to the gate below the barrier, electrons tunnel one at a time across the barrier.

A semiconductor tube may be fused to a conductor tube to produce a SET similar to an SET.

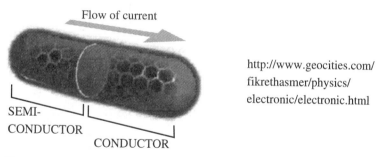

http://www.geocities.com/
fikrethasmer/physics/
electronic/electronic.html

23.1.1 X-rays can ionize materials through which they pass, generating X-ray photoelectrons. This process is the basis of X-ray photoelectron spectroscopy (in which the photoelectrons are detected and analyzed); however, in diffraction studies, it is an undesired side effect. X-ray photoelectrons are typically ejected from inner electron shells, but valence electrons can "fall" into the vacant energy level, with a possible disruption of bonding. Thus, radicals (atoms or molecules with unpaired valence electrons) can be formed in the wake of photoionization by losing a valence electron into the newly formed vacancy, by breaking a bond in a photoionized molecule, or by capture of a photoelectron generated from another species. Electron capture can also reduce the capturing species.

The effects of high energy X-ray radiation can be illustrated by discussing the protein modifications that can occur during protein crystallography with a synchrotron source. Protein crystals have high water content in channels throughout the crystal. X-ray radiolysis of water generates hydroxyl radicals and hydrated electrons.

$$H_2O + \text{ionizing radiation} \longrightarrow H_2O^{\cdot+} + e_{aq}^- \xrightarrow{H_2O} H_3O^+ + \cdot OH + e_{aq}^-$$

The hydrated electrons react with any oxygen that is present to produce superoxide anions and hydroperoxyl radicals.

$$e_{aq}^- + O_2 \longrightarrow O_2^{\cdot-} \xrightarrow{H_2O} HO_2 \cdot + OH^-$$

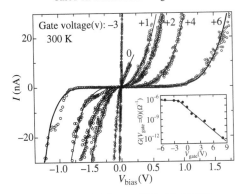

single-molecule transistor at room temperature
based on a semiconducting nanotube

Observations:

- For positive V_{gate}, a gap appears
- For negative V_{gate}, the conductance saturates at the metallic value
- Conductance modulation by 6 orders
- Gain ~ 0.35: improvements well possible

Tans et al., *Nature*, **393**, 49 (1998)

The radicals attack amino acid residues, cause cross-linking between adjacent protein molecules in the crystal, and may damage the protein backbone. Time-resolved macromolecular crystallography avoids this damage with a technique of shock freezing that rapidly plunges the crystal in liquid nitrogen or propane before exposure to synchrotron radiation. The following mechanism scheme (S.D. Maleknia, M. Brenowitz, and M.R. Chance, *Anal. Chem.*, **71**, 3965 (1999)) demonstrates side chain modification of aromatic residues by the hydroxyl radical.

23.2.1 The relationship between critical temperature and critical magnetic field is given by

$$H_c(T) = H_c(0)\left(1 - \frac{T^2}{T_c^2}\right).$$

Solving for T gives the critical temperature for a given magnetic field:

$$T = T_c \left(1 - \frac{H_c(T)}{H_c(0)}\right)^{1/2} = (7.19\,\text{K}) \times \left(1 - \frac{20 \times 10^3\,\text{A m}^{-1}}{63901\,\text{A m}^{-1}}\right)^{1/2} = \boxed{6.0\,\text{K}}.$$

24.1.1 Dry atmospheric air is 78.08% N_2, 20.95% O_2, 0.93% Ar, 0.03% CO_2, plus traces of other gases. Nitrogen, oxygen, and carbon dioxide contribute 99.06% of the molecules in a volume with each molecule contributing an average rotational energy equal to kT (eqn 20.30). The rotational energy density is given by

$$\rho_R = \frac{E_R}{V} = \frac{0.9906\,N\langle \epsilon^R \rangle}{V} = \frac{0.9906 \langle \epsilon^R \rangle p N_A}{RT}$$

$$= \frac{0.9906\,kT\,p N_A}{RT} = 0.9906 p$$

$$= 0.9906(1.013 \times 10^5\,\text{Pa}) = 0.1004\,\text{J cm}^{-1}$$

The total energy density (translational plus rotational) is

$$\rho_T = \rho_K + \rho_R = 0.15\,\text{J cm}^{-3} + 0.10\,\text{J cm}^{-3}$$

$$\rho_T = 0.25\,\text{J cm}^{-3}$$

24.2.1 The maximum flux in mediated transport is achieved at very high concentrations of the transported species. Under such conditions, the transported species A flood the carrier species C, pushing practically all of the latter into the form of the AC complex. (The mathematical condition for saturation of the flux at J_{max} is that $[A] \gg K$, the equilibrium constant for dissociation of the AC complex; this condition puts practically all C into the complex, regardless of its inherent stability.) The value of J_{max} depends on the concentration of carrier species, $[C]_0$. For a given value of $[C]_0$, J_{max} represents the transport capacity of the "fleet" of carriers. The oversupply of A keeps the carriers transporting at full capacity.

25.1.1 Analysis of NMR lineshapes can be used to infer time scales of protein folding or unfolding steps. Protons (or other nuclei, for that matter) that have different chemical shifts in folded and unfolded proteins will yield a single peak if the time scale for interconversion (*i.e.*, for folding or unfolding) is comparable to or less than the reciprocal of the two peaks' frequency difference. Monitoring the change from two peaks (indicating that a sample contains both folded and unfolded proteins, which might be observed at one temperature) to a broad single peak (indicating fast interconversion, which might be the case at a higher temperature) can allow the determination of the time constant for the conversion. One advantage of NMR over vibrational or electronic spectroscopy is that the radiation used to probe the system is much less energetic, and therefore much less likely to alter the folding or unfolding process it is designed to investigate. The lineshape strategy cannot be used to investigate processes as fast as those accessible by electronic or vibrational spectroscopy. (Cf. example 18.2.)

26.1.1 (a) $k_2 = 6.2 \times 10^{-34}\,\text{cm}^6\,\text{molecule}^{-2}\text{s}^{-1}$ (assuming M is dioxygen)

$$k_4 = 8.0 \times 10^{-15}\,\text{cm}^3\,\text{molecule}^{-1}\text{s}^{-1}$$

The concentration of atomic oxygen will be very, very small making a binary collision between atomic oxygen extremely unlikely. In fact, the reaction

$$O + O + M \rightarrow O_2 + M \qquad v = k_5[O]^2[M]$$

is ternary which makes it even less likely. Rate terms in k_5 may be safely omitted from consideration.

(b) For all practical purposes $d[O_2]/dt = 0$ because very little dioxygen reacts to form either atomic oxygen or ozone. Using "molecules per cm^3", or simply cm^{-3}, as the concentration unit, we find that

$$[O_2] = \frac{N_A p}{RT} = \frac{N_A(10 \text{ Torr})}{R(298 \text{ K})} = 3.239 \times 10^{17} \text{ molecules cm}^{-3}$$

$$\frac{d[O]}{dt} = 2k_1[O_2] - k_2[O][O_2]^2 + k_3[O_3] - k_4[O][O_3]$$

$$= a_1 - a_2[O] + a_3[O_3] - a_4[O][O_3]$$

$$\frac{d[O_3]}{dt} = k_2[O][O_2]^2 - k_3[O_3] - k_4[O][O_3]$$

$$= a_2[O] - a_3[O_3] - a_4[O][O_3]$$

where
$$a_1 = 2k_1[O_2] = 6.478 \times 10^9 \text{ s}^{-1} \text{ cm}^{-3}$$
$$a_2 = k_2[O_2]^2 = 65.036 \text{ s}^{-1}$$
$$a_3 = k_3 = 0.016 \text{ s}^{-1}$$
$$a_4 = k_4 = 8.0 \times 10^{-15} \text{ cm}^3 \text{ s}^{-1}$$

(c) We break the time period in two. The early period encompasses the first 0.05 s with the initial conditions $[O]_0 = [O_3]_0 = 0$. The second period covers the remainder of the 4 hours with the initial conditions provided by the $[O]_{0.05 \text{ s}}$ and $[O_3]_{0.05 \text{ s}}$ values of the early period. Numerical integration of the coupled differential equations yields the concentrations of the following graphs.

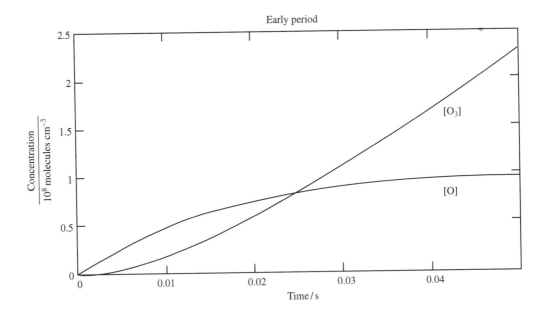

During the early period ($t < 0.05$ s) UV radiation causes the formation of a small amount of atomic oxygen (less than 10^8 molecules cm^{-3}). A steady-state is approached in which a near balance is established between production of atomic oxygen by dioxygen dissociation and usage of atomic

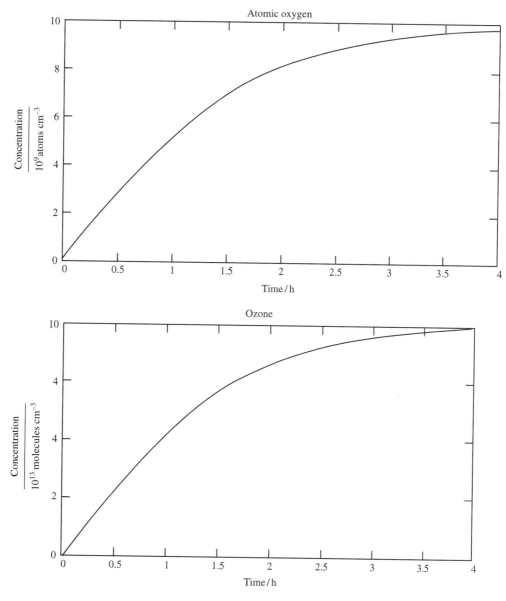

Atomic oxygen

Ozone

oxygen to produce ozone. There is, however, a 100-fold growth in the atomic oxygen concentration over the next 4 h so it is not a perfectly steady state.

After 4 h of photochemistry the percentage ozone is 0.0123%.

$$\text{percentage ozone} \sim \left(\frac{3.97 \times 10^{13} \text{ cm}^{-3}}{3.24 \times 10^{17} \text{cm}^{-3}} \right) 100 \sim 0.123\%$$

Ozone production does not require low pressure in the Chapman model. Changing the oxygen pressure to 100 Torr gives 0.025% ozone after 4 h but the increased collision rate reduces atomic oxygen to 1/5'th the value at 10 Torr. However, a pressure increase may require the inclusion in the mechanism of the step $O_3 + M \rightarrow O + O_2 + M$. This would reduce ozone production.

26.2.1
$$C + Q \xrightarrow{h\nu} C^* + Q \xrightarrow[\text{transfer}]{\text{electron}} C^+ + Q^-$$
chlorophyll quinone

Direct electron transfer from the ground state of C is not spontaneous. It is spontaneous from the excited state. The difference between the ΔG's of the two processes is given by the expression:

$$\Delta\Delta G = \Delta G_{c^*} - \Delta G_c \sim U_c - U_{c^*} \sim -(U_{\text{LUMO}} - U_{\text{HOMO}})$$

where U_{LUMO} and U_{HUMO} are energies of the LUMO and HOMO of chlorophyll. Since $\Delta\Delta G < 0$, we see that electron transfer is exergonic and spontaneous when the electron is transferred from the excited state of chlorophyll.

27.1.1
Figure 2 of the Zewail article on femtochemistry mentioned in *Further Reading* gives time scales for a large number of physical, chemical, and biological processes on the atomic and molecular scale. Radiative decay of excited electronic states can range from about 10^{-9} s to 10^{-4} s – even longer for phosphorescence involing "forbidden" decay paths. Molecular rotational motion takes place on a scale of 10^{-12} to 10^{-9} s. Molecular vibrations are faster still, about 10^{-14} to 10^{-12} s. The mean time between collisions in liquids is similarly short, 10^{-14} to 10^{-13} s. Proton transfer reactions occur on a timescale of about 10^{-10} to 10^{-9} s, about 100 times slower than "harpoon" reactions (10^{-12} to 10^{-10} s). Box 17.1 describes several events in vision, including the 200-fs photoisomerization that gets the process started. Box 25.1 discusses helix-coil transitions, including experimental measurements of time-scales of tens or hundreds of nanoseconds (10^{-8} to 10^{-7} s). Box 26.2 lists time scales of several energy-transfer and electron-transfer steps in photosynthesis. Initial energy transfer (to a nearby pigment) has a time scale of around 10^{-13} to 10^{-11} s, with longer-range transfer (to the reaction center) taking about 10^{-10} s. Immediate electron transfer is also very fast (about 3 ps), with ultimate transfer (leading to oxidation of water and reduction of plastoquinone) taking from 10^{-10} to 10^{-3} s.

27.2.1
The Rb atom must hit the I side of CH_3I in order to produce $RbI + CH_3$. The orientation of CH_3I can be controlled by exciting rotations about the CI axis with linearly polarized light; the optimal orientation aims the I side of CH_3I at the direction of approach of the beam of Rb atoms. Two possible alignments of the reactant beams are shown in the figure. In the top depiction, the beams are antiparallel, thereby maximizing the likelihood of collision and the volume within which collision can occur (but also putting each beam source in the path of the other beam). In the lower depiction, the beam paths are at right angles, thereby minimizing the region in which the beams collide, but facilitating the study of that well-defined collision volume by a "probe" laser at right angles to both beams.

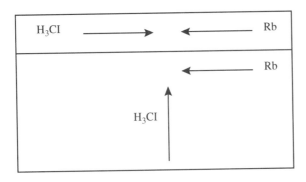

28.1.1 Refer to the figure below.

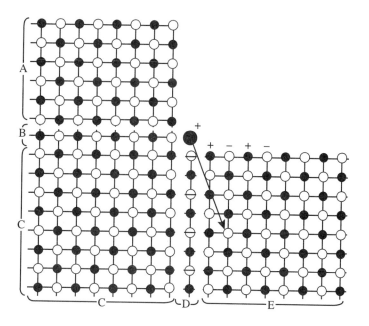

Evaluate the sum of $\pm\dfrac{1}{r_i}$, where r_i is the distance from the ion i to the ion of interest, taking $+\dfrac{1}{r}$ for ions of like charge and $-\dfrac{1}{r}$ for ions of opposite charge. The array has been divided into five zones. Zones B and D can be summed analytically to give $-\ln 2 = -0.69$. The summation over the other zones, each of which gives the same result, is tedious because of the very slow convergence of the sum. Unless you make a very clever choice of the sequence of ions (grouping them so that their contributions almost cancel), you will find the following values for arrays of different sizes

10×10	20×20	50×50	100×100	200×200
0.259	0.273	0.283	0.286	0.289

The final figure is in good agreement with the analytical value, $0.289\ 259\ 7\ldots$

(a) For a cation above a flat surface, the energy (relative to the energy at infinity, and in multiples of $\dfrac{e^2}{4\pi\varepsilon r_0}$ where r_0 is the lattice spacing (200 pm)), is

$$\text{Zone C} + \text{D} + \text{E} = 0.29 - 0.69 + 0.29 = \boxed{-0.11}$$

which implies an attractive state.

(b) For a cation at the foot of a high cliff, the energy is

$$\text{Zone A} + \text{B} + \text{C} + \text{D} + \text{E} = 3 \times 0.29 + 2 \times (-0.69) = \boxed{-0.51}$$

which is significantly more attractive. Hence, the latter is the more likely settling point (if potential energy considerations such as these are dominant).

28.2.1

$$R_{eq} = R_{max} \left(\frac{a_0 K}{a_0 K + 1} \right)$$

Taking the inverse of the above equation and multiplication by a_0 gives:

$$\frac{a_0}{R_{eq}} = \frac{1}{R_{max} K} + \frac{a_0}{R_{max}}$$

This working equation predicts that a plot of a_0/R_{eq} against a_0 should be linear if the model is applicable to the experimental data. The slope of a linear regression fit to the data gives the value of $1/R_{max}$ or $R_{max} = 1/\text{slope}$. Likewise, the regression intercept equals $1/R_{max} K$ or $K = \text{slope/intercept}$.

Cover design: *Cambraia Fonseca Fernandes*
Cover image: *Image of a DNA molecule obtained by scanning tunneling microscopy (STM), a technique that uses the principles of quantum theory to visualize atoms and molecules on surfaces. The STM technique is one of many discussed in the text that demonstrate the link between recent work in physical chemistry and all aspects of chemistry, including materials, environmental, and biological chemistry. The experiment was performed at the Ion Beam Materials Research Laboratory, Sandia National Laboratories, New Mexico, USA. The photograph was provided by Dr. John Cunningham, Visuals Unlimited, New Hampshire, USA.*

W. H. FREEMAN AND COMPANY
41 Madison Avenue
New York, NY 10010
Houndmills, Basingstoke RG21 6XS, England

ISBN 0-7167-4388-4

9 780716 743880   90000